Communications
in Computer and Information Science   727

Commenced Publication in 2007
Founding and Former Series Editors:
Alfredo Cuzzocrea, Xiaoyong Du, Orhun Kara, Ting Liu, Dominik Ślęzak,
and Xiaokang Yang

More information about this series at http://www.springer.com/series/7899

Beiji Zou · Min Li
Hongzhi Wang · Xianhua Song
Wei Xie · Zeguang Lu (Eds.)

Data Science

Third International Conference
of Pioneering Computer Scientists,
Engineers and Educators, ICPCSEE 2017
Changsha, China, September 22–24, 2017
Proceedings, Part I

 Springer

Editors
Beiji Zou
Central South University
Changsha
China

Min Li
Central South University
Changsha
China

Hongzhi Wang
Harbin Institute of Technology
Harbin
China

Xianhua Song
Harbin University of Science
and Technology
Harbin
China

Wei Xie
Harbin University of Science
and Technology
Harbin
China

Zeguang Lu
Sciences of Country Tripod Institute
of Data Science
Harbin
China

ISSN 1865-0929   ISSN 1865-0937 (electronic)
Communications in Computer and Information Science
ISBN 978-981-10-6384-8   ISBN 978-981-10-6385-5 (eBook)
DOI 10.1007/978-981-10-6385-5

Library of Congress Control Number: 2017953389

Printed on acid-free paper

This Springer imprint is published by Springer Nature
The registered company is Springer Nature Singapore Pte Ltd.
The registered company address is: 152 Beach Road, #21-01/04 Gateway East, Singapore 189721, Singapore

Preface

As the general and program co-chairs of the Third International Conference of Pioneer Computer Scientists, Engineers, and Educators 2017 (ICPCSEE 2017, originally ICYCSEE), it is our great pleasure to welcome you to the conference, which was held in Changsha, China, 22–24 September 2017, hosted by Central South University and the Computer Education Committee of the Hunan Higher Education Federation. The goal of this conference is to provide a forum for computer scientists, engineers, and educators.

The call for papers of this year's conference attracted 420 paper submissions. After the hard work of the Program Committee, 112 papers were accepted to appear in the conference proceedings, with an acceptance rate of 26.7%. The main topic of this conference is data science. The accepted papers cover a wide range of areas related to basic theory and techniques for data science including mathematical issues in data science, computational theory for data science, big data management and applications, data quality and data preparation, evaluation and measurement in data science, data visualization, big data mining and knowledge management, infrastructure for data science, machine learning for data science, data security and privacy, applications of data science, case study of data science, multimedia data management and analysis, data-driven scientific research, data-driven bioinformatics, data-driven healthcare, data-driven management, data-driven eGovernment, data-driven smart city/planet, data marketing and economics, social media and recommendation systems, data-driven security, data-driven business model innovation, and social and/or organizational impacts of data science.

We would like to thank all the Program Committee members, 216 coming from 147 institutes, for their hard work in completing the review tasks. Their collective efforts made it possible to attain quality reviews for all the submissions within a few weeks. Their diverse expertise in each individual research area helped us to create an exciting program for the conference. Their comments and advice helped the authors to improve the quality of their papers and gain deeper insights.

Great thanks should also go to the authors and participants for their tremendous support in making the conference a success. We thank Lanlan Chang and Jian Li from Springer, whose professional assistance was invaluable in the production of the proceedings.

Besides the technical program, this year the ICPCSEE offered different experiences to the participants. We welcome you to the Central South China to enjoy the beautiful summer in Changsha. We hope you enjoy the conference proceedings.

July 2017

Min Li
Fangxiang Wu
Qilong Han
Ronghua Shi

Organization

The Third International Conference of Pioneering Computer Scientists, Engineers, and Educators (ICPCSEE 2017, originally ICYCSEE) –http://2017.icpcsee.org– was held in Changsha, China, September 22–24, 2017, and hosted by Central South University and the Computer Education Committee of the Hunan Higher Education Federation.

ICPCSEE 2017 Steering Committee

Yaoxue Zhang	Central South University, China
Jianer Chen	Central South University, China
Yi Pan	Central South University, China
Jianxin Wang	Central South University, China

General Chair

Beiji Zou	Central South University, China

Program Chairs

Min Li	Central South University, China
Fangxiang Wu	Central South University, China
Qilong Han	Harbin Engineering University, China
Ronghua Shi	Central South University, China

Organization Chairs

Kehua Guo	Central South University, China
Xiaoning Peng	Huaihua University, China
Junfeng Man	Hunan University of Technology, China
Zeguang Lu	Sciences of Country Tripod Institute of Data Science, China

Publication Chairs

Hongzhi Wang	Harbin Institute of Technology, China
Guanglu Sun	Harbin University of Science and Technology, China
Weipeng Jing	Northeast Forestry University, China

Publication Co-chairs

Xianhua Song	Harbin University of Science and Technology, China
Wei Xie	Harbin University of Science and Technology, China

Yong Wang   Central South University, China
Liangwu Shi   Hunan University of Commerce, China

Education Chairs

Jiawei Huang   Central South University, China
Minsheng Tan   University of South China, China

Industry Chair

Yue Shen   Hunan Agricultural University, China

Demo Chairs

Jiazhi Xia   Central South University, China
Ying Xu   Hunan University, China

Panel Chairs

Jiawei Luo   Hunan University, China
Shaoliang Peng   National University of Defense Technology, China

Registration/Financial Chairs

Ya Huang   Central South University, China
Chengzhang Zhu   Central South University, China

Post/Expo Chair

Renren Liu   Xiangtan University, China

ICYCSEE Steering Committee

Jiajun Bu   Zhejiang University, China
Jian Chen   PARATERA, China
Xuebin Chen   North China University of Science and Technology, China
Wanxiang Che   Harbin Institute of Technology, China
Tian Feng   Institute of Software, Chinese Academy of Sciences, China
Qilong Han   Harbin Engineering University, China
Yiliang Han   Engineering University of CAPF, China
Yinhe Han   Institute of Computing Technology,
                          Chinese Academy of Sciences, China
Weipeng Jing   Northeast Forestry University, China
Hai Jin   Huazhong University of Science and Technology, China
Wei Li   Central Queensland University, Australia

Program Committee Members

Lei Deng	Central South University, China
Vincenzo Deufemia	University of Salerno, Italy
Xiaofeng Ding	Huazhong University, China
Jianrui Ding	Harbin Institute of Technology, China
Qun Ding	Heilongjiang University, China
Xiaoju Dong	Shanghai Jiao Tong University, China
Hongbin Dong	Harbin Engineering University, China
Zhicheng Dou	Renmin University of China, China
Jianyong Duan	North China University of Technology, China
Xiping Duan	Harbin Normal University, China
Lei Duan	Sichuan University, China
Junbin Fang	Jinan University, China
Xiaolin Fang	Southeast University, China
Guangsheng Feng	Harbin Engineering University, China
Jianlin Feng	Sun Yat-Sen University, China
Weisen Feng	Sichuan University, China
Guohong Fu	Heilongjiang University, China
Jing Gao	Dalian University of Technology, China
Dianxuan Gong	North China University of Science and Technology, China
Yu Gu	Northeastern University, China
Yuhang Guo	Beijing Institute of Technology, China
Jiafeng Guo	Institute of Computing Technology, Chinese Academy of Sciences, China
Meng Han	Georgia State University, USA
Qi Han	Harbin Institute of Technology, China
Xianpei Han	Chinese Academy of Sciences, China
Zhongyuan Han	Harbin Institute of Technology, China
Tianyong Hao	Guangdong University of Foreign Studies, China
Shizhu He	Chinese Academy of Sciences, China
Jia He	Chengdu University of Information Technology, China
Qinglai He	Arizona State University, USA
Liang Hong	Wuhan University, China
Zhang Hu	Shanxi University, China
Chengquan Hu	Jilin University, China
Wei Hu	Nanjing University, China
Hao Huang	Wuhan University, China
Lan Huang	Jilin University, China
Shujian Huang	Nanjing University, China
Ruoyu Jia	Sichuan University, China
Bin Jiang	Hunan University, China
Jiming Jiang	King Abdullah University of Science & Technology, Kingdom of Saudi Arabia
Wenjun Jiang	Hunan University, China
Feng Jiang	Harbin Institute of Technology, China
Weipeng Jing	Northeast Forestry University, China
Shenggen Ju	Sichuan University, China

Hanjiang Lai	Sun Yat-Sen University, China
Wei Lan	Central South University, China
Yanyan Lan	Institute of Computing Technology, Chinese Academy of Sciences, China
Min Li	Central South University, China
Mingzhao Li	RMIT University, Australia
Zhixu Li	Soochow University, China
Rong-Hua Li	Shenzhen University, China
Zhixun Li	Nanchang University, China
Xiaoyong Li	Beijing University of Posts and Telecommunications, China
Jianjun Li	Huazhong University of Science and Technology, China
Peng Li	Shaanxi Normal University, China
Qiong Li	Harbin Institute of Technology, China
Zhenghua Li	Soochow University, China
Qingliang Li	Changchun University of Science and Technology, China
Chenliang Li	Wuhan University, China
Xuwei Li	Sichuan University, China
Moses Li	Jiangxi Normal University, China
Mohan Li	Jinan University, China
Hua Li	Changchun University, China
Hui Li	Xidian University, China
Zheng Li	Sichuan University, China
Guoqiang Li	Norwegian University of Science and Technology, Norway
Xiaofeng Li	Sichuan University, China
Yan Liu	Harbin Institute of Technology, China
Yong Liu	HeiLongJiang University, China
Guanfeng Liu	Soochow University, China
Hailong Liu	Northwestern Polytechnical University, China
Yang Liu	Tsinghua University, China
Bingqiang Liu	Shandong University, China
Yanli Liu	Sichuan University, China
Shengquan Liu	XinJiang University, China
Ming Liu	Harbin Institute of Technology, China
Wei Lu	Renmin University of China, China
Binbin Lu	Sichuan University, China
Junling Lu	Shaanxi Normal University, China
Zeguang Lu	Sciences of Country Tripod Institute of Data Science, China
Jizhou Luo	Harbin Institute of Technology, China
Zhunchen Luo	China Defense Science and Technology Information Center, China
Jiawei Luo	Hunan University, China
Jianlu Luo	Officers College of PAP, China
Huifang Ma	NorthWest Normal University, China
Yide Ma	Lanzhou University, China
Hua Mao	Sichuan University, China
Xian-Ling Mao	Beijing Institute of Technology, China

Jun Meng	Dalian University of Technology, China
Tiezheng Nie	Northeastern University, China
Haiwei Pan	Harbin Engineering University, China
Jialiang Peng	Norwegian University of Science and Technology, Norway
Wei Peng	Kunming University of Science and Technology, China
Xiaoqing Peng	Central South University, China
Fei Peng	Hunan University, China
Yuwei Peng	Wuhan University, China
Jianzhong Qi	University of Melbourne, Australia
Shaojie Qiao	Southwest Jiaotong University, China
Zhe Quan	Hunan University, China
Yingxia Shao	Peking University, China
Qiaomu Shen	The Hong Kong University of Science and Technology, Hong Kong, China
Hongwei Shi	Sichuan University, China
Hongtao Song	Harbin Engineering University, China
Wei Song	North China University of Technology, China
Xianhua Song	Harbin Institute of Technology, China
Yanan Sun	Sichuan University, China
Chengjie Sun	Harbin Institute of Technology, China
Guanglu Sun	Harbin University of Science and Technology, China
Minghui Sun	Jilin University, China
Xiao Sun	Hefei University of Technology, China
Guanghua Tan	Hunan University, China
Wenrong Tan	Southwest University for Nationalities, China
Jintao Tang	National University of Defense Technology, China
Dang Tang	Chengdu University of Information Technology, China
Binbin Tang	Works Applications, Japan
Xifeng Tong	Northeast Petroleum University, China
Yongxin Tong	Beihang University, China
Vicenc Torra	Högskolan i Skövde, Sweden
Leong Hou U	University of Macau, China
Chaokun Wang	Tsinghua University, China
Chunnan Wang	Harbin Institute of Technology, China
Dong Wang	Hunan University, China
Hongzhi Wang	Harbin Institute of Technology, China
Jinbao Wang	Harbin Institute of Technology, China
Xin Wang	Tianjin University, China
Yunfeng Wang	Sichuan University, China
Yingjie Wang	Yantai University, China
Yongheng Wang	Hunan University, China
Zhifang Wang	HeiLongJiang University, China
Zhewei Wei	Renming University, China
Zhongyu Wei	Fudan University, China
Yan Wu	Changchun University, China
Zhihong Wu	Sichuan University, China

Huayu Wu	Institute for Infocomm Research, China
Rui Xia	Nanjing University of Science and Technology, China
Min Xian	Utah State University, USA
Tong Xiao	Northeastern University, China
Yi Xiao	Hunan University, China
Degui Xiao	Hunan University, China
Sheng Xiao	Hunan University, China
Minzhu Xie	Hunan Normal University, China
Jing Xu	Changchun University of Science and Technology, China
Jianqiu Xu	Nanjing University of Aeronautics and Astronautics, China
Dan Xu	University of Trento, Italy
Ying Xu	Hunan University, China
Yaohong Xue	Changchun University of Science and Technology, China
Mingyuan Yan	University of North Georgia, USA
Bian Yang	Norwegian University of Science and Technology, Norway
Yajun Yang	Tianjin University, China
Gaobo Yang	Hunan University, China
Lei Yang	HeiLongJiang University, China
Ning Yang	Sichuan University, China
Bin Yao	Shanghai Jiao Tong University, China
Yuxin Ye	Jilin University, China
Minghao Yin	Northeast Normal University, China
Dan Yin	Harbin Engineering University, China
Zhou Yong	China University of Mining and Technology, China
Jinguo You	Kunming University of Science and Technology, China
Lei Yu	Georgia Institute of Technology, USA
Dong Yu	Beijing Language and Culture University, China
Ye Yuan	Northeastern University, China
Kun Yue	Yunnan University, China
Lichen Zhang	Shaanxi Normal University, China
Yongqing Zhang	Chengdu University of Information Technology, China
Meishan Zhang	Singapore University of Technology and Design, Singapore
Xiao Zhang	Renmin University of China, China
Huijie Zhang	Northeast Normal University, China
Kejia Zhang	Harbin Engineering University, China
Yonggang Zhang	Jilin University, China
Jiajun Zhang	Institute of Automation, Chinese Academy of Sciences, China
Yu Zhang	Harbin Institute of Technology, China
Haixian Zhang	Sichuan University, China
Yi Zhang	Sichuan University, China
Boyu Zhang	Utah State University, USA
Wenjie Zhang	The University of New South Wales, Australia
Xiaowang Zhang	Tianjin University, China
Tiejun Zhang	Harbin University of Science and Technology, China
Dongxiang Zhang	University of Electronic Science and Technology of China, China

Liguo Zhang	Harbin Engineering University, China
Yingtao Zhang	Harbin Institute of Technology, China
Jian Zhao	ChangChun University, China
Xin Zhao	Renmin University of China, China
Qijun Zhao	Sichuan University, China
Bihai Zhao	Changsha University, China
Hai Zhao	Shanghai Jiao Tong University, China
Wenping Zheng	Shanxi University, China
Jiancheng Zhong	Hunan Normal University, China
Changjian Zhou	Northeast Agricultural University, China
Fucai Zhou	Northeastern University, China
Jinghua Zhu	Harbin Institute of Technology, China
Yuanyuan Zhu	Wuhan University, China
Min Zhu	Sichuan University, China
Zede Zhu	Hefei Institutes of Physical Science, Chinese Academy of Sciences, China
Quan Zou	Tianjin University, China
Wangmeng Zuo	Harbin Institute of Technology, China

Contents – Part I

Contents – Part II

A Fine-Grained Emotion Analysis Method for Chinese Microblog

Rui Zhou[✉], Hu-yin Zhang, and Gang Ye

Computer School of Wuhan University, Wuhan, Hubei, China
122191865@qq.com

Abstract. The Chinese microblog text is short, full of noise data and emoticons, and the words are often irregularly. For these characteristics, we proposed a fine-grained emotion analysis method. Combined with TF-IDF and variance statistics, we realized a method of calculating multi-class feature selection. We judge the text whether it is positive or negative first, then choose the fine-grained emotional tendency. And we get good result with the test using COAE data set. Compared with other method for feature selection and other emotional library, we did better.

Keywords: Fine-grained emotion analysis · Chinese microblog · Feature extraction

1 Introduction

With the development and advancement of the Internet, there are a lot of data spread on the Internet, which contain great value. Microblog, as the typical represent of the social network, is attracting more and more users to discuss hotpots and publish views in it. It will result in great commercial and political value to analysis the user's emotions fast and accurate.

Microblog text has many special characteristics, such as large noise, short text, irregularly words and contains plenty of emoticons, these characteristics make it difficult to get good emotional analysis result by traditional method. On the other hand, most of research stay on classify positive and negative emotion. In this paper, we proposed a fine-grained emotion analysis method, take the emoticons and multiple emotional vocabulary and repeated punctuation as features, using TF-IDF combine variance to choose features and calculate weights. When in the fine-grained emotional classification, we firstly classify the text into subject and object, then divided them into positive and negative class, and finally mark it as the emotion which has biggest emotional value.

2 Related Work

The emotion analyzing of text is also known as opinion mining, is the process of analysis, processing, induction and reasoning for subjective text with emotional color [1]. Zhao sort the emotional information classification into two missions, Subjective

B. Zou et al. (Eds.): ICPCSEE 2017, Part I, CCIS 727, pp. 1–11, 2017.
DOI: 10.1007/978-981-10-6385-5_1

and objective binary classification and emotional classification of subjective information [2]. But there is no concrete conclusion about how many categories of emotions should be divided, Pang divided the level of praise and grade into three [3, 4]. However Goldberg divided them for four level. Furthermore, there are some people divided the emotion into four roughly class [5, 6], then divided them into forty fine-grained emotion classes [7, 8]. In the research of fine-grained emotion for Chinese microblog, the emotional ontology library constructed by Dalian University of technology is mostly used, which divided the emotions into seven classes, including happy, like, surprise, sad, disgust, anger and fear [9, 10].

The two main methods of emotion analysis are using emotional dictionary and the machine learning [11, 12]. The first one is to add up the emotional words and calculate the emotional value of them [13, 14]; the second one is to train the marked trained text and build classification model through SVM, ME or KNN and so on, then classify them with the classify model [15, 16]. It is a little late when we start to do research about emotional analysis in our country, so it is lack of corpus about marked Chinese microblog text [17, 18]. Wang divided the text into positive and negative classes, based on dictionary and rule set, using the expression symbol as the auxiliary element, and get 77.3% accuracy [19, 20]. Zhang also based on dictionary, using expression picture and emotional words, get nearly 80% accuracy [21]. Shen get a good result based on dictionary and machine learning, using SVM, divided the text into seven fine-grained emotion [22].

As for feature extraction, the mostly used methods are IG, DF, MI and CHI. Yang think CHI and IG are the best after analyzing and comparing these methods [23]. Xiong improved the recognition effect by consider the word frequency and other information to overcome the shortage of CHI [24]. In this paper, we combine TF-IDF with variance to extract features.

3 Feature Extraction

The purpose of feature extraction is to reduce the dimension, that to avoid unnecessary calculations coursed by too much features. We choose the features which contain strong emotional tendencies to be the basis of emotional classification, so that we can reduce the calculation and improve the accuracy of classification.

3.1 Weight Calculating

TF-IDF is the most popular method to calculate the weight of features, its advantage is to consider the relationship between word frequency and anti-document frequency in the document. In this paper, we used TF-IDF to calculate weight.

Let D to represent the document collection, and it contains k classes, so there is $D = \{D_1, D_2, D_3 \ldots D_k\}, k \in K$, D_n represent the count that belongs to number k class in the whole documents. The Feature term set of feature item is represented by T. $T = \{t_1, t_2, t_3 \ldots t_i\}, i \in N$.

Definition 1 TF (Term Frequency): tf_{ik} is represent the amount of feature word t_i show in document D_k.

Definition 2 IDF (Inverse Document Frequency): idf_{ik} is represent the inverse frequency of t_i, d_{ik} is the number of t_i shown is the documents.

$$idf_{ik} = \log \frac{D_k}{d_{ik}} \tag{1}$$

To ensure the formula 1 is smooth, and not appear extreme value, we set a constant value in the formula. So the formula 1 is to be

$$idf_{ik} = \log \left(\frac{D_k}{d_{ik} + 0.5} + 0.5 \right) \tag{2}$$

Then we can calculate the weight of t_i in the document D_k

$$W_{ik} = tf_{ik} \times idf_{ik} = tf_{ik} \times \log \left(\frac{D_k}{idf_{ik} + 0.5} + 0.5 \right) \tag{3}$$

We can get the importance of the features for a class by the weight calculating used TF-IDF, but not the difference of the weight between classes. And based on experience, when the feature word has outstanding contribution to a class, it can be the consult of the class.

Considering above, we can calculate the variance of the feature's word frequency in each class, based on the weight calculated by IF-IDF, combined with variance. That the bigger the variance is, the more important for the feature word to the class.

$$D(x) = E(x^2) - E(x)^2 = \frac{1}{N} \left(\sum_{i=1}^{N} x_i^2 - N_x^{-2} \right) \tag{4}$$

If we calculate the variance of the word term frequency of feature word x_i, that is:

$$D(tf(x_i)) = \frac{1}{K} \left(\sum_{K=1}^{K} td_{ik}^2 - K\overline{tf_i}^2 \right) \tag{5}$$

$\overline{tf_i}$ is the average word frequency of the characteristic in each category, $\overline{tf_i} = \frac{\sum_{k=1}^{K} tf_{ik}}{K}$

We can also get the variance of weight that we get in formula 4.

$$D(tf_i df(x_i)) = \frac{1}{K} \left(\sum_{k=1}^{K} W_{ik}^2 - K\overline{W_i}^2 \right) \tag{6}$$

$\overline{W_i}$ is the average weight of the feature word in each category, and $\overline{W_i} = \frac{\sum_{k=1}^{K} W_{ik}}{K}$.

Then we sort the result of the variance of weights, choose the forefront of words, so that we can get the feature word which will be used to classify the text. These features are affected only by the frequency of words and the number of documents which contains the features.

3.2 Normalized Treatment

The result value we get above should be normalized further, change the weight into [0, 1]:

$$f(x_{ik}) = \frac{x_j}{\sum_{k=1}^{K} x_{ik}} \tag{7}$$

In order to filter out the noise that may appear in the corpus, we set a threshold. We take the value in statistics only by the value is bigger than the threshold. The value of the threshold is the arithmetic mean of the weight that the feature in each class.

$$\overline{x_i} = \frac{\sum_{k=1}^{K} x_{ik}}{K} \tag{8}$$

$$f(x) = \begin{cases} \dfrac{x_i}{\sum_{k=1}^{K} (x_{ik}|x_{ik} \geq \overline{x_i})}, x_{ik} \geq \overline{x_i} \\ 0, x_{ik} < \overline{x_i} \end{cases} \tag{9}$$

3.3 The Method of Choosing Features

To calculate the emotional value of the text, we choose the emoticons, high emotional tendency plural word combination and repetitive punctuation to be the emotional features. Compared with ordinary text, microblog contains lots of emoticons, these emoticons make it more directly for users to express themselves. In addition, there will be some new words in the microblog text, combination of network buzzwords, but can't be found by recognition tools. Some of them also contains a wealth of emotions; and the microblog users always don't follow the strict syntax and grammar rules, so there will be a lot of repetitive punctuation and tone words. Based on the above considerations, we add the three features above up to the emotional library.

In the previous section, we proposed a method that combining TF-IDF and variance to calculate the weight value of a feature. So in this section, we will extract emoticons, plural word combination and repetitive punctuation and tone words by the method. And sort the emotional tendency weight, get the higher ranking features.

3.3.1 Emoticons

There are 2 types of the emoticons in microblog. The first one is become from the microblog own expression library. It will be a picture shown in microblog, but when it comes to the text, it is the text like "[/难过]" (sad). So we can easily get them by regular expression, and almost all of the words are recorded in the emotional ontology library. The second one is composed of symbolic combination of emoticons.

Emoticons are in the form of network text, and often with strong emotion factors in. And it's difficult for segmentation tool to extract the emoticons, for they are made by some irregular symbols. By statistics, most of the emoticons can be expressed by a combination of 3 to 7 symbols. Just like "O.O" represent surprise, and "^.^" represent happy, ">.<" represent sad. That we carry out the expression symbol extraction through three to five window.

We extract all of the combinations of symbols through the window. Because all the train texts have been marked, we can get the weight of these combinations to each class based on the method we proposed above. Then we sort those weights, and get the top 50 combinations, and add them up to the emotional feature library. Some of the link is shown in Table 1.

Table 1. The top ten features of the emoticons, phrase and new words, repeated punctuation and modal words

type	Top 1 to 10
emoticons	O.O , ^.^ , ^-^ , >.< , ^_^||| , @.@ , ╯＿╰ , #^_^# , ≧ω≦ , →_→
phrase and new words	"尼玛"(like fuck) , "开森"(happy) , "给力"(wonderful) , "厉害了"(good done) , "坑爹"(cheating),"你妹"(like fuck) , "惊呆了"(be shocked),"喜大普奔"(news so exhilarating that everyone is celebrating and spreading it around the world), "屌丝"(loser) , "逆袭"(Counter attack)
repeated punctuation and tone words	"哈哈哈"(hahaha) , "!!!" , "???" ,"。。。", "啊啊啊"(au,au,au), "啦啦啦" (lalala), "呵呵"(hehe) , "33", "嘻嘻嘻"(xixixi) , "~~~"

3.3.2   Emotional Plural Word Combination

The text will be the single word one by one after segmentation. As we can see, the single word may have no emotional value, but it will be different when some of the words combined together. For example, in Chinese, "开森" (happy) will be "开" (open), "森" (forest) after segmentation. "开森" is a phrase with positive emotion, but the two single words are not. In addition, there are lots of network words in microblog. These words are difficult to identify using segmentation tools.

To solve this question, we use N-Gram and set N to two to five. We get all the phrase which combined by two to five words after segmentation. And then we calculate the weight of each phrase at the same way. Finally, we get the top 100 phrase and add them up to the emotional feature library, and most of them are new vocabulary in the network. Some of the link is shown in Table 1.

3.3.3   Repetitive Punctuation and Tone Words

We often use repetitive punctuation and tone words to rich our expression, such as "!!!", "???". These repetitive punctuation also have very strong emotional tendency inside. We count the repeated punctuation and modal words appearing continuously in the microblog text, when they appear three times or more, we marked them. Then we get the high weight ones to add up to emotional feature library using the method we propose above.

4   Emotion Analysis

We divide the emotional category into seven categories in this paper, which are happy, like, surprise, sad, disgust, anger and fear. Happy and like are the positive emotion, negative emotion includes sad, disgust, anger and fear. Surprise is not belong to positive or negative. The whole process is shown in Fig. 1.

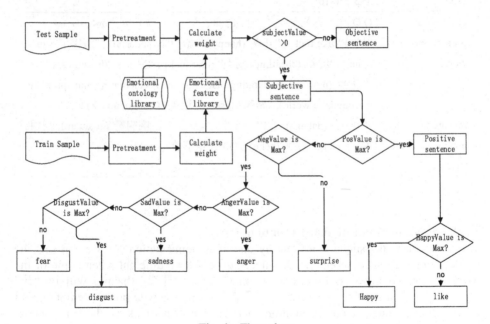

Fig. 1.   Flow chart

4.1   Pretreatment

There are many noisy data in the microblog text, just like URL, "@someone", "#-topic#" and so on. They are useless for emotion analysis, and even may make trouble with segmentation. So we remove them from the microblog text, and separate the word from the sentences using ICTCLAS, which is proposed by Chinese Academy of Sciences.

4.2 Weight Calculating

The emotional ontology library provided by Dalian University of Technology has scored emotional value for each kind of emotional tendency of the emotional word, these values are between 0 and 1, so they can be the weight of classification. These values, together with the feature words we have get in Sect. 3, constituted the emotional feature library. Then we will judge the emotional tendencies by sentence.

For each sentence in microblog, we extract every word, phrase and emoticons which is in the emotional feature library, and add up the feature's emotional score for each emotional tendency, that we can get the distribution of the sentences on each emotional category. We use 'AngerValue', 'DisgustValue', 'FearValue', 'HappyValue', 'LikeValue', 'SadValue' and 'SurpriseValue' to represent the seven emotional tendencies. And we use 'SubjectValue' to represent the whole emotional value, and 'PosValue' to represent the positive value, 'NegValue' to reprent the nagtive value. So there are:

$$
\begin{aligned}
\text{SubjectValue} &= \text{PosValue} + \text{NegValue} + \text{SurpriseValue} \\
&= (\text{HappyValue} + \text{LikeValue}) \\
&\quad + (\text{AngerValue} + \text{DisgustValue} + \text{FearValue} + \text{SadValue}) \\
&\quad + \text{SurpriseValue}
\end{aligned}
$$

Considering that the negative words in the sentence will flip the emotional tendencies of the whole sentence, we collected the dictionary of negative words. When the emotional word appears after the negative word and the surrounding window is within the distance of 2, we will reverse the emotional tendencies. As for there are seven tendencies, we will only reverse in the PosValue or NegValue.

4.3 Sub-Level Emotional Tendency Judging

We will take coarse-grained judgment first, and then fine-grained analysis of the approach. As we can see in Fig. 1. The judgment will be made by flow steps:

Step 1 : Judge whether a sentence is a subjective sentence or an objective sentence through SubjectValue. the If the SubjectValue > 0, it is a subjective sentence. Go to step 2.

Step 2 : Judge whether a subjective sentence is positive or not according to PosValue, NegValue and SurpriseValue. If PosValue is the biggest, it is a positive sentence, go to step 4.

Step 3 : Judge whether a subjective sentence is negative or surprise according to NegValue and SurpriseValue. If it's negative, go to step 5, else it is surprise and the process end.

Step 4 : Choose the biggest one in HappyValue and LikeValue, and it will be the final tendency of the sentence. The process go to end.

Step 5 : choose the biggest one in AngerValue, DisgustValue, FearValue and SadValue, and it will be the final tendency of the sentence. The process go to end.

4.4 Emotional Analysis for the Whole Microblog

We can get the emotional tendency for each sentence after the above treatment, then we will judge the emotion tendency for the whole microblog. We will set the emotional tendency as the one which are the most in sentences. When there are some same count of the categories, we will calculate the whole emotional tendency value of them, set the biggest one as the final emotional tendency. But if there is still no result, we will count the number of positive, negative and neutral sentences in the microblog, select the largest number of categories, and then identify the biggest emotional value of the fine-grained emotional category.

5 Experiment and Analysis

5.1 The Data of Experiment

We take the COAE for our experiment, and extracts some of the data of task 3 and task 4. The data of task 3 is regardless of field, and the data of task 4 is main about cars, food, phone and finance. Training data contains 4000 microblogs, and we get 14526 sentences after the clause and remove the repetition. Test data contains 5000 micro-blogs with 16785 sentences. These microblogs have been marked with emotional tendencies, all the microblog text are from Sina Microblog. The distribution of the number of emotional categories is shown in Table 2.

Table 2. Each emotional microblog count

Emotion type	Happy	Like	Anger	Disgust	Fear	Sadness	Surprise	No emotion
Train mount	371	597	235	425	49	388	112	1832
Test mount	558	779	218	467	51	372	168	2437

5.2 Evaluation Standard

We take accuracy, recall and F values as the evaluation criteria, the correct rate is the proportion of the text that matches the results of the test and the results of the artificial standard, and the recall rate is the number of relevant documents retrieved and the ratio of all relevant documents in the document library, F is the relationship between recall and correct. In addition, since there are a plurality of emotion categories referred to in this paper, the macro averages are used as the evaluation criteria. The formula is as follows:

$$\text{Macro Correct rate:} \quad p_{macro} = \frac{1}{7}\sum_{i=1}^{7} p_i$$

$$\text{Macro Recall rate:} \quad r_{macro} = \frac{1}{7}\sum_{i=1}^{7} r_i$$

$$\text{Macro F} \quad F_{macro} = \frac{2*p_{macro}*r_{macro}}{p_{macro} + r_{macro}}$$

5.3 Experimental Comparison and Analysis

5.3.1 Different Emotional Library

We experimented with the emotional ontology library of Dalian Polytechnic and our extended emotional feature library. The result is shown in Table 3, as we can see, the correct rate is much better to use our emotional feature library. That is to say, we can effectively improve the accuracy of emotional analysis by emotional feature library which we get through the feature we selected.

Table 3. The result of correct rate using Dalian University of Technology Emotional ontology library and our emotional feature library.

Emotion type	Happy	Like	Anger	Disgust	Fear	Sadness	Surprise	No emotion
Emotional ontology library	0.2425	0.1953	0.4161	0.1363	0.0941	0.3281	0.2552	0.505
Emotional feature library	0.6033	0.6541	0.6812	0.4553	0.3721	0.5832	0.4215	0.6788

5.3.2 Different Feature Selection Methods

The weights will be different if we use different feature selection method, and the result of classify will be different too. We compared with several other feature selection method, including CHIMAX-TF, TF-TF, TF-TFIDF, D(TF)-TF, all of them are improved well. As we can see in Fig. 2, the D (TFIDF)-TFIDF selected in this paper has the highest accuracy.

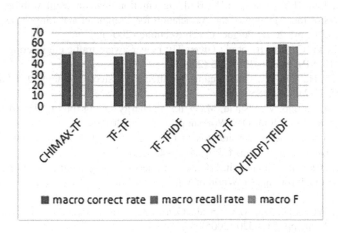

Fig. 2. The result of different feature choosing

6 Conclusion

After comparing the test, we can conclude that we achieve good fine-grained emotion Classification effect by the use of emotional feature library, which contains expression, the multiple emotional vocabulary and the repetitive punctuation selected by feature selection based on TF-IDF combined variance. In relation to other features selection methods and machine learning way to solve the problem of emotional classification, the method we proposed has certain advantages. However, in general, there is much to do to improve the accuracy of fine-grained emotional analysis, and the calculation of the emotional value of the unrecorded word is also an area for improvement in this paper.

References

1. Zhao, Z.-Y., Qing, B., Liu, T.: Sentiment analysis. J. Softw. **21**(8), 1834–1848 (2010)
2. He, J.-F., He, Y.-X., Liu, N.: A microblog short text oriented multi-class feature extraction method of fine-grained sentiment analysis. Acta Scicentiarum Nat. Univ. Pekin. **50**(1), 48–54 (2014)
3. Pang, B., Lee, L.: Seeing stars: exploiting class relationships for sentiment categorization with respect to rating scales. In: Meeting on Association for Computational Linguistics, pp. 115–124 (2005)
4. Bianzino, A., Chaudet, C., Rossi, D., et al.: A survey of green networking research. IEEE Commun. Surv. Tutor. **14**(1), 3–20 (2012)
5. Goldberg, A.B., Zhu, X.: Seeing stars when there aren't many stars. In: Textgraphs: the First Workshop on Graph Based Methods for Natural Language Processing. pp. 45–52 (2006)
6. Even, S., Itai, A., Shamir, A.: On the complexity of time table and multi-commodity flow problems. In: Foundations of Computer Science Annual Symposium on IEEE, pp. 184–193 (1975)
7. Yang, C., Lin, H.Y., Chen, H.H.: Building emotion lexicon from weblog corpora. In: Proceedings of the, Meeting of the Association for Computational Linguistics, ACL 2007, DBLP, 23–30 June 2007, Prague, Czech Republic (2007)
8. Zhang, M., Yi, C., Liu, B., et al.: Green TE: power-aware traffic engineering. In: Network Protocols (ICNP), pp. 21–30. IEEE (2010)
9. Xu, L., Lin, H., Pan, Y.: Constructing the affective lexicon ontology. J. China Soc. Sci. Techn. Inf. **27**(2), 180–185 (2008)
10. Cianfrani, A., Eramo, V., Listanti, M., et al.: An energy saving routing algorithm for a green OSPF protocol. In: INFOCOM Workshops, pp. 1–5 (2010)
11. Wu, W., Xiao, S.: Sentiment analysis of Chinese micro-blog based on multi-feature and combined classification. Chin. J. Eng. **4**, 39–45 (2013)
12. Amaldi, E., Capone, A., Gianoli, L.G., et al.: Energy management in IP traffic engineering with shortest path routing. In: World of Wireless Mobile and Multimedia Networks, pp. 1–6 (2011)
13. Ku, L.W., Wu, T.H., Lee, L.Y., et al.: Construction of an evaluation corpus for opinion extraction. Ntcir. pp. 513–520 (2005)
14. Cuomo, F., Cianfrani, A., Polverini, M., et al.: Network pruning for energy saving in the internet. Comput. Netw. **56**(10), 2355–2367 (2012)

15. Dasgupta, S., Ng, V.: Mine the easy, classify the hard: a semi-supervised approach to automatic sentiment classification. In: Proceedings of the, Meeting of the Association for Computational Linguistics and the, International Joint Conference on Natural Language Processing of the Afnlp, ACL 2009, DBLP, 2–7 August 2009, Singapore, pp. 701–709 (2009)
16. Shi, D., Ruixuan, L., Xiaolin, L.: Energy efficient routing algorithm based on software defined data center network. J. Comput. Res. Dev. **52**(4), 806–812 (2015)
17. Li, T., Ji, D.: Chinese journal of engineering. Appl. Res. Comput. **32**(4), 978–981 (2015)
18. Yoon, M., Kamal, A.: Power minimization in fat-tree SDN datacenter operation. In: Global Communications Conference, pp. 1–7 (2015)
19. Wang, Z., Yu, Z., Guo, B., Lu, X.: Sentiment analysis of Chinese micro blog based on lexicon and rule set. Comput. Eng. Appl. **51**(8), 218–225 (2015)
20. Mahadevan, P., Sharma, P., Banerjee, S., Ranganathan, P.: A power benchmarking framework for network devices. In: Fratta, L., Schulzrinne, H., Takahashi, Y., Spaniol, O. (eds.) NETWORKING 2009. LNCS, vol. 5550, pp. 795–808. Springer, Heidelberg (2009). doi:10.1007/978-3-642-01399-7_62
21. Zhang, S., Yu, L., Hu, C.: Sentiment analysis of Chinese micro-blogs based on emoticons and emotional words. Comput. Sci. **39**(s3), 146–148 (2012)
22. Shen, L.: The research on Chinese microblog sentiment analysis based on rules and machine learning methods. Anhui University (2015)
23. Yang, Y., Pedersen, J.O.: A comparative study on feature selection in text categorization. In: Fourteenth International Conference on Machine Learning. Morgan Kaufmann Publishers Inc. pp. 412–420 (1998)
24. Xiong, Z., Zhang, P., Zhang, Y.: Improved approach to CHI in feature extraction. Comput. Appl. **28**(2), 513–514 (2008)

Research of Detection Algorithm for Time Series Abnormal Subsequence

Chunkai Zhang$^{(\boxtimes)}$, Haodong Liu, and Ao Yin

Department of Computer Science and Technology, Shenzhen Graduate School,
Harbin Institute of Technology, Shenzhen, China
ckzhang812@gmail.com, haodong.1994@qq.com, yinaoyn@126.com

Abstract. The recent advancements in sensor technology have made it possible to collect enormous amounts of data in real time. How to find out unusual pattern from time series data plays a very important role in data mining. In this paper, we focus on the abnormal subsequence detection. The original definition of **discord** subsequences is defective for some kind of time series, in this paper we give a more robust definition which is based on the k nearest neighbors. We also donate a novel method for time series representation, it has better performance than traditional methods (like PAA/SAX) to represent the characteristic of some special time series. To speed up the process of abnormal subsequence detection, we used the clustering method to optimize the outer loop ordering and early abandon subsequence which is impossible to be abnormal. The experiment results validate that the algorithm is correct and has a high efficiency.

Keywords: Time series representation · Abnormal subsequence · K nearest neighbor

1 Introduction

A **time series** is a series of data points indexed (or listed or graphed) in time order. Most commonly, a time series is a sequence taken at successive equally spaced points in time. Abnormal data refers to some data which is significantly different from the others, anomaly detection is to discover the most unusual pattern in the time series data by algorithms. Generally speaking, data is usually generated by a single or a number of system processes, which can reflect the status of the system. When the process is not normal, it will lead to the production of abnormal data. Abnormal pattern means or contains important information of the system, anomaly detection has important application in data mining field, including improving the quality of clustering, data cleaning, summarization [2,5].

Keogh et al. is the first one who gave the definition of discord subsequence in his paper and he also proposed a very efficiently method to detect the time series discord (HOT SAX) [5]. Izakian and Pedrycz proposed fuzzy C-means clustering method to find out the abnormal subsequences [6]. Li et al. proposed Finding

© Springer Nature Singapore Pte Ltd. 2017
B. Zou et al. (Eds.): ICPCSEE 2017, Part I, CCIS 727, pp. 12–26, 2017.
DOI: 10.1007/978-981-10-6385-5_2

Time Series Discord Based on Bit Representation Clustering [8]. Liu et al. proposed the isolation forest [14,15] concept to detect anomalies. Besides, Liu et al. also proposed SCIForest [16] to detect cluster anomalies and the SCIForest is based on the Isolation Forest. Notice that the Isolation Forest is a state-of-art method to anomaly detection, it has high accuracy and very fast. But it still has probability of error, and in some case we can't allow the mistake. For example, the anomaly detec-tion in medical field. The distance-based method of anomaly detection can always find out the anomalies if you use appropriate parameters, it is slow but reliable. So in this paper, we used the distance-based method to discovery abnormal subsequence in time series.

The core contribution of this work is showed as follows:

- We indicate the definition of abnormal subsequence which is more robust than the discord [5] definition
- We propose a new representation for time series which is very easy to compute the similarity between subsequences
- Compared with the brute force algorithm, our new algorithm has greatly improved the efficiency.

The rest of this paper is organized as follows, In Sect. 2 we will introduce the necessary background materials and definitions. In Sect. 3 discuss a new representation for time series. In Sect. 4 we will introduce our detection algorithm. In Sect. 5 plenty datasets will be used to test our algorithm and shows its implications for many time series data mining tasks. In Sect. 6 concludes the paper.

2   Definition

Now we begin by defining the data type of interest.

Definition 1 (Euclidean distance). Given two time series Q and C of length n, the Euclidean distance between them is defined as:

$$Dist(Q, C) = \sqrt{\sum_{i=1}^{n}(q_i - c_i)^2}$$

While the definition of Euclidean distance is obvious and intuitive, we need it to exclude trivial matches. In general, the best matches to a subsequence (apart from itself) tend to be located one or two points to the left or the right of the subsequence in question (Fig. 1). Such matches have previously been called trivial matches [7–9]. As we shall see, it is critical when finding discords to exclude trivial matches; otherwise, almost all real datasets have degenerate and unintuitive solutions. We will therefore take the time to formally define a non-self match.

Definition 2 (Non-self match). Provided a time series T of length n, C is subsequence of T and the length is $m(m \leq n)$ and its start position is p, M is another subsequence and their start positions is q, when $|p - q| \geq m$, said subsequence M is non-self match of C.

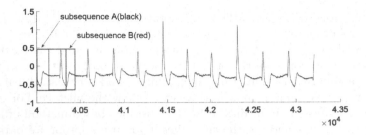

Fig. 1. Two trivial matches subsequences (A and B)

Definition 3 (Time series discord). Given time series T, the subsequence A of length m is said to be the discord subsequence of T if A has a largest distance to its nearest non-self match. Any subsequence C of length m, M_C is the non-self match of C, M_A is the non-self match of A, it satisfy:

$$\forall C \in T, min(DIST(A, M_A)) \geq min(DIST(C, M_C))$$

Suppose there are two similar abnormal subsequence in the time series, in that situation we can not use Definition 3 to find abnormal subsequences. Just consider the situation in Fig. 2, subsequence M and Q are two "discord" subsequences, But M and Q are very highly similar and they are non-self match to each other, so their distance is very small, we can't find Q or M as abnormal subsequence under Definition 3.

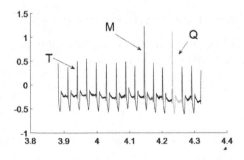

Fig. 2. M and Q has a very short distance, so neither M nor Q doesn't be considered as discord

A new abnormal subsequence definition is given as follows.

Definition 4 (Abnormal subsequences). Given time series T, in all subsequences of T, if subsequence M has maximum average distance to its K nearest non-self match (we call it the "abnormal degree"), we say that M is abnormal subsequence. The formula is expressed as follows:

$$M = argmax_{seq}(1/k \sum_{i=1}^{k}(seq, seq_i))$$

$$seq_i \in non\_self\_match\_knn(seq)$$

It is very clear that Definition 4 is more robust than Definition 3, and it can always find the most abnormal subsequences in the time series.

3 A New Representation for Time Series

As with most problems in computer science, the suitable choice of representation greatly affects the ease and efficiency of time series data mining. With this in mind, a great number of time series representations have been introduced, including the Discrete Fourier Transform (DFT) [9], the Discrete Wavelet Transform (DWT) [10], Piecewise Linear, and Piecewise Constant models (PAA) [11], (APCA) [11,12], and Singular Value Decomposition (SVD) [11].

The Symbolic Aggregate approXimation (SAX) [13] is the most popular way to presentation the time series data, it allows the transformation of a real-valued time series of any length n to a discrete string of length l, where $l << n$. We denote a time series T, reduced to cardinality c and dimensionality d, as $SAX(T)_{c,d}$. as shown in Fig. 3. After the PAA representation is obtained, it is then discretized into the symbolic representation.

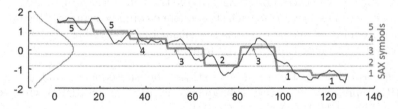

Fig. 3. Representation of time series T (blue) in PAA (green/bold) converted into a SAX word, SAX(T)5, 8 = 5, 5, 4, 3, 2, 3, 1, 1, with c = 5 and d = 8. (Color figure online)

There is an deficiency in the SAX representation, it's based on the PAA method, and PAA use the average value to represent a subsequence, it may ignore many important message of the subsequence. Sometimes subsequences has the same average value but they didn't similar at all. And the length of the subsequence is fixed, every subsequence just represent by a few words or numbers.

We propose a novel method for time series representation–**TSMBRB** (time series minimum bounding rectangle binarization) which is based on **MBR** (the minimum bounding rectangle), also known as bounding box or envelope (Fig. 4), is an expression of the maximum extents of a 2-dimensional object (e.g. point, line, polygon) or set of objects within its 2-D (x, y) coordinate system.

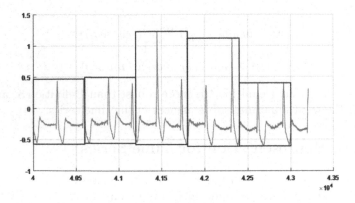

Fig. 4. Minimum boundary rectangle representation of time series

The following steps are given to extract all subsequence and transform them into **TSMBRB**:

(1) For a given time series T of length n, all the sub sequences are extracted from the start to the end in a fixed size length m. $C_{sub} = T_{1,m}, T_{2,m}, \ldots, T_{n-m+1,m}$

(2) For each subsequence c divide it into J part. We call every part as a segment c_k

(3) According to the time series range, divide the time series range into N parts, then it becomes a grid, and each subsequence could be represent by a matrix. Each item in the matrix only has two situations, covered or not covered. For covered area mark 1, not covered mark 0.

Figure 5(up) shows the result of extract the subsequence and convert to TSMBRB representation, every subsequence convert to TSMBRB will become a matrix. $A = \begin{pmatrix} 0 & 0 & 1 & 0 & 0 \\ 0 & 0 & 1 & 0 & 0 \\ 0 & 0 & 1 & 0 & 1 \\ 0 & 0 & 1 & 0 & 1 \\ 1 & 1 & 1 & 1 & 1 \end{pmatrix}$, $B = \begin{pmatrix} 0 & 0 & 0 & 0 & 1 \\ 0 & 0 & 0 & 0 & 1 \\ 0 & 1 & 0 & 0 & 1 \\ 1 & 1 & 1 & 0 & 1 \\ 1 & 1 & 1 & 1 & 1 \end{pmatrix}$. Figure 5(down) shows that how to extract all subsequence and convert them into TSMBRB representation, set a slide window with length m, from the beginning to the end of the time series data, every subsequence extract from the slide window, we convert it into TSMBRM representation. So finally there is $n-m$ subsequence totally, for each subsequence, we are just concerned about the non-self match subsequence. The pseudo code of extract all subsequences and convert to TSMBRB representation is shown in Algorithm 1.

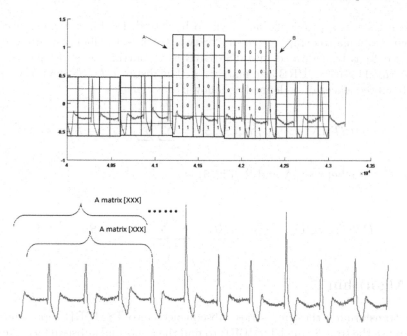

Fig. 5. An example of TSMBRB representation, subsequence A and B convert to TSMBRM

Algorithm 1. convertIntoTSMBRB(T, m, J, N)

---

input : time series T, subsequence length m, the length of segment of
subsequence J, the number of separate the time series range N
output: all subsequence represent by $TSMBRB$

1 C ← $(T_{1,m}, T_{2,m}, \ldots, T_{n-m+1,m})$
2 **foreach** $c \in C$ **do**
3   **foreach** $i \in range(1,N)$ **do**
4   **foreach** $k \in range(1,m/J)$ **do**
5   **if** c *has been covered* **then**
6   row.append(1)
7   **end**
8   **else** row.append(0);
9   **end**
10   tsm.append(row)
11   **end**
12   TSM.append(tsm)
13 **end**
14 **return** TSM

Through the above process, all the subsequences of time series are extracted and transformed into TSMBRB representation. How to measure the similarity between them is a problem that needs to be solved. In this paper we used the following formula to compute the similarity between subsequences, suppose subsequence S_1 is transformed into TSMBRB as a matrix S_A, subsequence S_2 is transformed into TSMBRB as a matrix S_B. The distance $DISTANCE_{TSMBRB}$ of two subsequences of is defined as follows:

$$DISTANCE_{TSMBRB}(S_1, S_2) = \sum_{i=1}^{N} \sum_{j=1}^{N} (S_{A_{i,j}} - S_{B_{i,j}})^2$$

so the distance between A and B (Fig. 5) is

$$DISTANCE_{TSMBRB}(A, B) = \sum_{i=1}^{5} \sum_{j=1}^{5} (S_{A_{i,j}} - S_{B_{i,j}})^2 = 9$$

4   Algorithm

After transforming the time series subsequences into TSMBRB representation, we can use the transformed TSMBRB to find the abnormal subsequence. According to definition, a subsequence will be considered as an abnormal subsequence if it has maximum average distance to its K-nearest non-self match. Obviously, the brute force algorithm for finding abnormal is simple and easily bethink. We simply take each possible subsequence and find the average distance to their K nearest non-self match. The subsequence that has the greatest such value is the abnormal. The pseudo code of brute force algorithm is shown in Algorithm 2.

Algorithm 2. The brute force algorithm for abnormal subsequence detection

---

input : all subsequences TSMBRB representation, subsequence length m.
output: The location of abnormal subsequence

1 MAX_DEGREE = INF
2 **foreach** $A \in TSM$ **do**
3 | MIN_DIS = 0
4 | **foreach** $B \in TSM$ **do**
5 | | **if** A,B is non-self match **and** $B \in KNN(A)$ **then**
6 | | | MIN_DIS += $DISTANCE_T SMBRB$(A,B)
7 | | **end**
8 | **end**
9 | **if** $MAX\_DEGREE < MIN\_DIS$ **then**
10 | | MAX_DEGREE = MIN_DIS
11 | | $position$ = A.location
12 | **end**
13 **end**
14 **return** $position$

Apparently the complexity brute force algorithm is not acceptable. If we carefully observe the algorithm, there are two point we can optimize in Algorithm 2

- Line 4, the inner loop, we do not need to calculate the true "K-nearest non-self match" distance, consider about that we select "K-random non-self match" subsequences of a particular subsequence(assume the subsequence m_1), if the distance between m_1 and its K-random non-self match subsequence is smaller than current "K-nearest non-self match" distance, then subsequence m_1 can't be abnormal subsequence, because its "K-nearest non-self match" distance only be smaller than "K-random non-self match".
- Line 2, the outer loop, we don't need to search every subsequence to find the abnormal one, if we change the order of search, make the most likely to be abnormal subsequence visited first, we can get a relatively large distance at the begining, so a lot of normal subsequence which has small "K-nearest non-self match" distance won't be considered in the next time, thus we early abandon a lot of normal subsequence in the inner loop.

The pseudo of our algorithm is shown in Algorithm 3. We call it KTSAD algorithm (KNN Time Series Anomaly Detection).

Algorithm 3. KNN Time Series Anomaly Detection algorithm **KTSAD**(TSB, k)

---

    input : all subsequences TSMBRB representation , the K value of
              non-self match KNN distance
    output: the location of abnormal subsequence.

1   cls $\leftarrow$ clustering(TSB)
2   $Candidate \leftarrow$ orderBy(cls)
3   **foreach** $A \in Candidate$ **do**
4   KRN = K random non-self match of A
5   KRN_DIS = $\sum_{i=1}^{k} DISTANCE_{TSMBRB}(A, KRN_i)$
6   **if** $MAX\_DEGREE < DIS$ **then**
7   KNN = K nearest non-self match of A
8   KNN_DIS = $\sum_{i=1}^{k} DISTANCE_{TSMBRB}(A, KNN_i)$
9   **if** $MAX\_DEGREE < KNN\_DIS$ **then**
10   MAX_DEGREE = KNN_DIS
11   position = i
12   **end**
13   **end**
14 **end**
15 return position

For each subsequences, we used the "K-random non-self match" distance instead of "K-nearest non-self match" distance. Thus we just need to calculate K times euclidean distance, not calculate all non-self match euclidean distance and find the K small-est distance. Besides, we used the clustering method, if two non-self match subsequences are highly similar then they are more likely in the same cluster after clustering. Obviously the subsequence in a cluster which has fewer subsequences is more likely to be abnormal than the other subsequences in a cluster which has more subsequences. So our algorithm started at the most likely to be abnormal subsequences, we will get a relatively large distance at the beginning, and early stop inner loop for plenty normal subsequences.

5 Experiment

We begin by showing some utility examples of time series abnormal subsequence discover, and prove that our algorithm speeds up the process.

5.1 Abnormal Detection in Space Shuttle Data

The sensors data is also extremely common, detecting the abnormal subsequence of sensors data is quite important. In the following experiment, we use our KTSAD algorithm to detect the subsequence of space shuttle data. In Fig. 6, we see an a Space Shuttle Marotta Valve time series that was annotated by a NASA engineer, it has two phase, the energize/de-energize and an abnormal which called "Poppet pulled significantly out of the solenoid before energizing". By using KTSAD algorithm we can easily find out the abnormal subsequence (the red bold one in Fig. 6 the location about 4300). According to the engineer experience we extract the subsequence with length 100. And the numbers of K-nearest neighbors is 3.

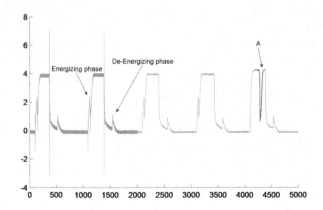

Fig. 6. Space Shuttle Marotta Valve time series, energizing/de-energizing and poppet pulled significantly out of the solenoid before energizing(A) (Color figure online)

5.2 Abnormal Detection in ECG Data

ECG data is the process of recording the electrical activity of the heart over a period of time using electrodes placed on the skin. These electrodes detect the tiny electrical changes on the skin that arise from the heart muscle's electrophysio-logic pattern of depolarizing during each heartbeat. It is a very commonly performed cardiology test [15]. Figure 8(left) is the interception of ECG data in a section of ECG data chf13_45590.

We use the KTSAD algorithm to detect abnormal subsequence for the time series. The length of the subsequence is 150, and the number of K-nearest neighbors is 3. By calling the algorithm and observing the results, it is obviously that the algorithm finds out the position of the abnormal subsequence, which is about in position of 2800. Figure 7(right) is the KTSAD algorithm to detect the abnormal subsequences and calculate the subsequence of the KNN abnormal degree. Figure 7(right) shows that the algorithm detects the abnormal subsequences located about 2800 of the position.

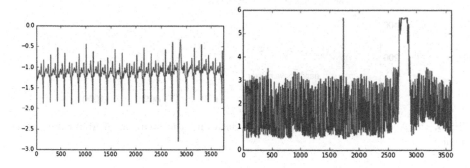

Fig. 7. ECG chf13_45590

5.3   Abnormal Detection in GUN Point Data

The 2D time series was extracted from a video of an actor performing various actions with and without a replica gun. The film strip above illustrates a typical sequence. The two time series measure the X and Y coordinates of the actor's right hand. The actor draws a replica gun from a hip mounted holster, aims it at a target, and returns it to the holster. Watching the video we discovered that at about ten seconds into the shoot, the actor misses the holster when returning the gun. An off-camera (inaudible) remark is made, the actor looks toward the video technician, and convulses with laughter. At one point (frame 450), she is literally bent double with laughter. See Fig. 8.

For using guns, we think it's dangerous when you are not aiming at your target or the gun is not in the holster (The red bold line in Fig. 8). So we need to find out the sequence by using our abnormal detection algorithm. In this situation we just consider about single coordinate(X or Y both works, we take X).

We also use the KTSAD algorithm to detect abnormal subsequence for the time series. The length of the subsequence is 150, and the number of K-nearest neighbors is 5. The result of KTSAD algorithm shown in Fig. 9.

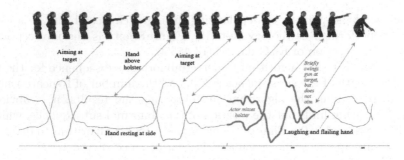

Fig. 8. The GUN point data (Color figure online)

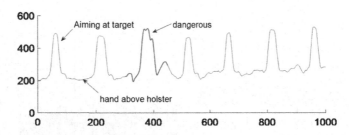

Fig. 9. the X coordinate of the gun point data and the location of abnormal subsequence

This time we not only just find out the abnormal subsequence location, but also compared with other algorithm. The "distance calculation" function is necessary for every abnormal detection algorithm. Because the distance-based method require to know the distance between subsequences. If an algorithm can find out the abnormal subsequence and access fewest "distance calculation" function, of course it's the good detection algorithm. So in the following experiment, we experiment with brute force, HOT SAX, BitClusterDiscord and our KTSAD algorithm. For each algorithm, we record the times to access the "distance calculation" function. Figure 10 denote the times of each algo-rithm call the "distance calculation" function.

According to Fig. 10, we can find that the BitClusterDiscord algorithm has smallest numbers of call the "distance calculation" function, but the BitCluster-Discord is based on PAA, in Sect. 3 we had discussed the disadvantage of PAA presentation. Although BitClusterDiscord perform better at this dataset, but our algorithm still has it's own advantage and it also perform better than brute force algorithm and HOT SAX algorithm.

5.4   A Special Situation

In Sects. 3 and 5.3 we had discussed the disadvantage of PAA presentation for time series, we just say that the PAA is based on the average value and it unseemliness for some kind of time series, but we didn't give an example about which

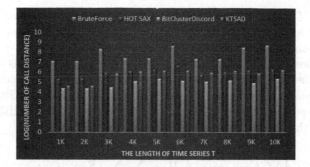

Fig. 10. the number of each algorithm calls distance function with different length time series

time series the PAA presentation can't work. So next time we will introduce a special time series which is not appropriate to use PAA/SAX representation to detect abnormal subsequence.

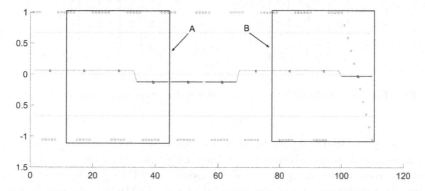

Fig. 11. Subsequence A and B are exactly the same by using the SAX representation, but obviously the subsequence B is abnormal one. So you will never find the most abnormal/discord subsequence as long as its based on the PAA/SAX representation

Consider about the situation of Fig. 11, it is a scatter map of square-wave from 1 to -1, at the end of the square-wave there are ten point between 1 and -1. By using SAX representation, we can see that the subsequence A and the subsequence B are exactly the same('CCB'), but obviously the subsequence B is the abnormal subsequence in the current time series. So if you use the SAX representation it can't find out the subsequence B is the most discord one, because subsequence B has the same "K nearest non-self" distance with subsequence A, as we discussed in part 3 the PAA/SAX is based on the average value, and sometimes two subsequences has same average value but their shape or the information they contained is totally different, so PAA/SAX representation is not appropriate for some kind of time series.

How's our TSMBRB representation performance in that special situation? Fig. 12 give us an answer. For same dataset, the TSMBRB representation accurately reflects the characteristic of the time series, for the square-wave part, the TSMBRM representation is same. And for the last ten point, It is totally different between the TSMBRB representation and PAA/SAX representation. And we select the same subsequence A and B as Fig. 11 does, PAA/SAX says their distance is 0(because their words is exactly same). but the TSMBRB representation denote their distance is

$$DISTANCE_{TSMBRB}(A, B) = \sum_{i=1}^{5} \sum_{j=1}^{5} (S_{A_{i,j}} - S_{B_{i,j}})^2 = 5$$

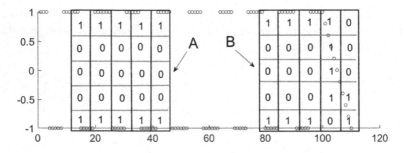

Fig. 12. The TSMBRM representation for the subsequences in the Fig. 11

By using TSMBRB representation, we can easily find out the most abnormal subsequence in this situation. We have explained that the algorithm can't detect the abnormal subsequence which based on the PAA/SAX representation. Through the experiment we can prove that our TSMBRB representation for time series is more effective, more accurate and more reliable for the detection of abnormal subsequence.

6   Conclusion

In this work, we give the definition of abnormal subsequence based on K-nearest neighbor. We also propose a new representation for time series data. Based on the new definition and new representation, we propose an efficiently abnormal subsequence detection algorithm KTSAD. Through the analysis of the definition of abnormal subsequence, we can figure out that our abnormal subsequence definition is more robust than **discord**. The experiment shows that our new representation method perform better than traditional method like PAA/SAX in some special time series, and our algorithm can always find out the abnormal subsequence with high-efficiency and high-accuracy.

There is also many future work we should focus on.

- The algorithm we proposed is for single time series data, but in reality, a lot of time series is multi-dimensional, how to analyze the relationship between multi-dimensional time series and abnormal detection is a very significance direction.
- The algorithm we proposed is only for static time series, but in reality, the online detection is more significant. So we need to do more research on this domain.

Acknowledgments. This work is supported by National High Technology Research and Development Program of China (No. 2015AA016008).

References

1. Hawkins, D.M.: Identification of Outliers. Chapman and Hall, London (1980). pp. 1–12
2. Chandola, V., Banerjee, A., Kumar, V.: Anomaly detection: a survey. ACM Comput. Surv. (CSUR) **41**(3), 75–79 (2009)
3. Bentley. J.L., Sedgewick. R.: Fast algorithms for sorting and searching strings. In: Proceedings of the 8th Annual ACM-SIAM Symposium on Discrete Algorithms, pp. 360–369 (1997)
4. Duchene, F., Garbayl, C., Rialle, V.: Mining Heterogeneous Multivariate Time-Series for Learning Meaningful Patterns: Application to Home Health Telecare. Laboratory TIMC-IMAG, Facult'e de m'edecine de Grenoble, France (2004)
5. Keogh, E., Lin, J., Fu, A.: Hot sax: efficiently finding the most unusual time series subsequence. In: Fifth IEEE International Conference on Data Mining. IEEE (2005)
6. Izakian, H., Pedrycz, W.: Anomaly detection in time series data using a fuzzy C-means clustering. In: 2013 Joint IFSA World Congress and NAFIPS Annual Meeting (IFSA/NAFIPS). IEEE (2013)
7. Chen, Z., Fu, A.-W.C., Tang, J.: On complementarity of cluster and outlier detection schemes. In: Kambayashi, Y., Mohania, M., Wöß, W. (eds.) DaWaK 2003. Lecture Notes in Computer Science, vol. 2737, pp. 234–243. Springer, Heidelberg (2003). doi:10.1007/978-3-540-45228-7_24
8. Li, G., Bräysy, O., Jiang, L., et al.: Finding time series discord based on bit representation clustering. Knowl.-Based Syst. **54**, 243 254 (2013)
9. Faloutsos, C., Ranganathan, M., Manolopoulos, Y.: Fast subsequence matching in time-series data-bases. In: Proceedings of the ACM SIGMOD International Conference on Management of Data, 24–27 May, Minneapolis, MN, pp. 419–429 (1994)
10. Chan, K., Fu, A.W.: Efficient time series matching by wavelets. In: Proceedings of the 15th IEEE International Conference on Data Engineering, 23–26 March, Sydney, Australia, pp. 126–133 (1999)
11. Keogh, E., Chakrabarti, K., Pazzani, M., Mehrotra, S.: Locally adaptive dimensionality reduction for indexing large time series databases. In: Proceedings of ACM SIGMOD Conference on Management of Data, 21–24 May, Santa Barbara, CA, pp. 151–162 (2001)
12. Ulanova, L., Begum, N., Keogh, E.: Scalable clustering of time series with U-shapelets. In: Proceedings of the 2015 SIAM International Conference on Data Mining (2015)

13. Lin, J., et al.: A symbolic representation of time series, with implications for streaming algorithms. In: Proceedings of the 8th ACM SIGMOD workshop on Research Issues in Data Mining and Knowledge Discovery. ACM (2003)
14. Liu, F.T., Ting, K.M., Zhou, Z.-H.: Isolation forest. In: 2008 Eighth IEEE International Conference on Data Mining, ICDM 2008. IEEE (2008)
15. Liu, F.T., Ting, K.M., Zhou, Z.-H.: Isolation-based anomaly detection. ACM Trans. Knowl. Discov. Data (TKDD) 6(1), 3 (2012)
16. Liu, F.T., Ting, K.M., Zhou, Z.-H.: On detecting clustered anomalies using SCi-Forest. In: Balcázar, J.L., Bonchi, F., Gionis, A., Sebag, M. (eds.) ECML PKDD 2010. LNCS, vol. 6322, pp. 274–290. Springer, Heidelberg (2010). doi:10.1007/978-3-642-15883-4_18

An Improved SVM Based Wind Turbine Multi-fault Detection Method

Shiyao Qin[1], Kaixuan Wang[2(✉)], Xiaojing Ma[1], Wenzhuo Wang[1], and Mei Li[2]

[1] Renewable Energy Research Center, China Electric Power Research Institute,
Beijing, China
[2] School of Information Engineering, University of Geosciences (Beijing),
Beijing 100083, China
wangkaixuan@outlook.com

Abstract. A fault detection method bases on wind turbines (WTs) supervisory control and data acquisition (SCADA) is proposed, principal component analysis (PCA) was used to reduce the dimension of target features to 1-D, so that PCA output 1-D data can be used as label of support vector machine (SVM). Thus on the premise of not losing the prediction correctness, one model can detect the fault of 2 to 4 features, largely reduce the complexity of model building. Different experiments are present to show the effectiveness of the proposed method.

Keywords: Wind turbine · Fault detection · Principal component analysis · Support vector machine · SCADA data

1 Introduction

The market forecasts that by 2018 the wind energy cumulative gigawatts (GW) will be 43% higher than of 2015's GW [1]. For WTs the maintenance costs are high because of their remote location, and this can amount to as much as 25 to 30% of the total energy production [2]. The most effective way to reduce maintenance costs is to continuously monitor generators status and predict the malfunction of WT. Then the system degradation problems can be found and responded in time. Maintenance can be carried out ahead of time before the system crash and excess maintenance is avoided. Therefore, fault prediction could maximize the normal production of wind power plant and greatly reduce maintenance costs.

The main methods of WT fault diagnosis include these types such as the fault mode analysis based on statistical data, the fault diagnosis based on time series prediction, model-based fault diagnosis of control system, the fault diagnosis based on vibration analysis, and other auxiliary diagnosis methods like acoustic emission technique and ultrasonic capacitance liquid level detection [3]. Schlechtingen has proposed an approach for WT supervisory control and data acquisition (SCADA) data for condition monitoring purposes [4]. They apply adaptive neuro-fuzzy interference system (ANFIS) models to a wide range of different SCADA signals.

© Springer Nature Singapore Pte Ltd. 2017
B. Zou et al. (Eds.): ICPCSEE 2017, Part I, CCIS 727, pp. 27–38, 2017.
DOI: 10.1007/978-981-10-6385-5_3

Since SCADA signals are recorded with a long interval, which was initially not for the purpose of condition monitering and fault detection (CMFD) [4], most dynamical features of WT faults that are useful for CMFD are lost. Therefore, the detailed information f most WT faults cannot be diagnosed by using SCADA signals via frequency or time frequency SCADA signals have been mainly used by model-based methods and prediction methods for WT CMFD and prognosis [4,5]. Schlechtingen's successive work give detailed examples of the application of their method [6]. Many other data mining methods are utilized in fault prediction, for instance, SVM-based solution by Santos et al. [7], GA optimization method by Odofin et al. [8], probabilistic neural network by Malik and Mishra [9], k-means and neural net by Liu et al. [10].

The proposed method in this paper aims to simplify the model building process. Using SCADA data one has to build many models to make fault detection, for example in [5] they have build $45 * 18 = 810$ models, which will cost a lot manual work. Using PCA to process training labels, we can Less build 2/3 models.

2   Principles and Methods

2.1   The Principal Component Analysis (PCA)

PCA is a dimension reduction method. [11] Take WT as example, suppose we have $x_j^i; (i = 1, 2, \cdots, n; j = 1, 2, \cdots, m)$ data set containing m features of WT. The gearbox oil temperature $x_a, (j = a)$ and gearbox bearing temperature $x_b, (j = b)$ may have the following relation shown in Fig. 1.

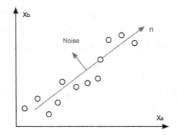

Fig. 1. Possible relation of gearbox oil temperature x_a and gearbox bearing temperature x_b

The two features x_a and x_b are strongly correlated, posit that they are along some diagonal axis u, with a little noise. Thus the data lies approximately on an $m1$ dimensional subspace. PCA can be used to detect, and perhaps remove, this redundancy. Before running PCA, we need to pre-process the data to normalize its mean and variance:

1. $\mu = \frac{1}{m} \sum_{i=1}^{m} x^i$

2. Replace each x^i with $x^i - \mu$
3. Let $\sigma^2 = \frac{1}{m} \sum_{i=1}^{m} (x_j^i)^2$
4. Replace each x_j^i with x_j^i / σ_j

First two step is to zero out the mean of the data and rest rescale each coordinate to have unit variance. Consider the following dataset (Fig. 2(a)), on which we have already carried out the normalization steps:

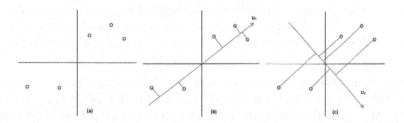

Fig. 2. (a) Example data set (b) Wanted direction u_1 with large projection, (c) u_2 direction with small projection for contrast.

The u direction will be find when the projection of data x onto vector u have the max variance. Since the length of the projection of x onto u is given by $x^T u$, we can maximizing the following subjected to $u^T u = 1$

$$\frac{1}{m} \sum_{i=1}^{m} ((x^i)^T u)^2 = \frac{1}{m} \sum_{i=1}^{m} u^T x^i (x^i)^T u = u^T \left(\frac{1}{m} \sum_{i=1}^{m} x^i (x^i)^T \right) u \qquad (1)$$

Let $\Sigma = \sum_{i=1}^{m} x^i (x^i)^T$, we use Lagrange multiplier method to realize the maximization.

1. $L(u, \lambda) = u^T \Sigma u - \lambda(u^T u - 1)$
2. Set $\nabla L(u, \lambda) = \Sigma u - \lambda u = 0$

This actually gives the principal eigenvector of Σ which is also empirical covariance matrix of the data. Thus, if we wish to project our data into a k-dimensional subspace ($k < n$), we should choose $u_1, ..., u_k$ to be the top k eigenvectors of U. The u_is now form a new, orthogonal basis for the data. Then, to represent x^i in this basis, we need only compute the corresponding vector.

$$y^i = \begin{bmatrix} u_1^T x^i \\ u_2^T x^i \\ ... \\ u_k^T x^i \end{bmatrix} \in R^k \qquad (2)$$

Thus, whereas $x^i R^n$, the vector y^i now gives a lower, k-dimensional, representation for x^i. The vectors $u_1, \cdots, u_k$ are called the first k principal components of the data. The typical goal of a PCA is to reduce the dimensionality of the original feature space by projecting it onto a smaller subspace, where the eigenvectors will form the axes. The eigenvectors with the lowest eigenvalues bear the least information about the distribution of the data.

2.2   Necessary Background of SVM

In the field of machine learning, SVM is a supervised learning model, which is usually used for pattern recognition, classification and regression analysis. In our work, we utilize SVM to train a normal behavior model and then give the prediction. A simple introduction to understand the principle of SVM here.

Logistic Regression. Popular speaking, SVM is a two class classification model, the basic model is a maximum linear classifier on feature space, the learning strategy is the maximum distance, and can be transformed into solving a convex quadratic problem [12]. To understand SVM, we must first understand a concept, linear classifier. Given a number of data points X, they belong to two different classes Y (Y can take 1 or -1, representing the different classes), and now to find a linear classifier to divide the data into two categories. The goal of a linear classifier is to find a hyperplane in the n-dimensional data space. The equation of this hyperplane can be expressed as the following:

$$w^T x + b = 0 \tag{3}$$

The purpose of logistic regression is to learn a 0/1 classification model from the features, the model is a linear combination of features as independent variables. Because the range of the independent variable is negative infinity. Therefore, the logistic function (or sigmoid function) is used to map the independent variables to (0,1), and the value of the mapping is considered to be the probability of belonging to y = 1. The following function g is the logistic function, where x is an n-dimensional feature vector (Fig. 3):

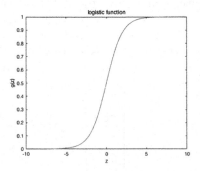

Fig. 3. Plot of logistic regression

$$h_\theta(x) = g(\theta^T x) = \frac{1}{1 + e^{-\theta^T x}} \qquad (4)$$

And its figure is

As you can see, the function maps infinity to (0,1). The function is the probability when $y = 1$

$$\begin{cases} P(y = 1|x; \theta) = h_\theta(x) \\ P(y = 0|x; \theta) = 1 - h_\theta(x) \end{cases} \qquad (5)$$

Thus, when we want to identify a new feature that belongs to which class, only needs solve $h_\theta(x)$, if $h_\theta(x)$ greater than 0.5 is the 'y = 1' class, otherwise the 'y = 0' class. Next, try to make a change in logistic regression. First change the result label $y = 0$ and $y = 1$ to $y = -1$ and $y = 1$. And then set θ_0 in $\theta^T x = \theta_0 + \theta_1 x_1 + \theta_2 x_2 + \cdots + \theta_n x_n (x_0 = 1)$ to b, and lastly $\theta_0 + \theta_1 x_1 + \theta_2 x_2 + \cdots + \theta_n x_n (x_0 = 1)$.

That is to say, in addition to the Y from $y = 0$ to $y = -1$, the linear classification function have no difference with the formal representation of logistic regression. Furthermore, the G (z) in the hypothesis function can be simplified to the $y = -1$ and $y = 1$. The mapping relationships are as follows:

$$g(z) = \begin{cases} 1, z \geqslant 0 \\ -1, z < 0 \end{cases} \qquad (6)$$

The next question is, how to determine the hyperplane? Intuitively, this hyperplane should be the best fit for separating two types of data. The criterion for the 'best fit' is the largest interval between the straight line and the straight line. So, we have to look for the super plane with the largest distance.

Functional Margin and Geometrical Margin. When we have a certain hyperplane $w^T x + b$, $|w^T x + b|$ can represent the distance from the point x to the hyperplane. And through observation the sign of $w^T x + b$ whether or not is consistent with the sign of y to determine the correctness the classification. In this way, we derive the concept of functional margin $\hat{\gamma}$:

$$\hat{\gamma} = y(w^T x + b) = yf(x) \qquad (7)$$

And the minimal functional margin of all the sample point (x_i, y_i) of training data set T in hyperplane (w, b) is the functional margin of training data set T:

$$\hat{\gamma} = min\hat{\gamma}_i \quad (i = 1, \cdots, n) \qquad (8)$$

But there is a problem with the defined functional margin, that is, if proportional change the value w and b, the functional margin become twice of the Original, so only the functional margin is not enough. In fact, we can add some constraints to the normal vector w, which leads to the definition of the distance between the real definition of the point to the hyperplane - geometrical margin.

Suppose that for a point x, the corresponding point of the vertical projection to the hyperplane is x_0, w is a vector perpendicular to the hyperplane, the distance between the sample x and the hyperplane, as shown in the following figure (Fig. 4):

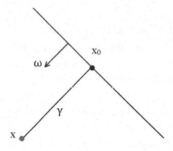

Fig. 4. Distance

According to the knowledge of plane geometry,

$$x = x_0 + \widehat{\gamma}\frac{w}{||w||} \tag{9}$$

And since x_0 is the point on the hyperplane, which satisfies $f(x_0) = 0$, substitute in the equation of hyper plane $w^T x + b = 0$, get $w^T x_0 + b = 0$ i.e. $w^T x_0 = -b$. Multiply both sides simultaneously $x = x_0 + \widehat{\gamma}\frac{w}{||w||}$ with w^T, according to $w^T x_0 = -b$ and $w^T w = ||w||$ we have:

$$\gamma = \frac{w^T x_0 + b}{||w||} = \frac{f(x)}{||w||} \tag{10}$$

It can be seen from the definition that the geometric margin is the functional margin divided by $||w||$ and the functional margin $y*(wx+b) = y*f(x)$ is in fact $|f(x)|$, it is only a measure of the margin, and the geometric margin $|f(x)|/||w||$ is the distance from the point to the hyperplane.

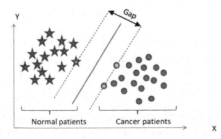

Fig. 5. Maximum margin classifier

Maximum Margin Classifier. The classification of a data point, when the greater the distance from the data points to hyperplane, the greater confidence. So, in order to make sure that the classification confidence is as high as possible, it is necessary to maximize the value of this margin by selecting the hyperplane, which is half of the Gap in the figure below (Fig. 5). By the previous analysis: Function margin $\widehat{\gamma}$ is not suitable to maximize the interval value, so the "gap" in the maximum interval hyperplane to be found here refers to the geometric margin $\widetilde{\gamma}$. The function of maximum margin classifier can be defined as: $max\widetilde{\gamma}$ Also, some conditions must be met:

$$y_i(w^T x_i + b) = \widehat{\gamma}_i \geqslant \widehat{\gamma} \quad i = 1, \cdots, n \tag{11}$$

As the following figure shows, the line in the middle is to find the optimal hyperplane, its distance to the two dotted lines are equal. The distance is the geometric margin $\widetilde{\gamma}$, the distance between the two dotted lines is equal to $2\widetilde{\gamma}$ The point on the boundary of the dotted line is the support vector. Because these support vectors are just on the dotted line boundary, so they satisfy $y(w^T + b) = 1$, for all points that are not support vectors, $y(w^T + b) > 1$. The actual SVM algorithm used is more complicated, we give only an introduction to SVM to help readers to understand the basic principle. In our work we use LibSVM toolbox to preform training and prediction (Fig. 6).

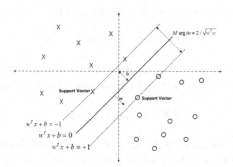

Fig. 6. Margin and support vectors

2.3   The Overall Process

Data Preprocessing Null Value: Generally can be divided into two types of null values: one is the value of the incomplete, and the other is the value is empty i.e., null values. The data in SCADA system are up from the WT number sensor, because the operating state of the machine often intermittent shutdown and other unstable state, data often appear empty values. Null data in the SCADA system can last for a period of time, for such data can be directly removed or set to 0. In the comparison experiment of the training model, the results of the direct removal of the null value points are better than the 0 processing. An other case is that the existence of a null value is occasionally appears, set these data to zero.

Outliner. Any of a large data set will have outliers, wind data set of the abnormal value is mainly manifested in some uncontrollable parameters such as wind speed, air temperature is higher than the abnormal range, the wind speed is less than 0, the temperature is less than 0, the wind speed is greater than the upper limit of history, air temperature is greater than the upper limit for the history of abnormal value, this kind of data is recorded should be deleted. In addition there are some performance parameters, such as speed, speed is less than 0 of the data records should be deleted. For other abnormal values, in the statistical analysis of the historical value of the fan, you can delete the data points deviate significantly from the normal value.

Normalization. Because of the different units, the degree of variation is different, in order to eliminate the influence of the dimension and the size of the variables, for different units or orders of magnitude can be compared and weighted, normalization must be applied. The normalization of the data is to scale the data so that it falls into a small specific interval. The most typical ways are 0-1 normalization and Z-score.

0-1 normalization applies linear transformation to the original data, so that the results fall to the [0,1] as the follows equation shows, where max is the maximum of the sample data, min is the minimum.

$$X^* = fracx - minmax - min \tag{12}$$

Z-score normalization process data to conform to the standard normal distribution, the mean value is 0 and the standard deviation is 1:

$$X^* = fracx - \mu\delta \tag{13}$$

where μ is the mean for all sample data, δ is standard deviation for all sample data. This experiment uses the Z-score normalization method.

Normal Behavior Model Training Data: According to fault event log, we choose SCADA data of one WT in good state for training. 50000 data (from 2015/12/9 to 2016/6/9) point of 6 features were selected.

PCA Process: Use PCA to reduce the dimension of target features to 1-D, the output 1-D data is set as the label of SVM. For example, if we want to predict gearbox bearing temperature, there are 3 relevant features: gearbox high speed bearing temperature, gearbox low speed bearing temperature and gearbox oil temperature. Notice that the PCA process will change data scale, so that the PCA output data should again applies z-score normalization.

Error: The SVM output is the predict value of the label feature. Suppose A to be actual value of the label, P to be the prediction value of the label data, thus define error:

$$e = A - P \tag{14}$$

The training data is when WT running in good state, so that the prediction value will have the similar form. When the related components of WT have fault, actual value as well as error will rise, thus error can be used to detect fault.

3   Experiments and Result

3.1   Experiment on Gearbox Bearing Temperature

Applying the model on the gearbox bearing temperature, three features - gearbox high speed bearing temperature, gearbox low speed bearing temperature and gearbox oil temperature were processed by PCA to make the label for training and prediction. The experiment result was shown in Fig. 7, we can see in Fig. 7(a) that the error goes up several times and parallel to that of Fig. 7(b), the actual value of generator bearing temperature.

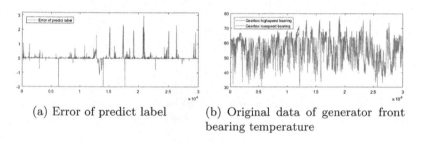

(a) Error of predict label   (b) Original data of generator front bearing temperature

Fig. 7. Error and original data

3.2   Experiment on Generator Bearing Temperature

Applying the model on the generator bearing temperature, two features - generator front bearing temperature and generator back bearing temperature were processed by PCA to make the label for training and prediction. The experiment result was shown in Fig. 8, in Fig. 8(a) the error raise and last for some time. In fact it is the cooling fan failure and the certain turbine had to working in power-limiting mode during that time.

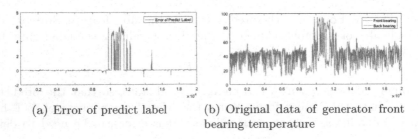

(a) Error of predict label

(b) Original data of generator front bearing temperature

Fig. 8. Error and original data

3.3 Set the Warning Line

According to steady state working Error of proposed PCA-NN method, we set warning lines as in Because we mainly concerns the case when temperature is too high, only two warning lines were set. The blue line and purple line stands for upperbound warning and upperbound alarming respectively, if the error keeps exceeding for a set time, the machine will halt. We can also find the error goes beyond the up alarm line for several times, this are some impulse which disappear quickly so that the alarm wont be triggered (Fig. 9).

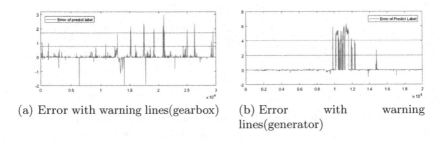

(a) Error with warning lines(gearbox)

(b) Error   with   warning lines(generator)

Fig. 9. Error and original data (Color figure online)

3.4 Discussion

The error of models build without PCA are shown in Fig. 10. Compare Fig. 10(a) with Fig. 7(a), Fig. 10(b) with Fig. 8(a), with PCA the fluctuation in error is large, but the high error section have large discrimination to the low error part the prediction result won't be affected. That is by applying PCA we can get the same result with 2 models as that with 5 models.

From Fig. 7(b), gearbox high speed bearing temperature and gearbox low speed bearing temperature has basically the same waveform, and another feature, gearbox oil temperature also takes the same trend. But in that of Fig. 8(b) we can find the generator front bearing temperature and generator back bearing temperature are not so parallel(especially data point 10000 to 12000). In this case PCA can still keeps the temperature raising trend as Fig. 8(a) show. That the information lost in PCA process won't affect the prediction result.

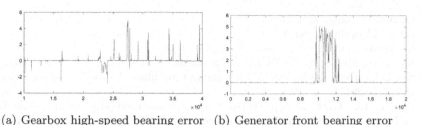

(a) Gearbox high-speed bearing error   (b) Generator front bearing error

Fig. 10. Error without PCA process

4 Conclusions

SCADA signals have been used by model-based methods and prediction methods for WT CMFD and prognosisUsing. After data preprocessing PCA is applied to process several features to an 1-D label, we can Less build 2/3 models that is saving 2/3 of model building time without losing the prediction correctness. SVM as a good non-linear classifier is used for setting up normal behavior models. The error together with the warning liner can give alarms when the abnormal state last for a certain time.

Acknowledgments. This paper is supported by Renewable Energy Research Center of China Electric Power Research Institute of STATE GRID's science and technology project: *Research on Key Technologies of condition monitoring and intelligent early detection of wind turbine based on big data.*

References

1. de Azevedo, H.D.M., Arajo, A.M., Bouchonneau, N.: A review of wind turbine bearing condition monitoring: state of the art and challenges. Renew. Sustain. Energy Rev. **56**, 368–379 (2016)
2. Aziz, M.A., Noura, H., Fardoun, A.: General review of fault diagnostic in wind turbines, pp. 1302–1307 (2010)
3. Amirat, Y., Benbouzid, M.E.H., Al-Ahmar, E., Bensaker, B., Turri, S.: A brief status on condition monitoring and fault diagnosis in wind energy conversion systems. RSER **13**, 2629–2636 (2009)
4. Schlechtingen, M., Santos, I.F., Achiche, S.: Wind turbine condition monitoring based on SCADA data using normal behavior models. Part 1: system description. Appl. Soft Comput. **14**, 447–460 (2013)
5. Yang, W., Court, R., Jiang, J.: Wind turbine condition monitoring by the approach of SCADA data analysis. Renew. Energy **53**(9), 365–376 (2013)
6. Schlechtingen, M., Santos, I.F.: Wind turbine condition monitoring based on SCADA data using normal behavior models. Part 2: Application examples. Appl. Soft Comput. **14**, 447–460 (2014). Elsevier Science Publishers B.V
7. Santos, P., Villa, L.F., Reones, A., Bustillo, A., Maudes, J.: An SVM-based solution for fault detection in wind turbines. Sensors **15**, 5627–5648 (2015)

8. Odofin, S., Gao, Z., Sun, K.: Robust fault estimation in wind turbine systems using GA optimisation. In: 2015 IEEE 13th International Conference on Industrial Informatics (INDIN), pp. 580–585. IEEE (2015)
9. Malik, H., Mishra, S.: Application of probabilistic neural network in fault diagnosis of wind turbine using FAST, TurbSim and Simulink. Procedia Comput. Sci. **58**, 186–193 (2015)
10. Liu, X., Li, M., et al.: A predictive fault diagnose method of wind turbine based on k-means clustering and neural networks. JIT **17** (2016). doi:10.6138/JIT.2016. 17.7.20151027i
11. Jolliffe, I.: Principal Component Analysis. Wiley Online Library (2002)
12. Statnikov, A., et al.: A Gentle Introduction to Support Vector Machines in Biomedicine. World Scientific, Singapore (2014)

GPU Based Hash Segmentation Index for Fast T-overlap Query

Lianyin Jia[1,2], Yongbin Zhang[1], Mengjuan Li[3(✉)], Jiaman Ding[1], and Jinguo You[1]

[1] Faculty of Information Engineering and Automation,
Kunming University of Science and Technology, Kunming 650500, China
[2] Yunnan Provincial Key Lab of Computer Technologies Application,
Kunming University of Science and Technology, Kunming 650500, China
[3] Library, Yunnan Normal University, Kunming 650500, China
lmjlykm@163.com

Abstract. T-overlap query is the basis of set similarity query and has been applied in many important fields. Most existing approaches employ a pruning-and-verification framework, thus in low efficiency. Modern GPU has much higher parallelism as well as memory bandwidth than CPU and can be used to accelerate T-overlap query. In this paper, we use hash segmentation to divide inverted lists into segments, then design an efficient inverted index called GHSII on GPU using hash segmentation. Based on GHSII, a new segmentation parallel T-overlap algorithm, GSPS, is proposed. GSPS uses segment at a time to scan segments and uses shared memory to decrease the number of accesses to device memory. Furthermore, an optimized algorithm called GSPS-TLLO using a heuristic query order is proposed to solve the problem of load imbalance. Experiments are carried out on two real datasets and the results show that GSPS-TLLO outperforms the state-of-the-art GPU parallel T-overlap algorithms.

Keywords: T-overlap query · Hash segmentation · Segment at a time · Shared memory · GHSII · GSPS-TLLO

1 Introduction

Set similarity query is a research focus in Database [1–3], Data Mining [4, 5], Natural Language Processing [6], Data Cleaning [7, 8], Information Retrieval [9] and many other fields. To compute similarity of sets, a certain similarity predicate is needed. Although there are many similarity predicates exists, such as Jaccard, Cosine and Dice, etc., most of them can be represented equivalently by T-overlap [10], so T-overlap query can be viewed as the foundation of set similarity query.

There are many T-overlap query algorithms emerged recent years, such as ProbeCount and MergeOpt [11], ScanCount and MergeSkip [12]. Most of these algorithms usually employ a pruning-and-verification framework and thus reduce the query efficiency. To overcome this shortage, paper [13] proposes trie-based index and algorithms which can find exact results directly by converting T-overlap query to finding query nodes with query depth equaling to T.

© Springer Nature Singapore Pte Ltd. 2017
B. Zou et al. (Eds.): ICPCSEE 2017, Part I, CCIS 727, pp. 39–51, 2017.
DOI: 10.1007/978-981-10-6385-5_4

However, the algorithms mentioned above are all CPU-based and not efficient when there are a large number of queries arrived nearly at the same time. Using modern hardware, e.g. GPU, provides an alternative. GPU, which is used initially in image processing and games, now is widely used in general purposed computing. Compared with CPU, GPU has much higher parallelism and much higher memory bandwidth and can be used to accelerate T-overlap query. Paper [14] proposes a GPU based T-overlap query algorithm GS-Parallel-Group which use ScanCount as the underlying algorithm. GS-Parallel-Group divides queries into groups and introduces both inter-query parallelism and intra-query parallelism to accelerate T-overlap query, it outperforms many other CPU based algorithms. However, the efficiency of GS-Parallel-Group can still be improved for the following reason: it uses element at a time (EAAT) to scan inverted lists and increases the counts of RIDs in device memory (DM). This may cause a large number of uncoalesced accesses to DM, thus deteriorate the performance.

To solve this problem, in this paper, we use hash segmentation to divide inverted lists into segments and design a GPU based hash segmentation inverted index (GHSII). Based on GHSII, an efficient T-overlap query algorithm, GPU segmentation parallel ScanCount (GSPS), is proposed. GSPS uses segment at a time (SAAT) strategy to scan segments and uses shared memory(SM) to decrease the number of accesses to DM. Furthermore, an optimized algorithm called GSPS-TLLO using a heuristic query order is also proposed to solve the problem of load imbalance. GSPS-TLLO can significantly reduce the number of accesses to DM and has a good characteristic of load balance, so it has a much higher efficiency than GS-Parallel-Group and GSPS.

There are many other segmentation approaches [15–18] proposed on CPU or GPU platform. But most of these approaches mainly focus on list intersection, none of them are designed for T-overlap query and none of them are designed to fully exploit the power of SM.

The remainder of this paper is organized as follows. Section 2 discusses the preliminaries. Section 3 describes hash segmentation approach and GHSII structure. In Sect. 4, we report GSPS and GSPS-TLLO algorithm in detail. Section 5 gives the experimental results and analysis.

2 Preliminaries

2.1 Basic Definition

Given a finite universe U of a set of elements, a database D is a set of records R, where each record R comprises a set of elements of U. Each record has a unique record ID (RID) from 0, 1, 2, $\cdots$, $|D|$-1, where $|D|$ is the number of records in dataset D.

T-overlap Query: Given a database D and a query $Q = \{e_0, e_1, e_2, \cdots e_{|Q|-1}\}$ where e_i is an element of U, identifies all records R in D having at least T common elements with Q, that is $|R \cap Q| \geq T$.

In this paper, we use inverted index, the most widely used data structure in text search engines, as the underlying data structure. It is made up of two components: the directory and inverted lists. The directory comprises all distinct elements in D. For each element e in the directory, there is a pointer pointing to an inverted list (list in short), l_e, which is made up of the RIDs of records containing e. For ease of query, we keep l_e in ascending order. In the following description, we will partition inverted lists into segments, so we use l_e^j to denote the j-th segment of l_e.

2.2   SCANCOUNT Algorithm

In this paper, we use SCANCOUNT algorithm as the underlying algorithm, so we discuss it in brief. A simple description of SCANCOUNT is shown in Fig. 1.

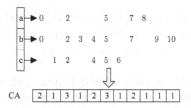

Fig. 1. SCANCOUT algorithm

SCANCOUNT is quite straightforward. It works on inverted index and execute in a scan-and-count manner. It scan relevant lists and count the occurrence of each record identifier(RID) encountered in these lists. An array named CA with size equaling to |D| is introduced to store the occurrences of each RID. For each RID encountered, we increase its count in corresponding position of CA by 1. After all lists counted, if the count of a certain position of CA is no less than T, then the corresponding RID is a qualified result.

2.3   CUDA Architecture

In this section, we give a brief overview of current GPU programming techniques used in our work.

The programming model we used is NVIDIA's Compute Unified Device Architecture (CUDA). For programmers, CUDA is a programming interface to the parallel architecture of NVIDIA GPU for general purpose computing. Thread is the basic execution unit in CUDA. Threads are organized into three levels: thread, block and gird. Multiple threads are organized to a block. Multiple blocks are organized to a grid. Usually each block can support up to 1024 threads. CUDA executes in SIMD mode, threads in a grid execute the same instruction but operate on different data. Massive threads can be executed in parallel to take full use of GPU's computing power.

There are two important storage devices in GPU: DM and SM. DM is large but slow. It usually takes hundreds of cycles to read or write a single data from DM [19]. Threads of different blocks can exchange information through DM. On the contrary, SM is small but much faster than DM. It takes only one cycle to access a single data. Threads in a block can exchange information through SM.

Note that when multiple threads in a block access consecutive memory addresses, these memory accesses can be grouped into a single access, this is so called coalesced access.

Many primitives have been proposed in CUDA, such as Parallel Scan and Map [20], Conditional scatter and Block Reduce [14], etc. These primitives use massive thread parallelism, inter-processor communication through SM and coalesced access, they are useful for our GPU based T-overlap query.

3   GHSII: GPU Hash Segmentation Inverted Index

3.1   Hash Segmentation Approach

As mentioned in Sect. 1, the count operation of GS-Parallel-Group may cause a large number of uncoalesced accesses to DM, thus cause high latency. Intuitively, SM is much faster than DM, if we can move the count operation to SM, then much higher speed can be expected.

Moving count step to SM is not a trivial job since SM is relatively small and CA can't be fully loaded in SM in most cases. So to fully utilize SM to execute T-overlap query, we partition lists into segments so that each segment can be counted directly in SM.

For ease of description, we assume each segment in a list has a unique segment identifier (SID) starting from 0. We use SEGs to denote the segments with the same SID, and j-SEGs[1] to denote the segments with SID equaling to j.

A naive segmentation strategy is to partition each list into fixed-width segments as shown in Fig. 2. But in this way, we may not use SM to accelerate T-overlap query, since the range from the minimum RID to the maximum RID in j-SEGs may be larger than M, the maximum number of RIDs that can be stored in SM.

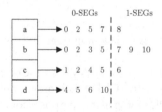

Fig. 2. Fixed-width segmentation

To overcome the shortage of fixed segmentation, in this paper, a new hash segmentation approach is proposed.

Hash Segmentation Approach: Given a hash function h, partition each inverted list into buckets, where $x \in l_e$ is put into bucket b, where $b = h(x)$.

To execute count operation in SM, we choose a simple hash function $h(x) = \lfloor x/P \rfloor$, here hash parameter P is a fixed integer satisfying P not greater than M. The hash function divide each inverted list into $\lceil |D|/P \rceil$ segments, RIDs in the j-th segment are restricted in

---

[1] When there is no ambiguity, we also use j-SEGs to denote all the j-th segments of query Q related lists.

the range from j * P to (j + 1) * P−1. Hash segmentation ensure that each segment can be counted directly in SM. An example of Hash segmentation is illustrated in Fig. 3.

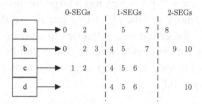

Fig. 3. Hash segmentation

3.2 GHSII Architecture

GPU is apt at array processing, so we create GHSII using 4 arrays. We first organize inverted index into 2 arrays: EA and DA. EA stores all distinct elements in the whole database and DA stores inverted lists sequentially for each distinct element.

Since we use hash segmentation to divide each list into segments, so to retrieve segments efficiently, another 2 arrays: segment length (SL) and segment offset (SO) are also been introduced. SL stores segment length for all elements sequentially and SO stores segment offsets in DA.

The GHSII created for inverted index of Fig. 2 is shown in Fig. 4.

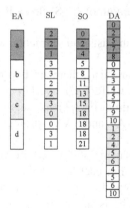

Fig. 4. GHSII structure

4 GSPS Algorithm

4.1 GSPS Algorithm

In this section, we discuss our GPU based T-overlap query algorithm in detail. We first discuss our parallel mode, then design a ScanCount kernel to execute T-overlap query, finally, we discuss approaches to gather the final results.

Parallel Model: For CUDA, threads are grouped in blocks and synchronization between threads of different blocks is expensive, so an intuitive idea is to assign each query to a unique block. In this way, to process n queries, we need allocate a count array in DM (DCA) with size n * |D| which has a large memory overhead when n is too large. To decrease the memory overhead, in this paper, we divide n queries into n/g groups, where g represents the group size. g queries in a group are processed in parallel by g blocks, while groups are processed serially. DCA can be used to store the results of a group and be reused for the next group after a group is processed, so the size of DCA can be decrease to g * |D|.

The ScanCount Kernel: After discussing the parallel model, we turn to process a query Q in a block.

Firstly, given n queries, to retrieve the corresponding lists and segments efficiently, a GPU based Cuckoo hashing (G-Cuckoo) is introduced. G-Cuckoo is implemented in cudpp 2.0, which is a parallel GPU version of Cuckoo hashing [21]. We create G-Cuckoo by elements of EA as hash keys and the locations of elements in EA as hash values. Using G-Cuckoo, we can get locations for all elements of n queries in parallel. For an element e in query Q, we can get its location loc in EA by G-Cuckoo, then locate l_e and l_e^j by the following formulas.

$$\text{The off set of } l_e \text{ in DA: SO}[\text{loc} * P] \tag{1}$$

$$\text{The off set of } l_e^j \text{ in DA: SO}[\text{loc} * P + j] \tag{2}$$

$$\text{The length of } l_e^j : \text{SL}[\text{loc} * P + j] \tag{3}$$

After getting the corresponding lists and segments, we start a SCANCOUNT kernel to execute scan-and-count task.

When scanning and processing lists, there are two popularly strategies used in IR, TAAT and DAAT [24]. Similar to IR, for T-overlap query, we consider two strategies to scan these lists: element at a time (EAAT) and segment at a time (SAAT). Different with TAAT and DAAT, the EAAT and SAAT strategies we proposed take advantage of massive thread parallelism.

EAAT can be seen as the byname of TAAT and works in lists perspective, that is lists are processed one by one. For each list, threads in a block scan and count it in parallel. GS-Parallel-Group algorithm is an example of using EAAT approach. The EAAT strategy for Q = {a,b} is shown in Fig. 5.

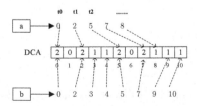

Fig. 5. EAAT strategy (For ease of description and understanding, we draw Figs. 5 and 6 based on inverted index, not GHSII).

EAAT strategy has one main shortage: despite scanning a list in parallel, the count operations to DM are uncoalesced and can't be in parallel in fact since adjacent RIDs are counted to nonadjacent positions in DCA as shown in Fig. 5. Assume the total of list lengths of query Q is qt, we should access DM qt times.

This shortage may cause a large number of uncoalesced accesses to DM which is orders of magnitude slower than SM and may reduce the performance. To overcome this shortage, we propose SAAT strategy based on hash segmentation and use SM to accelerate T-overlap query as shown in Fig. 6.

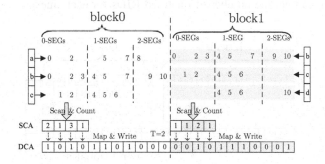

Fig. 6. SAAT strategy for queries {a,b,c} and {b,c,d}

SAAT first processes SEGs one by one which is different from DAAT and can be viewed as the combination of EAAT and DAAT. For j-SEGs, we process it using the following 2 steps:

Scan and Count: Threads in a block scan and count a segment of j-SEGs in parallel, then to the next segment. For each RID encountered, we don't count it in DCA directly, instead, we count it in SCA, a SM array. The size of SCA is equal to hash segmentation parameter P, and a RID r will be counted in position r%P of SCA.

Map and Write: For T-overlap query, given a query Q, what we need is not the counts themselves, but the qualified RIDs with overlap threshold greater than T. So we use a map primitive to mapping the counts in SCA to binary flags 0 or 1, then we write these flags back to DCA.

When processing j-SEGs for the i-th queries in current group, the k-th count in SCA will be map and write to DCA using Formula 4:

$$DCA[i * |D| + j * P + k] = \begin{cases} 1 & SCA[k] \geq T \\ 0 & otherwise \end{cases} \qquad (4)$$

Note that SCA can be used for (j + 1)-SEGs after j-SEGs processed. Also note that the write operations from SM to DM are coalesced accesses and execute much faster than uncoalesced ones. For j-SEGs we need to access DM P/TPB times where TPB represents the number of threads in per block. For query Q, the total number of DM accesses is ⌈|D|/P⌉ * P/TPB, so the accesses to DM can be greatly decreased.

Results Process in GPU: In order to get all RIDs satisfying threshold T, the following three primitives need to be executed as shown in Fig. 7:

(1) Parallel scan: By executing parallel scan primitive on DCA, we get a position array (PA) which points out the output locations for all 1 s in DCA.
(2) Conditional scatter: We execute conditional scatter primitive by DCA as input array and PA as location array. This primitive can map positions with value equaling to 1 of DCA to RIDs and scatter the these RIDs to the front of DCA.
(3) Block reduce: Since the results of all queries are put to the front of the DCA, we need to split the results for each query. By executing Block reduce primitive on DCA, we can get the number of qualified RIDs for each query.

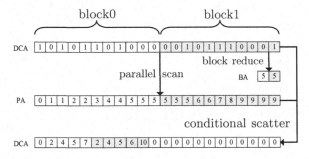

Fig. 7. Getting final results for queries {a,b,c} and {b,c,d}

Finally, we can transfer the exact final results back to main memory. The detailed description of GSPS is depicted in Algorithm 1

GSPS Algorithm

1. copy queries to DM
2. get locations in EA for elements of all queries **in parallel**
3. for each group
4.   Q← the query processed by current block
5.   for j=0 to $\lceil |D|/P \rceil$ -1
6.   initialize SCA to 0 **in parallel**
7.   use the locations in step 2 to retrieve the segments of j-SEGs of Q
8.   use SAAT to scan j-SEGs of Q and count RIDs to SCA **in parallel**
9.   map counts in SCA to binary flags and write these flags to DCA **in parallel**.
10.   execute the three primitives to get final results in DCA.
11.   copy results of current group back to MM.

Algorithm 1. GSPS Algorithm

4.2   Optimization

Queries are executed asynchronously on GPU which exposes a characteristic of load imbalance. In a group, some queries may be executed very quickly, while the others may be very slowly. The overall performance of a group is decreased since it is determined by the slowest query.

This observation motivates us using a new heuristic query order to reorder queries to get better performance. Since GSPS need to scan relevant lists, the overall performance of a query is tightly relevant with the total lists length of the query. So we first sort queries in total lists length order (TLLO), then we group queries using TLLO and call GSPS with TLLO GSPS-TLLO.

4.3   Complexity Analysis

Firstly, we analyze the storage overhead of GHSII which is made up of four arrays: EA, DA, SL and SO. Supposed that the total number of distinct elements in database D is td and the total length of all lists is tl, the size of EA, DA, SL and SO are td, tl, $td * \lceil |D|/P \rceil$, $td * \lceil |D|/P \rceil$ respectively. The total space overhead of GHSII is $td + 2 * td * \lceil |D|/P \rceil + tl$.

Then we analyze the storage overhead requirement for the algorithm. The major memory cost for GSPS-TILO is DCA and PA. The overall storage overhead of these two arrays is $2 * g * |D|$.

For each group, the average time complexity is avg/TPB where avg is the average number of RIDs in corresponding lists of a query. Since queries of a group executes in parallel, so for n queries, the average time complexity is $(n/g) * (avg/\text{TPB})$.

5   Experiments

5.1   General Setup

Environment: To evaluate the performance of our GPU based algorithms, we have performed extensive experiments on two real datasets to compare GSPS with GS-Parallel-Group. All experiments were carried on a Intel Core i7-4790 @3.60 GHz with 8 GB memory running windows7 64 bit, using Microsoft Visual Studio C++ 2010 as compiler. The GPU card we used is NVIDIA GeForce GTX 770 and its main parameters are shown in Table 1.

Table 1. Main parameters of GTX770

Parameters	Values
Cuda cores	256
Processor clock	1163 MHz
Multi processors	8
DM size	2048 MB
DM bandwidth	224.3 GB/s
SM size-per-block	48 KB

Datasets: The first real dataset is MSWEB [22] of UCI KDD Archive. MSWEB is a one-week log tracing of the virtual areas that users visited in the web portal www. microsoft.com. Each record corresponds to a user session and the set value comprises the areas visited. MSWEB is a small dense dataset and there are 32 K records in total and the vocabulary of the data set contains 294 distinct elements.

The second dataset is compiled from a snapshot of the DBLP data as described in [23]. Our snapshot consisted of almost 800,000 set on average. This dataset is a large sparse dataset and the total number of distinct elements is 530,000.

Queries: When generating queries, there is meaningless for a query Q returning an empty answer set, so we select 1000 records randomly with cardinality no less than T as queries for each threshold T = 2, 5, 10, 15, 20. We compute total running times for different algorithms. When performing queries for a certain threshold T, we use the corresponding queries generated for this threshold. In the experiments below, when not pointed out explicitly, we use queries for T = 2.

5.2   Compared with GS-Parallel-Group

To evaluate the efficiency of GSPS and GSPS-TILO, we compare them with the state-of-the-art algorithm, GS-Parallel-Group. Here we fix hash parameter P to 12288 (since GTX 770 has a 48 KB shared memory, if a integer occupies 4 Bytes, the maximum RID can be stored in SM is 12288) and TPB to 512 and then report the total elapsed time in Fig. 8a and b on MSWEB and DBLP respectively. We can see that both GSPS and GSPS-TILO run much faster than GS-Parallel-Group for all threshold T and GSPS-TILO is the fastest among the three algorithms for these two datasets. GSPS-TILO has an average performance gain about 30% than GS-Parallel-Group for MSWEB and DBLP respectively. For MSWEB, it reaches a maximum gain more than 40% when T is 10.

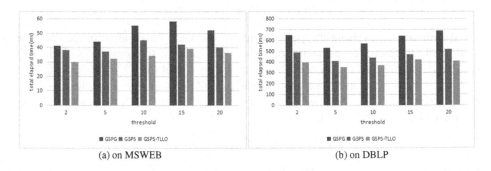

(a) on MSWEB   (b) on DBLP

Fig. 8. Total elapsed time

5.3   The Effect of P for GSPS

To examine the effect of the hash segmentation parameter P, we fix TPB to 512 and give the trends for P increasing from 4096 to 12288 on MSWEB and DBLP as shown in Fig. 9a and b. From these two figures, we can observe that the performance can be improved rapidly with the increase of P. The reason is that the larger segments can improve the parallelism as more RIDs can be processed in parallel. This observation indicates the effectiveness of GSPS and also indicates that the larger SM is, the higher performance gain.

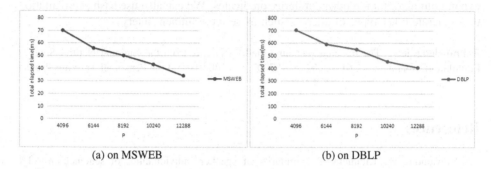

(a) on MSWEB   (b) on DBLP

Fig. 9. The effect of P

5.4   The Effect of TPB for GSPS

To examine the effect of TPB for GSPS, we selects queries of T = 2 in the MSWEB and DBLP and set P as 12288, then calculates the running time of the algorithm under different TPB. As shown in Fig. 10a and b, we can find that when TPB < 512 the running speed of GSPS algorithm is improved significantly, but when TPB > 512, the increase of TPB has no obvious effect.

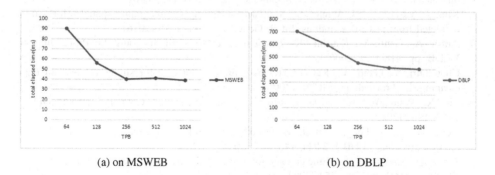

(a) on MSWEB   (b) on DBLP

Fig. 10. The effect of TPB

An interesting phenomenon we observed is that the curves become much flatter when TPB is greater than 256. The reason is that the RIDs' distribution in each segment is relatively sparse in MSWEB and DBLP, so when the TPB exceeds a certain threshold, the increase of TPB can't bring the increase of parallelism.

6 Conclusion

In this paper, to accelerate traditional T-overlap query using SM, a new GPU based hash segmentation index, GHSII, is designed. Based on GHSII, GSPS and an optimized version, GSPS-TLLO are proposed. These algorithms exploit shared memory and have a much higher efficiency than GS-Parallel-Group. In the next step, we try to expand our algorithm to other similarity predicates. We can also use hash segmentation to accelerate other types of queries, such as set containment query.

Acknowledgment. The research was supported by Grants from the National Natural Science Foundation of China (No. 61562054, 51467007, 61462050), and the Personnel Training Project of Yunnan Province (No. KKSY201603016).

References

1. Kawamoto, H., Kitamura, T.: Similarity of speaker individualities of sentence in ATR speech database set C. In: Proceedings of IEICE Technical Report Speech, pp. 33–34 (2013)
2. Hadjieleftheriou, M., Chandel, A., Koudas N., et al.: Fast indexes and algorithms for set similarity selection queries. In: Proceeding of IEEE 24th International Conference on Data Engineering (ICDE 2008), pp. 267–276 (2008)
3. He, B., Ke, F., Fang, R., et al.: Relational joins on graphics processors. In: Proceeding of ACM SIGMOD International Conference on Management of Data, pp. 511–524 (2007)
4. Shalom, S.A.A., Dash, M., Tue, M.: Efficient K-means clustering using accelerated graphics processors. In: Song, I.-Y., Eder, J., Nguyen, T.M. (eds.) DaWaK 2008. LNCS, vol. 5182, pp. 166–175. Springer, Heidelberg (2008). doi:10.1007/978-3-540-85836-2_16
5. Punronen, S., Terziyan, V.: A similarity evaluation technique for data mining with an ensemble of classifiers. In: Proceeding of IEEE Computer Society, International Workshop on Database & Expert Systems Applications, pp. 1155–1160 (2000)
6. Kim, J., Vasardani, M., Winter, S.: Similarity matching for integrating spatial information extracted from place descriptions. Int. J. Geogr. Inf. Sci. 1–25 (2016)
7. Arasu, A., Ganti, V., Kaushik, R.: Efficient exact set-similarity joins. In: Proceeding of VLDB, pp. 918–929 (2006)
8. Chaudhuri, S., Ganti, V., Kaushik, R.: A primitive operator for similarity joins in data cleaning. In: Proceeding of ICDE (2006)
9. Lin, X., Wang, W.: Set and string similarity queries a survey. Chin. J. Comput. **34**(10), 1853–1862 (2012)
10. Deng, D., Li, G., Feng, J., et al.: A unified framework for approximate dictionary-based entity extraction. VLDB J. **24**(1), 143–167 (2014)
11. Sarawagi, S., Kirpal, A.: Efficient set joins on similarity predicates. In: Proceeding of SIGMOD 2004, pp. 743–754 (2004)
12. Li, C., Lu, J., Lu, Y.: Efficient merging and filtering algorithms for approximate string searches. In: Proceeding of ICDE 2008, pp. 257–266 (2008)
13. Jia, L., Xi, J., Li, M., et al.: ETI: an efficient index for set similarity queries. Front. Comput. Sci. **6**(6), 700–712 (2012)
14. Li, M., Jia, L., You, J., et al.: Fast T-overlap query algorithms using graphics processor units and its applications in web data query. World Wide Web-internet Web Inf. Syst. **18**(2), 1–17 (2013)

15. Tatikonda, S., Junqueira, F., Cambazoglu, B.B., et al.: On efficient posting list intersection with multicore processors. In: Proceeding of ACM SIGIR 2009, pp. 738–739 (2009)
16. Ding, B., Nig, A.: Fast set intersection in memory. In: Proceeding of the VLDB Endowment 2011, pp. 255–266 (2011)
17. Ao, N., Zhang, F., Wu, D., et al.: Efficient parallel lists intersection and index compression algorithms using graphics processing units. In: Proceeding of the VLDB Endowment, pp. 470–481 (2011)
18. Ding, S., He, J., Yan, H.: Using graphics processors for high-performance IR query processing. In: Proceeding of WWW, pp. 1213–1214 (2008)
19. Programming of shared memory GPUs shared memory systems. http://site.uottawa.ca/ ~mbolic/ceg4131/CUDA_Report.pdf. Accessed Jan 2016
20. Wu, D., Zhang, F., Ao, N., et al.: Efficient lists intersection by CPU-GPU cooperative computing. In: Proceeding of 2010 IEEE International Symposium on Parallel & Distributed Processing, Workshops and Phd Forum (IPDPSW), pp. 1–8. IEEE (2010)
21. Pagh, R., Rodler, F.: Cuckoo hashing. J. Algorithms 51(2), 122–144 (2004)
22. Bay, S., Kibler, D., Pazzani, M., et al.: The UCI KDD archive of large data sets fordata mining research and experimentation. ACM SIGKDD Explor. Newsl. 2(2), 14–18 (2002)
23. Bayardo, R., Ma, Y., Srikant, R.: Scaling up all pairs similarity search. In: Proceeding of International Conference on World WideWeb 2007, pp. 71–81 (2007)
24. Broder, A.Z., Carmel, D., Herscovici, M., et al.: Efficient query evaluation using a two-level retrieval process. In: Proceeding of Twelfth International Conference on Information and Knowledge Management, pp. 426–434. ACM (2003)

A Collaborative Filtering Recommendation Algorithm Based on the Difference and the Correlation of Users' Ratings

Zhao-hui Cai$^{(\boxtimes)}$, Jing-song Wang, Yong-kai Li, and Shu-bo Liu

Computer School of Wuhan University, Wuhan, Hubei, China
zhcai@whu.edu.cn

Abstract. The traditional similarity algorithm in collaborative filtering mainly pay attention to the similarity or correlation of users' ratings, lacking the consideration of difference of users' ratings. In this paper, we divide the relationship of users' ratings into differential part and correlated part, proposing a similarity measurement based on the difference and the correlation of users' ratings which performs well with non-sparse dataset. In order to solve the problem that the algorithm is not accurate in spare dataset, we improve it by prefilling the vacancy of rating matrix. Experiment results show that this algorithm improves significantly the accuracy of the recommendation after prefilling the rating matrix.

Keywords: Collaborative filtering · Difference · Correlation · Prefill

1 Introduction

With the rapid development of information technology, the amount of information resources on the Internet is expanding exponentially. Faced with the huge amount of information resources, users need to spend a lot of time and energy to find what they need, eventually lost in the ocean of information. User personalized recommendation system can solve the contradiction between information overload [1] and user preferences in the internet. Now, personalized recommendation system has been widely applied in e-commerce [2], film and video sites, digital library [3], news websites [4], personalized music sites and other systems.

The technique of collaborative filtering is especially successful in generating personalized recommendations [5, 6]. The main idea is to recommend the views of other users who are similar to the target user. For example, users who are similar to target user have similar ratings on the same item in the rating system.

In the collaboration filtering recommendation system, there is no doubt that the key of the whole system is how to get the similar users of the target user. Therefore, the accuracy of similarity measurement between users is directly related to the quality of the whole recommendation system. The traditional similarity measurement methods include Cosine similarity, Adjusted Cosine similarity, Pearson correlation coefficient similarity, and the improved algorithm of the above algorithms [7]. In essence, the traditional similarity algorithm can only express the similarity or correlation, but not the difference of the interests between users, so it is difficult to fully reflect the relation

B. Zou et al. (Eds.): ICPCSEE 2017, Part I, CCIS 727, pp. 52–63, 2017.
DOI: 10.1007/978-981-10-6385-5_5

of interests between users, which will lead to the decline of the quality of the recommendation system. In the view of the above problem, a similarity measurement algorithm based on the difference and the correlation (DC) was proposed in this paper, which takes into account the correlation part as well as the difference part between users' ratings. However, the sparsity of the dataset is fatal to the accuracy of the DC algorithm, a prefilled DC algorithm (PDC) is proposed to improve the DC algorithm.

2 Related Work

There are three steps to generate recommendations for target user with traditional collaborative filtering algorithm [8, 9]:

(1) Data Representation

This step is mainly to collect the information which can reflect user interests, such as user ratings, user purchase information. Given a set of users $U = \{u_1, u_2, \cdots, u_m\}$ and a collection of items, the user interest model can be expressed as the user-item rating matrix R where the number m, n respectively denotes the number of users and items. As shown in formula (1), $R_{u,i}$ denotes that the user u is rating on the item i. When $R_{u,i} = 0$, it means that the user u doesn't rating on the item i.

$$R = \begin{bmatrix} R_{1,1} & R_{1,2} & \cdots & R_{1,n} \\ R_{2,1} & R_{2,2} & \cdots & R_{2,n} \\ \vdots & \vdots & & \vdots \\ R_{m,1} & R_{m,2} & \cdots & R_{m,n} \end{bmatrix} \tag{1}$$

(2) Find k Nearest Neighbors

We primarily should compute the similarity between target user and others based on the collected user interests model with a similarity measurement algorithm, and we could get the k nearest neighbors with the target user according to the similarity in descending order.

(3) Generate Recommended Items

We can predict a user's rating on an item with k nearest neighbors' ratings on the item based on the common sense that similar users are interested in the same project. If we define the collection of the user u's k nearest neighbors as N_u and the user u's prediction rating on the item i as $r_{u,i}$, we could get the user u's prediction rating according to his k nearest neighbors' rating by weighted average. As shown in formula (2):

$$r_{u,i} = \overline{R_u} + \frac{\sum\limits_{v \in N_u} sim(u, v) \times (R_{v,i} - \overline{R_v})}{\sum\limits_{v \in N_u} |sim(u, v)|} \tag{2}$$

In above, the $\overline{R_u}$, $\overline{R_v}$ respectively denotes the user u, v's average rating of all the items that the user has rated on. The $sim(u, v)$ represents the similarity between the user *u* and *v*. Finally, we could obtain the *N* items that has maximum *N* prediction rating as recommendation set.

We can get the key of the recommendation system is how to find the *k* nearest neighbors from the above three steps. The accuracy of the similarity measurement directly determines the quality of recommendation system.

2.1 Traditional Similarity Measurement Methods

Traditional similarity measurement methods have Cosine, Adjusted Cosine and Pearson correlation coefficient.

(1) Cosine

The rating of each user is viewed as an n-dimensional vector. And the relationship between the two users is treated as the cosine value which is the angle of two vectors. If the user does not rating on an item, the value will be 0. The rating of user "*u*" was seen as vector "**u**", and the rating of user "*v*" was seen as vector "**v**". Then the similarity measurement between user *u*, *v* can be expressed as the formula (3):

$$sim(u,v) = \cos(\mathbf{u},\mathbf{v}) = \frac{\mathbf{u} \bullet \mathbf{v}}{\|\mathbf{u}\| \times \|\mathbf{v}\|} = \frac{\sum_{c=1}^{n} R_{u,c}R_{v,c}}{\sqrt{\sum_{c=1}^{n} R_{u,c}^2}\sqrt{\sum_{c=1}^{n} R_{v,c}^2}} \qquad (3)$$

In the formula (3), the $R_{u,c}$, $R_{v,c}$ respectively denotes the rating of user *u*, *v*, which is on item *c*.

(2) Adjusted Cosine

The Cosine ignores different users have different scales of evaluation, so the Adjusted Cosine is used to modify the user's evaluation scale by subtracting the average rating of the user. If the I_{uv} denotes the collection of items that the user *u* and *v* both has rated on and the I_u, the I_v respectively denotes the collection of items that the user *u*, *v* has rated on, the similarity measurement between the user *u*, *v* can be expressed as the formula (4):

$$sim(u,v) = \frac{\sum_{c \in I_{uv}} (R_{u,c} - \overline{R_u})(R_{v,c} - \overline{R_v})}{\sqrt{\sum_{c \in I_u} (R_{u,c} - \overline{R_u})^2}\sqrt{\sum_{c \in I_v} (R_{v,c} - \overline{R_v})^2}} \qquad (4)$$

In the formula (4), the $\overline{R_u}$, $\overline{R_v}$ respectively denotes the average of the ratings on the items that the user *u* and *v* has rated on.

(3) Pearson correlation coefficient

The Pearson similarity measurement analyses the users' correlation in statistical methods. It only considers the items that two user both have rated on, which is suitable for two vectors that has fixed distance. If the I_{uv} denote the collection of items that the user *u* and *v* both has rated, the similarity measurement between the user *u*, *v* can be expressed as the formula (5):

$$sim(u,v) = \frac{\sum\limits_{c \in I_{uv}} (R_{u,c} - \overline{R_u})(R_{v,c} - \overline{R_v})}{\sqrt{\sum\limits_{c \in I_{uv}} (R_{u,c} - \overline{R_u})^2} \sqrt{\sum\limits_{c \in I_{uv}} (R_{v,c} - \overline{R_v})^2}} \tag{5}$$

2.2 The Analysis of Traditional Similarity Measurement Methods

The different algorithms have different drawbacks among traditional similarity algorithms. As shown in the following user-item rating matrix Table 1, the similarity between u_1 and u_3 is higher than that of u_1 and u_2 in Cosine. But in fact, both u_1 and u_2 like the items i_1 and i_2, while u_3 doesn't like i_1 and i_2. We can see that u_1 is more similar to u_2 than u_3. It's not consistent with the actual situation that the similarity between u_1 and u_4 is higher than that of u_1 and u_5, because u_1, u_4 and u_5 have the same ratings on the item i_2 and i_3 which can't compare who is more similar to u_1. The similarity computed in Pearson between u_3 and u_6 is -1, however they are very similar. What's more, the similarity between u_7 and others can't be computed in Adjusted Cosine and Pearson.

Table 1. User-item rating.

	i1	i2	i3	i4	i5
u1	4	5	4		
u2	4	4	1		
u3	1	2	1	2	1
u4		5	4	2	1
u5		5	4	5	4
u6	2	1	2	1	2
u7	4	4	4	4	4

Cosine evaluates the similarity between users by compute the cosine of the angle between two users' rating vectors, which only considers the aspect of users' interests, without considering the difference of users' interests.

$$\cos(\mathbf{u}, \mathbf{v}) = \frac{\mathbf{u} \bullet \mathbf{v}}{\|\mathbf{u}\| \times \|\mathbf{v}\|} = \mathbf{e_u} \bullet \mathbf{e_v} \tag{6}$$

As shown in the formula (3), the cosine of vector u and v can be viewed as the cosine of the vector $\mathbf{e}_u$ and $\mathbf{e}_v$ that are the unit vector of the vector u and v in the formula (6), which means that the difference of the vector u, v can't be reflected in Cosine. The Adjusted Cosine subtracts the average rating of the user compared with Cosine where the Pearson retains only the common rating items, discarding others. That's to say, the Adjusted Cosine and the Pearson has the same drawback as the same as the Cosine, which will influence the accuracy of similarity measurement.

For this reason, the researchers put forward different solutions. For example, in the paper [10], the author held that there is a negative correlation between user similarity and Euclidean distance. The paper [11] proposed a rating error filter with the Mean

Square Difference to exclude dissimilar users. In above two papers, they only take into account the difference but not the correlation between users' interests. Liu [12] combined information entropy with compressive distance weight based on the probability distribution of rating distance. Other researchers also try to improve the traditional similarity measurement which consider the difference in users' rating. Such as some researcher define a threshold to filter out the user whose rating is far from the rating of the target user.

3 The Difference and the Correlation Similarity

Through the analysis above, this paper proposed a similarity algorithm based on the difference and the correlation (DC) of users' rating. The algorithm has a good performance in non-spare dataset, however, it has a minor bad performance in spare dataset. In paper,we also proposed an improved DC algorithm with prefilling rating (PDC).

3.1 The Difference and the Correlation Similarity Measurement (DC)

Despite the correlation as a type of relationship between the users, the difference is also a type of relationship between users. For example, if the user u and v while they respectively has rating vector $(4, 4, 4)$ and $(1, 1, 1)$, the value of similarity of user u and v equals 1 in Adjusted Cosine and Pearson similarity algorithm, in fact they have different interests. The reason is that the two algorithms only take into account the correlation between users' ratings, neglecting the difference of uses' ratings. In this paper, we divide the relationship of users' ratings into different part and correlational part.

Definition 1: The difference of users' ratings. If the user u and v respectively rating on the item c with $R_{u,c}$ and $R_{v,c}$, we can express the difference of the item c as the $|R_{u,c} - R_{v,c}|$. What's more, the rating of each user is viewed as an n-dimensional vector, we can define the difference of users' ratings as the $(\mathbf{u} - \mathbf{v})^2$.

Definition 2: The correlation of users' ratings. If the user u and v respectively rating on the item c with $R_{u,c}$ and $R_{v,c}$, we can express the correlation of the item c as the $\min(R_{u,c}, R_{v,c})$. Similarly, we can define the correlation of users' ratings as the $\min(\mathbf{u}, \mathbf{v})^2$

The Definitions 1 and 2 describe quantitatively the difference and the correlation of users' ratings, the similarity measurement of the user u and v can be expressed as the formula (7):

$$sim(u, v) = \frac{\min(\mathbf{u}, \mathbf{v})^2}{(\mathbf{u} - \mathbf{v})^2 + \min(\mathbf{u}, \mathbf{v})^2} \tag{7}$$

In the formula (7), the vector $\boldsymbol{u}$, $\boldsymbol{v}$ respectively denote the user u, v's rating vector.

In practice, the matrix of users' ratings is very sparse. Each user has many items that have not been rated on, while the usual method is to set the value of rating to 0.

With the Definitions 1 and 2, we know that there is a calculation error in the similarity measurement. More concretely, the $(\mathbf{u} - \mathbf{v})^2$ will increase to $\max(\mathbf{u}, \mathbf{v})^2$ and the $\min(\mathbf{u}, \mathbf{v})^2$ will reduce to 0. In paper, we will give the formula (7) two specific solutions.

(1) The solution 1 (the Difference and the Correlation, DC)

To solve the above weakness, we can only take the common items that rated on into account. The solution will eliminate the influence of the error calculation which shown as formula (8):

$$sim(u, v) = \frac{\sum\limits_{c \in I_u \cap I_v} \min(R_{u,c}, R_{v,c})^2}{\sum\limits_{c \in I_u \cap I_v} (R_{u,c} - R_{v,c})^2 + \sum\limits_{c \in I_u \cap I_v} \min(R_{u,c}, R_{v,c})^2} \tag{8}$$

In the formula (8), the I_u, the I_v respectively denote the collection of items that the user u and v has rated on and the $R_{u,c}$, $R_{v,c}$ respectively denote that the user u, v's rating on item c.

(2) The solution 2 (the Prefilled Difference and Correlation, PDC)

The solution 1 discards the rating information of the item that the user u has rated on but the user v has not rated and otherwise, though it solve the matter of error calculation. In spare dataset, the count of both rated item will be less, which will influence the measurement of similarity.

To solve the sparse problem of dataset, the method of filling the rating matrix is proposed by many researchers. The solution to set the unrated-item to 0 is a kind of prefilling method, but this solution provides a bad value to rating-item matrix. It is obvious that giving a good prefilling method will improve the performance of the algorithm (7). It's easiest to fill the vacancy of matrix with the average rating of user. The paper [13] also proposed a trust propagation method to fill the vacancy. The paper [14] filled the blank with the theory of cloud model. In addition, a combination method of user preference on item properties and item popularity ratings is also proposed to solve the sparse problem in the paper [15].

Therefore, we can prefill the rating matrix to optimize the DC algorithm (PDC). If we define the $r_{u,i}$ as the prefilled rating that the user u rating on item i, the prefilled matrix R' can be expressed as formula (9):

$$R' = \begin{cases} R_{u,i} & \text{if user u rated item i} \\ r_{u,i} & \text{if user u not rated item i} \end{cases} \tag{9}$$

Now, we can measure the similarity of users with the prefilled rating matrix. What's more, we can take into account the all items that two users have rated on, as show in the following formula (10):

$$sim(u, v) = \frac{\sum_{c \in I_u \cup I_v} \min(R'_{u,c}, R'_{v,c})^2}{\sum_{c \in I_u \cup I_v} (R'_{u,c} - R'_{v,c})^2 + \sum_{c \in I_u \cup I_v} \min(R'_{u,c}, R'_{v,c})^2} \qquad (10)$$

3.2 Recommendation Algorithm

In paper, we proposed DC algorithm and PDC algorithm which have similar recommended steps. Here, we just describe the PDC algorithm which has a better performance.

Algorithm 1: PDC similarity recommendation algorithm

Input: user-item rating matrix R, prefilled user-item matrix R' ,need to be recommended user u, the number of nearest neighbors k, the number of items need to be recommended N

Output: the collection of items need to be recommended I

1 Begin

2   Set $I = \varnothing$;the collection of similarity S=$\varnothing$;the collection of prediction rating $RP = \varnothing$;the collection of users $U = \{u_1, u_2, \cdots, u_m\}$, the collection of items $I = \{i_1, i_2, \cdots, i_n\}$;the collection of items that the user u has rated on I_u ; the average rating of each user $\{\overline{R_1}, \overline{R_2}, \ldots \overline{R_m}\}$;

3   For each $v \in U - \{u\}$:

4

$$sim(u, v) = \frac{\sum_{c \in I_u \cup I_v} \min(R'_{u,c}, R'_{v,c})^2}{\sum_{c \in I_u \cup I_v} (R'_{u,c} - R'_{v,c})^2 + \sum_{c \in I_u \cup I_v} \min(R'_{u,c}, R'_{v,c})^2} \qquad ;$$

5   S=S$\cup$\{sim(u,v)\} ;

6   End For

7      $N_u = TOP(S, K)$

8   For each $i \in I - I_u$:

9           $r_{u,i} = \overline{R_u} + \dfrac{\sum_{v \in N_u} sim(u, v) \times (R_{v,i} - \overline{R_v})}{\sum_{v \in N_u} |sim(u, v)|}$;

10          $RP = RP \cup \{r_{u,i}\}$;

11   End For

12   For each $i \in I - I_u$:

13   IF $i \in TOP(RP, N)$:

14               $I = I \cup \{i\}$;

15   End IF

16   End For

17End

4 Experiment and Analysis

4.1 Dataset

In this paper, the experiment use the dataset provided by the MovieLens site founded in 1997. Its dataset is obtained from movies' recommendation website and widely used for the research of recommendation system. Now, the site has more than 71,000 users and over 10,000 movies which has been rated on. There are three different scale dataset and this paper choose the dataset that has 943 users, 1682 films and 100,000 ratings. The experiment will performed by dividing the dataset into training subset and test subset, 80% of the data is used as the training subset and the 20% data as the test subset.

4.2 Evaluation Metrics

In this paper, the experiment will use the Root Mean Square Error (RMSE) as evaluation metrics. Given the collection of prediction rating $\{p, p_2, \ldots, p_N\}$ and the collection of real rating $\{p_1, p_2, \ldots, p_N\}$, the RMSE is expressed as formula (11):

$$RMSE = \sqrt{\frac{\sum_{i=1}^{N} |p_i - q_i|^2}{N}} \tag{11}$$

4.3 Experimental Scheme

The experiment is conducted in following three parts. Firstly, we should evaluate the relationship between the accuracy of the DC algorithm and the sparsity of dataset; Secondly, we also should verify the fact that the accuracy of the DC algorithm is related to the differences of dataset. Finally, we will prove that the PDC algorithm has a better accuracy compared with the traditional similarity measurement.

(1) Experiment One

In this part of Experiment, we are able to obtain different sparse dataset by sorting the dataset according to the times of user rating and then respectively taking the first $\{100, 200, \ldots, 800, 900, 940\}$ users as different dataset. The Figs. 1 and 2 show the RMSE of Cosine, Pearson and DC algorithm with different number of neighbors and different sparse dataset.

The Figs. 1 and 2 show that the sparsity of dataset has a significant impact on the Pearson and DC algorithm which both only consider that common rating items. The In overall view, DC has a better RMSE value than Pearson. Compared with Cosine, when the sparsity of dataset is less than 0.92, the DC algorithm has better performance.

(2) Experiment Two

Compared with the traditional similarity algorithm, The DC algorithm takes into account the difference of dataset which mainly reflects on the rating on item. We can obtain a couple of difference dataset by sorting the dataset according to the standard

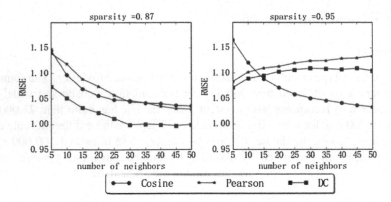

Fig. 1. Fixed sparsity with 0.87 and 0.95, compared to different recommendation algorithms

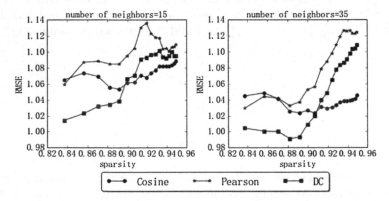

Fig. 2. Fixed number of neighbors with 15 and 35, compared with different recommendation algorithms

deviation of item rating and then respectively taking the first $\{100, 200, \ldots, 1500, 1600\}$ users as different dataset. The Figs. 3 and 4 show the RMSE of Cosine, Pearson and DC algorithm with different number of neighbors and different variance dataset.

In Figs. 3 and 4, they show that the accuracy of this three algorithm are related with the variance of dataset, however the DC algorithm performs better when standard deviation of dataset is bigger.

(3) Experiment Three

The experiment part one and part two analyze the DC Algorithm in sparsity and difference of dataset, which means we can improve the performance of recommendation system in this two aspects. In this paper, the PDC algorithm try to prefill the vacancy of rating matrix to solve the problem of sparsity of dataset. Specifically, we experiment with the PDC algorithm with two prefilling methods. The first method is to fill the vacancy with the average rating of each user, and second method is to fill the vacancy with combining the ratings and the properties of items as shown in Fig. 5.

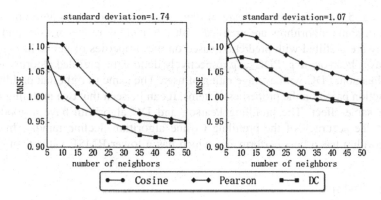

Fig. 3. Fixed standard deviation with 1.74 and 1.07, compared to different recommendation algorithms

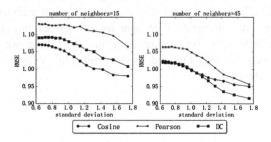

Fig. 4. Fixed number of neighbors with 15 and 45, compared with different recommendation algorithms

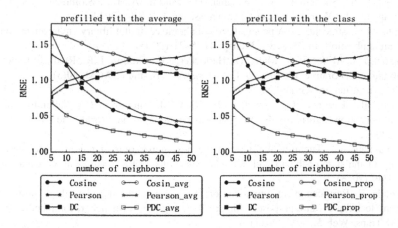

Fig. 5. Comparison of PDC algorithm with other recommendation algorithms prefilled with average rating of user on the left, and prefilled with prediction based on the properties of items on the right.

The Fig. 5 show that the compare of PDC algorithm and others algorithms, while on the left, some algorithms are prefilled with the average rating of user, and on the right, they are prefilled with prediction based on the properties of items. In Figure, the Cosine_avg, Pearson_avg, PDC_avg respectively denote the prefilled algorithms of the Cosine, Pearson, PDC with average rating of user. The same to right that prefilled with the prediction based on the properties of items. It can be seen that two prefilling method have the same effect. The prefilling Pearson and DC algorithm have a good effect, however, the accuracy of the prefilling Cosine algorithm decline rapidly. In all, the PDC algorithm has a best performance which has a lower RMSE value than others.

5 Conclusions

In this paper, we propose a similarity algorithm based on the difference and the correlation of users' ratings (DC). To solve the problem of sparsity of dataset in the DC algorithm, we improve it by prefilling the vacancy of rating matrix (PDC). The result show that the PDC algorithm has a better recommendation effectiveness than other traditional similarity measurements. The sparsity of dataset has an impact on DC algorithm, but has little effect on Cosine algorithm. In future, we can study a fuse similarity measurement jointing DC and Cosine algorithm which is able to adapt to changes in dataset density.

References

1. Eppler, M.J., Mengis, J.: The concept of information overload: a review of literature from organization science, accounting, marketing, MIS, and related disciplines. Inf. Soc. **38**, 325–344 (2004)
2. Sivapalan, S., Sadeghian, A., Rahnama, H., Madni, A.M.: Recommender systems in e-commerce. In: World Automation Congress, pp. 179–184 (2014)
3. Renda, M.E., Straccia, U.: A personalized collaborative digital library environment: a model and an application. Inf. Process. Manag. **41**, 5–21 (2005)
4. Konstan, J.A., Miller, B.N., Maltz, D., Herlocker, J.L., Gordon, L.R., Riedl, J.: GroupLens: applying collaborative filtering to Usenet news. Commun. ACM **40**(3), 77–87 (2000). Article (Konstan2000GroupLens)
5. Sarwar, B., Karypis, G., Konstan, J., Riedl, J.: Item-based collaborative filtering recommendation algorithms. In: International Conference on World Wide Web, pp. 285–295 (2001)
6. Adomavicius, G., Tuzhilin, A.: Toward the next generation of recommender systems: a survey of the state-of-the-art and possible extensions. IEEE Trans. Knowl. Data Eng. **17**, 734–749 (2013)
7. Cacheda, F., Formoso, V.: Comparison of collaborative filtering algorithms: limitations of current techniques and proposals for scalable, high-performance recommender systems. ACM Trans. Web **5**, 1–33 (2011)
8. Breese, J.S., Heckerman, D., Kadie, C.: Empirical analysis of predictive algorithms for collaborative filtering, pp. 43–52 (1998)
9. Su, X., Khoshgoftaar, T.M.: A survey of collaborative filtering techniques. Adv. Artif. Intell. **2009**, 4 (2009)

10. Singh, A., Yadav, A., Rana, A.: K-means with three different distance metrics. Int. J. Comput. Appl. **67**, 13–17 (2013)
11. Shardanand, U.: Social information filtering for music recommendation. Massachusetts Institute of Technology, pp. 74–81 (1994)
12. Liu, Y., Feng, J., Lu, J.: Collaborative filtering algorithm based on rating distance. In: International Conference on Ubiquitous Information Management and Communication, p. 66 (2017)
13. Jang, S., Yang, J., Kim, D.K.: Minimum MSE design for multiuser MIMO relay. IEEE Commun. Lett. **14**, 812–814 (2010)
14. Eldar, Y.C.: Universal weighted MSE improvement of the least-squares estimator. IEEE Trans. Sig. Process. **56**, 1788–1800 (2008)
15. Han, Y., Cao, H., Liu, L.: Collaborative filtering recommendation algorithm based on score matrix filling and user interest. Comput. Eng. (2016)

Research on Pattern Matching Method
of Multivariate Hydrological Time Series

Zhen Gai[(⊠)], Yuansheng Lou, Feng Ye, and Ling Li

College of Computer and Information, Hohai University, Nanjing, China
381498928@qq.com

Abstract. The existing pattern matching methods of multivariate time series can hardly measure the similarity of multivariate hydrological time series accurately and efficiently. Considering the characteristics of multivariate hydrological time series, the continuity and global features of variables, we proposed a pattern matching method, PP-DTW, which is based on dynamic time warping. In this method, the multivariate time series is firstly segmented, and the average of each segment is used as the feature. Then, PCA is operated on the feature sequence. Finally, the weighted DTW distance is used as the measure of similarity in sequences. Carrying out experiments on the hydrological data of Chu River, we conclude that the pattern matching method can effectively describe the overall characteristics of the multivariate time series, which has a good matching effect on the multivariate hydrological time series.

Keywords: Hydrology · Multivariate time series · Pattern matching · Dynamic time warping

1 Introduction

With the development of information technology, the types and quantities of hydrological data have increased dramatically, which gradually shows the characteristics of diversity, mass and polymorphism. The effective handling and utilization of hydrological data has become one of the key issues in the informatization of water conservancy [2]. Pattern matching is a similarity measuring method, the initial work of time series mining, as well as one of the core technologies of time series mining, which holds the basic position in time series analysis and processing [3].

At present, the pattern matching methods and techniques of single time series are becoming more and more mature, while the research of pattern matching in multivariate time series develops rather slowly due to the complicate relations between different dimensions. Such as rainfall, water level, evaporation and flow in the hydrological data, we cannot simply treat the multivariate hydrological time series as a superposition of the unitary time series because of the certain relations among them. Moreover, the general study of hydrological information is not only on the same site, but for the overall situation of a watershed. The multivariate time series contains more information than the single ones, and the mining of the data based on the multivariate can vividly reflect the characteristics. Therefore, the research on the pattern matching method of multivariate time series is of great significance and application prospect [1].

© Springer Nature Singapore Pte Ltd. 2017
B. Zou et al. (Eds.): ICPCSEE 2017, Part I, CCIS 727, pp. 64–72, 2017.
DOI: 10.1007/978-981-10-6385-5_6

2 Related Work

Nowadays, the research production of multivariate time series pattern matching are still less than that of single ones. The characterization of things is often described by multiple variables, among which there is usually certain relationship. Thus, we cannot consider the multivariate time series as a simple superposition of many single ones. Two important problems that need to be solved in pattern matching of multivariate time series are pattern representation and similarity measurement.

Pattern representation is the abstract and generalized representation of the time series. It is a re-description of the time series at a higher level, which is helpful to improve the efficiency of calculation and storage of the data [5]. There exist many methods of pattern representation. Principal Component Analysis (PCA), proposed by Krzanowski, takes the multivariate time series as a matrix, and extracts the first few main components to represent the pattern [6], which brings a high computation complexity due to the mining on original data. Guan Heshan extracts the local important points of the multivariate time series, and constructs the feature pattern vector according to the distribution feature of the important points [8]. It describes the original sequence and has a good effect on the small-scale multivariate time series. Korn uses the singular value decomposition method (SVD), regarding the recorded values at each time point as the sample points of random variables, using the correlation coefficient matrix as the base of feature extraction, and representing the pattern with singular value vector and the corresponding orthogonal matrix [7]. However, this method ignores the temporal relationship among data.

There exist many methods of similarity measurement as well. Eros is based on extended Frobenius norm. It first uses the PCA on the time series matrix to produce the principal components and the corresponding eigenvalues, and then calculates the similarity of multivariate time series [20]. SVD regards the variables in the time series as random ones, and constructs the pattern matching model according to the coordinate transformation principle in the linear space [7]. But its essence is based on statistics, can not reflect the temporal relationship of data, which means the possibility of misjudgment. PD extracts the local important points of the multivariate time series in the three-dimensional space as the feature, and measures the similarity based on the distribution characteristics of the important points [8]. However, it does not take into account the dimension and feature differences among different variables in multivariate time series.

In view of the problems above and the characteristics of multivariate hydrological time series, this paper first divides the time series into different segments, and uses the value of each segment as a feature to express the original data. Then, the PCA is performed on the feature sequence. Finally, we use the weighted DTW distance to measure the similarity between sequences, and prove the validity of the method by experiments.

3 Multivariate Time Series Pattern Representation

Definition 1 (Multivariate time series). A series of observations $x_i(j)$ is called multivariate time series, in which i $(i = 1, 2, \ldots, n)$ denotes the i th point in chronological order, j $(j = 1, 2, \ldots, m)$ denotes the j th variable, and $x_i(j)$ denotes the observed value of the j th variable at the i th point. A multivariate time series can be represented by a n $\times$ m matrix, in which n is the length of time, and m is the number of variables.

Definition 2 (Isomorphic multivariate time series). Isomorphic multivariate time series needs to satisfy the following conditions:

1. In multivariate time series data sets, the number and meaning of variables in each sample in the data set should be same, and the variables should correspond to each other.
2. For one sample, observations of different dimensions share same time.
3. To eliminate the impact of different units, we need to standardize the data. In this paper, we use Z-score standardized method as follow:

$$X_i = (X_i - \mu_i)/\sigma_i \tag{1}$$

μ_i and σ_i respectively denote the mean value and standard deviation of individual variables.

3.1 Piecewise Aggregation Approximation (PAA)

In order to reduce the time dimension of multivariate time series to improve the computational efficiency of data mining, it is necessary to represent the feature of multivariate time series. An important principle of feature representation is that features after the dimensionality should reflect the characteristics of the original time series as much as possible. Because of the correlation between the variables of multivariate time series, the ideal method is to segment all the relevant variables at the same time, so we use the PAA method to reduce the dimension of multivariate time series.

PAA is used to divide the time series of length n into w segments, in which each sequence has the same length k, and the mean value of each sequence is used to approximate the segment [12]. This method is a data dimension reduction process whose compression ratio is k = n/w. As shown in Figs. 1 and 2, in which Fig. 1 is the original data, Fig. 2 is the data after dimension reduction. This time the PAA process uses a compression rate of 75%, which keeps the overall trend of the original data and the correlation between variables.

3.2 Principle Component Analysis (PCA)

PCA is the most frequently used method of dimension reduction in multivariate time series. There is a certain degree of correlation between multiple variables, so it is

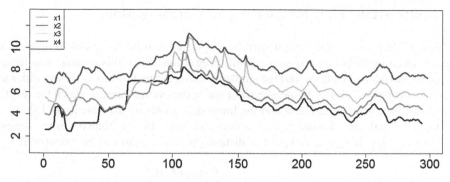

Fig. 1. The graph of original data

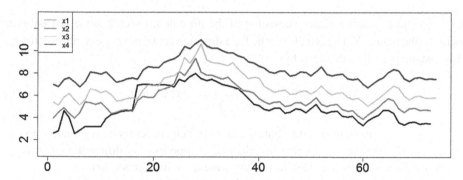

Fig. 2. The graph of the data after dimension reduction

possible to extract information from these indicators as much as possible through a certain linear combination. PCA can not only reduce the data dimension, but also retains most part of the original data. This method regards the original data as a whole and projects data from the higher-dimensional space to the lower-dimensional space in order to reduce dimensions, which fully reflects the correlation between the variables [14].

In this paper, we use PCA to deal with the data after feature representation. According to the PCA principle, we need first to calculate the covariance between the multivariate time series variables to obtain a covariance matrix $S_{m\times m}$, and then to carry out the eigenvalues and eigenvectors decomposition of $S_{m\times m}$ by singular value decomposition. We select the first k eigenvalues whose cumlative contribution has reached to a certain value (70%–90% is generally selected [15], here we select 90% as an indicator), and then combine the corresponding eigenvectors into a characteristic matrix $U_{m\times k} = [u_1, u_2, \ldots, u_k]$. Moreover, the contribution of the k eigenvectors will be used as the weight for the following distance measure. Finally, we project the feature sequence to the eigenvectors to get the principal component data, $Y_{n\times k} = X_{n\times m} \times U_{m\times k}$. Compared with the original multivariate time series $X_{n\times m}$, the dimension of feature sequence $Y_{n\times k}$ is reduced for $k < m$.

4 Multivariate Time Series Pattern Matching Model

DTW is widely used in the field of speech recognition to solve the problem of uneven speech speed in the process of single word recognition. Then, this method was introduced by Berndt and Clifford into the similarity measurement of time series in order to solve the problem of similarity measurement under the circumstance in which the length of the two time series is different and the linear drift exists on the time axis [13].

Let A and B denote two matrices of pattern-represented time series, $A[a_1, a_2, \ldots, a_n,]$, $B[b_1, b_2, \ldots, b_m]$. The distance between them can be calculated:

$$D(A, B) = D_{base}(a_1, b_1) + \min \begin{cases} D_{dtw}(A, B[2:-]); \\ D_{dtw}(A[2:-], B); \\ D_{dtw}(A[2:-], B[2:-]). \end{cases} \qquad (2)$$

$X[i:-]$ denotes a subsequence consisting of the ith column vector to the last column vector of the matrix X. $D_{base}(a_i, b_j)$ is the base distance between vector a_i and vector b_j. This distance can be calculated by:

$$D_{base}(a_i, b_j) = \frac{w_i + w_j}{\sum_1^n w_i + \sum_1^m w_i} DTW(a_i, b_j) \qquad (3)$$

w_i, w_j, the weight here, is the contribution rate of the eigenvectors corresponding to the vectors a_i, b_j. For the importance of principal components is different in matching multivariate time series, we need to use the weight to distinguish them.

The pattern matching method proposed in this paper is based on PAA and PCA, with the combination of DTW distance metric. We call it PP-DTW. The steps of the PP-DTW algorithm are as follows:

Input: original multivariate hydrological time series: X; the subsequence that require pattern matching: s; the length of segment: r; threshold: th.

Output: pattern matching result: R.

Step1: $L_{n \times m}$ = Z-score (X) // Preprocess the original multivariate time series and use Z-score to normalize it to obtain $L_{n \times m}$

Step2: $T_{u \times m}$ = PAA(r, $L_{n \times m}$) // $T_{u \times m}$ is the feature sequence, u = n/r

Step3: PCA($T_{u \times m}$) → $U_{m \times k}$, w; $P_{u \times k} = T_{u \times m} \times U_{m \times k}$ // $U_{m \times k}$ is a characteristic matrix combine by eigenvalues, w is the contribution rate, $P_{u \times k}$ is the result of projecting $T_{u \times m}$ to $U_{m \times k}$

Step4: S = segment ($P_{u \times k}$) // S is the segmented multivariate time series set

Step5: while (D(s, S[i]) < th), R.add(S[i]) // D is the distance between s and S[i] calculated by formulas (2) and (3)

Step6: return R

5 Experiment and Discussions

5.1 Experimental Environment and Datasets

Experiment software environment: R 3.31, Windows7 Ultimate operating system. Hardware environment: PC. Configuration parameter: P320 AMD Athlon (tm)IIP320 Dual-Core Processor, CPU 2.10 GHz, RAM 4 GB.

We choose Chuhe River as the research object to verify the method proposed above. The four sites of Chuhe River are: Sanchawan (station code: 62914600), Hongshanyao (station code: 62915000), Huazikou (station code: 62917000), Getang (station code: 62916400). We use the average water level of these four stations as the datasets and extract the data from 2010 to 2016, a total of 10228 records.

5.2 Analysis of Dimension Reduction

Firstly, we verify whether the proposed method can effectively reduce the dimension and express the characteristics of the original sequence. In this experiment, we randomly select the original time series, and carry out the pattern representation including the characteristic representation and principal component analysis on it. One variable of the original data and its corresponding pattern representation data are chosen to be drawn into a line chart, as shown in Figs. 3 and 4:

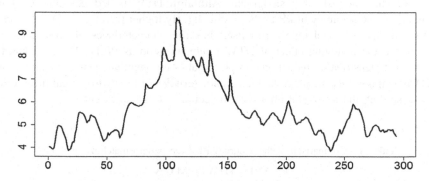

Fig. 3. The graph of original data

It can be seen that the method proposed in this paper reflects the trend and overall the characteristics of the original data accurately. What's more, the amount of data is greatly reduced. So this method is suitable for data processing of multivariate hydrological time series.

5.3 Accuracy Experiment

In this experiment, the matching effects of SVD, DTW and PP-DTW are compared. First, we classify the datasets after pattern representation, randomly select a sequence from the data set as an input, and use these three methods to match the k sequences

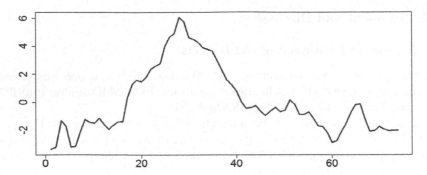

Fig. 4. The graph of the data after pattern representation

with the smallest distance to the input sequence, in this experiment, the value of k is 1, 5, 10. Suppose that there are N sequences in the k which is in the same class with the input sequence, we define the accuracy rate: $e = N/k$.

To reduce the error, we repeat the above steps. The average of the accuracy rate e^* is taken as an indicator of the pattern matching effect. The bigger the value of e^* is, the better the effect of pattern matching is. The results are as follows:

In Table 1, SVD based on the statistical method is the worst of the three methods, for that it cannot reflect the temporal relationship of the data when matching the time series, and has the risk of misjudgment. Although DTW is a direct mining of the original data, the accuracy of PP-DTW is still slightly higher than that of DTW. This is because the hydrological time series itself has the characteristics of noise and randomness, so the matching effect of DTW is not as good as PP-DTW which has the process of dimensionality reduction and principal component analysis. It is shown that PP-DTW can not only maintain the main characteristics of the hydrological time series, but also smooth the noise and improve the accuracy to some extent.

Table 1. Comparison of the accuracy of three pattern matching methods

k	SVD	DTW	PP-DTW
1	0.720	0.931	0.940
5	0.657	0.864	0.917
10	0.562	0.829	0.894

5.4 Time Analysis

We randomly select a period of time series, and use those three methods for pattern matching in different sizes of data sets which are hydrological data from one year to six years. The average value t^* of the pattern matching consumption time is used as an index, and Fig. 5 shows the comparison of time consumption of three different methods. It can be seen from the figure that the computational efficiency of SVD is the lowest in the three methods. Compared with DTW method the PP-DTW method has the process of dimensionality reduction, so the computational efficiency is better than the DTW method, which is the most efficient method in the three methods.

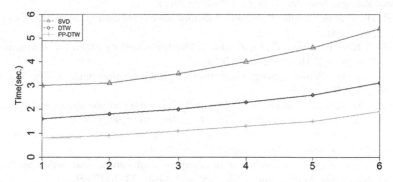

Fig. 5. The comparison of matching time of three methods

6 Conclusions

This paper proposes a pattern matching method PP-DTW for multivariate hydrological time series, the effectiveness of which has been verified by experiments. Compared with other methods, PP-DTW has the following advantages: (1) Multivariate time series are regarded as a whole and all the variable dimensions are segmented at the same time, which preserved the isomorphism and correlation of multivariate time series. (2) PP-DTW allows the match of multivariate time series with unequal length, and supports amplitude translation and stretching, time axis stretching and bending of multivariate time series. (3) Compared with DTW, PP-DTW reduces the dimension of multivariate time series in time and variable dimensions and improves the computational efficiency. (4) Compared with SVD and other statistical based methods, PP-DTW can preserve the temporal relationship of data, and reduce the risk of miscarriage of justice.

Using the mean value to characterize the original multivariate time series may smooth out some important points, such as extreme points, turning points, etc. Therefore, the following research should focus on how to better characterize the local features of multivariate time series.

Acknowledgement. This work is supported by (1) National Natural Science Foundation of China (61300122); (2) A Project Funded by the Priority Academic Program Development of Jiangsu Higher Education Institutions; (3) Water Science and Technology Project of Jiangsu Province (2013025).

References

1. Li, S., Zhu, Y., Zhang, X., Wan, D.: BORDA counting method based similarity analysis of multivariate hydrological time series. J. Hydraul. Eng. **40**(3), 378–384 (2009)

2. Mo, R., Ai, P., Wu, L., Yue, Z., Feng, P.: A fundamental frame of water resources data center supporting big data. Water Resour. Informatiz. **3**, 16–20 (2013)
3. Li, Z., Zhang, F., Li, K.: DTW based pattern matching method for multivariate time series. Pattern Recogn. Artif. Intell. **24**(3), 425–430 (2011)
4. Jia, P., He, H., Liu, L., Sun, T.: Overview of time series data mining. Appl. Res. Comput. **24** (11), 15–19 (2007)
5. Wu, H., Zhang, F., Zhang, C., Li, Z., Du, J.: Matching similar patterns for multivariate time series. J. Appl. Sci. **31**(6), 643–649 (2013)
6. Krzanowski, W.: Between-groups comparison of principal components. J. Am. Stat. Assoc. **74**(367), 703–707 (1979)
7. Korn, F., Jagadish, H., Faloutsos, C.: Efficiently supporting ad hoc queries in large datasets of time sequences. ACM SIGMOD Rec. **26**(2), 289–300 (1997)
8. Guan, H., Jiang, Q., Wang, S.: Pattern matching method based on point distribution for multivariate time series. J. Softw. **20**(1), 67–79 (2009)
9. Liu, Z., You, Z., Zou, Z.: Efficient matching of multivariate time series by wavelets. J. Southwest Chin. Normal Univ. (Nat. Sci. Ed.) **34**(4), 73–76 (2009)
10. Huang, H., Shi, Z., Zheng, Z.: Similarity search based on shape k-d tree for multidimensional time sequences. J. Softw. **17**(10), 2048–2056 (2006)
11. Yang, K., Shahabi, C.: A PCA-based similarity measure for multivariate time series. In: ACM International Workshop on Multimedia Databases, pp. 65–74. ACM (2004)
12. Cao, D., Liu, J.: Research on dynamic time warping multivariate time series similarity matching based on shape feature and inclination angle. J. Cloud Comput. **5**(1), 11 (2016)
13. Berndt, D., Clifford, J.: Using dynamic time warping to find patterns in time series. In: Working Notes of the Knowledge Discovery in Databases Workshop, pp. 359–370 (1994)
14. Liang, S., Zhang, Z., Cui, L., Zhong, Q.: Dimensionality reduction method based on PCA and KICA. Syst. Eng. Electron. **33**(9), 2144–2148 (2011)
15. Jolliffe, I.: Principal Component Analysis, pp. 55–58. Springer, New York (2002)

Further Analysis of Candlestick Patterns' Predictive Power

Tao Lv$^{(\boxtimes)}$ 🆔 and Yongtao Hao 🆔

College of Electronics and Information Engineering,
Tongji University, Shanghai 200092, China
superlvatao@163.com

Abstract. Since the candlestick patterns were mined, there is a contentious dispute on whether the candlestick patterns have predictive power in academia. To help resolve the debate, this paper uses the data mining methods of pattern recognition, pattern clustering and pattern knowledge mining to research the predictive power of candlestick patterns. In addition, we propose the similarity match model and nearest neighbor-clustering algorithm to solve the problem of similarity match and clustering of candlestick series, respectively. The experiment includes testing the predictive power of the Morning Star pattern and Evening Star pattern with the testing dataset of the candlestick series data of Shanghai 180 index component stocks over the latest 10 years. Experimental results show that (1) There have some spurious patterns in the existing candlestick patterns. However, after further classification of a spurious pattern based on its shape feature, those patterns with special shapes still have predictive power. (2) Some patterns do have the predictive power. (3) As there is no precise mathematical definition to describe the existing patterns' predictive power, it is essential to give the mathematical formula for improving the candlestick patterns' prediction performance.

Keywords: Candlestick chart · Candlestick series · Candlestick pattern · Similarity match · Cluster

1 Introduction

A time series is a series of observations listed in time order. The mathematical model of the time series is a matrix, as shown in formula (1).

$$\begin{bmatrix} x_1(t) & x_1(t+1) & \cdots & x_1(t+m) \\ x_2(t) & x_2(t+1) & \ldots & x_2(t+m) \\ \vdots & \vdots & \ddots & \vdots \\ x_i(t) & x_i(t+1) & \cdots & x_i(t+m) \end{bmatrix}, (i = 1, 2, \cdots, n; t, m \in N^+) \quad (1)$$

where $x_i(t)$ indicates the i-th observation at time t, n indicates the number of observations in the time series. If $n = 1$, it is called univariate time series (UTS); if $n \geq 2$, it is called multivariate time series (MTS). The MTS can easily be found anywhere in our daily life. For example, the time series of stock prices (stock time series for short)

© Springer Nature Singapore Pte Ltd. 2017
B. Zou et al. (Eds.): ICPCSEE 2017, Part I, CCIS 727, pp. 73–87, 2017.
DOI: 10.1007/978-981-10-6385-5_7

consisting of stock price observations, the time series of personal health consisting of the observation of blood pressure, temperature and white corpuscle, etc.

The time series have two import characteristics. One is the historical information will affect the future trend [1]. That is the historical values of observations will exert an influence on the future values in the time series. The other is history repeats itself [2]. That is to say, some special time sub-series will repeat in the entire time series. Because of the two features, various time series predictions have become a research hot spot, one of which is the prediction of stock time series, stock prediction for short.

Candlestick technical analysis is the best known technical analysis method about stock prediction. In the stock market, people create a candlestick chart (also called K-line in Asia) as the graphical representation of stock time series. Taking a daily candlestick chart for example (All candlestick charts shown in the paper refer to daily candlestick charts, unless otherwise stated), a candlestick chart indicates the fluctuation of stock prices during a day. It not only shows the close price, open price, high price and low price for the day, but reflects the difference and size between any two prices. If the candlestick charts of a stock list in time order, then a series used to reflect the fluctuation of the stock price for some time can be formed, which can be called candlestick series. Because each candlestick chart consists of four prices, the essence of candlestick series is stock series with four observations.

In candlestick series, the candlestick pattern series (candlestick pattern for short) refers to a candlestick sub-series containing some knowledge used to predict stock prices. For example, if the appearance of a sub-series can often cause the rising or falling of the stock price, this sub-series will be a typical pattern series. The essence of candlestick technology analysis lies in stock prediction using candlestick patterns. Therefore, how to mine the candlestick patterns and how to make use of these patterns for predicting are main research contents of candlestick technology analysis.

By the artificial method of observing the candlestick series of stock market (or Japanese rice market), people (the leading character is the founder of the candlestick chart, Munehisa Honma, who was a Japanese rice trader in the 18th century) have found many candlestick patterns. The literatures [3, 4] introduce the existing patterns and their features in detail, such as Morning Star (MS), Evening Star (ES) and Doji, etc. Some papers [5–8] conclude from the experiment that the existing candlestick patterns have a good forecasting capability for forecasting stock trends. Some other papers [9–13] have studied the stock prediction based on these patterns, and achieved some research results. However, there are also a number of papers [4, 14–16] challenging these patterns' predictive power. They argue that candlestick technology analysis violates the efficient market hypothesis, so it is not feasible for stock investment based on candlestick patterns. They also did some experiments, which show that the existing candlestick patterns have no predictive power.

Through reviewing the relevant literatures, this paper considers that the first reason why there are two different positions regarding the patterns' predictive power is that the existing candlestick patterns are lack of rigorous mathematical definition. For example, the shadow length and body size are not defined clearly in the definition of candlestick patterns, which means a candlestick pattern has many different shapes. Because the predictive power of a pattern may vary a lot for different shapes. If we ignore the shape difference and research the predictive power of a pattern by taking all patterns with various shapes as a whole instead of classifying the pattern further based on its shape feature, then the study result of candlestick patterns' predictive power may produce deviations. For instance, a MS pattern has three shapes: shape A, shape B and shape C, as shown in Fig. 1, where shape A is the generic form of MS pattern, and shape B and C are the infrequent form of MS pattern. Suppose that shape A has predictive power, and shape B and C don't have predictive power. When studying the predictive power of MS pattern, if we ignore the shape difference between the three patterns, and research them as a whole, then we will come to the wrong conclusion that MS pattern has no predictive power. However, if the three patterns are classified further based on shape features and researched separately, then we can get the correct conclusion that MS pattern has predictive power only at shape A.

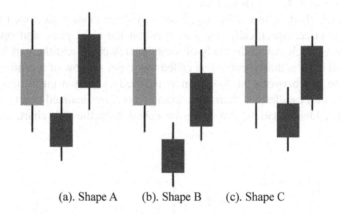

(a). Shape A   (b). Shape B   (c). Shape C

Fig. 1. Three kinds of shapes of morning star pattern

The second reason is that there also have no mathematical definitions about the existing patterns' predictive powers, which means that it is lack of standard evaluation indicators to evaluate the patterns' predictive power. That will also affect the evaluation of candlestick patterns' predictive power.

The third reason is that as the existing candlestick patterns are mined by artificial means, there may be some spurious pattern in them.

In order to resolve the debate and verify the above inference, this paper present the research of candlestick patterns' predictive power using the data mining related method, such as pattern recognition, pattern clustering, statistical analysis based on evaluation indicators, etc. The rest of this paper is organized as follow. In Sect. 2, we firstly introduce the candlestick chart, candlestick technology analysis and candlestick patterns in short. Then we define the similarity match model and nearest neighbor clustering algorithm of candlestick series. In Sect. 3, we define the analysis method of patterns' predictive power, including the evaluation indicators. Section 4 presents the experimental result and discussion. Section 5 concludes the paper.

2   Candlestick and Candlestick Series Clustering

Firstly, we give the mathematic definition of candlestick series. Let KS^i denotes the i-th candlestick series of any stock, then $KS^i = \left[D_1^i, D_2^i, \cdots, D_{|KS^i|}^i \right]$, $(|KS^i| \in N)$, where $|KS^i|$ is the number of elements in KS^i, which is also called the length of KS^i, $D_t^i (1 \leq t \leq |KS^i|)$ is the t-th candlestick chart in KS^i. $D_t^i = \{ C_t^i, O_t^i, H_t^i, L_t^i \}$, where C_t^i, O_t^i, H_t^i and L_t^i are the t-th day's close price, open price, high price and low price in KS^i, respectively.

2.1   Candlestick

2.1.1   Candlestick Chart Definition

The candlestick chart is drawn by four basic elements: close price, open price, high price and low price. Specifically, the part between the close price and open price is drawn into a rectangle called the body of a candlestick chart, and the part between the high price and body is drawn into a line called the upper shadow of a candlestick chart. Moreover, the part between the lower price and body is drawn into a line called the lower shadow of a candlestick chart. This kind of very personalized chart consisting of upper shadow, lower shadow and body are called a candlestick chart, as shown in Fig. 2.

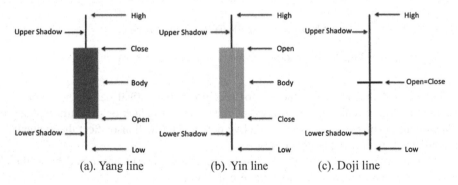

(a). Yang line   (b). Yin line   (c). Doji line

Fig. 2.   Candlestick chart (Color figure online)

In the candlestick chart, if the open price is lower than the close price, then the candlestick chart is often called Yang line, and the body is usually filled with white or green color, as shown in Fig. 2(a). And if the open price is higher than the close price, then the candlestick chart is often called Yin line, and the body is usually filled with black or red color, as shown in Fig. 2(b). Moreover, if the open price is equal to the close price, the candlestick chart is often called Doji line, and the body will collapse into a single horizontal line, as shown in Fig. 2(c).

2.1.2 Candlestick Technology Analysis

We introduce some key concepts of candlestick technology analysis in this section. Let D_t represents the t-th day's candlestick chart of any stock.

(1) Moving average: it is the average of stock price for some time. The three-day moving average at time D_t is defined by:

$$M_{avg}(D_t) = \frac{1}{3}\{C_{t-2} + C_{t-1} + C_t\} \tag{2}$$

where C_t denotes the close price of D_t.

(2) Candlestick trend: it is used to describe the trend of a candlestick chart, including uptrend and downtrend. D_t is said to be a downtrend if

$$M_{avg}(D_{t-6}) > M_{avg}(D_{t-5}) > \cdots > M_{avg}(D_t) \tag{3}$$

with at most one violation of the inequalities. Uptrend is defined analogously.

2.1.3 Candlestick Patterns

Many candlestick patterns have been mined up to now. Limited by space, only the patterns of MS and ES will be introduced in the below. Let $KS = [D_t, D_{t+1}, D_{t+2}]$ represents a three-day candlestick series.

The conditions of KS becoming the MS pattern are as follows: (1) D_t is a downtrend, and $C_t < O_t$. (2) $|C_{t+1} - O_{t+1}| > 0$, $C_t > C_{t+1}$ and $C_t > O_{t+1}$. That is the second day $t+1$ must be gapped from the first day and can be of either color. (3) $C_{t+2} > O_{t+2}$, $C_{t+2} > (C_t + O_t)/2$. That is the third day $t+2$ is Yang line and ends higher than the midpoint of the first day. A standard MS pattern is shown in Fig. 3(a).

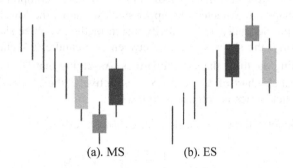

(a). MS   (b). ES

Fig. 3. The standard candlestick series chart of MS and ES

The predictive power of MS pattern from the existing literature is that: MS is a trend reversal pattern, which gives the bullish market signal. This means when the MS pattern appears, the stock prices will be likely to be transferred from downtrend into uptrend, or the stock market would may be changed from bearish market to bullish market, and the stock prices would be rise gradually.

The conditions of *KS* becoming the ES pattern are as follows: (1) D_t is an uptrend, and $C_t > O_t$. (2) $|C_{t+1} - O_{t+1}| > 0$, $C_t < C_{t+1}$ and $C_t < O_{t+1}$. That is the second day $t+1$ is gapped from the first day and can be of either color. (3) $C_{t+2} < O_{t+2}$, $C_{t+2} < (C_t + O_t)/2$. That is the third day $t+2$ is Yin line and ends lower than the midpoint of the first day. A standard ES pattern is shown in Fig. 3(b).

The predictive power of ES pattern from the existing literature is that: ES is a trend reversal pattern, which gives the bearish market signal. That means after the ES pattern appears, the stock prices will be likely to be transferred from uptrend into downtrend,or the stock market would may be changed from bullish market to bearish market, and the stock prices would be fall gradually.

2.2   Similarity Match of Candlestick Series

The similarity between two candlestick series is composed of two parts. (1) One is the shape similarity between candlestick charts, which is the similarity of shape features between the corresponding candlestick charts in the two candlestick series; (2) The other is the position similarity between candlestick charts, which is the similarity of position features between the corresponding candlestick charts in the two candlestick series. Therefore, this paper will define candlestick series' shape similarity model and position similarity model of candlestick series, respectively. Then based on these two kinds of similarity match models, the final similarity match model of candlestick series could be built. Let KS^i and KS^j represents two candlestick series, where $KS^i = KS^j$, $Sim^{i,j}$ indicates the similarity between KS^i and KS^j. So $Sim^{i,j}$ can be defined as follows.

2.2.1   The Shape Similarity of Candlestick Series
According to the shape feature of the candlestick chart, this paper proposes using the shape distance to measure the shape similarity between two candlestick charts. Firstly, based on the shape structure of the candlestick chart, three components of the candlestick chart's shape are extracted: the upper shadow shape, the lower shadow shape and the body shape. Secondly, the similarity match method of three shapes is defined, respectively. Finally, the shape similarity between two candlestick charts can be calculated by summing the three shapes' similarities. Assuming that D_t^i and D_t^j denote the t-th day's candlestick chart of KS^i and KS^j, respectively, the shape similarity match model of candlestick series is defined as follows:

(1)   Let $US^i[t]$ denotes the upper shadow length of D_t^i, as defined by:

$$US^i[t] = \begin{cases} \frac{H_t^i - O_t^i}{C_{t-1}^i * 0.1}, & O_t^i \geq C_t^i \\ \frac{H_t^i - C_t^i}{C_{t-1}^i * 0.1}, & O_t^i < C_t^i \end{cases} \tag{4}$$

where $C_{t-1}^i * 0.1$ is used to normalize the upper shadow length. According to the related regulations of Chinese A-share market, the range of daily fluctuations of stock prices cannot exceed 10% of the previous day's close price. Therefore, $C_{t-1}^i * 0.1$ can be used to normalize the length of the candlestick chart's upper shadow, lower shadow and body.

Let $Sim_{US}^{i,j}(t)$ denotes the upper shadow similarity between D_t^i and D_t^j, as defined by:

$$Sim_{US}^{i,j}(t) = \begin{cases} 0, & US^i[t] * US^j[t] = 0, US^i[t] \neq US^j[t] \\ \frac{Min(US^i[t],US^j[t])}{Max(US^i[t],US^j[t])}, & US^i[t] * US^j[t] > 0 \\ 1, & US^i[t] = US^j[t] = 0 \end{cases}$$
(5)

(2) Let $LS^i[t]$ denotes the lower shadow length of D_t^i, as defined in formula (6):

$$LS^i[t] = \begin{cases} \frac{C_t^i - L_t^i}{C_{t-1}^i * 0.1}, & O_t^i \geq C_t^i \\ \frac{O_t^i - L_t^i}{C_{t-1}^i * 0.1}, & O_t^i < C_t^i \end{cases}$$
(6)

Let $Sim_{LS}^{i,j}(t)$ denotes the lower shadow similarity between D_t^i and D_t^j, as defined by

$$Sim_{LS}^{i,j}(t) = \begin{cases} 0, & LS^i[t] * LS^j[t] = 0, LS^i[t] \neq LS^j[t] \\ \frac{Min(LS^i[t],LS^j[t])}{Max(LS^i[t],LS^j[t])}, & LS^i[t] * LS^j[t] > 0 \\ 1, & LS^i[t] = LS^j[t] = 0 \end{cases}$$
(7)

(3) Let $B^i[t]$ denotes the body length of D_t^i, as defined in formula (8):

$$B^i[t] = \frac{C_t^i - O_t^i}{C_{t-1}^i * 0.1}$$
(8)

Let $Sim_{Body}^{i,j}(t)$ denotes the body similarity between D_t^i and D_t^j, as defined by

$$Sim_{Body}^{i,j}(t) = \begin{cases} 0, & B^i[t] * B^j[t] < 0 \\ 0, & B^i[t] * B^j[t] = 0, |B^i[t]| \neq |B^j[t]| \\ \frac{Min(|B^i[t]|,|B^j[t]|)}{Max(|B^i[t]|,|B^j[t]|)}, & B^i[t] * B^j[t] > 0 \\ 1, & B^i[t] = B^j[t] = 0 \end{cases}$$
(9)

(4) Let $Sim_S^{i,j}(t)$ denotes the shape similarity between D_t^i and D_t^j, as defined by

$$\begin{cases} w_{Body} + w_{LS} + w_{US} = 1 \\ w_{Body} \geq 0 \\ w_{US} \geq 0 \\ w_{LS} \geq 0 \\ Sim_S^{i,j}(t) = w_{Body} * Sim_{Body}^{i,j}(t) + w_{US} * Sim_{US}^{i,j}(t) + w_{LS} * Sim_{LS}^{i,j}(t) \end{cases}$$
(10)

where w_{Body}, w_{US} and w_{LS} represent the weight of $Sim_{Body}^{i,j}(t)$, $Sim_{US}^{i,j}(t)$ and $Sim_{LS}^{i,j}(t)$, respectively.

(5) Let $SSim^{i,j}$ denotes the shape similarity between KS^i and KS^j, as defined by

$$SSim^{i,j} = \sum_{t=1}^{n} Sim_S^{i,j}(t) * w_S^t, \left(n = |KS^i|, \sum_{t=1}^{n} w_S^t = 1\right) \quad (11)$$

where w_S^t represent the weight of $Sim_S^{i,j}(t)$.

2.2.2   The Position Similarity of Candlestick Series

For computing the similarity between two candlestick series, we not only consider the shape similarity between candlestick series but also the position similarity. If we only consider the shape similarity, then it will cause the problem that two candlestick series having same shape features but different position features will be mistaken for being identical. For example, supposing that Fig. 2(a) and (b) represents the candlestick series chart of KS^i and KS^j, respectively, we can see that according to the shape feature definition of the candlestick chart, all of the corresponding candlestick charts in KS^i and KS^j have the same shape features. These mean that KS^i and KS^j have identical shape features, i.e., $SSim^{i,j} = 1$. However, as is vividly shown in Fig. 2, the relative position of D_2^i and D_2^j are different though D_1^i and D_1^j have the same relative position in the candlestick series. Therefore, the candlestick chart's trend between KS^i and KS^j are not identical, i.e., $Sim^{i,j} < 1$. If we only consider the shape similarity, we will draw the wrong conclusion that $SSim^{i,j} = Sim^{i,j} = 1$ (Fig. 4).

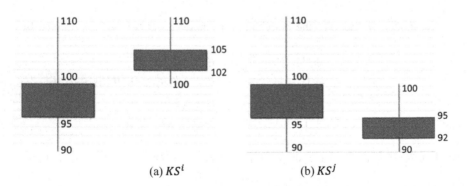

(a) KS^i (b) KS^j

Fig. 4. Candlestick series chart

To solve this problem, the concept of candlestick coordinate is introduced hoping to implement the position match between candlestick charts by defining the candlestick chart's coordinate in the candlestick series. In this paper, the sequence of a candlestick chart in the candlestick series is called x coordinate of the candlestick chart; the increase range of close price is called y coordinate of the candlestick chart. In addition, the first candlestick chart's y coordinate is set to 1 in the candlestick series. Therefore, the position similarity model of candlestick series based on the candlestick coordinate is defined as follows:

(1) Let (x_t^i, y_t^i) denotes the coordinate of D_t^i, as defined by

$$x_t^i = t,\ y_t^i = \begin{cases} 1, & t = 1 \\ (C_t^i - C_{t-1}^i)/C_{t-1}^i * 0.1, & t > 1 \end{cases} \tag{12}$$

Let $Sim_P^{i,j}(t)$ denotes the position similarity between D_t^i and D_t^j, as defined by

$$Sim_P^{i,j}(t) = \begin{cases} 0, & y_t^i * y_t^j = 0, y_t^i \neq y_t^j \\ \frac{Min(y_t^i, y_t^j)}{Max(y_t^i, y_t^j)}, & y_t^i * y_t^j > 0 \\ 1, & y_t^i = y_t^j = 0 \end{cases} \tag{13}$$

(2) Let $PSim^{i,j}$ denotes the position similarity between KS^i and KS^j, as defined by

$$PSim^{i,j} = \sum_{t=1}^{n} Sim_P^{i,j}(t) * w_P^t, \left(n = |KS^i|, \sum_{t=1}^{n} w_P^t = 1\right) \tag{14}$$

where w_P^t represent the weight of $Sim_P^{i,j}(t)$.

2.2.3 The Similarity of Candlestick Series

Finally, basing on the shape similarity and position similarity, the similarity between two candlestick series could be obtained. Therefore, the similarity match model between KS^i and KS^j is defined by

$$Sim^{i,j} = SSim^{i,j} * w_S + PSim^{i,j} * w_P \tag{15}$$

where w_S and w_P represent the shape similarity weight and position similarity weight of candlestick series, respectively.

2.3 Cluster of Candlestick Series

More accurate classification results of candlestick patterns can be got by clustering them using the nearest neighbor-clustering algorithm based on the similarity match model of candlestick series. The candlestick series' nearest neighbor clustering (CNNCA) algorithm is described as follows:

Input:
   $KSet = \{KS^1, KS^2, \cdots, KS^n\}$ // the data set of candlestick series
   θ // Similarity threshold
Output:
    $CSet$ // the set of clusters
CNNCA Algorithm:
   Assign initial value for parameters: $w_S, w_P, w_{Body}, w_{US}, w_{LS}, w_S^t, w_P^t$;
     $m=1$;
     $Q^m = \{KS^1\}$; // Q^m represents the m-th cluster
     $CSet = \{Q^m\}$;
     FOR i=2 **TO** n **DO**
     {
       $SimMax = 0$;
       FOR EACH Q^{item} **IN** $CSet$
         FOR EACH KS^j **IN** Q^{item}
        Get $Sim^{i,j}$ based on formula 15;
          If $(Sim^{i,j} > SimMax)$
           {
             $SimMax = Sim^{i,j}$;
             $f = item$; // f represents the ID of a cluster whose element is most similar
  to KS^i
           }
         End
        End
       IF $(SimMax > \theta)$ **THEN**
         $Q^f = Q^f \cup KS^i$;
       ELSE
       {
        $m = m + 1$;
        $Q^m = \{KS^i\}$;
       }
     }
     $CSet = \{Q^1, Q^2, \cdots, Q^m\}$;

In addition, $|Q^m|$ represents the number of elements in Q^m. As each candlestick series will be matched once with all of the candlestick series stored in the cluster, the time complexity and space complexity are both $O(n^2)$.

3   The Analysis of Patterns' Predictive Power

We analyze the patterns' predictive power by the following steps:

(1) Pattern recognition. Based on the definition of candlestick patterns, we identify all the candlestick series belonging to a pattern (such as MS or ES), and then they form a set (*KSet*).

(2) Pattern clustering. We use the CNSSC algorithm to cluster *KSet*, then the set of clusters (*CSet*) can be got, in which different clusters represent the same pattern's different shapes.

(3) Defining the evaluation indicators. Although the literatures [3, 4] introduce the existing patterns and their features in detail, they do not give the evaluation indicator for evaluating the candlestick pattern's predictive power. So we will define some evaluation indicators in this paper.

The pattern's predictive power is got primarily by analyzing the trend of the pattern's consequent candlestick series. In addition, according to the candlestick technology that in the consequent candlestick series, the trend of a candlestick, which is closer to pattern series, will be more affected by the pattern series. Therefore, we mainly analyze the next day's candlestick trend to evaluate the pattern's predictive power. There are four prices to describe the next day's candlestick trend, which are the open price, close price, high price and low price. However, it is normal to see that the high/low price is bigger/smaller than the previous day's close price. So only the next day's close price and open price will be used to evaluate the pattern's predictive power.

Let $KS = [D_{t-3}, D_{t-2}, D_{t-1}]$ denotes a three-day candlestick pattern, D_t denotes the next day's candlestick of the candlestick pattern. The evaluation indicators are defined as follows:

(a) Let C_t denotes the close price of D_t, $P_U(C_t)$ denotes the probability of the trend of C_t is uptrend, $P_D(C_t)$ denotes the probability of the trend of C_t is downtrend. $P_U(C_t)$ and $P_D(C_t)$ are calculated by

$$P_U(C_t) = |Q_{UC}^{m,t}|/|Q^m| \tag{16}$$

$$P_D(C_t) = |Q_{DC}^{m,t}|/|Q^m| \tag{17}$$

where $|Q_{UC}^{m,t}|$ represents the number of patterns meeting the condition of $C_t \geq C_{t-1}$ in Q^m, $|Q_{DC}^{m,t}|$ represents the number of patterns meeting the condition of $C_t \leq C_{t-1}$ in Q^m, C_{t-1} represents the close price of D_{t-1}. A higher value of $P_U(C_t)$ or $P_D(C_t)$ indicates a higher probability for predicting the next day's uptrend or downtrend of close price.

(b) Similarly, let O_t denotes the open price of D_t, $P_U(O_t)$ denotes the probability of the trend of O_t is uptrend, $P_D(O_t)$ denotes the probability of the trend of O_t is downtrend. $P_U(O_t)$ and $P_D(O_t)$ are calculated by

$$P_U(O_t) = |Q_{UO}^{m,t}|/|Q^m| \tag{18}$$

$$P_D(O_t) = |Q_{DO}^{m,t}|/|Q^m| \tag{19}$$

where $|Q_{UO}^{m,t}|$ represents the number of patterns meeting the condition of $O_t \geq O_{t-1}$ in Q^m, $|Q_{DO}^{m,t}|$ represents the number of patterns meeting the condition of $O_t \leq O_{t-1}$ in Q^m, O_{t-1} represents the open price of D_{t-1}. A higher value of $P_U(O_t)$

or $P_D(O_t)$ indicates a higher probability for predicting the next day's uptrend or downtrend of open price.

(4) Analysis. Based on the evaluation indicators, we analyze the pattern's predictive power.

4   Experiment and Result Analysis

We select the candlestick series data of Shanghai 180 index component stocks over the latest 10 years (from 2006-01-04 to 2016-08-24) as the test data. Limited by space, only the MS and ES pattern's predictive power will be analyzed in the experiment. And the parameters of CNSSC algorithm are set as follows: $\theta = 0.75$, $w_S = 0.2$, $w_T = 0.8$, $w_{Body} = 0.6$, $w_{US} = 0.2$, $w_{LS} = 0.2$, $w_S^t = w_P^t = 1/|KS^i|$, $(t = 1, 2, \cdots, |KS^i|)$.

4.1   Experiment One

The aim of the first experiment is to analyze the MS pattern's predictive power based on the method defined in Sect. 3. Firstly, based on the definition of MS, 1164 MS patterns are identified from the test data. Then we cluster these patterns using the CNSSC algorithm, and finally 451 clusters are obtained. We choose the top 10 clusters with the most elements to conduct statistical analysis, as shown in Table 1.

Table 1. The experiment result of MS

Q^m	Q^0	Q^1	Q^2	Q^3	Q^4	Q^5	Q^6	Q^7	Q^8	Q^9	Q^{10}
$|Q^m|$	1164	10	9	9	9	9	8	8	8	8	8
$P_U(C_t)$	0.52	0.10	0.70	0.44	0.56	0.56	0.38	0.38	0.63	0.50	0.25
$P_U(O_t)$	0.50	0.20	0.80	0.33	0.44	0.33	0.50	0.38	0.63	0.63	0.25

In Table 1, Q^0 represents the cluster composed of 1164 MS patterns. Its $P(C_1)$ and $P(O_1)$ are both approximately equal to 0.5, which means its predictive power is not obvious no matter for the next day's close price or open price. Therefore, it may be a spurious pattern to predict bullish market. However, after further classifying the MS patterns, we can see that (1) Q^2 and Q^8 have a good predictive power for predicting the next day's close price and open price, because both of their $P(C_1)$ and $P(O_1)$ are above 0.5. (2) Q^1, Q^3, Q^6, Q^7 and Q^{10} have a bad predictive power for predicting the next day's close price and open price, because both of their $P(C_1)$ and $P(O_1)$ are below 0.5. (3) Q^4 and Q^5, have a good predictive power for predicting the next day's close price, as their $P(C_1)$ is above 0.5. (4) Q^9 has a good predictive power for predicting the next day's open price, as its $P(O_1)$ is above 0.5.

The result of experiment one shows that (1) MS patterns do not have the predictive power for predicting the next day's close price and open price. Therefore, we can consider that MS patterns are definitely a spurious pattern. However, some MS patterns with special shapes also have a good predictive power for predicting the next day's close price and open price. (2) The predictive power of MS varies a great deal for different shapes, which is consistent with the expected analysis.

4.2   Experiment Two

The aim of the second experiment is to analyze the ES pattern's predictive power based on the method defined in Sect. 3. Firstly, based on the definition of ES, 1413 ES patterns are identified from the test data. Then we cluster these patterns using the CNSSC algorithm, and finally 537 clusters are obtained. We choose the top 10 clusters with the most elements to conduct statistical analysis, as shown in Table 2.

Table 2. The experiment result of ES

Q^m	Q^0	Q^1	Q^2	Q^3	Q^4	Q^5	Q^6	Q^7	Q^8	Q^9	Q^{10}
$|Q^m|$	1413	51	20	14	13	10	9	9	9	7	7
$P_U(C_t)$	0.45	0.43	0.55	0.57	0.46	0.70	0.56	0.44	0.78	0.33	0.86
$P_U(O_t)$	0.70	0.69	0.70	0.71	0.62	0.70	0.44	0.78	0.78	0.89	0.57

In Table 1, Q^0 represents the cluster composed of 1516 ES patterns. Its $P(C_1)$ is 0.45, while the $P(O_1)$ is 0.70, which means it has a good predictive power for predicting the next day's open price. After further classifying the TIU patterns, we can see that (1) Q^2, Q^3, Q^5, Q^8 and Q^{10}, etc., have a good predictive power for predicting the next day's close price and open price, because both of their $P(C_1)$ and $P(O_1)$ are above 0.5. (2) Q^6 has a good predictive power for predicting the next day's close price, as its $P(C_1)$ is above 0.5. (3) Q^1, Q^4, Q^7 and Q^9 only have a good predictive power for predicting the next day's open price, as their entire $P(O_1)$ are above 0.5.

The result of experiment two shows that (1) ES patterns have a good predictive power for predicting the next day's open price. However, some ES patterns with special shapes also have a good predictive power for predicting the next day's close price. (2) The predictive power of ES patterns varies a great deal for different shapes, which is consistent with the expected analysis.

4.3   Experiment Conclusion

Through the above experiment, we can draw the following conclusion: (1) The MS pattern is a spurious pattern, and the ES pattern only has the predictive power of open price. (2) The predictive power of a pattern varies a great deal for different shapes. Take MS and ES for example, some shapes' MS/ES patterns have a good predictive power for predicting the close price and open price, while some others do not have the predictive power.

The above two points have resulted in the dispute on whether the candlestick patterns have predictive power in academic. Therefore, to analyze the predictive power of a pattern, we should make a concrete analysis of concrete shapes. Each of the existing candlestick patterns requires further classification based on the shape feature. Meanwhile the definition of candlestick patterns and their predictive power should be more rigorous and precise. If so, it will be improved a lot for the performance of stock prediction based on candlestick patterns.

5   Conclusion

As a primary technology analysis method of stock prediction, there is different option on the stock price prediction based on candlestick patterns in the academic world, though it is widely used in reality. We consider that one reason for the debate is that the definition of candlestick patterns and their predictive power is more open and lack of mathematical rigor. The other reason is that there are some spurious patterns in the existing candlestick patterns. To verify the these inferences, this paper uses the data mining method, such as pattern recognition, similarity match, cluster and statistical analysis, etc., to study the predictive power of candlestick patterns. The experiment results show that our inferences are all right. Therefore, the best method to solve the debate is that the researchers should redefine the candlestick patterns' definition and their predictive power with mathematical method.

Acknowledgment. The Key Basic Research Foundation of Shanghai Science and Technology Committee, China (Grant No.14JC1402203) and the Science and Technology Support Program of China (Grant No. 2015BAF10B01) financially supported this work.

References

1. Li, L., Tian, X., Yang, H., et al.: Financial time series forecasting based on SVR. Comput. Eng. Appl. **41**(30), 221–224 (2005)
2. Edwards, R.D., Magee, J., Bassetti, W.H.C.: Technical Analysis of Stock Trends, 10th edn. CRC Press, Boca Raton (2012)
3. Nison, S.: Japanese Candlestick Charting Techniques: A Contemporary Guide to the Ancient Investment Technique of the Far East. New York Institute of Finance, New York (1991)
4. Marshall, B.R., Young, M.R., Rose, L.C.: Candlestick technical trading strategies: can they create value for investors? J. Bank. Finan. **30**(8), 2303–2323 (2006)
5. Caginalp, G., Laurent, H.: The predictive power of price patterns. Appl. Math. Finan. **5**(3–4), 181–205 (1998)
6. Lee, K.H., Jo, G.S.: Expert system for predicting stock market timing using a candlestick chart. Expert Syst. Appl. **16**(4), 357–364 (1999)
7. Goswami, M.M., Bhensdadia, C.K., Ganatra, A.P.: Candlestick analysis based short term prediction of stock price fluctuation using SOM-CBR. In: IEEE International Advance Computing Conference, pp. 1448–1452 (2009)
8. Lu, T.H., Shiu, Y.M.: Tests for two day candlestick patterns in the emerging equity market of Taiwan. Emerg. Mark. Finan. Trade **48**(1), 41–57 (2014)
9. Li, H., Ng, W.W.Y., Lee, J.W.T., et al.: Quantitative study on candlestick pattern for Shenzhen stock market. In: 2008 IEEE International Conference on Systems, Man and Cybernetics, SMC 2008, pp. 54–59. IEEE, Piscataway (2008)

10. Wei, X., Ng, W.W.Y., Firth, M., et al.: L-GEM based MCS aided candlestick pattern investment strategy in the Shenzhen stock market. In: Proceedings of the 2009 International Conference on Machine Learning and Cybernetics, ICMLC 2009, vol. 1, pp. 243–248. IEEE, Piscataway (2009)
11. Kamo, T., Dagli, C.: Hybrid approach to the Japanese candlestick method for financial forecasting. Expert Syst. Appl. **36**(3), 5023–5030 (2009)
12. Kimiagari, A.M., Jasemi, M., Kimiagari, S.: A modern neural network model to do stock market timing on the basis of the ancient investment technique of Japanese candlestick. Expert Syst. Appl. **38**(4), 3884–3890 (2011)
13. Barak, S., Dahooie, J.H., Tichy, T.: Wrapper ANFIS-ICA method to do stock market timing and feature selection on the basis of Japanese candlestick. Expert Syst. Appl. **42**(23), 9221–9235 (2015)
14. Zwergel, B., Klein, C., Fock, H.: Performance of candlestick analysis on intraday futures data. J. Deriv. **13**(1), 28–40 (2005)
15. Young, M.R., Marshall, B., Cahan, R.: Are candlestick technical trading strategies profitable in the Japanese equity market? Rev. Quant. Finan. Acc. **31**(2), 191–207 (2008)
16. Horton, M.J.: Stars, crows, and doji: the use of candlesticks in stock selection. Q. Rev. Econ. Finan. **49**(2), 283–294 (2009)

Partial Least Squares (PLS) Methods
for Abnormal Detection of Breast Cells

Yuchen Zhu⑩, Shanxiong Chen^(⊠), Chunrong Chen, and Lin Chen

School of Computer and Information Science, Southwest University,
Chongqing 400715, China
csxpml@163.com

Abstract. Breast cancer is one of the malignant tumors having high incidence in women, the incidence of breast cancer has increased in all parts of the world since twentieth century, but its etiology is not yet completely clear, so it is very important to detect breast cells. In this paper, we built a regression model to detect breast cells, and generated a method for predicting the formation of benign and malignant breast cells by training the model, then we used the 10 features of breast cells to predict it, the results reaching upto 93.67% accuracy, it was very effective to predict and analyse whether the breast cells getting cancer, It had an important role in the diagnosis and prevention of breast cancer.

Keywords: Partial least squares · Multivariate analysis · Breast cancer · Prediction

1 Introduction

Breast cancer is a malignant tumor that occurs in the glandular epithelium of the mammary gland, the female breast is composed of skin, fibrous tissue, mammary glands and fat. The breast is not an important organ in maintaining human life, the breast cancer in situ is not fat but breast cancer cells lost their normal character, and loosely connected with the surrounding cells, easy to fall off. Once cancer cells are loss, the free cancer cells can spread all over the body with the blood or lymph metastasis formation, endanger life. At present breast cancer has become a common cancer that threat to women's physical and mental health. Early-stage breast cancer often does not have typical symptoms and sign, and people usually difficult to pay attention, Usually the most of the breast cancer is painless mass, only a few of it have different degrees of pain or tingling. Currently, there are many ways to detect breast cancer, such as breast X-ray mammography, ultrasound and CT examination [1]. However, many methods need to be further discussed, some methods may not even be used as the main method to detect breast cancer. For patients, early detection of breast cancer is the key to reduce the disease, Symptoms of the early-stage breast cancer is changing in the shape or producing lumps [2], and the 80% patients who usually through physical examination or breast cancer screening are diagnosed breast lumps the first time, So we can judge whether there is a breast lump by detecting the breast cells, Therefore, it is an important means to detect the existence of breast cancer cells and to prevent the spread of breast cancer cells.

© Springer Nature Singapore Pte Ltd. 2017
B. Zou et al. (Eds.): ICPCSEE 2017, Part I, CCIS 727, pp. 88–99, 2017.
DOI: 10.1007/978-981-10-6385-5_8

Partial Least-Squares Regression is an advanced method of multivariate analysis, It consists of multiple linear regression analysis, canonical correlation analysis and principal component analysis [3], and is widely used in many fields. [4–9], at the same time, it has achieved good results in multivariate modeling and prediction, and the prediction model it established has high stability, accuracy and anti-noise ability [10, 11], So it is a good method to solve the problems of the breast cancer which have many factors, and it is also the motivation to use PLS method to solve the problems. In this paper, we selected the medical detection results of benign and malignant breast cancer cells, and adopted partial least square method to establish a regression model, the test data were predicted by choosing appropriate threshold as the criterion of the detection result. It was found that the prediction model based on partial least squares method has the advantages of fast detection ability and high detection accuracy.

This paper provides an idea and a basic approach to the detection of abnormal breast cells, which may be a revelation for future research. It makes the establishment of the model more convenient and flexible that the detection process is recyclable.

1.1 Partial Least Squares Regression

1. Standardized treatment of variables:

$$x_{ij}^* = \frac{x_{ij} - \overline{x_j}}{s_j}, (i = 1, 2, \ldots m; \; j = 1, 2, \ldots, n) \qquad (1)$$

$$y_{ij}^* = \frac{y_{ij} - \overline{y_j}}{s_y}, (i = 1, 2, \ldots m; \; j = 1, 2, \ldots, n) \qquad (2)$$

The data matrix is recorded as $E_0 = \left(E_{01}, E_{02}, \ldots, E_{0p}\right)_{n \times p}$ after X standardized processing, the data matrix is recorded as $F_0 = \left(F_{01}, F_{02}, \ldots, F_{0q}\right)_{n \times q}$ after Y standardized processing.

2. Extracting principal component and stepwise regression. t_1 is the first component of $E_0, t_1 = E_0 w_1, w_1$ is the first axis of E_0, It is a unit vector, i.e. $\|w_1\| = 1$. Note u_1 is the first component of $F_0, u_1 = F_0 c_1, c_1$ is the first axis of F_0, It is a unit vector, i.e. $\|c_1\| = 1$. In order to make the t_1 and u_1 relate to the greatest extent, according to the principal component analysis, it should be:

$$Var(t_1) \rightarrow max \qquad (3)$$

$$Var(u_1) \rightarrow max \qquad (4)$$

On the other hand, due to the need for regression modeling, it is asked that t_1 has the greatest explanatory power to u_1, According to the canonical correlation analysis, the correlation degree of t_1 and u_1 should reach the maximum value:

$$r(t_1, u_1) \rightarrow max \tag{5}$$

Together, the covariance of the t_1 and the u_1 is required to achieve the maximum value, and thus to solve the following optimization problems:

$$max <E_0 w_1, F_0 c_1> \tag{6}$$

$$s.t \begin{cases} w_1^T w_1 = 1 \\ c_1^T c_1 = 1 \end{cases} \tag{7}$$

That is, under the condition of $\|w_1\| = 1$ and $\|c_1\| = 1$, solving the maximum of $w_1^T E_0^T F_0 c_1$.

By using the Lagrangian algorithm,

$$s = w_1^T E_0^T F_0 c_1 - \lambda_1 \left(w_1^T w_1 - 1\right) - \lambda_2 \left(c_1^T c_1 - 1\right) \tag{8}$$

and find the partial derivative of w_1 and c_1 and λ_1 and λ_2 for s,

$$\frac{\partial s}{\partial w_1} = E_0^T F_0 c_1 - 2\lambda_1 w_1 = 0 \tag{9}$$

$$\frac{\partial s}{\partial c_1} = F_0^T E_0 w_1 - 2\lambda_2 c_1 = 0 \tag{10}$$

$$\frac{\partial s}{\partial \lambda_1} = -\left(w_1^T w_1 - 1\right) = 0 \tag{11}$$

$$\frac{\partial s}{\partial \lambda_2} = -\left(c_1^T c_1 - 1\right) = 0 \tag{12}$$

From (9) to (12), we can deduce that:

$$2\lambda_1 = 2\lambda_2 = w_1^T E_0^T F_0 c_1 = <E_0 w_1, F_0 c_1>$$

Then write θ_1 as $2\lambda_1 = 2\lambda_2 = w_1^T E_0^T F_0 c_1$, by the literature [3], $\theta_1 = w_1^T E_0^T F_0 c_1$, $E_0^T F_0 F_0^T E_0 w_1 = \theta_1^2 w_1$, $F_0^T E_0 E_0^T F_0 c_1 = \theta_1^2 c_1$, so w_1 is the characteristic vector of the matrix $E_0^T F_0 F_0^T E_0$, and the corresponding characteristic value is θ_1^2. θ_1 is the objective function, if we will solve the maximum value of it, that is to solve the feature vector w_1 what is the maximum eigenvalue of a matrix $E_0^T F_0 F_0^T E_0$ corresponding, then, to solve the components t_1 and residual matrix E_1:

$$t_1 = E_0 w_1 \tag{13}$$

$$E_1 = E_0 - t_1 p_1^T \tag{14}$$

And also:

$$u_1 = F_0 c_1 \tag{15}$$

$$F_1 = F_0 - t_1 r_1^T \tag{16}$$

$$F_1^* = F_0 - u_1 q_1^T \tag{17}$$

In the formula, $p_1 = \frac{E_0^T t_1}{\|t_1\|^2}, q_1 = \frac{F_0^T u_1}{\|u_1\|^2}, r_1 = \frac{F_0^T t_1}{\|t_1\|^2}$. By the same way [12]:

$$t_2 = E_1 w_2 \tag{18}$$

$$E_2 = E_1 - t_2 p_2^T \tag{19}$$

$$u_2 = F_1 c_2 \tag{20}$$

$$F_2 = F_1 - t_2 r_2^T \tag{21}$$

and $p_2 = \frac{E_1^T t_2}{\|t_2\|^2}, r_2 = \frac{F_1^T t_2}{\|t_2\|^2}$.

So calculated, we will finally get:

$$E_0 = t_1 p_1^T + t_2 p_2^T + \cdots t_A p_A^T \tag{22}$$

$$F_0 = t_1 r_1^T + t_2 r_2^T + \cdots t_A r_A^T \tag{22}$$

We return Eq. (22) to the regression equation of $x_1^* = E_{0j}$ for $y_k^* = F_{0k}$:

$$y_k^* = a_{k1} x_1^* + a_{k2} x_2^* + \ldots + a_{kp} x_p^* + F_{Ak} \quad (k = 1, 2, \cdots, q) \tag{23}$$

if the rank of A is X.

1.2   Cross-validation

In general, in the partial least squares regression model, it is only necessary to extract a part of the component, but we can observe the new component whether it can improve the model's prediction after adding it. Extracting h components after removing a sample point i from the sample y, and fitting a regression equation, then we bring i into the regression equation, we can get the fitted value $\hat{y}_{hj(-i)}$, so we define $S_{PRESS,hj}$ as the prediction error sum of squares of y_i:

$$S_{PRESS,h} = \sum\nolimits_{i=1}^{n} \left(y_{ij} - \hat{y}_{hj(-i)} \right)^2 \tag{24}$$

And $S_{SS,hj}$ as the error sum of squares of y_i:

$$S_{SS,h} = \sum_{j=1}^{n} \left(y_{ij} - \hat{y}_{hji}\right)^2 \qquad (25)$$

For the sample, the disturbance error of regression equation that has h components can to a certain extent less than the fitting error of regression equation that has $h - 1$ components, it is believed that adding a component t_h will improve the accuracy of prediction. Therefore, the value of $S_{PRESS,h}/S_{SS,h-1}$ the smaller, the better. We define the cross-validation of composition of t_h:

$$Q_h^2 = 1 - \frac{S_{PRESS,h}}{S_{SS,h\ 1}} \qquad (26)$$

When $Q_h^2 \geq 0.0975$, the contribution of t_h is obvious.

1.3   Specific Point in the Sample

We can find the specific point in the sample according to the contribution rate of the i sample point to t_h that of the component h. Define contribution rate as T_{hi}^2:

$$T_{hi}^2 = \frac{t_{hi}^2}{(n-1)s_h^2} \qquad (27)$$

In the formula: s_h^2 is the variance of t_h. So that it can be measured cumulative contribution rate of sample point i to $t_1, t_2, \ldots, t_m$:

$$T_i^2 = \frac{1}{n-1} \sum_{h=1}^{m} \frac{t_{hi}^2}{s_h^2} \qquad (28)$$

In the software, you can draw the elliptic map of T^2, it is believed that these sample points are specific if the sample point falls outside the ellipse, We can remove these specific points and re fit, we will repeat this process until there is no specific point in the sample.

1.4   Variable Importance in Projection

We can use the Variable Importance in Projection VIP_j to measure the importance of the role of x_i in the interpretation of Y, the value of VIP_j the bigger, the x_i has a more important role. The definition of VIP_j is as follows [3]:

$$VIP_j^2 = \frac{p \sum_{h=1}^{m} Rd(Y; t_h) w_{hj}^2}{\sum_{h=1}^{m} Rd(Y; t_h)} \qquad (29)$$

And $Rd(Y; t_h) = r^2(y, t_h)$; $r(y, t_h)$ is the correlation coefficient of y and t_h, w_{hj}, is the first j component of w_h.

1.5   The Anomaly Detection Model Based on PLS

The anomaly detection procedure based on PLS method and the flow chart (Fig. 1) are as follows:

1. Import the data set for the establishment of the model and set the dependent variables and independent variables, after extracting principal component from standardized datas, we fit and build PLS model. The number of principal components is the number of attributes that have an effect on the results
2. Observe T^2 ellipse and identify outliers, Then we remove outliers from data and fit again until there is no exception, and in this process human intervention is needed; Get the parameters and the formula of y.

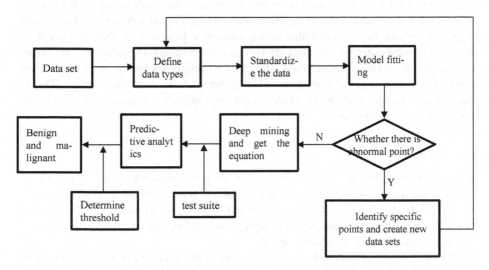

Fig. 1. Flow chart of anomaly detection based on PLS

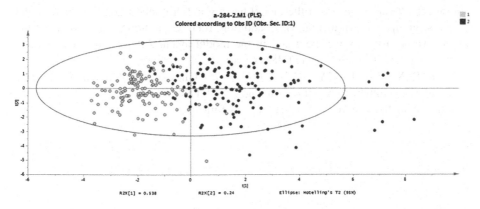

Fig. 2. The first fit of the T^2 elliptic

3. Enter the data set to be measured and use the Eq. (23) to calculate to get the predictive value of y, and judge whether the predictive value is benign or malignant according to the determined threshold.

2 Experimental Verification

2.1 Introduction to Data Sets

The data set of this experiment comes from Wisconsin Diagnostic Breast Cancer (WDBC) that was created by Dr. William H. Wolberg (General Surgery Dept. University of Wisconsin), W. Nick Street (Computer Sciences Dept. University of Wisconsin), Olvi L. Mangasarian (Computer Sciences Dept. University of Wisconsin) in November 1995 and donated by Nick Street [13]. The data has 569 cases of cell biopsy, each case has 32 attributes which contain a patient's number and a cancer diagnosis and the other 30 attributes are real measured values. In attributes, "B" stands for benign, "M" stands for malignant, The other 30 properties are formed by the mean, standard deviation and the maximum value of the 10 features of the nucleus. The 10 features are: radius, texture, perimeter, area, smoothness, compactness, concavity, symmetry, fractal dimension.

2.2 Data Processing

In this experiment, we set "B" to 0, and set "M" to 1 as dependent variables, and the ten feature attributes are used as independent variables, Select half of the data (284 samples) to establish the model, the remaining half (285 samples) used to verify, and the data are divided into two groups, the benign group and malignant group, after that, we import processed data into SIMCA-P 13.

We get three main components after the data is processed by principal component analysis (PCA), R2X is The fraction of the variation of the X variables explained by the model. R2Y is the fraction of the variation of the Y variables explained by the model, and Q2 is cross-validation. We draw its T^2 elliptic graph based on the 3 principal components, It can be seen that the model has a good distinction between benign and malignant groups (shown in Fig. 3). In addition, there are a number of outliers in the sample, we need to remove them and fit the model again. After remove the outliers several times, the T^2 elliptic is as follows (Fig. 4).

Then, R2X = 0.744, R2Y = 0.757, Q2 = 0.75. Click the "Coefficient Plot" view the regression coefficient (As shown in Fig. 2 and Table 1).

The standardized regression equation is as follows:

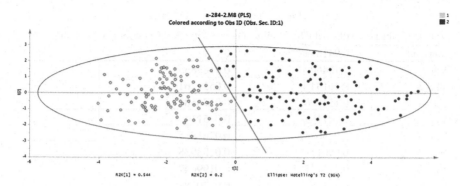

Fig. 3. Multiplex fitting

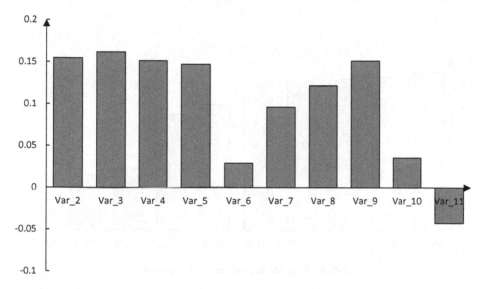

Fig. 4. The histogram of each variable coefficient

$$y = 0.941699 + \text{radius} * 0.154609 + \text{texture} * 0.16134 + \text{perimeter} * 0.154876$$
$$+ \text{area} * 0.147183 + \text{smoothness} * 0.0290967 + \text{compactness}$$
$$* 0.0961712 + \text{concavity} * 0.121727 + \text{concave points}$$
$$* 0.15144 + \text{symmetry} * 0.0355873 - \text{fractal dimension}$$
$$* 0.0424221$$

In VIP diagram (Fig. 5) we can see that the cell concave points, radius, perimeter, area and concavity is important to explain whether the cell is cancerous.

After data processing is completed, you can view the prediction results. In order to easily to view, we sort the data into form (Table 2).

Table 1. Regression coefficient

Var ID (Primary)	M8.CoeffCS [2] (Var_1)	1.89456* M8.CoeffCS [2] (Var_1) cvSE
Sonstant	0.941699	0
Var_2	0.154609	0.0182833
Var_3	0.16134	0.0531261
Var_4	0.154876	0.0176588
Var_5	0.147183	0.025114
Var_6	0.0290967	0.0261805
Var_7	0.0961712	0.0188938
Var_8	0.121727	0.0199647
Var_9	0.15144	0.0220582
Var_10	0.0355973	0.0575215
Var_11	−0.0424221	0.0398328

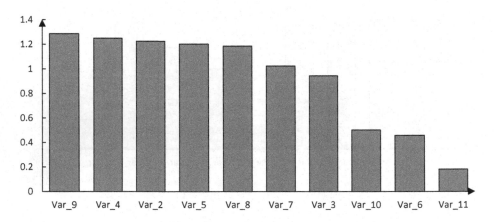

Fig. 5. Variable Importance in Projection

Here we set 0.5 as threshold, if the predictive value is greater than 0.5, it is considered as a malignant cell, if the predicted value is less than 0.5, it is considered as a benign cell. Thus we predict 339 benign cells in the 357 benign cells and 194 malignant cells in the 212 malignant cells, so the results reaching upto 93.67% accuracy, it can better predict whether the cell cancer.

3   Result Analysis

Partial least square method can be used in many fields and the model it established has better accuracy and availability. Breast cancer is a disease with high incidence but it has good effect in early diagnosis, so observing breast cells has become an important means to prevent breast cancer. In this paper, we use the partial least squares method to establish regression model for the breast cells that have characteristic variables, and get

Table 2. Results after sorting

Class	Predictive values	Actual values	Radius	Texture	Perimeter	Area	Smoothness	Compactness	Concavity	Concave points	Symmetry	Fractal dimension
1	0.295967838	0	13.54	14.36	87.46	566.3	0.09779	0.08129	0.06664	0.04781	0.1885	0.05766
1	0.267255496	0	13.08	15.71	85.63	520	0.1075	0.127	0.04568	0.0311	0.1967	0.06811
1	-0.173891653	0	9.504	12.44	60.34	273.9	0.1024	0.06492	0.02956	0.02076	0.1815	0.06905
1	0.166668372	0	13.03	18.42	82.61	523.8	0.08983	0.03766	0.02562	0.02923	0.1467	0.05863
1	-0.224458058	0	8.196	16.84	51.71	201.9	0.086	0.05943	0.01588	0.005917	0.1769	0.06503
1	0.144558497	0	12.05	14.63	78.04	449.3	0.1031	0.09092	0.06592	0.02749	0.1675	0.06043
1	0.389824863	0	13.49	22.3	86.91	561	0.08752	0.07698	0.04751	0.03384	0.1809	0.05718
1	0.110002247	0	11.76	21.6	74.72	427.9	0.08637	0.04966	0.01657	0.01115	0.1495	0.05888
1	0.127394918	0	13.64	16.34	87.21	571.8	0.07685	0.06059	0.01857	0.01723	0.1353	0.05953
...												
2	1.248139169	1	20.47	20.67	134.7	1299	0.09156	0.1313	0.1523	0.1015	0.2166	0.05419
2	1.425508788	1	20.55	20.86	137.8	1308	0.1046	0.1739	0.2085	0.1322	0.2127	0.06251
2	0.679341203	1	14.27	22.55	93.77	629.8	0.1038	0.1154	0.1463	0.06139	0.1926	0.05982
2	1.194395267	1	15.22	30.62	103.4	716.9	0.1048	0.2087	0.255	0.09429	0.2128	0.07152
2	1.736706578	1	20.92	25.09	143	1347	0.1099	0.2236	0.3174	0.1474	0.2149	0.06879
2	1.528192491	1	21.56	22.39	142	1479	0.111	0.1159	0.2439	0.1389	0.1726	0.05623
2	1.30694358	1	20.13	28.25	131.2	1261	0.0978	0.1034	0.144	0.09791	0.1752	0.05533
2	0.842298291	1	16.6	28.08	108.3	858.1	0.08455	0.1023	0.09251	0.05302	0.159	0.05648
2	1.915359342	1	20.6	29.33	140.1	1265	0.1178	0.277	0.3514	0.152	0.2397	0.07016

a good result, the results reaching upto 93.67% accuracy. it can be seen that the cell radius, texture, concave points and perimeter and area are positively correlated with the cell carcinoma, and the fractal dimension is negatively correlated; As can be seen from the VIP_j, concave points, perimeter, radius and area and concavity have a great contribution for the predicted value, but the contribution of cell symmetry, smoothness and fractal dimension to the predictive value is small, we can sometimes give up the independent variables when we choose the regression variables, But the conclusion that about VIP_j index analysis is still qualitative, we can only state that the independent variable plays a more important role in it, and there are some limitations of the VIP method, it can not be said that the number of variables is the best choice when the contribution of the independent variable is very large, sometimes we also consider the correlation between variables to choose [14]. In the choice of threshold, we have choosen an intermediate value, its persuasion is not too strong and we need to test and fit many times to improve.

Acknowledgments. This study was supported by the National Natural Science Foundation of China (61170192, 41271292), China Postdoctoral Science Foundation (No. 2015M580765), Chongqing Postdoctoral Science Foundation (No. Xm2016041), Scientific and Technological Research Program of Chongqing Municipal Education Commission (KJ1603109, KJ1503207), the Fundamental Research Funds for the Central Universities (XDJK2014C039, XDJK2016C045), Doctoral Fund of Southwestern University (swu1114033).

References

1. Wang, X., Qiu, W., Jia, Z.: The diagnostic actuality and latest progress of breast cancer. Clin. Med. China **28**(8) (2012)
2. Li, J.: Early diagnosis and treatment of breast cancer. Chin. J. Gener. Pract. **06**, 336–338 (2005)
3. Wang, H.: Linear and Nonlinear Methods for Partial Least Squares Regression. National Defence Industry Press, Beijing (2006)
4. Paulo, J.M.D., Barros, J.E.M., Barbeira, P.J.S.: A PLS regression model using flame spectroscopy emission for determination of octane numbers in gasoline. Fuel **176**, 216–221 (2016)
5. Bernardes, C.D., Figueiredo, M.C.P.D., Barbeira, P.J.S.: Developing a PLS model for determination of total phenolic content in aged cachaças. Microchem. J. **116**, 173–177 (2014)
6. Krishnan, A., Williams, L.J., Mcintosh, A.R., et al.: Partial least squares (PLS) methods for neuroimaging: a tutorial and review. Neuroimage **56**(2), 455–475 (2011)
7. Gan, X.S., Duanmu, J.S., Wang, J.F., et al.: Anomaly intrusion detection based on PLS feature extraction and core vector machine. Knowl.-Based Syst. **40**(1), 1–6 (2013)
8. Halstensen, M., Amundsen, L., Arvoh, B.K.: Three-way PLS regression and dual energy gamma densitometry for prediction of total volume fractions and enhanced flow regime identification in multiphase flow. Flow Meas. Instrum. **40**, 133–141 (2014)
9. Bo, L., Castillo, I., Chiang, L., Edgar, T.F.: Industrial PLS model variable selection using moving window variable importance in projection. Chemometr. Intell. Lab. Syst. **135**, 90–109 (2014)
10. Ji, S., Li, S.: Anomaly detection based on partial least squares. Chin. Bus. **9**, 138–139 (2009)

11. Luo, L., Bao, S., Mao, J., et al.: Quality prediction and quality-relevant monitoring with multilinear PLS for batch processes. Chemometr. Intell. Lab. Syst. **150**, 9–22 (2015)
12. Xue, Y., Chen, Q.: The application of partial least square method in the evaluation of customer satisfaction. Coop. Econ. Sci. Technol. **7**, 28–29 (2006)
13. The UCI Machine Learning Repository. http://archive.ics.uci.edu/ml/datasets/Breast+Cancer +Wisconsin+(Diagnostic). Accessed 27 June 2016
14. Zhou, Q., Ouyang, Y., Xuegang, H.: Using partial least-squares regression to found abnormal data in data-mining. Microelectron. Comput. **22**(01), 25–27 (2005)

Desktop Data Driven Approach to Personalize Query Recommendation

Xiao-yun Li and Ying Yu[✉]

School of Computer Science and Technology, University of South China,
Hengyang, Hunan, China
lxy.yy@yeah.net, yyingu@sina.com

Abstract. Query recommendation is an effective method to help users describe their search intentions. In a personalized system, cold-start and the data sparsity were unavoidable, which directly lead to deficient performance of personalizing. As a significant part of a user's personal information space, a personal computer owns lots of documents relevant to his or her interest. Therefore, desktop data was introduced to construct a user's preference model. Furthermore, considering the variety of desktop data, relationship between search task and work task was simultaneously exploited to predict a user's specific information need. Ten volunteers joined experiments to evaluate the potential of desktop data. A series of experiments were conducted and the results proved that desktop data greatly contributed to providing effective personalized reference words. Besides, the results demonstrated that a user's long-term interest model performed steadier than work task context, but the most valuable words were the top-3 words extracted from the work context.

Keywords: Query recommendation · Desktop data · User model · Work task · Search task

1 Motivations

Undoubtedly, explosive increase of various digital resources makes people feel extremely convenient to access any information. However, too much data is also confusing. Even with information services offered by many general search engines, like Google, Bing, Baidu, people still feel lost when they confront thousands, even millions of returned items. The prime reason is that search query cannot accurately describe an information seeking intention. Sometimes, people even do not realize what do they really need. Research found that there were only 24.5% user know exactly what they were looking for, and there were up to 68.4% user did not have explicit user goal [1]. Besides, a statistic report showed that people would rather use short query (1–3 words) than input more accurate words to describe their needs. This kind of short query went against satisfying a user's information need [2]. Therefore, inaccuracy and inadequacy of search query have become the most important reasons for deficient performance of current personalized information systems. Discovering more appropriate keywords without burdening users to depict search goals has become one of the hottest research issues.

© Springer Nature Singapore Pte Ltd. 2017
B. Zou et al. (Eds.): ICPCSEE 2017, Part I, CCIS 727, pp. 100–109, 2017.
DOI: 10.1007/978-981-10-6385-5_9

Query recommendation, as one of the solutions, has been proven effective in many information retrieval tasks. This method worked on offering additional relevant keywords to refine the description of search query. In prior researches, social annotations, query logs, query frequency, click-through data, query sessions and interaction context have been commonly exploited to facilitate query suggestion. The main problem of these approaches was that all necessary data resources relied on different internet servers for numerous services, which directly led to scarcity of useful data and irrelevancy of recommended items for a specific user.

However, plenty of valuable documents that highly relevant to a user's interest preference were stored on his or her personal computer. All these documents were regarded desktop documents in this paper relative to online documents. When people browsed through some online resources, they might save them on local disk for re-exploration. Compared to the limitless space on the internet, space of local disk was extremely rare and valuable. Therefore, it was reasonable to believe that desktop documents owned higher relevance to meet a user's specific needs. Based on these considerations, a desktop data driven approach was proposed to present useful query terms.

2 Related Works

Performance of a personalized query recommendation system relied on predicting accurate reference keywords to describe a user's specific need. Lots of resources have been exploited to achieve this goal. Based on data sources' type, all research methods were classified into three kinds.

The first one was based on documents. For example, some researchers analyzed all document in the corpus, and then used the space vector model to calculate a cosine similarity to find those valuable words as expanded keywords [3]. Some research works only exploited a part of relevant documents, and used a high-frequent vocabulary as expansion words [3], such as Top-N returned items or those items that a user had browsed during one session. Research of Verberne et al. [4] demonstrated that words selected from those clicked documents could improve performance of query suggestion in the interactive academic search. Besides, some corpuses, like dictionary, WordNet, Wikipedia, were introduced to analyze a query's morphology, syntax or semantics, and then to help understand the meaning of this query in different contexts [5, 6]. Moreover, social tags were found more valuable as a term resource than those words from feedback documents serving as expansion words [7].

The second type of method was based on search logs. To reach some specific search intentions, people always modify their queries during one whole search task. Some researchers regarded these multiple sessions as one session, and used the keywords that high frequent co-occur in a same session as reference words [3]. Some researchers focused on a user's behavioral signals to understand a user's true needs. Borisov et al. [8] proposed a context-aware time model by utilizing dwell times between user operations, and the experimental results showed that this model could obtain better performance on constructing ranking task than the results that based on the mean values. And two different queries were regarded as related if they led to a same link. Field and Allan [9] considered two types of queries: those that related to current task (regarded as on-task context) and those that did not (regarded as off-task context). Their research works illustrated on-task context were more useful, but were also easily affected by the off-task context, and experimental results showed that automatic identification of search task could reduce the impact of off-task queries in an interleave scenario. Furthermore, even an implicit negative feedback, defined by the behavior when users did not choose those recommended terms, was proven consistently and effectively boosted the performance of query auto-completion [10].

The third one was spatial-temporal context. For example, "百度" and "莆田系" were highly relevant during a particular period, and few people would link them together out of that special period. The temporal information was demonstrated valuable to offer nearby point-of-interest recommendations [11]. Association of season and topic were found useful to offer better recommendation service in a Community Question Answering system, such as users would expect different tour programs in different seasons [12]. Besides, with the fast increasing of the Internet of Things technology and the popularizing of smart terminals, people could be located immediately. And a location where a user started a seeking tasks could provide valuable information for a system to recommend personalized commercial information nearby that location. Besides, geographical information, like latitude, longitude and altitude, have been proven significantly valuable to capture a user's moving patterns, which was useful to improve point-of-interest recommendation [13]. Fang et al. used users' behavior and corresponding local items' information to facilitate local visiting and new location exploration [14]. Zhang and Chou's research works achieved significant improvement recommendation quality by exploiting geographical correlations, categorical correlations and social correlations [15]. Research of Zhang and Wang proposed a new model focusing on utilizing current location, time of visiting, social correlation and collaboration simultaneously, to satisfy a user's need, and their experiments achieved 47.3% growth in Recall@5 and 46.6% improvement in Precision@5 than the other best methods [16].

Undoubtedly, these works to some extent improved user experience. However, most researches were related to common users. It was difficult to offer a personalized suggestion to meet one user's specific information requirement. Of course, a user's search logs and search context have been introduced to predict this user's personal need in some research works. Research [17] showed that 35.7% search sessions were multi-task session tasks when people used Taobao, and 20% tasks were sibling and hierarchical search tasks, which were quite useful for product suggestion. However, cold-start and the data sparsity still were inevitable problems. Some researchers have noticed the value of desktop documents. Frenud [18] exploited a user's current task context to find what kind of document this user really need, and used the genre information to help next search task. Experimental results of Teevan et al. [19] proved that desktop information was valuable for improving search quality. Belkin [20] worked on the relationship between work task and search task. And our work focused on the combination of a user's desktop data and current work task simultaneously.

3   Desktop Data Driven Query Recommendation Model

This research assumed that a search user would use his or her personal computer to start a search task, and there were abundant digital documents related to this user's interest preference stored on this computer.

Rich desktop data could help avoid the problems of cold-start and data sparsity. However, a specific search task directly reflected a user's information need, which was generated by the current working scenario. Therefore, desktop data was reasonably regarded as a model that consisted of two parts: the user preference model and the work task context model, which served as the sources of providing suggestion words, as shown in Fig. 1.

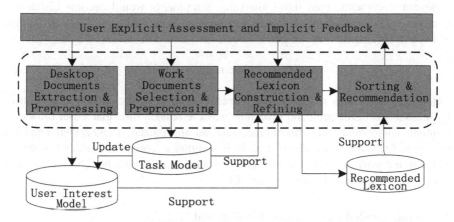

Fig. 1. Query recommendation model based on desktop context

3.1   Desktop Documents Extraction and Preprocessing

Experiments focused on text documents, which were regarded as a corpus. Word segmentation used NLPIR[1] (Institute of Computing Technology, Chinese Lexical Analysis System). Statistical analysis showed that verbs and nouns were the most valuable to represent the content of a document, so only top-20 highest frequency verbs and nouns were reserved. And the Space Vector Model (SVM) was introduced to represent the model of a document. The vector of a document d was represented as a group of two-tuple: $V_d = (<kw_1, wt_1>, \ldots, <kw_i, wt_i>, \ldots, <kw_m, wt_m>)$. Here kw_i was the keyword i, and w_{ti} represented the weight of this keyword in document d. The weight was achieved by the TF-IDF weighting algorithm, as in Eq. (1).

$$w_{i,d} = \frac{n_{i,d}}{\sum n_{k,d}} \cdot \log\left(\frac{N}{n}\right) \qquad (1)$$

---

[1] http://ictclas.nlpir.org/.

Here $n_{i,d}$ was the number of term i appears in document d. $\sum n_{k,d}$ represented the number of all terms appeared in this document. $\log(N/n)$ was the inverse document frequency. N was the total number of documents in this corpus. n was the number of term i appears in the corpus.

Then each keyword could be represented as a group of two-tuple: V_{kw} = (<d$_1$, wt$_1$>,..., <d$_i$, wt$_i$>,..., <d$_m$, wt$_m$>). This model was regarded as a user's interest preference model.

3.2   Work Documents Selection and Preprocessing

It was assumed that work tasks would directly lead to search tasks. Many factors could affect the conducting of a specific work task. If all the opened documents were regarded as relevant documents, then those low-value documents would become distracting. Therefore, high-frequent operation and session time were taken into consideration to select work documents. The weight of document i could be achieved by Eq. (2).

$$weight_d = A \cdot operation + B \cdot time \qquad (2)$$

Here *operation* was a summary of several operations, including input, remove, copy, mouse scroll, file print and save. *time* was the session time that a user browsed a document. These values were fuzzy processed and normalized. A and B were empirical parameters. Only those documents that their weights exceeded a threshold would be selected to the work context. And then the work task model could be achieved by using the same method as description in Sect. 3.1.

3.3   Recommended Lexicon Construction and Refining

A document consisted of many words, but only a few words owned high value to represent this document. Those documents opened before a user started the search task might be more related to a user's current work task. However, it was difficult to guarantee the quantity and quality of the documents in a task context, and then the problems of cold-start and the data sparsity were still inevitable. It was reasonable to choose both task context and user interest model as the sources to generate the most suitable recommended lexicon.

All the words in both task context model and user interest model was compared with user query. The similarity was defined as the cosine value of the angle between two vectors, as shown in Eq. (3).

$$sim(q, w) = \cos(v_q, v_w) = \frac{\sum\limits_{i=1}^{n} w_i \times q_i}{\sqrt{\sum\limits_{i=1}^{n} w_i^2 \cdot \sum\limits_{i=1}^{n} q_i^2}} \qquad (3)$$

Here w_i and q_i separately represented the weight of this word in document i.

3.4 Sorting and Recommendation

Two different sequences were achieved separately by task context and user model. With the combination of two models shown in Eq. (4), a final score of each word could be achieved. Based on this value, Top-10 highest scoring terms were presented to the search user.

$$score_w = \alpha \cdot wt_u + \beta \cdot wt_t \tag{4}$$

Here wt_u was the word score in user interest model, and wt_t was the word weight in task context model. α and β were experience parameters.

4 Experiments Evaluation

4.1 Experiments Design

The prime goal of this paper was to demonstrate the potential of desktop data serving as a valuable expansion words source in a personalized system. Ten volunteers were asked to join the experiments. Participants had different research directions. An application was asked to install in their personal computer. And this application obtained permission to analyze a folder in their local disks, where stored lots of digital documents related to the owner's interests. The experimental application could supervise the system and all the volunteers had to use the experimental system to search, which used Baidu as the meta-search engine.

Volunteers were asked to give a score to each top-10 document in the returned list: "2" representing the document was highly relevant to the information need, "1" meaning relevant and "0" indicating irrelevant. The Discounted Cumulative Gain (DCG) was introduced to evaluate the performance of our approach [9]. DCG was a performance measure for an algorithm in information retrieval. It had two assumptions: ① highly relevant (assigned to "2") documents were more useful when appearing earlier in a search engine result list, and ② highly relevant documents were more useful than relevant documents (assigned to "1"), which were in turn more useful than irrelevant documents (assigned to "0"). Then the DCG was defined as Eq. (5).

$$DCG(i) = \begin{cases} G(1), & if \quad i = 1 \\ DCG(i-1) + \frac{G(i)}{\log_2(i)}, & otherwise \end{cases} \tag{5}$$

Here $G(i)$ was the graded relevance of the result at position i.

To further assess the average effectiveness of a search system, the DCG was normalized to a value between "0" (the worst possible DCG given the ratings) and "1" (the best possible DCG given the ratings) when averaging across queries. The normalized DCG values could be achieved by Eq. (6).

$$nDCG = DCG/iDCGZ \tag{6}$$

Here *iDCG* represented the value of ideal DCG that was achieved by sorting documents according to their relevance to produce the maximum possible *DCG*.

4.2 Results Analysis

About the parameters' value in Eq. (3), user interview investigation results showed that high-frequency operation and browsing time owned equal value. Hence the value of A and B were both assigned to 0.5 in the experiments. And to achieve the parameters in formula (4), a series of experiments had been conducted. But the manual assessment results showed that different work scenario would lead to different dependence level on long-term user interest model. Therefore, the same weight was given to α and β at present.

Firstly, to prove the value of desktop data, we chose one user's evaluation results, as shown in Fig. 2. This task scenario was "*training of risk thinking ability*". Server documents discussing "*risk thinking*" were opened, which was regarded as the task context. General recommended terms for the query "思维" would be "导图", "训练", or "能力". However, there were lots of documents related to "*thinking*" stored in this user's personal computer, such as training of computational thinking, programming thinking, system thinking, and recently saved some documents about risk thinking. Thus, the experimental results were totally different. And the experimental results showed in Fig. 2 used the combination of the user interest model and the work task context model.

Baidu Original Result Query:思维	Gain	Experimental Results						
		The first recommended word: 风险	Gain	The second recommended word: 程序设计	Gain	The third recommended word: 计算	Gain
baike.baidu.com/link?url=z	0	wenku.baidu.com/link?ur	2	zhihu.com/question/278813	1	baike.baidu.com/link?url-	0
image.baidu.com/search/in	0	book.douban.com/subjec	1	wenku.baidu.com/link?url=p	1	blog.sina.com.cn/s/blog_	1
baike.baidu.com/link?url=>	1	kcf.com.cn/info.asp?id=9	0	blog.sina.com.cn/s/blog_49	0	douban.com/note/24276:	0
xuexila.com/naoli/siwei/5(	0	blog.sina.com.cn/s/blog_	1	blog.jobbole.com/67886/	1	product.dangdang.com/2:	0
zhihu.com/question/19599	0	xueshu.baidu.com/s?wd=	2	blog.csdn.net/robotcsdn/arti	2	zybang.com/question/03:	0
wenku.baidu.com/link?url=	0	xueshu.baidu.com/s?wd=	2	book.douban.com/subject/1	0	icourse163.org/course/h	1
tieba.baidu.com/f?kw=%C:	0	news.xinhuanet.com/fort	1	vdisk.weibo.com/s/z4DS9ol	2	huizhi123.com/view/247	0
wiki.mbalib.com/wiki/%E6	0	foodmate.net/zhiliang/is(	1	dangdang.com/?_ddclickuni(	0	baike.baidu.com/link?url-	1
yuedu.baidu.com/ebook/38	0	financialnews.com.cn/yh	0	igeekgroup.com/Content/ht	0	bookdao.com/book/1838	0
baike.baidu.com/link?url=:	1	bank.jrj.com.cn/2015/04	1	zhihu.com/question/198314	1	mooc.guokr.com/course.	0
DCG	1.4	DCG	6.126	DCG	4.375	DCG	1.72
Normalized DCG	0.5	Normalized DCG	0.843	Normalized DCG	0.735	Normalized DCG	0.654

Fig. 2. Comparison of evaluation results between the experimental system and the Baidu

The results clearly illustrated that the returned items for the first three recommended words in the experimental results all achieved high assessment scores. Furthermore, explicit desktop analysis found that the first two words were extracted from the work task context, and the third one came from the user interest model. The results showed that the user gave a relative better evaluation to the experimental system than the Baidu. Actually, Baidu also suggested 5 words, but they were more distracting than the original query, which was the reason that those results were not shown in this list.

To further evaluate the average performance of those selected words and to compare the value of user interest model and work task context, each volunteer was asked to evaluate top-10 returned items for top-10 presented words, and the average evaluation results were shown in Fig. 3.

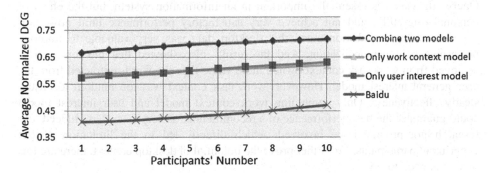

Fig. 3. Average performance of top-10 recommended words

The results showed that the combination of the user interest model and the work context model provided the best service as the expansion lexicon than the others. There were two very adjacent lines in the middle of the figure, representing the experimental system relied only on user interest model or only on the work task model. The results demonstrated that the average performance of top-10 recommended words for each user had small difference. But the performance of these two models had different value for the words with different ranked place, as shown in Fig. 4.

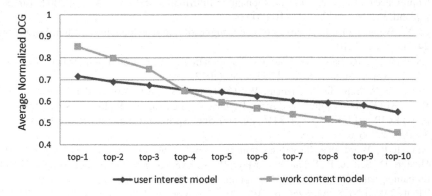

Fig. 4. Average performance for top-10 words

According to Fig. 4, it was obvious that top-3 words extracted from the work task context were more valuable than those from the user interest model. However, the line representing work context model had bigger slope than another one, which meant that the user interest model owned more reliable effectiveness, especially for cold-start scenario.

5 Conclusion

Query suggestion is essentially important in an information system, but the effect of personalizing still could not achieve very satisfactory performance until now. Our experimental results demonstrated that desktop data was very valuable to improve performance of personalization. And the results also showed that the top-3 terms achieved from work task context owned more value than the words extracted from the user general interest model. However, work task context was too random to ensure steady effectiveness. Only combining work context model and user interest model could guarantee the best performance of a recommendation service system. Indeed, our research got privacy issue involved, which directly led to the limitation of our experiment participants. To further prove the potential of desktop context, there are lots of works need to do.

Acknowledgements. This work is supported by the National Natural Science Foundation of China (No. 61402220), Scientific Research Project of Education Bureau of Hunan Province, China (Grant No. 15C1186), the Construct Program for the Key Discipline in University of South China (No. NHxk02), the Construct Program for Innovative Research Team in University of South China.

References

1. Broder, A.: A taxonomy of web search. In: 25th Annual International ACM SIGIR Conference on Research and Development in Information Retrieval, August 2002, Tampere, Finland, vol. 36, no. 2, pp. 3–10 (2002)
2. Li, Y., Wang, B., Li, J.: A survey of query suggestion in search engine. J. Chin. Inf. Process. 24(6), 75–84 (2010)
3. Carpineto, C., Romano, G.: A survey of automatic query expansion in information retrieval. ACM Comput. Surv. (CSUR) 44(1), 1–50 (2012)
4. Verberne, S., Sappelli, M., Kraaij, W.: Query term suggestion in academic search. In: de Rijke, M., Kenter, T., de Vries, Arjen P., Zhai, C., de Jong, F., Radinsky, K., Hofmann, K. (eds.) ECIR 2014. LNCS, vol. 8416, pp. 560–566. Springer, Cham (2014). doi:10.1007/978-3-319-06028-6_57
5. Navigli, R.: Word sense disambiguation: a survey. ACM Comput. Surv. 41(2), 1–69 (2009)
6. Hienert, D., Schaer, P., Schaible, J., Mayr, P.: A novel combined term suggestion service for domain-specific digital libraries. In: Gradmann, S., Borri, F., Meghini, C., Schuldt, H. (eds.) TPDL 2011. LNCS, vol. 6966, pp. 192–203. Springer, Heidelberg (2011). doi:10.1007/978-3-642-24469-8_21

7. Lin, Y., Lin, H., Jin, S., et al.: Social annotation in query expansion a machine learning approach. In: Proceedings of 34th Annual ACM SIGIR Conference, Beijing, China, July 2011, pp. 405–414 (2011)

8. Borisov, A., Markov, I., de Rijke, M., et al.: A context-aware time model for web search. In: Proceedings of 39th International ACM SIGIR Conference on Research and Development in Information Retrieval, 17–21 July 2016, Pisa, Italy, pp. 205–214 (2016)

9. Field, H., Allan, J.: Task-aware query recommendation. In: Proceedings of 36th International ACM SIGIR Conference on Research and Development in Information Retrieval, 28 July–1 August 2013, Dublin, Ireland, pp. 83–92 (2013)

10. Zhang, A., Goyal, A., Kong, W., et al.: adaQAC: Adaptive query auto-completion via implicit negative feedback. In: Proceedings of 38th International ACM SIGIR Conference on Research and Development in Information Retrieval, 9–13 August 2015, Santiago, Chile, pp. 143–152 (2015)

11. Yuan, Q., Cong, G., Ma, Z., et al.: Time-aware point-of-interest recommendation. In: Proceedings of 36th International ACM SIGIR Conference on Research and Development in Information Retrieval, 28 July–1 August 2013, Dublin, Ireland, pp. 363–372 (2013)

12. Otsuka, A., Seki, Y., Kando, N., et al.: QAque: faceted query expansion techniques for exploratory search using community QA resources. In: Proceedings of 21st International Conference on World Wide Web, April 2012, Lyon, France, pp. 799–806 (2012)

13. Cheng, C., Yang, H., King, I., et al.: A unified point-of-interest recommendation framework in location-based social networks. ACM Trans. Intell. Syst. Technol. (TIST) 8(1), 10 (2016)

14. Fang, Q., Xu, C., Hossain, M.S., et al.: STCAPLRS: a spatial-temporal context-aware personalized location recommendation system. ACM Trans. Intell. Syst. Technol. (TIST) - Special Issue on Crowd in Intelligent Systems, Research Note/Short Paper and Regular Papers, 7(4), 59 (2016)

15. Zhang, J., Chou, C.: GeoSoCa: exploiting geographical, social and categorical correlations for point-of-interest recommendations. In: Proceedings of 38th International ACM SIGIR Conference on Research and Development in Information Retrieval, 09–13 August 2015, Santiago, Chile, pp. 443–452 (2015)

16. Zhang, W., Wang, J.: Location and time aware social collaborative retrieval for new successive point-of-interest recommendation. In: Proceedings of 24th ACM International on Conference on Information and Knowledge Management, 18–23 October 2015, Melbourne, Australia, pp. 1221–1230 (2015)

17. Zhou, X., Zhang, P., Wang, J.: Examining task relationships in multitasking consumer search sessions: a query log analysis. In: Proceedings of 79th ASIS&T Annual Meeting: Creating Knowledge, Enhancing Lives through Information & Technology, Copenhagen, Denmark, 14–18 October 2016. Article No. 102

18. Freund, L.S.: Exploiting task-document relations in support of information retrieval in the workplace. Doctoral dissertation, University of Toronto (2008)

19. Teevan, J., Dumais, S.T., Horvitz, E.: Potential for personalization. ACM Trans. Comput.-Hum. Interact. (TOCHI) 17(1), 4 (2010)

20. Belkin, N.J.: Helping people find what they don't know. Commun. ACM 43(8), 58–61 (2000)

Disease Prediction Based on Transfer Learning in Individual Healthcare

Yang Song[1], Tianbai Yue[2], Hongzhi Wang[1(✉)], Jianzhong Li[1], and Hong Gao[1]

[1] Massive Data Computing Research Center, Harbin Institute of Technology,
P.O. Box 750, 150001 Harbin, China
16s003050@stu.hit.edu.cn, {wangzh,hzwang}@hit.edu.cn
[2] Heilongjiang University, Harbin, China
letianbai1005@qq.com

Abstract. Nowadays, emerging mobile medical technology and disease prevention become new trends of disease prevention and control. Based on this technology, we present disease prediction models based on transfer learning. Breast cancer disease data has been used to build our model. According to the neural networks, the basic model has been provided. With unlabeled data, transfer learning is a appropriate way to revise the module to increase accuracy. The test results show that the algorithm is suitable for data classification, especially for unlabeled health data.

Keywords: Individual healthcare · Transfer learning · Neural networks · Disease prediction · Unlabeled data

1 Introduction

In recent years, the importance of individual health care is of increasing priority. There are more and more applications combining hospital treatment data, the user's medical information, and patient hospital information, to achieve a full range of medical monitoring. Monitoring the health condition using mobile technology is effective in increasing the accuracy of disease prediction.

For disease prediction, many methods have been used to provide relatively accurate prediction results. For example, the decision tree, naive Bayes, and neural networks are used for heart disease prediction [1]. However, some problems in disease prediction remain unsolved.

First of all, there is a lack of valid data. Hospital data is private, therefore, data is not easy to access and analyze, creating many challenges for disease prediction. Second, given that the health data is diversified with different formats, such data is difficult to utilize. Third, the most important problem is that with the popularity of mobile phones, more and more data comes in with no labels. Thus, how to classify unlabeled data is an urgent problem.

Therefore, for efficient and accuracy disease prediction, it is necessary to handle unlabeled data efficiently. To achieve high efficiency and accuracy, algorithms such as neutral networks and transfer learning [12] could be adopted.

© Springer Nature Singapore Pte Ltd. 2017
B. Zou et al. (Eds.): ICPCSEE 2017, Part I, CCIS 727, pp. 110–122, 2017.
DOI: 10.1007/978-981-10-6385-5_10

For disease prediction, many approaches have been proposed. Cardiovascular disease prediction [2] can be predicted with lipid-related markers. However, this is dependent on the doctor's experience. Decision trees and Bayesian classification have been tested to have good performance in accurate heart disease prediction [3]. However, the other methods such as KNN and classification based on clustering have not performed well. The association rule discover [4] is shown accurate results in disease prediction. The feature subset selection [5] also shows a good performance in heart disease prediction. However, all of these methods disregard the fact that the dynamic data may have unlabeled data.

Recently, an increasing amount of methods such as neural network [6] successfully assist in medical diagnoses. Artificial Neural Networks, Decision Tree and Naive Bayes [7] are employed in different systems to predict different diseases. Furthermore, based on ranking and learning techniques [8], image-based diseases can be diagnosed as well. These methods benefit the progress of individual healthcare. However, so far, a dynamic data based algorithm to handle both labeled and unlabeled data has not yet been created.

Thus, to support efficient analytics on individual healthcare, we have proposed an optimization algorithm for disease prediction. Given that existing disease prediction methods, unlabeled data or data lacking classification do not contribute to the accuracy, the goal of the optimization algorithm is to combine unlabeled data to improve the accuracy of disease prediction.

To achieve this goal, we developed disease prediction algorithm which combines a supervised study and unsupervised study [11] to improve the accuracy of disease prediction. We utilized transfer learning to improve the common model and enhance performance. To address this problem, we proposed the supervised study to be the basis of our algorithm and used the unsupervised study to be part of transfer learning. Then, we revised the present prediction and reconstructed the model. Based on these algorithms, we have provided several patterns suitable for different situations including iteration.

In summary, this paper makes the following contributions.

1. First, we propose an optimization algorithm for disease prediction on individual healthcare. As we know, this is an innovative strategy that takes unlabeled data into account on individual healthcare.
2. The second contribution is using transfer learning in disease prediction. As the disease data is dynamic and hard to find the pattern, we employ transfer learning to solve this problem. With transfer learning, a supervised study and an unsupervised study have been combined to achieve higher performance.
3. Finally, we employ the algorithm with R. Our evaluation shows that the algorithm significantly improves the correlation of the prediction by approximately 30%.

The remainder of this paper is organized as follows. We first overview the algorithm in Sect. 2. The detailed disease predict algorithm is then proposed in Sect. 3. The experimental evaluation is reported in Sect. 4. Finally, we conclude and discuss future work in Sect. 5.

2   Overview of Disease Prediction

In this section, we give an overview of our proposed algorithm. We introduce the design of the algorithm in the first section. After that, we discuss the choice of the present machine learning algorithms.

2.1   Algorithm Design

The algorithm is designed into 4 forms suitable for different needs. As we know that we take unlabeled data into account, we provide different forms to suit different situations, and combinations of data form, shown below in Fig. 1.

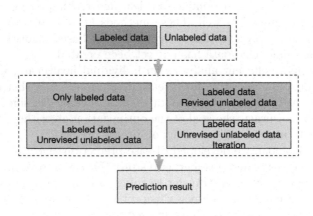

Fig. 1. The design of the algorithm. The algorithm is composed with 4 parts.

The detailed process for the design is shown as follows.

1. Only labeled data (OLD). This algorithm id built for the case that there is a sufficient amount of labeled data. A supervised method is used in this situation.
2. Labeled data and unrevised unlabeled data (LDUUD). This algorithm uses labeled data first. Then, we use the model to obtain a prediction result, After which, we use the result to train the model again. This method is suitable for data with less unlabeled data, normally less then 1%.
3. Labeled data and revised unlabeled data (LDRUD). The same process as above except that the prediction result will be revised with an unsupervised algorithm. This method is for a large amount of unlabeled data. In this way, the unlabeled data can be divided into clusters.
5. Labeled data, revised unlabeled data and iteration (LDRUDI). The same process as the above algorithm, but this method, will revise and train a small part of unlabeled data each time. This method is particularly for a massive amount of unlabeled data, as shown in Fig. 2.

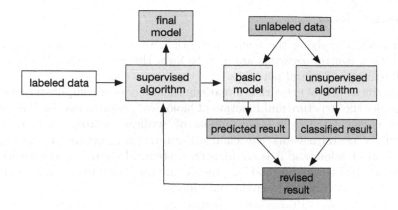

Fig. 2. The process of the last version of algorithm.

In this paper, we will introduce the main idea of the last version of the algorithm, for the others are simplified versions of the last one. In addition, we present all conducted experiments to testify about the performance of our proposed methods.

2.2   Algorithm Choice

Supervised methods and unsupervised methods contribute to the whole algorithm. To make transfer learning a more optimized algorithm, we selected our algorithm carefully.

For the unsupervised method, we compare different methods and ultimately use neutral networks [10] for its flexibility and accuracy. For the supervised method, we propose simple methods like K-means [9] to achieve effective and efficient outcomes.

This is the overview of the disease prediction algorithm. The detailed design is shown in the following section.

3   Disease Prediction Algorithm

This section introduces the entire algorithm and detailed methods including finding the optimized parameters. First, we will introduce our basic model using neural networks. Next, we present methods to process unlabeled data. The core idea is to combine the cluster result to adjust the prediction result. To achieve this goal, we use the iteration formula to find the optimal parameter. Finally, we present our final model with improvements in accuracy and correlation. The detailed design will be presented as follows.

3.1 Basic Model

Basic model is applying the neural networks. As we have a small amount of labeled data, neutral networks are used to form the basic model. We provide a brief illustration of neural networks.

An artificial neural network is a mathematical model or computational model that mimics the structure and function of biological neural networks. The neural network is calculated by a large number of artificial neurons. In most cases, the artificial neural network can change the internal structure on the basis of external information, and it is an adaptive system. Modern neural networks are a nonlinear statistical data modeling tool, commonly used to construct complex relationship modeling or to explore data models.

We use the Back Propagation Neutral Network. The learning consists of two forward processes, the forward propagation of information and the reverse propagation of errors. Three layers are essential in this algorithm.

1. **Input layer.** The input layer of neurons are responsible for receiving input from the outside world and sent processed input to the middle layer of neurons.
2. **Hidden layer.** The middle layer is the internal information processing layer, responsible for information transformation. According to the needs of information processing ability, the middle layer can be designed as a single hidden or multi-hidden layer structure. The last hidden layer sends information to the output layer. After further processing, the learning forward communication process is completed.
3. **Output layer.** As the name suggests, the output layer outputs the processing results to the outside world.

In our algorithm, we considered using a 5 perception machine in the hidden layer. As the picture shows in Fig. 3, circles present perception machines, which are a functions to calculate the input data to give a classification. Different weights are given to each perception machine.

The basic model has the following process. First, we normalize the given labeled data. Then, we initialize the BP neural network with hidden layer and sigmoid function. After determining that the sample is convergent, we adjust the weights according to the error. Finally, we arrive at a classification result.

3.2 Data Handle

After achieving a basic model, we still do not satisfy requirements for accuracy and correlation. Thus, the next step is to employ transfer learning to mark and categorize unlabeled data. We also give a brief introduction of transfer learning.

Data calibration will require a significant amount of manpower and material resources. The goal of transfer learning is to use the knowledge learned from an environment to help learn tasks in a new environment. There is only a small amount of new data, but there are a large amount of usable previously marked

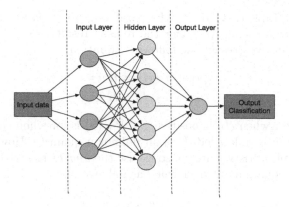

Fig. 3. The structure of Back Propagation Neutral Network algorithm.

data, and even other types of valid data. Thus, by selecting effective data from this usable data pool and adding it to the current training data, it is possible to train the new model.

With proper methods, unlabeled data can be used as labeled data to improve accuracy, as shown in Fig. 4. Thus, we provide a unsupervised algorithm to process unlabeled data.

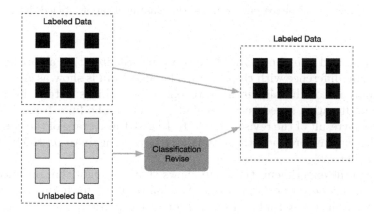

Fig. 4. The basic idea of transfer learning.

There are few steps to process unlabeled data.

The first step is to use the basic model to get a prediction result. As we know the prediction result may have some errors, we use the next step to revise the model.

Secondly, we use the unsupervised algorithm to give a classification to unlabeled data. By choosing K-means algorithm, we can then divide the data into two parts. After classifying data, we give a revision of the formal prediction.

As shown in Table 1, we define the parameter to be used in the following
algorithm. In the revision, we prefer the revised result to be close to the real
classification. Thus we present the following equation.

$$pre + a * clu = tru \tag{1}$$
$$pre - a * clu = tru \tag{2}$$

Equation (1) is for when clu is positive and Eq. (2) is for when clu is negative.

Given that tru is unlabeled. It is an unknown quantity. However, the data
will have the same statistic pattern, so we use the $avgres$ to arrive at an answer.
In this way, the equation is transformed as follows.

$$avg((pre_i - a * clu_i)or(pre_i + a * clu_i)) = avgres \tag{3}$$

Table 1. Unlabeled data parameter table

Parameters	Meaning
pre	The prediction result from the basic model
clu	The cluster result from k-means
a	The parameter used to revise the prediction result
tru	The real classification of unlabeled data
$avgres$	The average result of the classification of labeled data

The optimized of a is prioritized, thus, we use the iteration algorithm to get
the minimum difference with the $avgres$. The algorithm is show as Algorithm 1.

The purpose of the algorithm is to optimize a. Lines 1 to 6 initiate the
parameters. Iteration speed is represented e. And for the lines from 7 to 14, it
calculates the result of the revise. Line 15 to 19 give the average of the calculation
result. The end part produces the optimal a or iterates again to get the best result
from lines 20 to 26.

Through the experiment, the result shows that this algorithm produces optimal a and give a better performance to the final result. After finding the optimal
a, we can obtain revised labeled data through calculation. Thus, we require a
lager amount of labeled data. And then we take the next step.

3.3 Model Revise

With all the labeled data, we run the neutral network again. This time, all
labeled data is taken into account.

When working with a massive amount of unlabeled data, the prediction will
result in incorrect classifications that hinder prediction performance. In order
to solve this problem, we use an iteration method to arrive at a result. In our
experiment, we use half of the unlabeled data to train for the first time and

---

Algorithm 1. Optimal a

Input: pre_i,clu_i,$avgres$
Output: a

```
 1  a=1;
 2  e=0.01;
 3  tmpᵢ= new array();
 4  sum = 0;
 5  count = 0;
 6  avgtmp = 0;
 7  for each i do
 8  |   if cluᵢ is negative then
 9  |   |   tmpᵢ =preᵢ-a*cluᵢ;
10  |   end
11  |   else
12  |   |   tmpᵢ=preᵢ+a*cluᵢ;
13  |   end
14  end
15  for each i do
16  |   sum+=tmpᵢ;
17  |   count=count+;
18  end
19  avgtmp= sum/count;
20  if avgres equal to avgtmp then
21  |   return a;
22  end
23  else
24  |   a=a-e;
25  |   return to line 7;
26  end
```

---

the other unlabeled data for the second time. The two rounds of remodeling give a better performance in contrast to a one-round method. Therefor, when manipulating a massive dataset, the iteration method is recommended.

4   Evaluation

We evaluated disease prediction through extensive experiments on disease datasets. We conducted experiments to test the following points.

1. What unlabeled data can be used in the algorithm?
2. Does the proposed algorithm lead to an improvement in accuracy?
3. What is the general performance of the proposed algorithm?

4.1   Experimental Setup

For our evaluation, we tested various disease datasets. We ran all experiments 10 times and reported the average and standard deviation of the results.

Testbeds: We implemented our program with R and ran the program on R 3.3.0.

Dataset: For evaluation of disease prediction algorithms, we used datasets of UCI. We handled the dataset to imitate real life application of disease predictions. We use the breast cancer dataset [13] as an example. The data consists of 14 attributes, and are divided into 4 part to compare results.

- OLD: 100 labeled data only to get the prediction model. More than 300 data points for experimental verification.
- LDUUD: 100 labeled data and 200 unrevised unlabeled data for training. More than 300 data points for experimental verification.
- LDRUD: 100 labeled data and 200 revised unlabeled data for training. More than 300 data points for experimental verification.
- LDRUDE: 100 labeled data and 200 unrevised unlabeled data for training. More than 300 data points for experimental verification.

We compare the accuracy, error rate, precision, recall and correlation of the final prediction model. The detailed definitions are show in Table 2.

Table 2. Detailed definition of evaluation parameter.

Name	Definition
Accuracy	The predicted positive cases/all the cases
Error rate	The predicted negative cases/all the cases
Precision	The real positive cases/the predicted positive cases
Recall	The right predicted positive cases/the real positive cases
Correlation	The similarity of predicted case and real case

4.2 Result Evaluation

We implemented the 4 version of the experiment to evaluate breast cancer data. From the neural networks from Figs. 5, 6, 7 and 8, we find that as the dataset grows in size, the error rate increases. The detailed difference will be given as follows.

OLD. This method gives a relatively accurate performance of cancer disease prediction with around 94.5%. The prediction result has an 82% correlation to the real classification, which is acceptable for this size of data. Therefore, this method can be chosen when labeled data comprises the majority of the dataset.

LDUUD. This method gives the least accurate performance. The reason lies in the management of unlabeled data. The unlabeled data is marked by the basic model, so it will cause the reconstructed model to be over-fitted, leading to poor performance. This shows that the method should be revised. However, this method can still be used for a small amount of unlabeled data.

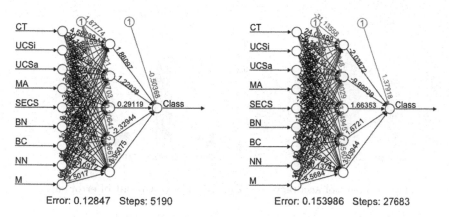

Error: 0.12847 Steps: 5190   Error: 0.153986 Steps: 27683

Fig. 5. The neutral networks of OLD.   **Fig. 6.** The neutral networks of LDUUD.

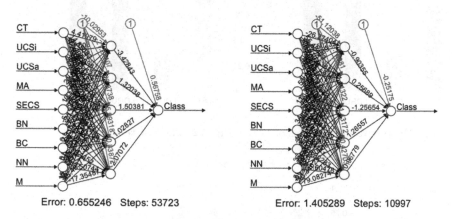

Error: 0.655246 Steps: 53723   Error: 1.405289 Steps: 10997

Fig. 7. The neutral networks of LDRUD. **Fig. 8.** The neutral networks of LDRUDI.

LDRUD. This method gives a relatively accurate result with revision of the basic model. We come up with the best parameter a around 0.6. The result shows that it has a similar level of accuracy and correlation. This indicates that the revision greatly improves the number of right classifications, thus showing that our algorithm effectively improves the model by 30% in contrast with the LDRUD.

LDRUDI. The last method give the best performance of the cancer prediction algorithm. It has the best percentage in accuracy, error rate, precision, recall and correlation, especially in correlation with an improvement of 10%. This shows that the iteration method is effective and can improve accuracy as well.

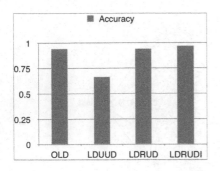

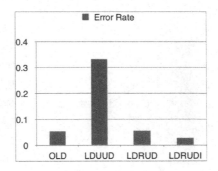

Fig. 9. The test result of accuracy.   **Fig. 10.** The test result of error rate.

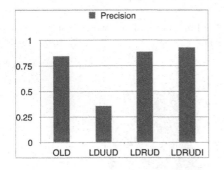

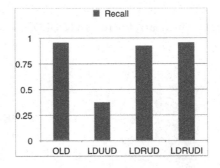

Fig. 11. The test result of precision.   **Fig. 12.** The test result of recall.

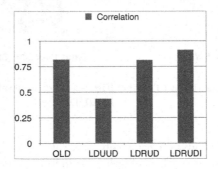

Fig. 13. The test result of correlation.

OVERVIEW. The comparison of all 4 methods is shown in Figs. 9, 10, 11, 12 and 13. We can see that with our revision algorithm, the accuracy improves significantly. With unlabeled data, we can still improve the accuracy of our prediction model. Transfer learning makes great contributions to disease prediction.

5   Conlusion

In this paper, we propose a disease prediction algorithm as a method to process dynamic unlabeled data for individual healthcare. To achieve this goal, we designed the algorithm with transfer learning. The whole process is composed with neural networks and K-means. The basic model is comprised of neural networks. After K-means algorithm, we give a revision of the prediction and provide a new model with all the labeled data. As a result, the prediction accuracy is improved significantly. We implement the algorithm into R. Finally, we have also designated 4 versions of algorithm to process data under different circumstances. From the experimental results, the proposed approach leads to significant accuracy improvement in disease data.

For future work, we plan to extend the design of disease prediction in transfer learning. As previously mentioned, transfer learning is a method to combine the experienced or formal model with new data even from another domain. Therefore, the next step for our work is to identify the common features of different diseases and make a model according to these similarities. In this way, making slight modification to a model of heart disease can predict kidney disease with accurate results. The solution needs to be verified in the future work.

Acknowledgments. This paper was partially supported by National Sci-Tech Support Plan 2015BAH10F01, NSFC grant U1509216,61472099, the Scientific Research Foundation for the Returned Overseas Chinese Scholars of Heilongjiang Province LC2016026 and MOE-Microsoft Key Laboratory of Natural Language Processing and Speech, Harbin Institute of Technology.

References

1. Palaniappan, S., Awang, R.: Intelligent heart disease prediction system using data mining techniques. IEEE/ACS International Conference on Computer Systems and Applications, pp. 108–115. IEEE (2008)
2. Di, A.E., Gao, P., Pennells, L., et al.: Lipid-related markers and cardiovascular disease prediction. J. Am. Med. Assoc. (JAMA) **307**(23), 2499–506 (2012)
3. Soni, J., Ansari, U., Sharma, D., et al.: Predictive data mining for medical diagnosis: an overview of heart disease prediction. Int. J. Comput. Appl. **17**(8), 43–48 (2011)
4. Ordonez, C.: Association rule discovery with the train and test approach for heart disease prediction. IEEE Trans. Inf Technol. Biomed. **10**(2), 334 (2006)
5. Anbarasi, M., Anupriya, E., Iyengar, N.C.S.N.: Enhanced prediction of heart disease with feature subset selection using genetic algorithm. Int. J. Eng. Sci. Technol. **2**(10), 5370–5376 (2010)
6. Weng, C.H., Huang, C.K., Han, R.P.: Disease prediction with different types of neural network classifiers. Telemat. Informat. **33**(2), 277–292 (2016)
7. Chadha, R., Mayank, S.: Prediction of heart disease using data mining techniques. CSI Trans. ICT 1–6 (2016)
8. Huang, W., Zeng, S., Wan, M., et al.: Medical media analytics via ranking and big learning: a multi-modality image-based disease severity prediction study. Neurocomputing **204**, 125–134 (2016)

9. Hartigan, J.A., Wong, M.A.: Algorithm AS 136: a K-Means clustering algorithm. Appl. Stat. **28**(1), 100–108 (1979)
10. Haykin, S.: Neural networks: a comprehensive foundation. Neural Netw. Compr. Found. 71–80 (1994)
11. Hastie, T., Tibshirani, R., Friedman, J.: The elements of statistical learning. J. Roy. Stat. Soc. **167**(1), 192 (2001)
12. Pan, S.J., Yang, Q.: A survey on transfer learning. IEEE Trans. Knowl. Data Eng. **22**(10), 1345–1359 (2010)
13. Breast Cancer Dataset. https://archive.ics.uci.edu/ml/datasets/Breast+Cancer+Wisconsin+(Diagnostic)

Research on Fuzzy Matching Query Algorithm Based on Spatial Multi-keyword

Suzhi Zhang$^{(\boxtimes)}$, Yanan Zhao, and Rui Yang

School of Computer and Communication Engineering,
Zhengzhou University of Light Industry, Zhengzhou 450002, Henan, China
zhsuzhi@zzuli.edu.cn

Abstract. With the rapid growth of spatial data, POI (Point of Interest) is becoming ever more intensive, and the text description of each spatial point is also gradually increasing. The traditional query method can only address the problem that the text description is less and single keyword query. In view of this situation, the paper proposes an approximate matching algorithm to support spatial multi-keyword. The fuzzy matching algorithm is integrated into this algorithm, which not only supports multiple POI queries, but also supports fault tolerance of the query keywords. The simulation results demonstrate that the proposed algorithm can improve the accuracy and efficiency of query.

Keywords: Spatial data · Multi-keyword search · Approximate query algorithm · RB-tree

1 Introduction

In recent years, more and more search engines have begun to support spatial keyword query [1], so that more and more web resources begin to have location attributes, for example: Through twitter users could publish their geographic location information; Restaurants and tourist attractions' web pages usually included their location of the information and check-in function. With the rapid development of Location Based Service, spatial keywords are gradually emerging at the forefront of technology.

Spatial Keyword Query is one of the most important methods to cope with spatial data. Most of the existing search engines are beginning to support spatial data query, by importing text data and spatial constraints as query conditions to return the results that satisfy the spatial information and text data, and display the results according to the corresponding sorting algorithm. In the past, the query algorithm had been impossible to meet all needs of users. How to locate the analogous results in the massive spatial data has become the hot-spot of the current research.

In real life, the user may query multiple keywords to get more abundant information, how to solve problems that the corresponding geographical location in the spatial data to meet the needs of users, and the text data of one POI can not fully include the user input information, for example users want to buy goods, eat foods, sing songs, thus it may require multiple POIs to solve the corresponding problem; and at time of inquiry, each different keywords needs different corresponding edit distance threshold.

© Springer Nature Singapore Pte Ltd. 2017
B. Zou et al. (Eds.): ICPCSEE 2017, Part I, CCIS 727, pp. 123–133, 2017.
DOI: 10.1007/978-981-10-6385-5_11

2 Formal Description of Problem

2.1 Source of the Problem

The query of spatial keywords has just started and some achievements have been made: MHR-tree proposed by Yao et al. [2], previously, the index structure only supports exact match query, and MHR-tree index will approximate string matching for the first time; And string matching is introduced into the spatial query, and the min-wise signature is added to each tree node to prune the query result yet, but while the signature data are not guaranteed, resulting in a smaller return rate. Alsubaiee et al. [3] proposed the LBAK tree memory index structure, which ensures the balance between the index cost and the query cost on the basis of Yao. However, the problem of node limitation can not be solved by adding the inverted index into the index node; Wang et al. proposed RB-tree [4] to introduce the bitmap into the index results to improve the query rate and the accuracy of the results; Hu et al. [5] applied the approximate matching to the surrounding query. However, only a single text keyword in the POI point is studied, and there is less research on the multi-text keywords for spatial location points. In the application of multi-keyword query, Zhang et al. [6] proposed the MCK query, that is, the distance between the two most distant POIs of the keyword query as the area diameter, but they only proposed the idea, and not really applied to the index; Cao et al. [7] developed the concept and designed the corresponding cost function, but they did not deal with the text relevance problem.

For keywords approximate matching, the study above are only for single keyword research. In the study of multi-spatial keywords, only the distance between the farthest and the closest points are considered, and how to solve the edit distance in multi-keyword of spatial database is not studied; The spatial attributes are considered much more in multi-spatial keyword query, less consideration is given to textual information. Aiming at the existing problems, this paper proposes a spatial multi-keyword query (SMQ) algorithm, which supports multi-spatial keyword approximation, and combines the spatial correlation and the text correlation effectively. This algorithm introduces the cost coefficients of multi-keyword query into the spatial index structure to support multi-spatial keywords fuzzy matching. In order to satisfy the user's ability to find all the information about the location information associated with the keywords, the SMQ algorithm should do something as follows:

1. The query results contain all the query keywords (including some complex keywords such as "Seven Days Hotel"), and support multi-keyword approximation query.
2. SMQ algorithm is oriented around the range query algorithm, so the query location should be closest to the P position.
3. The query POI point set is relatively nearest.

   Based on the above statements, the main contributions of this paper are as follows:

1. Firstly, Building index structure based on RB-tree to support multi-keyword approximate matching.
2. According to the index structure, an efficient approximate query algorithm is proposed.

3. By using the real data to carry out the simulation experiment, and using the experimental data to prove the superiority and feasibility of the approximation.

2.2   Basic Concepts

In a given set of spatial data, the POI points p in the spatial data are represented $O = \{p1, p2, p3, \dots, pn\}$, $p = \{o, t\}$, then $O = \{(o_1, t_{1.1}, t_{1.2}, \dots t_{1.w}) \dots (o_n, t_{n.1}, \dots t_{n.w})\}$, as showed in Fig. 1, the space collection contains multiple POI, o_i said POI's location, o_t said the corresponding keywords set. Set spatial keyword query SKQ: $q_t = \{(t_1, \tau_1), \dots (t_m, \tau_m)\}$, $\forall_i \in [1, m]$, $\tau_i \in N$, q_l represents a region or a discrete point in Euclidean space, q_t represents the text attribute of SKQ, and t represents the keyword, τ represents the corresponding edit threshold.

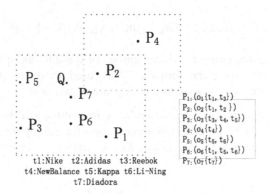

Fig. 1.   The relationship between spatial information and textual information.

2.3   Algorithm Preparation

Definition 1: Spatial Multi-keywords Query Set SMS:

$$\{t_1, t_2, t_3 \dots t_m\}, 1 \le |m| \le |q_t| \tag{1}$$

The textual relevance of SMS include two aspects: keyword weight and edit distance. In different situations, POI has the same keyword information in many cases. As showed in Fig. 1: P1, P2 and P6 have the keyword t_1 "Nike", since the algorithm is oriented around the query, so sometimes only one spatial location information can meet the needs of user, however, each POI corresponding to different weight value of text information, such as: large supermarkets sell many different attributes of goods, but compared with the store, the store goods style and reputation will be better. For this purpose, the corresponding weight value w(t) is set for each keyword. And according to the different needs of users to set the corresponding coefficient.

In this paper, we use the classical TF-IDF model [8] to compute the weight w(t) of a single keyword, and optimize it on the basis of the original formula:

$$w(t) = \frac{\sum (tf_{w,d} \cdot idf_{w,D,S})}{w_{max}} \tag{2}$$

Among them, $tf_{w,d}$ indicates the number of queries that appear in the document D, the greater the number the greater the weight value, $idf_{w,d,s}$ denotes the reciprocal of the number of query words in the query region |Ds| only. w_{max} represents the maximum weight value of the keyword.

In the past, for a single keyword approximation query, q-gram index is introduced into R-tree, and query words were fuzzy matched. If the distance between the query word t_i and the target word o_t in the database is less than the given threshold, the matching succeeds; Otherwise, it fails. And each value is related to t_i, and each size threshold associated with t_i, in the spatial query, the following formula is generally used for pruning:

$$\text{if } \varepsilon (t_i, o_{t,i}) = \tau, \text{ then } Gt_i \cap Go_{t,i}| \geq \max(|t_i|, |o_{t,i}|) - 1 - (\tau - 1) * q \tag{3}$$

Here, Gm denotes the number of index entries of q-gram, and q denotes the index entry length. In this paper, the formula (2) is used to query pruning rules for multi-spatial keywords, and q is set to 2 during the experiment.

Definition 2: Formal definition of text-similarity in multi-spatial keywords query. In the formula, the weight W is taken into account, and the editing distance is set as the influence factor adding weight calculation. Key words text correlation attributes calculation formula:

$$Q_T(t, o_{t,i}) = \frac{w(o_{t,i})}{(1 + \max(\tau^*))^\theta} \tag{4}$$

W(o) denotes the weight value of the keyword in the spatial data which have the least distance from the ti $\in$ SMS edit distance. In order to realize the value range of [0, 1], the formula (2) has been adjusted to show the edit distance; τ^* represents the edit distance, θ is to adjust the specific gravity between the edit distance τ and Weight w, initially set to 2.

Definition 3: The query diameter D of multiple spatial keywords. The minimum spatial query range includes the corresponding POI points of all the keywords q_t in the SMS set. Since q_t and q_i are not one-to-one correspondence, the expression D is divided into two cases, let the length of QRS (Query Result Set) be the number of POI returned, then the diameter D (o_i, o_j) expression is as follows:

$$\left\{ \begin{array}{ll} 1 \leq i < j \leq m & \\ D(o_i, o_j) = 0 & \text{if } |QRS| = 1 \\ \text{else} & \\ D(o_i, o_j) = \max(dist(o_i, o_j)) & 1 < |QRS| \leq m \end{array} \right\} \tag{5}$$

The normalization of Eq. (5) is performed, in this paper, the SMQ query algorithm is oriented to the surrounding query algorithm. the user query point is p, then the point p is the center, the distance from the farthest point of the p point in the QRS result set is taken as the radius, Let the radius Max(dist(p, om)), the diameter Dmax = 2Max(dist (p, o_m)), the modified cost of the query diameter D(SMS) is:

$$
\left\{
\begin{array}{l}
D = 0 \\
\text{else} \\
D = \dfrac{D(o_i, o_j)}{2\mathrm{Max}(\mathrm{dist}\,(p, o_m)) + D(o_i, o_j)} \quad 1 \le i < j < m \le n
\end{array}
\right\}
\tag{6}
$$

Definition 4: Farthest radius R of Multi-spatial Keyword Query. R is the distance from the query point p to the farthest POI in QRS. The expression is as follows:

$$
R(p, o_m) = \mathrm{max}(\mathrm{dist}(p, o_m)), \ o_m \in QRS
\tag{7}
$$

The formula (7) was normalized [9].The formula (7) is set into the normalized linear transformation formula, R as the most distant radius of the nearest neighbor query. In the neighborhood query N(q) [8], Min(R) = max(dist(p, o_i)), Max(R) = max(dist(p, o_i)) + maxdist$(o_i, o_j), o_i, o_j \in$ N(q), after finishing the formula:

$$
R = \frac{R(p, o_m) - \mathrm{Max}(\mathrm{dist}(p, o_i))}{\mathrm{Max}(\mathrm{dist}(o_i, o_j))} \quad 1 \le i < j \le n
\tag{8}
$$

Therefore, the spatial correlation of the set SMS is given by formula (6) and (8):

$$
Q_L(p, o_{i,j,m}) = \alpha \frac{D(o_i, o_j)}{2\mathrm{max}(dist(p, o_m)) + \mathrm{max}(dist(o_i, o_j))} + \beta \frac{R(p, o_m) - \mathrm{max}(dist(p, o_i))}{\mathrm{max}(dist(o_i, o_j))}
\tag{9}
$$

α, β are the correlation coefficients.

The spatial correlation formula (4) and the text correlation formula (9) of spatial keyword approximation are combined to obtain the cost formula:

$$
(\cos t)_{t,o} = \lambda \cdot Q_T(t, o) + (1 - \lambda) \cdot Q_L(t, o), \ (\alpha + \beta + \lambda = 1)
\tag{10}
$$

Define the problem: A query location P, a set of multi-spatial keyword query SMS $(o_1, o_2, o_3, \ldots o_m)$, the collection of all objects through the SMS(t, o) function filter, in the spatial database to find the corresponding POI point collection, and according to the needs of users to change the corresponding coefficient. If asked the required distance is closest, set the value of λ smaller, and if asked the text related, set λ some larger.

Case 1: a query set in SMS keywords (t1, t2, t3, t4), as shown in Fig. 1, Q is the position of the user, and seven points are given to choose, in the approximate match, depending on the user's requirements, the corresponding results are given:

When the user needs for the most recent query, the results are as follows: QRS {(p2, p3), (p1, p2, p3), (p2, p3, p4)}.

When the demand for the main keyword weights, the results are as follows: {(p1, p3, p4), (p2, p3, p4), (p2, p3)}.

In this paper, we consider both textual and distance-dependent, and give the corresponding algorithm.

3   Query Algorithm for Spatial Multi-keyword Based on RB-Tree

The most straightforward method for querying spatial keywords is tantamount to search for matches according to the edit distance threshold [9] and the similarity of the keywords, and then sort according to the spatial distance. The original spatial keyword query stores the latitude and longitude of the spatial data separately, and the query needs to traverse all the data to find the corresponding position. The RB-tree, which supports spatial keyword approximation query, is invoked as the index structure of the algorithm. The index tree is similar to IR-tree, and combines spatial and textual attributes effectively.

3.1   RB-Tree Index

RB-tree is based on the R-tree by adding two bitmap length bitmap LB and Gram bitmap to support the spatial keyword approximation matching. In the index tree, the keywords of SMS set, we can further pruning on the basis of LB and calculating the Gram bitmap of the keywords in the node by formula (3). Once less than the threshold, it is directly thrown; In the RB-tree with N as the root node, but also an inverted file in the node for the calculation of the weight of the keyword, such as Table 1, the file is a set of keywords arranged in the relevant order by the word frequency matrix in each document.

In the RB-tree, each non-child node N_i stores (P, N_i.LB, N_i.GB, MBR, N_i.DI, N_i.CS). P is the location of nodes corresponding to the disk, N_i.LB indicates the length bitmap of the node; N_i.GB represents the Gram bitmap; MBR represents the minimum matrix area; N_i.DI represents the node inverted file; N_i.CS represents the child node related information. The leaf node N_j contains the contents (P, MBR, N_j.DI). In the index tree, the leaf node's bitmap is obtained by the keyword set, and the parent node's bitmap is derived from the sub-node bitmap (Figs. 2 and 3).

Table 1. Keyword frequency in inverted document

Inverted text attributes	Inverted text attributes
$O_1[t_1, 12], [t_3, 7]$	$O_6[t_3, 20], [t_4, 8]$
$O_2[t_2, 12], [t_1, 7]$	$O_7[t_7, 15], [t_{10}, 13]$
$O_3[t_3, 12]$	$O_8[t_8, 17], [t_{11}, 6]$
$O_4[t_4, 10], [t_1, 9], [t_7, 8]$	$O_9[t_9, 18]$
$O_5[t_5, 21], [t_9, 10]$	

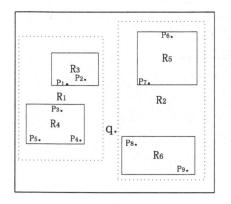 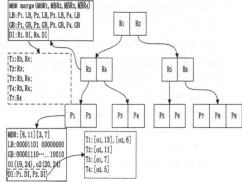

Fig. 2. Spatial location   **Fig. 3.** RB-tree structure

3.2 An Approximate Algorithm Based on RB-Tree

For multi-keyword approximation query algorithm, in the circle with the query point p as the center and R as the radius, QRS is matched with the SMS set. All the POIs in the open circle are saved by the sequence U. In this paper, the query factor has two parts: spatial correlation and textual relevance. The spatial correlation is as described in Eq. (2), including the farthest diameter D of the query keywords and the farthest query radius R; The text relevant includes the edit distance threshold and the keyword weight.

The approximate query algorithm is divided into the following steps:

1. Querying the point o_m in the U, which satisfies the text-dependent and distance-dependent, the dist (o_m, p) is set to the farthest radius (the farthest radius is temporarily set), and placing the spatial position o_m into the QRS collection.
2. Querying the remaining POI points in the set U, the iterative process is as follows: Suppose that the point POI, which is finally added to QRS, is o_i, then the $(i + 1)$ th POI point needs to satisfy the following two formulas, the spatial correlation formula:

$$M(cost)_{i,i+1} = \frac{D(o_i, o_{i+1})}{2\max(dist(p,o)) + \max(dist(o_i, o_{i+1}))} \quad (11)$$

The text-related formulas:

$$M(cost)_{t,o_t} = \frac{w\left(\sum_{m=1}^{n} o_{t.i}\right) \quad t.i \in o_{i+1}}{(1 + \max(\tau^*))^\theta} \quad (12)$$

Combined to find the minimum set with value point o_t coexist in QRS, becoming the new contrast point, until all the query keywords are iterated.
3. During the query, if the value of M(cost) is smaller than the initial value. The initial value is updated.

4. Repeating the above steps until the query end. If the keywords in the SMS are not matched, the loop will be skipped and the process will end.

Algorithm 1: Get the initial set N (q), and the initial cost value

Input: RB-tree, query point P, SMS set $\{t_1, t_2, t_3 \ldots\ldots t_m\}$

Output: $N(q)$, $(cost)_{N(q)}$

1. $N(q)$=Null, V=Null
2. Initialization sequence Se
3. Se.Enqueue(RB-tree Node,0)
4. V=SMS
5. While Se!=Null{
6. w=Se.Dequene()
7.   If w is a POI {
8.   o=w
9. $N(q)=N(q) \cup \{o\}$
10. $V=V—\{o_t\}$
11.   If(v=null)
12.   Break
13.   }
14.   else
15.   for $w_i \in$ w{
16.   if $(o_i==w_i)$
17.   {
18.   if (o_i is a leaf)
19.   Se.Enqueue(o_i,dist(o_i,p));
20.   else   Se.Enqueue(o_i,Mindist(o_i,p))
21.   }
22.   }
23.}
24. The minimum circle and the associated radius and center point of the set N(q) are calculated and overwritten
26.Return $N(q)$,$(cost)_{N(q)}$

In Algorithm (1), from the first row to the fourth row initialize the sequence, we put the multi-spatial keywords into the set V, enter the RB-tree into the sequence Se, and from fifth line to twenty-fourth line, we use the best first traversal algorithm to query RB-tree to meet the conditions of the object, and they are saved to the collection N(q), set n for the number of cycles in the query process, the time complexity of the best query algorithm is O(|SMS| * log(O)). Since the number of query POI is less than the number of query keywords, and the time complexity of Algorithm (1) is O(n * |SMS| log (O)).

Algorithm 2: Based on the results of the algorithm (1) as a parameter, the (cost)N(q) is compared, and the results of QRS are optimized.

Input: RB-tree, query point P, SMS set $\{t_1, t_2, t_3 \dots t_m\}$, parameters α, β, γ, r=Mindist(o_i,p).

Output: QRS,(cost)$_{QRS}$

  1.Se.Enqueue(RB-tree Node,0)

  2.Call the algorithm (1) to get N(q), (cost)$_{N(q)}$

  3.QRS=N(q)，M=(cost)$_{N(q)}$,V=SMS

  4.While Se!=Null{

  5.   w=Se.Dequene()

  6.   if w is a POI{

  7.   o=w

  8.   If V!=Null

  9.   QRS=QRS $\cup$ {o}

10.   if V=Null

11.   break

12.   }

13. else

14.   for $w_i \in$ w{

15. $o_{i+1}=w_i$

16.   if (o_{i+1} is a leaf) {

17.   if($o_i > o_{i+1}$)

18.   Se.Dequene(o_{i+1},(cost)o_{i+1});

19.   else

20.   Se.Dequene(o_i,(cost)o_{i+1})

21.   }

22. }

23.   if ((cost)o_{i+1}<M)

24.   M=(cost)o_{i+1};

25. return QRS,M

In Algorithm (2), we first get the results by using Algorithm (1), then compare them with formula (10) in 23th row. If the new cost value is smaller, replace the original value, and get the set of values as the results of multi-spatial keyword query. The time complexity is closely related to the number of query results obtained by Algorithm 1 above, and its value is O(|SMS| * log(O)), so it shows better robustness.

4 Experimental Results and Analysis

The experimental environment consists of Inter (R) Core (TM) i5-4590CPU, 4 GB installed memory and WIN8 operating system; the data set is crawled through the micro-blog API, and each micro-blog contains the content of the publication and where it is published. The experiment consisted of 11,243 spatial data sets of shopping centers and restaurants, and 396 different keyword sets, each line consists of a POI and 20 keywords. The location of the query and keywords are randomly selected, each calculation result is based on the average of multiple results as a reference value to use.

Figure 4 shows that the consumption time decreases from the initial 500 ms to 480 ms with the increase of the value, and then the time increases slowly. When the value is 0.35, the consumption time is the smallest because the text correlation is added in the query function to provide some pruning strategy, but with the increasing of the value, the relative weight of the spatial is reduced, which makes the query radius increase, and finally, the query result is increased, resulting in a certain redundancy, and these are not conducive to the query. In the experiment, set the number of query keywords to 3.

Figure 5 shows the comparison of the different number of spatial keywords n, the spatial multi-keyword approximate matching algorithm and the previous spatial matching query are compared, the former has a greater advantage, with the query number increases, the query time consumption linear growth. And add text-related attributes, making the query pruning to play a better effect.

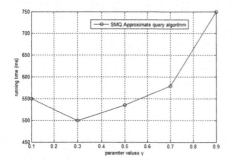

Fig. 4. Effect of parameter value on query time

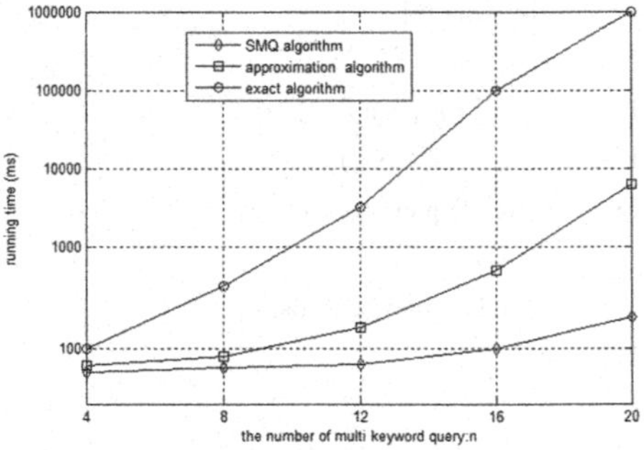

Fig. 5. The influence of the number of keywords on the time

Figure 6 shows that, compared with the exact query, the SMQ query algorithm is more accurate than the previous query algorithm. When the number of query keywords is small, the ratio of the approximation algorithm to the exact algorithm is close to 1, and with the increase of the number of queries showing a linear growth, when the number of queries is 10, still close to 1.

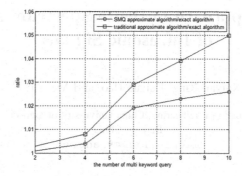

Fig. 6. The ratio between the approximation algorithm and the exact algorithm

5 Conclusion

In this paper, an approximate query algorithm based on RB-tree is proposed to support multi-spatial keywords fuzzy matching. In the algorithm, we firstly define the correlation function, in which the edit distance, weight, spatial distance and other elements are introduced. Compared with the traditional single query, the algorithm is more able to meet the needs of users. Finally, the optimal coefficient values of the relevant elements are given by experiment and then compared with the traditional approximate algorithm, the results show that the algorithm has better performance in terms of query efficiency and precision.

References

1. Liu, X.P., Wan, C.X., Liu, D.X.: Survey on spatial keyword search. J. Softw. **27**(2), 329–347 (2016)
2. Yao, B., Li, F.F., Hadjieleftheriou, M.: Approximate string search in spatial databases. In: Proceedings of ICDE, pp. 545–556. IEEE, Washington (2010)
3. Alsubaiee, S., Behm, A., Li, C.: Supporting location-based approximate-keyword queries. In: Proceedings of SIGSPATIAL GIS, pp. 61–70. ACM Press, New York (2010)
4. Wang, J.B., Gao, H., Li, J.Z.: An index supporting spatial approximate keyword search on disks. J. Comput. Res. Dev. **49**(10), 2142–2152 (2012)
5. Hu, J., Fan, J., Li, G.L., et al.: Top-k fuzzy spatial keyword search. Chin. J. Comput. **35**(11), 2237–2246 (2012)
6. Zhang, D.X., Chen, Y.M., Mondal, A., et al.: Keyword search in spatial databases: towards searching by document. In: Proceeding of 2009 IEEE International Conference Data Engineering, pp. 688–699. IEEE, Computer Society, Washington, DC (2009)
7. Cao, X., Cong, G., Ji, L., et al.: Collective spatial keyword querying. In: Proceedings of 2011 ACM SIGMOD, International Conference on Management of Data, pp. 373–384. ACM, New York (2011)
8. Long, C., Wong, R.C., Wang, K., et al.: Collective spatial keyword queries: a distance owner-driven approach. In: Proceedings of 2013 ACM SIGMOD International Conference on Management of Data, pp. 689–700. ACM, New York (2013)
9. Fan, J., Li, G.L., Zhou, L.Z., et al.: Seal: spatio-textual similarity search. Proc. VLDB Endow. **5**(9), 824–835 (2012)

A New Approach to Dense Spectrum Analysis of Infrasonic Signals

Kaiyan Xing, Kaixue Hao, and Mei Li[(⊠)]

School of Information Engineering, China University of Geosciences (Beijing),
Beijing, China
maggieli@cugb.edu.cn

Abstract. Spectrum analysis is very important in geological hazards of infrasonic signal observation systems. The spectrum of infrasonic signal is a dense spectrum which can leads to potential erroneous spectrum analysis. Hereby we propose a dense spectrum analysis algorithm combining all phase Fast Fourier Transform (apFFT) and Chirp Z-transform (CZT) to analyse dense low frequency signal. This is called all phase Chirp Z transform (apCZT). The apFFT spectrum analysis can reduce spectrum leakage, but does not enhance resolution while the CZT vice versa. The novel algorithm apCZT can suppress spectral leakage and improve the resolution at the same time. Simulation results demonstrate that the apCZT algorithm can distinguish the frequencies whose intervals are less than the ordinary frequency resolving power of Discrete Fourier Transform (DFT) and apFFT. The apCZT it is not only suitable for infrasonic signals but also in other dense spectrum analysis applications, such as voice, vibration, noise, electrocardiography, radar signals, power system harmonics and other engineering practice.

Keywords: Infrasonic signal · apFFT · CZT · apCZT

1 Introduction

In geological hazards of infrasonic signal observation system, we monitor the infrasonic signals and predict the occurrences of geological hazards according to their spectrum characteristics. It is essential to estimate the frequency, amplitude and phase of sampling signals by transforming that includes Discrete Fourier Transform (DFT), Laplace transform, Z-transform, etc. Accurate frequency resolution is required in order to obtain the perfect characteristic of the spectrum. Thus, dense spectrum analysis is of great significance because the frequency intervals of these infrasonic signals are very small.

A finite length signal obtained by truncating produces spectral leakage causing the frequency, amplitude, and phase to be grossly inaccurate. Therefore many improved algorithms are put forward in order to obtain high resolution and accurate characteristic of the spectrum, including interpolated fast Fourier transform method [1, 2], analytical leakage compensation [3, 4] and all phase Fast Fourier Transform (apFFT) spectrum analysis [5] and so on. These spectrum analysis methods improve resolution and accuracy through spectrum correction or refinement, but they are not applicable unsuitable for multi and dense frequency signals.

© Springer Nature Singapore Pte Ltd. 2017
B. Zou et al. (Eds.): ICPCSEE 2017, Part I, CCIS 727, pp. 134–143, 2017.
DOI: 10.1007/978-981-10-6385-5_12

A novel algorithm, which is called all phase Chirp Z-transform (apCZT) is put forward in this paper to analyse dense low frequency such as infrasonic signals. As a pre-processing step, the apFFT algorithm can greatly reduce spectral leakage and preserve the phase value [6]. On the other hand, CZT algorithm can be utilized to obtain a higher resolution. The proposed algorithm has the advantage of both apFFT and CZT.

2 Methods

In this section, apFFT, CZT and the novel algorithm apCZT are described. This section will focus on the basic principles and calculation methods of apCZT, while formulas and figures will briefly described apFFT and CZT methods.

2.1 Traditional Methods of Dense Spectrum Signal Analysis

2.1.1 apFFT Algorithm–Reduce Spectral Leakage

The apFFT algorithm is put forward to improve the method of estimation of phase and amplitude [5]. The apFFT improves the truncating way of DFT and greatly reduces spectral leakage. All segments, which include an arbitrary point of sampling sequence, are considered. It is assumed that there is a sampling sequence with a length of $2N-1$, including $x(0)$, $x(1)$, $x(2)$, $x(3)$, $x(4)$, $x(5)$, $x(6)$, and the apFFT algorithm truncates the sequence with a rectangular window. All of the segments with length N, which contain $x(3)$, are listed as follows.

$$
\begin{aligned}
&\text{Segment 1}: & x(3) \quad x(4) \quad x(5) \quad x(6) \\
&\text{Segment 2}: & x(2) \quad x(3) \quad x(4) \quad x(5) \\
&\text{Segment 3}: & x(1) \quad x(2) \quad x(3) \quad x(4) \\
&\text{Segment 4}: & x(0) \quad x(1) \quad x(2) \quad x(3)
\end{aligned}
$$

Periodic extension is then applied to each segment with the next step is truncating the rectangular window with length N. Finally, add the values corresponding position of all segments as a new sampling sequence with length N. As shown in the following block diagram, we can see that all of the segments with length N contain sampling point $x(3)$:

$$
\begin{aligned}
&\ldots\ldots x(4) \quad x(5) \quad x(6) \quad \mathbf{x(3)} \quad x(4) \quad x(5) \quad x(6) \ldots\ldots \\
&\ldots\ldots x(4) \quad x(5) \quad x(2) \quad \mathbf{x(3)} \quad x(4) \quad x(5) \quad x(2) \ldots\ldots \\
&\ldots\ldots x(4) \quad x(1) \quad x(2) \quad \mathbf{x(3)} \quad x(4) \quad x(1) \quad x(2) \ldots\ldots \\
&\ldots\ldots x(0) \quad x(1) \quad x(2) \quad \mathbf{x(3)} \quad x(0) \quad x(1) \quad x(2) \ldots\ldots
\end{aligned}
$$

Summary: $4x(3)\ 3x(4) + x(0)2x(5) + 2x(1)x(6) + 3x(2)$

The sampling sequence $x(n)$ is processed by apFFT algorithm using rectangular window, the new sequence is $y(n)$,

$$y_N(n) = \left[\frac{N-n}{N}x(n) + \frac{n}{N}x(n-N)\right]u_N(n), n = 0, 1, \cdots, N-1 \tag{1}$$

Assume that each segment is $y_i(i = 0, 1, \cdots, N-1)$, corresponding to the cycle shift segment. The Eq. (2) represents the relationship between the DFT's y_i of and y_i',

$$y_i'(k) = Y_i(k)e^{j\frac{2\pi}{N}ki} \tag{2}$$

We can obtain $Y_N(k)$, which is the DFT of $y(n)$, according to the linearity of the Fourier transform [7],

$$Y_N(k) = \frac{1}{N}\sum_{i=0}^{N-1}Y_i'(k) = \frac{1}{N}\sum_{i=0}^{N-1}Y_i(k)e^{j\frac{2\pi}{N}ki} = \frac{1}{N}\sum_{i=0}^{N-1}\left[\sum_{m=0}^{N-1}x(m-i)e^{-j\frac{2\pi km}{N}}\right]e^{j\frac{2\pi ki}{N}} \tag{3}$$

Equation (3) can be rewritten as Eq. (4).

$$\begin{aligned}Y_N(k) &= \frac{A}{N}\sum_{i=0}^{N-1}\left[\sum_{m=0}^{N-1}e^{j\left(\frac{2\pi k_0(m-i)}{N}+\phi_0\right)}e^{-j\frac{2\pi km}{N}}\right]e^{j\frac{2\pi ki}{N}}\\[2mm] &= \frac{A}{N}e^{j\phi_0}\frac{1-e^{-j\frac{2\pi}{N}(k_0-k)N}}{1-e^{-j\frac{2\pi}{N}(k_0-k)}}\cdot\frac{1-e^{-j\frac{2\pi}{N}(k_0-k)N}}{1-e^{-j\frac{2\pi}{N}(k_0-k)}}\\[2mm] &= \frac{A}{N}e^{j\phi_0}\frac{\sin^2(\pi(k_0-k))}{\sin^2(\pi(k_0-k)/N)}\end{aligned} \tag{4}$$

Normalizing Eq. (4).

$$Y_N(k) = \frac{A}{N^2}\left|\frac{\sin(\pi(k_0-k))}{\sin(\pi(k_0-k)/N)}\right|^2 e^{j\phi_0} \tag{5}$$

Equation (5) indicates that the squared term $|\sin(\pi(k_0-k))/\sin(\pi(k_0-k)/N)|^2$ makes apFFT algorithm reduce spectral leakage, while the phase of apFFT is the prime phase of the signal.

The apFFT considers the circumstances of all segments and suppresses spectral leakage well. Methods based on the apFFT approach are widely used in spectrum analysis, but it is unavailable to enhance resolution.

2.1.2 CZT Algorithm–Improve Resolution

CZT algorithm is the generalization of the discrete Fourier transform and widely used in the field of spectrum analysis, such as frequency obtaining [8], image processing [9, 10], and power system measurements [11]. CZT can significantly enhance the frequency resolution without increasing the number of sampling data, while less calculation is required, compared with other spectrum refinement methods, which are quite suitable for short signals [12]. However, CZT refinement involves significant leakage of the signal spectrum. For dense frequency signals, the result of the CZT refinement may be inaccurate.

The CZT algorithm refinement principles are as follows:

The Z-transform result for sample sequence $x(n)$ with length N, $0 \leq n \leq N-1$ is Eq. (6).

$$X(z) = \sum_{n=0}^{N-1} x(n) z^{-n} \tag{6}$$

While the DFT samples the z-plane at uniformly spaced points along the unit circle, the CZT algorithm samples along spiral arcs in the z-plane, corresponding to straight lines in the s-plane [13]. Suppose sampling point is z_k in z-plane.

$$z_k = AW^{-k}, \quad k = 0, 1, \cdots, M-1 \tag{7}$$

where M is an arbitrary integer, both A and W are arbitrary complex numbers as follows.

$$A = A_0 e^{j2\pi\theta_0} \tag{8}$$

$$W = W_0 \cdot e^{j2\pi\phi_0} \tag{9}$$

When $k = 0, z_0 = A_0 e^{j2\pi\theta_0}$, the amplitude is A_0, and the amplitude angle is θ_0. When $k = 1, z_1 = A_0 W^{-1} e^{j(\theta_0 + \phi_0)}$, $A_0 W^{-1}$ is amplitude, and $(\theta_0 + \phi_0)$ is amplitude angle. If the sampling sequence has N points, the output is a sequence of M points. With k changing, $X(z_k)$ presents in the form of spiral contours. Since ϕ_0 is arbitrary, the frequency resolution can be adjusted. In other words, the property of CZT, which has assumed considerable importance, is that the frequency resolution can be arbitrary, while the frequency resolution of traditional DFT is a constant value.

CZT performs well on a single-frequency or partial multi-frequency dense frequency signal with a relatively large interval, and the resolution is much higher than the DFT's. It has been proven that CZT can obtain frequencies with values is less than f_s/N. Therefore, in practical applications, CZT algorithm spectral refinement method has limitations, since it cannot reduce spectral leakage.

2.2 New Method of Dense Spectrum Signal – apCZT Algorithm

The spectrum analysis is not satisfied performing CZT or apFFT. A novel algorithm, called all phase Chirp Z-transform (apCZT) is put forward to analyze dense low frequency signals. This is based on apFFT and CZT since the apFFT algorithm can reduce spectral leakage and CZT can obtain high resolution.

The first step of apCZT is pre-process of windowing similar to apFFT algorithm. This improves the traditional input data truncating way before transforming and can reduce spectral leakage as already stated in Sect. 2.1. Then perform CZT to obtain the amplitude and phase of signals.

The specific processes to perform apCZT include following steps:

1. Divide the sampling sequence into segments

$$x_1(n), x_2(n), \cdots, x_i(n), \cdots, x_N(n)$$

Each segment $x_i(n)$ includes $x(m)$ and has N points totally.

2. Extend all segments periodically and add corresponding values of each segment.
3. Choose L, the smallest integer greater than or equal to N + M − 1.
4. Form an L points sequence $y(n)$ from $x'(n)$ by

$$y(n) = \begin{cases} A^{-n} W^{n^2/2} \cdot x'(n), & n = 0, 1, \cdots, N-1 \\ 0, & n = N, N+1, \cdots, L-1 \end{cases} \tag{10}$$

A and W can be computed as Eqs. (8) and (9), Fig. 2 indicates the contour in z-plane [6].

5. Compute the L points DFT of $y(n)$.

$$Y(r) = \sum_{n=0}^{L-1} y(n) e^{-j\frac{2\pi}{L}rn}, r = 0, 1, \cdots, L-1 \tag{11}$$

6. Define an L points sequence $h(n)$ by the equation

$$h(n) = \begin{cases} W^{-\frac{n^2}{2}}, & 0 \leq n \leq M-1 \\ W^{\frac{-(L-n)^2}{2}}, & L-N+1 \leq n < L \\ arbitrary, & other\ n,\ if\ any \end{cases} \tag{12}$$

7. Compute L points DFT of $h(n)$.

$$H(r) = \sum_{n=0}^{L-1} h(n) e^{-j\frac{2\pi}{L}rn}, r = 0, 1, \cdots, L-1 \tag{13}$$

8. Multiply $Y(r)$ and $H(r)$ point by point.

$$G(r) = Y(r) \cdot H(r) \tag{14}$$

9. Compute the L points IDFT of $G(r)$, which is $g(k)$, $k \leq M$.
10. Multiply $g(k)$ and $W^{k^2/2}$.

$$X(k) = g(k) \cdot W^{\frac{k^2}{2}}, k = 0, 1, \cdots, M-1 \tag{15}$$

Each step of apCZT algorithm is as shown above, then we will test that the result of apCZT is better than traditional CZT's, FFT's, apFFT's and zoomFFT's by programming.

3 Simulation and Comparison

Assume that there is a low frequency signal $x(n)$, which contains three frequencies,

$$x(n) = \cos\left(\frac{2\pi \cdot f_1 \cdot n}{f_s}\right) + \sin\left(\frac{2\pi \cdot f_2 \cdot n}{f_s}\right) + \cos\left(\frac{2\pi \cdot f_3 \cdot n}{f_s}\right) \qquad (16)$$

where f_s is the sampling rate. FFT, apFFT, CZT and apCZT methods were applied respectively using MATLAB. Set value $f_1 = 21.12$, $f_2 = 21.13$, $f_3 = 21.3$, which are low frequencies. The value of f_s is 200 Hz and the sampling points N are 3999, respectively. The resolution of DFT is 0.05 Hz, but the interval between 21.12 Hz and 21.13 Hz is 0.01 Hz, which is less than 0.05 Hz.

The parameters of CZT functions are as follows: $f_0 = 21$, $f_1 = 22$, $f_s = 200$. $M = 128$. f_0 is starting frequency, f_1 is stopping frequency, f_s is sampling rate.

The formula of Hanning window is written as

$$w(n) = 0.5\left(1 - \cos\left(2\pi \frac{n}{N}\right)\right), 0 \leq n \leq N$$

Figure 1 represents magnitudes of FFT, apFFT and Fig. 2 shows magnitudes of apCZT.

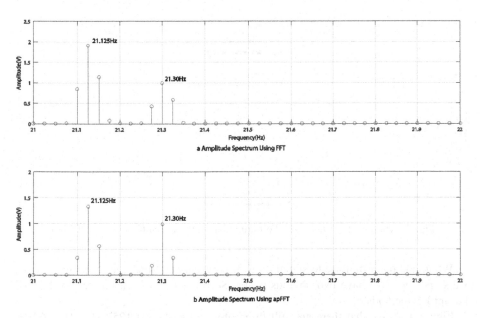

Fig. 1. Amplitude spectrum using traditional methods

Figure 1(a) and (b) indicate that it is hard to obtain the accurate frequencies. However, three peaks, 21.1250 Hz, 21.1563 Hz and 21.2969 Hz, appear in Fig. 2,

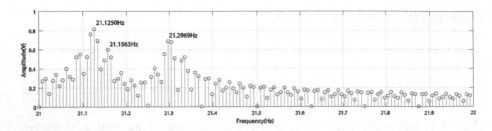

Fig. 2. Amplitude spectrum using new methods apCZT

which represents amplitude spectrum using apCZT, and the frequency errors compared with theoretical values are 0.02%, 0.12% and 0.01%.

Then we add random noise to x(n) and set Signal to Noise Ratio (SNR) to be −40 dB. Figure 3 indicates the effects of noise on the traditional transforming result. Figure 4 indicates the effects of noise on the apCZT result which is a new method.

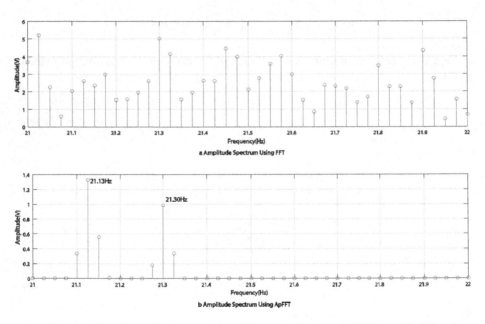

Fig. 3. The effects of noise on the traditional methods result (SNR = −40 dB)

When add random noise to signal, Fig. 3 represents that Fig. 3(a) have no obvious peaks and Fig. 3(b) have two obvious peaks. So noise has a great impact on the FFT and apFFT methods.

Figure 4 shows that there are still three obvious peaks, 21.1250 Hz, 21.1562 Hz and 21.2969 Hz when add random noise to signal, meanwhile, the error is very small.

Therefore, we can conclude that apCZT algorithm is more helpful to analyze dense frequency signals.

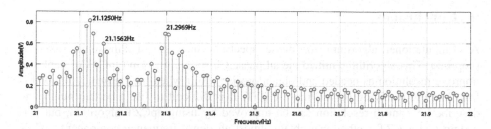

Fig. 4. The effects of noise on the new method result (SNR = −40 dB)

4 Application in Infrasonic Signal

After mathematical proof and simulation verification, we apply the apCZT algorithm to infrasonic signals. A sampling sequence x(n) is the infrasound signal of debris flow was obtained by Dongchuan Debris Flow Observation and Research Station of Chinese Academy of Sciences in Jiangjia Gully, Yunnan Province, China. According to literature [14], the infrasound signal bandwidth of debris flow is within the range of 4 to 15 Hz, which belongs to the low frequency, and apCZT algorithm is suitable to spectrum analysis of infrasound.

The sampling rate of infrasound sequence with 49999 points is 60 Hz, the apCZT algorithm was applied to perform the analyzation. According to sampling rate, it is easy to compute the resolution, which is approximately 0.0012 Hz, the bandwidth focus on is less than or equal to 20 Hz. We perform apCZT as steps in Fig. 3, values of parameters in Fig. 4 are as follows.

$$M = 256, f_0 = 1, f_1 = 20, f_s = 60$$

Figure 5 shows the spectrum of debris flow infrasound signal in 0.01–0.06 Hz in order to demonstrate the apCZT analysis results of dense low-frequency signal is superior.

Figure 5 already shows the dense frequency components. It is easy to find peaks.

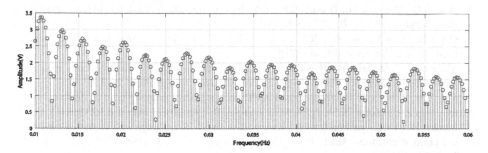

Fig. 5. Amplitude spectrum of infrasound signal x(n)

5 Conclusions

A computational algorithm for dense spectrum analysis of infrasonic signal is proposed. This algorithm entitled the all phase Chirp-Z transform (apCZT), enables to improve resolution as needed and obtain precise result when the frequency interval is less than the resolving power limit of DFT. It is difficult to use a certain number to describe how much resolution can be improved by using apCZT algorithm, but it is certain that apCZT method can obtain almost all dense frequency components, while traditional methods can only get some of the frequency components. The apCTZ algorithm is not only suitable in infrasonic signals but can also applied in other dense spectrum analysis applications, such as voice, vibration, noise, electrocardiography, radar signals, power system harmonics and other engineering practice.

Acknowledgement. This work is financially supported by the National Natural Science Foundation of China (Grant Nos. 41374185 and 41572347), the Fundamental Research Funds for the Central Universities (Excellent Instructors Fund, Grant No. 2652016139).

References

1. Jain, V.K., Collins, W.L., Davis, D.C.: High-accuracy analog measurements via interpolated FFT. IEEE Trans. Instrum. Meas. **28**, 113–122 (1979). doi:10.1109/TIM.1979.4314779
2. Chen, K.F., Li, Y.F.: Combining the Hanning windowed interpolated FFT in both directions. Comput. Phys. Commun. **178**(12), 924–928 (2008)
3. Renders, H., Schoukens, J., Vilain, G.: High-accuracy spectrum analysis of sampled discrete frequency signals by analytical leakage compensation. IEEE Trans. Instrum. Meas. **33**, 287–292 (1984). doi:10.1109/TIM.1984.4315226
4. Diao, R., Meng, Q.: Frequency estimation by iterative interpolation based on leakage compensation. Measurement **59**, 44–50 (2015)
5. Wang, Q., Xiao, Y., Kaiyu, Q.: Parameters estimation algorithm for the exponential signal by the interpolated all-phase DFT approach. In: IEEE 2014, pp. 37–41 (2014). doi:10.1109/ICCWAMTIP.2014.7073356
6. Rabiner, L.R., Schafer, R.W., Rader, C.M.: The chirp z-transform algorithm. IEEE Trans. Audio Electroacoust. **17**, 86–92 (1969). doi:10.1109/TAU.1969.1162034
7. Wu, G.Q., Wang, Z.H., Huang, X.H.: All phase correction method for discrete spectrum. Data Acquis. Process. **20**, 287–290 (2005)
8. Oppenheim, A.V., Schafer, R.W., Buck, J.R.: Discrete-Time Signal Processing, vol. 2. Prentice-Hall, Englewood Cliffs (1989)
9. Mostarac, P., Malarić, R., Hegeduš, H.: Adaptive chirp transform for frequency measurement. Measurement **45**, 268–275 (2012). doi:10.1016/j.measurement.2011.12.005
10. Ma, T.N., Takaya, K.: High-resolution NMR chemical-shift imaging with reconstruction by the chirp Z-transform. IEEE Trans. Med. Imaging **9**, 190–201 (1990). doi:10.1109/42.56344
11. Lanari, R.: A new method for the compensation of the SAR range cell migration based on the chirp Z-transform. IEEE Trans. Geosci. Remote Sens. **33**, 1296–1299 (1995). doi:10.1109/36.469496
12. Aiello, M., Cataliotti, A., Nuccio, S.: A chirp-Z transform-based synchronizer for power system measurements. IEEE Trans. Instrum. Meas. **54**, 1025–1032 (2005). doi:10.1109/TIM.2005.847243

13. Smith, J.O.: Mathematics of the Discrete Fourier Transform (DFT): With Audio Applications. Julius Smith (2007)
14. Kogelnig, A., Hübl, J., Suriñach, E., et al.: Infrasound produced by debris flow: propagation and frequency content evolution. Nat. Hazards **70**, 1713–1733 (2014). doi:10.1007/s11069-011-9741-8

Research on XDR Bill Compression Under Big Data Technology

Bing Zhao[1(✉)], Sining Zhang[1], and Jun Zheng[2]

[1] Department of Electronic Engineering, Heilongjiang University, Harbin, Heilongjiang, China
zb0624@163.com
[2] Vixtel Technologies Holdings Limited, Beijing, China

Abstract. Communication industry has been walking in front of the big data technology application. Some studies on precision-controlled compression method are carried on in this paper by using the big data technology based on the widely used XDR bill in communication industry. According to different application scenarios, this paper puts forward targeted compression strategy and technological implementation method and verify the high efficiency of associated method in practice. It solves the problem of occupying large storage and low analysis efficiency in areas like the storage of the massive XDR bill, pretreatment, aggregation. It provides valuable references for telecommunication-related researchers and engineering practitioner in respect of using the big data technology.

Keywords: Big data · XDR bill · Compression strategy

1 X Data Recording

In a communication system, XDR (X Data Recording) bill initially evolved from CDR (Call Data Recording). CDR is the record of the network key information during the call in the traditional communication network. XDR is an extension of CDR concept [1,2]. In this paper, it refers to the record of the key information of data traffic in Mobile Network and Hosted Network, namely, traffic log. It is based on the user session as a unit and each session forms a XDR record. XDR record each online behavior of each user. So it is also called Internet bill.

XDR bill has various detailed information of online behavior. It can capture much information by combining user information, device information and network topology, such as users behavior information, target site information, network resources information, network performance information and other information with application value and commercial value [3–5]. How to store, process and analyze XDR bill efficiently has become key research project of relevant telecommunication companies in the near future [6–8].

The locations of XDR bills collection (generation) are: user side, IDC export, access network, metropolitan area network, province network export, backbone network export, etc.

© Springer Nature Singapore Pte Ltd. 2017
B. Zou et al. (Eds.): ICPCSEE 2017, Part I, CCIS 727, pp. 144–152, 2017.
DOI: 10.1007/978-981-10-6385-5_13

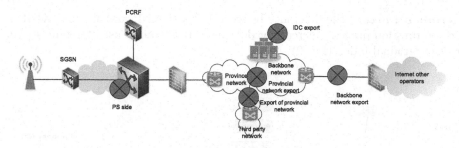

Fig. 1. Application scenario

The collection of XDR bill in every nodes involves huge data quantity. According to the past experience, every 1 million users can produce 50 billion XDR records per day. If each record is calculated in 100 bytes, every 1 million users data will take up 5T storage space (uncompressed) per day (Fig. 1).

Storing and processing massive XDR bill have very great difficulty. Traditional data compression technology has very limited effect on the processing of XDR bill. From the perspective of information processing and analysis, this paper puts forward the compression strategy of XDR bill and the relevant implementation method. It also proves the effectiveness of relevant methods and provides useful references and directions for follow-up similar work.

2 Basic Principles of XDR Bill Compression

It can be seen from each recorded information in XDR bill that there are large amount of self-similar, inefficient, redundant information. Many information can be compressed. Secondly, the original XDR bill retains accurate time information, which is usually accurate to seconds, milliseconds or even microseconds. But in the subsequent processing and statistics, it is usually aggregated into hours, days, weeks, months and other granularity. So time granularity can also be compressed. In addition, huge fluctuation of related index in XDR bill is rare less seen. In most cases, the index changes uniformly and slowly in a small time window. So the index can be approximate sampling compressed.

Overall, the possible strategies for XDR bill compression are as follows:

Summary: Calculate the arithmetic mean, geometric mean, mean square deviation of the parameters in the multiple XDR. Summary results become a new XDR bill and replace the original multiple XDR bill. It is mainly aimed at performance data, such as throughput, delay, error packets, etc.

Accumulation: Accumulate single field of multiple XDR bill. The cumulative results become a new bill and replace the original multiple XDR bill. It is mainly aimed at volume, duration, etc.

Sample: Extract a bill from multiple XDR bill to replace the original multiple bill. It is mainly used for XDR bill that has many interactions in a short time.

Increment: Record the difference between the last XDR and the first XDR in the compression process (increase or decrease). It is mainly used for time stamp or other gradual index (Fig. 2).

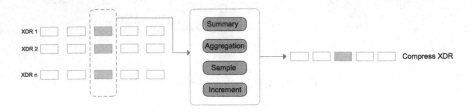

Fig. 2. XDR bill compression strategies

Compression process of XDR bill can introduce the following problem:

1. Introduce new delays. (Complaint handling sensitivity)
2. Miss some users operation content. (Behavior audit sensitivity)
3. Time and space granularity of analysis increase. (Network analysis sensitivity)

3   Rearrangement Before XDR Compression

Before being compressed, the original XDR bill should be rearranged. The bill is rearranged according to the source IP, destination IP and other key fields. After the rearrangement, the results are compressed. This process will increase the time-delay of XDR processing. So it is necessary to set the time window to ensure that only the XDR bill in the time window is rearranged (Fig. 3).

Timestamp	Source IP Address	Destination IP Address	XDR Key Field
T1	a	X	S1
T2	b	Y	S2
T3	c	Z	S3
T4	a	X	S4
T5	b	Y	S5
T6	c	Z	S6
T7	a	X	S7
T8	b	Y	S8
T9	c	Z	S9

Fig. 3. XDR bill before rearrangement

Subsequent compression processing can be carried out after rearrangement (Figs. 4 and 5).

Timestamp	Source IP Address	Destination IP Address	XDR Key Field
T1	a	X	S1
T4	a	X	S4
T7	a	X	S7
T2	b	Y	S2
T5	b	Y	S5
T8	b	Y	S8
T3	c	Z	S3
T6	c	Z	S6
T9	c	Z	S9

Fig. 4. XDR bill after rearrangement

Timestamp	Source IP Address	Destination IP Address	XDR Key Field
T7-T1	a	X	(S1+S4+S7)/3
T8-T2	b	Y	(S2+S5+S8)/3
T9-T3	c	Z	(S3+S6+S9)/3

Fig. 5. Compression processing

4   Example of XDR Bill Compression

A user accesses to www.baidu.com and this operation generate XDR data. The delay of page opening is as follows:

This user visit the same website many times and generate many XDR bills during this time. The parameter fluctuation of site access delay (in ms) is not very significant. It means also: During this period, the network communication quality between the user and the target server is stable. Concurrent access of the target server has no significant fluctuation and the parameters of network delay remain basically constant. We can compress the XDR data in this period.

Sample: Sample multiple data and choose one or several data as the representative.

Summary: Aggregate multiple data and calculate the mean, variance, maximum, minimum and other statistical indicators of time delay.

5   Basic Flow of XDR Compression Algorithm

Multiple bills are compressed into one bill, which is executed in the following three ways: Counter, Trigger, Timer.

Counter: When the statistical value of the counter reaches the set threshold, the compression process is triggered and the counter is reset. Counter includes (Figs. 6 and 7):

File Counter: It is used to compress cross-file XDR. When a user's XDR is spread out across a number of different files, this counter can control the span of compression.

WWW.BAIDU.COM page opens

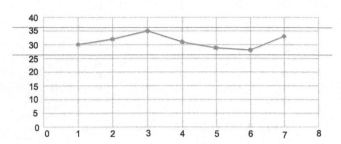

Fig. 6. Delay of page opening

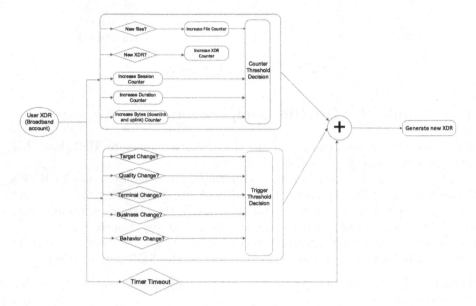

Fig. 7. Flow of XDR compression

Session Counter: It is used to control how many sessions are compressed at one time.

Duration Counter: It is used to control how long XDR should be compressed together. It can prevent the large time-delay when the user interaction is very small or the low compression ratio when the interaction is very dense.

Volume Counter: It is used to control the threshold number of accumulated total data to output a XDR. It is mainly used for compression control when the applications speed is high or the data interaction number is small.

Trigger: It make trigger judgment by using a certain condition. When some parameters mutate and do not match the similarity, it triggers a new XDR aggregate output. For example, a user accesses to all the XDR of one page and the page opening threshold is set to 10 ms. If the parameter is higher than the threshold, a new XDR is triggered. Triggers are used to ensure that the details of the XDR can be retained when the users behavior changes (Fig. 8).

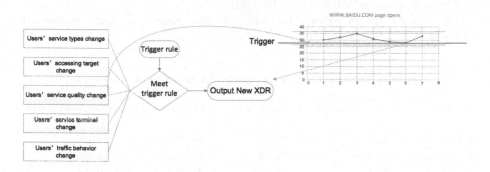

Fig. 8. Triggered flow

Trigger can be set by a single indicator, a combination of multiple indicators, or other complex logical combination.

Timer: Set time window. When the new XDR bill exceeds the time window, XDR compression should be executed. The purpose of the time window is to control the delay of the compression algorithm and prevent the XDR bill from being triggered when the trigger conditions do not permit. For example, a user accesses an address and does not access it again. The user's XDR fails to output because it can not reach the compressed counter, which leads to the disappearance of XDR. Through adjusting the parameters of counter, trigger, and timer, the compression ratio and the delay of the XDR bill can be adjusted. Compression feedback control chart is as follows (Fig. 9):

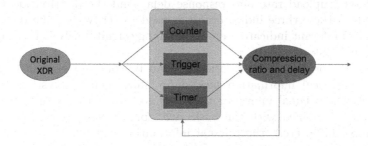

Fig. 9. Compression feedback control chart

6   Experiment Results

Take the XDR data of a province as an example. Select some days and test one hours data per day.

Timestamp range 2016082712–2016090112.

The results are as follows by using Apache Spark big data framework (Table 1).

Table 1. The results by using Apache Spark big data framework

Test Timestamp	Original number of lines	Number of lines after compression	Compression rate	Compression time
2016082712	3430673	1250479	36.40%	2.0 min
2016082812	3204621	1144552	35.70%	1.9 min
2016082912	3059840	1033240	33.70%	1.9 min
2016083012	3844542	1578274	41.10%	2.3 min
2016083112	3500261	1337467	38.20%	2.0 min
2016090112	3542573	1427836	40.30%	2.2 min

The experimental results show:

The compression rate in the experimental results is the ratio of the number of log records after compression to the original number of records. The compressed results can be further compressed by using the compression algorithms, such as gzip, Snappy, LZO and so on. This experiment is to compress the digital information in the original log. After compression, multiple logs merge into one log. Needful statistical information in the original log is reserved after sacrificing some useless details. And as can be seen from the experimental results that the information compression rate can reach about 40%.

After compressing log, we will sectionally count the average value of the download rate, upload rate and response delay and the distribution of effective records of these three indicators. It is used to verify that the distribution probability of relevant indicators does not change significantly before and after compression (Tables 2 and 3).

Overall, the original XDR information can be compressed on a large scale when the statistical information needed for subsequent analysis is basically unchanged. The original information can be compressed to about 40% of the original size. Combined with the basic text compression algorithms (such as gzip, Snappy, LZO, etc.), the original information can be compressed significantly. It provides great convenience for the storage, distribution and processing of XDR information.

There has not been a similar compression and processing based on statistical information for the logs of telecommunications industry such as XDR. This paper provides a good reference for similar work. Its basic ideas and used ideas can

Table 2. Download rate (Bytes/s)

Time stamp	Before compression	After compression	Difference (%)
2016082712	243286.8	242615.5	0.28
2016082812	283500.7	282621.7	0.31
2016082912	223745.8	223945.7	−0.09
2016083012	190767.6	189923.2	0.44
2016083112	202872.3	203168.6	−0.15
2016090112	263286.8	264286.5	−0.38

Table 3. Upload rate (Bytes/s)

Time stamp	Before compression	After compression	Difference (%)
2016082712	16127.38	16108.37	0.12
2016082812	18234.24	18302.83	−0.38
2016082912	15181.83	15139.18	0.28
2016083012	13834.75	13880.28	−0.33
2016083112	14824.38	14900.28	−0.51
2016090112	17389.27	17329.83	0.34

Table 4. Response delay (ms)

Time stamp	Before compression	After compression	Difference (%)
2016082712	109.02	109.21	−0.18
2016082812	108.29	108.37	−0.07
2016082912	110.73	110.79	−0.05
2016083012	109.21	109.01	0.18
2016083112	109.39	108.89	0.45
2016090112	107.36	108.02	−0.61

provide inspiration for the same type of work. Meanwhile, It proves the validity of the method from real experimental results (Table 4).

Acknowledgements. This work is supported by Heilongjiang Provincial Education Department Science and Technology Research Project (No. 12531492). Many thanks to the anonymous reviewers, whose insightful comments made this a better paper.

References

1. Jiang, G., Hu, F., Shi, L.: Urban functional area identification based on call detail record data. J. Comput. Appl. **36**(7), 2046–2050 (2016)

2. Li, B., Shen, L., Dai, P., Ren, X.: Methods of TD-LTE signaling data accuracy verification. Telecom Eng. Tech. Stand. (2), 22–27 (2017)
3. Xu, H., Xu, J.: Calculation of spatial position data based on mobile phone signaling work and live. Beijing Surv. Mapp. (6), 69–71 (2016)
4. Liu, G., Wang, X., Zhang, J., Li, S.: Study on intelligent management and control of tourist attraction based on mobile signaling data. J. Univ. Electron. Sci. Technol. China 44(5), 769–777 (2015)
5. Wu, S., Luo, J., Zhou, Y., Lin, J., Shu, Z.: Method of real-time traffic statistics using mobile network signaling. Appl. Res. Comput. 31(3), 776–779 (2014)
6. Du, C., Jiang, S.: Research on user ridership characteristic based on mobile signaling data. Mob. Commun. 39(23), 9–12 (2015)
7. Sui, Y., Shen, L., Tao, L., Dai, P., Wan, R., Wang, W.: Research on user location based on signaling big data. Telecommun. Sci. (s1), 197–201 (2016)
8. Li, X., Gong, Q.: Research on intelligent verification scheme of signaling XDR data quality. Shandong Commun. Technol. 36(4), 1–4 (2016)

The Scalability of Volunteer Computing for MapReduce Big Data Applications

Wei Li[✉] and William Guo

School of Engineering and Technology, Central Queensland University,
Rockhampton, QLD 4702, Australia
{w.li,w.guo}@cqu.edu.au

Abstract. Volunteer Computing (VC) has been successfully applied to many compute-intensive scientific projects to solve embarrassingly parallel computing problems. There exist some efforts in the current literature to apply VC to data-intensive (i.e. big data) applications, but none of them has confirmed the scalability of VC for the applications in the opportunistic volunteer environments. This paper chooses MapReduce as a typical computing paradigm in coping with big data processing in distributed environments and models it on DHT (Distributed Hash Table) P2P overlay to bring this computing paradigm into VC environments. The modelling results in a distributed prototype implementation and a simulator. The experimental evaluation of this paper has confirmed that the scalability of VC for the MapReduce big data (up to 10 TB) applications in the cases, where the number of volunteers is fairly large (up to 10K), they commit high churn rates (up to 90%), and they have heterogeneous compute capacities (the fastest is 6 times of the slowest) and bandwidths (the fastest is up to 75 times of the slowest).

1 Introduction

When the data sets of business transactions or social media become massive as termed as Big Data, the computational analysis of the big data, in order to predict business trends, deepen customer engagement and optimize operations, challenges the traditional data processing and demands newer parallel and distributed approaches and tools [16]. The issues, such as distributing the data, parallelizing the computation and synthesizing results, must be handled in a reasonable amount of time, in order to support timely smart decision. In this area, MapReduce [7] has been a successful programming paradigm developed by Google to process the large data sets, such as *crawled documents*, *inverted indices* and *web request logs*. Nowadays, MapReduce has been extensively used by the enterprises, such as Yahoo, Facebook and Microsoft, to process their enterprises big data.

MapReduce consists of 3 steps to process a big data set. In the *map step*, the original, big data set is divided into a number of small data sets, which are distributed onto a cluster of computers as *map tasks*. The data sets will be computed in parallel by the same *map function* by the entire cluster so that each computer will emit a number of *<key, value>* pairs at the end of this step. In the *shuffle step*, all the *<key, value>* pairs with the same key from the last step will be merged together in the form of *<key, a list of values>*. These pairs will be further sorted by the keys into a number of *reduce tasks*,

© Springer Nature Singapore Pte Ltd. 2017
B. Zou et al. (Eds.): ICPCSEE 2017, Part I, CCIS 727, pp. 153–165, 2017.
DOI: 10.1007/978-981-10-6385-5_14

which are redistributed onto the cluster. In the *reduce step*, the reduce tasks are computed in parallel by the same *reduce function* by the entire cluster so that each computer will emit a number of *<key, value>* pairs as the final results. The difference between the map step and the reduce step is that any 2 computers in the map step may emit the *<key, value>* pairs with the same key, but neither in the reduce step will emit a *<key, value>* pair with the same key. This is because the keys emitted from the reduce step are the same as the keys of the shuffle step that already merged by the same keys.

When MapReduce has succeeded for a variety of big data applications such as *inverted indices, k-means, classification* and more in cluster environments [1], it involves moving a large amount of data, particularly in the shuffle step, and therefore puts stress on network bandwidth and introduces runtime overhead. This concern was reflected in the research of [1], which proposed the MaRCO model to achieve a full overlap between the map computation and shuffle communication to speed up the overall performance, and [2], which proposed Meta-MapReduce to avoid big data set migration across remote sites and only transmit the very essential data for obtaining the result, in the standard venue of *grid (cluster) computing environments*. Coming to the scope of research of this paper, it remains open in the current literature whether MapReduce can be effective in the extended environment of Volunteer Computing (VC) as defined by [17]. From one aspect, VC makes use of the potential computing power from millions of volunteer computers from the Internet and has been successfully applied to large-scale scientific projects such as SETI@home [11], FiND@Home [8] and Climateprediction.net [4]. Thus instead of using expensive computing clusters (grids), it is inspiring to utilize the free volunteered computing power for MapReduce big data applications. From another aspect, MapReduce challenges VC in terms of whether the system is able to scale (1) when a cluster mainly needs data communication for the shuffle step only, VC needs data communication for all 3 steps of MapReduce; (2) when a cluster fails rarely, volunteers commit churn; (3) when a cluster consists of homogeneous machines, working in a high speed network, volunteers are heterogeneous in compute capacity, bandwidth and storage.

This paper is to extend VC to big data applications via MapReduce by proposing a DHT (Distributed Hash Table) based strategy for task scheduling and data migration, a prototype implementation to verify the functional correctness of the model, and a simulator to evaluate the scalability of the model. The evaluation aims at answering whether VC is an appropriate distributed model for big data applications like MapReduce.

The rest of this paper has been structured as: related work is reviewed in Sect. 2. Modelling MapReduce for VC environments is presented in Sect. 3. Section 4 describes the experimental settings for the evaluation of a virtual MapReduce job. Section 5 details the simulation results and analysis. Section 6 concludes the initial evaluation that VC scales for 10K peers in the opportunistic volunteer environments.

2   Related Work

Some works in the current literatures are worth to review in terms of promoting VC for MapReduce style big data applications. [6] presented a model BOINC-MR, which is based on BOINC but exploited a pull method to allow inter-peer data transfer for moving data between mappers and reducers to speed up the shuffle step data communication and reduce the burden on the central servers. There are 2 open issues with BOINC-MR. First, the MapReduce progress, depending on direct peer communication with each other rather than on a higher level overlay, would cause a halt when peers commit churn. Second, a hybrid structure, combing super nodes and P2P rather than pure P2P, has to be used because direction communication cannot go through firewalls. When the model tried to extend VC for MapReduce applications, their experimental results could not confirm whether VC is effective for MapReduce in that the model had no better performance (over the original BOINC), even on 1 GB data set and small number (40) of peers in a grid environment without churn.

VMR [5] is an extension to their previous work [6], aiming at the execution of MapReduce tasks on large scale VC resources from the Internet by direct peer communication, tolerating transient server failure, peer failure or byzantine behaviors. Their experimental results showed that VMR performed better (in terms of the number of map and reduce tasks and the replication factors) for the MapReduce applications like word count, inverted indices, N-Gram ans NAS EP for 50, 100, 200 peers. However, how peers committed churn and how the performance could be affected by the churn were not mentioned. The results still could not confirm VC for the Internet scope, where peers are in a large number (much more than 200) and commit churn frequently.

MOON [14] extended the MapReduce middleware Hadoop [9] for adaptive task and data scheduling to offer reliable MapReduce services to VC systems that were supported by a small set of dedicated nodes. MOON tried to confirm how well the existing MapReduce frameworks performed on VC environments. To cope with churn, MOON exploited data replication and task replication and proposed corresponding management and scheduling models. When the evaluation results of MOON were provided to compare its performance with Hadoop, they also showed the scalability of the model against churn rates. Their results, in a couple of real world applications and in a small cluster environment (60 volunteers plus 6 dedicated nodes), were somehow compliant with our simulation results in this paper.

P2P-MapReduce was proposed by [15] to provide a reliable middleware for MapReduce in dynamic cloud environments. The main idea was to use backup of data and tasks to cope with peer churn. Peers were treated differently as master nodes, slave nodes and user nodes, who performed different roles for the computation and backup, and might change their roles upon peer failure. The simulation results showed that P2P-MapReduce outperformed MapReduce in the centralized master/worker model, in terms of reliability and scalability with a large scale of network of 40,000 nodes and a small failure rate of up to 0.4%, measured by the number of failed jobs and the amount of data exchanged for the maintenance of the peer network.

The idea of [3] was similar to MOON in that it exploited peers to store and transfer data files so as to reduce the overhead on the central servers. In structure, it was a hybrid of SCOLARS (a BOINC based master/worker model) and BitTorrent (a P2P data share overlay) and used replication of input, intermediate and output data files to improve reliability. Comparing with BOINC and SCOLARS, the evaluation results showed that the proposed model was more efficient for data distribution and more scalable with varying file sizes and available bandwidths. The shortcoming of the evaluation was a small grid of 92 nodes without churn.

Based on the above review in this section, whether VC is scalable for big data applications remains open and makes the research of this paper necessary.

3   Modelling Volunteer Computing for MapReduce

The underlying P2P VC overlay to model MapReduce is the work of [12, 13], where a pure P2P is built on the Chord DHT protocol [18]. In the model, a central point is maintained for only advertising available VC projects and providing a bootstrap pool for a volunteer to join a project. When a project is of the interest of a newer coming volunteer, the volunteer will join the Chord overlay from one of the bootstrap nodes. Once joined the overlay, the peer keeps searching for and then doing a task. A peer can leave the overlay when there is no more tasks; a peer can leave or crash at any time even when doing a task. When a peer leaves, the unfinished task is check-pointed and will be picked up by another peer. However the unfinished task of a crashed peer must be totally redone by another peer. Built on Chord DHT protocol, the VC model takes the advantage of Chord in terms of the proved reliability, scalability and performance. In this paper, the remodeling of VC for MapReduce is based on the above fundamental overlay by adding newer components and coordination as illustrated in Fig. 1 and detailed in the following sections.

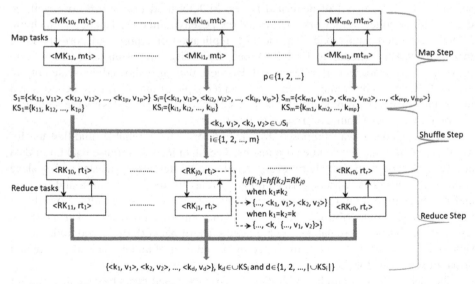

Fig. 1. The coordination of peers for MapReduce

3.1   The Map Step

The global object $<MK, m>$ represents the number of map tasks m that can be accessed by the key MK, which is known to every peer. Initially the m map tasks are: $<MK_{10}, mt_1>$, $<MK_{20}, mt_2>$, ..., $<MK_{m0}, mt_m>$, where 0 manes the task is available to download for execution by a peer. Once a map task $<MK_{i0}, mt_i>$ is put into execution by the peer, the task will change to $<MK_{i1}, mt_i>$, where 1 means it is now in execution and where $i \in \{1, 2, ..., m\}$.

Each peer looks up $<MK_{i0}, mt_i>$, where $i \in \{1, 2, ..., m\}$, for an available map task. If $<MK_{i0}, mt_i>$ is available, the peer changes it to $<MK_{i1}, mt_i>$ and then put it into execution. If the peer leaves before finishing the task, it will change it back to $<MK_{i0}, mt_i>$. When a map task is in execution, the peer will need to re-timestamp the task in a regular time interval ui. The ui is a predefined parameter, which can be accessed by the key UI (via $<UI, ui>$ on the overlay) that is known to every peer. It is worth to note that in a real implementation, it is unnecessary to re-timestamp the real task object mt_i. Instead, a simple timestamp mts_i is accessed by the key MK_{i11}. To check or update the state of a map task, a peer retrieves $<MK_{i11}, mts_i>$ more efficiently than retrieving the real task object mt_i. To access the real task object mt_i, a peer retrieves $<MK_{i1}, mt_i>$.

If failed with looking up $<MK_{i0}, mt_i>$, each peer looks up $<MK_{i1}, mt_i>$, where $i \in \{1, 2, ..., m\}$, for an available map task that satisfies the condition: (the current time - the timestamp mts_i of mt_i) $> ui$. Such a map task was in execution by a peer that is treated as crashed already.

A map task mt_i, where $i \in \{1, 2, ..., m\}$, is a self-satisfied object, which includes the executable code and data that are encapsulated in the data structures that are appropriate to a particular MapReduce application. A method call on mt_i such as $mt_i.execute()$ will perform the map task and emit p key-value pairs $S_i = \{<k_{i1}, v_{i1}>, <k_{i2}, v_{i2}>, ..., <k_{ip}, v_{ip}>\}$, where p is at least 1 in the general sense of MapReduce as defined by [7].

3.2 The Shuffle Step

The global object $<RK, r>$ represents the number of reduce tasks r that can be accessed by the key RK, which is known to every peer. Initially the r reduce tasks are: $<RK_{10}, rt_1>, <RK_{20}, rt_2>, ..., <RK_{r0}, rt_r>$, where 0 manes that the task is available to download for execution by a peer. A reduce task rt_j, where $j \in \{1, 2, ..., r\}$, is a self-satisfied object, which includes the executable code but its data set is initially empty and will be filled up by this shuffle step. The data structure of rt_j is in principle similar to a hash table, which will chain the values when keys clash. As a consequence, the data structure of rt_j will be treated as a hash table in the following description without losing generality.

For the p key-value pairs $S_i = \{<k_{i1}, v_{i1}>, <k_{i2}, v_{i2}>, ..., <k_{ip}, v_{ip}>\}$ that are emitted from the execution of a map task mt_i, where $i \in \{1, 2, ..., m\}$, we assume $KS_i = \{k_{i1}, k_{i2}, ..., k_{ip}\}$ and there is a hash function hf. For any $k \in \cup KS_i$, $hf(k) = RK_{j0}$, where $j = \{1, 2, ..., r\}$. For $<k_1, v_1>$ and $<k_2, v_2> \in \cup S_i$, we assume $hf(k_1) = hf(k_2) = RK_{j0}$ and thus $<RK_{j0}, rt_j>$ is retrieved. Under such a situation, if $k_1 \neq k_2$, $rt_j.put(<k_1, v_1>)$ and $rt_j.put(<k_2, v_2>)$ will store data v_1 and v_2 for the 2 different keys k_1 and k_2. If $k_1 = k_2$, $rt_j.put(<k_1, v_1>)$ and $rt_j.put(<k_2, v_2>)$ will store data v_1 and v_2 by chaining them for key k_1 or k_2. Then the reduce task $<RK_{j0}, rt_j>$ is stored back to the DHT ring. In this design, when the mapping step finishes, the shuffle step finishes in principle.

3.3 The Reduce Step

Each peer looks up $<RK_{j0}, rt_j>$, where $j = \{1, 2, ..., r\}$, for an available reduce task. If $<RK_{j0}, rt_j>$ is available, the peer changes it to $<RK_{j1}, rt_j>$ and then put it into execution. If the peer leaves before finishing the task, it will change it back to $<RK_{j0}, rt_j>$. When a reduce task is in execution, the peer will need to re-timestamp the task in a regular time interval ui, which was defined in Sect. 3.1.

   If failed with looking up $<RK_{j0}, rt_j>$, each peer looks up $<RK_{j1}, rt_j>$, where $j \in \{1, 2, ..., r\}$, for an available reduce task that satisfies the condition: (the current time - the timestamp rts_j of rt_j) > ui. Such a reduce task was in execution by a peer that is treated as crashed already.

   A method call on rt_j such as $rt_j.execute()$ will perform the reduce task and emit at least 1 key-value pair. When a peer emits a $<k, v>$, where $k \in \cup KS_i$, where $i \in \{1, 2, ..., m\}$, it will simply store it back to the DHT ring.

4 The Experimental Environment

We have implemented 2 versions of the proposed VC model for MapReduce. One is distributed version that has been implemented on the Open Chord platform [10]. This version of prototype implementation is to verify the functional correctness of the proposed model. In our experiments, the prototype was functionally correct for a 5-machines overlay, which was connected by a high speed Ethernet. This part is omitted from this paper as it is not the focus of scalability. The other version is a simulator, aiming at the evaluation of scalability of the model. When a large number of volunteer machines is unavailable to access, this version of simulator is able to evaluate the model on any numbers of virtual volunteer and map or reduce task. The compute load of tasks, the compute capacity and network speed of volunteers are all allowed to set for evaluation as detailed in this section.

   Instead of using any real MapReduce applications, a virtual MapReduce job is proposed to generalize any MapReduce jobs as long as they comply with the programming paradigm as defined by [7]. In other words, this virtual job is able to demonstrate the generality of the simulation for any MapReduce jobs in the volunteer computing environments. Thus the runtime behaviors of the job execution could be applied to a wide range of MapReduce applications.

   There is no restriction for the number of map tasks, reduce tasks or the number of peers involved in a job. There is no restriction for the number of peers that commit churn and when they commit churn. There is no restriction on the compute load of a map or reduce task.

   A peer needs a *search time* to look for an available map or reduce task. The *standard step* is used as a time unit to simulate a real world time unit such as a second, minute, hour or day etc. For example, if the search time is set as 100, it means that a peer needs 100 standard steps to find an available task or make sure that no any available tasks.

   A peer's *compute capacity* is the relative compute speed in standard steps. If a peer's capacity is 1, it can perform a standard step in one step. However, if a peer's capacity is ½, it needs 2 steps to perform a standard step. That is, the latter is 2 times slower that the former.

Each map or reduce task has a *compute load*, which is defined by standard steps as well. For example, if a map or reduce task's computing load is 1,000 steps, a peer of capacity of 1 will need 1,000 steps to finish the task, but a peer of capacity of ½ will need 2,000 steps to finish the task.

Each map or reduce task has a *download time* and an *upload time*, which are defined by standard steps as well. The download time simulates the network capacity that a peer accesses the network to download a map or reduce task, and similarly the upload time is the time to upload the result of a map or reduce task.

As illustrated in Fig. 2, the normal workflow for a peer to do a map or reduce task consists of 4 time slots: *search, download, compute* and *upload*. A peer can leave or crash at any time slot.

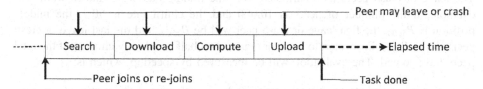

Fig. 2. The workflow of a peer in doing a map or reduce task

As our model is built on Chord, a safe assumption for the search time is naturally based on the proved efficiency of Chord lookup services, looking up a map or reduce task in $O(logN)$ messages [18], where N is the number of peers. If we assume that each message needs *3* standard steps (e.g. *3* s) for safety, the search time for a *10,000* peers overlay will be $3xlog(10,000) \approx 40$ when these messages are serially processed. The real situation could be better when these messages are somehow processed in parallel.

A safe assumption for the download time and upload time is based the use of ADSL2$^+$ as the volunteer internet connection, which is neither a high end nor a low end internet plan that can be possessed by most of the home internet users. The download speed of ADSL2$^+$ is about 24 Mbps (3 MB/s) and the upload speed is 1.4 Mbps (0.18 MB/s).

A safe assumption for the size of a map or reduce task or its result is 50 MB, which needs 50 MB ÷ 3 MB/s = 16.7 s to download a task and 50 MB ÷ 0.18 MB/s = 278 s to upload a result. Thus a safe assumption of download time is 17 steps and of upload time is 300 steps, provided the result of a map or reduce task is not expended or shrunk. If we assume that the entire job consists of 200,000 such size of map or reduce tasks, there will be 200,000 × 50 MB = 10 TB data set to be processed.

5   The Scalability Evaluation

In this section, the scalability of the proposed VC model in performing MapReduce tasks will be evaluated from 3 aspects when peers commit churn or have heterogeneous communication cost or compute capacity.

5.1 The Scalability Against Churn

The evaluation scenario is set as in Table 1, where the search time, download or upload time and the compute load are all set as standard steps. Particularly, the download and upload time are set as 17 and 300, corresponding to ADSL2$^+$ bandwidth minimum standards of 24 Mbps and 1.4 Mbps for downloading and uploading 50 MB data. The peers join the overlay in 20 (randomly chosen) standard steps interval, while peers leave or crash in 40 (randomly chosen) standard steps of interval. The numbers of peer are set as from 2,000 to 10,000, of which half of them have the compute capacity of 1, the other half have the capacity of ½, and the average capacity is 0.75. To evaluate the scalability against churn, the churn rate is set as from 10% to 90% of the total number of peers. For example, for 10,000 peers, the evaluation will be performed when there are 1,000 (10%); 3,000 (30%); 5,000 (50%); 7,000 (70%) and 9,000 (90%) peers to leave or crash respectively. For the churn peers, half leaves and the other half crashes. The leave or crash peers are distributed from the middle backward and forward. For example, if the number of peers is 10,000 and the churn rate is 50%, the middle position is P_{5000}, the first leave or crash peer will be P_{2500}, and the last leave or crash peer will be P_{7499}. Peers start to leave or crash when half (randomly chosen) of the total peers have joined. The evaluation will be measured by speedup, which is:

$$\frac{\text{the total standard steps to complete the entire job by a single peer of the average of capacity}}{\text{the total standard steps to complete the entire job by the overlay of peers}}.$$

The evaluation results in terms of speedup are showed in Fig. 3. The following 2 observations support the system scalability against churn in the dynamic, opportunistic VC environments.

1. At the same churn rate, the more peers the system has, the faster the overall computation is.
2. At the same peer number, the smaller the churn rate is, the faster the overall computation is.

Table 1. The experimental setting to evaluate scalability against churn

Scenario variable	Value
The number of map tasks	200,000
The number of reduce tasks	200,000
The compute load of a map or reduce task	8,000
The search time for a map or reduce task	40
The download time for a map or reduce task	17
The upload time for a map or reduce result	300
The number of peers	2,000; 4,000; 6,000; 8,000; 10,000
The compute capacity	50% of peers: 1; 50% of peers: ½
The churn rate	10%; 30%; 50%; 70%; 90%
The peer join interval	20
The peer leave or crash interval	40

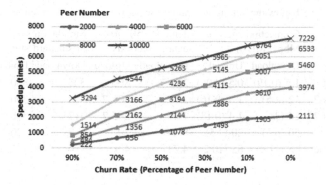

Fig. 3. The speedup against churn

There is 1 more observation about the speedup difference between neighbor churn rates as showed in Fig. 4, where a bigger churn rate affects the speedup more significantly than a smaller churn rate does.

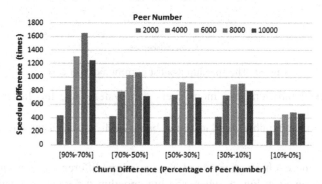

Fig. 4. The speedup difference against churn difference

5.2 The Scalability Against Communication Cost

The scenario setting of this evaluation is the same as those in Table 1 except that the churn rate is fixed as 30% and the download and upload time are set as in Table 2 to reflect the communication cost for the commonly available network bandwidth. The download and upload speed are chosen for the minimum standard of the service in each bandwidth. As showed in Table 2, the fastest connection is 75 (300/4) times (download) and 160 (800/5) times (upload) of the slowest connection.

The evaluation results in terms of speedup are showed in Fig. 5. The following 2 observations support the system scalability against the heterogeneous bandwidth of the commonly available internet connections for volunteers in the dynamic, opportunistic VC environments.

Table 2. The bandwidth setting to evaluate scalability against communication cost

	Connection speed in Mbps (MB/s)		Time for 50 MB data in seconds		Simulation setting in standard steps	
	Download	Upload	Download	Upload	Download	Upload
ADSL	1.5 (0.19)	0.5 (0.06)	266.7	800	300	800
ADSL2	12 (1.5)	1.3 (0.16)	33.3	307.7	35	300
ADSL2$^+$	24 (3)	1.4 (0.18)	16.7	285.7	17	300
NBN	50 (6.3)	20 (2.5)	8	20	8	20
Ethernet	95 (11.9)	82 (10.2)	4.2	4.9	4	5

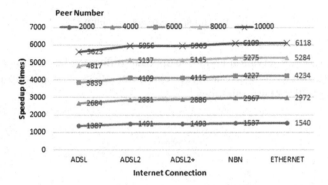

Fig. 5. The speedup against communication cost

1. At the same bandwidth, the more peers the system has, the faster the overall computation is.
2. At the same peer number, the larger the bandwidth is, the faster the overall computation is.

There is 1 more observation about the speedup difference between bandwidth differences as showed in Fig. 6, where the big speedup difference happened between the big bandwidth difference between ADSL and ADSL2 or between ADSL2$^+$ and NBN (National Broadband Network of Australia). That is, when bandwidth goes a big

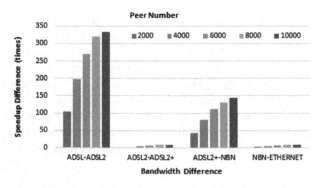

Fig. 6. The speedup difference against bandwidth difference

jump, the speedup does in the same way. There is no too much speedup difference happened between ADSL2 and ADSL2⁺ or between NBN and Ethernet.

5.3 The Scalability Against the Heterogeneity of Compute Capacity

The scenario setting of this evaluation is the same as those in Table 1 except that the churn rate is fixed as 30% and the compute capacities are set as in Table 3 to reflect the heterogeneity of compute capacity of peers.

The evaluation results in terms of speedup are showed in Fig. 7. The following 2 observations support the system scalability against the heterogeneity of compute capacity of volunteers in the dynamic, opportunistic VC environments.

1. At the same capacity variation, the more peers the system has, the faster the overall computation is.
2. At the same peer number, the smaller the capacity varies, the faster the overall computation is.

There is 1 more observation about the speedup differences between the capacity differences as showed in Fig. 8, where, the more the capacity varies between peers, the more the speedup drops (or the slower the overall computation is).

Table 3. The setting to evaluate scalability against the heterogeneous compute capacity

Setting	Compute capacity variation
C1	All peers: 1
C2	1/2 peers: 1, 1/2 peers: 1/2
C3	1/3 peers: 1, 1/3 peers: 1/2, 1/3 peers: 1/3
C4	1/4 peers: 1, 1/4 peers: 1/2, 1/4 peers: 1/3, 1/4 peers: 1/4
C5	1/5 peers: 1, 1/5 peers: 1/2, 1/5 peers: 1/3, 1/5 peers: 1/4, 1/5 peers: 1/5
C6	1/6 peers: 1, 1/6 peers: 1/2, 1/6 peers: 1/3, 1/6 peers: 1/4, 1/6 peers: 1/5, 1/6 peers: 1/6

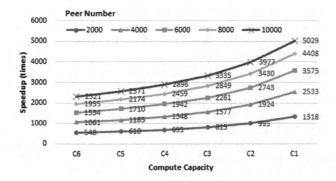

Fig. 7. The speedup against heterogeneous compute capacity

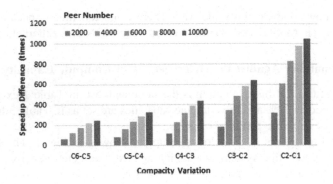

Fig. 8. The speedup difference against compute capacity variation

6 Conclusion

A DHT-based volunteer computing model is proposed for performing MapReduce big data applications and verified functionally correct by a prototype implementation on Chord protocol and Open Chord APIs. To evaluate the scalability of the model in the opportunistic volunteer computing environments, where a large number of peers have heterogeneous compute and network resources and commit churn, a simulator of the model has been implemented to test for an overlay of a large number of peers. The experimental evaluations showed that VC scales for a large number (up to 10,000) of peers, which commit high churn rate (up to 90%), have heterogeneous compute capacity (the fastest is 6 times of the slowest) and rely on different bandwidth (from the slow ADSL to high speed Ethernet). The results from this paper have confirmed that VC is suitable for big data applications like MapReduce when the model is built properly and the data splitting is appropriate. The future work will include the optimization of the simulator to verify the scalability of VC for even larger number of peers, who may commit more discrete churn.

References

1. Ahmad, F., Lee, S., Thottethodi, M., Vijaykumar, T.N.: MapReduce with communication overlap (MaRCO). J. Parallel Distrib. Comput. **73**(5), 608–620 (2013)
2. Afrati, F., Dolev, S., Sharma, S., Ullman, J.D.: Meta-MapReduce: a technique for reducing communication in MapReduce computations (2015). arXiv preprint arXiv:1508.01171
3. Bruno, R., Ferreira, P.: FreeCycles: efficient data distribution for volunteer computing. In: Proceedings of the Fourth International Workshop on Cloud Data and Platforms (2014)
4. Climateprediction.net (2016). http://www.climateprediction.net
5. Costa, F., Veiga, L., Ferreira, P.: Internet-scale support for map-reduce processing. J. Internet Serv. Appl. **4**, 18 (2013)
6. Costa, F., Silva, L., Dahlin, M.: Volunteer cloud computing: MapReduce over the Internet. In: Proceedings of IEEE International Symposium on Parallel and Distributed Processing Workshops and Ph.D. Forum (IPDPSW), pp. 1855–1862 (2011)

7. Dean, J., Ghemawat, S.: MapReduce: simplified data processing on large clusters. Commun. ACM **51**(1), 107–113 (2008)
8. FiND@Home (2016). http://findah.ucd.ie
9. Hadoop (2014). https://wiki.apache.org/hadoop/ProjectDescription
10. Kaffille, S., Loesing, K.: Open Chord Version 1.0. 4 User's Manual. The University of Bamberg, Germany (2007)
11. Korpela, E.J.: SETI@home, BOINC, and volunteer distributed computing. Annu. Rev. Earth Planet. Sci. **40**, 69–87 (2012)
12. Li, W., Franzinelli, E.: Decentralizing volunteer computing coordination. In: Che, W., et al. (eds.) ICYCSEE 2016. CCIS, vol. 623, pp. 299–313. Springer, Singapore (2016). doi:10. 1007/978-981-10-2053-7_27
13. Li, W., Guo, W., Franzinelli, E.: Achieving dynamic workload balancing for P2P volunteer computing. In: Proceedings of the 44th International Conference on Parallel Processing Workshops (ICPPW), pp. 240–249 (2015)
14. Lin, H., Ma, X., Archuleta, J., Feng, W.C., Gardner, M., Zhang, Z.: Moon: MapReduce on opportunistic environments. In: Proceedings of the 19th ACM International Symposium on High Performance Distributed Computing, pp. 95–106 (2010)
15. Marozzo, F., Talia, D., Trunfio, P.: P2P-MapReduce: parallel data processing in dynamic cloud environments. J. Comput. Syst. Sci. **78**(5), 1382–1402 (2012)
16. Oracle: An Enterprise Architect's Guide to Big Data - Reference Architecture Overview. Oracle Enterprise Architecture White Paper (2016)
17. Sarmenta, L.: Volunteer Computing. Ph.D., thesis, Massachusetts Institute of Technology (2001)
18. Stoica, I., Morris, R., Liben-Nowell, D., Karger, D.R., Kaashoek, M.F., Dabek, F., Balakrishnan, H.: Chord: a scalable peer-to-peer lookup protocol for Internet applications. IEEE/ACM Trans. Netw. (TON) **11**(1), 17–32 (2003)

An Improved FP-Growth Algorithm Based on SOM Partition

Kuikui Jia[(✉)] and Haibin Liu

China Aerospace Academy of Systems Science and Engineering,
Beijing 100048, China
jia_kuikui@163.com

Abstract. FP-growth algorithm is an algorithm for mining association
rules without generating candidate sets. It has high practical value in
many fields. However, it is a memory resident algorithm, and can only
handle small data sets. It seems powerless when dealing with massive
data sets. This paper improves the FP-growth algorithm. The core idea
of the improved algorithm is to partition massive data set into small data
sets, which would be dealt with separately. Firstly, systematic sampling
methods are used to extract representative samples from large data sets,
and these samples are used to make SOM (Self-organizing Map) cluster analysis. Then, the large data set is partitioned into several subsets
according to the cluster results. Lastly, FP-growth algorithm is executed
in each subset, and association rules are mined. The experimental result
shows that the improved algorithm reduces the memory consumption,
and shortens the time of data mining. The processing capacity and efficiency of massive data is enhanced by the improved algorithm.

Keywords: FP-growth · SOM · Data mining · Cluster · Partition

1 Introduction

Data mining called knowledge discovery in database is a process of extracting
useful knowledge for the decision-making process in a large number of data sets.
Data mining has been widely used in many fields since it was proposed in 90s.
Association rule mining is an important part of data mining. In 1993, Agrawal
[1] first proposed the concept and model of association rules, which described
a relation of one thing and other things. At present, there are two kinds of
association rule mining algorithms: Apriori, FP-growth. The Apriori algorithm
[2] proposed by Agrawal et al. in 1994 needs to scan the database many times and
generates a large number of candidate sets, so that it wastes a lot of time to test
these candidates. FP-growth algorithm [3] was proposed by Han in 2000. It is a
mining algorithm based on frequent pattern tree. The algorithm can efficiently
mine frequent patterns, and is about one order of magnitude faster than the
Apriori algorithm. However, with the increase of the amount of data and useful
information in the data set becomes more and more sparse, it will consume a

© Springer Nature Singapore Pte Ltd. 2017
B. Zou et al. (Eds.): ICPCSEE 2017, Part I, CCIS 727, pp. 166–178, 2017.
DOI: 10.1007/978-981-10-6385-5_15

lot of memory space in the establishment of FP-tree, so that the memory is not enough to complete the mining task [4]. Park et al. [5] proposed the use of systematic sampling method for data mining, however, it is easy to result in the deformity and inaccuracy of data mining. Therefore, the researchers began to consider the parallel computing environment to solve the above problems. In paper [6], the parallel algorithm based on multi thread is used, which reduces the storage and computation pressure, but the limitation of the memory resource restricts the expansion of the algorithm. In the paper of [7,8] the MPI parallel environment are described in detail, but inter process communication way is used to coordinate parallel computing, leading to parallel scalability problems of low efficiency. The parallel algorithm usually has large inter process scheduling and communication overhead, and it is difficult to segment the task of constructing FP-tree.

Based on the previous researches, an improved FP-growth algorithm is proposed. The core idea of the improved algorithm is to partition massive data set into small data sets, which would be dealt with separately. The main process of the algorithm is firstly to use clustering algorithm to partition the data set, and then FP-growth algorithm is implemented in each data subset. Therefore, the large memory that FP-growth algorithm needs is divided into small memory, and because the subset is partitioned according to the cluster result, association rule can be mined efficiently in each subset, rather than the random partition of the data set caused by mining uncertainty rules. Because of the large amount data, cluster analysis would consume a large amount of computing resources, so this paper use the method of systematic sampling to extract representative samples from large dataset, and then SOM algorithm is used to cluster analysis. The data set can be partitioned into several subsets by using cluster result, and then association rules would be mined in each subset by FP-growth algorithm. The improved algorithm not only improves the efficiency of mining, but also effectively solves the problem of insufficient memory, and the validity of the algorithm is verified by experiments.

2 Research on SOM Algorithm

Clustering analysis is one of the hot topics in data mining, which is to separate the data objects into several classes or clusters, so that the objects in the same cluster have high similarity, and have a big difference between different clusters. Clustering analysis is an unsupervised analysis method, and the classes to be classified are unknown in advance. The classical clustering algorithms such as Kmeans, CURE, neural network, etc., all need to specify the number of clusters in advance. Specifying the number of clusters is based on experience, and need repeated attempts to get correct classification, which is blindness, depressing overall efficiency and reliability of the clustering results. Self-organizing map is a special neural network algorithm proposed by Finland scholar Kohonen in 1981. SOM is used to adjust the high dimensional data points to the low dimensional grid, which is similar to the common neural network. SOM analysis have made many valuable achievements in the field of visual. With the help of high dimensional data in a low dimensional organizational ability, SOM classification, clustering and prediction in data mining field has many successful applications. In this paper, SOM clustering algorithm is used to cluster the data set, and the clustering model is used to partition data set, and FP-growth data mining is implemented in each subset.

SOM is a special kind of neural network, which is composed of input layer and output layer. The input layer has only one node corresponding to the input vector $x = [x_1, x_2, \ldots, x_d]$, where d is the dimension of input data. The output layer consists of a set of ordered nodes on a two dimensional grid, and each node corresponds to a weight vector $m = [m_1, m_2, \ldots, m_d]$. The basic SOM network training steps are described below:

(a) The initial weights are assigned to each node of the output layer. The training end condition is defined by using the weight error limit, or a predefined training length.

(b) A sample x is selected from the training data set, and the distance between the sample and each output node is calculated. The node that is nearest to the sample x is selected, which is called the best matching node (BMN) of the input sample, denoted as m_c.

$$||x - m_c|| = min||x - m_i|| \tag{1}$$

(c) According to the defined neighborhood function, the nodes in the BMN's neighborhood are determined, and the weights of BMN and neighbor nodes are adjusted by:

$$m_i(t + 1) = m_i(t) + a(t)h_{ci}(t)[x(t) - m_i(t)] \tag{2}$$

where $m_i(t)$ represents the weight of the i node of step t, and $a(t)$ is the learning rate of step t, and $h_{ci}(t)$ is a neighborhood function. The learning rate is usually reduced gradually with the training, which can be reduced by linear, exponential reduction, etc. Gauss function, bubble function etc. can be selected as the neighborhood function.

(d) If the training end condition is not satisfied, training returns to step (b) to continue.

3   Research on FP-Growth Algorithm

3.1   The Definition of Association Rules Mining

Association rule is a description of a thing and other things. Association rule mining is to find out the relationship between these data from a large number of data sets. Two metrics for measuring rules are support and confidence.

Definition 1. Rule $A \Rightarrow B$ is established in transaction set D with support $S(Support)$, which is the ratio of the number of transactions containing $A \cup B$

to all transactions in D, denoted as $Support(A \cup B)$. It is equal to the probability that the transaction contains the set A and B, that is:

$$support(A \Rightarrow B) = P(A \cup B) = D(X)/|D| \qquad (3)$$

where $D(X)$ is the number of transactions in database D containing X.

Definition 2. Rule $A \Rightarrow B$ confidence in the transaction set D is the ratio of the number of transactions containing A and B to the number of transactions containing A, which is the conditional probability $P(A|B)$, that is:

$$confidence(A \Rightarrow B) = P(A|B) = support(A \cup B)/support(A) \qquad (4)$$

Association rules mining must first find out all the itemset that satisfy the minimum support, and then calculate all the association rules with the confidence greater than the threshold value.

3.2   FP-Growth Algorithm

The basic idea of FP-growth algorithm is to compress the transactions with the tree structure, and to keep the relationship between the attributes in the transaction. The algorithm needs to scan the transactions set twice: the first scan is to find frequent 1-itemset, which is set in accordance with the support count in descending order; The purpose of the second scan is to build the FP-tree with the "Null" as the root node.

FP-growth algorithm mainly includes three modules: FP-growth module reflects the flow of FP-growth algorithm; The inserting module mainly completes the function of generating FP-tree; The searching module is used to obtain the conditional pattern base to facilitate the next recursive operation. In order to quickly traverse the FP-tree, header table T is created, each row of which represents a frequent item and it's frequency, and the pointer to the node is set corresponding to itself in FP-tree, and FP-tree is recursively mined to find all frequent itemset.

The following example illustrates how to build a FP-tree: a transaction set T is shown in Table 1, and min_sup $= 0.2$, so support count $= 0.2 * 10 = 2$ times. Firstly, the transactions set is traversed to count the support counts of all kinds of attributes, and the attributes whose support count is less than 2 are excluded. According to the support count, the attributes are sorted descending, and the frequent 1-itemset L_1 are generated, and $L_1 = [$B:8, A:7, C:5, D:2, E:2$]$.

Create the item header table, and then traverse a transaction set to generate the FP-tree. According to L_1, each transaction's items ranks according to L_1, and are inserted into the tree that NULL is used as the root node, meanwhile, the support count of each node is updated. According to the above principle, after traversing all transactions, the FP-tree is shown in Fig. 1. Therefore, the FP-growth algorithm is that eligible transaction items, counts of items and the relationship between items are compressed into a tree.

Table 1. Transaction collection T

Transactions	Items
T1	A, B, E
T2	B, D, F
T3	B, C
T4	A, B, D
T5	A, B
T6	B, C, G
T7	A, C
T8	A, B, E
T9	A, B, C
T10	A, C

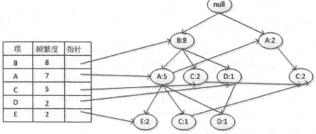

Fig. 1. FP-tree and item header table

4 FP-Growth Algorithm Based on SOM Partition

4.1 Algorithmic Process

FP-growth algorithm constructs the FP-tree to find the frequent patterns. However, when the data set is large, it is not realistic to construct FP-tree based on memory. In this paper, the FP-growth algorithm based on SOM partition can solve the problem. In this algorithm, SOM clustering method is used to get the subset of the data that have aggregated information, and then the FP-growth algorithm is implemented on each subset in parallel. The algorithm is divided into 5 steps: (a) Data standardization; (b) Extract Sampling data from data set according to systematic sampling; (c) SOM algorithm is used to cluster the sample data to get the classification model of the database transactions; (d) The transactions in the whole database is partitioned into several subsets according to the classification model, and the FP-growth mining is performed on the data subset in parallel; (e) Summary results. The program flow chart is shown in Fig. 2.

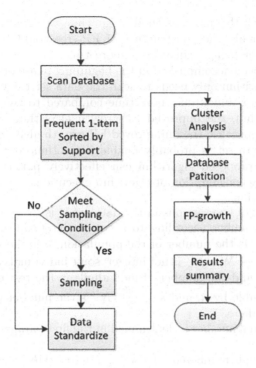

Fig. 2. Algorithmic program flow chart

(a) Data standardization

The items contained in each transaction in the database are discrete. In order to solve the Euclidean distance between transactions, it is necessary to change the item to digital representation. If one item is included in one transaction is denoted as 1, otherwise, denoted as 0. The specific process is as follows:

(1) The frequent 1-itemset L_1 in the database is obtained by scanning the database once.

(2) L_1 is sorted according to the support count in descending order, so a new sequence L_1' is obtained.

(3) For each transaction in dataset, the items contained in the L_1' are retained, not included in the L_1' being removed, and the items are sorted according to the order in L_1'.

(4) Standard data vector $x = x_1, x_2, \ldots, x_n$ is constructed, where n is equal to the number of elements in the L_1', and $x_1, x_2, \ldots, x_n$ corresponds to the corresponding location item of L_1'. For example, the transaction set T in Table 1, according to support set as 0.2, would get $L_1' = $ B, A, C, D, E, therefore T1 $= 1, 1, 0, 0, 1$, and so on until T10. It is shown in Table 2 after data standardization.

The algorithm that this paper proposed firstly scanned data sets to obtain the sorted list L_1', which is used for the standardization of entire data set in accordance with the above method, meanwhile, systematic sampling being completed. Finally the construction of FP-tree could be completed just via the data set being scanned one more time, for the frequent 1-itemset sequence has been accomplished in the beginning phase of the algorithm, so the entire algorithm only needs to scan the data set three times. Although the data set is scanned one more time compared to the unimproved FP-growth algorithm, the improved FP-growth algorithm is more useful for large data set, because the unimproved FP-growth algorithm for data mining of large data set is probably unable to continue owing to insufficient memory. The proposed algorithm can effectively partition data set, and reduce memory consumption of algorithm execution.

(b) Sample extraction

(1) The number of transactions in the data set D is set randomly.

(2) The entire number according to a certain interval (set to k) segment. When $\frac{N}{n}$ (N is the number of the population, n is the size of sample) is integer, $k = \frac{N}{n}$. When $\frac{N}{n}$ is not integer, some individuals are removed from population, and the number of individuals in the rest of the population can be divisible by n, and $k = \frac{N'}{n}$ (N' is the number of the rest of the population) here.

(3) The individual number is determined in the first paragraph using random sampling.

(4) The individuals numbered $l, l + k, \ldots, l + (n - 1)k$ are extracted.

(c) Cluster analysis

The SOM algorithm is used to cluster the sample data. After several iterations, the distance between the self-organizing nodes are obtained. The nodes that have a small distance belong to a same class, and that have a large distance belong to different classes. The distribution of neurons in the two-dimensional plane is easy to be distinguished. The number of categories can be determined according to the need, in general, the higher the degree of fineness, the more categories of data set. After determining the number of classification, the minimum circle cover method of point set [10] is used to determine the central position of each class, and the center set is as follows:

$$C = \{c_1, c_2, \ldots, c_n\} \tag{5}$$

where n represents the number of categories. Any element c_k in C is represented as follows:

$$c_k = \{c_{k1}, c_{k2}, \ldots, c_{km}\} \tag{6}$$

where m represents the dimension of the center coordinate point.

(d) FP-growth mining

By comparing each data with the central point in the dataset, the data has the minimum distance with the central point belongs to the corresponding category, that is:

$$||t_i - c_{min}|| = min\{||t_i - c_k||\} \tag{7}$$

where t_i represents any transaction data, and c_{min} represents the closest center point to t_i, and c_k represents any central point of C. Transactions are assigned to the corresponding flag according to the category that they belong to, and the data mining process is completed by computer on each subset. The data partition program is based on the Euclidean distance between the transaction data and the center point. After the data partition, the data set D is partitioned into s subsets, expressed as $D = D_1 \cup D_2 \cup \ldots \cup D_s$.

The FP-tree is built on each subset, and the construction process is based on the algorithm in Sect. 3.2. Since frequent 1-itemset sorted by large to small in accordance with the size of the support has been obtained before this phase, so the items can be added to the FP-tree directly.

(e) Results summary

(1) Based on the search algorithm in the Sect. 3.2, all the conditional pattern bases can be found on each subset.

(2) A conditional pattern base satisfying the preset support degree is selected to obtain frequent patterns. Frequent pattern's confidence is calculated, which is greater than the confidence threshold would be added into the data association rules. The solving process of support and confidence is as follows: Suppose X and Y are itemset in frequent patterns, and $X \cap Y = \emptyset$, so the association rule $(X \Rightarrow Y)$'s support $s(X \Rightarrow Y) = P(X \cup Y) = support(X \cup Y)$, and it's confidence $c(X \Rightarrow Y) = P(Y|X) = \frac{support(X \cup Y)}{support(X)}$, and refer to Sect. 2.1 for details.

(3) The association rules contained in the whole data set can be obtained through combining the association rules obtained in (2). Because of the previous SOM clustering analysis, each subset of data contains the high-density frequent items of corresponding categories, which contains the corresponding association rules. Assume that the data set D contains association rules set $R = \{r_1, r_2, \ldots, r_n\}$ (n represents the number of association rules contained in the data set, and r_i represents any association rule), and any data subset contains association rule set $R_{D_i} = \{r_{D_i 1}, r_{D_i 2}, \ldots, r_{D_i m}\}$, (i = 1, 2, \ldots, n, and m represents the number of association rules contained in a subset, and $r_{D_i j}$ represents any association rule in subset), then: $R_{D_1} \cup R_{D_2} \cup \ldots \cup R_{D_s} = R = \{r_1, r_2, \ldots, r_n\}$.

Therefore, the association rules in the data set D can be obtained by merging the association rules on each subset of data.

4.2 Algorithm Complexity Analysis

It is assumed that there are s association rules in a mining task and m items that meet the minimum support and n transactions in data set and a items contained in each transaction averagely that meet the minimum support. Then the classical algorithm to obtain s association rules, ignoring the cost of connecting of data set, the algorithm needs calculate $m \times n \times a + m \times s$ times. Since n and m dominate, the time complexity is $O(mn)$. Assume that the number of samples is b, and the number of clusters is c, then the time complexity of the improved algorithm is $a \times n \times a + m \times s + c \times b$. Because b and c are smaller than others, they can be ignored, so the time complexity is also $O(mn)$. Although the improved algorithm

Table 2. Standardized transaction set T

Transactions	Standardized data
T1	1 1 0 0 1
T2	1 0 0 1 0
T3	1 0 1 0 0
T4	1 1 0 1 0
T5	1 1 0 0 0
T6	1 0 1 0 0
T7	0 1 1 0 0
T8	1 1 0 0 1
T9	1 1 1 0 0
T10	0 1 1 0 0

and the original algorithm's time complexity is same, because the improved algorithm partition the data set into small subsets, which greatly reduces the space complexity, and reduces the memory occupation of program, and increases the possibility of massive data mining.

5   Experiment and Result Analysis

In this paper, the FP-growth algorithm based on SOM partition is verified by experiments. The experimental data sets are derived from Frequent Itemset Mining Dataset Repository.

5.1   SOM Cluster Analysis Result

In this experiment, the connect.data data set is used. Samples are extracted from the data set, and SOM clustering analysis is executed on the samples. After 100 iterations, the two-dimensional plane of 36 neurons appeared in clusters, whose similar neurons are close to each other, and different classes neurons are away from each other. Each neuron has a strong connection with some input points, and there is also a connection weight between neurons. The neurons with the small distance are classified into one class. The neurons are projected to the two-dimensional plane after the training of the weights, and the coordinates of the neurons are determined according to the distance between the sample points and neurons.

In order to highlight the effect of clustering, a shrinkage factor is added to each category, and the sample points are close to each data center under the action of the shrinkage factor. Different colors are marked in each category, for the classification result is observed more intuitively. After adjusting the parameters for many times, we can get the clustering result as shown in Fig. 3. Because of the complexity of the process of programming, Java programming are used to obtain the result of clustering.

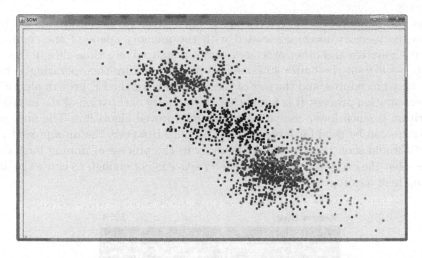

Fig. 3. Optimized clustering result (Color figure online)

5.2 Performance of FP-Growth Algorithm Based on SOM Partition Experiment

According to the clustering results in Sect. 5.1, the original data set is divided into 4 subsets, and FP-growth algorithm is executed in each subset in parallel. We can get the rule sets as shown in Table 3. Since all transaction items used in the experiment have been corresponding to numbers, rules are made up of numbers. A number in rules represents an item, and the last number of every rule is the frequent count of the rule. It can be seen from the table that the association rules can be mined efficiently from the database by using the algorithm proposed in this paper.

Table 3. Association rules

Rules	Support count
1 4 7 10 13 16 19 22	5520
4 7 10 13 16 19 23	4360
10 13 16 19 22 25 28 31	3346
1 4 7 10 28 31 34	3269
3 4 7 10 13 16 19 22	2594
2 6 7 10 13 16 19	2498
1 4 7 61 64 67	2280
4 7 10 13 31 34 37	2060

Memory usage is recorded during the execution of the algorithm. The (a) of Fig. 4 shows the memory usage of the improved algorithm, and the (b) of Fig. 4

shows the unimproved FP-growth algorithm's memory consumption. Figure 4 shows the memory usage associated with the memory space of the computer operating system and other processes, and is displayed as a time slice in 60 s. As can be seen from the figure 40% memory is occupied by the operating system and other procedures, and the rest of the memory is for the FP-growth algorithm implementation process. It is found that the memory occupation of the improved algorithm is much lower compared to the unimproved algorithm. The improved algorithm can be used for large data set mining. However, the unimproved FP-growth would soon occupy a lot of memory in the process of mining large data set, so that the computer physical memory space is not enough to cause the data mining task terminated.

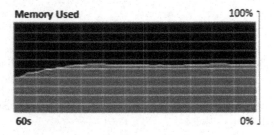

(a) Memory usage of improved algorithm

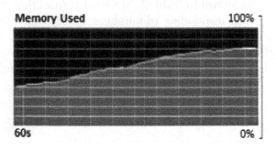

(b) Memory usage of unimproved algorithm

Fig. 4. Memory usage

The improved algorithm is compared with the Apriori algorithm and the FP-growth algorithm, and the operation time of each algorithm is shown in Fig. 5 with the support of 5%. As can be seen from the Fig. 5, with the increase of the amount of data, the time consumed by the three algorithms are increasing, however, the time consumption of the FP-growth algorithm is significantly lower than the Apriori algorithm, due to the increase of the amount of candidate items produced by the Apriori algorithm, which takes a lot of time in process. The improved FP-growth algorithm significantly reduces the computation time compared to the unimproved FP-growth algorithm. As can be seen from the figure, the time of the FP-growth algorithm consumes with the increase of the amount of data is growing rapidly, and the improved FP-growth algorithm presents a gentle growth trend. Therefore, the performance of the improved algorithm is obviously improved.

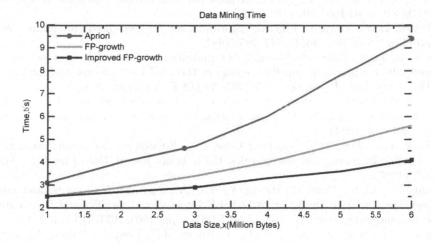

Fig. 5. Execution time comparison

6   Summary

This paper describes the process of the improved FP-growth algorithm. Firstly, SOM clustering algorithm is used to analysis the samples extracted from big data set, and according to the results of clustering analysis big data set is partitioned into several subsets with high density association rules, and finally the association rules contained in each subset are mined by FP-growth algorithm. The experimental results show that the improved algorithm can reduce the memory usage, and shorten the time of data mining.

References

1. Adnan, M., Alhajj, R.: A bounded and adaptive memory-based approach to mine frequent patterns from very large databases. IEEE Trans. Syst. Man Cybern. B Cybern. **41**(1), 154–172 (2011)
2. Agrawal, R., Srikant, R., et al.: Fast algorithms for mining association rules. In: Proceedings of 20th International Conference on Very Large Data Bases, VLDB, vol. 1215, pp. 487–499 (1994)
3. Aouad, L.M., Le-Khac, N.A., Kechadi, T.M.: Distributed frequent itemsets mining in heterogeneous platforms. J. Eng. Comput. Archit. **1**(2), 1–12 (2007)
4. Dean, J., Ghemawat, S.: MapReduce: simplified data processing on large clusters. Commun. ACM **51**(1), 107–113 (2008)

5. El-Hajj, M., Zaiane, O.R.: Parallel leap: large-scale maximal pattern mining in a distributed environment. In: 12th International Conference on Parallel and Distributed Systems, ICPADS 2006, vol. 1, 8-p. IEEE (2006)
6. Goethals, B., Zaki, M.J.: Advances in frequent itemset mining implementations: report on FIMI 2003. ACM SIGKDD Explor. Newsl. **6**(1), 109–117 (2004)
7. Han, J., Pei, J., Yin, Y.: Mining frequent patterns without candidate generation. ACM SIGMOD Rec. **29**, 1–12 (2000)
8. Hearn, D.W., Vijay, J.: Efficient algorithms for the (weighted) minimum circle problem. Oper. Res. **30**(4), 777–795 (1982)
9. Liu, L., Li, E., Zhang, Y., Tang, Z.: Optimization of frequent itemset mining on multiple-core processor. In: Proceedings of the 33rd International Conference on Very Large Data Bases, pp. 1275–1285. VLDB Endowment (2007)
10. May, R.: Mining association rules between sets of items in large database. In: Proceedings of ACM SIGMOD International Conference on Management of Data, pp. 207–216 (1993)
11. Park, J.S., Chen, M.S., Yu, P.S.: Using a hash-based method with transaction trimming for mining association rules. IEEE Trans. Knowl. Data Eng. **9**(5), 813–825 (1997)
12. Qiu, H., Gu, R., Yuan, C., Huang, Y.: YAFIM: a parallel frequent itemset mining algorithm with spark. In: 2014 IEEE International Parallel and Distributed Processing Symposium Workshops (IPDPSW), pp. 1664–1671. IEEE (2014)
13. Zhou, L., Zhong, Z., Chang, J., Li, J., Huang, J.Z., Feng, S.: Balanced parallel FP-growth with MapReduce. In: 2010 IEEE Youth Conference on Information Computing and Telecommunications (YC-ICT), pp. 243–246. IEEE (2010)
14. Zou, X., Zhang, W., Liu, Y., Cai, Q.: Study on distributed sequential pattern discovery algorithm. J. Softw. **16**(7), 1262–1269 (2005)

A Novel Recommendation Service Method
Based on Cloud Model and User Personality

Jing Yao[1], Zhigang Hu[1(✉)], Hua Ma[2], and Bingting Jiang[1]

[1] Central South University, Changsha 410075, China
{yaojing, zghu, jbeating}@csu.edu.cn
[2] Hunan Normal University, Changsha 410081, China

Abstract. The number of Internet Web services has become increasingly large recently. Cloud services consumers face a critical challenge in selecting services from abundant candidates. Due to the uncertainty of Web service QoS and the diversity of user characteristics, this paper proposes a Web service recommendation method based on cloud model and user personality (WSRCP), which employs cloud model similarity method to analyze the similarity of QoS feedback data among different users, to identify the user with high similarity to the potential user. Based on the QoS data of the users' feedback, Finally, user characteristic attribute Web service recommendation is implemented by personalized collaborative filtering algorithm. The experimental results on the WS-Dream dataset show that our approach not only solves the drawbacks of the sparse user service, but also improves the recommend accuracy.

Keywords: Cloud model · Personality · Cloud similarity algorithm · Services recommendation

1 Introduction

As the number of Internet services with much uncertainty grows, it becomes difficult to meet the needs of their own services easily for users in the service selection. It is an actual and challenging issue to choose a service to meet the needs of users [1–3]. Quality of Service (QoS) is often used to describe the non-functional features of Web services, Characterize the quality of Web services in a certain way, such as response time, throughput and cost. There are many similarities in many services, so you can use the appropriate similarity calculation method to find these services have similarities in the service, and then through a certain calculation to recommended to the user, to meet the needs of users. In recent years, rapid growth of service recommendation brings an increasing interest in collaborative filtering algorithm [4] (CF) and many service recommendation have used CF as an important tool [5] proposed a collaborative filtering algorithm based on cloud model is proposed to calculate the similarity of users' scoring, which avoids the problem that the traditional strict matching object is not enough. In [6], a collaborative filtering recommendation algorithm based on user characteristics and project attributes is proposed. In [7], proposed a personalized recommendation algorithm, which presents a user-oriented Web service QoS prediction method, UL-WSRec, which treats the user as a similar user, and selects the service by

© Springer Nature Singapore Pte Ltd. 2017
B. Zou et al. (Eds.): ICPCSEE 2017, Part I, CCIS 727, pp. 179–191, 2017.
DOI: 10.1007/978-981-10-6385-5_16

similar users proposed a collaborative filtering recommendation method based on user attribute and cloud model, but only considers the user's scoring characteristics. In this paper, the cloud model is introduced into the service recommendation process. We proposed the service recommendation method based on the cloud model, combined with the multi-attribute user characteristics and the similarity analysis algorithm. Under different application scenarios, this algorithm uses the real data set simulation experiment to characterize the service recommendation precision, and obtains the ideal recommendation result.

This paper describes the current situation and main problems of service selection. This paper proposes a method of cloud model service recommendation based on user characteristics, and then introduces the cloud model and cloud similarity algorithm. Combined with the service recommendation effect of cloud model, through the simulation experiment, to determine the various similarity algorithm in different application scenarios of the calculation accuracy; Finally, we conclude the paper.

2 Related Work

Web service discovery has become a research hotspot in recent years. How to find that the service to meet the needs of users quickly and accurately has become a key problem in service computing. Users are difficult to timely and effective choice for their own needs of the service in the face of a large number of services, thus affecting the user experience. Moreover, In a large number of networks and information resources, the recommended system is a very good tool, it can provide users with more convenient and personalized service. It through the different needs of users, the flexibility to adjust a variety of information services. To a certain extent, the recommendation system can provide the user with the service suitable for the user.

There are many uncertainties and things in the real world. How to express and deal with the phenomenon of uncertainty and things has been the focus of research, but also a difficult problem. The uncertainty of knowledge and reasoning is divided into two kinds of fuzziness and randomness. In recent years, some progress has been made in the study of fuzziness, and fuzzy set theory has been gradually established, and membership degree is used as the criterion of fuzziness. The membership degree is the exact value obtained by the membership function, which violates the fuzzy mathematics and returns to the scope of precise mathematics. In response to this problem, Academician Li put forward the cloud model theory [8]. The cloud model combines ambiguity and randomness to achieve the transformation of qualitative concepts and quantitative values.

According to the definition of the cloud, different parameters of the cloud can describe the same qualitative concept [9]. Different clouds to represent the quantitative characteristics of the same qualitative problem, the correlation between these clouds, that is, the degree of similarity between these clouds, can describe the difference between them. Because of the three digital features of the cloud to characterize the overall shape of the cloud, we can use the three digital features of the cloud to study the similarity between clouds. With the cloud model theory has been applied by other areas, cloud similarity measurement method is also endless.

Collaborative Filtering (CF) is one of the most commonly used technologies in e-commerce personalized recommendation system. It is mainly to compare and find the user preferences of the process, it needs a number of user information to predict the user's preferences. Collaborative filtering recommends that for a user, there must be a similar person in the user community, and the historical information of these people can be used to learn. Through these data, it can provide some reference for similar users' service recommendation.

In the field of service choice research, the user is faced with how to choose from a number of uncertain services for their own services. So this paper will be introduced to the cloud model of service selection process, while the multi-user attributes are also taken into account. Based on the characteristics of different users, We proposed a collaborative filtering algorithm based on cloud model. And we observe whether the method has improved the accuracy of the service in the real data set under the system simulation. cloud model and similarity algorithm.

3   Cloud Model and Similarity Algorithm

3.1   Cloud Model

On the basis of traditional fuzzy mathematics and probability statistics, Academician Li put forward a cloud model expressing qualitative and quantitative interchange [10]. The cloud model mainly reflects the ambiguity and randomness of things. By combining these two uncertainties, it is possible to express the meaning of a qualitative concept with a quantitative value or to describe a quantitative value in a qualitative language.

Definition 1. Ω is a quantitative field, T a qualitative value. The membership degree $C_T(x)(C_T(x) \in [0, 1])$ is a random number with a stable tendency. It describes the qualitative relationship between the elements x and T in Ω. The distribution of membership in the domain of the domain known as the subordinate cloud, referred to as the cloud. Thus, the cloud is a mapping from the domain Ω to the interval [0,1], that is, $x \rightarrow C_T(x), x \in \Omega$ sequence $(x, C_T(x))$ is called cloud [11].

A cloud droplet is a qualitative concept, in the number of a realization, a large number of cloud droplets composed of clouds. Although the cloud droplets are different at different times, the cloud's digital character determines the overall shape of the cloud. The numerical features of the cloud are represented by the expected Ex, entropy, and super-entropy He, which reflect the quantitative characteristics of the qualitative concept. The digital features of the cloud are shown in Fig. 1.

Definition 2. Expectation Ex: The point at which the qualitative concept is best embodied in the field is the best sample point for numericalizing the qualitative concept. The cloud is reflected in the cloud where the peak position.

Definition 3. Entropy En: indicates the size of the range of values that can be accepted by the qualitative concept in the domain space, that is, the degree of ambiguity, which is also a measure of qualitative and also. The cloud is reflected in the cloud width.

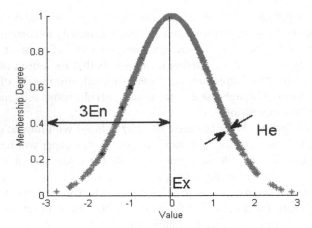

Fig. 1. Cloud digital features

Definition 4. The super-entropy *He*: the measure of the uncertainty of the entropy, reflecting the degree of discretization of the cloud droplets, representing the randomness of the cloud drops, revealing the association of ambiguity and randomness. The greater the super entropy, the greater the degree of cloud droplets, the greater the cloud thickness in the cloud. The three numerical features of the cloud are calculated as shown in Eqs. (1)–(3).

$$Ex = \overline{X} = \frac{1}{N}\sum\nolimits_{i=1}^{N} x_i \tag{1}$$

$$En = \sqrt{S^2 - \frac{1}{3}He} \tag{2}$$

$$He = \sqrt{\frac{\Pi}{2} \times \frac{1}{N}\sum_{i=1}^{N}|x_i - Ex|} \tag{3}$$

where S^2 is the sample variance and is expressed as follows:

$$S^2 = \frac{1}{N-1}\sum_{i=1}^{N}(x_i - \overline{X})^2 \tag{4}$$

3.2   Cloud Similarity Measure Algorithm

There are some similarities between two clouds or multiple clouds that describe the same qualitative concept, saying that these clouds are similar to each other or cloud, where 'certain similarity' refers to the similarity between clouds less than the given similarity threshold. In theory, the cloud is made up of numerous clouds. But in reality, we can only by generating a limited number of cloud droplets, to mark the overall shape of the cloud. Thus, due to the number of cloud droplets, even if the digital features are exactly same, they can only be very similar, not exactly the same. The greater the number of cloud droplets produced by the same cloud of two digital features, the greater the coverage of the cloud, the greater the calculated similarity we get. So the number of cloud drops has a significant effect on the calculation of similarity between clouds.

The earliest cloud model similarity algorithm [10] is achieved by the distance mean between the cloud droplets, and two cloud models randomly generate a certain number of cloud droplets. After filtering, the distance between the cloud drops is calculated, That is, the difference squared as the degree of similarity between the two cloud models.

The similarity calculation of cloud model based on cosine similarity [4] (CCM) takes the user score as a vector on the n-dimensional space. The similarity between users is measured by the cosine angle between vectors. Let the user u and user v in the n-dimensional project space on the score were expressed as the vector $\vec{u}$, $\vec{v}$, then the user $\vec{u}$ and the user $\vec{v}$ between the similarity can be calculated by (5).

$$sim(u, v) = \cos(\vec{u}, \vec{v}) = \frac{\vec{u} \cdot \vec{v}}{\|\vec{u}\| \cdot \|\vec{v}\|} \tag{5}$$

Cloud model using Euclidean distance similarity measurement method [12] (OCM) proposed a standardized multi-dimensional Euclidean similarity calculation method, by mapping three digital features into three-dimensional space, calculate the exponential function Euclidean similarity to characterize similarity. As shown in Eq. (6):

$$sim(u, v) = \frac{1}{e\sqrt{(Ex_u - Ex_v)^2 + (En_u - En_v)^2 + (He_u - He_v)^2}} \tag{6}$$

The normal cloud model similarity calculation method based on the expected curve [13] (ECM) can easily and effectively describe the general characteristics of the normal cloud because of the normalized cloud expectation curve with analytic form. Therefore, the similarity of the cloud model can be solved by means of the normal cloud expectation curve. The similarity between the two cloud models is expressed by solving the area S of the overlapping lines of the expected curves of the two cloud models. The area of the shaded parts reflects the similarity of the two cloud models.

The normal cloud similarity calculation method based on the expected curve is mainly to describe the similarity between different cloud models through the expected curve. The expectation curve can well reflect the overall characteristics of the cloud model. It is a kind of similarity from the perspective of the global geometric feature to study the similarity of the normal cloud model, and can ignore the super entropy description in the cloud model. In many cases, however, to use the normal cloud definition and the 3En rule of the positive norm cloud model, almost all of the cloud drops are under this maximum boundary curve, which is determined by the rules of the normal distribution. The calculation method is shown in Eqs. (7)−(9)

$$y = e^{-\frac{(x-Ex)^2}{2En^2}} \tag{7}$$

$$S = \int_{-\infty}^{+\infty} y \, dx \tag{8}$$

$$ECM(u, v) = \frac{2S}{\sqrt{2\pi}((En_u - En_v)} \in [0, 1] \tag{9}$$

The maximum boundary curve is a method of studying the geometric properties of cloud models from the local perspective of maximum cloud droplet values. Using the normal cloud similarity calculation method (MCM) based on the maximum boundary curve, let en = 3He + En obtain the maximum boundary curve resolution similar to the expected curve method. Similarly, by similar cloud model similarity based on the expected curve. The method is used to get the area of the overlapping part of the maximum boundary curve between the two cloud models, and the similarity of the two normal cloud models is obtained.

The formula is as follows:

$$MCM(u, v) = \frac{2S}{\sqrt{2\pi}((en_u - en_v)} \in [0, 1] \tag{10}$$

There are many similar methods to calculate the similarity degree in the cloud model. We list some of the commonly used methods above. And these methods proposed have a certain effect on our research, which solves the technical problems to a certain extent. However, these methods are still worthy of improvement, there is a large space for modification. In other words, in some scenarios, the accuracy of the calculated results is indeed not high, which follow the research to bring greater interference. To sum up, in different application scenarios, we can choose the right way to improve the accuracy of service recommendations.

4   Method

The main idea of the recommendation service method based on cloud model and user personality as follows: Firstly, the user's multi-feature attribute data is standardized, then employ the user's multi-attribute historical data and transform to cloud model to calculate each user as a "cloud", each user's historical data is regarded as a 'cloud drop', and represent as a cloud model. The similar user is presumed by data of these users' geographic information, then employ the multi-attribute personality data of these similar users, obtain the candidate service list after weighted synthesis, and finally recommend the corresponding service to user.

The algorithm is as follows:

Input: User characteristic attribute matrix $U_{u,attr}$, nearest neighbor set k, weighting factor λ.

Output: Predict the top-N recommended set $\{S1, S2, S3,..., Sn\}$ of the target user R_u's.

Step 1: Standardize multi-attribute data.

Step 2: The user is divided into similar user sets by user geographic information.

Step 3: verify the user's similarity and calculate $\text{sim}_{attr}(u, v)$.

Step 4: Use the historical data of similar users to calculate the service recommendation results, and use the formula of (11) obtain the top-N recommendation set as the prediction value.

Step 5: Calculate the average absolute error with the actual user's QoS value, and analyze the recommended accuracy.

We select the user characteristic attribute matrix $T_{u,j}$, $A_{u,j}$, respectively represent the user's response time and throughput, the matrix data is sparse, the goal is to select different users with common services data. Where u_i denotes the user i, j_n denotes the service n, $t_{ui,jn}$ denotes the response time under the n service of user i; $a_{ui,jn}$ denotes the throughput under n service of user i. all shown in (11) and (12).

$$T_{u,j} = \begin{bmatrix} t_{u1,j1} & \cdots & t_{u1,jm} \\ \vdots & \ddots & \vdots \\ t_{ui,j1} & \cdots & t_{ui,jm} \end{bmatrix} \tag{11}$$

$$A_{u,j} = \begin{bmatrix} a_{u1,j1} & \cdots & a_{u1,jm} \\ \vdots & \ddots & \vdots \\ a_{ui,j1} & \cdots & a_{ui,jm} \end{bmatrix} \tag{12}$$

First of all, it is necessary to normalize the data. In this paper, the Z-score normalization method is used to standardize the mean and standard deviation of the original data. After processing data meets the standard normal distribution, mean is 0, the standard deviation is 1, the transformation function is:

$$x^* = \frac{x - \mu}{\sigma} \tag{13}$$

The normalization is completed, the matrices of $T_{u,j}$ and $A_{u,j}$ are to integrate. The formula is calculated by the pre-set weighted ratio to obtain the matrix $B_{u,j}$. The formula is as follows:

$$B_{u,j} = T_{u,j} + A_{u,j} \tag{14}$$

$$B_{u,j} = \begin{bmatrix} b_{u1,j1} & \cdots & b_{u1,jm} \\ \vdots & \ddots & \vdots \\ b_{ui,j1} & \cdots & b_{ui,jm} \end{bmatrix} \tag{15}$$

$B_{u,j}$ denotes the characteristics of multi-attribute users. It is assumed that users with the same or similar geographical location information are similar users and get similar user sets. Then using the cloud similar method to perform similar user verifying the hypothesis. We use the cosine similarity algorithm as an example to calculate the similarity, using the formula (5), for each row of $B_{u,j}$ and the rest of each row. Select the top five services similarity into its corresponding similar user set. In order to avoid the differences in the service, the similarity value is too small, we set a threshold, if the similarity is less than the threshold, even if we do not reach the five similar users, we will also abandon it to ensure result accuracy.

$$sim_{attr}(u, v) = \begin{cases} value, & value \text{ is more than the threshold} \\ 0, & value \text{ is less than the threshold} \end{cases}$$

Through the similar result set, the service recommendation result is calculated by using the historical data of similar users, and the top-N recommendation set is obtained by using the formula (16) as the prediction value. Calculate the service recommendation value S_{Ui} (**u, v**). Combined with the similarity of user characteristics attribute obtain the results. The final service value is calculated as follows:

$$S(u, v) = x_1 S_{U1}(u, v) + x_2 S_{U2}(u, v) + \cdots + x_i S_{Ui}(u, v) \tag{16}$$

where x_i is the weighting factor and can be adjusted according to the different set of data x_i, $\sum x_i = 1$.

The error calculation is performed on the obtained $S_{Ui}(u, v)$, and the average absolute error between the QoS value and the actual potential user is calculated and analyze the recommendation accuracy.

5 Experiment

We developed the test program on MATLAB, compared the UL-WSRec method and the LICM method, and compared the experimental results of the multi-attribute service method in WS-Dream # 1 data set [14, 15], which geometrically located from different parts of the world with real real-time QoS data. To a certain extent, it can be assumed that the same geographical location and similar users of the computer's IP address can be approximated as similar users. After the analysis of QoS data of some similar users, the similarity calculation method is used to compare and analyze.

We use kinds of QoS experiments that can be regarded as similar users in geographically position. There are three kinds of different backgrounds (universities, government organizations, institutions) in the data. The data of each application background of simulation experiments were carried out for each five sets, results were compared and analyzed, one of sets data was extracted from Table 1.

Where User ID 45 as U1, User ID 46 as U2, User ID 45 as U3, User ID 45 as U4, and simulation experiment is carried out with different similarity algorithm, and data result is recorded in Table 2.

WS-Dream # 1 data set, each user has 5825 different services QoS data. This more than 5800 data is divided into two parts, select part of the data for user training, another part of the test. The data of the normalized data is modeled by the cloud model, and the

Table 1. WS-Dream # 1 user data

User ID	IP address	Country	IP no	AS	Latitude	Longitude
45	128.83.122.161	United States	2152954529	AS18	30.2961	−97.7369
46	128.83.122.162	United States	2152954530	AS18	30.2961	−97.7369
47	128.83.122.180	United States	2152954548	AS18	30.2961	−97.7369
48	128.83.122.181	United States	2152954549	AS18	30.2961	−97.7369

similarity degree is verified by the cosine similarity measure. Then the common index of the collaborative prediction method is the average absolute error MAE [16] to verify the recommended algorithm, Which measures the accuracy of the prediction by calculating the deviation of the predicted user score from the actual user score. Assuming that the user's predicted score is $\{pre_1, pre_2, ..., pre_N\}$, the corresponding actual score set is $\{R_1, R_2, ..., R_N\}$, the MAE can be calculated from the following expression:

$$MAE = \frac{\sum_{i=1}^{N} |pre_i - R_i|}{N} \tag{17}$$

Figure 2 is the summary of the MAE values of the multiple sets of data obtained after the service forecast. We can see that the historical data of the users are also increased with the increase of the number of training users, and the historical data of these similar users can be better recommended to the potential users of the corresponding services. Moreover, the method of this article has a significant advantage over other algorithms in recommendation results.

Similarity algorithm is mainly used for similar users to recommend the service, therefore, in order to better represent the accuracy of similar algorithms, we achieved several common similarity calculation method, by comparing the different similarity method to express WSRCP in different application scenarios recommended results.

Table 2 shows the different similarity calculation methods. Then, the MAE performance comparison chart of Fig. 3 is obtained by using the service recommendation method based on the user characteristic attribute proposed in this paper.

In the experiment, we can conclude that the traditional similarity calculation method (SCM) is uncertain and computationally high, and its accuracy is accompanied by the change of the number of generated clouds. It is not suitable as the similarity calculation method in the current application environment The algorithm based on cosine similarity measure does not take into the length and dimension of dimension account, ignoring the relationship between three important characteristic in cloud model, the relationship of expectation, entropy and super entropy, such as the different nature and weight of each digital feature, result to feature loss and the degree of distinction is too small. The similarity calculation method based on Euclidean distance is simple and straightforward and easy to understand, but the Euclidean measure needs to ensure that the dimension indicators are at the same scale level. In other words, the three numerical features of the cloud model are defined equally important. In the practical application is not consistent with the situation. ECM and MCM are the measurement methods of the expected curve and the maximum boundary curve of the

Table 2. WS-Dream # 1 QoS data similarity comparison

	SCM				ECM				MCM				OCM				CCM			
	U1	U2	U3	U4	U1	U2	U3	U4	U1	U2	U3	U4	U1	U2	U3	U4	U1	U2	U3	U4
U1	1	0.0236	0.0284	0.0258	1	0.9633	0.9543	0.9788	1	0.9944	0.9777	0.9826	1	0.9336	0.9109	0.9326	1	0.9993	0.9997	0.9993
U2	0.0236	1	0.0335	0.0191	0.9633	1	0.9208	0.9763	0.9944	1	0.9731	0.9782	0.9336	1	0.8578	0.9323	0.9993	1	0.9983	1
U3	0.0284	0.0335	1	0.0352	0.9543	0.9208	1	0.9436	0.9777	0.9731	1	0.9930	0.9109	0.8578	1	0.8969	0.9997	0.9983	1	0.9983
U4	0.0258	0.0191	0.0352	1	0.9788	0.9763	0.9436	1	0.9826	0.9782	0.9930	1	0.9326	0.9323	0.8969	1	0.9993	1	0.9983	1

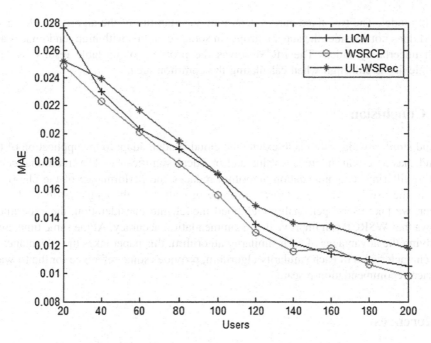

Fig. 2. WS-Dream # 1 university organization QoS data MAE performance comparison

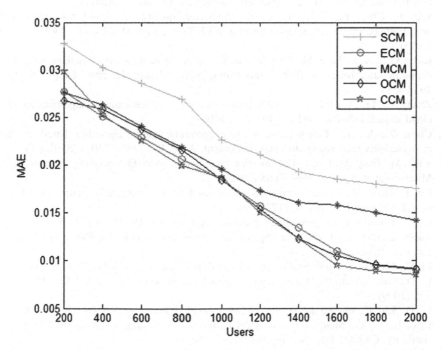

Fig. 3. WS-Dream # 1 data under different similarity MAE performance comparison chart

cloud model. The two fixed curves determine the stability of the algorithm. However, ECM does not consider the super entropy, in some scenarios with high randomness and high uncertainty index. The MCM solves the problem of unstable results, but the complexity is still high when calculating the common area.

6 Conclusion

Cloud similarity algorithm is to extend the cloud model, adapt to the application of the cloud, has a certain theoretical value and practical significance. The traditional cooperative filtering recommendation algorithm reduces the performance of the algorithm due to the user's own characteristic attribute or only take the single attribute. In this paper, we take users' personality and cloud model into consideration, the experiment shows that WSRCP can improve the recommendation accuracy. At the same time, after studying the advantages of each similarity algorithm, this paper takes the advantages of the characteristics of each similarity algorithm, provides some reference for the forward service recommendation research.

References

1. Zheng, Z., Ma, H., Lyu, M.R., et al.: QoS-aware web service recommendation by collaborative filtering. IEEE Trans. Serv. Comput. 4(5), 140–152 (2011)
2. Wu, J., Chen, L., Feng, Y., et al.: Predicting quality of service for selection by neighborhood-based collaboration filtering. IEEE Trans. Syst. Man Cybern. 43(2), 428–439 (2013)
3. Kang, G., Liu, J., Tang, M., Cao, B., Xu, Y.: An effective web service ranking method via exploring user behavior. IEEE Trans. Netw. Serv. Manag. 12, 554–564 (2015). ISSN 1932-4537
4. Zhang, G., Li, D., Li, P., et al.: Collaborative filtering recommendation algorithm based on cloud model. J. Softw. 18(10), 2403–2411 (2007)
5. Chen, Z., Li, Z.: Collaborative filtering recommendation algorithm based on user characteristics and project attributes. J. Comput. Appl. 07, 1748–1750, 1755 (2011)
6. Tang, M., Jiang, Y., Liu, J.: User location aware web service QoS prediction method. Small Micro-comput. Sys. 12, 2664–2668 (2012)
7. Liu, F., Hong, Y.: Communication algorithm based on user characteristic attribute and cloud model. Comput. Eng. Sci. 6, 1172–1176 (2014)
8. Li, D.: Uncertainty in knowledge representation. Eng. Sci. 10, 73–79 (2000)
9. Zadeh, L.A.: Fuzzy Logic with Engineering Applications. Publishing House of Electronics Industry, Beijing (2001)
10. Li, D., Liu, C.: The universal of normal cloud model. Eng. Sci. 6(8), 28–34 (2004)
11. Li, D., Liu, C., Gan, W.: A new cognitive model: cloud model. Int. J. Intell. Sys. 3(24), 357–375 (2009)
12. Li, D., Liu, C.: On the universality of cloud model. China Eng. Sci. 08, 28–34 (2004)
13. Liao, L., Li, C., Meng, X.: Free-model co-filtering algorithm based on Euclidean space similarity. Comput. Eng. Sci. 10, 1977–1982 (2015)
14. Li, H., Guo, C., Qiu, W.: Equivalent cloud model similarity calculation method. Acta Electron. Sin. 11, 2561–2567 (2011)

15. Zheng, Z., Zhang, Y., Lyu, M.R.: Investigating QoS of real-world web services. IEEE Trans. Serv. Comput. **7**(1), 32–39 (2014)
16. Zheng, Z., Zhang, Y., Lyu, MR.: Distributed QoS evaluation for real-world web services. In: Proceedings of the 8th International Conference on Web Services (ICWS 2010), Miami, Florida, USA, 5–10 July 2010, pp. 83–90 (2010)
17. Llinas, G.A.G., Nagi, R.: Network and QoS-based selection of complementary services. IEEE Trans. Serv. Comput. **9**(1), 79–91 (2015)

A Cooperative Abnormal Behavior Detection Framework Based on Big Data Analytics

Naila Marir$^{(\boxtimes)}$ and Huiqiang Wang

College of Computer Science and Technology,
Harbin Engineering University, Harbin, China
nailamarir@gmail.com, wanghuiqiang@hrbeu.edu.cn

Abstract. As cyber attacks increase in volume and complexity, it becomes more and more difficult for existing analytical tools to detect previously unseen malware. This paper proposes a cooperative framework to leverage the robustness of big data analytics and the power of ensemble learning techniques to detect the abnormal behavior. In addition to this proposal, we implement a large scale network abnormal traffic behavior detection system performed by the framework. The proposed model detects the abnormal behavior from large scale network traffic data using a combination of a balanced decomposition algorithm and an ensemble SVM. First, the collected dataset is divided into k subsets based on the similarity between patterns using a parallel map reduce k-means algorithm. Then, patterns are randomly selected from each cluster and balanced training sub datasets are formed. Next, the subsets are fed into the mappers to build an SVM model. The construction of the ensemble is achieved in the reduce phase. The proposed structure closely delivers a high accuracy as the number of iterations increases. Experimental results show a promising gain in detection rate and false alarm compared with other existing models.

Keywords: Support vector machines · Abnormal behavior detection · Big data · Cyber attacks · Ensemble classifier · MapReduce

1 Introduction

Prior to the twentieth century, the cyber attacks were still relatively random and undeveloped. However, and recently, due to the modem age of internet connectivity, cyber attacks are becoming increasingly intelligent, dynamic, and more sophisticated. They are a kind of long-term intrusion that can avoid all the existing security strategies on the target system. Their goal is to destroy, steal or modify confidential data and damage the entire target system. Many security systems offer the protection of computer users from malicious intrusion such as firewalls, antivirus and behavior analytics. According to various reports, most of the proposed systems and techniques are not sufficient to detect this type of attacks. Nowadays, security teams are increasingly turning to behavior based detection models to detect the cyber attacks. This type of intrusion detection

© Springer Nature Singapore Pte Ltd. 2017
B. Zou et al. (Eds.): ICPCSEE 2017, Part I, CCIS 727, pp. 192–206, 2017.
DOI: 10.1007/978-981-10-6385-5_17

system is based on the process of analyzing accumulated data from multiple sources (host, network or security logs). They have the ability to learn new patterns by constructing an analytical model of the system behaviors.

The identification of malicious behavior and the delineation of the boundaries between normal and abnormal behavior have been facilitated by machine learning and data mining techniques [1]. These are methods of data analysis that learn the correlations between the features and the labels to find hidden insights. The major advantage offered by these methods is their ability to recognize the unknown and zero-day attacks. Most of the time, researchers have developed classification models for abnormal behavior detection using single classifiers such as decision tree, support vector machine (SVM), artificial neural networks (ANNs), and K-nearest neighbors (KNN). However, with the increase of the events that need to be observed for the detection of cyber attacks, machine learning for Big Data technology has provided the possibility of collecting and analyzing a large volume of real-world data. According to recent researches [2], Big Data analytics can be used to improve cyber security and situational awareness. New Big Data technologies, such as databases related to the Hadoop ecosystem and stream processing, are allowing the storage and analysis of large heterogeneous data sets at an unparalleled scale and speed. These technologies could help transform security analytics.

In order to benefit from machine learning techniques using big data technologies to overcome the limitation in the detection of cyber attacks of current security systems, this paper proposes a cooperative large scale abnormal behavior detection framework based on big data analytics to help organizations to early detect new threats and react quickly before they propagate. In this research, we focus on the detection of the abnormal behavior in network traffic data. The motive of this research is to provide a new framework for an abnormal behavior detection system that processes network traffic and is intelligent enough to identify cyber attacks.

The abnormal behavior detection technique is based on ensemble SVM classifier. First, the network data flow is divided into multiple subsets based on map reduce k-means clustering algorithm and then we randomly regroup flows from each cluster to form balanced subsets that can represent the entire data set. The obtained subsets are trained separately in the map phase based on SVM classifier. Next, the output of the ensemble is combined in the reducer phase. The proposed system also provides a confidence measure for the decision of continuing iteration of ensemble layer learning. This means that, if the confidence and diversity level of the ensemble is insufficient, the system will repeat the training utilizing the previous layer projection data as an input to the next ensemble layer. The ensemble will terminate the training if the calculation of the confident and the diversity measure of the ensemble is good enough. The effectiveness of the proposed method is shown in this paper by the obtained simulation result. The proposed system is also compared with existing machine learning classifiers in terms of accuracy and effectiveness.

The rest of this paper is organized as follows. Section 2, provides an understanding of the background behind the detection of the abnormal behavior. In Sect. 3, we describe the structure of the proposed model. Simulation results are given in Sect. 4. Finally, the conclusion to our work is given in Sect. 5.

2 Background and Related Work

In this section, recent research on cyber attack detection based on big data technologies is first introduced. Then, recent work on ensemble classifiers in abnormal behavior detection is presented.

2.1 Cyber Attack Detection Based on Big Data Technologies

The construction of a system for detecting cyber attacks has always been an ambitious goal, not only for the highly complex pattern to detect, but it is also a big challenge to support the long-term and large scale analysis with the current traditional technologies. According to [3] traditional technologies failed to fill the gaps in cyber security and deal with recent and unknown threats. Recently, the evolution of data driven security has been enhanced through the use of big data technologies. Various researchers working in the field of cyber security have described intrusion detection system (IDS) as one of the most visible uses for big data analytics. An IDS is designed to monitor events that happen in a computer system or network, analyzing them for signs of possible incidents and frequently prohibiting the unauthorized access [4]. There are two approaches to distinguish between the normal and the malicious behavior of the system, which are the misuse detection approach and the anomaly detection approach. The First approach detects a series of actions as an attack if it matches a previously full description of the sequence of actions (signature) executed by the attacker. The other one finds patterns in data that do not conform to expected behavior. One of the major techniques that have proven to be of great practical value in the anomaly-based detection is machine learning (ML). The basic principle behind the ML techniques is learning and making prediction from the data [5]. Machine learning can be divided into two categories: approaches based on Artificial Intelligence (AI) techniques and approaches based on Computational Intelligence (CI) methods [6]. The decision trees, k-Nearest Neighbor (KNN), Multi-Layer Perception (MLP), k-means clustering and Support Vector Machines (SVM) are some examples of the AI techniques, whereas, genetic algorithms, artificial neural networks, fuzzy logic, and artificial immune systems are examples of the CI techniques. Big Data technology and machine learning techniques offer the immense opportunity to learn and adapt to the new requirements of data analysis for the cyber security. However, few work has been done in the realm of search in the area of cyber attack detection in big data environment using the ML approaches which have, for most of them, some limitation in the efficiency and the implementation. Suthaharan [7] discusses in his paper the issues related to combining supervised learning techniques and Big Data technologies such as

Hadoop, Hive and Cloud for solving network traffic classification problems. Lee and Lee [8] present a Hadoop-based traffic monitoring system that performs IP, TCP, HTTP, and Netflow analysis of multi-terabytes of Internet traffic in a scalable manner. Ahn et al. [9] propose a new model for unknown attacks detection based on big data analysis techniques while extracting information from various sources. Marchal et al. [10] have proposed an architecture based on big data for large-scale security monitoring. In addition, in [11] a real-time intrusion detection system includes four-layered IDS architecture for ultra-high-speed big data environment using Hadoop implementation is proposed.

2.2 Ensemble Classifier for Abnormal Behavior Detection

The application of ensemble techniques for the detection of abnormal behavior has recently witnessed a surge of research efforts. However, [12] showed theoretically and empirically that the accuracy of an ensemble classifier is better than one single classifier. The advantages of ensemble classifiers are particularly evident in the field of intrusion detection, as there are many different types of intrusions, and different detectors are needed to detect them [13]. In [14], an investigation of the possibility of using ensemble techniques to improve the performance of network intrusion detection systems is performed. An ensemble of different methods such as bagging, boosting and stacking have been applied, in order to ameliorate the accuracy and reduce the false positive rate. Naive Bayes, J48, JRip and iBK are the four base classifiers implemented for the ensemble learning. Folino and Pisani [15] proposed a method for a distributed intrusion detection that used genetic programming to generate decision-tree classifiers. These classifiers were then combined into an ensemble using AdaBoost.M2, a variant of AdaBoost. [16] also proposed a new ensemble construction method that uses PSO generated weights to create ensemble of classifiers with better accuracy for intrusion detection. Despite yielding significant results, these studies are not suitable for area demanding the analysis of large scale data.

3   A Cooperative Anomaly Behavior Detection Framework Based on Big Data Analytics

As has been mentioned, the collaboration of big data analytics and machine learning techniques can play a crucial role in the detection of unknown cyber attacks. This paper attempts to detect the abnormal behavior through big data analytics and a parallel learning process. Therefore, in this section, we first introduce a cooperative abnormal behavior detection framework based on big data analytics and ensemble machine learning techniques. Our framework consists of several layers that use different categories of data information for the detection of known and unknown attacks. We then focus on the network traffic abnormal behavior model. We will present our proposed large scale network traffic abnormal behavior detection method based on the combination of the k-means clustering algorithm for balanced decomposition and the parallel ensemble svm based on map reduce paradigm.

3.1   Architectural Overview of the Proposed Framework

The proposed framework aims to detect anomalous patterns from multiple data sources such as (network traffic data, host events data and security log data). It works by identifying reference points with behaviors that are very different from the expectation in high speed big data environments through a parallel ensemble machine learning techniques. The ensemble learning helps to improve the accuracy of the detection and lead to a remarkable reduction in the running time. The proposed framework helps to a deeper understanding of the cyber situational awareness.

3.1.1   Event Traffic Collection and Storage

The bottom-layer of our framework consists of two main tasks: gathering event data from multiple sources and storing that data. The network event data, host event data and security log sources are the three traditional input types considered in our framework. Each cluster node stores the large scale collected data from the sensors that are installed in the network into the cluster file system HDFS that's stands for Hadoop Distributed File System. The HDFS is designed for storing very large files and managing the storage across a network of machines.

3.1.2   Preprocessing Layer

This layer includes the ensemble of techniques that illustrate the ways for the preparation of the collected data. Preprocessing methods are applied to input data in order machine learning algorithm to produce more effective results and to reduce the calculation load of algorithms used. This stage handles the removal of noise and inconsistent data by cleaning the data, combines multiple sources of data, transforms the data into forms which are appropriate for specific machine learning tasks and reduces the data by the selection of both features and instances in a dataset. In order to achieve the best decomposition of the dataset, our framework provides different techniques: sampled based decomposition, resampling techniques and clustering based decomposition. All of these techniques are based on the map reduce paradigm, the hive and Hbase technologies.

3.1.3   Ensemble Learning

In order to find the abnormal behavior among the collected data, ensemble learning techniques are applied. The process of the ensemble learning is composed of several stages. During the building of the ensemble two main types of techniques are proposed. The dependent and the independent techniques:

- Dependent strategy: In this method we convert the weak classifiers to strong ones. The classifiers are trained in series with training instances having different weight distributions. In the training procedure, the weak learners will update their weight according to the training errors in each iteration. After a sufficient number of iterations, an ensemble of weak learner models is constructed.

- Independent strategy: In this method, each base classifier is trained with a different subset of the original training data. The manner of dividing the training dataset is very important to improve the accuracy, reduce the training time, and balance the load of data for the ensemble learning. After the partition, the classifiers are trained on the subsets and the models are constructed.

After the construction of the classifiers, the suitable result will be obtained in the combining stage. During this stage the decision of each base learner will be aggregated. Two methods for combining the base-learner outputs could be used: weighting methods and meta-learning methods.

3.1.4   Interpretation

It refers to the correlation of the alarm events generated by the ensemble learning layer and the visualization of the whole situation of the system over a period of time. The primary objective of this layer is to make the results meaningful to administrators and decision makers.

3.2   A Large-Scale Network Traffic Abnormal Behavior Detection Model

In this section, we elaborate the details of the proposed large scale network traffic abnormal behavior detection system. Conventionally, host event data, network traffic data and security log data are the three basic inputs for constructing our proposed framework. But considering that the cyber attacks in the new era of distributed information systems are rapidly growing and that numerous instances of malware through network occur over a long period of time, the detection of the abnormal behavior of the network traffic data at an early stage is our major concern. For the abnormal detection, an ensemble structure consisting of multiple SVMs is proposed. In order to apply ensemble method upon large scale data, we provide a partition algorithm for clustering the data into balanced subsets. Each subset is trained by an individual SVM in the map phase. The final obtained SVMs are combined with their estimations to detect the abnormal behavior. The overall structure is shown in Fig. 1.

3.2.1   Balanced Decomposition

The problem of the partition of the dataset, in our paper, is directed towards the accuracy of the SVM ensemble. Our aim is to achieve a balanced decomposition of the dataset, so that the combination of all SVMs that are built on each data subset, will achieve a lower generalization error and a lesser time compared to an SVM built on the entire data set. To address the above mentioned problem, a balanced decomposition based on map reduce is proposed. The decomposition is achieved by two main steps: the clustering of the data with K-means algorithm and the forming of the subsets based on random selection of the events from each

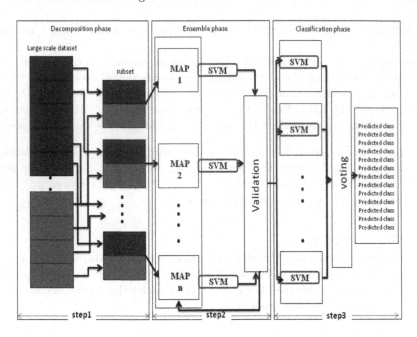

Fig. 1. Overall structure of large-scale network traffic abnormal behavior detection model

cluster that are obtained from the first step. The K-means is one of the scalable and simplest unsupervised learning algorithms that solve the well-known partitioning clustering problem. The main objective of clustering algorithm is to partition data into k clusters $C = \{C_1, ..., C_k\}$ based on a similarity measure between samples. In the parallel version of K-means, each iteration of the algorithm is formulated as a map reduce job. Firstly the dataset is randomly divided and broadcasted to all mappers. The k points are randomly selected from the sub datasets as an initial clustering centroids. In the map phase, we first calculate the Euclidian distance between the centroid and the other data points. Then we assign each data point to the nearest cluster. This process iterates until all the data points are processed. The output of the map phase is the pair of center cluster points and the corresponding data points. All the output pairs of all mappers are the input of the reducer phase. In the reduce phase, first we read the data and group the values with the same key into a set of clusters. Then the average of each cluster is calculated in order to determine the position of k centroids. After that, the difference between the new centroids and the original ones is calculated. If the difference is bigger than the threshold the output is fed back to the mappers for computing another set of centroid again. After the termination of the clustering process we obtain different similar clusters. We then reconstruct the train subsets by selecting random packets from each cluster, each subset has the same size and the same distribution of events. The main goal of this step is to yield a group of subsets that represent the whole dataset. This ensures the strongest of the constructed classifiers in the combining phase.

Algorithm 1. Balanced decomposition based on Map Reduce

Input: Train set $X = \{(x1, y1), (x2, y2), ..., (xi, yi)\}$
Output: Ensemble of k balanced clusters
1:Create random m split of the train dataset;
2:Begin MR Job;
3:MAP j: $\forall j \in \{1, ..., m\}$
Input: Global variable centers, m datasets
Output: Index of the center cluster point and pattern information
4:For each train pattern do;
5:Calculate the Euclidian distance between the centroid and the other train pattern;
6:Assign each train pattern to the nearest cluster;
7:End for;
8:REDUCE (R());
Input: Pair of center cluster points and corresponding data points
Output: Ensemble of clusters
9:Read the data and group the values with the same key into a set of clusters;
10:Calculate the average of each cluster;
11:Calculate the difference between the new centroids and the original ones;
12:If the difference > threshold repeat map job;
13:Create a random subsets from each of the obtained k clusters;
14:End;

Based on the algorithm proposed above, the network traffic data can be equiponderantly clustered into a number of sub clusters, which contains a form that represent the whole dataset. This design yields balanced partitions of the data while maintaining the same structure of the whole dataset, and also improves prediction accuracy of our proposed model.

3.2.2 Building Classifier

For large scale network traffic abnormal behavior detection, an iterative layered structure where each layer consists of ensemble SVM classifiers is proposed. We first give a brief description of the SVM classifier which is our base classifier. After that we give a detailed description of the proposed algorithm.

(a) SVM algorithm: Support vector machine is one of the most representative models in the machine learning field. It is based on the concept of decision planes that define decision boundaries. A decision plane is one that separates between a set of objects having different class memberships. In SVM, the problem of learning binary classification is given by

$$\min_{\beta, b, \xi} \left(\frac{1}{2}\beta'\beta + C \sum_j \xi_j\right) \tag{1}$$

such that

$$y_j f(x_j) \geq 1 - \xi_j$$
$$\xi_j \geq 0.$$

Where C represents the positive regulation constant indicating the tradeoff between the empirical error and the complexity term, and ζ_j refers to the slack variables for the j-th pattern indicating the tolerance of the margin for the separating hyper-plane. This problem can be considered as a dual problem to this soft-margin formulation. Using Lagrange multipliers [17]

$$\max_a \sum_j \alpha_j - \frac{1}{2} \sum_j \sum_k \alpha_j \alpha_k y_j y_k x'_j x_k \tag{2}$$

Subject to

$$\sum_j y_j \alpha_j = 0 \tag{3}$$

$$0 \leq \alpha_j \leq C, \tag{4}$$

the dual soft-margin problem can be solved by quadratic programming optimization problems. Then each non-zero α_i indicates that the corresponding sample is remain to form the solution of the hyper plane. Then, the final form of the hyper plane is determined by the equation:

$$f(X) = \sum_{i=1}^n \alpha_i y_i K(X_i, X) + b \tag{5}$$

Where i represents the index set of support vectors and b represents the solution of bias term.

(b) Calculate the confidence measure: In our algorithm, the degree of confidence for each classifier decision is calculated by the estimation of the class membership probabilities based on the prediction's distance to the hyper plane. However, basically, the output value of the SVM classifier is unassociated to the probability of the class. Platt's proposes a method to estimate the conditional class probability $P(y|x)$ as a function of the output of the SVM. In this method, the posterior class probability has the form of

$$P(y = 1|x) \approx P_{A,B}(f) \equiv \frac{1}{1 + \exp(Af + B)}, \tag{6}$$

where the parameters A and B represent variables trained from the output patterns of SVM and $f(x)$ is the decision function of the SVM. The best parameter setting $z^* = (A^*, B^*)$ is decided by solving the following regularized maximum likelihood problem of the training data, which is the entropy error function E

$$\min_{z=(A,B)} F(z) = -\sum_{i=1}^l (t_j \log(p_i) + (1 - t_i) \log(1 - p_i)) \text{for} \quad p_i = P_{A,B}(f_i), \tag{7}$$

$$\text{for} \quad p_i = P_{A,B}(f_i), t_j = \begin{cases} \dfrac{N_+ + 1}{N_+ + 2}, & \text{if} \quad y_i = +1, \\ \dfrac{1}{N_- + 2}, & \text{if} \quad y_i = -1, \end{cases} \quad i = 1, ..., l. \tag{8}$$

(c) **Ensemble diversity measure:** The idea to calculate the diversity among classifiers is based on the entropy measure. The highest diversity among classifiers for an instance $z_j \in Z$ shows $L/2$ of the votes in z_j with the same value (0 or 1). One possible measure of diversity based on this concept is:

$$E = \frac{1}{N} \sum_{j=1}^{N} \frac{1}{(L - [L/2])} \min\{l(z_j), L - l(z_j)\}. \tag{9}$$

E is between 0 and 1, where 0 indicates no difference and 1 indicates the highest possible diversity.

(d) **Proposed ensemble SVM based on the map reduce paradigm:** In our proposed system, an ensemble of SVM models is used to perform the classification decision for the detection of the abnormal behavior. Then, the decision is made by selecting the class with the majority vote. The first step is to assign to each data chunk, a sub dataset obtained from the balanced decomposition algorithm. The number of the produced map tasks is equal to the number of training subsets. At each layer, in the map stage we train SVM on the corresponding training sub dataset and generate a respective set of Support Vectors ($SV set_1, ..., SV set_n$). Then we classify the training data by the obtained set of Support Vectors in order to compute the probability output of each input pattern to determine the confidence measure by (8). If the given input is considered as an unconfident, that pattern is pushed to the reduce phase to construct the new train dataset as an input for the next layer. This projection is repeated in each layer in the map stage until the pattern meets the required confidence measures for our system. In reduce stage, the first step is to collect all the projected data from the mappers and build the training dataset. After that, we combine the support vectors of each map with a linear combination to form an ensemble of SVM. Then the obtained ensemble is examined by calculating the measure of diversity on the validate dataset. After that, we compare with the threshold: if the ensemble is regarded as inadequate by the diversity checker or if the constructed train data is enough, the system is initialed for another layer. Else the algorithm stops and the final output of the reducer from the final layer is the ensemble of SVMs that will be used in the classification phase. The construction of ensemble SVM by map and reduce function are given in Algorithm 2.

Algorithm 2. Construction of ensemble SVM based on map reduce

---

Input: k sub datasets;
Output: ensemble SVMs;
1:Assign each sub data to a map chunk;
2:Begin MR Job;
3:MAP j: $\forall j \in \{1, ..., k\}$;
Input:k train datasets
Output: Set of support vectors, train data;
4:Train SVM on k training datasets;
5:Obtain svm set for k training datasets
6:Construct the model;
7:Classify the train dataset;
8:Add the train input pattern to the train vector if it is considered unconfident;
Input: Set of support vectors, vector
of input train data;
Output: Ensemble of SVM classifiers;
9:Group the data obtained from all the mappers and construct the train dataset;
10:Classify the ensemble svm with the validate dataset;
11:Calculate the diversity measure of the ensemble
12:If the measure > threshold repeat the map reduce job;
13:Else construct the ensemble svm
14:End;

In our proposed algorithm, the ensemble is obtained only if the diversity measure reaches the required threshold. A problem occurs if the number of input pattern from the previous layers is insufficient for the next round, while the obtained diversity measure is not satisfied yet. To solve this problem, we reduce the threshold value of the confidence measure in the map phase. This simple method helps to increase the number of unconfident patterns obtained from the previous layer.

3.2.3   Pattern Classification

After the building of the ensemble SVM using the algorithms explained above, we need to combine the results of each SVM in order to detect the abnormal behavior. The majority voting algorithm is used for the final decision. Each SVM classifier will vote for a class and eventually the class with a higher number of votes will be reported as the final decision. The Algorithm 3 below summarizes the process of the classification of new pattern as a normal or an abnormal behavior.

Algorithm 3. Abnormal behavior classification

Step 1:For the given test pattern x, a series of svm output is obtained;
Step 2:Choose the class that receives the highest total vote as the final decision;

---

4   Evaluation and Experimental Result

This section includes the experimental results obtained during the simulation and the comparison between the implemented system and the machine learning classifiers. Using Map Reduce paradigm, we implemented the proposed ensemble SVM for detecting the abnormal behavior in large scale network traffic. The experiments were conducted on a virtual Hadoop cluster. We installed Hadoop version 2.7.3. The operating system was Ubuntu 14.0.4. Our implementation was done using eclipse. All experiments run on the same machine with 16 GB RAM and a 4 core processor.

(a) **Data collection:** In order to verify the efficiency of the proposed large scale abnormal behavior detection system, NSL-KDD cyber security dataset is used. NSL-KDD dataset contains approximately 4 GB network traffic data. Each connection instance is described by 41 attributes deploying different features of the network flow and a label assigned to each flow denoting its type either normal or some specific attack class. The dataset is selected because of its widespread use in the related work for testing abnormal detection models and it has a large number of instances and attributes. The details of the training and testing datasets used for the experiments are shown in Table 1.

Table 1. Distribution of NSL-KDD data set

Dataset	Records	Normal	Dos	Probe	$U2R$	$R2L$
KDD train+	125973	67343	45927	11656	52	995
KDD test+	22544	9711	7458	2421	200	2754

(b) **Simulation results:** The first section of this part, presents the impact of balanced decomposition technique on the accuracy of the ensemble SVM classifier. The second experience compares the proposed system with the state of art methods. The proposed system is evaluated for accuracy by considering true positive rate.

4.1   The Effect of Balanced Decomposition on Classification Performance of Ensemble SVM

Table 2 shows the performance of the proposed ensemble SVM using the proposed balanced decomposition algorithm. The decomposition implemented does not demand the training datasets to be sampled in any specific way. The ensemble method has been compared with the random data decomposition using the same base classifier. As can be seen from the Table 2, the ensemble SVM combined with the data balanced decomposition algorithm that we proposed, has made, at each iteration, a significant improvement compared to the unbalanced one, for both the training true positive rate and testing true positive rate. This means that the ensemble methods has raised the learning and prediction accuracy.

Table 2. Performance comparison of the proposed system using balanced and random decomposition

Iteration number	Random TPR		Balanced TPR	
	Training TPR	Testing TPR	Training TPR	Testing TPR
1	94.74	92.61	97.01	95.21
2	96.89	95.23	98.90	97.65
3	97.92	96.16	99.70	98.41
4	98.12	97.88	99.73	98.43

As illustrated in the Fig. 2, the rate of the true positive increases as the system iterations increase. This is due to the obtained unconfident patterns from the first iteration which need to be learnt more. In the overfitting phenomena, the accuracy of the confidante pattern could be decreased in each iteration layer. However, the diversity between the classifiers was calculated in each iteration and if there is a difference in the diversity between iterations, the system stops and recognition accuracies are safely maintained.

4.2 The Comparison with the State-of-the-Art Methods of Machine Learning

In this experiment, the proposed system is compared with machine learning methods, namely the K-nearest neighbors and Naive Bayes classifier. For the sake of comparison, we used the implementation of the sequential machine learning available in Weka library. The same feature set and training and test sets are used. Table 3 shows the comparison results between the proposed model, Naive Bayes classifier and KNN, while considering accuracy in terms of true positive rate. It is obvious from the results that the proposed model delivered a higher accuracies than both of the classifiers. If we consider the efficiency in terms of processing time, the results indicate that map reduce model plays great role in saving the training and test time. It is quite obvious that the ML classifier takes more modeling time as well as more decision-making time for the abnormal behavior detection. The evaluation showed that the proposed cooperative framework based on big data analytics is capable of detecting behavior anomalies in network traffic data in a very short time and with high accuracy.

Table 3. Comparaison results between the proposed method and the state of art methods.

Attack types	K nearest neighbor	Naive Bayes	Proposed system
DoS	99.97	99.98	99.994
Probe	97.88	96.27	99.09
R2L	88.59	90.372	90.53
U2R	80.76	63.46	84.61

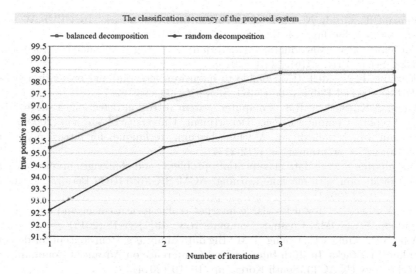

Fig. 2. The classification accuracy of the proposed system

5   Conclusion

In this paper, a cooperative abnormal behavior detection framework based on big data analytics and ensemble learning machine has been proposed. In addition, an iterative Map Reduce implementation of a balanced decomposition and a parallel ensemble SVM for the detection of the abnormal behavior in large scale network traffic data has been developed. The system uses a balanced decomposition algorithm which divides the whole dataset into several subsets to ensure the diversity among the base classifiers and to increase the detection accuracy. The SVMs are then trained on each subset. The efficiency of the proposed method, in terms of accuracy, has been illustrated with experimental results. The proposed system has shown high performances in the detection of abnormal behavior in a distributed way. As a next step, we plan to improve the performance and processing phase of the proposed system by employing an appropriate feature selection method. Another important area to be investigated is to find a suitable strategy to combine the predictions.

Acknowledgments. The National Natural Science Foundation of China under Grant Nos. 61370212, 61402127, 61502118; the Natural Science Foundation of Heilongjiang Province under Grant Nos. F2015029, F2016009; the Fundamental Research Fund for the Central Universities under Grant No. HEUCF100601.

References

1. Tsai, C.F., Hsu, Y.F., Lin, C.Y., Lin, W.Y.: Intrusion detection by machine learning: a review. Expert Syst. Appl. **36**(10), 11994–12000 (2009)

2. Janssen, T., Grady, N.: Big data for combating cyber attacks. In: CEUR Workshop Proceedings, Fairfax, vol. 1097, pp. 151–158 (2013)
3. Cardenas, A.A., Manadhata, P.K., Rajan, S.P.: Big data analytics for security. IEEE Secur. Priv. **11**(6), 74–76 (2013)
4. Scarfone, K.A., Mell, P.M.: Guide to intrusion detection and prevention systems (IDPS), Special Publication (NIST SP), pp. 800–894 (2007)
5. Jones, A.K., Sielken, R.S.: Computer system intrusion detection: a survey. Technical report, Computer Science Department, University of Virginia (2000)
6. Zamani, M., Movahedi, M.: Machine learning techniques for intrusion detection (2013). arXiv preprint arXiv:1312-2177
7. Suthaharan, S.: Big data classification: problems and challenges in network intrusion prediction with machine learning. ACM SIGMETRICS Perform. Eval. Rev. **41**(4), 70–73 (2014)
8. Lee, Y., Lee, Y.: Toward scalable internet traffic measurement and analysis with Hadoop. SIGCOMM Comput. Commun. Rev. **43**(1), 5–13 (2012)
9. Ahn, S.H., Kim, N.U., Chung, T.M.: Big data analysis system concept for detecting unknown attacks. In: 16th International Conference on Advanced Communication Technology (ICACT), South Korea, pp. 16–19 (2014)
10. Marchal, S., Jiang, X., State, R., Engel, T.: A big data architecture for large scale security monitoring. In: Proceedings of IEEE International Congress Big Data, Anchorage, pp. 56–63 (2010)
11. Rathore, M.M., Ahmad, A., Paul, A.: Real time intrusion detection system for ultra-high-speed big data environments. J. Supercomput. **72**(9), 3489–3510 (2016)
12. Dos Santos, E.M.: Static and dynamic overproduction and selection of classifier ensembles with genetic algorithms. Ecole de Technologie Superieure, Canada (2008)
13. Aburomman, A.A., Ibne Reaz, M.B.: A survey of intrusion detection systems based on ensemble and hybrid classifiers. Comput. Secur. **65**, 135–152 (2017)
14. Gaikwad, D.P., Thool, R.C.: Intrusion detection system using bagging ensemble method of machine learning. In: International Conference on Computing Communication Control and Automation, pp. 291–295. IEEE (2015)
15. Folino, G., Pisani, F.S.: Combining ensemble of classifiers by using genetic programming for cyber security applications. In: Mora, A.M., Squillero, G. (eds.) EvoApplications 2015. LNCS, vol. 9028, pp. 54–66. Springer, Cham (2015). doi:10.1007/978-3-319-16549-3_5
16. Aburomman, A.A., Ibne Reaz, M.B.: A novel SVM-kNN-PSO ensemble method for intrusion detection system. Appl. Soft Comput. **38**, 360–372 (2016)
17. Vapnik, V.: Statistical Learning Theory. Wiley-Interscience, New York (1998)

Composite Graph Publication Considering Important Data

Yuqing Sun, Hongbin Zhao$^{(\boxtimes)}$, Qilong Han, and Lijie Li

College of Computer Science and Technology, Harbin Engineering University,
Harbin 150001, China
yuqing_s@163.com, z-hb@vip.sina.com,
{hanqilong,lilijie}@hrbeu.edu.cn, hqlwxm2002@163.com

Abstract. Under the premise to protect the privacy of personal information, publishing valuable graph is a challenging issue in privacy research. Appling differential privacy in graph, most of the work focused on graph structure characteristic values, because the basic of differential privacy is data distortion, it's hard to get valuable composite graph if we add a large number of random noise into the raw data. In this article, we show the key that influence availability is whether the important data keep original value in a composite graph. We analysis the properties of important data of k triangle count, and provide a new method for synthesis graph publication. We show the application of this method in k triangle count, and the experimental results proved the accuracy of the method.

Keywords: Differential privacy · Composite graph · k triangle count

1 Introduction

In reality, a lot of information can use graph to represent the structure, such as social network, the spread of disease, traffic network, etc. The nodes in graph represent individuals, edges represent relationships between individuals. These data are seen as the main data source of data mining and analysis, if we release or processing them directly, it usually brings the risk of personal privacy leak. Many studies have also shown that simple anonymous methods could not prevent the attacker's diverse attack, and expose personal information.

The current study of privacy protection are based on differential privacy, it is established on solid mathematical foundation, the quantitative expression and proof of privacy risk are given. Differential privacy limits the impact of any one of the records on the published results, making it impossible for an attacker to determine the information of the target record. Differential privacy has a good performance on histogram release, however, there is a challenge of scalability and usability in applying graph data. Because the application of differential privacy to graph data easily leads to the destruction of the structure of the original graph data.

How to publish a composite graph that satisfies edge differential privacy and has exact structure attribute are the problem that we need to solve. The most important structural attribute data of the graph data is subgraph count, that is, the number of

© Springer Nature Singapore Pte Ltd. 2017
B. Zou et al. (Eds.): ICPCSEE 2017, Part I, CCIS 727, pp. 207–219, 2017.
DOI: 10.1007/978-981-10-6385-5_18

occurrences of a particular subgraph in the graph. k triangle count is the most important study of subgraph count because triangle is the smallest nontrivial subgraph. k triangle is a graph that contains k triangles and has a common edge. It is worth noting that we calculated a copy of the subgraph. For example, if there is a common edge in d triangles (d > k), there are C_d^k k triangles. In this paper we use $f_{k\triangle}(G)$ to represent the k triangle of graph G.

In this paper, we analyze the factors that influence the results of $f_{k\triangle}(G)$ in the process of synthesizing graphs and define them as important data. We generate the corresponding matrices based on the degree of nodes, so that we can identify them in matrix. Using a new partition method to explore the dense and sparse areas, so that important data in the reconstruction can be accurately restored, so as to obtain accurate $f_{k\triangle}(G)$.

1.1 Related Work

Publishing graph data under differential privacy usually uses some spatial decomposition method. Xiao et al. [1] proposed an adaptive standard spatial indexing method KD tree to provide differential privacy protection. The nodes in KD tree are iterated and the heuristic method is used to select the split points so that the two sub-regions segmented are as close as possible to uniform distribution. Cormode et al. [2] proposed a similar approach. In addition, the article also shows that KD hybrid method performs optimally. Many literatures have improved the method based on KD hybrid, including the constraint reasoning proposed in [3] and optimized privacy budget allocation scheme. Their experiments show that KD hybrid method is superior to traditional KD tree and the method in [1]. Qardaji and Li [4] proposed a standard iterative segmentation method for multidimensional data. When the method is applied to a two-dimensional data set, it is similar to a KD tree based on the number of noise values. The above partitioning method basically establishes a hierarchical structure on the representation of the data points. Some papers [3, 5, 6] improved the count query on the hierarchy.

Xiao et al. [5] proposed a Privlet method that supports histogram queries, which has the advantage that it can respond long range count queries more accurately. Zhang et al. [6] proposed a hierarchical decomposition to construct histogram, through the quantitative noise to determine whether a region needs to continue to divide, eliminating the effect of granularity and tree height on data availability. The work closest to ours is by Chen et al. [7]. They put forward a method in the case of correlated network data, which can still achieve differential privacy. However, the performance of the method on the subgraph count is not given, and this is the main measure of usability for our method.

1.2 Contribution

The main contributions of this paper are as follows:

We analyze the factors that affect the availability of composite graphs in non-interactive ways, and define important data for usability functions.

We propose a composite graph publishing method NoiseGraph for edge differential privacy, considering $f_{k\triangle}(G)$ important data, which enables the published composite graph to respond triangle count queries accurately of arbitrary constant k. As far as we know, this is the first time that a solution is proposed for the k triangle count of composite graph.

Using real datasets, we compared the usability of triangle count with different k values of the composite graph. The results show that NoiseGraph with low privacy budget can guarantee k triangle count of the composite graph is still accuracy.

2 Preliminaries

In this paper, we input simple graphs as usual, G = (V, E), we temporarily give a set of node label that is independent of the edge set (generate node tag randomly). The adjacency matrix under the label is expressed as $A(G) \to |V| \times |V|$ satisfies:

$$A(G)_{ij} = \begin{cases} 1 & \text{if} (v_i, v_j) \in E \\ 0 & \text{otherwise} \end{cases} \tag{1}$$

For any simple graph G, its adjacency matrix A(G) is also called the (0,1) matrix. It is clear that the (0,1) matrix is a graph if and only if it is self-symmetric and the diagonal is all zero. $A_{i,j}$ represents the elements in matrix A(G)(i, j ≤ |V|), and |V| represents the number of nodes in G. $A_{i,j} = A_{j,i}$, and $A_{i,i} = 0$. As shown in Fig. 1(a), the adjacency matrix A(G) is shown in Fig. 1(b).

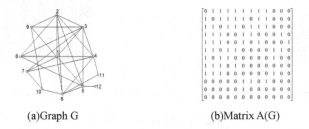

(a)Graph G (b)Matrix A(G)

Fig. 1. Example of social networking

2.1 Problem Definition

We consider the following problem, given a two-dimensional graph G, our goal is to publish the composite graph $\widehat{G}$. The general process is as follows. A group of node label L is obtained according to the important data characteristics, and the two-dimensional matrix A (G) of graph G is generated by L. The region is denoted by dom(A). Generate the ladder matrix B, with dom(B) denote the region of B. First, dom (B) is divided into N coarse-grained unit, for each coarse-grained unit with qiasi-lines to continue fine-grained division, and eventually forming M irregular sub-regions, C = {R_1, R_2, ..., R_M}, for any arbitrary $i \in [1, M]$, calculate the noise count of the sub-region R_i, that is, add the Laplace noise based on the count value of the region, and restore (0,1) distribution of region dom(R_i), finally reconstruct the (0,1) distribution of dom(B), which is obtained from the characteristics of the graph. The above process satisfies differential privacy.

Taking the network graph G shown in Fig. 2(a) as an example, we use the NoiseGraph method to get a node label L and generate the matrix A(G). We denote the point with 1 in matrix A(G) in Fig. 2(a). The small area consisting of four cells, centered on points, is dom (A_{ij}), and 0–12 represents the boundary of dom(A_{ij}). To divide the ladder region B into N coarse-grained region, N = 1, followed by fine-grained division using a quasi-lines (the polyline in Fig. 3(b)), each partition dividing the original area into two sub-regions.

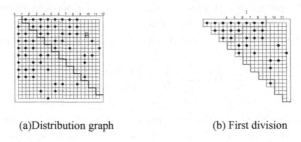

(a)Distribution graph   (b) First division

Fig. 2. Example of a division process

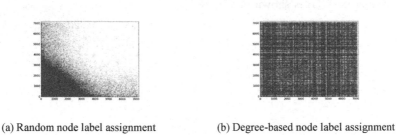

(a) Random node label assignment   (b) Degree-based node label assignment

Fig. 3. Matrix distribution

2.2 Source of Error

NoiseGraph introduces two kinds of errors in the original data set, noise error and reduction error. The noise error is due to we add the Laplace noise to each region to achieve the differential privacy, which has the same variance.

The reduction error is due to the fact that we use the noise count to randomly generate the (0,1) distribution of each region, and the reduction error have a correlation with the noise error. When the specific usability function is not taken into account, given the original data set G and the synthetic data set, the reduction error is defined as

$\sum_{i=1}^{|V|} \sum_{j=1}^{|V|} \left| A_{ij} - \widehat{A}_{ij} \right|$. The minimum of reduction error is equal to the added independent noise, and [7] also confirms that the reduction error depends on the size of the region density. When the noise error is small, the region is dense or sparse, and the distribution of the random distribution is similar with the original data distribution. Using the division can make the density of the region becomes sparse or dense, but with the increase in the number of division, the added noise is also increasing, so the size of the partition and the density of the region together determine the size of the reduction error.

The reduction error directly affects the usability of the data, but when the reduction error is min, the data availability may not be the biggest. Since the same reduction error value may correspond to a variety of different data distribution schemes, and there is a difference in the edge of the graph G being accurately restored, which also leads to different data availability. There are often many important edges in the real graph, which are included in some special subgraphs that determine the structure of the graph, such as triangles, stars, clusters, etc. These important edges exist or not, which significantly affect the subgraph count result. To ensure the structural characteristics of the graph, we hope that the important edges of the usability function in the graph data can be accurately restored.

3 Graph Publication

We consider the following problem, given a two-dimensional graph data G, our goal is to publish the noise synthesis graph $\widehat{G}$, which can accurately respond to k triangle count query, k is an arbitrary positive integer. In this section we propose a NoiseGraph method that satisfies differential privacy, as shown in Algorithm 1.

Algorithm 1 NoiseGraph

Input: Raw graph G, privacy budget ε

Output: Composite graph $\widehat{G}$

---

1. $\varepsilon = \varepsilon_a + \varepsilon_b + \varepsilon_c$;
2. DDNL(G, ε_a)$\rightarrow$graph label L;
3. Generate the matrix A(G) based on L;
4. QPR(A(G), $\varepsilon_b + \varepsilon_c$)$\rightarrow$IRTree T;
5. Generate the matrix A($\widehat{G}$) based on T;
6. Return $\widehat{G}$;

3.1 Distribute Degree-Based Node Label

At present, it is difficult to publish an accurate composite graph under differential privacy of non-interactive environment, add noise to the original graph and the triangle count of graph have a great change because the original dataset has a great influence on the triangulation result. In the worst case, the addition or deletion of an edge causes the result of the 1-triangles to increase or decrease $|V| - 2$.

We found the process, original graph achieve differential privacy, will bring different impact on the results when different data changes, what's more, same data has different effects on different usability functions. For this we define the concept of important data. The important data of the usability function not only appears in the graph data, but also the important data of the median of the list data, the Top-k position of the trajectory data.

Definition 1. Important Data. Given graph data G, usability measure function f_u, change a record t in G to generate the data set G', set $I_{fu}(t) = f_u(G) - f_u(G')$, if $I_{fu}(t)$ β, then t is the important data.

β is constant and β $\geq$ 1, the size depends on the actual application of the graph. $I_{fu}(t)$ measures the importance of t to f_u. The greater the $I_{fu}(t)$, the greater the importance of t, but the important data is not unique, it is depending on the size of β. f_u is measure function. The tuple t may be an edge, a node, or a track, determined by the type of dataset to be published.

The effect of important data is reflected in the release of synthetic data, rather than for some direct release of statistical results in non-interactive or provide interactive query data release, because they are both calculate noise through the query function of the global sensitivity or local sensitivity, and added to the real statistical count, the sensitivity is determined by the maximum value of the query results, and this maximum value is not changed, so no need to define and identify important data.

The nodes in network data are often tightly connected. The true triangle count value is often a multiple of total number of edges in the graph, so at least one edge is included in multiple triangles. According to the power law distribution, the distribution of the edges must be nonuniform. Combining the definition of important data, we can analyze that there must be important data.of triangle count.

In the process of adding noise to original graph G to obtain synthesized graph $\widehat{G}$, the more important data is retained, the more accurate the query results. According to the type of f_u and G, we can easily determine the existence of important data in G, for which we focus on the analysis of the character of important data and identify them withou reveal the privacy. So as to guide the process of data synthesis, and make important data as accurate as possible. Despite the diversity of f_u, we find that there is a correlation in important data for a certain type of usability measure.

Theorem 1. The important data of $f_{k\Delta}$ exists or not must change the query results of $f_{m\Delta}$ ($1 \leq m \leq k$).

According to the definition of important data and $f_{k\Delta}$, it is easy to conclude that the above theorem is established.

Because we want to publish a composite graph to satisfie a series of $f_{k\Delta}$ quiries, so we need to restore important data accurately. Here we refer to the important data of $f_{k\Delta}$ as important edge. If the important edge of $f_{k\Delta}$ exists in matrix must exist in the node degree greater than k. Intuitively, the relationship between any node and the node with larger degree is more important than the node with smaller degree.

First of all, we have to solve the problem that find important edges in adjacent matrix, according to the above analysis, we need to arrange the edge according to the degree of the node. However, it will violate differential privacy if determine the relative position of nodes by comparing the magnitude of the degree directly. The DDNL method which we propose does not compare the true degree of the node $f(V_i)$, but to compare the size of the degree after adding noise, denoted by $Nd(V_i)$. The global sensitivity is 2, and the assigned privacy budget is ε_a.

The DDNL method is proposed to identify the important edges of triangle count in matrix, but we find that it can also make the edge 1 better together, forming a more dense region so that provide support for our next division. For example, for the wiki-vote dataset used in this paper, we randomly generate a set of node labels to generate adjacency matrices and represent their distributions with scatterplots. As shown in Fig. 3(a), the points in the graph represent edge 1. We use DDNL generate a set of node label and generate matrix, the distribution of the matrix shown in Fig. 3(b), we can clearly see the aggregation effect of the method.

3.2   Quasi-line Partition and Reconstruction

There are often some dense clusters in real graph, such as communities in social networks, corresponding to the adjacency matrix, that is, there are a large number of 1 in some regions. It is possible to reduce the reduction error by relying on the original data, rationally dividing the matrix to obtain dense and sparse areas, and designing appropriate stopping conditions to limit the injection of noise.

Algorithm 2   Quasi-line Partition and Reconstruction

Input: Adjacency matrix A(G), privacy budget ε_b and ε_c
Output: IRTree T
1.   Generate the ladder region B;
2.   Partition B to N sub-regions set $\Re$ with vertical lines;
3.   for each $R_i \in \Re$ do
4.   Initialize a BinaryTree T with a root node $V_{i,}$;
5.   Set dom(V_i)=R_i and mark V_1 as unvisited;
6.   i=1;
7.   while there exits an unvisited node V do
8.   Subregion V_l, V_r ←Partition$(V,\varepsilon_b/2^i)$;
9.   Mark V, V_l as visited;
10.   Caculate $\hat{c}$ = count(V_l) + Lap$(1/\varepsilon_c)$;
11.   If$(\hat{c}/$dom$(V_l) > \delta)$
12.   Mark V_r as unvisited;
13.   Add V_l and V_r to T;
14.   i++;
15. Retuen T;

Currently, although there are dependent data partitioning methods, but original data space is divided into regular area. While ignoring the true distribution of data irregularities, thereby reducing the accuracy of the division. In order to solve the above problem, we propose a quasi-line partition method, as shown in Algorithm 2. The time complexity of Algorithm 2 is chosen by the exponential mechanism to determine the quasi-line.

Theorem 2. Given $(0,1)$ distribution area R and all possible segmentation lines set L. The two regions separated are expressed as R_1 and R_2. Set $l \in L$, and $f(l) = max|Den(R_1) - Den(R_2)|$, there must be a quasi-line, so that $\mathbf{f}(l_q) \geq \mathbf{f}(l)$.

Proof: Let $Den(R_1) = C_1/S_1$, $Den(R_2) = C_2/S_2$, and $Den(R_1) > Den(R_2)$, obviously it must exist with 0 cells or 1 cells of public side e, we are expressed as $l \cap Cell_0$ or $l \cap Cell_1$.

(1) If $e \in l \cap Cell_0$, move m $Cell_0$ $(m \geq 1)$ from R_1 to R_2
   hence $f(l_q) = \left|\dfrac{C_1}{S_{1-m}} - \dfrac{C_2}{S_{2+m}}\right| > f(l)$

(2) if $e \in l \cap Cell_1$, move m $Cell_1$ $(m \geq 1)$ from R_2 to R_1
   hence $f(l_q) = \left|\dfrac{C_1+m}{S_{1+m}} - \dfrac{C_{2-m}}{S_{2-m}}\right| > f(l)$
   Since $l \in l_q$
   We get $f(l_q) \geq f(l)$

Then we consider the application of exponential mechanism to select the quasi-line, the first time on the division of B, there are a total of $(|V| - 1)!$ possible quasi-lines. The time complexity is exponential order. So, we first divide B into N sub regions by coarse grain, the size of N also affects the number of fine-grained, that is, the number of sub-region. The coarse-grained partition is decomposed by straight line, taking into account the data distribution characteristics of the matrix generated by DDNL. The larger the nodes label, the larger coarse-grained region. We use the noise degree of the matrix A(G) as a reference for the distribution of the nodes, and determine the coarse-grained partition by the difference of the area head node and the tail node noise level. Detailed process such as Algorithm 3.

Algorithm 3   Coarse Partition

Input: Ladder region B, each vertex noisydegree $Nd(V_i)$

Output: Sub-regions set $\mathfrak{R}$

1.   Initialize $\mathfrak{R} = \emptyset$;
2.   Caculate $Avg = \sum_{i=1}^{|V|} Nd(V_i)$, add B to $\mathfrak{R}$;
3.   for each $\mathfrak{R}_i \in \mathfrak{R}$ do
4.   if(dif(R_i)>Avg)
5.   search the boundary which left node V_j>Avg and right node V_{j+1}<Avg, Partition R_i to R_1 and R_2;
6.   Add R_1 and R_2 to $\mathfrak{R}$,delete R from $\mathfrak{R}$;
7.   else
8.   break;
9.   Return $\mathfrak{R}$;

IRTree has the following characteristics. IRTree is a binary tree structure. Each node V_x records two parts, namely, the left and right borders of dom (V_x), and they are composed of a set of vectors of length $|V|$. The left child of all non-leaf nodes are leaf nodes. We use the privacy budget allocation scheme proposed in [8], dividing the assigned privacy budget by half of the last spent privacy budget.

Although we reduce the number of possible quasi-lines of the area to be partitioned after coarse-grained partitioning, it is still exponential time complexity. We use heuristic method to find the best dividing line $CV_{i,j}$ for each row based on greedy idea. The quasi-line is connected by each $CV_{i,j}$, and the local optimal is used instead of the global optimal. Different dom(V) division produces different sub-regions, due to the influence of important edges, we hope that the regions that they exit are more dense or sparse. At the same time, in order to the overall effect, there is a need to distinguish between dense and sparse areas, so we designed the following scoring function.

For unvisited node V, if the row of dom(V) contains S cells, the row has a total of S possible dividing lines. We use R to represent all possible lines in each row, and the scoring function q (R, p) is used to pick the split line p in R.

$$q(R, p) = \left| \mathrm{Den}(R_1) - \frac{1}{2} \right| \times |\mathrm{Den}(R_1) - \mathrm{Den}(R_2)| \tag{2}$$

Since the scoring function is defined by the density, and the maximum density of any region is 1 and the minimum value is 0, the global sensitivity of q is 1/2. The arbitrary division line p_i of the region R is outputted with the following probability.

$$\frac{\exp\left(\frac{\varepsilon}{2GS(q)} q(R, p_i) \right)}{\sum_{p_j \in P} \exp\left(\frac{\varepsilon}{2GS(q)} q\left(R, p_j \right) \right)} \tag{3}$$

Although we have chosen the best division of each row in dom(V) through the exponential mechanism to form a relatively sparse or dense area, there is a problem that we can not guarantee each row is divided left node dom(V_l) is sparse, so we still need to make a judgment, first find each row of dom (V_l) noise count $= c + \mathrm{lap}(1 / \varepsilon)$, by $\hat{c}/\mathrm{dom}(V_l)$ to find the density, we use y to represent the region, when $\hat{c}/\mathrm{dom}(V_l) \geq \delta$, y is 1, and vice versa y is 0, if the number of 1 is large, then divide each row of c/dom $(V_l) \geq 0.5$, and vice versa is divided into less than 0.5.

The most critical issue of regional division is to determine the stop conditions, a small number of division will introduce too much reduction error, on the contrary, will introduce a large number of noise errors. Therefore, we set the stop conditions as V_l is less than or equal to a certain threshold, because the data set itself is very sparse and the value of threshold is generally small. In the following operation, we set it to 0.5%.

The overall privacy budget of the QPR algorithm is divided into three parts, ε_a, ε_b and ε_c, for three phases, respectively, ε_a used to obtain the node label, ε_b used to select the division of each class, ε_c used to calculate the noise count for each sub-region. First we have to determine these three values. In general, we have to allocate a relatively large privacy budget for ε_b and ε_c, and between ε_b and ε_c, we will give more privacy budgets to ε_c. Because it is difficult to quantify the size of these values theoretically, we have chosen the ratio for them in the experiment.

Once the value of ε_a, ε_b and ε_c are determined, we apply the following scheme to assign the privacy budget for each node. In order to obtain the noise count for each sub-region, we assign $\mu\varepsilon_b/2^i$ to select the dividing line for the ith division of each row, $(1 - \mu)\varepsilon_b/2^i$ to determine the sparseness of the region.

The ith divided area corresponds to the node in ith layer of the tree (the root node is layer 0). Each node consumes a privacy budget of $\varepsilon_b/2^i$. Finally, assign the ε_c privacy budget for each leaf node to find the noise count for each sub-region.

Theorem. NoiseGraph satisfies ε differential privacy.

Proof. NoiseGraph consists of three parts. DDNL obviously satisfies the ε_a difference privacy. In the Quasi-line Partition and Reconstruction method, the coarse partition does not involve the original data set, so it does not involve the privacy leak. Next the quasi-line division form the IRTree T, the data in the same layer of T is not intersecting, so the parallel composition is satisfied, in other words, the path of each of the paths from the root node to the leaf node is independent of its privacy. Any one of the paths, except for the leaf nodes, we have performed a three-step differential privacy operation: using the exponential mechanism to select the best segmentation for each row, using the laplace mechanism to determine which segmentation we currently need, to judge whether meet the stop condition. Each path satisfies the sequence composition and the privacy budget is less than ε_b. Finally, we use ε_c privacy budget and application of the laplace mechanism to compute the noise count of IRTree leaf node. The three parts apply the sequence combination again, and the NoiseGraph satisfies the ε differential privacy.

Table 1. Datasets scale

Datasets	|V|	|E|	Density	$\triangle$	$2\triangle$	$3\triangle$
GrQc	5,242	14,485	0.1054%	48,260	2,041,499	23,754,699
Wiki-Vote	7,115	100,762	0.3981%	608,389	40,544,543	1,153,145,067
HepTh	9,877	25,975	0.0533%	28,339	429,013	2,906,030
HepPh	12,008	118,491	0.1644%	3,358,499	936,890,335	69,737,335,659

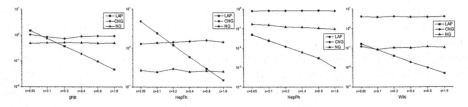

Fig. 4. Triangle count

4 Experimental Evaluation

In this part, we have experimentally evaluated the performance of the NoiseGraph method (NG) in terms of data availability. As a reference, we compared three methods (1) Laplace [9], to inject the synthetic graph directly into the original graph. Expressed as LAP; (2) we consider a changed NoiseGraph method, ignoring the presence of important edges, using a random node label instead of the degree-based node label, assigning this portion of the consumed privacy budget to other steps, while the scoring function corresponding to $q(R,p) = |\text{Den}(R_1) - \text{Den}(R_2)|$, we call this method CNG. We measure the three methods by the average relative error, $\text{MRE} = \left|f\left(\widehat{G}\right) - f(G)\right| / f(G)$. Which for the synthesis of the query results for the real query results, Where $f\left(\widehat{G}\right)$ is the query result of the composite graph, and $f(G)$ is the real query result.

We used four real data sets: GrQc, HepPh, HepTh, wiki-Vote. The first three datasets are arXiv's paper relations, covering general relativity, high-energy physics and high-energy physics theory. If two authors co-authored a paper, there is a relationship between them. The wiki-Vote dataset records the voting relationship of the Wikipedia administrator, and if a person casts to someone or votes by them, there is a relationship between the two. The following Table 1 shows the properties of the dataset and the results of the four triangle count queries.

The first query we evaluated is $f_\triangle$, i.e. $k = 1$, as shown in Fig. 4, in general, NG's performance on all datasets is superior to the CNG method, but its MRE is often higher than LAP. This is because the MRE of the LAP method is determined by the noise error and decreases strictly with the increase of the privacy budget. However, the error result of the NG method is not strictly reduced with the increase of NG, because there are two kinds of error sources, not entirely determined by the privacy budget. The results of NG method and CNG method is different not only because the CNG method does not consider the important data, but also through the DDNL method, can make the matrix 1 more closely than before, so as to get a better division effect. From the figure we can see that, when ε is smaller, the error of HepTh is relatively large, mainly because the data set is too sparse.

The experimental results of $f_{2\triangle}$ are similar to $f_\triangle$, but the LAP error of $f_{2\triangle}$ is significantly larger than $f_\triangle$, because $f_{2\triangle}$ is more sensitive than $f_\triangle$. And our NG method on the application of $f_{2\triangle}$ is not affected, and the error fluctuation range than $f_\triangle$, because relative to the triangle count the double triangle is a combination of the selection process. Compared with $f_\triangle$, the presence or absence of some data is further amplified.

As can be seen from Figs. 4 and 5, it can be seen that the performance of the synthetic method generated by the NG method is not necessarily optimal in a single usability measure, but the NG method is to publish a method that satisfies the accuracy of the triangulation at different values In this paper, we compare the NG method with the CNG method, the error mean $(\text{MRE}(f_\triangle) + \text{MRE}(f_{2\triangle}) + \text{MRE}(f_{3\triangle}))/3$, and the LAP method has a large error in all the experimental results at $k = 2$ and 3, and the published data is lost on data availability Meaning, so we show only the performance of the remaining two methods in Fig. 6. It can be seen from Fig. 6 that the global

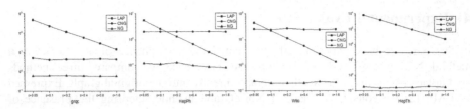

Fig. 5. Triangle count

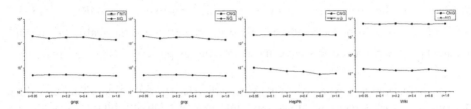

Fig. 6. Comprehensive triangle count

k-triangler counts on the four data sets show a lower average relative error (k = 1, 2, 3) compared to the CNG method, It is shown that the method proposed in this paper can improve the accuracy of the k triangle count of the composite graph.

5   Conclusions

In this article, we present a more interesting idea of analyzing the usability function itself to discover the meaning of important data, thereby guiding the process of distorting the entire data. We have also experimentally proved that the accurate reduction of important data can ensure the accuracy of usability results to a large extent. However, we mainly analyze the k triangle query in this paper. In the future work, we can consider other usability metrics, and make the composite graph meet the accuracy of multiple availability metrics.

Acknowledgments. This article is partly supported by the National Natural Science Foundation of China under Grant No. 61370084, and the China Numerical Tank Project.

References

1. Xiao, Y., Xiong, L., Yuan, C.: Differentially private data release through multidimensional partitioning. In: Jonker, W., Petković, M. (eds.) SDM 2010. LNCS, vol. 6358, pp. 150–168. Springer, Heidelberg (2010). doi:10.1007/978-3-642-15546-8_11
2. Cormode, G., Procopiuc, C., Srivastava, D., Shen, E., Yu, T.: Differentially private spatial decompositions. In: 2012 IEEE 28th International Conference on Data Engineering, pp. 20–31. IEEE Computer Society, Virginia (2011)

3. Hay, M., Rastogi, V., Miklau, G., et al.: Boosting the accuracy of differentially private histograms through consistency. J. Proc. Vldb Endow. **3**(1–2), 1021–1032 (2010)
4. Qardaji, W., Li, N.: Recursive partitioning and summarization: a practical framework for differentially private data publishing. In: ACM Symposium on Information, Computer and Communications Security, pp. 38–39. ACM, Raleigh (2012)
5. Xiao, X., Wang, G., Gehrke, J.: Differential privacy via wavelet transforms. IEEE Trans. J. Knowl. Data Eng. **23**(8), 1200–1214 (2010)
6. Zhang, J., Xiao, X., Xie, X.: PrivTree: a differentially private algorithm for hierarchical decompositions. In: Proceedings of the ACM SIGMOD International Conference on Management of Data, pp. 155–170. ACM, San Francisco (2016)
7. Chen, R., Fung, B.C., Yu, P.S.: Correlated network data publication via differential privacy. VLDB J. **23**(4), 653–676 (2014)
8. Dwork, C.: A firm foundation for private data analysis. Commun. ACM **54**(1), 86–95 (2011)
9. Dwork, C., McSherry, F., Nissim, K., Smith, A.: Calibrating noise to sensitivity in private data analysis. In: Halevi, S., Rabin, T. (eds.) TCC 2006. LNCS, vol. 3876, pp. 265–284. Springer, Heidelberg (2006). doi:10.1007/11681878_14

Hierarchical Access Control Scheme of Private Data Based on Attribute Encryption

Xi Lin[1] and Yiliang Han[1,2(✉)]

[1] Department of Electronic Technology,
Engineering University of PAP, Xi'an 710086, China
hanyil@163.com
[2] College of Information Science and Technology,
Northwest University, Xi'an 710127, China

Abstract. To solve the problems of data sharing in social network, such as management of private data is too loose, access permissions are not clear, mode of data sharing is too single and soon on, we design a hierarchical access control scheme of private data based on attribute encryption. First, we construct a new algorithm based on attribute encryption, which divides encryption into two phases, and we can design two types of attributes encryption strategy to make sure that different users could get their own decryption keys corresponding to their permissions. We encrypt the private data hierarchically with our algorithm to realize "precise", "more accurate", "fuzzy" and "private" four management modes, then users with higher permissions can access the private data inferior to their permissions. And we outsource some complex operations of decryption to DSP to ensure high efficiency on the premise of privacy protection. Finally, we analyze the efficiency and the security of our scheme.

Keywords: Data sharing · Hierarchical access control · Attribute encryption · Outsourcing decryption

1 Introduction

With the rapid development of mobile Internet and varies of social software appear, users will produce lots of private data in daily life and work, which is casually released in mobile social network. In these social software, we can easily know a lot of private data about the user as soon as we exist in the huge "friend" list. Such as, we can speculate the user's name, job, address and so on just from the track he released day after day. So, user's private data is in a huge risk while he is enjoying the service provided by social software. Especially, with the function is more and more rich, the share is more and more convenient and the increasing operating costs, at present most of the social software in pursuit of a good user experience, often choose to sacrifice part of the security. Therefore, as for the management of user's private data, there are widely existing the problems such as management of private data is too loose, access permissions are not clear, mode of data sharing is too single and soon on [1, 2]. In fact, these social softwares often only provide users with two choices, "visible" and "invisible" to manage their data. However, users often need to share only a part of information rather than expose themselves completely.

© Springer Nature Singapore Pte Ltd. 2017
B. Zou et al. (Eds.): ICPCSEE 2017, Part I, CCIS 727, pp. 220–230, 2017.
DOI: 10.1007/978-981-10-6385-5_19

With the increasing of people's self-protection, the security requirements of their own private data have become more and more urgent. At present, as for the management of private data, most solutions are going to define roles for the users to realize fine-grained access control [3–5]; Lv et al. [6] use CP-ABE to realize access control of ciphertext; Xiong et al. [7] design a scalable access control model based on double-tier role and organization; Hao [8] layered the traditional RBAC to realize mass authorized hierarchical management. In our paper, we construct a new algorithm based on attribute encryption to realize private data hierarchical access control. Firstly, it divides encryption into two phases and we can encrypt the data hierarchically. Then, it also provides "precise", "more accurate", "fuzzy" and "private" four modes of data sharing, then users with higher permissions can access the private data inferior to their permissions. Finally, we outsource some complex operations of decryption to DSP to ensure high efficiency on the premise of privacy protection.

2 Preliminary

2.1 Bilinear Maps

There are two multiplicative cyclic groups G_1, G_2, whose prime order is p and if they meet the requirements: (1) Bilinearity. If $\forall u, v \in G_1$ and $\forall a, b \in Z_p$, we can get $e(u^a, v^b) = e(u, v)^{ab}$. (2) Non-degeneracy. $\exists u, v \in G_1$, it has $e(u, v) \neq 1$. (3) Calculability. $\forall u, v \in G_1$, $e(u, v)$ can be calculated out in a polynomial time. We think $e : G_1 \times G_1 \to G_2$ as the bilinear map [9].

2.2 Ciphertext-Policy Attribute-Based Encryption (CP-ABE)

As for the traditional CP-ABE, A is regarded as the attribute set in system, S is the set of users' attributes and P is the attribute strategy. Then they often use policy based on the access tree structure [10] or linear secret sharing [11, 12] to control the access. Our scheme adopts linear secret sharing to make it: (1) Use (M, ρ) to express P, where M is a $l \times h$ matrix and ρ is an injective function. When $i = 1, \cdots, l$, we regard $\rho(i)$ as the ith row associated attributes in M. (2) If users' attribute set S can meet the attribute strategy P, $I = \{i \mid \rho(i) \in S\}$, then we can find a constant coefficient group $\{\theta_i \in Z_p\}$ to satisfy $\sum_{i \in I} \theta_i \vec{M}_i = \{1, 0, \ldots, 0\}$, where $\vec{M}_i$ is the ith row vector in M. (3) Hide the secret we need to share. When $s \in Z_p$ is the secret to share, then we randomly select $v_2, v_3, \ldots, v_h \in Z_p$ to contribute $\vec{v} = (s, v_2, \ldots, v_h)$ and figure out the sharing value $\lambda_i = \vec{M}_i \vec{v} (i = 1, 2, \ldots, l)$.

3 Our Construction

3.1 System Mode

Figure 1 is the private data management system of our scheme. Attribute authorities AA are responsible for distributing unique private key to users U and managing the

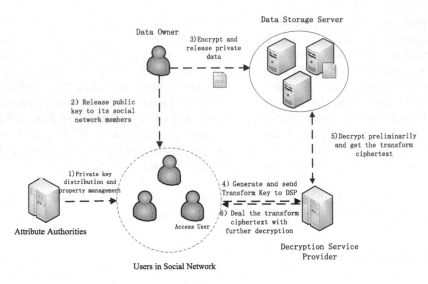

Fig. 1. Private data management system

attributes. Outsourcing decryption service provider DSP goes to provide access user AU with decryption service. When it encrypts, data owner DO uses the hierarchical attribute encryption we designed to encrypt the private data and release it to data storage server DSS.

When access user AU goes to access the private data in DSS, AU generates transform key TK according to his private key SK and sends it to DSP through secure channel, then DSP decrypts preliminarily and get the transform ciphertext. Then, DSP returns the result of preliminary decryption and AU goes to complete the further decryption. If the attributes of AU can satisfy the main encryption strategy, AU could complete the first phase of decryption and get fuzzy private data m_1. If the attributes of AU can still satisfy the secondary encryption strategy, AU could go on decryption and get more accurate private data $m_1 \| m_2$ or precise private data $m_1 \| m_2 \| m_3$.

3.2   Definitions

Definition 1. Private data. We divide the private data m into three parts m_1, m_2, m_3. And we use m_1 to constitute the fuzzy private data, $m_1 \| m_2$ to constitute the more accurate private data and $m_1 \| m_2 \| m_3$ to constitute the precise private data.

Definition 2. Attribute. There are two types of attributes in our scheme: main attribute and secondary attribute. We define U as the attribute in our system. When U is a main attribute, we use $h_1, h_2, \ldots, h_u \in G_1$ to express it and use A to represent the set of main attributes. When it is a secondary attribute, we use $n_1, n_2, \ldots, n_u \in Z_p$ to express it and use A' to represent the set of secondary attributes (any sum of two in $n_1, n_2, \ldots, n_u$ is out of the set A'). We define the main attribute set of users' as S and the secondary attribute set as S'. We assume that u_1 has $S = \{h_1, h_2\}$, $S' = \{\}$, u_2 has $S = \{h_1, h_2\}$, $S' = \{n_3\}$, u_3 has $S = \{h_1, h_2\}$, $S' = \{n_3, n_4\}$.

Definition 3. Hierarchical attribute encryption. Different from the method referred in [10], we use two types of strategies to complete the encryption, where the main strategy P_1 requires the main attributes and secondary strategies P_2, P_3 requires the secondary attributes. When it goes to encrypt, it can encrypt a few files at one time. We take three files as an example. Firstly, we divide the private data into m_1, m_2, m_3, then we use main strategy P_1 to encrypt the first part m_1 with the traditional attributes encryption and use secondary strategies P_2, P_3 to encrypt m_2, m_3 further. Therefore, if the user's main attribute set can meet P_1, he can complete the first phase of the decryption and get the first part data m_1. Then, if his secondary attribute set can meet P_2, P_3, he can complete the further decryption and get the data m_2, m_3.

3.3 Our Algorithm

(1) Setup ()

When it initializes, AA chooses a multiplicative group G_1 whose order is P and generator is g. There is existing the mapping $e : G_1 \times G_1 \to G_2$. We define the main attribute set of system as $A = \{h_1, h_2, \ldots, h_u\}$ and secondary attribute set as $A' = n_1, n_2, \ldots, n_u$, where $h_1, h_2, \ldots, h_u \in G_1$ and $n_1, n_2, \ldots, n_u \in Z_p$. Randomly choose $\alpha, \beta, n_0 \in Z_p$ and compute the public key of system

$$PK = <g, e(g,g)^\alpha, g^\beta, g^{n_o}, h_1, h_2, \ldots, hu, g^{\alpha n_1/n_0}, g^{\alpha n_2/n_0}, \ldots, g^{\alpha n_u/n_0} >$$

Then, compute the private key of system

$$MSK = g^\alpha$$

(2) Encrypt ()

Divide user's private data m into three parts m_1, m_2, m_3 and complete the encryption. Generate the main strategy P_1, secondary strategies P_2, P_3 to control the access: user with main attributes h_1, h_2 can meet P_1 and decrypt to get m_1; User with main attributes h_1, h_2 and secondary attribute n_3 can meet P_2 and decrypt to get m_2; User with main attributes h_1, h_2 and secondary attributes n_3, n_4 can meet P_3 and decrypt to get m_3.

First of all, randomly generate three groups of symmetric key k_1, k_2, k_3 and encrypt private data m_1, m_2, m_3 with symmetric encryption algorithm (expressed as E below) respectively to get $E_{k_1}(m_1)$, $E_{k_2}(m_2)$, $E_{k_3}(m_3)$. Then output the ciphetext of data file CM:

$$CM = <E_{k_1}(m_1), E_{k_2}(m_2), E_{k_3}(m_3) >$$

Second, encrypt k_1, k_2, k_3 with attribute encryption. We use the methods referred in [12] to construct LSSS matrix of main strategy P_1 and (M, ρ) is used to express it, where M is a matrix of $l \times h$ and ρ is an injective function. Then, randomly choose h dimension vector $\overrightarrow{V} = (s, v_2, \ldots, v_h) \in Z_p$. From $i = 1$ to l, compute $\lambda_i = \overrightarrow{v} \cdot M_i$, where M_i expresses the ith row vector of M. Randomly choose $r_1, r_2, \ldots, r_i \in Z_p$ and compute

$$\tilde{C}_1 = k_1 e(g,g)^{\alpha s} \quad C' = g^s$$

$$\left(C_1 = g^{\beta \lambda_1} h_{\rho(i)}^{-r_1}, D_1 = g^{r_1}\right), \ldots, \left(C_l = g^{\beta \lambda_l} h_{\rho(l)}^{-r_l}, D_l = g^{r_l}\right)$$

Above is the first phase of our attribute encryption. Then, it goes to encrypt k_2, k_3 and compute as follows according to g^{n_o} in PK and $g^{\alpha n_1/n_0}, g^{\alpha n_2/n_0}, \ldots, g^{\alpha n_u/n_0}$:

$$\tilde{C}_2 = k_2 e(g^{n_0 s}, g^{\alpha n_3/n_0}) = k_2 e(g,g)^{\alpha s n_3}$$

$$\tilde{C}_3 = k_3 e(g^{n_0 s}, g^{\alpha n_3/n_0} g^{\alpha n_4/n_0}) = k_3 e(g,g)^{\alpha s (n_3 + n_4)}$$

Output the ciphertext of symmetric key CT

$$CT = <\tilde{C}_1, \tilde{C}_2, \tilde{C}_3, C', (C_i, D_i)_{\{i \in l\}} >$$

(3) KeyGen ()

Input master private key MSK, main attribute set S and secondary attribute set S', then generate user's own private key. Randomly choose $t \in Z_p$ and compute SK

$$SK = <K_S = g^\alpha \cdot g^{\beta t}, K_{S'} = g^{\alpha/t} \cdot g^{\beta t}, L = g^t,$$
$$\{K_x = h_x^t\}_{\forall x \in S}, \{K_{x'} = n_{x'} \cdot t\}_{\forall x' \in S'} >$$

(4) KeyGen_out ()

Access user randomly choose $z \in Z_p$, compute transform key TK according to the user's SK

$$TK = <K_S' = g^{\alpha/z} \cdot g^{\beta t/z}, L' = g^{t/z}, \{K_x' = h_x^{t/z}\}_{\forall x \in S} >$$

(5) Transform_out ()

Access user sends CT and TK to the Decryption Service Provider, then DSP complete preliminary decryption

$$\prod_{i \in I} \left(e(C_i, \mathrm{L}) \cdot e\left(D_i, K'_{\rho(i)}\right) \right)^{w_i}$$
$$= \prod_{i \in I} e(g, g)^{t\beta\lambda_i w_i / z}$$
$$= e(g, g)^{t\beta s / z}$$

$$e(C', K'_S) / \left(\prod_{i \in I} \left(e(C_i, \mathrm{L}) \cdot e\left(D_i, K'_{\rho(i)}\right) \right)^{w_i} \right)$$
$$= e(g^s, g^{\alpha/z} \cdot g^{\beta t/z}) / e(g, g)^{t\beta s / z}$$
$$= e(g, g)^{\alpha s/z} \cdot e(g, g)^{\beta s t/z} / e(g, g)^{t\beta s / z}$$
$$= e(g, g)^{\alpha s/z}$$

DSP returns the transform ciphertext $CT' = <\tilde{C}_1, \tilde{C}_2, \tilde{C}_3, e(g, g)^{\alpha s/z}, e(g, g)^{t\beta s/z}>$

(6) Decrypt ()

If user's main attribute set S can meet the main strategy P_1 expressed by (M, ρ), let $I \in \{1, 2, \ldots, l\}$, $I = \{i : \rho(i) \in S\}$ and then begin the first phase of decryption. $\{w_i \in Z_p\}_{i \in I}$ is a set of constant, $\{\lambda_i\}$ is the effective part of s hidden in M and we can get $\sum_{i \in I} w_i \lambda_i = s$.

The first phase of decryption:

Access user executes KeyGen$_{out}$ () to compute transform key TK according to the user's SK and send CT and TK to DSP. Then, DSP executes Transform$_{out}$ () to complete the preliminary decryption of CT to get the transform ciphertext CT'.

Access user gets the transform ciphertext CT' and compute

$$k_1 = \tilde{C}_1 / (e(g, g)^{\alpha s/z})^z$$
$$m_1 = D_{k_1}(E_{k_1}(m_1))$$

The second phase of decryption:

If user's secondary attribute set can meet secondary P_2, P_3, compute

$$e(C', K_{S'}) / \left(\prod_{i \in I} \left(e(C_i, \mathrm{L}) \cdot e\left(D_i, K_{\rho(i)}\right) \right)^{w_i} \right)$$

According to Transform$_{out}$ (), we can easily know that

$$\prod_{i \in I} \left(e(C_i, \mathrm{L}) \cdot e\left(D_i, K_{\rho(i)}\right) \right)^{w_i} = \left(\prod_{i \in I} \left(e(C_i, \mathrm{L}) \cdot e\left(D_i, K'_{\rho(i)}\right) \right)^{w_i} \right)^z, \qquad \text{then}$$

compute

$$e(C', K_{S'}) / \left(\prod_{i \in I} \left(e(C_i, \mathrm{L}) \cdot e\left(D_i, K_{\rho(i)}\right) \right)^{w_i} \right)$$
$$= e(g^s, g^{\alpha/t} g^{\beta t}) / (e(g, g)^{t\beta s/z})^z$$
$$= e(g, g)^{\alpha s/t} e(g, g)^{s\beta t} / e(g, g)^{t\beta s}$$
$$= e(g, g)^{\alpha s/t}$$

According to $\{K_{x'} = n_{x'} \cdot t\}_{\forall x' \in S'}$ in SK, we can get

$$e(g,g)^{\alpha s n_3} = (e(g,g)^{\alpha s/t})^{n_3 t}$$
$$e(g,g)^{\alpha s(n_3 + n_4)} = (e(g,g)^{\alpha s/t})^{n_3 t + n_4 t}$$
$$k_2 = \tilde{C}_2 / e(g,g)^{\alpha s n_3}$$
$$k_3 = \tilde{C}_3 / e(g,g)^{\alpha s(n_3 + n_4)}$$

Use symmetric key k_2, k_3 and we can get

$$m_2 = D_{k_2}(E_{k_2}(m_2))$$
$$m_3 = D_{k_3}(E_{k_3}(m_3))$$

3.4 Our Scheme

System Initialization and User Registration. At the beginning, system runs Setup () to initialize itself and we can get PK and MSK according to the system parameter 1^λ. Then, release PK and preserve MSK. According to the main attribute set $A = \{h_1, h_2, \ldots, h_u\}$ and the secondary attribute set $A' = \{n_1, n_2, \ldots, n_u\}$ in the system, Attribute authorities AA goes to set user's main attribute set S and secondary attribute set S'. Then, it runs KeyGen () to generate user's own private key SK.

Private Data Encryption and Release. Data owner DO devides the private data into three parts, sets the access permissions and generates the corresponding main strategy and secondary strategy. Run Encrypt () to combine symmetric encryption and CP-ABE to complete hierarchical encryption. Then, upload the encrypted data.

Private Data Access and Outsourcing Decryption. When access user AU goes to access the private data of DO, if his SK can satisfy the DO's main encryption strategy, AU runs KeyGen$_{\text{out}}$ () to generate transform key TK and sends TK and ciphertext CT to the decryption service provider DSP. DSP can conduct preliminary decryption of CT and return transform ciphertext CT'. Then, AU can do further decryption of CT' and get the private data of the corresponding rank. For example, u_1 with $S = \{h_1, h_2\}$, $S' = \{\}$ can satisfy main strategy P_1, then he can decrypt and get fuzzy private data m_1; u_2 with $S = \{h_1, h_2\}$, $S' = \{n_3\}$ can satisfy main strategy P_1 and secondary strategy P_2, then he can get m_1, m_2 to construct more accurate private data $m_1 \| m_2$; u_3 with $S = \{h_1, h_2\}$, $S' = \{n_3, n_4\}$ can satisfy main strategy P_1 and secondary strategies P_2, P_3, then he can get m_1, m_2, m_3 to construct precise private data $m_1 \| m_2 \| m_3$.

4 Security Analysis

4.1 Confidentiality of Data

In our scheme, attribute encryption is divided to two phases to realize the hierarchical access control of private data. In fact, the first phase can be converted to the traditional algorithm of CP-ABE [10] which is judgmental PBDHE mathematical problems and proved to be secure under the standard model. The second phase is based on the result of first phase and its security depends on NPC problem of discrete logarithm decomposition in Elgamal algorithm. Assuming the attacker can decrypt the first phase and get the result $e(g,g)^{\alpha s}$ in some way, he still needs to get n_3, $n_3 + n_4$ to compute $k_2 = \tilde{C}_2/e(g,g)^{\alpha s n_3}$, $k_3 = \tilde{C}_3/e(g,g)^{\alpha s(n_3 + n_4)}$. It means he have to calculate out n_3, n_4 from $g^{\alpha n_1/n_0}, g^{\alpha n_2/n_0}, \ldots, g^{\alpha n_u/n_0}$ in PK. However, it is the NPC problem of discrete logarithm decomposition.

Our scheme outsources some complex operations so that we have to provide DSP with transform key TK. However, even though TK is generated by SK, attackers could never get the SK from TK because we randomly choose $z \in Z_p$ and add it to TK. It means if the attackers want to get SK, they have to recover z from K'_S, L' or K'_x, which is also the NPC problem of discrete logarithm decomposition. Therefore, the hierarchical attribute encryption adopted in our scheme is secure.

4.2 Privacy Protection in Outsourcing Decryption

In the traditional outsourcing decryption of attribute encryption, decryption service provider DSP could not know the private key SK of user's but it can get the attribute encryption strategies from the ciphertext. In other word, DSP can easily estimate the attributes of decryption, which leads to the leakage of users' privacy.

Our hierarchical attribute encryption algorithm divides the encryption into two phases and adopts main encryption strategy and secondary strategy to encrypt the private data. When it decrypts, we outsource the complex part of the first phase to DSP, which includes a lot of bilinear pairings computation. DSP uses the transform key provided by users to decrypt the ciphertext preliminarily, then the user completes the remain operations. Transformation key TK is generated by the user independently to make sure the security of private key SK. The second part only need a few simple operation and it is entirely completed by the user so that we only outsource the first phase and TK only includes the main attribute of decryption. Therefore, the secondary attribute of decryption is completely confidential to decryption service provider DSP. User can flexibly design the two kinds of strategies to encrypt so that we can not only realize the hierarchical access control of private data but also reduce the leakage of user's privacy in some ways.

4.3 Resist Collusion Attack

In our system, there may be two kinds of collusion attack forms: collusion between the users and collusion between user and DSP. Now we are divided into two cases prove the security.

Collusion Between the Users. Assuming that there exists collusion between user A and user B. In the first phase of decryption, if the user wants to get k_1, he needs to get $e(g,g)^{\alpha s}$. Normally, A and B can't recover s from $\prod_{i \in I} \left(e(C_i, L) \cdot e(D_i, K_{\rho(i)}) \right)^{w_i}$ if both of they can't satisfy main attributes of decryption, but it is possible to expand their permissions through the exchange of main attribute parameter $\{h_x\}_{\forall x \in S}$. Therefore, we add random number t into the user's SK to let $\{h_x\}_{\forall x \in S}$ exist as $\{K_x = h_x^t\}_{\forall x \in S}$ so that the collusion users A, B cannot decrypt the data through combining their private keys. In the second phase of decryption, if the user wants to get k_2, k_3, he needs to get $e(g,g)^{\alpha s n_3}$, $e(g,g)^{\alpha s(n_3+n_4)}$. Assuming that he has recovered $e(g,g)^{\alpha s}$ in the first phase, he still needs to get n_3, n_3+n_4. Therefore, our scheme releases secondary parameters $g^{\alpha n_1/n_0}, g^{\alpha n_2/n_0}, \ldots, g^{\alpha n_u/n_0}$ instead of releasing parameters $n_1, n_2, \ldots, n_u$ directly. Normally, A and B can't recover n_3, n_3+n_4, which are existing in users' private key if both of they can't satisfy the secondary strategy. But it is possible to expand their permissions through though the exchange of parameters $n_1, n_2, \ldots, n_u$ they have. Therefore, we also add the random number t into $\{K_{x'} = n_{x'} \cdot t\}_{\forall x' \in S'}$ so that the collusion users A, B cannot decrypt the data through combining their private keys. In conclusion, our scheme can resist collusion between users.

Collusion Between User and DSP. Assuming that there exists collusion between user B and DSP. When it decrypts, user A add random number $z \in Z_p$ to generate transform key TK, then DSP uses TK to decrypt the data preliminarily. Normally, there is no main attribute parameter $\{h_x\}_{\forall x \in S}$ needed in the private key of user B if he cannot satisfy the encryption strategy. Now, we assume that the collusion DSP has the main attribute parameter needed and user B maybe try his best to combine the TK of DSP to expend his access permission. In our scheme, we add $z \in Z_p$ to let main attribute parameters $\{h_x\}_{\forall x \in S}$ exist as $\{K'_x = h_x^{t/z}\}_{\forall x \in S}$ in TK. If they want to recover $\{h_x\}_{\forall x \in S}$ they need to solve the NPC problem of discrete logarithm decomposition. Therefore, user B cannot conspire with DSP to expend the permission and our scheme can resist the collusion between user and DSP.

5 Efficiency Analysis

In our paper, we have proposed a hierarchical attribute encryption algorithm. It divides attribute encryption into two phases and uses two strategies to realize hierarchical access control of private data. When it encrypts a single file, our scheme only needs to complete the first phase and goes to be converted into CP-ABE automatically. When it encrypts multiple files, the first phase of encryption is also similar to CP-ABE and the efficiency is improved little. In the second phase of decryption, it can reuse $\tilde{C}_1$, C', (C_i, D_i) and only need once bilinear pairings computation to compute $\tilde{C}_2$, $\tilde{C}_3$. When it goes to decryption, we use the method of outsourcing decryption. We outsource most complex bilinear pairings computation in decryption to DSP and it can greatly reduce the local computation. In the first phase of decryption, it just like the traditional outsourcing decryption algorithm based on attribute encryption. In the second phase of decryption, it can reuse the previous result $\prod_{i \in I} \left(e(C_i, L) \cdot e(D_i, K_{\rho(i)}) \right)^{w_i} = e(g,g)^{t\beta s}$ and only need to complete once bilinear pairings computation $e(g^s, g^{\alpha/t} g^{\beta t})$. Therefore, the attribute encryption algorithm proposed in our paper can reduce lots of bilinear pairings computation and improve the efficiency greatly when users need to use attribute encryption many times. And we outsource mass complex bilinear pairings computation to DSP so that users only need to complete little necessary computation locally. Tables 1 and 2 show the efficiency analysis of two times attribute encryption

Table 1. Efficiency analysis of two times attribute encryption

Scheme	Access structure	Number of attribute	Encryption	Decryption	Local decryption
[10]	Access tree	$a + b$	$(4a + 4b + 2)$ $e + 2p$	$(2a + 2b)$ $e + (2a + 2b + 4)p$	$(2a + 2b)$ $e + (2a + 2b + 4)$ p
[12]	LSSS	$a + b$	$(6a + 6b + 2)$ $e + 2p$	$(2a + 2b)$ $e + (4a + 4b + 2)p$	$(2a + 2b)$ $e + (4a + 4b + 2)$ p
[13]	LSSS	$a + b$	$(6a + 6b + 2)$ $e + 2p$	$(4a + 4b + 8)$ $e + (4a + 4b + 2)p$	$(2a + 2b + 8)e$
Ours	LSSS	$a + b$	$(3a + 1)$ $e + 2p$	$(2a + 6)e + (2a + 2)$ p	$(a + 6)e + p$

Table 2. Efficiency analysis of three times attribute encryption

Scheme	Access structure	Number of attribute	Encryption	Decryption	Local decryption
[10]	Access tree	$a + b$	$(6a + 6b + 3)$ $e + 3p$	$(3a + 3b)$ $e + (3a + 3b + 6)p$	$(3a + 3b)$ $e + (3a + 3b + 6)$ p
[12]	LSSS	$a + b$	$(9a + 9b + 3)$ $e + 3p$	$(3a + 3b)$ $e + (6a + 6b + 3)p$	$(3a + 3b)$ $e + (6a + 6b + 3)$ p
[13]	LSSS	$a + b$	$(9a + 9b + 3)$ $e + 3p$	$(6a + 6b + 12)$ $e + (6a + 6b + 3)p$	$(3a + 3b + 12)e$
Ours	LSSS	$a + b$	$(3a + 1)$ $e + 3p$	$(2a + 7)e + (2a + 2)$ p	$(a + 7)e + p$

and three times attribute encryption respectively. In these tables, the bilinear pairings computation is expressed by p, exponent arithmetic is expressed by e and the numbers of main attributes and secondary attributes associated are expressed by a, b respectively.

6 Conclusions

To solve the problems of data sharing in social network, such as management of private data is too loose, access permissions are not clear, mode of data sharing is too single and soon on, we design a private data hierarchical access control scheme based on attribute encryption, which divides encryption into two phases, and we can design two types of attributes encryption strategy to realize "precise", "more accurate", "fuzzy" and "private" four modes of private data management. We outsource some complex operations of decryption to DSP to ensure high efficiency on the premise of privacy protection.

Finally, we analyze the efficiency and the security of our scheme. It shows that our scheme has the advantages of confidentiality, collusion attack resistance and high privacy in outsourcing decryption. Compared with the traditional encryption algorithms, our scheme works on a high efficiency if it needs to use two or more times of attributes encryption. Compared with the traditional outsourcing decryption, our scheme only outsources part phase of decryption to DSP so that we can hide part attributes of user's, which improves the privacy of outsourcing decryption and confidentiality of key. However, there are still some questions need to solve, such as, reveal of attributes and verification of result in outsourcing decryption. Therefore, we are going to improve our scheme further to ensure a higher efficiency on the premise of privacy protection.

Acknowledgments. This work is supported by National Natural Science Foundation of China (61572521), Project funded by China Postdoctoral Science Foundation (2014M562445, 2015T81047).

References

1. Zhu, Y.Q., Li, J.H., Zhang, Q.H.: A new dynamic hierarchical RBAC model for web services. J. Shanghai Jiaotong Univ. **41**(5), 783–787 (2007)
2. Zhao, J.: Research on improved access control model based on T&RBAC. J. Yanshan Univ. **34**(4), 331–335 (2010)
3. Li, H.M., Wang, H.J., Fu, L.: Organization-based access control model for web service. Comput. Eng. **40**(11), 65–70 (2014)
4. Xiong, Z., Wang, P., Xu, J.Y., et al.: Attribute based access control strategy for enterprise cloud storage. Appl. Res. Comput. **30**(2), 513–517 (2013)
5. Wang, X.W., Zhao, Y.M.: A task-role-based access control model for cloud computing. Comput. Eng. **38**(24), 9–13 (2012)
6. Lv, Z.Q., Zhang, M., Feng, D.G.: Cryptographic access control scheme for cloud storage. J. Front. Comput. Sci. Technol. **5**(9), 835–844 (2011)
7. Xiong, H.R., Chen, X.Y., Zhang, B., et al.: Scalable access control model based on double-tier role and organization. J. Electron. Inf. Technol. **37**(7), 1612–1619 (2015)
8. Hao, X.L.: Application of improved RBAC model in grid video monitoring platform. Comput. Technol. Dev. **12**, 212–215 (2014)
9. Boneh, D., Franklin, M.: Identity-based encryption from the weil pairing. In: Kilian, J. (ed.) CRYPTO 2001. LNCS, vol. 2139, pp. 213–229. Springer, Heidelberg (2001). doi:10.1007/3-540-44647-8_13
10. Bethencourt, J., Sahai, A., Waters, B.: Ciphertext-policy attribute-based encryption. In: IEEE Symposium on Security and Privacy, pp. 321–334. IEEE Computer Society, Berkeley (2007)
11. Beimel, A.: Secure Schemes for Secret Sharing and Key Distribution. Israel Institute of Technology, Haifa (1996)
12. Waters, B.: Ciphertext-policy attribute-based encryption: an expressive, efficient, and provably secure realization. Publ. Key Crypt. **6571**, 53–70 (2011)
13. Green, M., Hohenberger, S., Waters, B.: Outsourcing the decryption of ABE ciphertexts. In: Usenix Conference on Security, pp. 34–34. USENIX Association, San Francisco (2011)

Secret Data-Driven Carrier-Free Secret Sharing Scheme Based on Error Correction Blocks of QR Codes

Song Wan$^{(\boxtimes)}$, Yuliang Lu, Xuehu Yan$^{(\boxtimes)}$, Hanlin Liu, and Longdan Tan

Hefei Electronic Engineering Institute, Hefei 230037, China
wsong1031@163.com, publictiger@126.com

Abstract. In this paper, a novel secret data-driven carrier-free (semi structural formula) visual secret sharing (VSS) scheme with $(2, 2)$ threshold based on the error correction blocks of QR codes is investigated. The proposed scheme is to search two QR codes that altered to satisfy the secret sharing modules in the error correction mechanism from the large datasets of QR codes according to the secret image, which is to embed the secret image into QR codes based on carrier-free secret sharing. The size of secret image is the same or closest with the region from the coordinate of $(7, 7)$ to the lower right corner of QR codes. In this way, we can find the QR codes combination of embedding secret information maximization with secret data-driven based on Big data search. Each output share is a valid QR code which can be decoded correctly utilizing a QR code reader and it may reduce the likelihood of attracting the attention of potential attackers. The proposed scheme can reveal secret image visually with the abilities of stacking and XOR decryptions. The secret image can be recovered by human visual system (HVS) without any computation based on stacking. On the other hand, if the light-weight computation device is available, the secret image can be lossless revealed based on XOR operation. In addition, QR codes could assist alignment for VSS recovery. The experimental results show the effectiveness of our scheme.

Keywords: Visual secret sharing · QR code · Error correction blocks · Carrier-free · Big data · Data-driven · Multiple decryptions

1 Introduction

A secret sharing scheme is a method that encoding a secret image into n noise-like shares where each share reveals no information about the secret image. The secret can only be recovered when qualified shares combined [1]. Visual cryptography (VC), also called Visual Secret Share (VSS), is a kind of secret sharing scheme [2–4] where the secret image could be revealed by stacking the qualified number of shares based on the human visual system (HVS) without any computation. On the other hand, the secret image can be revealed by XOR

B. Zou et al. (Eds.): ICPCSEE 2017, Part I, CCIS 727, pp. 231–241, 2017.
DOI: 10.1007/978-981-10-6385-5_20

operation when light-weight computation device is available [5]. By splitting and encoding a secret into n noise-like shares, VSS can overcome the problem of storing a secret in a single information-carrier which would be damaged and lost easily. When less than the qualified number of shares, it can reveal nothing of the secret image by inspecting the shares.

However, since each share looks like a random pattern of pixels, it will raise suspicion and increase the likelihood of attracting the attention of potential attackers. Furthermore, the alignment can be an important issue for VSS recovery.

QR code [6] is a popularly used two-dimensional barcode recently with the advantages of the high information density, robustness and error correction capability. QR codes are widely being used for many applications such as advertisements and user authentication, because the data in QR codes can be read easily with the help of a QR code reader.

In addition, the appearance of QR codes is similar to the share of VSS. Based on the advantages above, the technology of combining VSS and QR codes can be applied in many scenes, such as transferring secret information through public channels.

Recently, many researchers have proposed many schemes combining the technologies of VSS and QR codes. Chow et al. [7] proposed a secret sharing scheme for $(n, n)(n \geq 3)$ threshold based on XOR operation. The idea is to distribute and encode the information of a QR code containing a secret message into a number of QR codes. Each QR code share is a valid QR code which can be decoded correctly. The secret could be revealed by first XORing the light and dark modules contained in the encoding region of all the QR code shares and adding the function patterns. Nevertheless, it needs a computational device with XOR ability for secret recovery as well as is only for cases (n, n) that n is equal to or greater than 3.

Wan et al. [8] proposed a VSS scheme for (k, n) threshold based on QR code with multiple decryptions. The proposed scheme based on QR code can visually reveal secret image with the abilities of stacking and XOR decryptions as well as scan every shadow image, i.e., a QR code, by a QR code reader. Nevertheless, the embedding capacity of secret image is low.

In this article, we propose a novel secret data-driven carrier-free (semi structural formula) VSS scheme with $(2, 2)$ threshold based on the error correction blocks of QR codes. We first introduce the scheme based on carrier-free secret sharing. The proposed scheme is to search two QR codes that altered to satisfy the secret sharing modules in the error correction mechanism from the large datasets of QR codes according to the secret image, which is to embed the secret image into QR codes based on carrier-free secret sharing. The size of secret image is the same or closest with the region from the coordinate of $(7, 7)$ to the lower right corner of QR codes. Based on this, the optimal combination of two QR codes that modified in the error correction mechanism can be searched from the large datasets of QR codes, which can improve the embedding capacity than the existing schemes. Each share is a valid QR code that can be scanned and decoded by a QR code reader. The secret image can be recovered by stacking shares without any computation based on HVS, and can also be lossless revealed based on XOR operation when the light-weight computation device is available. As a result, when recovering the secret image, two different ways can be chosen to reveal the secret due to the different scenarios. As each QR code share can be recognized, it means that the shares may not be suspected if distributed via public channels and can reduce the likelihood of attracting the attention of potential attackers.

The remainder of the paper is organized as follows. The introduction to QR codes and VSS are presented in Sect. 2. The secret data-driven carrier-free algorithm is described in Sect. 3. Section 4 demonstrates the simulation results and analyses. Finally, Sect. 5 concludes this paper.

2 Background

2.1 Visual Cryptography

VSS was first introduced by Naor and Shamir [2]. The idea is to split a secret image into n images that called shares and afterwards using the OR-operation to stack some of these shares, the original image will be revealed. In a general (k, n) threshold VC scheme, the secret image would be divided into n random shares which respectively reveals nothing about the secret information. Stacking any k or more shares can visually recover the secret image based on the human visual system (HVS), but any $k - 1$ or less shares can not reveal the secret. In $(2, 2)$, VSS, the secret is encrypted into two random shares that have twice size of the secret image. Although some contrast [9] loss appears, the revealed image is identified clearly. When light-weight computation device is available, the secret can be revealed by XOR operation. VSS can be applied in many scenes [4], such as, authentication and identification, social computing security, watermarking, information hiding and transmitting passwords etc.

In this paper, we use the so-called $(2, 2)$ scheme.

2.2 QR Codes

QR code is defined as a two-dimensional barcode, which was invented by the Denso [10] Incorporated in 1994. The standard [6] defines forty sizes of QR code versions which are range from version 1 to version 40. Each QR code is divided into a number of modules and each version has four modules more than the previous one. For example, version 2 is made up of 25×25 modules while version 3 is made up of 29×29 modules. The structure of a QR code version 7 is shown in Fig. 1. Each QR code has three Finder Patterns which are used to detect the position of the symbol and recognize the QR code. The three Finder Patterns are located in the upper left, lower left and upper right corner. Alignment Patterns which permit QR code readers to compensate for image distortion only occur from version two up to forty and the higher the version is, the more Alignment Patterns exist.

The data in QR codes is encoded into the binary numbers of 0 and 1 based on Reed-Solomon codes. There are four different error correction levels (L = 7%, M = 15%, Q = 25%, H = 30%). The error correction [11] is used for recovering the QR code when parts of the symbol are destroyed or dirty. So, the QR code can also be recognized when embedding other information into it. The recovery capacity of QR codes will be improved by using the higher error correction levels, but it would increase the amount of data to be encoded. It means that a larger QR code version may be required if using a higher error correction level to encode the same message. The codewords of QR codes are divided into a large number of error correction blocks and corresponding error correction codewords are generated for each block. The number of error correction blocks, error correction codewords and data codewords depend on the QR code version and error correction level.

In [6], the error correction codewords for each block is given as (c, k, r). Here, c is the total number of codewords, k is the number of data codewords and r is the error correction capacity which represents the maximum number of codewords that can be modified per block. It means that the QR code would not be decoded correctly if more than r codewords per block contain errors. To minimize the possibility that localized damage will cause the QR code to become undecodable, the codewords from the blocks are encoded in an interleaved manner, with the error correction codewords appended to the end of the data codewords sequence. The error correction codewords and data codewords arrangement for QR code version 7, with an error correction level of H, is shown in Fig. 1.

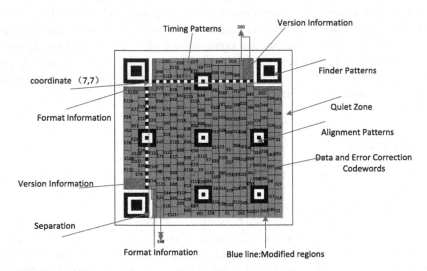

Fig. 1. The structure and codewords arrangement for QR code version 7 with error correction level H (Color figure online)

D1-D13	Data Block 1
D14-D26	Data Block 2
D27-D39	Data Block 3
D40-D52	Data Block 4
D53-D66	Data Block 5
E1-E26	Error Correction Block 1
E27-E52	Error Correction Block 2
E53-E78	Error Correction Block 3
E79-E104	Error Correction Block 4
E105-E130	Error Correction Block 5

3 Proposed Secret Data-Driven Carrier-Free Secret Sharing Scheme

In this section, a novel scheme which deeply integrates the theory of VSS with the error correction mechanism of QR codes based on carrier-free secret sharing is proposed. The errors can be corrected by the error correction codewords per block which would not be more than r, so that the QR code could be decoded correctly by a QR code reader while manipulated some of the codewords but still maintaining a QR code symbol. In this paper, the proposed scheme is $(2, 2)$ VSS threshold.

3.1 The Proposed Scheme

The idea behind our scheme is to search two QR codes that altered to satisfy the secret sharing modules in the error correction mechanism from the large datasets of QR codes according to the secret image, which is to embed the secret image into QR codes based on carrier-free secret sharing. The QR code version is selected based on the size of secret image from the large datasets of QR codes and the errors of each block are less than r to make sure that the QR code altered can be decoded correctly. In such a way, the optimal combination of two QR codes that modified in the error correction mechanism can be searched, which can improve the embedding capacity than the existing schemes. The secret image can be recovered by stacking two QR code shares visually when no light-weight computation device. On the other hand, if the light-weight computation device is available, the secret QR code can be lossless revealed based on XOR operation. In the proposed scheme, the large datasets of QR codes are to use the highest level of QR code error correction.

The QR code shares generation architecture of the proposed scheme is shown in Fig. 2, the corresponding algorithmic steps are described in detail in Algorithm 1.

The Algorithm 1 takes one secret image S and a large datasets of QR codes as the input and outputs two QR code shares SC_1, SC_2. Each QR code share is a valid QR code which can gain the original message when decoded by a QR code reader. The secret image recovery of the proposed scheme can be based

Algorithm 1. Secret data-driven carrier-free secret sharing
Input: A $M \times M$ binary secret image S, large datasets of QR codes
Output: QR code shares SC_1, SC_2
Step 1: Turn left and right the secret image S, then rotate it 90 degrees anticlockwise
Step 2: Search the QR codes whose regions are the same size or the closest with S from the coordinate of $(7,7)$ to the lower right corner of QR codes in the large datasets of QR codes, denoted $QR_1, QR_2 \ldots QR_N$, where N is a large number.
Step 3: Select two QR codes from the N QR codes randomly, for each position $(i,j) \in \{(i,j)\|1 \leq i \leq M; 1 \leq j \leq M\}$ repeat Steps 4–5.
Step 4: Let Q_1, Q_2 denote the regions of the two selected QR codes with the same size as S at the same locations from the coordinate of $(7,7)$ to the lower right corner of QR codes, where $Q_1(i,j)$ and $Q_2(i,j)$ indicates the bits corresponding to the two selected QR message.
Step 5: Check $Q_1(i,j)$ and $Q_2(i,j)$ to satisfy any one of the $S(i,j)$ sharing modules or not, if not, $Q_1(i,j) = \overline{Q}_1(i,j)$ or $Q_2(i,j) = \overline{Q}_2(i,j)$ alternately. If (i,j) belong to the data block and error correction block x, $count_x$++, which indicates the number of bits need to be changed in the data block and error correction block x.
Step 6: For each block, if $count_x > 2r$, go to Step 3; else go to Step 7.
Step 7: Output the two QR code shares SC_1, SC_2

on stacking or XOR operation, and the secret image can be recovered in the encoding regions of QR code shares.

From the above steps, according to the mechanism of QR codes, data generation when encoded before data mask is stored in an array module from left to right while the pixels of image showed are from top to bottom, so that the secret image preprocessing as Step 1 can be showed correctly while embedded into the cover QR codes. Step 2 search the QR codes whose regions are the same size or the closest with S from the coordinate of $(7,7)$ to the lower right corner of QR codes from the large datasets of QR codes, denoted $QR_1, QR_2 \ldots QR_N$. In Step 3, select two QR codes from the N QR codes randomly, for each position $(i,j) \in \{(i,j)\|1 \leq i \leq M; 1 \leq j \leq M\}$ repeat Steps 4–5. Let Q_1, Q_2 denote the regions of the two selected QR codes with the same size with S at the same locations from the coordinate of $(7,7)$ to the lower right corner of QR codes, where $Q_1(i,j)$ and $Q_2(i,j)$ indicates the bits corresponding to the two selected QR message in Step 4. As we all known, the codewords of QR codes are divided into a large number of error correction blocks and corresponding error correction codewords are generated for each block. The error correction codewords for each block is given as (c, k, r), r is the error correction capacity which represents the maximum number of codewords that can be modified per block. In Step 5, check $Q_1(i,j)$ and $Q_2(i,j)$ to satisfy any one of the $S(i,j)$ sharing modules or not, if not, we will flip $Q_1(i,j) = \overline{Q}_1(i,j)$ or $Q_2(i,j) = \overline{Q}_2(i,j)$ alternately. If (i,j) belong to the error correction block x of Q_1 or Q_2, we make $count_x$++, which indicates the number of bits need to be changed of the error correction block x. In Step 6, for the error correction blocks, if $count_x > 2r$, it means that the changed codewords are more than the maximum number of codewords that can be modified per block when the errors assigned to the two QR codes averagely.

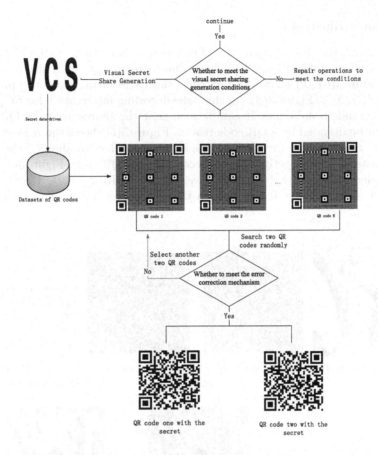

Fig. 2. The QR code shares generation architecture of the proposed scheme

So, it need to go to Step 3 to choose another two QR codes from the N QR codes to repeat Steps 4–5. If $count_x \leq 2r$, output the two QR code shares.

In such a way, the two QR code shares are modified in the range of error correction mechanism, which can be decoded correctly by any standard decoding software. The proposed approach which can satisfy the error correction mechanism in the QR code is applicable to all QR code versions. In addition, the secret can be revealed with the abilities of stacking and XOR decryptions.

4   Experimental Results and Analyses

In the experiments, QR code version 17 with error correction level H, original binary secret image1, with size of 78 × 78 pixels, as shown in Fig. 3(a), to test the efficiency of the proposed scheme.

4.1 Image Illustration

Figure 3 shows the simulation results of the seventeen version, 17-H, QR barcode with 85 × 85 modules. Figure 3(a) shows the secret image, with size of 78 × 78 pixels. Figure 3(b)–(c) show the QR code shares resulting from the proposed scheme, SC_1, SC_2. Figure 3(d)–(e) show the decoding information for SC_1, SC_2, which are random noise-like. It can be seen that the shares are valid QR codes which can be decoded by a QR code reader. Figure 3(f) shows the reconstructed QR code, S_1, which is revealed by stacking two QR code shares. The secret message can be seen directly from the reconstructed QR code. Figure 3(g) shows the revealed QR code, S_2, which is revealed by XORing two QR code shares with light-weight device. It can be seen that the secret image can be recovered lossless.

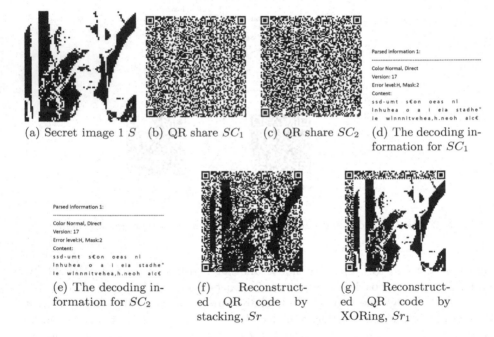

(a) Secret image 1 S (b) QR share SC_1 (c) QR share SC_2 (d) The decoding information for SC_1

(e) The decoding information for SC_2 (f) Reconstructed QR code by stacking, Sr (g) Reconstructed QR code by XORing, Sr_1

Fig. 3. The results of QR code version 17 with error correction level H by the proposed scheme.

4.2 Analysis

The encoded data before data mask is stored in an array module of QR code from left to right and the coordinate of $(0,0)$ represents the top left corner of the array module. The three identical finder patterns which are made up of 7 × 7 modules are used to recognize the QR code and to determine the rotational orientation of the symbol. In order to avoid changing the three patterns, we define the data which is blue regions as shown in Fig. 1 that could be modified is from the coordinate of $(7,7)$ to the lower right corner of QR codes.

In the experiments of QR code version 17 with error correction level H, the data regions which can be modified is a rectangle from the coordinate of $(7,7)$ to $(85,85)$ and The secret image is the same size with the rectangle from the coordinate of $(7,7)$ to $(85,85)$. According to the principle of the error correction mechanism in the QR code, the cover QR code could be decoded correctly if less than r codewords per block contain errors. In the Algorithm 1 above, we can search two QR codes from the large datasets of QR codes, which can satisfy the error correction mechanism when some bits of the two QR codes altered to match the secret sharing modules of $S(i, j)$. In this way, the QR code shares generated according to the proposed scheme as steps above can be recognized correctly by any standard decoding software. The secret image can be recovered by HVS without any computation based on stacking when no light-weight computation device On the other hand, if the light-weight computation device is available, the secret image can be lossless revealed based on XOR operation. Finally, the scheme can be applied to the general threshold.

As a result, in the experiments, through looking up Fig. 3(a), the following conclusions are obtained:

1. The secret message embedded into QR codes would be larger than the previous schemes with secret data-driven based on carrier-free secret sharing, which is the first investigated.
2. The QR code shares can be scanned and decoded correctly by any standard decoding software.
3. The secret image can be recovered by stacking two shares based on the human visual system without any computation and QR code readers. In addition, the secret image can also be lossless revealed based on XOR operation with light-weight computation device.
4. The proposed scheme can reach $(2,2)$ VSS threshold with the abilities of stacking and XOR lossless recovery.

4.3 Compared with Related Schemes

Recently, as the most similar and latest scheme with ours is only Wan et al.'s scheme [8], so we compare the proposed scheme with Wan et al.'s scheme in this section.

Wan et al. [8] proposed a VSS scheme for (k, n) threshold based on QR code with multiple decryptions. The proposed scheme based on QR codes can visually reveal secret image with the abilities of stacking and XOR decryptions as well as scan every shadow image, i.e., a QR code, by a QR code reader. Nevertheless, the embedding capacity of secret image is low. Figure 4 shows the results of QR code version 17 with error correction level H by Wan et al.'s scheme [8] proposed.

Compared with Wan et al.'s scheme, the core idea of our scheme is to search the two QR codes that altered to satisfy the secret sharing modules in the error correction mechanism from the large datasets of QR codes according to the secret image, which is to embed the secret image into QR codes based on carrier-free secret sharing. In this way, the secret image can be the maximum which is the same size with the blue regions that can be modified as shown in Fig. 1. Figure 3(a) show the secret image 1, with size of 78×78 pixels, Fig. 4(a) show the secret image 2, with size of 60×60 pixels. It can be seen that the rate of embedding capacity increased by 30% based on our scheme.

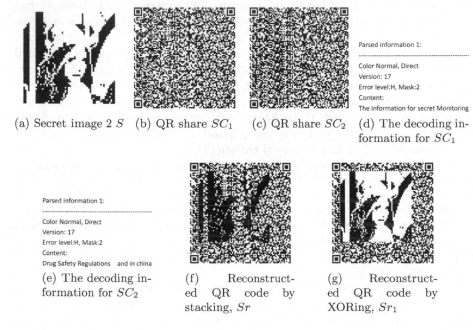

(a) Secret image 2 S (b) QR share SC_1 (c) QR share SC_2 (d) The decoding information for SC_1

Parsed information 1:

Color Normal, Direct
Version: 17
Error level:H, Mask:2
Content:
The information for secret Monitoring

Parsed information 1:

Color Normal, Direct
Version: 17
Error level:H, Mask:2
Content:
Drug Safety Regulations   and in china

(e) The decoding information for SC_2 (f)   Reconstructed QR code by stacking, Sr (g)   Reconstructed QR code by XORing, Sr_1

Fig. 4. The results of QR code version 17 with error correction level H by $(2,2)$ VSSQR proposed scheme.

5   Conclusion and Future Work

This paper presents a novel secret data-driven carrier-free (semi structural formula) visual secret sharing (VSS) scheme with $(2,2)$ threshold based on the error correction blocks of QR codes. The proposed scheme exploits the error correction capacity of each block in the QR code structure, to search two QR codes that altered to satisfy the secret sharing modules from the large datasets of QR codes according to the secret image, which is to embed the secret image into QR codes based on carrier-free secret sharing. In this way, the secret message embedded into QR codes can be larger than the previous schemes based on our scheme. Each share is a valid QR code which contains original information when scanned. So, it reduces the likelihood of attracting the attention of potential attackers if distributed via public channels. The secret image can be recovered by stacking QR code shares without any computation and can be also lossless revealed based on XOR operation if light-weight device is available. As the proposed scheme has the abilities of stacking and XOR operation, the proposed scheme could be applied in different scenarios whether the light-weight device is available or not. Experiments are conducted to show the efficiency of the proposed scheme.

References

1. Beimel, A.: Secret-sharing schemes: a survey. In: International Workshop, pp. 11–46 (2011)
2. Naor, M., Shamir, A.: Visual cryptography. In: Santis, A. (ed.) EUROCRYPT 1994. LNCS, vol. 950, pp. 1–12. Springer, Heidelberg (1995). doi:10.1007/BFb0053419
3. Wang, D., Zhang, L., Ma, N., Li, X.: Two secret sharing schemes based on boolean operations. Pattern Recognit. **40**(10), 2776–2785 (2007)
4. Weir, J., Yan, W.Q.: A comprehensive study of visual cryptography. In: Shi, Y.Q. (ed.) Transactions on Data Hiding and Multimedia Security V. LNCS, vol. 6010, pp. 70–105. Springer, Heidelberg (2010). doi:10.1007/978-3-642-14298-7_5
5. Yan, X., Wang, S., El-Latif, A.A.A., Niu, X.: Visual secret sharing based on random grids with abilities of AND and XOR lossless recovery. Multimed. Tools and Appl. **74**(9), 3231–3252 (2015)
6. Jtc1/Sc, I.: Information technology - automatic identification and data capture techniques - QR code 2005 bar code symbology specification (2006)
7. Chow, Y.-W., Susilo, W., Yang, G., Phillips, J.G., Pranata, I., Barmawi, A.M.: Exploiting the error correction mechanism in QR codes for secret sharing. In: Liu, J.K.K., Steinfeld, R. (eds.) ACISP 2016. LNCS, vol. 9722, pp. 409–425. Springer, Cham (2016). doi:10.1007/978-3-319-40253-6_25
8. Wan, S., Lu, Y., Yan, X., Wang, Y., Chang, C.: Visual secret sharing scheme for (k, n) threshold based on QR code with multiple decryptions. J. Real-Time Image Proc. 1–16 (2017)
9. Yan, X., Lu, Y., Huang, H., Liu, L., Wan, S.: Clarity corresponding to contrast in visual cryptography. In: Che, W., et al. (eds.) ICYCSEE 2016. CCIS, vol. 623, pp. 249–257. Springer, Singapore (2016). doi:10.1007/978-981-10-2053-7_23
10. Denso, W.I.: (2002). http://www.qrcode.com
11. Yan, X., Guan, S., Niu, X.: Research on the capacity of error-correcting codes-based information hiding. In: International Conference on Intelligent Information Hiding and Multimedia Signal Processing, IIHMSP 2008, pp. 1158–1161 (2008)

Template Protection Based on Chaotic Map for Face Recognition

Jinjin Dong[1], Xiao Meng[1], Meng Chen[1], Zhifang Wang[1(✉)],
and Linlin Tang[2]

[1] Department of Electronic Engineering,
Heilongjiang University, Harbin, Heilongjiang, China
xiaofang_hq@126.com
[2] Harbin Institute of Technology Shenzhen Graduate School,
Xili, Shenzhen, China

Abstract. With the widespread deployment of biometric recognition, personal data security and privacy are attracted more and more attentions. A crucial privacy issue is how to ensure the security of user template. This paper proposes a novel template protection algorithm for face recognition based on chaotic map. Each face template is corresponding to different chaotic sequence produced by system master key and user identification number. The order of chaotic sequence controls the substitution index of face template. Experiment results on facial FERET database show that our algorithm can significantly improve the recognition performance and ensure the security of face template.

Keywords: Template protection · Face recognition · Logistic map · Substitution index

1 Introduction

Biometric refers to identify or verify a person according to their physiological or behavioral characteristics such as fingerprint and handwriting. It is an automated process of identification and authentication, in which specific biometric traits of an individual are extracted during enrollment and stored as biometric templates. However, such authentication technology needs large-scale capture and storage of biometric data which leads to serious concern about leaking of privacy and identity theft. Unlike the traditional identification technology such as password or credit card, biometric characteristics are inherent to a person. Once the template is compromised, it is compromised forever because it cannot be revoked, reissued or even destroyed. Furthermore, since the same template might be used for various application and location, it would be possible to perform cross matching between them. In this way, the privacy of the user cannot be guaranteed.

Many template protection algorithms have been proposed. Jain divided these algorithms into two kinds: feature transformation and biometric encryption [1]. The former is a popular scheme in which the same transformation function respectively applies to register biometric characteristics and testing biometric characteristics in enrollment procedure and testing procedure. Then the transformed testing

© Springer Nature Singapore Pte Ltd. 2017
B. Zou et al. (Eds.): ICPCSEE 2017, Part I, CCIS 727, pp. 242–250, 2017.
DOI: 10.1007/978-981-10-6385-5_21

characteristics directly compares with the transformed template generated by register characteristics. The latter is firstly proposed to use biometric feature to encrypt key. It is deferent from feature transformation is that public information related to the biometric feature is stored in the database. The former security depends on the key safety and noninvertible transforms. And the latter depends on the safety of help data. This paper proposes a face template protection method based on chaotic map which belongs to the former.

Due to the high sensitivity of chaotic systems to parameters and initial conditions as well as the availability of many circuit realizations, chaos based algorithms are developed and studied as the core of encryption algorithms [2]. Chaotic image encryption refers to use discrete chaotic sequence to encrypt the image, and its essence is to play the characteristics of chaos to conceal the original face image and avoid the valuable information being achieved by other people even they get encrypted information [3, 4].

In this paper, an eigenvalues permutation algorithm for face template protection based on chaotic map is proposed. The original face template is extracted using principal component analysis (PCA) algorithm and disturbed based on chaotic logistic map. The rest of the paper is organized as follows: Sect. 2 introduces the existing biometric protection methods. Section 3 gives the proposed algorithm. Section 4 is devoted to the experiments and security analysis. At last, Sect. 5 concludes this paper.

2 Related Works

2.1 Biometric Protection Method

In recent years, there are increasing applications of biometric identification technology in the identity authentication, but also gradually expose its inherent weaknesses in some aspects of security and privacy, so higher requirements of the security of biometric template in the practical applications is raised. The template protection scheme should have properties like diversity, revocability, security and performance. The template protection schemes are broadly classified into two main types as function transformation approach and biometric cryptosystem based approach. Function transformation approach is further categorized as salting and non-invertible transformation method. Biometric cryptosystem based approach is further categorized as key binding and key generation methods.

A large number of famous biometric cryptosystems have been proposed such as fuzzy vault scheme [5–8], fuzzy commitment scheme [9], and fuzzy extractor [10]. Fuzzy vault is used to encrypt the biometric template which is described in the form of point sets, such as fingerprint minutiae sets. The fuzzy vault scheme proposed by Juels and Sudan [5–7] has become one of the most popular key-binding approaches as it provides effective and provable security for biometric template protection [11]. Since the fuzzy vault scheme is proposed, many biometric characteristics have been used to construct biometric cryptosystems based on fuzzy vault scheme, such as fingerprints, ear, and face.

Fuzzy commitment and fuzzy extractor are among the two most popular template protection schemes. Fuzzy commitment binds a binary key to a binary biometric representation and the key can only be recovered if a similar binary biometric re-presentation is presented. Fuzzy extractor directly transforms a binary biometric input into a stable binary string that can be used as encryption keys in cryptographic applications. Both these schemes take an ordered multi-biometric representation as input in our context. When modalities that are represented by un-ordered features are fused with modalities that are represented by ordered features, an unordered-to-ordered feature transformation is required.

3 Proposed Algorithm

A face recognition system firstly collects several face images of each legitimate user and extracts face features stored in the database corresponding with this user's ID number. One of the popular approaches for face recognition is eigenface method (Principal Component Analysis, PCA) [12]. The key idea is to find the best set of projection directions to span the feature space that will maximize the total scatter. Our proposed algorithm adopts PCA to produce face features for each user. So each user's face feature is a vector. Then we use image encryption for reference to displace the order of face feature vector by utilizing chaotic map. This section describes the generalized logistic map and the detail steps of our algorithm.

3.1 Logistic Map

The logistic map is an one-dimensional map and it is very simple and can produce fundamental results on non-linear dynamics, and it attracts many attention in image encryption lately. A simple chaotic logistic map is defined by Eq. (1):

$$x_{n+1} = \mu x_n (1 - x_n) \tag{1}$$

Where n is a non-negative integer, x_0 is the initial value of the logistic map, $x_0 \in [0, 1]$, μ is a control parameter. For a fixed value μ, we can get a certain sequence x_n by iteration with an initial value x_0.

Figure 1 shows the bifurcation diagram of the logistic map while μ belongs to (3, 4]. From the bifurcation diagram, we can see that the sequences generated by logistic map are chaotic sequences when $3.57 < \mu \le 4$.

Figure 2 shows the Lyapunov exponent curve of logistic and generalized logistic maps. Generally, the Lyapunov exponent is usually taken as an indication that the system is chaotic. If the value of Lyapunov exponent is positive number, the system is chaotic. It is noted that the logistic map has some drawbacks such as non-uniform behaviour and blank windows in the chaotic region as can be seen in Fig. 2(a), there are some areas where the Lyapunov exponent is either zero or negative.

To overcome the issue of the logistic map, we introduce a generalized logistic map [13] defined by Eq. (2)

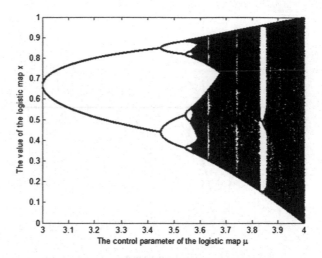

Fig. 1. Bifurcation diagram of the logistic map

$$x_{n+1} = \frac{4\mu^2 x_n (1 - x_n)}{1 + 4(\mu^2 - 1)x_n(1 - x_n)} \tag{2}$$

Where n is a non-negative integer, $-4 < \mu \leq 4$. The logistic map is in the chaotic region when $-4 < \mu \leq -2$ or $2 < \mu \leq 4$. Figure 2(b) shows the Lyapunov exponent of the generalized logistic map. We can see that there is no non-chaotic area in the chaotic region, and the uniform behavior of the generalized map is further proved.

The logistic map has been applied to image encryption since it satisfies the ergodicity, pseudorandom property and extreme sensitive to initial conditions and system parameters.

3.2 Our Algorithm

In this section, we presented our template protection algorithm for face recognition using a position permutation approach for face feature with the chaotic logistic sequences. Each user has its own specific ID, so we use the user ID and the master key of the system to determine the initial conditions of logistic map by hash function. The order of chaotic sequence controls the substitution index of face template. Figure 3 shows the generation process of the face template.

Figure 4 shows an example of the detailed generation flow of the substitution index. For example, the initial value of the logistic map is 0.9, the control parameter of the logistic map is 4, and we choose 10 numbers from the logistic sequence as an example to show the process of the permutation. The method is described in steps as follows:

(1) Encode the master key and user ID to a vector satisfied the input request of hash function, the output of hash function is taken as the initial value of logistic map. Then select an appropriate control parameter through Fig. 2(b), the chaotic

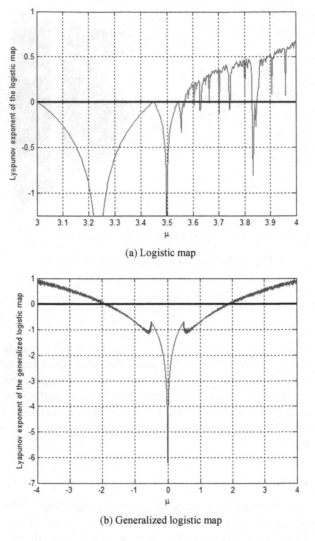

(a) Logistic map

(b) Generalized logistic map

Fig. 2. Lyapunov exponent curves of logistic and generalized logistic maps

logistic sequence is produced. Because different user has different ID, the initial
conditions of the chaos are not the same, so the chaotic sequences are different.

(2) Let $V = \{v_1, v_2, \ldots, v_n\}$ be a face feature of the user where n is the dimensionality
of eigenvectors, the dimension of chaotic sequence is often great than n. We select
the former n dimension of the logistic sequence gained by step (1) as the final
used chaotic sequence $L = \{l_1, l_2, \ldots, l_n\}$.

(3) Sort $L = \{l_1, l_2, \ldots, l_n\}$ in ascending order of size and obtain the new sequence
$L' = \{l'_1, l'_2, \ldots, l'_n\}$. Meanwhile, the substitution index $S = \{s_1, s_2, \ldots, s_n\}$ is
produced where s_i is the position in L of l'_i. In Fig. 4, 0.9 is the first dimension in
L, the corresponding index is seven.

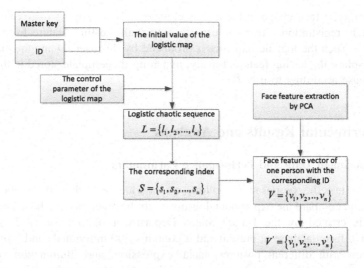

Fig. 3. The generation process of the face template

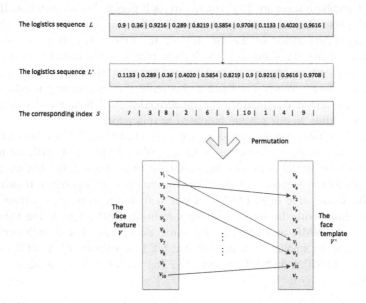

Fig. 4. An example of the detailed generation flow of the substitution index

(4) $V = \{v_1, v_2, .., v_n\}$ is replaced with $V' = \{v'_1, v'_2, \ldots, v'_n\}$ by the same rule of permutation, because the elements of the sequence $V = \{v_1, v_2, .., v_n\}$ and the sequence $L = \{l_1, l_2, \ldots, l_n\}$ are corresponding one by one. $V' = \{v'_1, v'_2, \ldots, v'_n\}$ is the face template.

(5) All the face features of every user are carried out the above four steps to get a new face template.

(6) In testing, a face image of the user is collected. The mapping matrix attained by PCA in registration process is used to extract the testing feature from the face image. Then the logistic sequence is produced by this user ID and the master key to displace the testing feature. Finally, matching the template stored in the system database according to user ID.

4   Experimental Results and Analysis

4.1   Experiment Data and Performance Parameters

In order to test the proposed algorithm, we conduct experiments on the FERET database, and compare the experimental results with the eigenface method. The FERET database is created by the United States Department of Defense, it is the most authoritative face database at present and it contains 200 individuals and 7 images for each person with different postures, facial expression, and illumination condition. Three samples of each person are used as training set, others are used for test.

The original template is the facial feature vector extracted using PCA algorithm. In the FERET database there are 200 persons, so each person has its own user ID, which means each person has its own corresponding logistic map since the initial value of the logistic map is determined by the user ID and the master key. Here we choose 200 logistic map, the initial conditions of the logistic map is determined by the user ID and the master key with an appropriate transformed function, the control parameter of the logistic map is fixed, we set $\mu = 4$ and at this point the logistic map is entirely in the state of chaos, and it can increase the randomness. And then the new template is generated according to our scheme proposed in this paper.

In our algorithm, we use False match rate (FMR) and False non-match rate (FNMR) to evaluate the performance of the algorithm. FMR and FNMR are the major two parameters of recognition algorithm performance evaluation. In addition, the equal error rate (EER) can measure the overall performance of an algorithm. It unifies FMR and FNMR at the same time. FMR increases with the increase of the threshold and FNMR decreases with the increase of the threshold. EER refers to the value of the intersection of FNMR and FMR in the same coordinate. For a high-performance algorithm, there is a smaller value of EER. The DET curve is similar to EER, x axis and y axis represent FMR and FNMR respectively. The Lower DET curve means the higher performance.

4.2   Experimental Results and Security Analysis

Figure 5 shows the DET curve with eigenface and the proposed algorithm. Experimental results show that the ERR of eigenface and the proposed algorithm are 14.2% and 8% respectively. From the graph, we can see more clearly and directly that the performance of our proposed algorithm using the logistic map is better than the original PCA algorithm.

In system, we store four data: the mapping matrix of PCA, user ID, the control parameter of logistic map and the face templates. The security of our algorithm depends

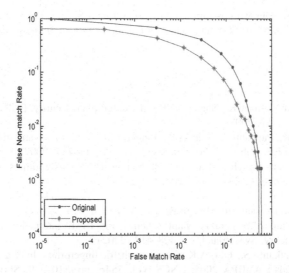

Fig. 5. The comparison of DET curve

on the logistic map. If the attacker doesn't know the logistic sequence, he can't recover the original face feature. The control parameter can't produce the logistic map without the initial condition. The master key and user ID conduct the initial condition by hash function. The irreversibility of hash ensures the security of the master key. The master key is the superlative key of the system, and it is used to protect the primary key and the encryption key. Generally speaking, it is very difficult to attack the master key directly, because the master key is usually stored in a dedicated cryptographic device, it is secure not only physically, but also logically. The master key is only mastered by a few security managers, these security managers are not directly contact to the user's key and plaintext data. And it is not possible for the user to obtain the master key that the security manager has, which is helpful to ensure the security of the key.

In our scheme, we can generate a new feature template by changing the control parameters of chaotic map or changing the master key. In this way, if the original template is destroyed or stolen, you can change the relevant parameters to generate a new template to replace the original template.

5 Conclusion

In this paper, we presented a template protection algorithm for face recognition using chaotic map. We use the properties of the logistic map, such as the ergodicity, pseudorandom property and extreme sensitive to initial conditions and system parameters to generate a chaotic sequence, and the original face feature can be displaced by the chaotic sequences. Since the representation of disturbed template features is the same as before, we can use the same matching scheme to measure the performance of the face recognition system. Experiment results show that our algorithm significantly improve the recognition performance and ensure the security of face template.

Acknowledgments. This work is supported by the National Science Foundation of China (nos. 61201399, 61501176, 61601174), and Startup Fund for Doctor of Heilongjiang University.

References

1. Jain, A.K., Nandakumar, K., Nagar, A.: Biometric template security. EURASIP J. Adv. Sig. Process. **2008**, 1–17 (2008)
2. Nandakumar, K., Pankanti, S., Jain, A.K.: Fingerprint-based fuzzy vault implementation and performance. IEEE Trans. Inf. Forensics Secur. **2**(4), 744–757 (2007)
3. Kocarev, L., Lian, S.: Chaos-Based Cryptography Theory. Algorithms and Applications. Springer, Heidelberg (2011). doi:10.1007/978-3-642-20542-2
4. Abd-El-Hafiz, S.K., Radwan, A.G., AbdElHaleem, S.H., Barakat, M.L.: A fractal-based image encryption system. IET Image Process **8**(12), 742–752 (2014)
5. Juels, A., Sudan, M.: A fuzzy vault scheme. In: IEEE International Symposium on Information Theory, Switzerland, pp. 408–413 (2002)
6. Uludag, U., Pankanti, S., Jain, A.K.: Fuzzy vault for fingerprints. In: Kanade, T., Jain, A., Ratha, N.K. (eds.) AVBPA 2005. LNCS, vol. 3546, pp. 310–319. Springer, Heidelberg (2005). doi:10.1007/11527923_32
7. Barakat, M.L., Mansingka, A.S., Radwan, A.G., Salama, K.N.: Hardware stream cipher with controllable chaos generator for colour image encryption. IET Image Process **8**(1), 33–43 (2014)
8. You, L., Wang, Y., Chen, Y., Deng, Q., Zhang, H.: A novel key sharing fuzzy vault scheme. KSII Trans. Internet Inf. Syst. **10**, 453–460 (2012)
9. Lafkin, M., Mikram, M., Ghouzali, S.: Biometric cryptosystems based fuzzy commitment scheme: a security evaluation. Int. Arab J. Inf. Technol. **13**(4), 443–449 (2016)
10. Dodis, Y., Ostrovsky, R., Reyzin, L., Smith, A.: Fuzzy extractors: how to generate strong keys from biometrics and other noisy data. SIAM J. Comput. **38**(1), 97–139 (2008)
11. Nagar, A., Nandakumar, K., Jain, A.K.: Multibiometric cryptosystems based on feature level fusion. IEEE Trans. Inf. Forensics Secur. **7**(1), 255–268 (2012)
12. Turk, M., Pentland, A.: Face recognition using eigenfaces. In: IEEE Transaction on Pattern Analysis and Machine Intelligence, pp. 586–591 (1991)
13. Song, X., Wang, S., El-Latif, A.A.A., Niu, X.: Quantum image encryption based on restricted geometric and color transformations. Quantum Inf. Process. **13**(8), 1765–1787 (2014)

A Fast and Secure Transmission Method Based on Optocoupler for Mobile Storage

Lu Zou[1], Dejun Zhang[1(✉)], Fazhi He[2], and Zhuyang Xie[1]

[1] College of Information and Engineering,
Sichuan Agricultural University, Yaan 625014, China
djz@sicau.edu.cn
[2] School of Computer Science, Wuhan University, Wuhan 430072, China

Abstract. This paper presents a one-way data transmission method in order to ensure the safety of data transmission from mobile storage to secure PC. First, an optocoupler is used to achieve the one-way transmission of physical channel, so that data can only be transmitted from mobile storage to secure PC, while the opposite direction is no physical channel. Then, a safe and reliable software system is designed which contains one-way communication protocol, fast CRC check method and packet retransmission algorithm together to ensure the safety of data transmission. After that, to obtain the maximum transmission rate, the frequency of data bus($slwr$) and the packet size(num) which effect on transmission rate are detailed analyzed. Experimental results show the proposed method is high-efficiency and safe.

Keywords: Data security · Mobile storage · Protocol · CRC check

1 Introduction

With the advent of global information and digital era, mobile storage technology has penetrated into every corner of the society, greatly improving the efficiency of people's work. However, the leakage events caused by the use of mobile storage [1] have posed a serious challenge to the national security system and corporate security system, which makes state secrets, financial information, business production and operation face a huge threat [2]. What's more, the recent outbreak of the Symantec event and PRISM [3] even exacerbate people's concerns about the safety of PC information. Although leakage events can be temporarily resolved by firewalls, proxy servers, intrusion detection or other security measures, it still cannot meet the requirements of government and sensitive departments for information security. In departments with high confidentiality requirements, the mobile storage even cannot be directly inserted to secure PC [4].

How to realize that not only can the data be transmitted from mobile storage to secure PC, but also prevent the data in the secure PC leaked to mobile storage. Hence, one-way transmission technology is one of the research hotspots in the field of information security in the world. To solve the leakage problems

© Springer Nature Singapore Pte Ltd. 2017
B. Zou et al. (Eds.): ICPCSEE 2017, Part I, CCIS 727, pp. 251–261, 2017.
DOI: 10.1007/978-981-10-6385-5_22

and ensure the safety of secure PC, this paper presents a one-way transmission method based on optocoupler for mobile storage to access to secure PC. In this way, data from mobile storage can be transmitted into secure PC, but that in secure PC cannot be leaked out.

The proposed one-way transmission system is efficient and safe with double guarantee. To achieve the goal, at hardware-level, the characteristics of optocoupler is well used to achieve the one-way physical channel. At software-level, a unique program flow and a new communication protocol is designed, which guarantees device at the bottom only send data out, so as to ensure the transmission between mobile storage device and secure PC is one-way, safe and fast.

2   Previous Work

At present, the main researches about one-way transmission can be classified into: software solution [5] and hardware solution [6–9].

At software-level, Kang et al. proposed pump technology [5], which uses a reverse validation mechanism to prevent the flow of data from the outside to the inside. The technology is one-way, but the corresponding protocol is bi-way, once the protocol exists loopholes, it can be used to fetch the data in secure PC. Wan et al. [7] developed an optical shutter data protocol suitable for one-way communication, of which the protocol is similar to TCP, but there is no connection, no cache and no variable distribution consumption as TCP in the establishment of the connection requires for. And in this way, data sending is direct and more efficient. Zhao [8] analyzed the shortcomings of one-way transmission device, and put forward a more secure and reliable solution which is mainly realized by modifying UDP.

In recent years, with the rapid development of security attack technology, software-level isolation is far from meeting the requirements of confidential information security, and hardware-level isolation with corresponding device came into being. It's said that hardware-level isolation can guarantee the unilateralism of data transmission physically [9].

At hardware-level, Li [10] proposed a new secure data transmission method based on the traditional isolation card which is called SGAP. Xiao et al. [11] integrated USB controller and SPI controller of MCU to realize one-way transmission in hardware-layer and firmware layer. Wang and Meng [12] proposed a fiber-based one-way data transmission system of which the receiving end is connected to mobile storage, while the sending end is connected to PC. This system makes full use of a dedicated hardware device whose sending end and receiving end are composed of optical fiber connection.

Most of the transmission methods above are "blindly sending", which means the one is only for sending while the other is only for receiving, and whether the received data is incorrect or complete is not concerned. Because of there is no interactive control protocol, it always leads to error seriously. Furthermore, the use of dedicated device (e.g. fiber optic transceivers) is costly and usually limited by the lack of flexibility. Apart from this, infrared communication technology is low transmission rate of which the highest transmission rate is 4 Mbps that is far from meeting the requirements of high-speed data transmission.

The proposed method in this paper is not only inexpensive, but also can ensure the safety of transmission between mobile storage and secure PC. In our method, an optocoupler is used as the one-way transmission medium, and then a safe transmission mechanism by software programming is designed to ensure the security and reliability of the data in the transmission channel.

3 Hardware Scheme and Overview

The one-way data transmission system aims at achieving a one-way data flow, specifically, it attempts to send data from the mobile storage to the secure PC. The overall framework of hardware component is shown in Fig. 1, which is divided into three parts: master controller, one-way physical channel and USB Device. (a) The master controller (ARM Cortex-A5) provides the USB host for reading the files of mobile storage. (b) The one-way physical channel is used to connect the master controller and USB Device. (c) The USB Device takes CY7C68013A as the core and it communicates with secure PC via USB bus.

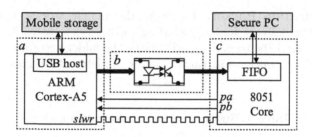

Fig. 1. The overall framework of hardware system.

From Fig. 1, both the USB host and the FIFO are set as input mode only which guarantees data security from hardware-level. In addition, an optocoupler is innovatively adopted in the one-way physical channel, which guarantees the data can only be transmitted from master controller to USB Device.

Data flow of the hardware-level is described as follows: (1) Since mobile storage has been inserted into USB host, the master controller reads basic information of the mobile storage. (2) Secure PC sends the transmission command to 8051 core via USB bus. (3) 8051 core parses the received command and triggers pa. (4) The master controller launches the one-way transmission procedure when it detects pa signal. (5) The one-way transmission procedure creates a transmission queue according to the file information of mobile storage. (6) The data packet is sent to optocoupler by master controller, and then passes the FIFO, eventually reaches to secure PC. (7) Secure PC parses the received packet, then recoveries the data packet.

4   One-Way Communication Protocol

Based on the hardware scheme, it is necessary to design a software system according to the one-way data transmission demand. Therefore, communication protocol for one-way transmission is the key and difficulty of this paper. UDP [13] based on the broadcast transmission, does not need to receive any confirmation message from the receiving end and supports for one-way transmission. However, UDP is generally used in the case where the amount of data for transmission is not large and the requirements of reliability is not high. Faced with large-capacity mobile storage, the problems of reliability must be solved. Therefore, this paper presents a new one-way communication protocol.

In the proposed protocol, the data packet required to send are defined as three types: *folder information packet, file information packet* and *file block packet*. In the one-way transmission process, each packet in the transmission channel is sent according to the order of *folder information packet, file information packet, file block packet 1, file block packet 2, · · · , file block packet N*. The packet in the one-way transmission channel is composed of data block and CRC block.

4.1   Packet Format

Folder Information Packet. Folder in the mobile storage carries various attributes of the delivery folder, including packet type, operation permission, pathname, etc. And the receiving end creates the folder according to the structure of the *folder information packet*. The complete information as shown in Table 1.

Table 1. Structure of folder information packet

Variable name	Variable type	Field description
PackType	int	Packet type
DesNum	int	File index
Mode	int	Operation permission
FoLength	int	Pathname length
Folder[*num*]	unsigned char	Pathname

File Information Packet. *File information packet* is the starting packet which notifies the receiving end of the starting of a new transmission process and it carries various attributes (packet type, file size, access time, etc.). The receiving end checks the validity of the packet and recoveries the file according to the attributes. The complete information as shown in Table 2.

Table 2. Structure of file information packet

Variable name	Variable type	Field description
PackType	int	Packet type
DesNum	int	File index
FileSize	int	File size
Access	time_t	Access time
Modify	time_t	Modify time
Mode	int	Operation permission
FiLength	int	Filename length
File[num]	unsigned char	Filename

File Block Information Packet. When one file transmits, it is divided into several blocks by a predetermined value (*num*). As a result, a specific packet is responsible for sending a specific block of file. *File block information packet* is composed of packet type, block length, block contents these three parts. The complete information as shown in Table 3.

Table 3. Structure of file block information packet

Variable name	Variable type	Field description
PackType	int	Packet type
DaLength	int	Block length
Data[num]	unsigned char	Block contents

4.2   16-Bit CRC Check

Due to the use of block strategy, block packet may loss at the receiving end, and it's difficult to be reorganized into a complete original file. Thus, append check information to the end of folder information packet, file information data packet and file block information packet is necessary. The CRC check theory [14] is employed in this paper, according to the content of the packet (n-bit), a 16-bit check redundancy code is generated, which is appended to the packet structure to form a new packet ($n + 16$ bit). The procedure of generating 16-bit check redundancy code is as follows.

The n-bit packet $D(X)$ to be sent is shifted 16-bit to the left and then divided by the polynomial $G(X)$, and the resulting remainder is the 16-bit check redundancy code. As shown in Eq. (1), which, $Q(X)$ is an integer.

$$\frac{D(X) \cdot 2^{16}}{G(X)} = Q(X) + \frac{R(X)}{G(X)} \tag{1}$$

The Mod-2 addition and subtraction used in Eq. (1) is bitwise with carry and borrow free. In fact, it is logical exclusive-OR operation. The multiplication and division operations conforms to the same rule. Thus, the polynomial of the 16-bit check redundancy code is shown below:

$$G(X) = X^{16} + X^{12} + X^5 + 1. \tag{2}$$

Ultimately, the received packet will be checked according to the rules of CRC at secure PC. Specifically, the received packet (including the packet information and CRC information) is divided by the polynomial. If the remainder is 0, it means there is no error, otherwise, there is error.

5   Structure of System

The one-way data transmission system is composed of bottom device and secure PC, as shown in Fig. 2, the bottom device is responsible for sending while the secure PC is responsible for receiving. These two components together to achieve the one-way transmission.

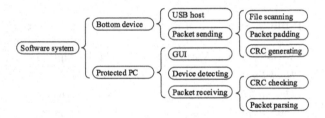

Fig. 2. The structure of software system.

The bottom device is divided into USB host module and packet sending module, and packet sending module consists of file scanning module, packet padding module, and CRC generating module. The USB host monitors the state of the USB host of the master controller in real-time, and when the state of the USB host is MOUNTED (see in Sect. 5.1), the packet sending module runs.

The secure PC is divided into GUI, device detecting module and packet receiving module, and packet receiving module consists of CRC checking module and packet parsing module. The device detecting module is for detecting whether the bottom device is connected or whether the optocoupler is normal.

5.1   USB Host

The status of USB host are described as three states: (a) no mobile storage inserted, denoted as DISKOUT, (b) mobile storage inserted but not mounted; in this situation, the contents of mobile storage cannot be read by OS(Ubuntu

14.04); denoted as IN, (c) mobile storage inserted and mounted; in this situation, the contents of mobile storage can be read; denoted as MOUNTED.

The USB host runs automatically with OS until the bottom device is powered off. Figure 3 shows the flow chart based on the three states of the USB host. The steps as follows: (1) Detecting the status of USB host. (2) If mobile storage is detected, mount it to the directory /mnt/usb in OS, otherwise, continue to detect. (3) Sending 2048 bytes of zero to test the physical channel and check whether there is hardware (optocoupler) failures, according to the status of pa. (4) Starting the one-way data sending process. (5) Unmount mobile storage, return to step (1).

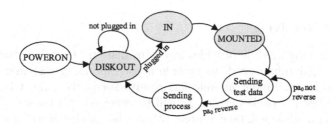

Fig. 3. The state machine of USB host.

5.2 Packet Sending

The packet sending module performs two deep recursive scans on the directory /mnt/usb. The first scan records the total number of leaf nodes whose type value is the initial value of $DesNum$ in the packets of $FOLDER$ or $FILE$. The second scan reads the leaf node and encapsulates it according to the three packet structures (*file information packet, folder information packet*, and *file block information packet*), and then appends the 16-bit CRC code to the packet to send. After each $FOLDER$ or $FILE$ has been transmitted, then $DesNum$ is decremented by one to zero. Since the communication protocol proposed in this paper is unreliable, a retransmission algorithm is designed to ensure the packet can be retransmitted in time. The retransmission algorithm is shown as below, when one packet is sent, the status of pa and pb is read and compared with sa and sb, respectively. If pa is reversed, it indicates that the secure PC has received correct packet. If pb is reversed, it indicates that the secure PC has received incorrect packet, and the retransmission algorithm is executing.

```
int sa=1,sb=1;
for(i=1;i<N;i++)
{
    SendingPacket(i);
    while(1)
    {
        if(pa!=sa)
```

```
        {
            sa=1-sa;
            SendingPacket(i);
        }
        elseif(pb!=sb)
        {
            sb=1-sb;
            break;
        }
    }
}
```

5.3 Packet Receiving

The packet sending module provides a retransmission algorithm, therefore, the packet receiving module needs to provide a corresponding response algorithm. When the packet receiving module has received a packet, the CRC checking module executes: (1) packet errors, *pa* will be reversed; (2) packet corrects, the packet parsing module will be executed; (3) another packet is waiting to receive, *pb* will be reversed.

The steps of packet parsing as follows: (1) identify the type of the packet. (2) type value is *FILE*, parse the structure of *file information packet* and create a new file; type value is *FOLDER*, parse the structure of *folder information packet* and create a new folder; type value is *DATA*, parse the structure of *file block information packet* and write the binary stream into the created file. (3) since the file is completed recovered, the write stream is closed.

6 Implementation and Performance

In this section, to further analyze various characteristics of our method, we discussed the effect of *slwr* signal frequency and *num* value on the transmission rate. The program of bottom device was implemented in C and performed on Raspberry Pi 3 Model B, running Ubuntu14.04. The program of secure PC was implemented in C# 4.0 and performed on a computer with Inter (R) core (TM) i7-4470 (3.4 GHz) and 8 GB of Main Memory, running Win7.

The master controller (ARM Cortex-A53), as shown in Fig. 1, provides *slwr* signal which generates write clock to (falling-edge enable) the FIFO of USB Device [15]. Therefore, the *slwr* signal indirectly affects the transmission rate. In its best case, the shorter the period of *slwr* is, the faster the transmission rate is. However, in practical applications, when the main controller generates excessive frequency of GPIO, packet loss is seriously due to the problem of soft delay in Ubuntu, which affects the transmission rate of the system.

The relationship between *slwr* and transmission rate is shown in Fig. 4. The transmission rate of the system is improved with the increasing of the period of

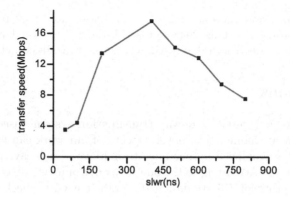

Fig. 4. The relationship between *slwr* and transmission rate.

slwr, and the maximum transmission rate is obtained when *slwr* is increased to 2.5 MHz. Moreover, with a higher period of *slwr*, the transmission rate is decreased.

Apart from the period of *slwr*, the size of the packet also plays a critical role in the transmission rate. In its best case, when there is no packet loss, *num* is not the bigger the better. For example, when mobile storage contains files of different sizes, set *num* as 100 Mbytes or larger, a large number of invalid data will be filled, which seriously affects the transmission rate of the system. Meanwhile, when there is packet loss, *num* is not the smaller the better. For example, set *num* as 256 bytes, as for the identical file, the rate of packet loss will be great with the increasing of *num*, which effectively reduce the rate of transmission loss of the whole system.

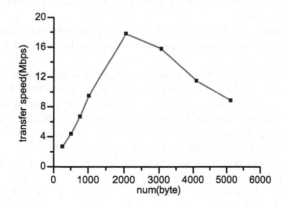

Fig. 5. The relationship between *num* and transmission rate.

The relationship between *num* and transmission rate (set *slwr* = 2.5 MHz) is shown in Fig. 5. The transmission rate of the system is improved with the increasing of num, and the maximum transmission rate of up to 17.8 Mbps is obtained when num is set as 2048 bytes. Moreover, with a larger num, the transmission rate is decreased due to the waste of transmission channel.

7 Conclusions

In this paper, we proposed a one-way transmission system based on optocoupler as the physical channel. The novel aspects of our work can be summarized as follows: (1) We employed optocoupler to achieve a one-way transmission of physical channel. (2) According to the one-way environment, we construct a new communication protocol. (3) We designed an efficient CRC check method and a reliable retransmission algorithm. When packet transmission failures or packet losses, the algorithm ensures the reliability of transmission. (4) We improved the transmission rate up to 17.5 Mbps, basically meet the requirements of using mobile storage.

Acknowledgments. This work is supported by the National Science Foundation of China (Grant No. 61472289) and Hubei Province Science Foundation (Grant No. 2015CFB254).

References

1. Reddy, S.T., Lakshmi, D.L., Deepthi, C., et al.: USB_SEC: a secure application to manage removable media. In: 2016 10th International Conference on Intelligent Systems and Control, pp. 1–4. IEEE (2016)
2. Walker, S.J.: Big data: a revolution that will transform how we live, work, and think. Am. J. Epidemiol. **17**(9), 181–183 (2014)
3. Guardian, T.: NSA collecting phone records of millions of Verizon customers daily. Commun. ACM (2013)
4. Shah, N.N., Kumar, G.N., Raval, J.A.: Web based framework for data confidentiality in removable media ensuring safe cyber space. Int. J. Sci. Eng. Technol. Res. **4**(5) (2015)
5. Kang, M.H., Moskowitz, I.S.: A pump for rapid, reliable, secure communication. Proc. ACM Conf. Comput. Commun. Secur. **39**(7), 119–129 (1993)
6. Prost, W., Auer, U., Tegude, F.J., et al.: Tunnelling diode technology. In: International Symposium on Multiple-Valued Logic, pp. 49–58. IEEE (2001)
7. Wan, Y.L., Zhu, H.J., Liu, H.Z., et al.: Reliability of one-way transmission system base on optical shutter. Netinfo Secur. **12**, 25–27 (2010)
8. Zhao, B.: Study and design of safe one-way information transmission equipment. Comput. Appl. Softw. **27**(6), 98–99 (2010)
9. García-Dorado, J.L., Mata, F., Ramos, J., Santiago del Río, P.M., Moreno, V., Aracil, J.: High-performance network traffic processing systems using commodity hardware. In: Biersack, E., Callegari, C., Matijasevic, M. (eds.) Data Traffic Monitoring and Analysis. LNCS, vol. 7754, pp. 3–27. Springer, Heidelberg (2013). doi:10.1007/978-3-642-36784-7_1
10. Li, G.: Implementation of file transfer system based on physical isolation. Comput. Eng. Appl. **18**, 166–168 (2004)

11. Xiao, Y.J., Fang, Y., Zhou, A.M., et al.: Design of data unilateral transmission based on USB 2.0. J. Comput. Appl. **26**(6), 1481–1490 (2006)
12. Wang, H.Y., Meng, F.Y.: Design and implementation of one-way data transmission system based on optical fibre. Netinfo Secur. **3**(1), 68–72 (2011)
13. Singh, R., Tripathi, P., Singh, R.: A survey on TCP (transmission control protocol) and UDP (user datagram protocol) over AODV routing protocol. J. Appl. Physiol. **31**(1), 63–9 (2014)
14. El-Khamy, M., Lee, J., Kang, I.: Detection analysis of CRC-assisted decoding. IEEE Commun. Lett. **19**(3), 483–486 (2015)
15. Sowmya, M.S., Shwetha, H.N., Savitha, A.P.: USB interface with FPGA for data acquisition system using in X-ray applications. Int. J. Sci. Technol. **2**(5), 270 (2014)

Android Malware Detection Using Local Binary Pattern and Principal Component Analysis

Qixin Wu[1,2], Zheng Qin[1,2(✉)], Jinxin Zhang[1,2(✉)], Hui Yin[1,2], Guangyi Yang[3], and Kuangsheng Hu[3]

[1] College of Computer Science and Electronic Engineering,
Hunan University, Changsha, Hunan, China
{zqin,zhangjixin}@hnu.edu.cn
[2] Hunan Key Laboratory of Big Data Research and Application,
Changsha, Hunan, China
[3] Hunan Institute of Metrology and Test, Changsha, China

Abstract. Nowadays, analysis methods based on big data have been widely used in malicious software detection. Since Android has become the dominator of smartphone operating system market, the number of Android malicious applications are increasing rapidly as well, which attracts attention of malware attackers and researchers alike. Due to the endless evolution of the malware, it is critical to apply the analysis methods based on machine learning to detect malwares and stop them from leakaging our privacy information. In this paper, we propose a novel Android malware detection method based on binary texture feature recognition by Local Binary Pattern and Principal Component Analysis, which can visualize malware and detect malware accurately. Also, our method analyzes malware binary directly without any decompiler, sandbox or virtual machines, which avoid time and resource consumption caused by decompiler or monitor in this process. Experimentation on 5127 benigns and 5560 malwares shows that we obtain a detection accuracy of 90%.

Keywords: Android malware detection · Binary texture feature · Local binary pattern · Principal component analysis

1 Introduction

In the coming era of big data, analysis methods based on big data have been widely used in research and application areas. With the explosively growing of malware, it is critical to apply the analysis methods based on machine learning to the field of malware detection, especially in Android malware detection. In the third quarter of 2016, International Data Corporation stated that Android operating system dominated the smartphone market with a share of 86.8% [1]. As the most widely used smart phone platform in the world, Android platform has attracted the attention of many malware developers, who tempt users to

© Springer Nature Singapore Pte Ltd. 2017
B. Zou et al. (Eds.): ICPCSEE 2017, Part I, CCIS 727, pp. 262–275, 2017.
DOI: 10.1007/978-981-10-6385-5_23

download malicious or vulnerable applications and install on Android devices [2–7] without their knowledge, which may make their personal privacy information such as account details, photographs, contact information, etc. easy to leakage.

In response to the damages where malware attackers have brought, analysis methods based on big data, especially machine learning, have played a more and more important role in detecting malicious variants on smartphone systems. Comparing with the traditional detection methods, analysis methods based on machine learning can detect unknown malwares, which could defend the zero-day attack. Besides, analysis methods based on machine learning can also prevent attacks like code-morphism, which allows the malware to evolve into many variants and bypass traditional code feature based detection systems.

During the past decades, some techniques were proposed to prevent, thwart and detect Android malware. Basically they can be categorized into static analysis and dynamic analysis. The former is the method of analyzing code without actually running it. Moser et al.'s work [8] tried to find out the control flow of an application to figure out its behavior. It often applied for finding programming faults and vulnerabilities in benign applications, and was furthermore used to detect malicious behavior without having to run an application. Dynamic analysis, on the contrary, captures apps' behavior by running its code on virtual machines or in a sandboxed environment. However, both static analysis and dynamic analysis have some shortages. For examples, static analysis needs to decompile apps to get its features and dynamic analysis needs to monitor the behavior of app and trace what an app really does as well, which are all time consuming. Also, static analysis may occupy local resources while decompiling and dynamic analysis may also occupy network resource when uploading applications on the cloud. Moreover, there do exist some scenarios without any decompiler, sandbox or virtual machines, etc., which cannot detect Android malware any more. Hence, we want to propose a novel method which can detect Android malware without any decompiler, sandbox or virtual machines, etc.

Proposing a novel malware detection method without any decompiler, sandbox or virtual machines, etc. does have some challenges. Since previous work, static analysis or dynamic analysis, have to use decompiler, sandbox or virtual machines to get malware's system calls, opcodes and so on. In our approach, we extract binaries from malware as feature, which do not need those tools any more. However, it is not easy to detect malware by binaries since the binary sequence is composed of 0 and 1, which is more sophisticated to distinguish malicious applications from benign ones using common techniques. On the other hand, binary sequence is fine enough to cause high detection complexity when applying image recognition detection method in practice.

In this paper, we propose a novel Android malware detection method based on binary texture feature recognition by Local Binary Pattern (LBP) and Principal Components Analysis (PCA), an intuitive way by extracting binary texture feature and converting to an image, which do not need to decompile or run .apk suffix file during detection and can visualize it as well. Nataraj et al.'s work [10] proposed to extract binary texture features by GIST [11], however, was only used to detect malware with the .exe suffix and the detection time cost of GIST is too high. Our approach targets the .apk suffix file and constructs binary image by [10], extracts Android malware texture features based on LBP, reduces the dimension by PCA, and then recognizes malicious and benign applications based on kNN algorithm with texture features. Our experiment illustrates that our method reduce time cost because of running without any decompiler or monitor, and improve the detection accuracy up to 90%.

Our approach can fit this problem pretty well is that there do exist similarities in various kinds of Android malware binaries since the binary segments we extracted from Android malware samples are similar. Also, the analogous texture features can we obtained when we convert similar binary segments to image texture features. Since the textures of target samples are analogous, we can easily find out the most similar samples by searching the most analogous texture features with kNN in our approach. Therefore, we consider the target sample as malicious application if the target sample is sufficiently similar to training samples. What PCA used in our method is to reduce dimensions while retaining accuracy.

The main contributions of our work are summarized as follows. (1) We propose a novel approach for Android malware detection based on binary texture feature recognition. (2) Our approach can efficiently detect Android malware without any decompiler, sandbox or virtual machines. (3) By using LBP texture feature extraction, PCA dimension reduction and kNN classifier, the experiments demonstrate that our approach can achieve high detection accuracy and low detection time cost.

The remainder of this paper is structured as follows. Section 2 reviews the previous work in Android malware detection. Section 3 states the problem to be addressed in this work and Sect. 4 introduces our method in detail. In Sect. 5, we discuss our experiment and conclude our work in Sect. 6.

2   Related Work

In the past years, mobile malware detection, especially Android malware detection, has become a research hot area and lots of techniques have been proposed to conquer the growing amount and sophistication of Android malware. Typically, those techniques can be summarized as static analysis and dynamic analysis. We review some of them below.

2.1   Static Analysis

Static analysis is a technique which extracts static malware features to classify malware without running code. The static malware features often include permissions [12,13,16], operation code [14,17,29], and PE header fields [15].

Permission based detection tools Kirin [16] was one of the most representative methods in the early times, which detected malicious permission combinations by defining a group of rules, however, with high positive false rate. RiskRanker

[18] detected high and medium risk apks by several predetermined features, such as the presence of native code, the use of functionality which costs users money without his/her permission, the dynamic loading of code which is stored encrypted in the app, etc. DroidAPIMiner [19] extracted features from a number of malicious and benign apps to detect malwares during static analysis. Kong et al. [20] classified malware instance into different malware families by structural information. Zhang et al. [17, 29] proposed a malware detection method based on opcode image recognition, which disassemble binary executables into opcodes sequences and then converts the opcodes into images. Only when decompiler work can opcode be obtained, however.

Texture feature extraction and image recognition are widely used in static analysis. Several detection techniques have been proposed before. Nataraj et al. [9, 10] proposed to extract binary texture features by GIST and visualize malwares, however, the .apk suffixes files are not included and the detection time cost of GIST is too high.

2.2 Dynamic Analysis

The other Android malware detection method is the dynamic analysis which monitors the behavior of applications and collects the methodology calls, operations and raised exceptions of malwares on sandbox or virtual machine environment during runtime. Enck et al. [21] proposed TaintDroid strategy which focused on taint analysis to dynamically monitor applications in a protected environment. Crowdroid [22] gathered information from smart phones and transferred them to the cloud for malware classification with k-means algorithm. Kwong and Yin [23] presented DroidScope platform which collected detailed execution traces of application by rebuilding the semantic information of both OS-level and Java-level, however, only two malicious apks were analyzed. Some limitations still exist in the aforementioned dynamic analysis, though. For examples, cloud servers are needed in some methods to classify malwares, which is time consuming and occupies too much resource. Also, some malicious apks would not perform aberrant behavior unless get specific instructions, etc. Under these circumstances, it is difficult to monitor the behavior of malicious applications.

3 Problem Statement

As Android has the most popular market share in the world, more and more Android malwares have been discovered. More than 650,000 individual pieces of malware for Android have been discovered by May 2014 [24]. Due to the endless evolution of the malware, it is critical to detect malware variants and stop them from leakaging our privacy information and disturbing our lives.

Existing detection schemes, either static analysis or dynamic analysis, all have some common shortcomings as follows.

(1) Time consumption. As is known, the first step of static analysis is to decompile .apk files into Java or jar files using tools, which costs a lot of time and different decompiler may cause different structure during this process. Dynamic analysis, analogously, monitors the behavior of application on sandbox or virtual machine environment during runtime, which is also time consuming.

(2) Resource occupation. When decompiling .apk file in static analysis, it is natural to occupy local resource to run decompile tools. In dynamic analysis, tracing what an application really does on sandbox or virtual machine environment is also a cost of local resource. Moreover, Crowdroid [22] leverages crowdsourcing to gather information from smart phones and transfers them to a cloud server for malware classification, which occupies too much network resource and cloud resource, etc.

(3) Do not fit for some specific situation. There are always some scenarios without decompile tools, sandbox or virtual environment, for examples, decompile tool may not work in some systems because it has strong dependency on system environment, and there also exists devices which do not have enough RAM to implement sandbox or virtual machines, which can no longer detect Android malware by dynamic analysis.

With all these shortages, we propose a new Android malware detection approach based on binary texture feature recognition by LBP and PCA, which can better solve these limitations aforementioned. In our method, we analyze malware binary directly without decompiler, sandbox or virtual machines, which avoid time and resource consumption caused by decompiler or monitor in this process and overcome those disadvantages in the above mentioned techniques. Also, our method visualize malware and convert malware detection problem to image recognition problem, which can efficiently classify malware variants. The detailed methodology will be elaborated in the next section.

4   Methodology

In this section, we first give a series of definitions below.

Definition 1. *Let M be the training data sets, $M = m_1, m_2$, where m_1 represents the malware data set, m_2 represents the benign data set.*

Definition 2. *Let x_j be a binary, x_j belongs to m_1 if x_j is a malware binary, otherwise, x_j belongs to m_2.*

Definition 3. *Let Tex be the texture feature, which is extracted from binary image.*

Definition 4. *Let $T(x_j)$ be the texture feature column vector generated from x_j, which is calculated by LBP algorithm. It is easy to know that $T(x_j)$ is a vector, which has 256 rows and 1 column.*

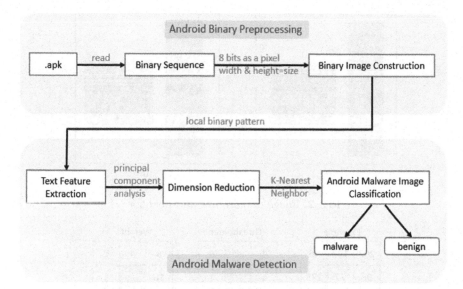

Fig. 1. The workflow of our methodology

Four steps are included in our approach. We first read the Android binary directly and convert it to image representation which is used for LBP feature extraction. To reduce the time cost and retain accuracy, we use the PCA to reduce the dimension of Android binary texture features. And then, we classify malware and benign by kNN. The work flow of our methodology is shown in Fig. 1.

Step 1: Binary Image Construction. In this step, we read Android binary directly and convert to an image representation which is used for obtaining texture based feature. The binary is read as a 1D array (vector) of 8 bit unsigned integers. The width of the matrix is fixed depending on the file size and the height varies according to the file size [25]. We give a few binary images of malwares (shown as Fig. 2(a)) and benigns (shown as Fig. 2(b)).

Step 2: Texture Feature Extraction. As Android binary has already converted to image, the texture based feature can be computed on the image to characterize the Android binary. The texture feature we used here is the LBP feature, which is often used to describe local texture features of an image. We introduce the LBP feature of binary image next.

LBP, short for local binary pattern, is an operator that describes the local texture features of an image. We now give a simple example to illustrate how to calculate LBP feature of Android binary image. Consider a local 3×3 neighborhood of an Android binary texture (shown as Fig. 3), each value in the 3×3 neighborhood is correspond to the gray levels of each cell in Android binary image.

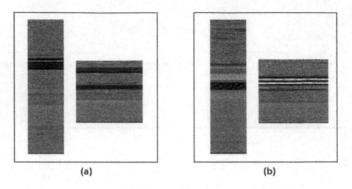

(a)	(b)

Fig. 2. Image of malwares and benigns

Example				Thresholder				Weights		
202	192	46		1	1	0		8	4	2
96	77	178		1		1		16		1
47	18	6		0	0	0		32	64	128

Fig. 3. Calculation of LBP

It is easy to see that the gray value of the center cell is 77, we compare this pixel to each of its 8 neighbors, following the pixels along a circle, i.e. clockwise or counter-clockwise, along clockwise as an example, we consider 77 to compare with 6, 18, 47, 96, 202, 192, 46, 178, if the center pixel's value is greater than the neighbor's value, write "0", otherwise, write "1". This gives an 8-bits, shown as Thresholder in Fig. 3. Thus, along the clockwise direction, the bits of the center pixel is 00011101 and the corresponding decimal is 29, which is the pixel value of the center cell. Now we give a common formula of LBP [28]:

$$LBP_p = \sum_{p=1}^{P-1} s(g_p - g_c)2^p \tag{1}$$

where

$$s(x) = \begin{cases} 0, & x \geq 0 \\ 1, & x < 0 \end{cases} \tag{2}$$

where g_p represents the gray value of the surrounding pixel, g_c represents the gray values of the center pixel, P is the number of the surrounding pixels.

Step 3: Dimension Reduction. In our approach, we use PCA, short for principal component analysis, to reduce dimension. PCA is a common method of data analysis and has been widely used to extract the key features from data and reduce the dimension as well. Since we have extracted the texture features from images, we consider the texture features from an image as an input and obtain the feature vector as an output. Note that the feature vector we get is calculated by the matrix multiplication of the variance matrix and the eigenvectors, where the eigenvectors are calculated by the covariance matrix of the texture features. We introduce the calculation of PCA method in detail blow.

Initially, we calculate the average texture feature vector according to T (x_j). Let $T_{average}$ be the average feature texture vector according to Eq. (3), where n is the number of the training feature texture vectors T (x_j).

$$T_{average} = \frac{\sum_{j=1}^{n} T(x_j)}{n} \tag{3}$$

Secondly, we calculate the variance vector according to the average feature texture vector $T_{average}$. Let $V(x_j)$ be the variance vector according to the Eq. (4).

$$V(x_j) = T(x_j) - T_{average} \tag{4}$$

Thirdly, we calculate the covariance matrix and eigenvectors. We define Cov as the covariance matrix, which is according to Eq. (5).

$$Cov = \frac{\sum_{j=1}^{n} V(x_j)V(x_j)^T}{n} = \frac{\sum_{j=1}^{n} AA^T}{n} \tag{5}$$

Let G be the column eigenvectors according to AA^T, G_i be the i-th eigenvector in eigenvectors G order by their eigenvalues λ_i from large to small, according to the Eqs. (6) and (7).

$$|AA^T - \lambda E| = 0 \tag{6}$$
$$AA^T \cdot G_i = \lambda_i \cdot G_i \tag{7}$$

To reduce the dimension, we choose k eigenvectors $G_{i<k}$ with the top k largest eigenvalues. What mentioned before is that the eventual feature vector we get is calculated by the matrix multiplication of the variance vector and the eigenvectors and the feature vector is used for our next step, the Android binary image classification. We define Vec(x_j) as the eventual feature vector, which can be calculated according to Eq. (8).

$$Vec(x_j) = V(x_j)^T \cdot G_{i<k} \tag{8}$$

Step 4: Android Binary Image Classification. The method we used here for Android binary image recognition is kNN (k-Nearest Neighbor), a common algorithm used in classification.

By setting k = 1 in kNN algorithm, if the feature vector Vec(x_j), obtained from step 3, is sufficiently similar to any previous malware feature vector in our training data set, we say this feature vector represents the Android malware image and vice versa. Note that the distance function used here is Euclidean distance.

5   Experiment

In this section, we provide several experiments to show the performance of our approach. At the beginning, we present the experiment setup and the data set. And then, we illustrate the accuracy and the detection time cost of our experiment.

5.1   The Experiments Setting

Experiment Setup. We implement our method on one computer. The version of the CPU is AMD A6-3420M @ 1.50 GHz, the RAM is 4.0 GB and the operation system is Windows 7. Our approach is developed by Java programming language.

Data Set. Two different data sets are considered in our approach to test Android malware detection methodology: the malware binary data set and the benign binary data set. The benign binary data set we used in our experiment is downloaded from Google Play. The malware samples we used in our experiment are collected from Drebin [26]. The samples have been collected in the period of August 2010 to October 2012 and were made available to us by the MobileSandbox project [27]. Our final dataset contains 5127 benign apps and 5560 malware samples.

As is shown in Table 1, we provide an overview of the top malware families in our dataset, which contains several popular families in current application markets.

5.2   The Approaches for Comparison

We use two approaches to distinguish malwares from benigns and make a comparison between them to demonstrate the efficiency of our methodology. The first

Table 1. Top 20 malware families in our dataset

Malware family	Number	Malware family	Number
FakeInstaller	925	Adrd	91
DroidKungFu	667	DroidDream	81
Plankton	625	ExploitLinuxLotoor	70
Opfake	613	MobileTx	69
GingerMaster	339	Glodream	69
BaseBridge	330	FakeRun	61
Iconosys	152	SendPay	59
Kmin	147	Gappusin	58
FakeDoc	132	Imlog	43
Geinimi	92	SMSreg	41

method we used here is kNN and the second is kNN based on PCA (PCA+kNN), where PCA is used for dimension reduction while achieves a predictive accuracy here.

Five training data sets are used in our experiment, which are 600, 700, 800, 900 and 1000 respectively. And we also implement our method in 5 different dimensions which are Dim(10), Dim(15), Dim(20), Dim(25) and Dim(30) to find out the appropriate dimension with comparatively higher accuracy in the second method.

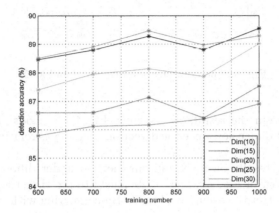

Fig. 4. Detection accuracy in different dimensions

5.3   Accuracy Comparision

Three kinds of metrics are used to evaluate the accuracy of our proposed methodology, which are TPR (true positive rate), FNR (false negative rate) and accuracy. Followings are the equations of these metrics:

$$TPR = \frac{TP}{TP + FN} \tag{9}$$

$$FNR = \frac{FN}{TP + FN} \tag{10}$$

$$accuracy = \frac{TP + TN}{TP + TN + FP + FN} \tag{11}$$

where TP (true positive) is defined as the number of malwares which are correctly detected, FN (false negative) is the number of malwares which are misrecognized, FP (false positive) is the number of benign applications which are misrecognized, TN (true negative) is the number of benign applications which are correctly detected.

At the beginning, we implement our method on 5 different dimensions and 5 different training data sets to find out a suitable dimension with relatively

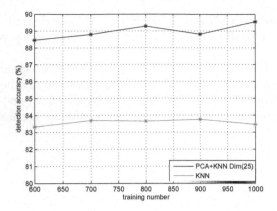

Fig. 5. The accuracy of our method and kNN

high accuracy. From the results as shown in Fig. 4, we choose Dim(25) as the most appropriate dimension since the detection accuracy is relatively higher than others.

The accuracy results on kNN and our method under different data sets in Dim(25) are shown in Fig. 5, which is obviously to see that the larger training data set is, the better the accuracy obtains. By comparing with kNN in different training data sets, the accuracy of our approach achieves from 88.46% to 89.55% while kNN achieves from 83.32% to 83.46%.

The results of TPR and FNR in kNN and our method are separately shown in Fig. 6, from which we can find out our method improves the TPR and reduce the FNR significantly. The experiment shows that the TPR of our approach grows from 86.98% to 89.03% and the FNR drops from 10.06% to 9.93% when the training data sets turn 600 to 1000. And in kNN, TPR grows from 77.76% to 78.63% and the FNR drops from 11.72% to 11.02%.

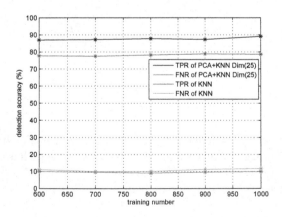

Fig. 6. TPR and FNR in our method and kNN

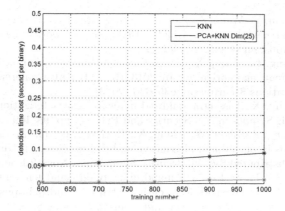

Fig. 7. The detection time cost of our method and kNN

5.4   Detection Time Cost Comparision

The results of detection time cost between our method and kNN are presented in this subsection. As is shown in Fig. 7, our method is a little slower than kNN since we used PCA before to reduce the dimension, but the detection accuracy of our method increases substantially on the other side.

6   Conclusion

In this paper, we propose a novel Android malware detection method based on binary texture feature recognition by Local Binary Pattern and Principal Component Analysis, which do not need any decompiler, sandbox or virtual machines. Since this detection scheme can only classify Android malware variants and benigns, but cannot explain what the texture is. So in the future, we will consider if we can translate the binary texture feature to semantic information for binary behavior analysis.

Acknowledgement. This work is partially supported by the National Science foundation of China under Grant No. 61472131, No. 61300218 and No. 61472132. The Natural Science Foundation of Hunan Province under Grant No. 2017JJ2292 and Outstanding Youth Research Project of Provincial Education Department of Hunan under Grant No. 17B030. Science and Technology Key Projects of Hunan Province (2015TP1004, 2015SK2087, 2015JC1001, 2016JC2012).

References

1. IDC: Smartphone OS Market Share (2016) Q3. http://www.idc.com/promo/smartphone-market-share/os
2. Felt, A.P., Finifter, M., Chin, E., Hanna, S., Wagner, D.: A survey of mobile malware in the wild. In: ACM Workshop on Security and Privacy in Smartphones and Mobile Devices, pp. 3–14 (2011)

3. Grace, M., Zhou, Y., Wang, Z., Jiang, X.: Systematic detection of capability leaks in stock android smartphones. In: Nineteenth Annual Network & Distributed System Security Symposium Ndss12 Isoc (2012)
4. Wang, R., Xing, L., Wang, X., Chen, S.: Unauthorized origin crossing on mobile platforms: threats and mitigation. In: 2013 ACM SIGSAC Conference on Computer & Communications Security, pp. 635–646 (2013)
5. Zhou, Y., Jiang, X.: Dissecting android malware: characterization and evolution. In: 2012 IEEE Symposium on Security and Privacy, pp. 95–109 (2012)
6. Zhou, Y., Wang, Z., Zhou, W., Jiang, X.: Hey, you, get off my market: detecting malicious apps in official and alternative android markets. In: Nineteenth Annual Network & Distributed System Security Symposium Ndss12 Isoc (2012)
7. Wei, F., Roy, S., Ou, X., Robby: Amandroid: a precise and general inter-component data flow analysis framework for security vetting of android apps. In: 2014 ACM SIGSAC Conference on Computer and Communications Security, pp. 1329–1341 (2014)
8. Moser, A., Kruegel, C., Kirda, E.: Limits of static analysis for malware detection. In: Twenty-Third Annual Computer Security Applications Conference (ACSAC 2007), pp. 421–430 (2007)
9. Nataraj, L., Manjunath, B.S.: SPAM: signal processing to analyze malware. J. IEEE Sig. Process. Mag. **33**(2), 105–117 (2016)
10. Nataraj, L., Yegneswaran, V., Porras, P., Zhang, J.: A comparative assessment of malware classification using binary texture analysis and dynamic analysis. In: 4th ACM Workshop on Security and Artificial Intelligence, pp. 21–30 (2011)
11. Olivia, A., Torralba, A.: Modeling the shape of a scene: a holistic representation of the spatial envelope. J. Comput. Vis. **42**(3), 145–175 (2001)
12. Rashidi, B., Fung, C., Vu, T.: RecDroid: a resource access permission control portal and recommendation service for smartphone users. In: ACM MobiCom Workshop on Security and Privacy in Mobile Environments, pp. 13–18 (2014)
13. Felt, A.P., Chin, E., Hanna, S., Song, D., Wagner, D.: Android permissions demystified. In: 18th ACM Conference on Computer and Communications Security, pp. 627–638 (2011)
14. Anderson, B., et al.: Improving malware classification: bridging the static/dynamic gap. In: 5th ACM Workshop on Security and Artificial Intelligence, pp. 3–14 (2012)
15. Raman, K., et al.: Selecting features to classify malware. In: InfoSec Southwest, pp. 1–5 (2012)
16. Enck, W., Ongtang, M., McDaniel, P.: On lightweight mobile phone application certification. In: 16th ACM Conference on Computer and Communications Security (CCS), pp. 235–245 (2009)
17. Zhang, J., et al.: Malware variant detection using opcode image recognition with small training sets. In: 2016 25th International Conference on Computer Communication and Networks (ICCCN), pp. 1–9 (2016)
18. Grace, M., et al.: Riskranker: scalable and accurate zero-day android malware detection. In: 10th International Conference on Mobile Systems, Applications, and Services (MOBISYS), pp. 281–294 (2012)
19. Aafer, Y., Du, W., Yin, H.: DroidAPIMiner: mining API-level features for robust malware detection in android. In: Zia, T., Zomaya, A., Varadharajan, V., Mao, M. (eds.) SecureComm 2013. LNICSSITE, vol. 127, pp. 86–103. Springer, Cham (2013). doi:10.1007/978-3-319-04283-1_6
20. Kong, D., et al.: Discriminant malware distance learning on structural information for automated malware classification. In: 19th ACM SIGKDD International Conference on Knowledge Discovery and Data Mining, pp. 1357–1365 (2013)

21. Enck, W., et al.: Taintdroid: an information-flow tracking system for realtime privacy monitoring on smartphones. In: 9th USENIX Symposium on Operating Systems Design and Implementation (OSDI), pp. 393–407 (2010)
22. Burguera, I., Zurutuza, U., Nadjm-Tehrani, S.: Crowdroid: behavior-based malware detection system for android. In: 1st ACM Workshop on Security and Privacy in Smartphones and Mobile Devices, pp. 15–26 (2011)
23. Kwong, L., Yin, Y.H.: Droidscope: seamlessly reconstructing the OS and Dalvik semantic views for dynamic android malware analysis. In: 21st USENIX Conference on Security Symposium USENIX Association, pp. 1–16 (2012)
24. Sophos mobile security threat report, mobile world congress. https://www.sophos.com/en-us/press-office/press-releases/2014/02/mobileworldcongress2014.aspx
25. Nataraj, L., Karthikeyan, S., Jacob, G., Manjunath, B.: Malware images: visualization and autmatic classification. In: International Symposium on Visualization for Cyber Security, pp. 1–7 (2011)
26. Arp, D., Spreitzenbarth, M., Huebner, M., Gascon, H., Rieck, K.: Drebin: efficient and explainable detection of android malware in your pocket. In: 21th Annual Network and Distributed System Security Symposium (NDSS) (2014)
27. Spreitzenbarth, M., Echtler, F., Schreck, T., Freling, F., Hoffmann, J.: Mobile-Sandbox: looking deeper into android applications. In: 28th International ACM Symposium on Applied Computing (SAC) (2013)
28. Ojala, T., Pietikäinen, M., Mäenpää, T.: Gray scale and rotation invariant texture classification with local binary patterns. In: Vernon, D. (ed.) ECCV 2000. LNCS, vol. 1842, pp. 404–420. Springer, Heidelberg (2000). doi:10.1007/3-540-45054-8_27
29. Zhang, J., et al.: Malware variant detection using opcode image recognition. In: 2016 IEEE 22nd International Conference on Parallel and Distributed Systems (ICPADS), pp. 1175–1180 (2016)

Elderly Health Care - Security and Privacy Issue

Kaiyu Wan[1], Vangalur Alagar[2(✉)], and Peter Oyikanmi[2]

[1] Xi'an Jiaotong-Liverpool University, Suzhou, People's Republic of China
Kaiyu.Wan@xjtlu.edu.cn
[2] Concordia University, Montreal, Canada
alagar@cse.concordia.ca

Abstract. The current trend in offering universal health care based on open distributed health care network which will provide health care to elderly, socially and economically weak population, and physically challenged patients has lead to rapid digitization of health care data, and wireless medium for data communication. Although it has many advantages and has immense potential to improve health care availability, it has created many challenges in maintaining health care services, in particular in offering security and privacy of the most vulnerable members of society who are the elderly. In this paper we identify the current status of elderly care, their vulnerabilities, and challenges faced to offer them health care in total privacy and dignity.

Keywords: Health care · Elderly · Security · Privacy · Big data

1 Introduction

The word *"Elderly"* is descriptive of someone who has achieved old age, and most developed countries accept the range of 60 to 65 as the inception of old age. In Canada, this is the age when many people are eligible for social services such as government pensions. The U.S Census in 2010 reported that there were 40.3 million elderly people in the United States which is a sharp increase from the 3.1 million reported in 1900. In 2012, the population of the elderly made up 11% of the world's population and was projected to reach 22% by 2050, with 68% of the world's population over 80 living in Asia and Latin America and the Caribbean [23]. At the moment, one in six Europeans can be regarded as elderly [50]. In Australia, older people currently make up 13.6% of the population [42]. According to [34], Japan is the demographically oldest country in the world with 20% of its population aged 65 and above. Population in China has grown substantially over the last 20 years. In 2000, population aged 65 and older in China was almost 90 million. It is estimated that by 2020 China will have 230 million elderly people and this number will increase to well over 300 million by 2050, thereby making China the largest population of elderly worldwide. Due to the projected growth in the population of the elderly, it is expected that the proportion of care givers assisting the elderly will not be sufficient and the cost of giving care for the elderly will also be on the increase. An effective solution to this

© Springer Nature Singapore Pte Ltd. 2017
B. Zou et al. (Eds.): ICPCSEE 2017, Part I, CCIS 727, pp. 276–291, 2017.
DOI: 10.1007/978-981-10-6385-5_24

problem is *health care service automation*, which can provide services whenever and wherever they are demanded thereby maximizing health service utilization and minimizing service cost. This paper gives an account of the current status of technology and tools used in health care automation, and raises awareness on its consequences on the privacy and dignity of the elderly.

The elderly community form an important focus area due to the numerous *vulnerabilities* and health care needs they face especially personal care in daily living which could range from taking care of personal hygiene to feeding themselves. A health condition that can result out of poor nutrition is diabetes which has been found to lead to higher rates of premature death. Sedentary lifestyle, social isolation, loneliness, or depression can lead to malnourishment, and depression medications can also change how nutrients are absorbed or how food tastes, thus making the dietary care of elderly paramount. Mental illness is a very difficult issue to address as many seniors are either unwilling or unable to report their situations. Most elderly people lack knowledge about the causes or symptoms behind their health problems and assume their health problems are simply due to their aging [13]. The elderly also go through *physical, psychological* and *financial* vulnerabilities [11], become dependent on others which exposes them to societal dangers [30] such physical and psychological assaults, and financial exploitation [51]. The different vulnerable situations and the health conditions discussed above can be related to specific contexts thereby creating a need for technology-enabled context-aware solutions. In this paper we bring out the flaws and inadequacies in existing technology-enabled solutions that adversely affect the privacy, safety, and dignity of the elderly patients.

2   Vulnerabilities of Elderly

The elderly in society suffer different kinds of vulnerabilities, although most of them don't report them due to guilt, embarrassment and shame [15]. The three most cited types of abuse old people pass through are *physical, psychological* and *financial* [11]. In order to improve the quality of care of the elderly it is essential to identify the *contexts* in which these vulnerable situations are likely to arise, and set checks and balances at each context that will ensure patient safety.

Older people are vulnerable to physical assaults ranging from care givers to strangers due to the frailty they suffer [44]. Frailty can be defined as experiencing a progressive physiologic decline in multiple body systems, loss of body functions, loss of physiologic reserve, and increased vulnerability to disease and death. They experience hardship during and after disasters because of diminished sensory awareness [30] and lower rescue aids than their younger counterparts. It is also likely that old people have limited mobility in dangerous situations thereby exposing them to physical harm. Hence, they become dependent on others for special needs such as protection from abuse and exploitation resulting in frailty which further exposes them to societal dangers [30].

Old people are often isolated from the rest of the society, especially in areas with linguistic barriers, which lead to loneliness [52]. A survey revealed that on the average, 32.2% of elderly people feel lonely while only 69.5% of elderly have some recreational activity outside the home [18]. Lonely elderly person can have additional health complication such as depression and anxiety. It is of no surprise that on the average 55.1% of elderly have sad attitude towards their lives [18].

The elderly are often targets from fraudulent business practices due to their lack of market knowledge as most of the elderly don't have clear guidance on the appropriate channels to contact for financial advice [27]. In addition to this, the frail elderly can also be more vulnerable to property damage owing to lack of insurance [43]. According to the reports [16,51], 20% of the of elderly in USA were victims of financial exploitation, and more than 50% of all reported cases of elder abuse in Canada involved material exploitation [16].

3   Health Care Legal Acts for Protection and Privacy

In many countries health care institutions and government agencies have provided a set of legal guidelines to follow in offering health care services. Below we review two of them and emphasize that for elderly health care more stringent measures should be implemented.

3.1   USA Health Care Act

HIPAA [19] was enacted by the United States Congress in 1996 to protect health insurance coverage for medical workers and establish specific guidelines to follow when processing electronic medical information especially on mobile devices due to the sensitivity of the data being measured and processed in the health care industry and the risks involved in any probable breaches. HIPAA dictates that "protected health information (PHI) such as name, phone number, email address, photograph, payment information and any other information that can be linked back to the patient should never be disclosed to or tampered by third party without the express permission of the patient". It also provides certain guidelines on the availability (disclosure) of information to authorized personnel and the use of access controls to safeguard medical equipments.

3.2   Canadian Health Information Protection Acts

The Ontario Health Information Protection Act [14] has suggested that physicians should obtain either express or implied consent before disclosing personal health information to external parties. It enforces the concept of *lockouts* which is a situation where the patient can explicitly restrict the physician from disclosing personal health information even to others directly involved in providing health care services to the patient [49]. In addition, the Act provides recommendations for information disclosure relating to family member and the police. Saskatchewan Health Information Protection Act (SHIPAC) [41] is the standard developed and adopted by the Government of Saskatchewan. It guards the rights of individuals and specifies the obligations of *trustees* regarding the collection, storage, use, access and disclosure of personal health information in the province of Saskatchewan. According to the Act, personal health information can be defined as information with respect to the physical or mental health of the individual or about any health service provided to the individual. A *trustee* includes government institutions, community clinics, district health boards, regional health authorities, ambulance operators and health professionals. SHIPAC upholds the right of the individual to have full control over the use or disclosure of personal health information about himself or herself and trustees must obtain such consent voluntarily and purposefully from the individual before using or disclosing such information. The Act further dictates that a trustee in charge of personal health information must establish written actionable policies and procedures to ensure security and privacy of such information. Alberta Health Information Protection Act (AHIPC) [40] is the standard developed by the Government of Alberta guiding health professionals on handling electronic health information in the province of Alberta. The Act states that custodians of electronic health information must submit a privacy impact assessment to, meet any security requirements established by the department of the Provincial Commissioner and obtain approval for access to the provincial Electronic Health Record (EHR) from the Department. It stipulates that a custodian must ensure its electronic health record information system creates and maintain logs containing the essential information about system access for security audit.

4   United Nations Proposed Health Care Systems for Assisting Elderly

While multiple authors have attempted to classify health care systems into different research areas, this section discusses the seven key themes proposed by United Nations [53]. Collectively they provide a framework for raising patient awareness, training health care givers in ethical practice, and providing timely health care service for patients in total privacy. We emphasize that none of these systems can protect the privacy of patients. Moreover, there exists no one system that includes all the features.

1. Education and Awareness: These kind of health care systems are used to raise public awareness such as educating people on important health topics such as HIV/AIDS. This category of systems makes use of text messaging for health promotion or to alert target groups of health campaigns. These programmes can be set up either as one-way alerts or interactive tools and can often be downloaded to patients' mobile phones or sent as a series of text messages. Using SMS messages for health campaigns is preferred due to low cost, confidentiality and broad reach. However, there have been certain barriers to using mobile technology for education and awareness such as SMS length restrictions, language barriers, illiteracy, shortage of skilled technical personnel and lack of technical support in rural areas.

2. Helpline: These systems use telecommunication technology to provide a range of health care services to patients and health professionals. Usually, trained health care assistants are available as call center agents to answer queries from patients. This system is currently being capitalized by low income countries to overcome health care challenges created by shortage of health professionals, cost of service, lack of transportation in remote areas, and lack of sources that can provide reliable information. However, due to mounting infrastructure and data management costs, patients in many low income countries are unable to afford this service. Moreover, such systems cannot guarantee availability, reliability, and security of services.

3. Diagnostic and Treatment Support: Health care systems under this category attempt to achieve treatment compliance, disease eradiation, and overcome challenges through *reminder systems,* usually by voice or SMS. These systems can be used to support patients with conditions such as diabetes and tuberculosis. Patients can use this category of health care systems to schedule appointments or receive medical advice remotely through voice or SMS. However, the use of these systems does not automatically translate to neither the improvement of health service metrics nor patient satisfaction.

4. Communication and Training: Health care projects within this category provide access to information on health science publications from government agencies and medical research institutions, and to databases of health care workers such as nurses and care givers. This is useful in cases where medical support is scare and in home-care services are demanded. It also promotes timely communication between different health units to improve patient care services. For example, a patient can use such systems during an emergency to check bed availability in a local clinic. Health Canada is currently supporting two PDA-related projects in Saskatchewan.

5. Diseases and Epidemic Outbreak Tracking: This category of health care systems focus on collecting and transmitting information about incidences of communicable diseases thereby helping the containment of such outbreaks. The information collected usually includes the location and levels of such diseases that health care authorities can use to target medical assets towards affected areas. It is particularly important in developing countries where collecting localized medical data is hard because a lot of affected patients often don't have access to hospitals that can measure and collect their health information in order to help health care authorities develop and gauge medical policies.

6. Remote Monitoring: This category of health care systems makes use of medical information exchanged via electronic mobile communications to monitor and improve patients' health conditions usually with the help of medical sensors. Remote monitoring systems read physiological information about chronic health conditions such as diabetes from the patient, and send them through mobile devices to a centralized repository hosted by health care providers for analysis.

The information gathered can be used for the early detection of extreme critical conditions that can lead to loss of life. However, remote monitoring is usually very costly to maintain.

7. Remote Data Collection: This category of health care systems collect data for use by health care authorities and health care providers which can help in the development and improvement of policies, especially in situations where information is needed in real-time. This is very important for conducting statistical analysis from surveyed data in developing countries. Such systems involve the use of mobile devices by health workers to collect and enter health data after which the information is transmitted to a centralized database for analysis. Nevertheless, systems designed for remote data collections are usually very costly to implement because of the complexity in integrating data which have the Big Data characteristics, such as *Volume, Velocity, Variety*, and *Veracity*.

5   Current Research Projects and Systems in Health Care

We confine here to discussing the merits and inadequacies of context-aware health care systems, because health care monitoring of elderly patients use sensor networks and wearable devices. The context-awareness behavior emerges in determining adaptations and timely reactions that are to be associated with the timely observations of patient behavior. Below we provide a review of some existing health care systems for the elderly and point out their inadequacies.

Codeblue [32], developed by the Harvard Sensor Network Lab, involves the use of multiple medical sensors for the tracking of patient's health information by end-user devices through wireless communication. In other words, each patient has a channel that sensors can publish data to and end-users such as doctors and nurses can subscribe to through their hand-held devices. This project seems to suffer from numerous attacks [25] such as *denial-of-service attacks*, and routing loop attacks and *grey-hole attacks*. *UbiMon*, developed by Imperial College, London, is aimed at capturing detectable and predictable life threatening abnormalities in elderly patients [36]. An analysis reported in [29] suggests that *UbiMon* has not met up the security requirements for wireless health care monitoring.

Many research projects have started using patient-wearable medical devices and sensors in the environment to monitor elderly patients. We survey some of them below.

In [12] the mental health of the elderly, especially those suffering from bipolar disorder (BP), is monitored in a two-level architecture. Personal Ambient Monitoring Infrastructure, known as PAM-I, is one level and Personal Ambient Monitoring Programming Architecture, known as PAM-A, is another level. PAM-I consists of medical sensors implanted into individual's body and environmental sensors placed in the home environment. The mobile phones are responsible for aggregating information from the sensors using rule-oriented applications and sending it to the personal computer for storage. The information measured by environmental sensors is transmitted to the personal computer through common communication protocols for example bluetooth. The PAM-A module, contains applications for handling inter-device network communication. The main focus of this project was to provide an acceptable level of health care solution for users, and privacy issues were completely ignored.

Alarm-Net [29], developed at the University of Virginia, is targeted at assisting elderly patient in their daily living. It makes use of wearable sensors to measure individual physiological data. The observed information transferred along with its source address, ID and sensor type are transferred in real-time for remote analysis. Network security is achieved in Alarm-Net through the use of secure remote password (SRP) protocol. Further, its sensors make use of AES encryption scheme to protect information. The major drawback of this approach is that there is no method given to decrypt the information while in communication.

MobiCare [9] was designed in 2006 and funded by the European Commission. The main motivation behind the development of MobiCare is to improve the quality of life for the elderly and in essence save their lives. The architecture consists of a sensor network containing wearable sensors, a BSN manager and a back-end infrastructure known as the MobiCare Server. The sensors are responsible for gathering information and broadcasting it to the MobiCare client which makes use of application layer standard HTTP POST protocol to store the information in the MobiCare Server. The security of patient information and level of security of sensor network and risk was not addressed.

MEDiSN [28], developed by John Hopkins University, contains several battery powered physiological monitors and medical sensors that measure physiological health information of a patient. The monitors are responsible for temporarily storing health information, encrypting and signing it and then transmitting it to relay points (RPs). These relay points are implemented as bidirectional routing tree.

Another wearable sensor-based mobile health care system that reads context-based information through the use of small and battery-operated beacons such as motion and location of an elderly is *CarePredict* [20]. It transmits data through wireless communication service to remote servers from where experts can examine the information in order to detect any acute deviation, such as restless sleep patterns and changes in eating patterns of the elderly which can then be isolated and investigated further. *CarePredict* allows care-givers and family members to track information about their loved ones from any computer and subscribers can grant or revoke access at any time they want.

In the above research access control methods for resource and information access by different subjects have not been discussed. The two recent works [2, 39] propose smart-home architectures with different objectives. The focus in [2] is to estimate loneliness of elderly patients in a senior home, whereas the focus in [39] is to monitor and assist elderly patients in a senior home. Both these works use a network of body-ware sensors (medical devices such as blood sugar monitors) and environmental sensors to identify, monitor, and measure the parameters of interest. Both works have not addressed security of sensor networks, and there is a potential for misusing the data collected by the sensors. Another important issue is data integrity, which is not addressed in these works.

In summary, existing systems are more focused on providing specific care and information to patients and aid health care staff in a limited manner. They neither provide any meaningful feedback to patients nor assist the health care staff to share knowledge in a secure manner. Another drawback is that the user interfaces for most of these systems are very complex thereby making interaction with such systems difficult for elderly patients. Some of the systems are targeted towards only one specific health condition. Consequently, elderly patients who suffer multiple medical problems need to use more than one system, which compounds the complexity of interface usage.

6   Current State of Security and Privacy in mHealth

For us to have a complete health care architecture, it is important to consider the serious challenge of protecting the information stored on these systems (security) and ensuring the information is used for appropriate or specified purposes (privacy). Most health care systems use wearable or implantable medical sensors that measure physiological information about patients and send them to a remote server through electronic devices. This architecture opens door to multiple entry points for attackers to compromise the system ranging from threats that focus on eavesdropping on information while in transit to threats that targeted towards making the system unavailable for usage. Over 18 million patients had their health records breached between 2009 and 2011 with an average of 49,000 records are breached per incident, thereby making the impact of breaches significant [45]. However, a report identified that there are more data breaches due to loss or theft than hacking [5].

The consequences of information breach on health care architectures are overwhelming. Health care providers can lose data, revenue, trust, customers and opportunities as a result of having their system compromised. In fact, for health care providers to benefit from the 2009 American Recovery and Reinvestment Act (ARRA), they must ensure that the health information collected from patients are stored with appropriate security and privacy measures and also meaningfully used [47]. A stolen personal health record is estimated to worth $50 on the black market which is much more than the cost of a stolen credit card with CCV estimated to worth between $1 and $6 [45]. It was further pointed that a US health insurance firm recently incurred fines and penalties worth $250,000 as a result of a recent breach of 1.5 million medical records. Apart from the financial consequences, many users usually lose faith in health care systems once there has been at least one case of breach reported. Therefore, both economic and ethical considerations require us to consider both security and privacy as "essential requirements" in developing health care systems.

6.1   Common Issues for Security Breach

The term mHealth is used for health care systems that allow its users use mobile devices for interacting with the system. In addition the system will use wireless communication technology for communication and data transfer across the health care network. Surveys revealed that a lot of users of health care systems are unaware that they are targets of cyber-attacks and what they can do to mitigate such threats. For example, as at 2012, 44% of adult mobile users are not aware of the vulnerabilities and solutions for their mobile systems [48]. According to the report, the number grew substantially to 57% in 2013. Usually, this lack of awareness is enhanced by the fact that hackers can easily cover their tracks in the mobile ecosystem as mobile devices have restricted user interfaces and low resources especially in the mHealth domain as most of such systems usually collect large amounts of sensitive personalized information about subscribers and this can lead to adverse consequences if such information is made available to unauthorized or unavailable to authorized users. An unauthorized user can range from an attacker that attempting to view information being sent over a public network to a third party accessing information from the mobile device if it is lost by the user.

However, most existing mHealth systems do not incorporate security. A recent survey, only 38% of mobile health care systems have clearly defined mobile privacy policies that they follow in ensuring patient's privacy [3]. Another study of mobile health and fitness applications reported that only 25% of free mHealth applications and 48% of paid mHealth applications informed their users about the existence of such policies [33]. Some of these applications also share personalized health information with third party systems without informing users [22].

In addition, because most mHealth systems make use of sensors to transmit sensitive detailed health information about patients over a wireless or wired network, an attacker can install a malicious node within the range of the network and intercept information which is referred to as passive eavesdropping or even take the attack further by using the malicious node to grab information via sending queries to an existing transmitter which is referred to as active eavesdropping [31]. Such information may contain additional information about patients' habits, location and movements and therefore unintended disclosure of such information pose huge risk to patients for example an attacker can use the compromised message to physically harm the patient [21].

Most mHealth systems store information about patients on cloud servers susceptible to various common internet attacks [55]. The recent attack on healthcare.gov website was due to the fact that the appropriate security measures for example, changing manufacturer's default password, were not implemented on the cloud servers used for the processing and storage of personal health information and in 2014, attackers targeted health care accounts more than the business sector [4].

Another important security and privacy issue for mHealth systems is the situation where patients lose their mobile devices without implementing appropriate authentication mechanisms such as passwords on such devices [26]. This also applies to health care providers implementing the *BYOD* paradigm where employees bring their own personal devices for use at work. 2% of all mobile users in the UK reported theft in 2009 [21]. In addition, security issues could also occur in cases of ownership changes where the application stores information locally and the user forgets to wipe his health information before decommissioning.

Usually, most of the security and privacy issues existing in mHealth systems occur because they are no standard development procedures [26]. This serves as an hindrance to scalability and integration with other mobile services such as mobile banking [38]. The lack of a standardized process means that ensuring security and privacy is the responsibility of the development team which might not even health professionals. Unintended data exposure can lead to loss of trust among clinicians and patients especially in cases of applications that additional services for example credit card processing.

6.2   The IAS-Octave Model for Software Security

In this section, we discuss the IAS-Octave Model proposed by the Information Assurance and Security in 2013, which states the requirements for implementing software security [10]. It is essential to integrate these requirements in an automated health care system. The following features are part of the IAS-Octave model.

- *Confidentiality:* Confidentiality refers to keeping the information private from unauthorized users. It can also be seen as the right of an individual to have personal information kept private [17]. If this right is violated, it could lead to grave consequences. Generally, confidentiality is achieved through cryptography. Whereas privacy is patient-centric, confidentiality is system-centric, in the sense that the checks and balances in the system must keep information confidential in order to meet the privacy requirements of its clients.
- *Integrity:* Integrity refers to keeping data as accurate and consistent as possible during its lifecycle [6]. In other words, information about users can only be modified by authorized parties especially in the health care industry where the personal health information about the patient is expected to be recent and only susceptible to changes by the patient himself. To ensure data integrity, database updates must synchronize with data modifications initiated by patients or their health care staff who are authorized to do so.
- *Availability:* A secure system is expected to be available to users at all times. Information is unavailable not only when it is lost or destroyed, but also when access to the information is denied or delayed (usually referred to *Denial Of Service* attacks which is very common nowadays). In health care, "availability" is a very important property. A typical scenario is patient monitoring where the system is expected to provide real time updates to health care providers. If information becomes unavailable because of device or network failure in such a system, it could lead to disastrous outcomes.
- *Accountability:* Accountability refers to the system feature that ensures that every action/has a source and target, and the transaction of an entity can uniquely be traced to its source. For example, the system should be able to link data violations leading to "compromising confidentiality" to appropriate actors in the system. It is necessary to impose appropriate sanctions/punishments on violators of security. In essence, accountability is answerability with sanctions [8].

- *Auditability:* Auditability is the capacity of a system to follow all activities associated with an information resource. Usually, it involves a backward trace of activities to ensure accountability. Usually auditability is achieved through the use of audit logs or trails.
- *Authenticity:* Authenticity refers to genuineness of information and those working with the information trust it. In other words, authenticity involves all parties involved provide a form of identity for authentication and authorization. There are three ways of authenticating a user which are things the user knows (for example, username and password combination), things the user has (for example, ID card) and things the user is or does (for example, fingerprints) [7]. Authorization is usually achieved through role-based access control.
- *Non-repudiation:* Non-repudiation is a legal concept which translates to the assurance that all parties involved in a digital transaction cannot deny their involvements. For example, non-repudiation can be used to guarantee that a sender of a message cannot deny having sent it and the recipient cannot deny receiving it.

6.3   Prototype Architecture

The security of the context-aware health care architecture [37] is built around the above requirements. This architecture specializes the context-aware framework architecture [1] and personalizes it to patients. Personalization is achieved by enabling patient interaction with system through body-ware sensors and the use of personalized hand-held devices, such as cell phone or laptop with fingerprint authentication mechanism. Data transmission to Cloud happens through a middleware where data security and integrity principles are enforced by the system. Health care actors and members/friends related to patients are assigned "roles", and assigned access rights based on these roles. In addition, access control methods make use of context information in determining the extent of access to patient data for actors. Thus, a physician may have access to a patient under her care only when either the patient is physically present in her office or the patient has initiated an on-line session for consultation. At other times, the physician cannot access the data relevant to patients under her care. This is just an example to illustrate the importance of context in assigning access to patient data. In principle, roles, contexts, and context-dependent access controls are to be determined jointly by patients, medical experts, and health counselors. A prototype of such a system under this architecture has been implemented. The platform used for the implementation is Java using the Java Development Kit(JDK) which includes a virtual machine to compile source code into a working application. This specialized architecture can be further enriched with the design pattern architecture [39] in which we can embed mechanisms for ensuring patient-centric security and privacy.

7   Conclusions

We have provided an informal in-depth review of elderly health care, its requirements, current status and its inadequacies in offering patient-centric care in total privacy. Overcoming these inadequacies is a primary motivation for our ongoing work on a new context-aware architecture in which IAS-Octave model security requirements are integrated. Given that health care systems reviewed by us are being largely deployed worldwide, there is definitely a need for an improved architecture that can be used as the skeleton for such systems. Our ongoing research is expected to result in a robust system. Our approach based on Component-based Software Engineering (CBSE) approach [35] has the following virtues.

- We use CBSE principles in constructing the architecture. It is an assemblage of components, where each component provides/receives services through its well-defined interfaces. Two components are connected only if services provided by one can be received by the other, subject to security and privacy constraints.
- Each component can be tested both individually as well as in conjunction with the components with which it interacts. The functionality at its interfaces, and the conformance of policies for invoking the functionality can be tested.
- By enriching an interface specification and by adding new interfaces the architecture is allowed to grow. That is, new components can be added to the current architecture in order to accommodate new requirements. As an example, "heath insurance entity", if required, can be added to the architecture without violating the existing architectural connections. Thus, the system grows "conservatively" meaning the overall behavior of the extended system absorbs (without violation) the behavior of the original architecture.
- Situation and Workflow languages are also extendable, in the sense new operations can be added within their current grammatical structures. We can add macros, pre-compiled predefined functions, to speed up the triggering of reactions.
- We have defined actuator types in order that any number of actuators within each type can be attached to the system. At run time the actuator controller will choose one actuator that is available to execute reactions. Moreover, it will choose actuators necessary to execute multiple actions as specified in the WPL language. This is possible because we can define a mapping from WPL expressions to actuator types.

Another important feature that we are investigating to integrate in our architecture is the *Autonomic Computing System* (ACS) principles. The importance of Autonomic and Cognitive sensing is emphasized by Yang [54]. The eight characteristics enumerated by Yang are (1) Self-management, (2) Self-configuration, (3) Self-integration, (4) Self-protection, (5) Self-optimization, (6) Self-healing, (7) Self-adaptation, and (8) Self-scaling. We adapt these eight characteristics of autonomic computing, first introduced by Horn [24], and subsequently strengthened by engineering informatics community [46] to health care as follows.

- ACS has to *know itself*. Since the system can exist in many layers, detailed knowledge of each layer, its components, and their relationships to other layers must be self-recorded. Such layers in health care are (1) environmental sensor layer, (2) body sensor layer, (3) patient layer (which might include social circles, friends and relatives), (4) layer of medical experts, (5) care givers and clinical layer, (6) service provision layer, and (7) data/knowledge base layer.
- ACS must *self-configure*. It refers to an automated configuration of components and subsystems according to well-defined policies, which will allow dynamic adaptation to changing environmental conditions. The environment includes a host of elderly homes, out-patient clinics, pharmacists, and institutional managers together with the devices/instruments they use to connect with the system.
- ACS has to *optimize*. Health care work flows that demand resources must be executed to minimize cost and maximize utility.
- ACS must *self-heal*. It must repair itself and recover from unusual event activity, such as network failure, external hacking, and lack of resources. Health care systems should recognize smart solutions and discover alternative ways of functioning in order to keep the availability of the system.
- ACS must *protect* itself. It must continuously monitor itself in order to "know who does what in the system", and prevent unauthenticated users from entering the system. It must protect confidentiality of data while authorized health care staff are actively sharing information, as well as when data/information is disclosed to outside of system.
- ACS must be *context-aware*. It has to know its environment which dynamically might change. It has to have a sufficient set of rules to interact with its neighbors.
- ACS must be both *open* and *closed*. It must use open standards to maximize its utility, while applying strict rules to prevent unauthorized access.
- ACS must *self-optimize*. It has to predict the optimal set of resources for accomplishing the current tasks and improve its performance. In health care the system should be able to function without delay when a large number of patients from remote areas access it with different service requests.

There is no system yet in which the above characteristics are integrated. Our ongoing research investigates techniques for constructing loosely coupled hierarchical control systems that are to be interconnected with our context-aware health care architecture to achieve privacy preserving personalized patient care with dependable timely adaptation.

References

1. Alagar, V., Mohammad, M., Wan, K., Hnaide, S.A.: A framework for developing context-aware systems. Trans. Context-aware Syst. Appl.: ICST **1**, 1–26 (2014). Springer
2. Austin, J., Dodge, H.H., Riley, T., Jacobs, P.C., Thielke, S., Kaye, J.: A smart-home system to unobtrusively and continuously assess lonliness in older adults. IEEE J. Transl. Eng. Health Med. **4**, 1–11 (2016)

3. Bari, F., Mark, F.: Security and privacy in mobile health. CIO J. (2013)
4. Benton, L.: Dissecting the mhealth storage dilemma: Cloud or on-site? October 2014. http://www.mhealthnews.com/blog/dissecting-mhealth-storage-dilemma-cloud-or-site. Accessed 29 Jan 2015
5. Bitglass: The 2014 bitglass healthcare breach report (2014)
6. Boritz, J.E.: IS practitioners' views on core concepts of information integrity. Int. J. Acc. Inf. Syst. 6(4), 260–279 (2005)
7. Braithwaite, W.R.B.: Why two-factor authentication in healthcare? (2009). https://www-304.ibm.com/partnerworld/gsd/showimage.do?id=25727. Accessed 29 Jan 2015
8. Brinkerhoff, D.: Accountability and health systems: Overview, framework and strategies, January 2003. http://www.who.int/accountability/Accountability HealthSystemsOverview.pdf. Accessed 29 Jan 2015
9. Chakravorty, R.: A programmable service architecture for mobile medical care, pp. 532–536. IEEE (2006)
10. Cherdantseva, Y., Hilton, J.: A reference model of information assurance and security. In: 2013 Eighth International Conference Availability, Reliability and Security (ARES), September 2013
11. Claudette, D.: Aboriginal Elder Abuse in Canada. Aboriginal Healing Foundation, Ottawa (2002)
12. Curtis, D.W., Pino, E.J., Bailey, J.M., Shih, E.I., Waterman, J., Vinterbo, S.A., Stair, T.O., Guttag, J.V., Greenes, R.A., Ohno-Machado, L.: SMART—an integrated wireless system for monitoring unattended patients. J. Am. Med. Inform. Assoc. 15(1), 44–53 (2008)
13. David, B., Alistair, T.: Community care of vulnerable older people: cause for concern. Br. J. Gen. Prac. 63(615), 549–550 (2013)
14. Debbie, T.: Overview of the personal health information protection act. Law and Governance, LegalFocus on Healthcare and Insurance, vol. 8, no. 4, June 2014. http://www.weirfoulds.com/overview-of-the-personal-health-information-protection-act
15. Elizabeth, C., Barbara, B., Lawson, W.: Reporting elder mistreatment. J. Gerontological Nurs. 23(7), 24–32 (1997)
16. Elizabeth, P.: National survey on abuse of the elderly in Canada. J. Elder Abuse Negl. 4(1–2), 5–58 (1993)
17. Fallon, L.F.: Patient confidentiality (2008). http://www.surgeryencyclopedia.com/Pa-St/Patient-Confidentiality.html. Accessed 29 Jan 2015
18. Goel, P., Garg, S., Singh, J., Bhatnagar, M., Chopra, H., Bajpai, S., et al.: Unmet needs of the elderly in a rural population of Meerut. Indian J. Commun. Med. 28, 165–166 (2003)
19. Greene, A.: When hipaa applies to mobile applications. Mobile Health News, June 2011. http://mobihealthnews.com/11261/when-hipaa-applies-tomobile-applications
20. Group, C: Empowering independent living for seniors. Technical report, Carepredict (2015). Accessed 29 Jan 2015
21. Group, V: Evaluating mhealth adoption barriers: Privacy and regulation. Technical report, Vodafone Global Enterprise (2013). Accessed 29 Jan 2015
22. HealthCareBusinessTech: Mobile health apps create privacy risk, study says. Technical report, Healthcare Business Technologies (2014). Accessed 29 Jan 2015
23. Hope, P., Bamford, S., Beales, S., Brett, K., Kneale, D., Macdonnell, M., McKeon, A.: Creating sustainable health and care systems in ageing societies. Technical report, Ageing Societies Working Group (2012). Accessed 29 Jan 2015

24. Horn, P.: Autonomic computing: IBM's perspective on the state of information technology (2001). http://www.research.ibm.com/autonomic/manifesto/ autonomiccomputing.pdf
25. Kambourakis, G., Klaoudatou, E., Gritzalis, S.: Securing medical sensor environments: the codeblue framework case. In: The Second International Conference on Availability, Reliability and Security, ARES 2007, pp. 637–643. IEEE (2007)
26. Kharrazi, H., Chisholm, R., VanNasdale, D., Thompson, B.: Mobile personal health records: an evaluation of features and functionality. Int. J. Med. Inform. **81**(9), 579–593 (2012). Accessed 29 Jan 2015
27. Kim, E., Geistfeld, L.: What makes older adults vulnerable to exploitation or abuse? (2008)
28. Ko, J., Lim, J.H., Chen, Y., Musvaloiu-E, R., Terzis, A., Masson, G.M., Gao, T., Destler, W., Selavo, L., Dutton, R.P.: Medisn: medical emergency detection in sensor networks. ACM Trans. Embed. Comput. Syst. (TECS) **10**(1), 11 (2010)
29. Kumar, P., Lee, H.J.: Security issues in healthcare applications using wireless medical sensor networks: a survey. Sensors **12**(1), 55–91 (2011)
30. Lauren, F., Deana, B., Chien-Chih, L., Samuel, B., Joseph, B.: Frail elderly as disaster victims: emergency management strategies. Prehosp. Disaster Med. **17**, 67–74 (2002)
31. Madhukar, A., Zachary, I., Insup, L.: Quantifying eavesdropping vulnerability in sensor networks. In: Proceedings of the 2nd International Workshop on Data Management for Sensor Networks, DMSN 2005, pp. 3–9. ACM, New York (2005). http://doi.acm.org/10.1145/1080885.1080887
32. Malan, D., Fulford-Jones, T., Welsh, M., Moulton, S.: Codeblue: an ad hoc sensor network infrastructure for emergency medical care. In: International Workshop on Wearable and Implantable Body Sensor Networks, vol. 5 (2004)
33. McCarthy, M.: Experts warn on data security in health and fitness apps. BMJ **347**, 1 (2013)
34. McDaniel, S.A.: The conundrum of demographic aging and policy challenges: a comparative case study of Canada, Japan and Korea. Can. Stud. Popul. **36**(1–2), 37–62 (2009)
35. Mohammad, M., Alagar, V.: A formal approach for the specification and verification of trustworthy component-based systems. J. Syst. Softw. **84**(1), 77–104 (2011)
36. Ng, J.W., Lo, B.P., Wells, O., Sloman, M., Peters, N., Darzi, A., Toumazou, C., Yang, G.Z.: Ubiquitous monitoring environment for wearable and implantable sensors (ubimon). In: International Conference on Ubiquitous Computing (Ubicomp). Citeseer (2004)
37. Oyekanmi, T.P.: A context-aware healthcare architecture for the elderly. Master of Computer Science thesis, Concordia University, Canada (2015)
38. Payne, J.: The state of standards and interoperability for mhealth among low- and middle-income countries. mHealth Alliance, March 2013
39. Periyasamy, K., Wan, K., Alagar, V.: Health care design patterns: an internet of things approach. In: Proceedings of the International Conference on Computers and their Applications, CATA 2017. IEEE (2017)
40. Province of Alberta: Alberta health information protection act. Technical report, The Government of Alberta (2010). http://www.qp.alberta.ca/documents/Regs/ 2010_118.pdf
41. Province of Saskatchewan: Health information protection act. Technical report, The Government of Saskatchewan (2003). http://www.health.gov.sk.ca/hipa

42. Renehan, E., Dow, B., Lin, X., Blackberry, I., Haapala, I., Gaffy, E., Cyarto, E., Brashe, K.: Healthy ageing literature review 2012. Technical report, State of Victoria, Department of Health (2012). Accessed 29 Jan 2015

43. Robert, B., Daniel, K.: Response of the elderly to disaster: an age-stratified analysis. Int. J. Aging Hum. Dev. **16**(4), 283–296 (1982)

44. Ronet, B., Heather, D., Mark, L.: Violence against the elderly a comparative analysis of robbery and assault across age and gender groups. Res. Aging **20**(2), 183–198 (1998)

45. SecureWorks: Healthcare security breaches: secure patient information and avoid financial penalties. Technical report, Secure Works (2015). Accessed 29 Jan 2015

46. Sterrit, R.M., Parashar, H., Tianfield, R., Unland, A.: A concise introduction to autonomic computing. Adv. Eng. Inform. **19**, 181–187 (2005)

47. Symantec: Security and privacy for healthcare providers. Best Practices Series For Healthcare (2009)

48. Symantec: Internet security threat report 2014. Technical report, Symantec (2014). Accessed 29 Jan 2015

49. The Council of Physicians and Surgeons: Confidentiality of personal health information. Technical report, College of Physicians and Surgeons of Ontario, April 2006. https://www.cpso.on.ca/uploadedFiles/policyitems/Confidentiality.pdf

50. The Economist Intelligence Unit Limited: Healthcare strategies for an ageing society. Technical report, Economist Intelligence Unit (2009)

51. Tueth, M.: Exposing financial exploitation of impaired elderly persons. Am. J. Geriatr. Psychiatry **8**(2), 104–111 (2000)

52. Way, U.: Seniors vulnerability report (2011). http://www.theprovince.com/pdf. Accessed 29 Jan 2015

53. WHO: mhealth: new horizons for health through mobile technologies. Technical report, World Health Organization, June 2011. Accessed 29 Jan 2015

54. Yang, G.: Body sensor networks - research challenges and applications (power point presentation) (2014)

55. Zhang, R., Liu, L.: Security models and requirements for healthcare application clouds. In: 2010 IEEE 3rd International Conference on Cloud Computing (CLOUD), pp. 268–275, July 2010

Secure Multi-party Comparison Protocol and Application

Jing Zhang[1,2,3(✉)], Shoushan Luo[2], and Yixian Yang[1,2]

[1] School of Computer and Information Technology,
Benjing Jiaotong University, Beijing 100444, China
zj_jsj@126.com
[2] Information Security Center,
Beijing University of Posts and Telecommunications, Beijing 100876, China
[3] College of Computer Science and Technology, Henan Polytechnic University,
Jiaozuo 454000, Henan, China

Abstract. The problem of information comparison is always an important field of SMC. In order to effectively solve the fully equal problem of multi-data for all information, a secure two-party multi-data comparison protocol for equality (STMC) is proposed with the aid of the NTRU encryption. The protocol converts multi-data comparison problem for equality to polynomials comparison for equality. Analysis shows that the protocol is correct and security in semi-honest model. Being STMC as basic building block, a secure multi-party multi-data comparison protocol for equality (SMMC) is proposed. SMMC provides a solution which n participants hope to determine the equality of their private input sets, on the condition of no information leaked. This protocol is proved to be collusion-resistance security. The last, computational complexity and communication complexity of the two protocols are analyzed. It is shown that new protocols have low complexity. We also give applications in the secure multi-party information comparison problem and secure multi-party polynomial comparison problem.

Keywords: SMC · Secure multi-party comparison problem · NTRU encryption

1 Introduction

The research of Yao's Millionaires [1] problem is always an important branch of Secure Multi-party Computation (SMC). The problem mainly solves that on the premise of not leaking their private information, two untrusted participants work together to compare the size of data. In Ref. [1], Yao suggested three solutions by one-way functions, but he did not give the systemic proof. Then Goldreich etc. proved that all secure multi-party computation problems are solvable in theory in Ref. [2], and suggested the general solution for secure multi-party computation in Ref. [3]. With the further development, the Millionaires problem has been extended to secure multi-party data equal comparison, secure multi-party intersection, security information sorting and security vector equal etc. In Ref. [4], a protocol of secure multi-party data comparison

© Springer Nature Singapore Pte Ltd. 2017
B. Zou et al. (Eds.): ICPCSEE 2017, Part I, CCIS 727, pp. 292–304, 2017.
DOI: 10.1007/978-981-10-6385-5_25

has been proposed, which based on the F function and semantic addition homomorphic encryption; Based on full homomorphic encryption, a two-party data comparison protocol was suggested in Ref. [5]; The research in Ref. [6] was an analysis method about privacy compared, which based on quantum check model; In Ref. [7], under the assumption of LWE in the lattice, the comparison problem was converted to a decryption ability of a random string. The method solved the secure two-party computation such as the relationship of an element and a set, set-intersection, and set equation etc.; In Ref. [8], Hazay study the two fundamental functionalities Oblivious Polynomial Evaluation committed oblivious PRF evaluation, and proved that the POE protocol were readily used for secure set-intersection; In Ref. [9], Neugebauer, etc. showed a new protocols, which solve the problem of fair and privacy-preserving ordered set reconciliation in the malicious model, based on a variety of novel non-interactive zero-knowledge-proofs; Luo proposed a security multi quantum sort agreement based on d dimension entanglement state in Ref. [10]; Dan etc. described the design and implementation of four different oblivious sorting algorithms in Ref. [11]; Among four algorithms, The first two were designs based on sorting networks and quicksort with the capability of sorting matrices, and the other two were designed with a low round count and an oblivious radix sort algorithm; In Ref. [12], Zhang etc. focused on the problem of verifiable privacy preserving multiparty computation, and design a series of protocols for dot product, ranging and ranking, which are proved to be privacy preserving and verifiable. In Ref. [13], a simple and efficient sorting algorithm was proposed for secure multi-party computation. The algorithm was efficient when the number of parties and the size of the underlying field were small; Geng etc. constructed a security equal protocol in Ref. [14] using secure scalar product protocol. In Ref. [15], López-Alt etc. show how on-the-fly MPC can be achieved using a new type of encryption scheme- multikey FHE scheme based on NTRU. Here we list contributions and defects about equal comparison protocol in recent years (Table 1).

Table 1. The contributions and defects of secure multi-party comparison protocol for equality

Reference	Main contribution	Defect
Ref. [4]	Comparing the multi-party's secret information and calculating the number of information	Higher computing complexity
Ref. [5]	Constructing a scheme to compare two private numbers	More comparison times; Can't ensure privacy value fully equal
Ref. [7]	Proving a two-party set equal problem with resisting quantum attack	Higher computing complexity
Ref. [14]	Solving vector comparison	Only solve vector comparison; No extensibility

In this paper, we study protocol of security two-party multi-data comparison for equality (STMC) and introduce a new technique for designing efficient secure protocols for these problems in the presence of semi-honest. We describe the execution of the protocol, and prove the correctness and security in detail. Being STMC as basic building block, we give a protocol of secure multi-party multi-data comparison for equality (SMMC). we make sure that SMMC is security against collusion through security proof. The result of analysis shows that our protocol has lower computation complexity than that of protocol in Refs. [4] and [7]. On the other hand, our SMMC can ensure privacy value fully equal. We further demonstrate that our technique is useful for various search functionalities.

2 Preliminaries

2.1 NTRU Encryption Algorithm [16]

NTRU is a cryptosystem based on the polynomial ring. The security of cryptosystem is associated with the shortest vector problem (SVP) of lattice solution [17]. So the encrypt scheme of NTRU is better than many public encryption systems such as the discrete logarithm, the factorization algorithm and so on. In addition, NTRU algorithm has the ability of resistance to quantum computing attack. Moreover, NTRU have many advantages, such as reasonably short, easily created keys, high speed, and low memory requirements, etc. [18]. We design our two protocols of comparison for equality based on the NTRU cryptosystem.

NTRU is a cryptosystem based on the polynomial ring $R_q = \mathbb{Z}[x]/(X^N - 1)$, where N denotes security coefficient and q denotes the prime integer. The polynomial ring is defined as $f(x) = f_0 + f_1 x + \ldots + f_{N-1} x^{N-1} \in R_q$ or the row vector $f = [f_0, f_1, \ldots, f_{N-1}]$. Define $(R_q, *)$ as the convolution product such as $(f * g)(x) = \sum\limits_{k=0}^{N-1} \sum\limits_{i+j=k(\mathrm{mod}N)}$ $f \cdot g_j \cdot x^k$, where $k \in [0, N], f, g \in R_q$. We have $L(d_1, d_2) = F \in \mathbb{Z}[x]/(X^N - 1)$.

Parameter. Assume that p and q are two prime integers with $p > q$, and select $d_j, d_g, d_r \in (0, N)$ by specified encoding rules, plaintexts space is

$$M = \left\{ a_0 + a_1 x + \ldots + a_{N-1} x^{N-1} \in \mathbb{Z}[x]/(X^N - 1) \mid -\frac{q}{2} < a_i < \frac{q}{2}, i = 0, \ldots, N - 1 \right\}$$

Key Creation. To create private key, we randomly choose the polynomial $f \in L_f(d_f, d_{f-1})$. The polynomial f must satisfy the requirement that there are two polynomials f_p, f_q make the identities true, that is $f_p * f \equiv 1 (\mathrm{mod}\, p), f_q * f \equiv 1 (\mathrm{mod}\, q)$. f, f_p are kept secret. We choose the polynomial $g \in L_g(d_g, d_{g-1})$ randomly, and compute $h = pf_q * g (\mathrm{mod}\, q)$. We public $PK = (h)$ and keep g, f_q secret.

Encryption. Suppose that the plaintexts is $m(x) \in M$, we randomly choose a polynomial $r(x) \in L_r(d_r, d_{r-1})$ with small coefficients to attack encrypted ciphertexts, and compute the result of ciphertexts e. where $e = r * h + m (\mathrm{mod}\, q)$.

Decryption. In order to decrypt e, we firstly compute $a \equiv f * e (\mathrm{mod}\, q)$, where we chose the coefficients of a in the interval from $(-\frac{q}{2}, \frac{q}{2})$. We recover the message by computing $d \equiv a * f_p (\mathrm{mod}\, p)$ based on the private key f. The result of computation d is decryption plaintexts.

NTRU have two features, that is $D(E(m_1) + E(m_2)) = m_1 + m_2, D(bE(m)) = bm.$

2.2 Security Theorem in Semi-honest Model

In the semi-honest model, both participants are strictly enforced the rules of the protocol without halfway forcibly exiting or malicious mixing with false data. However, each of parties could keep all process information that he get during the execution of protocol. The security proof in semi-honest model is implicit in the following description. There a probability polynomial time algorithm (sometimes referred to as simulator). With the help of the simulator, semi-honest participants can simulate the execution process of the protocol separately and obtain any information of the execution process, through directly using their own input and the final result of protocol.

Definition 1 (The security of secure two-party computations in semi-honest model). Let $f : \{0,1\}^* \times \{0,1\}^* \to \{0,1\}^* \times \{0,1\}^*$ be a functionality, where $f_i(x,y), i = 1,2$ is the i-th element of $f(x,y)$ and Π be a two-party protocol for computing f (denoted f^2). The view of the i-th party during an execution of Π on (x,y), denoted $VIEW_i^\Pi(x,y)$, is $(x, r, m_1, \ldots, m_t)$, where r represents the outcome of the i-th party's internal coin tosses, and m_j represent the j-th message it has received. Let $OUTPUT_i^\Pi(x,y)$ be the i-th output result. There exist probabilistic polynomial-time algorithm, denoted S_1 and S_2, making the Eqs. (1) and (2) workable, then it is sure for Π to the secure calculation f.

$$\{S_1(x, f_1(x,y)), f_2(x,y)\}_{x,y \in \{0,1\}^* s.t. |x|=|y|}$$
$$\overset{c}{\equiv} \{VIEW_1^\Pi(x,y), OUTPUT_2^\Pi(x,y)\}_{x,y \in \{0,1\}^* s.t. |x|=|y|} \tag{1}$$

$$\{f_1(x,y), S_2(y, f_2(x,y))\}_{x,y \in \{0,1\}^* s.t. |x|=|y|}$$
$$\overset{c}{\equiv} \{OUTPUT_1^\Pi(x,y), VIEW_2^\Pi(x,y)\}_{x,y \in \{0,1\}^* s.t. |x|=|y|} \tag{2}$$

Definition 2 (The security of secure multi-party computations in semi-honest model). Let $f : \{0,1\}^* \times \ldots \times \{0,1\}^* \to \{0,1\}^* \times \ldots \times \{0,1\}^*$ be an n-ary functionality, where $f_i(x_1, x_2, \ldots x_n), i = 1,2,\ldots n$ is the i-th element of $f(x_1, x_2, \ldots, x_n)$ and Π be multi-party protocol for computing f (denoted f^n). For a subset of participants $I = \{i_1, i_2, \ldots i_t\}$, let $VIEW_I^\Pi(\overline{x}) = \left(I, VIEW_{i_1}^\Pi(\overline{x}) \ldots, VIEW_{i_2}^\Pi(\overline{x}) \right)$ denote the message sequence during an execution of the I-th subset, where $f_I(\overline{X}) = \{f_{i_1}(x_1, x_2, \ldots x_n), f_{i_2}(x_1, x_2, \ldots x_n), \ldots, f_{i_t}(x_1, x_2, \ldots x_n)\}$. If there exist probabilistic polynomial-time algorithm, denoted S, such that the formulations (3) is workable, it is sure for Π to the secure calculation n-ary function f.

$$\{S(I, (X_{j_1}, \ldots, X_{j_s}), f_I(\overline{X}))\}_{\overline{X} \in (\{0,1\}*)^n} \overset{c}{\equiv} \{(view_I^\Pi(\overline{X}))\}_{\overline{X} \in (\{0,1\}*)^n} \tag{3}$$

2.3   The Problems

This paper studies the following two problems. Their applications will then be shown.

Problem 1. Secure two-party multi-data comparison for equality (STMC). Alice and Bob have private data sets $S_A = \{m_{A1}, m_{A2} \ldots, m_{An}\}$ and $S_B = \{m_{B1}, m_{B2} \ldots, m_{Bn}\}$ respectively, and they want to compare mutually the secure data on the condition of no private information leakage, and get whether the two-party private data sets are all equal.

Problem 2. Secure multi-party multi-data comparison for equality (SMMC). There are n participants. And they have respectively private set $S^{(i)} = \left\{m_1^{(i)}, m_2^{(i)}, \ldots, m_{k_1}^{(i)}\right\}, i = 1, 2, \ldots, n$ $S^{(i)} \subseteq \{1, 2, \cdots, N\}$. They hope to determine the equality of their private input sets, on the condition of no information leaked.

3   Solutions

3.1   Secure Two-Party Multi-data Comparison for Equality (STMC)

We consider two parties, Alice and Bob, who rank their respective collection elements in the order of smallest to largest in size. Without loss of generality, we assume $m_{A1} < m_{A2} < \ldots < m_{An}$ and $m_{B1} < m_{B2} < \ldots < m_{Bn}$. The public key and private key of protocol are generated by Alice, which are expressed as $SK = (f, f_p)$ and $PK = (h)$.

Protocol 1: secure two-party multi-data comparison for equality
Inputs: Alice inputs private sets $S_A = \{m_{A1}, m_{A2} \ldots, m_{An}\}$, and Bob inputs another private set $S_B = \{m_{B1}, m_{B2} \ldots, m_{Bn}\}$.
Output: Alice gets $F(x)$. If $F(x) = 0$, $S_A = S_B$ or $S_A \neq S_B$.

Step 1. Alice and Bob construct corresponding polynomials respectively, as follow:

$$M_i(x) = m_{i1} + m_{i2}x + \ldots + m_{in}x^{N-1}, (M_i \in L_M), (i = A, B) \tag{4}$$

Step 2. Alice selects a random polynomial $r_A(x)$ with small coefficient, calculates $E_{pk_A}(M_A(x))$ and sends it to Bob.
Step 3. Bob picks $b \neq 0$ randomly and calculates $bE_{pk_A}(M_A(x))$ after he receives $E_{pk_A}(M_A(x))$. Also, Bob selects a random polynomial $r_B(x)$ with small coefficient, calculates $E_{pk_A}(b(M_A(x) - M_B(x))) = bE_{pk_A}(M_A(x)) + E_{PK_A}(-bM_B(x))$, and sends it to Alice.
Step 4. Alice decrypts $D_{pk_A}(E_{pk_A}(b(M_A(x) - M_B(x))))$ and obtains a polynomial $F(x) = b(M_A(x) - M_B(x))$. Alice gets whether their data are all equal according to the result of $F(x)$. If $F(x) = 0$, then $S_A = S_B$; if not, $S_A \neq S_B$. And Alice sends the computing result to Bob.

Theorem 1: In semi-honest model, STMC is private.

Proof: The security of the protocol based on one of the NTRU encryption system. For the private key $SK = (f, f_p)$ of Alice, the adversaries construct a CS-lattice with 2 N dimension, which is shown as follow:

$$L_{CS} = \begin{bmatrix} \alpha I_N & M(h) \\ 0 & qI_N \end{bmatrix} = \begin{bmatrix} \alpha & 0 & \cdots & 0 & h_0 & h_1 & \cdots & h_{N-1} \\ 0 & \alpha & \cdots & 0 & h_{N-1} & h_0 & \cdots & h_{N-2} \\ \cdots & \cdots & \cdots & \cdots & \cdots & \cdots & \cdots & \cdots \\ 0 & 0 & 0 & \alpha & h_1 & h_2 & \cdots & h_0 \\ 0 & 0 & \cdots & 0 & q & 0 & \cdots & 0 \\ 0 & 0 & \cdots & 0 & 0 & q & \cdots & 0 \\ \cdots & \cdots & \cdots & \cdots & \cdots & 0 & \cdots & 0 \\ 0 & 0 & \cdots & 0 & 0 & 0 & \cdots & q \end{bmatrix} \quad (5)$$

The shortest vector $(\alpha f, g)$ with 2N dimension in L-lattice is obtained from the public key $PK = (h) = p \cdot f^{-1} \cdot g \pmod{q}$ of Alice. According to Ref. [14], the aim vector $\tau(\alpha f, g)$ of lattice can be solved, where $\tau = \sqrt{\frac{2\pi e \|f\|_2 \|g\|_2}{Nq}}$ is the length ratio between the aim vector and the shortest vector. If N is large enough, this problem of security is a NP problem of polynomial. It is almost impossible to decrypt the NTRU using lattice-reduction algorithm under the existing computational capabilities. Therefore, Bob cannot decrypt radically the private data of Alice using message sequence that he has acquired. We demonstrate the security of the protocol through constructing two simulators S_1 and S_2.

1) Simulating Bob's view through simulator S_2

Based on the inputs $B = \{r_B(x), b, M_B(x)\}$ and $f_2(A, B)$, S_2 proceeds as follows:

(1) S_2 first constructs $M_B(x) = m_{B1} + m_{B2}x + \ldots + m_{Bn}x^{N-1}$ ($M_B \in L_M$), and computer $E_{pk_A}(-bM_B(x))$.

(2) S_2 chooses randomly a set $S'_A = \{m'_{A1}, \ldots, m'_{An}\}$, $m'_{A1} < \ldots < m'_{An}$, and constructs $M'_A(x) = m'_{A1} + m'_{A2}x + \ldots + m'_{An}x^{N-1}$ ($M'_A \in L_M$); Then S_2 selects a polynomial $r'_A(x)$ with small coefficient and computes:

$$E_{pk_A}(M'_A(x)), bE_{pk_A}(M'_A(x)),$$
$$E_{pk_A}(F'(x)) = bE_{pk_A}(M'_A(x)) + E_{PK_A}(-bM_B(x)) = E_{pk_A}(b(M'_A(x) - M_B(x))).$$

(3) S_2 outputs

$$S_2(f_2(A, B), B) = \{r_B, b, M_B(x), E_{pk_A}(M'_A(x)), bE_{pk_A}(M'_A(x)), E_{pk_A}(-bM_B(x)), E_{pk_A}(F'(x)), F'(x)\}$$

Because of the choice of S'_A, it must hold that

$$M'_A(x) \overset{C}{\equiv} M_A(x), E_{pk_A}(M'_A(x)) \overset{C}{\equiv} E_{pk_A}(M_A(x)), bE_{pk_A}(M'_A(x)) \overset{C}{\equiv} bE_{pk_A}(M_A(x)),$$
$$F'(x) \overset{C}{\equiv} F(x), E_{pk_A}(F'(x)) \overset{C}{\equiv} E_{pk_A}(F(x))$$

Note that in this protocol: $output_1^{\pi} = output_2^{\pi} = \{F(x)\}$

$VIEW_2^{\Pi}(A, B)$
$= \{r_B(x), b, M_B(x), E_{pk_A}(M_A(x)), bE_{pk_A}(M_A(x)), E_{pk_A}(-bM_B(x))E_{pk_A}(F(x)), F(x)\}$

Therefore, we have:

$$\{f_1(A, B), S_2(B, f_2(A, B)),\} \overset{C}{\equiv} \{VIEW_2^{\Pi}(A, B), output_1^{\Pi}(A, B)\}.$$

2) Simulating Alice's view through simulator S_1, homoplastically.

3.2   Secure Multi-party Multi-data Comparison for Equality (SMMC)

As the extension of STMC, we consider the following problem: the n participants have respectively private set $S^{(i)} = \{m_1^{(i)}, m_2^{(i)}, \ldots, m_{k_1}^{(i)}\}, i = 1, 2, \ldots, n$ $S^{(i)} \subseteq \{1, 2, \cdots, N\}$. They hope to determine the equality of their private input sets, on the condition of no information leaked. If $F_i(x) = 0, M_n(x) = M_i(x)$ where $i = 1, 2, \ldots, n - 1$. Obviously, $M_1(x) = M_2(x) = \ldots = M_n(x)$. So we can demonstrate $S^{(1)} = S^{(2)} = \ldots = S^{(n)}$.

Protocol 2: Secure multi-party comparison protocol for equality.

Preliminary: The participant $P_i(i = 1, \ldots, n)$ ranks his set elements in ascending ranking sequence of $m_1^{(i)} < m_2^{(i)} < \ldots < m_{k_i}^{(i)}, i = 1, 2, \ldots, n$, and construct his polynomial $M_i(x) = m_1^{(i)} + m_2^{(i)}x + \ldots + m_{k_i}^{(i)}x^{k_i}$, where $i = 1, 2, \ldots, n$.

Inputs: each participant $P_i(i = 1, \ldots, n)$ inputs own polynomial $M_i(x)$ where $i = 1, 2, \ldots, n$.

Outputs: $F_i(x) = 0, i = 1, 2, \ldots, n$ or not.

Step 1. The participant P_n selects a random polynomial $r_n(x)$ with small coefficient, calculates his public key cryptosystem $E_{pk_n}(M_i(x))$ and sends it to the other participants.

Step 2. The participants $P_i(i = 1, \ldots, n - 1)$ randomly picks $b_i \neq 0$ and calculates $b_i E_{pk_n}(M_n(x))$ after they received $E_{pk_n}(M_n(x))$. Then each participant selects random polynomials $r_i(x), i = 1, 2, \ldots n - 1$ with small coefficient, calculate $r_i(x) \cdot h(x) - b_i M_i(x)$, which is denoted by $E_{PK_i}(-b_i M_i(x))$, and sends it to P_n.

Step 3. Received $E_{pk_n}(F_i(x))$, the participant P_n obtains a polynomial $E_{pk_n}(F_i(x))$ and computes $D_{pk_n}(E_{pk_n}(b_i(M_n(x) - M_i(x))))$. He gets the result, $S^{(1)} = S^{(2)} = \ldots = S^{(n)}$, by checking $F_i(x) = 0, i = 1, 2, \ldots, n$.

Theorem 2: SMMC is collusion-resistance security for semi-honest participants.

Proof: In the semi-honest model, the security of SMMC problem is that each participant cannot obtain any input-data from the other participants, since he has many information such as his own input, output of SMMC and the intermediate results that he has collected. The security of our protocol is considered from the following cases.

1) P_n is not corrupted.

In our protocol, all of the participants $P_i(i = 1, 2, \ldots, n-1)$ compute result with ciphertext. Namely they can only obtain ciphertext or intermediate result in the form of ciphertext. According to Theorem 1, it is difficult for the corruptors to get the private data of some participants in the case of unknowing private key. Because they have to solve the problem of shortest vector on lattice, which is non-deterministic polynomial problem.

Supposing corruptors' set is $P = \{P_1, P_2, \ldots, P_t\}$, where $P_i = \{r_i(x), M_i(x)\}$, $i = 1, \ldots, t$. The input of corruptors is $\{I, f_I(\overline{P})\}$ $I = \{1, 2, \ldots, t\}$, where $f_I(\overline{P}) = \{f_i(P_1, P_2, \ldots, P_n)\}$.

We construct the simulator S_1 to implement the view of corruptors:

(1) I and $f_I(P_1, P_2, \ldots, P_n)$ represent the input of S_1, and calculate $E_{pk_n}(M_i(x))$, where $i = 1, 2, \ldots, t$.

(2) S_1 chooses randomly sets $\{S'_{t+1}, S'_{t+2}, \ldots, S'_n\}$, where $S'_i = \left\{m_1^{(i)'}, \ldots, m_{k_i}^{(i)'}\right\}$, $m_1^{(i)'} < \ldots < m_{k_i}^{(i)'}$. The aim of S_1 is comparing equal or not, so no matter what S_1 chooses, the ideal result is only one result, yes or no. Obviously, the computation is indistinguishable, that is $\{S_1, \ldots S_t, S'_{t+1}, S'_{t+2}, \ldots, S'_n\} \overset{C}{\equiv} \{S_1, \ldots S_t, S_{t+1}, S_{t+2}, \ldots, S_n\}$.

(3) S_1 constructs $M'_i(x) = \sum\limits_{j=1}^{k_i} \left(m_j^{(i)'} \cdot x\right)$ $(M_i \in L_\mathbf{M}), i = t+1, t+2, \ldots, n$. Then S_1 selects randomly polynomial sequences $r'_i(x), i = t+1, \ldots, n$ with small coefficient and calculates $E_{pk_n}\left(M'_i(x)\right)$.

(4) S_1 selects randomly $b'_i \neq 0$, and simulate the implementation of protocol. S_1 obtains the information sequence as follows:

$$E_{pk_n}\left(M'_n(x)\right), b'_i E_{pk_n}\left(M'_n(x)\right), i = 1, 2, \ldots, t, E_{pk_n}\left(b'_i M_i(x)\right), i = 1, 2, \ldots, t,$$
$$E_{pk_n}\left(F'_i(x)\right) = E_{pk_n}\left(b'_i (M'_n(x) - M_i(x))\right), i = 1, \ldots, t.$$

(5) S_1 outputs $S(I, (P_1, \ldots, P_k), f_I(\overline{P})) = \{\{1, 2, \ldots, t\}, \{r_i(x), M_i(x)\}, \{E_{pk_n}\left(M'_n(x)\right), b'_i E_{pk_n}(M_n(x)), E_{pk_n}\left(F'_i(x)\right)\}\}(i = 1, 2, \ldots, t)\}$.

During the real implementation of the protocol, the corruptors' view is the following sequences.

$$VIEW_I^\Pi(\overline{P}) = (I, VIEW_1^\Pi(\overline{P}) \ldots, VIEW_k^\Pi(\overline{P}))$$

Where

$$VIEW_i^\Pi(\overline{P}) = \{\{r_i(x), M_i(x)\}, E_{pk_n}(M_i(x)), E_{pk_n}(M_n(x)), b_i E_{pk_n}(M_n(x)), E_{pk_n}(F_i(x))\}i$$
$$= 1, 2, \ldots, k$$

According to computational indistinguishable, we obtain:

$$b_i' \overset{C}{\equiv} b_i, i = 1, 2, \ldots, t, M_n'(x) \overset{C}{\equiv} M_n(x), E_{pk_n}(F_i'(x)) \overset{C}{\equiv} E_{pk_n}(F_i(x)), i = 1, 2, \ldots, t.$$

Therefore

$$\{S(I, (P_1, \ldots, P_k), f_I(\overline{P}))\}_{\overline{P}\in(\{0,1\}*)^n} \overset{C}{\equiv} \{VIEW_I^\Pi(\overline{P}))\}_{\overline{P}\in(\{0,1\}*)^n}.$$

2) P_n is corrupted

According to the description, we know that P_n has private key. So all $F_i(x) = b_i(M_n(x) - M_i(x)), i = 1, 2, \ldots, n$ are obtained by P_n. In fact, P_n cannot obtain $M_i(x), i = 1, 2, \ldots, n-1$ only if he knows the exact value about b_i. On the other hand, the corruptor can only obtain his or the other corruptors' intermediate result from the implementation of the protocol. Therefore it is impossible for corruptor group to obtain private information about other participants. Without loss of generality, we suppose corruptors' set is $P = \{P_1, P_2, \ldots, P_t, P_n\}$, where $P_i = \{r_i(x), M_i(x)\}, i = 1, \ldots, t$. The initial input of corruptors is $\{I, f_I(\overline{P})\}$, where

$$I = \{1, 2, \ldots, t, n\}$$
$$f_I(\overline{P}) = \{f_1(P_1, P_2, \ldots, P_n), f_2(P_1, P_2, \ldots, P_n), \ldots, f_k(P_1, P_2, \ldots, P_n), f_n(P_1, P_2, \ldots, P_n)\}.$$

Next, we construct the simulator S_1 to implement the view of corruptors.

(1) S_1 inputs I and $f_I(P_1, P_2, \ldots, P_n)$, and calculates $E_{pk_n}(M_i(x)), i = 1, 2, \ldots, t, n$.
(2) S_1 chooses randomly sets $\{S_{t+1}', S_{t+2}', \ldots, S_{n-1}'\}$, where $S_i' = \left\{m_1^{(i)'}, \ldots, m_{k_i}^{(i)'}\right\}$. $m_1^{(i)'} < \ldots < m_{k_i}^{(i)'}$. The aim of S_1 is comparing equal or not, so no matter what S_1 chooses, the ideal result is only one result, yes or no. Obviously, $\{S_1, \ldots S_t, S_{t+1}', S_{t+2}', \ldots, S_n\}$ is computational indistinguishable with $\{S_1, \ldots S_t, S_{t+1}, S_{t+2}, \ldots, S_n\}$.
(3) S_1 constructs $M_i'(x) = \sum_{j=1}^{k_i}\left(m_j^{(i)'} \cdot x\right)(M_i \in L_M), i = t+1, t+2, \ldots, n-1$. Then S_1 selects randomly polynomial sequences $r_i'(x), i = t+1, \ldots, n-1$ with small coefficient and calculates $E_{pk_n}(M_i'(x))$.
(4) S_1 selects randomly $b_i' \neq 0, i = 1, 2, \ldots, n$, and simulate the implementation of protocol. S_1 obtains the information sequence as follows:

$$E_{pk_n}(M_n'(x)); b_i' E_{pk_n}(M_n(x)), i = 1, 2, \ldots, t; E_{pk_n}(b_i' M_i(x)), i = 1, 2, \ldots, t;$$
$$E_{pk_n}(F_i'(x)) = E_{pk_n}(b_i'(M_n'(x) - M_i(x))), i = 1, \ldots, t;$$

$$F'(x) = {}^{(1)}F'_i(x) \cup {}^{(2)}F'_i(x), \text{ where } {}^{(1)}F'_i(x) = b'_i(M_n(x) - M_i(x)), i = 1, 2, \ldots, t, n$$
$${}^{(2)}F'_i(x) = b'_i(M_n(x) - M'_i(x)), i = t+1, t+2, \ldots, n-1$$

(5) S_1 outputs

$$S(I, (P_1, \ldots, P_k), f_I(\overline{P})) =$$
$$\{\{1, 2, \ldots, t\}, \{r_i(x), M_i(x)\}, \{E_{pk_n}(M'_n(x)), b'_i E_{pk_n}(M_n(x)), E_{pk_n}(F'_i(x))\} (i = 1, 2, \ldots, t), F'(x)\}$$

The information sequences which are obtained by corruptors in the real implementation of the protocol, are described as follow:

$$VIEW_I^\Pi(\overline{P}) = (I, VIEW_1^\Pi(\overline{P}) \ldots, VIEW_k^\Pi(\overline{P}), VIEW_n^\Pi(\overline{P}))$$

Where

$$VIEW_i^\Pi(\overline{P}) =$$
$$\{\{r_i(x), M_i(x)\}, E_{pk_n}(M_i(x)), E_{pk_n}(M_n(x)), b_i E_{pk_n}(M_n(x)), E_{pk_n}(F_i(x))\}, i = 1, 2, \ldots, k$$
$$VIEW_n^\Pi(\overline{P}) = \{\{r_n(x), M_n(x)\}, F(x)\};$$

According to computational indistinguishable, we have

$$b'_i \overset{C}{\equiv} b_i, i = 1, 2, \ldots, n, M'_i(x) \overset{C}{\equiv} M_i(x), i = t+1, \ldots, n-1$$
$$E_{pk_n}(F'_i(x)) \overset{C}{\equiv} E_{pk_n}(F_i(x)), i = 1, 2, \ldots, n; F'(x) \overset{C}{\equiv} F(x);$$

Therefore, $\{S(I, (P_1, \ldots, P_k), f_I(\overline{P}))\}_{\overline{P} \in (\{0,1\}*)^n} \overset{C}{\equiv} \{VIEW_I^\Pi(\overline{P}))\}_{\overline{P} \in (\{0,1\}*)^n}$

From what has been discussed above, we prove the protocol is correctness and collusion-resistance security for the semi-honest participants.

4 Complexity Analysis

In the area of SMC, many of the protocols that have been given have been called upon to invoke existing SMC algorithms. So there is no specific algorithm that is explicitly used. In order to calculate, we analyse complexity of our protocol in computational complexity and communication round.

In STMC, NTRU encryption is performed for one time; NTRU decryption is performed for two times; Multiplication is performed one time, and addition is performed for two times; So ignoring multiplication and addition, the computing complexity is $O(3N \log N)$ according to the computational complexity of NTRU encryption. Because Alice need sends ciphertext to Bob, and Bob sends the calculation results back after he received ciphertext. While Calculation operations of protocol in Ref. [7] include hash, collections, sorting, vector linear operation (L), number and a vector of sampling arithmetic (expressed in D and G), respectively. The hash functions and collections are expected to be sorted, so don't worry about the complexity. So the total complexity is (3 k + 1) L + mD + kG. In protocol, the sampling run time of number is O (n), And the sampling run time of vector can be reduced to input the length of the O (n^2/P) (P represents the number of processors). Communicational rounds is for two times. In order to more clearly comparison. Table 2 list computational complexity and communication rounds of the two protocols.

Table 2. Comparision of complexity

Paper	Computational complexity	Communication rounds
STMC	$O(3N \log N)$	2
Ref. [7]	$O\left(mN + k\frac{N^2}{P}\right)$	2

In SMMC, 2n − 1 times NTRU encryption and NTRU decryption are performed in the SMMC; n times multiplication are performed. So ignoring multiplication, computational complexity is $O(2nN \log N)$, and communication rounds is for 2(n − 1) times. While nL + 1 times homomorphic encryption and homomorphic decryption are performed in the Ref [4]. Multiplication are performed for n times. So computational complexity is $O(LnN \log N)$, and communication rounds is for kLn times. We list computational complexity and communication rounds of the two protocols in Table 3.

Table 3. Comparision of complexity

Paper	Encryption	Decryption	Multiplication	Computational complexity	Communication rounds
SMMC	n	n − 1	n	$O(2nN \log N)$	$2(n − 1)$
Ref. [4]	nL	1	n − 2	$O(LnN \log N)$	kLn

It can be seen that the computational complexity of our STMC is lower than that of Ref. [7], when solving two-party multi-data equal comparison; And the computational complexity and communication rounds of our SMMC are better than that of Ref. [4], when solving multi-party multi-data equal comparison. In addition, NTRU encryption system itself has a higher operation speed, therefore, under the same condition it can greatly reduce protocol encryption and decryption operation time, improve the operation speed of the protocol.

5 Application

In reality, there are a lot of use about secure multi-party multi-data comparison problem for equality, for example:

(1) Secure multi-party information comparison problem for equality

Assuming that $P_1, P_2, \ldots, P_n$ are participants who engage in the computation which can compare private information for equality without any information leaked, and each of them has a private information, $m_i, i = 1, 2 \ldots, n$. In order to solve this problem, each of participants covert private data into binary code. Then they construct their own polynomial, which puts code as its coefficients according to the order from low to high. For example, we suppose $m_1 = 5, m_2 = 2, m_3 = 5, m_4 = 3$ is respectively the private information which each of participants has, then we get the follow mapping relations $f(x) : 5 \rightarrow 101, 2 \rightarrow 010, 5 \rightarrow 101, 3 \rightarrow 011$, and follow polynomials: $M_1(x) = 1 + x^2$; $M_2(x) = x$; $M_3(x) = 1 + x^2$; $M_1(x) = 1 + x$. At last, we can use the new protocol to solve the problem directly. Need to point out that the protocol proposed in Ref. [4] is able to calculate the number of information which is not equal. About this problem, we can solve through expanding our protocol from unilateral to multilateral. The number of non-zero polynomial that each party gets is the number of unequal information. And the computational complexity and communication complexity of this protocol is superior to the Ref. [4].

(2) Secure multi-party polynomial comparison problem for equality

Assuming each of participants, $P_1, P_2, \ldots, P_n$, has a private polynomial with n degree, $f_i(x), i = 1, 2 \ldots, n$. They want to engage in the computation of polynomials comparing equal without any private information leaked. We define this problem as secure multi-party polynomials comparison problems for equality. Although our protocol mainly solve multi-party multi-information comparison problem, we achieve the final result by polynomials comparison. So we can use the new protocol to solve multi-party polynomials compare equal problem directly.

In addition, our protocol can solve the problem such as secure multi-party vector comparison problem for equality and secure multi-party tuples comparison problem for equality which are proved in Ref. [19].

6 Conclusion

In this paper, we study secure multi-party multi-data comparison problems for equality. We introduce NTRU encryption system for designing efficient secure protocols. Our starting point is the protocol for secure two-party comparison, whose security base on the difficulty of the shortest path on the lattice. The protocol has a lower computational complexity. Using STMC as basic building block, we present a secure multi-party multi-data comparison protocol for equality. We prove the protocol has collusion-resistance security for semi-honest participants. In addition, our protocols have good scalability, which used widely in secure multi-party computation. It should be pointed out that we implement the protocol in semi-honest model, we hope in the future work on secure multi-party comparing computation in the malicious model.

Acknowledgements. We would like to thank the anonymous reviewers. This work is supported by Asia 3 Foresight Program of the National Natural Science Foundation of China (Grant No. 61411146001).

References

1. Yao, A.C.: Protocols for secure computations. In: 23rd Annual Symposium on Foundations of Computer Science, SFCS 1982, pp. 160–164. IEEE (1982). doi:10.1016/0022-2836(81)90087-5
2. Goldreich, O., Micali, S., Wigderson, A.: How to play any mental game. In: Proceedings of the Nineteenth Annual ACM Symposium on Theory of Computing, pp. 218–229. ACM (1987)
3. Goldreich, O.: Secure multi-party computation [EB/OL] (1998). http://www.wisdom.weizman.ac.il/∼oded/pp.html
4. Liu, W., Wang, Y.B.: Secure multi-party comparing protocol and its applications. Acta Electronica Sinica 40(5), 871–876 (2012). (in chinese)
5. Tang, Q.Y., Chuan-Gui, M.A., Yan, G.: Comparing private numbers based on fully homomorphic encryption. J. Inf. Eng. Univ. 13(6), 654–657 (2013). (in chinese)
6. Yang, F., Yang, G., Hao, Y., et al.: Security analysis of multi-party quantum private comparison protocol by model checking. Modern Phys. Lett. B 29(18), 1550089 (2015)
7. Feng, X.: Secure two-party computation for set intersection and set equality problems based on LWE. J. Electron. Inf. Technol. 34(2), 462–467 (2012). (in chinese)
8. Hazay, C.: Oblivious polynomial evaluation and secure set-intersection from algebraic PRFs. In: Dodis, Y., Nielsen, J.B. (eds.) TCC 2015. LNCS, vol. 9015, pp. 90–120. Springer, Heidelberg (2015). doi:10.1007/978-3-662-46497-7_4
9. Neugebauer, G., Brutschy, L., Meyer, U., Wetzel, S.: Privacy-preserving multi-party reconciliation secure in the malicious model. In: Garcia-Alfaro, J., Lioudakis, G., Cuppens-Boulahia, N., Foley, S., Fitzgerald, W.M. (eds.) DPM/SETOP-2013. LNCS, vol. 8247, pp. 178–193. Springer, Heidelberg (2014). doi:10.1007/978-3-642-54568-9_12
10. Luo, Q.B., Yang, G.W., She, K., et al.: Multi-party quantum private comparison protocol based on d-dimensional entangled states. Quantum Inf. Process. 13(10), 2343–2352 (2014)
11. Bogdanov, D., Laur, S., Talviste, R.: A practical analysis of oblivious sorting algorithms for secure multi-party computation. In: Bernsmed, K., Fischer-Hübner, S. (eds.) NordSec 2014. LNCS, vol. 8788, pp. 59–74. Springer, Cham (2014). doi:10.1007/978-3-319-11599-3_4
12. Zhang, L., Li, X.Y., Liu, Y., et al.: Verifiable private multi-party computation: ranging and ranking. In: IEEE Proceedings of INFOCOM 2013, pp. 605–609. IEEE (2013)
13. Hamada, K., Ikarashi, D., Chida, K., et al.: Oblivious radix sort: an efficient sorting algorithm for practical secure multi-party computation. IACR Cryptol. ePrint Archive 2014, 121 (2014)
14. Geng, T., Li, H.C., Luo, S.S., et al.: A privacy-preserving dynamic point distance determination protocol and its extension. J. Beijing Univ. Posts Telecommun. 35(3), 47–51 (2012). (in chinese)
15. López-Alt, A., Tromer, E., Vaikuntanathan, V.: On-the-fly multiparty computation on the cloud via multikey fully homomorphic encryption. In: Forty-Fourth ACM Symposium on Theory of Computing, pp. 1219–1234. ACM (2012)
16. Hoffstein, J., Pipher, J., Silverman, J.H.: NTRU: a ring-based public key cryptosystem. In: Buhler, J.P. (ed.) ANTS 1998. LNCS, vol. 1423, pp. 267–288. Springer, Heidelberg (1998). doi:10.1007/BFb0054868
17. Babai, L.: On Lovász' lattice reduction and the nearest lattice point problem. Combinatorica 6(1), 1–13 (1986)
18. Hermans, J., Vercauteren, F., Preneel, B.: Speed records for NTRU. In: Pieprzyk, J. (ed.) CT-RSA 2010. LNCS, vol. 5985, pp. 73–88. Springer, Heidelberg (2010). doi:10.1007/978-3-642-11925-5_6
19. Li, S.D., Wang, S.D., Dai, Y.Q., Luo, P.: Multiparty secure computation for comparing two sets. Sci. China Ser. F: Inf. Sci. 39(3), 305–310 (2009). (in chinese)

Security Analysis of Secret Image Sharing

Xuehu Yan[✉], Yuliang Lu, Lintao Liu, Song Wan, Wanmeng Ding,
and Hanlin Liu

Hefei Electronic Engineering Institute, Hefei 230037, China
publictiger@126.com

Abstract. Differently from pure data encryption, secret image sharing (SIS) mainly focuses on image protection through generating a secret image into n shadow images (shares) distributed to n associated participants. The secret image can be reconstructed by collecting sufficient shadow images. In recent years, many SIS schemes are proposed, among which Shamir's polynomial-based SIS scheme and visual secret sharing (VSS) also called visual cryptography scheme (VCS) are the primary branches. However, as the basic research issues, the security analysis and security level classification of SIS are rarely discussed. In this paper, based on the study of image feature and typical SIS schemes, four security levels are classified as well as the security of typical SIS schemes are analyzed. Furthermore, experiments are conducted to evaluate the efficiency of our analysis by employing illustrations and evaluation metrics.

Keywords: Secret image sharing · Shamir's polynomial-based secret image sharing · Visual cryptography · Security analysis · Linear congruence

1 Introduction

Traditional image encryption protects the image security with no loss-tolerant. Secret image sharing (SIS) for (k, n) threshold encrypts the secret image into n noise-like shadow images (also called shares or shadows), and then distributes the shadow images among multiple participants. The secret can be recovered by collecting k or more authorized shadow images while less than k shadow images reveal nothing of the secret. SIS may be applied in not only information hiding, but also watermarking, access control, authentication, and transmitting passwords etc. visual secret sharing (VSS) [1] also called visual cryptography scheme (VCS) and Shamir's polynomial-based scheme [2], are the main branches in SIS.

In VSS [3–9] with (k, n) threshold, the generated n shadow images are first printed onto transparencies and then distributed to n associated participants. The key advantage of VSS lies in, the secret image can be reconstructed by superposing any k or more shadow images and human eyes with no cryptographic computation. Less than k shares will give no clue about the secret even if infinite

B. Zou et al. (Eds.): ICPCSEE 2017, Part I, CCIS 727, pp. 305–316, 2017.
DOI: 10.1007/978-981-10-6385-5_26

computation power is available. After Naor and Shamir's original work, the associated VSS problems and its physical properties are extensively researched, such as contrast [10,11], pixel expansion [4,5,12,13] threshold [14], multiple secrets [15], noise-like patterns [7,16–19], and so on [20,21].

In Shamir's original polynomial-based SIS [2] with (k, n) threshold, the secret image is encrypted into the constant coefficient of a random $(k-1)$-degree polynomial to obtain n shadow images, which are then also distributed to n associated participants. The secret image can be reconstructed with high-resolution based on Lagrange interpolation when collecting any k or more shadow images. Following Shamir's scheme and employing all the coefficients of the polynomial for embedding secret, Thien and Lin [22] reduced shadow image size $1/k$ times to the original secret image. Inspired by Thien and Lin's research, some Shamir's polynomial-based schemes [15,21,23,24] were proposed to obtain more features. The advantage of Shamir's polynomial-based scheme is the secret can be recovered with high quality. Although Shamir's polynomial-based SIS only needs k shadow images for reconstructing the distortion-less secret image, while it requires more complicated computations, i.e., Lagrange interpolations, in the recovery phase and known order of shares.

Besides the above mentioned two primary branches, some other SIS schemes [25–27] were introduced as well to achieve different features.

However, as the basic research issues, the security analysis and security level classification of SIS are rarely discussed in the previous studies, which will be considered in this paper. The contributions of this paper lie in, first four security levels of typical SIS are classified based on the study of image feature and typical SIS schemes. Second, the security levels corresponding typical SIS schemes are theoretically analyzed. Finally, experimental results of typical SIS schemes further show the effectiveness of our work.

The rest of the paper is organized as follows. Section 2 introduces some basic requirements for the proposed method. In Sect. 3, four security levels are presented in detail. Section 4 gives the security analyses of typical SIS schemes. Section 5 is devoted to experimental results. Finally, Sect. 6 concludes this paper.

2   Preliminaries

In this section, we give some preliminaries as the basis for our work. In (k, n) threshold SIS, the original secret image S is shared among total n shadow images $SC_1, SC_2, \cdots SC_n$, while the reconstructed secret image S' is reconstructed from t $(k \leq t \leq n, t \in \mathbf{Z}^+)$ shadow images. $S(i, j) \in [0, P-1]$, where $[0, P-1]$ denotes the pixel values range and P indicates the maximum possible pixel value. For instance, in VSS $P = 2$ for binary secret image and $P = 251,256$ or a suitable prime number in gray image sharing or colour image sharing.

s is the original secret pixel $S(i, j)$ or a secret region. s'_{k-1} denotes the recovered secret pixel $S'(i, j)$ or region by $k-1$ or less shadow images. In general, for (k, n) threshold SIS, security is defined as that s'_{k-1} gives no clue about the secret.

In what follows, symbols $\oplus$ and $\otimes$ denote the Boolean XOR and OR operations.

2.1   VSS

In RG-based VSS [28], 0 denotes white pixel, 1 denotes black pixel. The generation and recovery phases of original $(2,2)$ RG-based VSS are described below.

Step 1: Randomly generate 1 RG SC_1.

Step 2: Compute SC_2 as in Eq. (1).

Recovery: $S' = SC_1 \otimes SC_2$ as in Eq. (2). If a certain secret pixel $s = S(i,j)$ of S is 1, the recovery result $SC_1 \otimes SC_2 = 1$ is always black. If a certain secret pixel is 0, the recovery result $SC_1 \otimes SC_2 = SC_1(i,j) \otimes \overline{SC_1(i,j)}$ has half chance to be black or white since SC_1 are generated randomly.

$$SC_2(i,j) = \begin{cases} SC_1(i,j) & if\ S = 0 \\ \overline{SC_1(i,j)} & if\ S = 1 \end{cases} \tag{1}$$

$$\begin{aligned} S'(i,j) &= SC_1(i,j) \otimes SC_2(i,j) \\ &= \begin{cases} SC_1(i,j) \otimes SC_1(i,j) & if\ S(i,j) = 0 \\ SC_1(i,j) \otimes \overline{SC_1(i,j)} = 1 & if\ S(i,j) = 1 \end{cases} \end{aligned} \tag{2}$$

The above approach can be extended to (k,n) threshold scheme by applying the above process repeatedly for the first k bits and generating the last n - k bits to be equal to subset of the first k bits or even to 0. one (k,n) RG-based VSS example [29] is described as follows:

Algorithm 1. (k,n) RG-based VSS
Input: A $M \times N$ binary secret image S, the threshold parameters (k,n)
Output: n shadow images $SC_1, SC_2, \cdots SC_n$
Step 1: For each position $(i,j) \in \{(i,j) \| 1 \le i \le M, 1 \le j \le N\}$, repeat Steps 2–4
Step 2: To obtain $b_1, b_2, \cdots b_k$ from $S(i,j)$ in parallel one by one repeatedly using Eq. (1) where $\tilde{b}_x$ denotes the temporary pixel
Step 3: Compute $N_k = \lfloor n/k \rfloor$, set $b_{k+1} = b_1, b_{k+2} = b_2, \cdots b_{2k} = b_k, b_{2k+1} = b_1, \cdots b_{N_k \times k} = b_k$ and $b_{N_k \times k+1} = b_{N_k \times k+2} = b_n = 0$
Step 4: Randomly rearrangement $b_1, b_2, \cdots b_n$ to $SC_1(i,j), SC_2(i,j), \cdots SC_n(i,j)$
Step 5: Output the n shadow images $SC_1, SC_2, \cdots SC_n$

2.2   Shamir's Polynomial-Based SIS Scheme

We take a secret pixel value s as the currently processing gray value of the secret image, and then to split s into n pixels corresponding to n shadows by original Shamir's polynomial-based SIS scheme. The specific scheme is listed as follows:

Algorithm 2. Original Shamir's polynomial-based SIS

Input: A $M \times N$ binary secret image S, the threshold parameters (k, n)

Output: n shadow images $SC_1, SC_2, \cdots SC_n$

Step 1: A prime number P is selected. For each position $(i, j) \in \{(i, j)|1 \leq i \leq M, 1 \leq j \leq N\}$, repeat Steps 2–4

Step 2: For the current processing pixel value $s = S(i, j)$, to divide into pieces sc_i, we generate a $k - 1$ degree polynomial

$$f(x) = (a_0 + a_1 x + \cdots + a_{k-1} x^{k-1}) \bmod P \qquad (3)$$

in which $a_0 = s$, and a_i is random, $i = 1, 2, \cdots k - 1$.

Step 3: Compute

$$sc_1 = f(1), \cdots, sc_i = f(i), \cdots, sc_n = f(n). \qquad (4)$$

Step 4: Arrangement $sc_1, sc_2, \cdots sc_n$ to $SC_1(i, j), SC_2(i, j), \cdots SC_n(i, j)$

Step 5: Output the n shadow images $SC_1, SC_2, \cdots SC_n$

In the recovery phase of original Shamir's polynomial-based SIS, given any k pairs of these n pairs $\{(i, sc_i)\}_{x=1}^{n}$, where i serves as an identifying index or a order label corresponding to the ith participant, we can obtain the coefficients of $f(x)$ by the Largrange's interpolation, and then evaluate $s = f(0)$. And the process repeats until all pixels of the secret image are finished. And the secret image S cannot be revealed without k or less shadow images.

Following Shamir's scheme and utilizing all coefficients of the polynomial for embedding secret, i.e., in Eq. (3) a_i, $i = 0, 1, 2, \cdots k - 1$, is sequentially picked up from not-shared-yet pixels of the encrypted secret image S_1, Thien and Lin [22] reduced share size $1/k$ times to the secret image. The differences between original Shamir's polynomial-based SIS and Shamir's polynomial-based SIS with smaller share size (namely improved Shamir's polynomial-based SIS) lie in:

1. Only the first coefficient a_0 in Eq. (3) covers the secret pixel in original Shamir's polynomial-based SIS, while all the k coefficients $a_0, a_1, \cdots a_{k-1}$ cover the secret pixels in improved Shamir's polynomial-based SIS. Thus, share size of improved Shamir's polynomial-based SIS is reduced $1/k$ times to that of original Shamir's polynomial-based SIS.
2. Before the really sharing processing, the secret image S is encrypted to obtain secret image S_1 in improved Shamir's polynomial-based SIS. Then the encrypted secret image S_1 is severed as the secret image for sharing.

3   SIS Security Analysis

Here, in order to give the security analysis and security level classification of SIS, in this section, based on the study of image feature and typical SIS schemes, four security levels are classified.

3.1   The Feature Analysis of Image

Digital image is different from pure electronic data. The image is composed of pixels, and there is a certain correlation between pixels, such as texture, structure, edge and other related information. In the local region of an image, the pixel value of one pixel is close to its adjacent pixels value. Therefore, the security of the image needs to consider two aspects, i.e., corresponding pixel security and region security.

3.2   Security Level Classification and Analysis

Based on the study of image feature and typical SIS schemes, four security levels of SIS are classified as follows:

1. s'_{k-1} is random, i.e., $\Pr ob\left(s'_{k-1} = i\right) = \frac{1}{P}$ for any $i \in [0, P-1]$.
2. s'_{k-1} is random or $s'_{k-1} = c$, where c indicates a constant number with finite certain values or guided by another input as well as c has no relation with the original secret s.
3. $s'_{k-1} = f(s_1)$, where $f(x)$ means a relation function of x and s_1 denotes the encrypted result of s.
4. $s'_{k-1} = f(s)$.

For **Level 1**, since s'_{k-1} is random, SIS in **Level 1** has similar security to "one-time pads", the most secure encryption method known.

If $s'_{k-1} = c$ for **Level 2**, c has no relation with the original secret s, thus s'_{k-1} of SIS in **Level 2** gives no clue about the secret s as well.

As a result, s'_{k-1} of SIS in both **Level 1** and **Level 2** can not disclose the secret s.

For **Level 3**, s'_{k-1} has a relation with the secret encrypted result, such as, linear function. If there is no encryption technique, s'_{k-1} may give clue about the secret s. Hence, the security is guaranteed in a degree by another encryption technique other than only SIS itself.

s'_{k-1} has a relation with the secret in **Level 4**. There is information leakage for SIS, which may be designed for special applications.

4   Security Analysis of Typical SIS Schemes

In this section, the security of typical SIS schemes will be analyzed.

4.1   VSS

In fact, Eq. (1) is equal to $sc_2 = sc_1 \oplus s$ or $s = sc_1 \oplus sc_2$. Since if $s = 0 \Rightarrow sc_2 = sc_1 \oplus 0 \Rightarrow sc_2 = sc_1$, and if $s = 1 \Rightarrow sc_2 = sc_1 \oplus 1 \Rightarrow sc_2 = \overline{sc_1}$.

Since sc_1 is random and $sc_2 = sc_1 \oplus s$, sc_2 is random as well. In traditional RG (2,2) VSS, we have $k = 2, n = 2$. When $k - 1 = 1$ or less shadow image is stacked, $s'_{k-1} = sc_2$ or sc_1, hence s'_{k-1} is random. As a result, traditional RG (2,2) VSS belongs to **Level 1**.

The equation $s = sc_1 \oplus sc_2$ in traditional RG (2,2) VSS could be extended to $s = sc_1 \oplus sc_2 \oplus \cdots \oplus sc_k$, which is applied in the (k, n) RG-based VSS shown in Algorithm 1.

In step 2 of (k, n) RG-based VSS, the k bits are utilized in parallel to gain threshold mechanism, i.e., when less than k shadow images are collected, the secret will not be revealed due to $s = b_1 \oplus b_2 \oplus \cdots \oplus b_k$ and $s'_{k-1} = b_1 \oplus b_2 \oplus \cdots \oplus b_{k-1}$. In Step 3, it sets the next $N_k \times k - k + 2$ bits to be distinct one of the first k bits so that every bit of the next $N_k \times k - k + 2$ bits is random. On the other hand the last $n - N_k \times k - 1$ are set to be 0 (here c = 0) in (k, n) RG-based VSS. As a result, the (k, n) RG-based VSS belongs to **Level 2**.

4.2 Shamir's Polynomial-Based SIS

In original Shamir's polynomial-based SIS, knowledge of just $k-1$ of these values does not suffice in order to calculate s since Largrange's interpolation. In other words, from Eq. (3), for the first k shadow images we have

$$
\begin{aligned}
f(1) &= (a_0 + a_1 + \cdots + a_{k-2} + a_{k-1}) \bmod P \\
f(2) &= (a_0 + 2a_1 + \cdots + 2^{k-2}a_{k-2} + 2^{k-1}a_{k-1}) \bmod P \\
&\cdots \\
f(k-1) &= (a_0 + (k-1)a_1 + \cdots + (k-1)^{k-2}a_{k-2} \\
&\quad + (k-1)^{k-1}a_{k-1}) \bmod P \\
f(k) &= (a_0 + ka_1 + \cdots + k^{k-2}a_{k-2} + k^{k-1}a_{k-1}) \bmod P
\end{aligned}
\tag{5}
$$

Since $a_1, \cdots, a_{k-2}, a_{k-1}$ all are random, from the first k-1 equations, we can only gain $s'_{k-1} = c_1 \times a_{k-1} + c_2$, where c_1, c_2 indicate two constant numbers. Due to a_{k-1} is random, s'_{k-1} is random. Finally, original Shamir's polynomial-based SIS belongs to **Level 1**.

All the k coefficients $a_0, a_1, \cdots a_{k-1}$ cover the secret pixels in improved Shamir's polynomial-based SIS, and the pixel value of one pixel is close to its adjacent pixels value, thus we have $a_0 \approx a_1, \cdots \approx a_{k-1}$. Then, from $f(1) = (a_0 + a_1 + \cdots + a_{k-2} + a_{k-1}) \bmod P$, we know $s'_1 = \frac{f(1)}{k} \approx a_0$. Hence, $s'_1 \approx f(s_1)$. s'_1 has a relation with the secret encrypted result. s'_1 will give clue about the secret s without encryption technique. As a result, improved Shamir's polynomial-based SIS belongs to **Level 3**.

5 Experimental Results and Analyses

In this section, experiments and analysis are conducted to evaluate the effectiveness of our methods and analysis.

Simulation result by original RG-based (2, 2) VSS [28] is presented in Fig. 1. The binary secret image S is illustrated in Fig. 1(a). Two generated shares SC_1

and SC_2 which are noise-like are demonstrated in Fig. 1(b)–(c). The stacking result denoted as S' by the two shares is shown in Fig. 1(d), which reveals the secret. Thus, $S'_1 = SC_1$ or $S'_1 = SC_2$ is random, i.e., $\Pr{ob}\left(s'_{k-1} = i\right) = \frac{1}{2}$ for any $i \in [0, 1]$. The result validates our analysis, i.e., traditional RG $(2, 2)$ VSS belongs to **Level 1**.

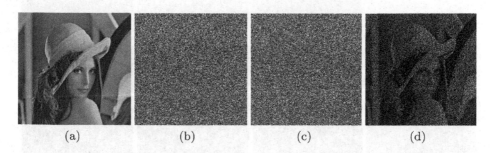

(a)   (b)   (c)   (d)

Fig. 1. Simulation result of traditional RG-based $(2, 2)$ VCS [28]. (a) The secret image S; (b)–(c) two random shares SC_1 and SC_2; (d) stacking result S' by two shares (SC_1 and SC_2).

Figure 2 illustrates the test of (k, n) RG-based VSS [29], where $k = 3, n = 4$ and the binary secret image is in Fig. 2(a), 2(b–e) show the 4 shadow images SC_1, SC_2, SC_3 and SC_4, which are noise-like. Figure 2(f–j) show the recovered secret images with any 3 or 4 shadow images with stacking recovery, from which the secret image recovered from $k = 3$ or more shadow images could be recognized based on superposition. Figure 2(k–p) show the recovered secret image with any less than $k = 3$ shadow images based on stacking recovery, from which there is no information could be recognized. Since s'_{k-1} is random or $s'_{k-1} = c$, the (k, n) RG-based VSS belongs to **Level 2**.

We will consider original Shamir's polynomial-based SIS with $(2, 2)$ threshold and gray secret image in Fig. 3(a). As shown in Fig. 3(b) and (c), two shadow images SC_1 and SC_2 are generated, which are noise-like. The secret image revealed by the two shadow images is exhibited in Fig. 3(d). Thus, $s'_1 = sc_1$ or sc_2 is random. Finally, original Shamir's polynomial-based SIS belongs to **Level 1**.

Figure 4 shows the experiment of improved Shamir's polynomial-based SIS with secret image encryption for $(2, 4)$ threshold. Figure 4(a) presents the grayscale secret image whose encrypted result is given in Fig. 4(b) by XOR encryption. The corresponding four shadow images SC_1, SC_2, SC_3 and SC_4 generated by improved Shamir's polynomial-based SIS from Fig. 4(b) are illustrated in Fig. 4(c)–(f) which are noise-like image. The result revealed by the first two shadow images is shown in Fig. 4(g) and its decryption secret image is in Fig. 4(h).

Furthermore, Fig. 5 gives the experiment of improved Shamir's polynomial-based SIS without secret image encryption for $(2, 4)$ threshold. Figure 4(a)

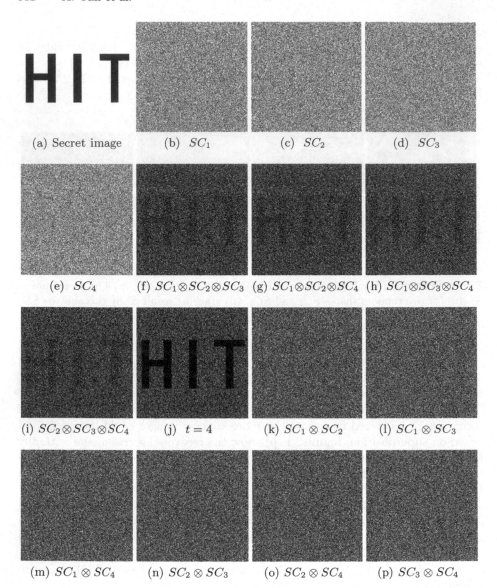

Fig. 2. Experimental example of (k, n) RG-based VSS, where $k = 3, n = 4$

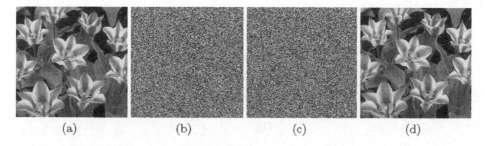

Fig. 3. Simulation results of original Shamir's polynomial-based SIS method, where $k = 2, n = 2$. (a) The gray secret image; (b)–(c) two original shadow images SC_1, SC_2; (d) revealing result by SC_1 and SC_2.

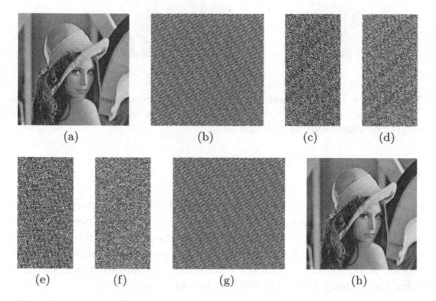

Fig. 4. Simulation results of improved Shamir's polynomial-based SIS with secret image encryption, where $k = 2, n = 2$. (a) the secret image; (b) encrypted secret image by XOR encryption on (a); (c)–(f) four shadow images; (g) revealing result by SC_1 and SC_2; (h) decryption secret image

presents the grayscale secret image whose corresponding four shadow images SC_1, SC_2, SC_3 and SC_4 generated by improved Shamir's polynomial-based SIS without secret image encryption are exhibited in Fig. 4(b)–(e). The result revealed by the first two shadow images is shown in Fig. 4(f). From the four shadow images, we can see that each shadow image gives clue about the secret image, hence improved Shamir's polynomial-based SIS belongs to **Level 3**.

(a)   (b)   (c)   (d)

(e)   (f)

Fig. 5. Simulation results of improved Shamir's polynomial-based SIS without secret image encryption, where $k = 2, n = 2$. (a) The secret image; (b)–(e) four shadow images; (f) revealing result by SC_1 and SC_2

6   Conclusion

In this paper, first based on the study of image feature and typical secret image sharing (SIS) schemes, four security levels of typical SIS are classified. Second, the security levels corresponding typical SIS schemes are theoretically analyzed. Finally, experimental results of typical SIS schemes further show the effectiveness of our work. Quantificationally evaluating and its evaluation metrices for analyzing the security of typical SIS will be our future work.

Acknowledgement. The authors would like to thank the anonymous reviewers for their valuable comments. This work is supported by the National Natural Science Foundation of China (Grant No. 61602491).

References

1. Naor, M., Shamir, A.: Visual cryptography. In: Santis, A. (ed.) EUROCRYPT 1994. LNCS, vol. 950, pp. 1–12. Springer, Heidelberg (1995). doi:10.1007/BFb0053419
2. Shamir, A.: How to share a secret. Commun. ACM **22**(11), 612–613 (1979)
3. Yan, X., Wang, S., El-Latif, A.A.A., Niu, X.: Visual secret sharing based on random grids with abilities of AND and XOR lossless recovery. Multimedia Tools Appl. **74**(9), 3231–3252 (2015). doi:10.1007/s11042-013-1784-2
4. Yang, C.N.: New visual secret sharing schemes using probabilistic method. Pattern Recogn. Lett. **25**(4), 481–494 (2004)

5. Cimato, S., De Prisco, R., De Santis, A.: Probabilistic visual cryptography schemes. Comput. J. **49**(1), 97–107 (2006)
6. Wang, D., Zhang, L., Ma, N., Li, X.: Two secret sharing schemes based on Boolean operations. Pattern Recogn. **40**(10), 2776–2785 (2007)
7. Wang, Z., Arce, G.R., Di Crescenzo, G.: Halftone visual cryptography via error diffusion. IEEE Trans. Inf. Forensics Secur. **4**(3), 383–396 (2009)
8. Weir, J., Yan, W.Q.: A comprehensive study of visual cryptography. In: Shi, Y.Q. (ed.) Transactions on Data Hiding and Multimedia Security V. LNCS, vol. 6010, pp. 70–105. Springer, Heidelberg (2010). doi:10.1007/978-3-642-14298-7_5
9. Yang, C.N., Sun, L.Z., Yan, X., Kim, C.: Design a new visual cryptography for human-verifiable authentication in accessing a database. J. Real-Time Image Process. **12**(2), 483–494 (2016)
10. Wu, X., Sun, W.: Improving the visual quality of random grid-based visual secret sharing. Sig. Process. **93**(5), 977–995 (2013)
11. Yan, X., Liu, X., Yang, C.N.: An enhanced threshold visual secret sharing based on random grids. J. Real-Time Image Process. 1–13 (2015)
12. Guo, T., Liu, F., Wu, C.: Threshold visual secret sharing by random grids with improved contrast. J. Syst. Softw. **86**(8), 2094–2109 (2013)
13. Fu, Z., Yu, B.: Visual cryptography and random grids schemes. In: Shi, Y.Q., Kim, H.-J., Pérez-González, F. (eds.) IWDW 2013. LNCS, vol. 8389, pp. 109–122. Springer, Heidelberg (2014). doi:10.1007/978-3-662-43886-2_8
14. Yan, X., Wang, S., Niu, X.: Threshold construction from specific cases in visual cryptography without the pixel expansion. Sig. Process. **105**, 389–398 (2014)
15. Li, P., Ma, P.J., Su, X.H., Yang, C.N.: Improvements of a two-in-one image secret sharing scheme based on gray mixing model. J. Vis. Commun. Image Represent. **23**(3), 441–453 (2012)
16. Yan, X., Wang, S., Niu, X., Yang, C.N.: Generalized random grids-based threshold visual cryptography with meaningful shares. Sig. Process. **109**, 317–333 (2015)
17. Zhou, Z., Arce, G.R., Di Crescenzo, G.: Halftone visual cryptography. IEEE Trans. Image Process. **15**(8), 2441–2453 (2006)
18. Liu, F., Wu, C.: Embedded extended visual cryptography schemes. IEEE Trans. Inf. Forensics Secur. **6**(2), 307–322 (2011)
19. Yan, X., Wang, S., Niu, X., Yang, C.N.: Halftone visual cryptography with minimum auxiliary black pixels and uniform image quality. Digit. Sig. Process. **38**, 53–65 (2015)
20. Ateniese, G., Blundo, C., De Santis, A., Stinson, D.R.: Visual cryptography for general access structures. Inf. Comput. **129**(2), 86–106 (1996)
21. Li, P., Yang, C.N., Wu, C.C., Kong, Q., Ma, Y.: Essential secret image sharing scheme with different importance of shadows. J. Vis. Commun. Image Represent. **24**(7), 1106–1114 (2013)
22. Thien, C.C., Lin, J.C.: Secret image sharing. Comput. Graph. **26**(5), 765–770 (2002)
23. Yang, C.N., Ciou, C.B.: Image secret sharing method with two-decoding-options: lossless recovery and previewing capability. Image Vis. Comput. **28**(12), 1600–1610 (2010)
24. Bs, S., Atrey, P.K., Khabbazian, M.: On the semantic security of secret image sharing methods. In: Proceedings - 2013 IEEE 7th International Conference on Semantic Computing, ICSC 2013, vol. 13, no. 4, pp. 302–305 (2013)
25. Liu, L., Lu, Y., Yan, X., Wan, S.: A progressive threshold secret image sharing with meaningful shares for gray-scale image. Accepted by 12th International Conference on Mobile Ad-hoc and Sensor Networks, pp. 1–12 (2016)

26. Yan, X., Wang, S., El-Latif, A.A.A., Sang, J., Niu, X.: A novel perceptual secret sharing scheme. In: Shi, Y.Q., Liu, F., Yan, W. (eds.) Transactions on Data Hiding and Multimedia Security IX. LNCS, vol. 8363, pp. 68–90. Springer, Heidelberg (2014). doi:10.1007/978-3-642-55046-1_5

27. Yan, X., Lu, Y., Chen, Y., Lu, C., Zhu, B., Liao, Q.: Secret image sharing based on error-correcting codes. Accepted by 3rd IEEE International Conference on Big Data Security on Cloud, pp. 1–12 (2017)

28. Kafri, O., Keren, E.: Encryption of pictures and shapes by random grids. Opt. Lett. **12**(6), 377–379 (1987)

29. Yan, X., Lu, Y., Liu, L., Wan, S.: Random grids-based threshold visual secret sharing with improved visual quality. In: Shi, Y.Q., Kim, H.J., Perez-Gonzalez, F., Liu, F. (eds.) IWDW 2016. LNCS, vol. 10082, pp. 209–222. Springer, Cham (2017). doi:10.1007/978-3-319-53465-7_16

PRS: Predication-Based Replica Selection Algorithm for Key-Value Stores

Liyuan Fang, Xiangqian Zhou, Haiming Xie, and Wanchun Jiang[✉]

School of Information Science and Engineering,
Central South University, Changsha 410083, Hunan, China
{lyfang,jiangwc}@csu.edu.cn

Abstract. The tail latency of end-user requests, which directly impacts the user experience and the revenue, is highly related to its corresponding numerous accesses in key-value stores. The replica selection algorithm is crucial to cut the tail latency of these key-value accesses. Recently, the C3 algorithm, which creatively piggybacks the queue-size of waiting keys from replica servers for the replica selection at clients, is proposed in NSDI 2015. Although C3 improves the tail latency a lot, it suffers from the timeliness issue on the feedback information, which directly influences the replica selection. In this paper, we analysis the evaluation of queue-size of waiting keys of C3, and some findings of queue-size variation were made. It motivate us to propose the Prediction-Based Replica Selection (PRS) algorithm, which predicts the queue-size at replica servers under the poor timeliness condition, instead of utilizing the exponentially weighted moving average of the state piggybacked queue-size as in C3. Consequently, PRS can obtain more accurate queue-size at clients than C3, and thus outperforms C3 in terms of cutting the tail latency. Simulation results confirm the advantage of PRS over C3.

Keywords: Prediction · Replica selection · Tail-latency · Key-value stores

1 Introduction

In distributed key-value stores, a single end-user request can generate a large number of keys, which need tens or hundreds of servers to serve them [1,2, 5]. Consequently, the response time of each end-user request depends on the tail latency of these key-value access. Actually, every latency of these key-value accesses may dominate the end to end latency. Because the long response time of end-user requests have a great negative impact on user experience and revenue [3,4], cutting tail latency in key-value stores is crucial. Recent research shows the replica selection algorithm in key-value stores has a large impact on the distribution of the tail latency [10].

Replica selection algorithm determines which replica server will be selected for each key at clients. Obviously, selecting the fastest server is one of the effective way to cut tail latency. In brief, the fastest server is the one who can process

© Springer Nature Singapore Pte Ltd. 2017
B. Zou et al. (Eds.): ICPCSEE 2017, Part I, CCIS 727, pp. 317–330, 2017.
DOI: 10.1007/978-981-10-6385-5_27

the key and returns the value in the shortest time. In this process, the latency of each key consists of three parts: the propagation delay, the queueing delay at replica server and the service time of this key. Among them, the propagation delay is in the order of millisecond even microseconds. Therefore, the queueing delay and the service time account for the main part of latency, especially the queueing delay may be very large when the performance of replica server is poor.

However, it is hard to determine which replica server is the fastest one for the following reasons. First, clients can't know the waiting queue-size timely, because the replica selection algorithms must be simple enough without large overhead in term of coordination and computation. Second, clients can't observe the behavior of other clients, and thus the concurrence of key-value access from different clients to a server may result in a dramatic growth of the waiting queue-size. In addition, the performance fluctuations of servers and the service rate at servers varies greatly across machines and times, and can be impacted by many factors like maintenance activities, periodic garbage collection and so on. Finally, if all clients send keys to one faster server together, the concurrence of key-value access may cause sharp performance degradation of current replica server. Correspondingly the performance of replica servers changes periodically due to the key-value accesses. This phenomenon is called the herd behavior, and should be avoided.

In current popular key-value stores, the replica selection algorithms are all very simple, without taking the performance of servers into consideration. Recently, C3, a replica selection algorithm proposed in NSDI [10], is designed to adapt the time-varying performance of servers and avoid the herd behaviors. Piggybacking the waiting queue-size and the service time from replica severs to guide the replica ranking at clients, C3 acquires significant improvement on the tail latency, i.e., up to 3 times at the 99.9^{th} percentile. However, the feedback information utilized by C3 may be of poor timeliness, as pointed in [11]. For example, the EWMA of piggybacked waiting queue-size may still be used for replica ranking hundreds of microseconds later in C3, during which the real queue-size may have changed greatly. In addition, we use τ_w denote the time interval between the receiving of the feedback and the utilization of it in replica ranking, there is a very large probability that $\tau_w \geq 100$ ms, as reported in Tars [11]. Tars, an improved algorithm of C3, points out this timeliness issue. It doesn't totally address this problem. As a result, Tars' improvement on tail latency is minor compared to C3.

In face of this timeliness issue, we proposed the PRS algorithm in this paper, predicting the queue-size used when $\tau_w \geq 100$ ms. Specifically, PRS observes that the trend of the queue-size variation would not change fast, and the future trend can be inferred by the variation of queue-size Δq_s and the difference of Δq_s piggybacked from replica server. PRS distinguishes three types of queue-size trends: stable, increase and decrease. The differentiation of different trend is executed on the receiving of each feedback information. Although mistake may be made on the differentiation of different trends, PRS can revise it as soon as the next feedback packet is received, and guarantee the correctness in most cases. When the trend is changed, PRS drops all the feedback information in history. When the trend is not changed, PRS collects all the recently received feedback queue-size for the following queue-size prediction. Under the increase or the decrease trend states, PRS utilizes the feedback queue-sizes to fit a line against time according to the least square method, and computes the predicted queue-size with this fitting function. Under the stable trend, PRS treats the EWMAs of the feedback queue-sizes as the predicted queue-size. In this way, the predicted queue-size is strictly in accordance with the current trend of queue-size variation, and thus has a large probability to be closer to the real queue-size, compared to the EWMAs of all the feedback queue-size in history utilized in C3. Consequently, PRS obtains more accurate queue-size than C3, and thus outperforms C3 in terms of cutting the tail latency. Simulation results confirm the advantage of PRS over C3.

The rest of this paper is organized as follows. The Sect. 2 introduces the background and related work. We show the motivation of this paper in Sect. 3, and Sect. 4 presents the PRS algorithm developed for replica selection. We evaluate the performance of the PRS algorithm with simulations based on the open source code of C3 in Sect. 5. Finally, Sect. 6 is the conclusion of this paper.

2   Background and Related Work

In this section, we first present the basic framework of replica selection in key-value stores, and then discuss the current replica selection algorithms.

2.1   Framework of Replica Selection

When an end-user request (e.g. a web page request) reaches a web server, which is also a client of key-value stores, hundreds of keys will be generated [5]. The values of these keys are distributed in different of a group replica servers, and the client needs to access these servers for the corresponding values, as shown in Fig. 1. When a key needs to be sent, the client has to choose one of this replica servers to serve the key. Which replica server will be selected is determined by the replica selection algorithm.

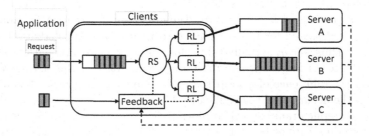

Fig. 1. Framework of replica selection in key-value stores.

A server can be chosen by a group of clients. When a key reaches, it may either be served immediately or enter the waiting queue, according to the status of this server, free or busy. The service time of each key changes with the performance fluctuation of servers. When a key is served, the corresponding value will be returned to the client, and the feedback information of server can be piggybacked within the value.

2.2 Replica Selection Algorithms

Follow above framework, a number of replica selection algorithms are developed for key-value stores. These replica selection algorithms are all quite simple for coordination and computation, as they all only make a decision at clients. For example, Riak [6] or Amazon ELB [7] use an external load balancer such as Nginx selects replica server by Least-Outstanding Requests (LOR) strategy. These algorithms fail to cut the tail latency due to unaware of the condition of servers, i.e., the queue-size of waiting keys.

The C3 algorithm, is designed to adapt the time-varying performance of servers and avoid the herd behaviors. The biggest innovation of C3 against other classic algorithms is that feedback information is piggybacked from servers. Specifically, the queue-size of waiting keys and the server's service time are returned back with the value. Utilizing these feedback information, C3 employs a scoring function to rank replica servers at clients. For each replica server s, clients compute score by following function:

$$\Psi_s = R_s - 1/\bar{\mu}_s + (\hat{q})^3/\bar{\mu}_s \tag{1}$$

where R_s is the response time which is witnessed by the client, $\bar{\mu}_s^{-1}$ is the EWMA of service time feedback return from the server s, and $\hat{q}$ is the estimated queue-size of the waiting queue of server s, $\hat{q}$ is defined by the following equation:

$$\hat{q} = 1 + \bar{q}_s + n * os_s \tag{2}$$

where $\bar{q}_s$ is the EWMA of feedback queue-size, n is the number of clients in the system, os_s is the outstanding keys from the client to server s. When a new key reaches, the client individually ranks the corresponding replica servers and selects the first server to serve this key.

In the scoring function, the estimation queue-size $\hat{q}$ is raised to cubic, this is because the queueing delay contributes the most to tail latency [10]. In addition, the $n * os_s$ is designed for load-balancing to insure each client doesn't burst to each server. Thus, the $\bar{q}_s$ actually takes effect on queue-size estimation of servers. The queue-size estimation is the most important part of the scoring function. Therefore, the difference of $\bar{q}_s$ and the real queue-size of server s has a directly impact on the accuracy of the replica ranking.

After the replica ranking, C3 employs a distribute rate control mechanism to limit the sending rate of keys at client to avoid the servers to be overwhelmed. C3 sets a limited rate for each client to individual server. Keys just can be sent when the server is within its rate limit (RL in Fig. 1). If all the replicas are limited, the key will be put in a backlog queue until at least one server is within its rate limit.

Tars [11] is an improved replica selection algorithm of C3. Tars indicates the timeliness issue of feedback in C3. Specifically, the delay time of feedback is composed of the single-pass network latency τ_d and the interval τ_w during receiving the response and using the feedback to score. Actually, the single-pass network latency τ_d is millisecond even microsecond, but the τ_w can be tens or hundreds of times larger as than τ_d. Large τ_w can be resulted from two aspects. First, in the link of one client to one server, the flow is inconsecutive. After a client receives a value from a server, the following keys arrive this client may not belong to this server group for a long time. Second, score of the replica server may be not dominant in the scoring, therefore, the client will not select this server for a long time until a worse server appears. In addition, the feedback is really square in some cases, which lead to a longer τ_w. For example, if we have a total of 600000 keys to be sent, there are 300 clients and 50 servers, for one pair of client and server, only 40 keys are assigned in average. If there are only 150 clients, the average number of feedback information will be doubled, and τ_w will be smaller. In general, after a long delay, the outdated feedback can not reflect the current queue-size and the service rate of the corresponding servers.

However, Tars doesn't solve this timeliness problem effectively. Tars follows the scoring idea of C3, realizes the timeliness issue, the queue-size estimation function are divided into two parts according to τ_w. When $\tau_w < 100\,\text{ms}$, Tars employs a flow rate idea to estimate queue-size. In fact, in this condition, even just using feedback queue-size to estimate can be probably accurate, other operations bring little effect. The other is $\tau_w \geq 100\,\text{ms}$, if LOR is 0, the queue-size estimation is set to 0. If LOR is not 0, Tars continue use the queue-size estimation of C3. In fact, when $\tau_w \geq 100\,\text{ms}$, Tars degenerates to LOR, doesn't solve the timeliness issue. Finally, in reducing tail latency, Tar is minor improved.

3   Motivation

As mentioned in above section, the queuing delay is the major part of tail latency, and the queue-size estimation is the core of scoring function. Obviously, the accuracy of estimated queue-size has a directly impact on replica ranking. Tars [11] points out that feedback information suffers a poor timeliness issue, when $\tau_w \geq 100\,\text{ms}$. In this poor timeliness condition, the estimated queue-size employed the EWMA of outdate feedback is unable to represent the queue-size of servers. As shown in Fig. 2, Q_s is the real queue-size of server, $\bar{q}$ is the EWMA of feedback queue-size. The difference among the $\bar{q}_s$ and the queue-size Q_s is large. Moreover, due to the lack of feedback, $\bar{q}$ keeps unchanged for a long time. However, after hundreds of milliseconds, the real queue-size of the server has already changed in a large range. Therefore, when $\tau_w \geq 100\,\text{ms}$, providing a better queue estimation method is an effective way to improve C3.

We reproduce the simulation results of C3, and some findings of queue-size variation were made under the poor timeliness condition. First, the queue-size

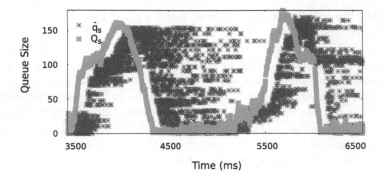

Fig. 2. Queue-size and its estimation in C3 with $\tau_w \geq 100\,\text{ms}$

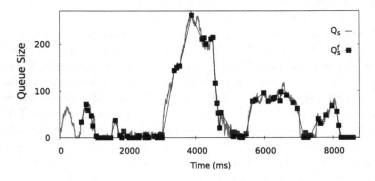

Fig. 3. Queue-size and feedback queue-size in C3

changes frequently at servers, but the variation of queue-size can maintain a trend for a long period, from hundreds to thousands of milliseconds. Second, although the poor timeliness feedback can't represent the present state of servers, it still can reflect the trends of queue-size varied recently. Finally, although the feedback between one server and one client is relatively sparse, the trends of feedback queue-size can represent the trends of queue-size in the server roughly. Figure 3 confirms these opinions, wherein Q_s^f is the feedback queue-size to a random client from a server. Follow these discoveries, we can infer the trends of queue-size variation by utilizing the feedback. According to the trends and taking the τ_w into consideration, it is probable to obtain a more accurate predicted queue-size than EWMA. Note that, the prediction may not absolutely accurate, but the prediction based on the right variation trends can increase the probability to close the reality. Therefore, the accuracy of trends is very important. How to recognize the changes of trends is the important and difficult points.

4   Design of PRS

In this section, we introduce the design of the PRS algorithm. Generally, the PRS algorithm is similar to C3, i.e., both of them use the EWMA of the piggybacked queue-size in replica ranking and use the same scoring function (1). The main difference are twofold. First, the PRS algorithm predicts the queue-size of replica servers when $\tau_w \geq 100\,\text{ms}$. Second, the rate control is not used in the PRS algorithm.

4.1 Feedback Information

Feedback information is still important in the PRS algorithm. Specifically, the queue-size Q_s^f and the service rate of respective key μ_s are still piggybacked as in C3. Both $\bar{q}_s$ and $\bar{\mu}_s$ are still the EWMA of Q_s^f and μ_s at clients. Excepting Q_s^f and μ_s, the variation of the waiting queue-size Δq_s, which is collected every 20 ms, is also piggybacked to clients in the PRS algorithm.

4.2 Trends of Queue-Size Variation

Once clients receive feedback, PRS makes decsions about the trends of current queue-size variation. In the PRS algorithm, three different trends of the queue-size variation are defined, as represented by the three states shown in Fig. 4. On the receiving of each feedback information, the PRS algorithm distinguishes these states according to Q_s^f and Δq_s. Specifically, the difference between Δq_s and last feedback Δq_s (denoted by $\Delta q_{s_{last}}$) is computed at first. The sign of Δq_s can reasonable represent the trends of the queue-size variation. However, the change of the sign of Δq_s is always slower than the change of the trends of the queue-size variation, as shown in Fig. 6. Combining Δq_s and $\Delta q_s - \Delta q_{s_{last}}$, PRS can know the trends of the queue-size variation earlier. Moreover, when the feedback queue-size is small enough, PRS treats it to be stable, i.e., the queue-size would vary just a little. To smooth the small oscillation in stable state, Q_s^f is the dominated value used to distinguish the stable state with other states.

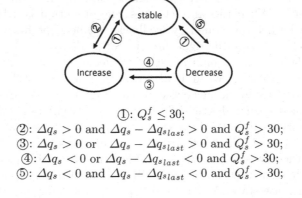

①: $Q_s^f \leq 30$;
②: $\Delta q_s > 0$ and $\Delta q_s - \Delta q_{s_{last}} > 0$ and $Q_s^f > 30$;
③: $\Delta q_s > 0$ or $\quad \Delta q_s - \Delta q_{s_{last}} > 0$ and $Q_s^f > 30$;
④: $\Delta q_s < 0$ or $\Delta q_s - \Delta q_{s_{last}} < 0$ and $Q_s^f > 30$;
⑤: $\Delta q_s < 0$ and $\Delta q_s - \Delta q_{s_{last}} < 0$ and $Q_s^f > 30$;

Fig. 4. State transition for the trends of queue-size variation

The initial state of the queue-size variation is the stable state in PRS. The states transition conditions are listed in Fig. 4. Any time when the condition ①, i.e., $Q_s^f \leq 30$, is satisfied, the future queue-size variation is considered to be stable, i.e., PRS enters into the stable state. Conversely, $Q_s^f \leq 30$ is one of the necessary condition that PRS deviates from the stable state and transfers to either the increase state or the decrease state.

② and ③ are the conditions that PRS enters into the increase state. If PRS enters into the increase state from the stable state, the queue-size increases rapidly and the sign of Δq_s changes rapidly. Thus, the condition ② includes both inequality $\Delta q_s > 0$ and $\Delta q_s - \Delta q_{slast} > 0$. If PRS enters into the increase state from the decrease state, either $\Delta q_s > 0$ or the decrease rate of queue-size gets slower indicates that the trend is going to change. Therefore, the condition ③ involves either $\Delta q_s > 0$ or $\Delta q_s - \Delta q_{slast} > 0$. In both cases, inequality $Q_s^f > 30$ should be satisfied.

④ and ⑤ are the conditions that PRS enters into the decrease state. These conditions are designed in a similar way with conditions ② and ③. Note that PRS can not directly transfers from the stable state to the decrease state in fact. But there may be no feedback information received during the increase state, and thus the state transformation with condition ⑤ must be added.

In some cases, PRS may fail to capture the trends correctly in time, but mistake can be revised once the subsequent feedback information is received.

4.3   Feedback Selection and Queue-Size Prediction

After the judgement of trends, PRS selects and stores the useful feedback queue-size and the time receiving them. Obviously, only the feedback queue-size in the same trend takes effect to prediction, and the feedback belongs to different state is useless or even counterproductive. However, the feedback at boundary of state transition which is also the last feedback is hard to define. It can not be dropped roughly, as more positive feedback can increase prediction accuracy. Therefore, when the state is unchanged, PRS just simply puts the newcome feedback into the feedback sets. When the state changes, the operation is a bit complicate, PRS desides whether the boundary feedback can be kept first, and drops all useless feedback to ensure the right direction of prediction.

Actually, there are only conditions that the feedback at the boundary should be remained. Specifically, when PRS transfers from the stable state to the increase state, the feedback queue-size at this boundary will be involved in the prediction of queue-size under the following increase trend, as shown in Fig. 5(a). When PRS transfers from the increase state to the decrease state, whether the feedback queue-size at this boundary will be kept depends on if it can be the start point of the new state. As shown in Fig. 5(b), when the queue-size is lager than the new feedback, it can be kept. When PRS transfers from the decrease state to the increase state. As shown in Fig. 5(c), the left point is smaller than the new feedback, the increase state can be considered start over here. Excepting above three conditions, all feedback in last state are dropped. With these operations, most of the useable feedback points are obtained.

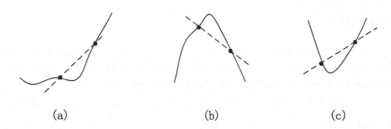

(a)   (b)   (c)

Fig. 5. The cases that points need to be remained in the next trend state

Once there is a new key coming, PRS provides two different methods to predict the queue-size using all the stored feedback according to their current state. Specifically, if the current state is stable, PRS just uses the EWMAs of the feedback queue-size collected under this trend as the predicted queue-size. On the other hand, if the current state is increase or decrease, PRS fits a line with the least square method by using these feedback queue-size collected within the current trend, and computes the queue-size under current time based on this fitted line.

In addition, when the same fitting function is used several times in the increase state, without new feedback for a period, we employs the following complementary mechanism. If the LOR is equal to 0 for 4 times, we set the predicted queue-size 0 in the fifth score, ensuring this replica will be selected. It is actually a trial. When the value of decrease state is smaller than 0, PRS just set it to 0.

In general, for each key, we can make a prediction for the queue-size of each replica. Following the trends, the predicted queue-size will be closer to reality in a high probability.

4.4 Discussion

The PRS method is composed of two parts, the trends judgement and the queue-size prediction. The correctness of the trends judgment directly affects the direction of the queue-size prediction. The difficulty of trends judgement is state transition. In this condition, there is a few feedback information can be used. For example, we know the queue-size is going to decrease, but we don't know how fast it decreases, and thus we can only wait for the next feedback. Therefore, the performance of PRS algorithm will drop down a bit when the states transfer.

The distribution of feedback have a decisive impact on the accuracy of prediction function. The more balance the feedback distribution, the more accuracy the prediction. The proportion of the number of clients and servers, link utilization, replica number, the performance of servers and so on all influence the distribution of feedback. Actually, the performance of servers is poor for a long time and the link utilization is high results in a long τ_w. When τ_w is large, the proportion of the number of clients and servers is influential.

5   Evaluation

In this section, we evaluation the performance of the PRS algorithm based on the open source code of C3 [12]. We first introduce the setup of experiment, second we present our simulation results, compared with C3.

5.1   Setup

As C3, the simulation is based on a discrete-event simulator. To mimic the arrival of user requests, workloads generate keys at clients according to Poisson process [5]. The service time of each key takes on the exponent distribution with a mean of 4 ms [13]. With the time-varying performance fluctuations in consideration, each server sets its mean service rate either μ or to $\mu * D$ with a uniform probability every fluctuation interval [14]. Here is 500 ms for mimic the situation that server maintain poor performance for a long time. D is set to 3 by default. In this simulation, the replication factor is 3. The single-pass network latency is set to 250 μs. The system uses 200 workload generators, 50 servers and 150 clients by default. Each experiment repeats 5 times using different random seeds. The total number of keys is 600000.

5.2   Simulation Results

We conduct simulations with PRS, Tars and C3 under the default configuration. First of all, we study the effect of PRS over queue-size estimation with the simulation of C3. One client and one server is selected to show the accuracy of predicted queue-size by PRS. As shown in Fig. 6, Q_s is the real queue-size of server, Q_s^f is the feedback queue-size, q_e is the EWMA queue-size of C3, q_p is the predicted queue-size of PRS, only the estimation of $\tau_w \geq 100$ are output. Obviously, the PRS queue-size is more accurate than C3, especially when the

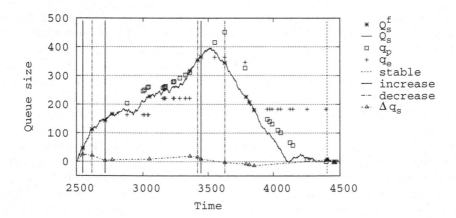

Fig. 6. The effect of predicted queue-size and the trends judgement

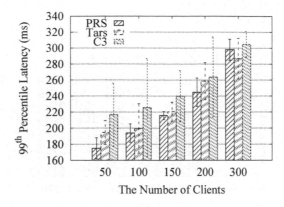

Fig. 7. Impact of number of clients

queue-size trend is not changed. For example, in the increase state, EWMA just stays unchanged, and PRS increases over time. Compared to EWMA, PRS is more aggressive and more accurate. The correctness of queue-size trend definitively influence the accuracy of prediction. The Fig. 6 also shows our foresight of queue-size trend. The three kinds of lines represent the corresponding states transition. The correctness of most of trend judgement can be guaranteed. Sometimes PRS makes a wrong decision, i.e., when the increase state changes to the decrease at 3500 ms. However, once a new feedback return, the trend is corrected. To be more convincing, we count up the difference between the real queue-size and PRS or EWMA, respectively. The simulation is repeated 5 times and taken average. As a result, PRS is better than C3 with **57%** of all points, these two algorithms are similar with **13%** of all points, and the rest **30%** of all points PRS is worse than C3. Note that we consider both methods are similar when the difference of two methods is smaller than or equal to 1. In total, PRS has more accurate queue-size estimation than the EWMA of C3.

Subsequently, we compare the performance of PRS, Tars and C3 under different number of clients. Figure 7 shows the 99th percentile tail latency when the number of clients is varied from 50 to 300. Obviously, the smaller the number of clients, the smaller the tail latency. This is expected, due to the following two reasons. One is the number of clients directly influence the degree of concurrency. The smaller proportion of the number of clients and servers doesn't lead to high concurrency. Another one is that the less the clients, the feedback is more intensive, and the more accurate the prediction, leads to better performance. Therefore, our PRS method is similar to C3 and Tars due to the lack of feedback. And in other cases, we are more outperformed than C3 and Tars.

We also study the effect of heavy demand skews, as many workloads are skewed in reality. The 80% of total keys are generated by 20% clients and 50% clients, respectively. Figures 8 and 9 shows the simulation results, respectively.

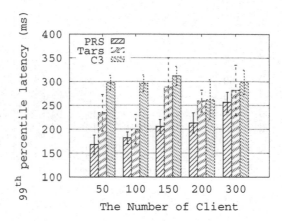

Fig. 8. Impact of the skewed client demands: 20% clients generate 80% of total keys.

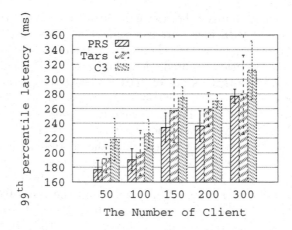

Fig. 9. Impact of the skewed client demands: 50% clients generate 80% of total keys.

Obviously, the tail latency of PRS is much smaller than C3 and Tars. In the heavily skewed traffic, the feedback is more frequent, makes the performance of PRS better.

Figure 10 shows the impact of the replica number. Here we vary the replica factor from 3 to 5. Due to the increase of replica number, it becomes more difficult to select the fastest replica server. Therefore, with the number of replicas increases, the tail latency increases. However, PRS still outperforms than C3 and Tars.

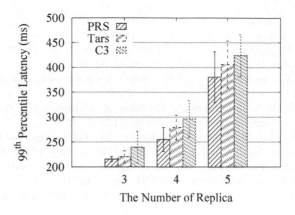

Fig. 10. Impact of the number of replica

6   Conclusion

The replica selection strategy in key-value stores influences the distribution of the tail latency directly. However, how to select a faster replica server is challenging. C3 is a replica selection algorithm proposed recently, and acquires significantly improvement on tail latency. But C3 suffers from the timeliness issue. In this paper, we proposed the PRS algorithm, designed for solve the timeliness issue of C3. We are aware that the queue-size variation can maintain a trend for a long period, when the feedback is poor timeliness. Dividing the feedback into different trend states, we can get the usable feedback and fit a function to predict current queue-size. Compared to C3, the PRS is more active and the predicted queue-size is more chose to real queue-size of servers. Through many simulation results under different scenarios, it is also confirmed that PRS outperforms a lot than C3 in cutting tail latency, especially in the case of feedback intensive.

References

1. Decandia, G., Hastorun, D., Jampani, M., Kakulapati, G., Lakshman, A., Pilchin, A., Sivasubramanian, S., Vosshall, P., Vogels, W.: Dynamo: amazons highly available key-value store. In: Proceedings of the SOSP (2007)
2. Jalaparti, V., Bodik, P., Kandula, S., Menache, I., Rybalkin, M., Yan, C.: Speeding up distributed request-response workflows. In: Proceedings of the SIGCOMM (2013)
3. Brutlag, J.: Speed Matters (2009). http://googleresearch.blogspot.com/2009/06/speed-matters.html
4. Rumble, S.M., Ongaro, D., Stutsman, R., Rosenblum, M., Ousterhout, J.K.: Its time for low latency. In: Proceedings of the HotOS (2011)
5. Nishtala, R., Fugal, H., Grimm, S., Kwiatkowski, M., Lee, H., Li, H.C., McElroy, R., Paleczny, M., Peek, D., Saab, P., Stafford, D., Tung, T., Venkataramani, V.: Scaling memcache at facebook. In: Proceedings of the NSDI (2013)

6. Riak Load Balancing and Proxy Configuration (2014). http://docs.basho.com/riak/1.4.0/cookbooks/Load-Balancing-and-Proxy-Configuration/
7. Amazon ELB (2014). http://docs.aws.amazon.com/ElasticLoadBalancing/latest/DeveloperGuide/TerminologyandKeyConcepts.html
8. Cassandra Documentation (2014). http://www.datastax.com/documentation/cassandra/2.0
9. Atikoglu, B., Xu, Y., Frachtenberg, E., Jiang, S., Paleczny, M.:. Workload analysis of a large-scale key-value store. In: SIGMETRICS (2012)
10. Suresh, L., Canini, M., Schmid, S., Feldmann, A.: C3: cutting tail latency in cloud data stores via adaptive replica selection. In: Proceedings of the NSDI (2015)
11. Jiang, W., Fang, L., Xie, H., Zhou, X., Wang, J.: Timeliness-aware adaptive replica selection for key-value stores. In: ICCCN, Tars (2017)
12. Suresh, L: Simulation Code of C3. https://github.com/lalithsuresh/absim/tree/table
13. Vulimiri, A., Godfrey, P.B., Mittal, R., Sherry, J., Ratnasamy, S., Shenker, S.: Low latency via redundancy. In: CoNEXT (2013)
14. Schad, J., Dittrich, J., Quiane-Ruiz, J.-A.: Runtime measurements in the cloud: observing, analyzing, and reducing variance. VLDB Endow. 3(1–2), 460–471 (2010)

A General (k, n) Threshold Secret Image Sharing Construction Based on Matrix Theory

Wanmeng Ding$^{(\boxtimes)}$, Kesheng Liu, Xuehu Yan, and Lintao Liu

Hefei Electronic Engineering Institute, Hefei 230037, China
wanmeng502@sina.com

Abstract. Shamir proposed a classic polynomial-based secret sharing (SS) scheme, which is also widely applied in secret image sharing (SIS). However, the following researchers paid more attention to the development of properties, such as lossless recovery, rather than the principle of Shamir's polynomial-based SS scheme. In this paper, we introduce matrix theory to analyze Shamir's polynomial-based scheme as well as propose a general (k, n) threshold SIS construction based on matrix theory. Besides, it is proved that Shamir's polynomial-based SS scheme is a special case of our construction method. Both experimental results and analyses are given to demonstrate the effectiveness of the proposed construction method.

Keywords: Secret image sharing · Shamir's polynomial · Vandermonde matrix · Matrix theory · Linear space · Threshold construction

1 Introduction

Shamir's polynomial-based schemes [1–5] and visual cryptographic schemes (VCS) [6–8] are the primary branches in secret image sharing. As a classic secret sharing scheme, Shamir's polynomial-based scheme is based on polynomial interpolation. The scheme encrypts secrets into the constant coefficient of a random $(k - 1)$-degree polynomial, and then the secret pixel can be reconstructed by Lagrange interpolation. In fact, the principle of the Lagrange interpolation is polynomial solving theory.

In Shamir's polynomial-based method, to divide the secret number S, a random $k - 1$ degree polynomial $f(x) = a_0 + a_1x + \ldots + a_{k-1}x^{k-1}$ in which $a_0 = S$, and evaluate: $S_1 = f(1), \ldots, S_i = f(i) \ldots S_n = f(n)$.

Given any subset of k of these S_i values, Shamir find the coefficients of f(x) by interpolation, and then S = f(0) is evaluated.

Actually, the substance is how to get the coefficients of f(x) by solving equations. Thus, in another point of view, we can introduce matrix theory to solve this problem, instead of the interpolation method. More importantly, based on matrix theory and linear space theory, a general (k, n) threshold secret image sharing construction method will be introduced.

© Springer Nature Singapore Pte Ltd. 2017
B. Zou et al. (Eds.): ICPCSEE 2017, Part I, CCIS 727, pp. 331–340, 2017.
DOI: 10.1007/978-981-10-6385-5_28

In this paper, a general (k, n) threshold secret image sharing construction based on matrix theory is proposed. Firstly, by analyzing Shamir's polynomial-based method, we summarize the condition of construct the polynomials, and then propose a general (k, n) threshold secret image sharing construction based on matrix theory. It is demonstrated that the proposed construction is a general method. Experimental results and analyses indicate the effectiveness of the proposed construction.

The rest of the paper is organized as follows. Section 2 introduces some preliminary techniques as the basis for the proposed construction. In Sect. 3, the proposed construction is presented in detail. Section 4 gives experimental results and analyses. Finally, Sect. 5 concludes this paper.

2   Preliminaries

In this section, firstly, Shamir's polynomial-based scheme is given. Then we introduce analysis of Shamir's polynomial based on matrix theory. And at the end of this section, by analyzing Shamir's polynomial, we propose the condition of our construction method.

2.1   Shamir's Polynomials-Based Scheme

Our goal is to share the secret image S into n shadow images $SC_1, SC_2, \ldots SC_n$ in such a way that: (1) knowledge of any k or more shares makes S easily computable; (2) knowledge of any $k - 1$ or fewer shares leaves S completely undetermined. Such a scheme is called a (k, n) threshold scheme. The scheme is based on polynomial interpolation: given k points in the 2-dimensional plane $(x_1, y_1), \ldots, (x_k, y_k)$ with distinct x_i, there is one and only one polynomial $f(x)$ of degree $k - 1$ such that $f(x_i) = y_i$ for all i. For example, We take a pixel value s as the gray value of the first secret pixel, and then to split s into n pixels corresponding to n shadows. The specific scheme is listed as follows:

1. In the sharing phase, given an pixel value s, we select a prime number p. To divide into pieces sc_i, we generate a $k - 1$ degree polynomial

$$f(x) = (a_0 + a_1 x + \cdots + a_{k-1} x^{k-1}) \bmod p \tag{1}$$

   in which $a_0 = s$, and then compute

$$sc_1 = f(1), \cdots, sc_i = f(i), \cdots, sc_n = f(n). \tag{2}$$

   and take (i, sc_i) as a secret pair, where i serves as an identifying indice or a order lable and sc_i serves as a pixel value.
   And the process repeats until all pixels of the secret image are finished. In the end, n shadow images are generated.
2. In the recovery phase, given any k pairs of these n pairs $\{(i, sc_i)\}_{i=1}^n$, we can obtain the coefficients of $f(x)$ by the Largrange's interpolation, and then evaluate $s = f(0)$. Knowledge of just $k-1$ of these values, on the other hand, does not suffice in order to calculate s.

2.2 Analysis of Shamir's Polynomials Based on Matrix Theory

In Shamir's polynomials shown in (1), equations in (2) are evaluated in the recovery phase. When given any subset of these S_i values, we can solve the coefficients of $f(x)$ by solving these k equations. Without loss of generality, we can assume that we are given the first k subset. Thus, we can get k equations as follows:

$$
\begin{cases}
f(1) = a_0 + a_1 + \cdots + a_{k-1} \\
f(2) = a_0 + 2^1 a_1 + \cdots + 2^{k-1} a_{k-1} \\
f(3) = a_0 + 3^1 a_1 + \cdots + 3^{k-1} a_{k-1} \quad (\text{mod } p) \\
\vdots \\
f(k) = a_0 + k^1 \cdot a_1 + \cdots + k^{k-1} \cdot a_{k-1}
\end{cases}
\tag{3}
$$

Here, the parameter k is fixed, and $a_0, a_1, ..., a_{k-1}$ are unknown. So Eq.(3) is a linear system with k equations and k unknowns. In another point of view, Eq.(3) is equivalent to the following vector where a deserves as a variable:

$$
\mathbf{Ka} = \mathbf{f} \tag{4}
$$

The system in matrix-vector form can also can be written as

$$
\begin{bmatrix}
1 & 1 & 1 & \cdots & 1 \\
1 & 2 & 2^2 & \cdots & 2^{k-1} \\
1 & 3 & 3^2 & \cdots & 3^{k-1} \\
\vdots & \vdots & \vdots & & \vdots \\
1 & k & k^2 & \cdots & k^{k-1}
\end{bmatrix}
\begin{bmatrix}
a_0 \\
a_1 \\
a_2 \\
\vdots \\
a_{k-1}
\end{bmatrix}
(\text{mod } p) =
\begin{bmatrix}
f(1) \\
f(2) \\
f(3) \\
\vdots \\
f(k)
\end{bmatrix}
(\text{mod } p)
\tag{5}
$$

Actually, linear Eq. (3) and vector Eq. (4) will be mixed without distinction, in addition, solution and solution vector are indiscriminate. Using the rank of the coefficient matrix $\mathbf{K}$ and the augmented matrix $(\mathbf{K}, \mathbf{f})$, we can easily discuss whether the linear system has a unique solution according to the following theorem.

Theorem 1. *k-variables linear equations* $\mathbf{Ka} = \mathbf{f}$. *The necessary and sufficient condition for a unique solution is:* $R(\mathbf{K}) = R(\mathbf{K}, \mathbf{f}) = k$.

To solve the Eq. (3) and then get the coefficients of $f(x)$, we must ensure that the rank of the coefficient matrix $\mathbf{K}$ is k.

In Shamir's polynomial, the coefficient matrix equations $S_1 = f(1), ..., S_i = f(i), ..., S_n = f(n)$ is :

$$
\begin{bmatrix}
1 & 1 & 1 & \cdots & 1 \\
1 & 2 & 2^2 & \cdots & 2^{k-1} \\
1 & 3 & 3^2 & \cdots & 3^{k-1} \\
\vdots & \vdots & \vdots & & \vdots \\
1 & k & k^2 & \cdots & k^{k-1} \\
\vdots & \vdots & \vdots & & \vdots \\
1 & n & n^2 & \cdots & n^{k-1}
\end{bmatrix}
\tag{6}
$$

The coefficient matrix is a Vandermonde matrix. Because a qualitative property of Vandermonde matrix is that the rank of k order Vandermonde matrix is k. Thus the polynomial interpolation problem is solvable with unique solution.

2.3 Conditions of Proposed Construction Method

In fact, Shamir's polynomial constructed by the Vandermonde matrix is only a special case of constructing a polynomial. In this paper, we proposed a general polynomial construction method and then proposed a general (k, n) threshold secret image sharing construction based on matrix theory.

  As shown in Sect. 2.2, it is not just the Vandermonde matrix, as long as the coefficient matrix satisfies the requirement of "any k-order matrix rank k", the coefficient matrix can be applied to construct polynomials. The problem now is how to qualify the coefficient matrix. We find that the coefficient matrix $\mathbf{K}$ should be qualified by such a condition:

Theorem 2. *In the row vector group of the coefficient matrix* $\mathbf{K}$*, any k vectors are linearly independent.*

3 The Proposed Construction Method

Our construction is based on matrix theory. In preparation for the secret sharing, we first construct a $n \times k$ matrix $\mathbf{K}$, $\mathbf{K}$ is constructed by a special matrix $\mathbf{A}$, which is qualified as follows: All the minors of matrix $\mathbf{A}$ are nonzero. Given k linearly independent k-dimensional row vector group $\alpha_1, \alpha_2, \cdots, \alpha_k$, which form a $k \times k$

matrix $\alpha = \begin{pmatrix} \alpha_1 \\ \alpha_2 \\ \vdots \\ \alpha_k \end{pmatrix}$. The $k \times k$ matrix $\mathbf{A}$ quality our condition mentioned above

is a coefficient matrix. By multiplying two matrices $\mathbf{A}\alpha = \beta$, we can get another

k linearly independent k-dimensional row vector group $\beta = \begin{pmatrix} \beta_1 \\ \beta_2 \\ \vdots \\ \beta_k \end{pmatrix}$. That is to

say, the row vector group of β can be expressed linearly by the row vector of α, and A is the coefficient matrix for this representation. That is,

$$\begin{pmatrix} \beta_1 \\ \beta_2 \\ \vdots \\ \beta_k \end{pmatrix} = \begin{pmatrix} a_{11} & a_{12} & \cdots & a_{1k} \\ a_{21} & a_{22} & \cdots & a_{2k} \\ \vdots & \vdots & \vdots & \vdots \\ a_{k1} & a_{k2} & \vdots & a_{kk} \end{pmatrix} \begin{pmatrix} \alpha_1 \\ \alpha_2 \\ \vdots \\ \alpha_k \end{pmatrix} \tag{7}$$

Then, we concatenate α and β to a new matrix $\mathbf{K}$:

$$\mathbf{K} = \begin{pmatrix} \alpha \\ \beta \end{pmatrix} = \begin{pmatrix} \alpha_1 \\ \vdots \\ \alpha_k \\ \beta_1 \\ \vdots \\ \beta_k \end{pmatrix} \tag{8}$$

and $\mathbf{K}$ is a matrix with size $2k \times k$. Due to the special properties of the matrix A, the matrix $\mathbf{K}$ has such a property: in the row vector group of matrix $\mathbf{K}$, any k vectors are linearly independent. In this way, we can get a (k, n_x) threshold construction, and n_x range from k to $2k$. In the following of our paper, we assume (k, n_x) as (k, n) threshold, which is enough for real applications.

3.1   The Sharing Phase

In what follows, the orginal gray secret image S, with size of $M \times N$. Without loss of generation, we take a pixel value s as the gray value of the first secret pixel, and then to split s into n pixels corresponding to n shadows images. In the sharing phase, given a matrix $\mathbf{K}$ constructed by specific way mentioned above, we select a prime number p. To divide the secret s into pieces sc_i, we generate a vector $\mathbf{a}(a_0, a_1, \cdots, a_{k-1})$ which is generated by k integer numbers $a_0, a_1, \cdots, a_{k-1}$, among them $a_0 = s$, and others are randomly generated. And then compute $\mathbf{Ka} = \mathbf{f}$ as follows:

$$\mathbf{Ka} = \begin{pmatrix} \alpha_1 \\ \vdots \\ \alpha_k \\ \beta_1 \\ \vdots \\ \beta_k \end{pmatrix} \bullet \begin{pmatrix} a_0 \\ \vdots \\ a_{k-1} \end{pmatrix} = \begin{pmatrix} sc_1 \\ \vdots \\ sc_i \\ \vdots \\ sc_n \end{pmatrix} (\bmod p) \tag{9}$$

And take $(\mathbf{k_i}, sc_i)$ as a secret pair, where α_i serves as an identifying indice or a key and sc_i serves as a pixel value. The steps are described in Algorithm 1.

In the end, n shadow images are generated by the construction mentioned in Algorithm 1. Thus, the n shadow images are generated from our general (k, n) threshold SIS construction based on matrix theory. Note that: we select the prime number p to be 251 which is the greatest prime number not larger than 255. Actually, since the gray pixel value of an image ranges from 0 to 256, our construction is not totally lossless. But it is still a high resolution SIS construction method.

Algorithm 1. The proposed general (k, n) threshold SIS construction by matrix theory for sharing phase

Input: The threshold parameters (k, n), a matrix $\mathbf{K}$ constructed by Theorem 3, a secret image S with size of $M \times N$ and a prime number p

Output: n shadow images $SC_1, SC_2, \cdots SC_n$

Step 1: For every secret pixel s in each position$(i, j) \in \{(i, j)|1 \leq i \leq M, 1 \leq j \leq N\}$, repeat Step 2-3.

Step 2: Generate a vector $\mathbf{a} = (a_0, a_1, \cdots, a_{k-1})$, set $s = a_0$, and generate others randomly in the finite domain $[0, p-1]$.

Step 3: Compute $\mathbf{f} = \mathbf{Ka}(\text{mod } p)$, where $sc_1(i, j) = f(1), \cdots, sc_n(i, j) = f(n)$.

Step 3: Output n shadow images $SC_1, SC_2, \cdots SC_n$

3.2   The Recovery Phase

In the recovery phase, given any k pairs $\{(\mathbf{k_{i_j}}, SC_{i_j})\}_{j=1}^{k}$, $(i_1, i_2, \cdots, i_k) \subset \{1, 2, \cdots, n\}$, we can put together k vectors $\mathbf{k_i}$ to a matrix $\mathbf{K_{mini}}$. Thus we can finally obtain the vector $\mathbf{a}(a_0, a_1, \cdots, a_{k-1})$ by solving the following linear equation:

$$
\begin{bmatrix} a_0 \\ a_1 \\ \vdots \\ a_{k-1} \end{bmatrix} = \mathbf{K_{mini}^{-1}} \bullet \begin{bmatrix} sc_{i_1} \\ sc_{i_2} \\ \vdots \\ sc_{i_k} \end{bmatrix} (\text{mod } p) \tag{10}
$$

thus we get the secret pixel $s = a_0$. Knowledge of just $k-1$ or fewer values does not suffice in order to calculate a_0. The recovery steps can be described in Algorithm 2.

Algorithm 2. The proposed general (k, n) threshold SIS construction in matrix method in recovery phase

Input: The k shadow images which are randomly selected from n secret shadow images $SC_1, SC_2, \cdots, SC_n$ and corresponding k vectors $\mathbf{k_{i_j}}$

Output: The original secret image S

Step 1: According to the k vectors $\mathbf{k_i}$, concatenate k vectors $\mathbf{k_i}$ to a matrix $\mathbf{K_{mini}}$.

Step 2: For each position$(i, j) \in \{(i, j)|1 \leq i \leq M, 1 \leq j \leq N\}$, repeat Step 3-4.

Step 3: According to the Eq.(4), construct linear system.

Step 4: Get the coefficient a_0 of $f(x)$ by computing linear system according to Eq.(10), and set the pixel $S(i, j) = a_0$.

Step 5: Output the secret image S.

---

4   Experimental Results and Analyses

4.1   Image Illustration

In this section, this article carries on the experiments and analyses of the proposed method. Experimental images demonstrate the effectiveness intuitively, while corresponding histograms draw the probability distribution of pixel values to prove features from the view of statistic analysis. In the experiments, image 1 as shown in Fig. 1(a), and image 2 as shown in Fig. 2(a) are used as the original gray secret images, with size of 256×256.

Simulation results of the proposed construction are presented in Fig. 1 and Fig. 2, where threshold is $(2, 4)$. As a special case of our construction method, Fig. 1 shows another experiment based on our construction, where the coefficient matrix is a 4×2 Vandermonde matrix:

$$
\begin{pmatrix}
1 & 1 \\
1 & 2 \\
1 & 3 \\
1 & 4
\end{pmatrix}
\tag{11}
$$

Four shadow images SC_1, SC_2, SC_3 and SC_4 are generated from $(2, 4)$ threshold construction, and the noise-like image Fig. 1(b)–(e) shows the four shares. The result revealed by the first two shadow images is shown in Fig. 1(f).

This is the construction of Shamir's polynomial-based scheme. And this it is proved that Shamir's polynomial-based scheme is a special case in our construction method.

(a)   (b)   (c)

(d)   (e)   (f)

Fig. 1. Simulation results of Shamir's method, where $k = 2, n = 4$. (a) The secret image; (b)–(e) four original shadow images SC_1, SC_2, SC_3 and SC_4; (f) revealing result by SC_1 and SC_2; The coefficient matrix is a Vandermonde matrix which is used in Shamir's polynomial-based scheme.

Next, in more general case, using the method mentioned in Sect. 3, we generate a random 4×2 matrix $\mathbf{K}$:

$$\begin{pmatrix} 7 & 6 \\ 3 & 3 \\ 69 & 63 \\ 43 & 39 \end{pmatrix}. \tag{12}$$

Then, as shown in Fig. 2(b), (c), (d) and (e), we generate four shadows images SC_1, SC_2, SC_3 and SC_4, which are noise-like, using Algorithm 1. In the recovery phase, Fig. 2(f) is a secret image revealed by SC_1 and SC_2.

(a)   (b)   (c)

(d)   (e)   (f)

Fig. 2. Simulation results of the proposed method, where $k = 2, n = 4$. (a) The secret image; (b)–(e) four original shadow images SC_1, SC_2, SC_3 and SC_4; (f) revealing result by SC_1 and SC_2, using our general (k,n) threshold secret image sharing construction based on matrix theory; The coefficient matrix is a general matrix that qualify our conditions.

In addition, corresponding histograms draw the probability distribution of pixel values to prove the security of the construction method from the view of statistic analysis, as shown in Fig. 3.

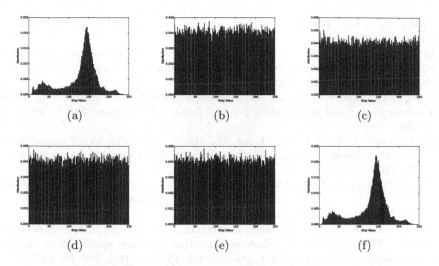

Fig. 3. The statistical histogram of each pixel value in the $(2, 4)$ threshold scheme. (a) the secret image S; (b) shadow image SC_1; (c) shadow image SC_2; (d) shadow image SC_3; (e) shadow image SC_4; (f) revealing result by SC_1 and SC_2

4.2 Brief Summary

Based on experimental results shown in Figs. 1, 2 and 3, we can conclude that:

1. The shares are noise-like, hence either every single original share gives no clue about the secret and there is no cross interference of secret image in the shares for the proposed construction.
2. The proposed construction achieves (k, n) threshold.
3. Our general (k, n) threshold secret image sharing construction based on matrix theory is effective, and Shamir's polynomial-based scheme which use Vandermonde matrix as coefficient matrix is a special case of our construction method.

5 Conclusion

This paper summarized the characteristics and revealed the principle of Shamir's polynomial-based scheme, and proposed a general (k, n) threshold secret image sharing construction based on matrix theory, which is more adaptive. Several experimental results and analyses are performed in this paper to evaluate the effectiveness of the proposed construction. Theory analysis and mathematical proof will be the future work.

References

1. Shamir, A.: How to share a secret. Commun. ACM **22**(11), 612–613 (1979)

2. Thien, C.C., Lin, J.C.: How to share a secret. Comput. Graph. **26**(5), 765–770 (2002)
3. Lin, S.J., Lin, J.C.: A two-in-one two-decoding-options image sharing method combining visual cryptography (VC) and polynomial-style sharing (PSS) approaches. Pattern Recogn. **40**(12), 3652–3666 (2007)
4. Yang, C.N., Ciou, C.B.: Image secret sharing method with two-decoding-options: lossless recovery and previewing capability. Image Vis. Comput. **28**(12), 1600–1610 (2010)
5. Li, P., Yang, C.N., Wu, C.C., Kong, Q., Ma, Y.: Essential secret image sharing scheme with different importance of shadows. J. Vis. Commun. Image Represent. **24**(7), 1106–1114 (2013)
6. Wang, Z., Arce, G.R., Di Crescenzo, G.: Halftone visual cryptography via error diffusion. IEEE Trans. Inf. Forensics Secur. **4**(3), 383–396 (2009)
7. Naor, M., Shamir, A.: Visual cryptography. In: Santis, A. (ed.) EUROCRYPT 1994. LNCS, vol. 950, pp. 1–12. Springer, Heidelberg (1995). doi:10.1007/BFb0053419
8. Yan, X., Wang, S., Niu, X.: Threshold construction from specific cases in visual cryptography without the pixel expansion. Sig. Process. **105**, 389–398 (2014)

A Real-Time Visualization Defense Framework for DDoS Attack

Yiqiao Jin, Qidi Liang, Jian Zhang$^{(\boxtimes)}$, and Ou Jin

College of Information Science and Engineering, Central South University,
South Lushan Road 932, Changsha 410012, China
391157404@qq.com, 273734434@qq.com, 308409399@qq.com, 715011769@qq.com

Abstract. In recent years, with the continuous development of DDoS attacks, DDoS attacks are becoming easier to implement. More and more servers and even personal computers are under the threat of DDoS attacks, especially DDoS flood attacks. Its main purpose is to cause the target host's TCP/IP protocol layer to become congested. In this paper, we propose a real-time visualization defense framework for DDoS attack. Our framework is based on spark-streaming so that it allows for parallel and distributed traffic analysis that can be deployed at high speed network links. Moreover, this framework includes a cylindrical coordinates Visualization Model, which enables users to recognize DDoS threats promptly and clearly. The experiments show that our framework is able to detect and visualize DDoS flooding attacks timely and efficiently.

Keywords: DDoS · Visualization · Attack detection · Kafka · Spark-streaming

1 Introduction

Distributed Denial of service attacks are one of the most widely used form of attacks by the attackers, hackers or international cyber terrorists for crippling network infrastructure. With the continuous development of DDoS attacks, DDoS attacks are becoming easier to implement. More and more servers and even personal computers are under the threat of DDoS attacks, especially DDoS flood attacks. Relevant investigation [1] pointed out that the game industry and Internet financial industry are the hardest-hit area. Moreover, UDP Flood and TCP Flood are still the biggest threat to websites. The survey also pointed out that over 80% of the DDoS attack is occurred within one hour, which reflected that First-Hour Defense is very crucial.

There are many destructive DDoS attacks [2–5]. Rai et al. [6] provided the classification of DDoS attacks and indicated that DDoS flood attack is one of the most destructive DDoS attacks. Therefore, DDoS protection generally includes two aspects: One is aimed at the development of the form of attack, can effectively detect the attack traffic; Another, the most importantly, is how to reduce the impact on the business system or the network, so as to ensure the continuity and availability of business systems. At present, there are many DDoS

© Springer Nature Singapore Pte Ltd. 2017
B. Zou et al. (Eds.): ICPCSEE 2017, Part I, CCIS 727, pp. 341–351, 2017.
DOI: 10.1007/978-981-10-6385-5_29

attack detection methods [7–12] in academic circles but few combine with the visualization technology.

In this paper, we built a real-time visualization defense framework for DDoS attack, which includes parallel processing module, mitigation module and visualization module. Especially, we focus on the DDoS attack traffic visualization module and propose a cylindrical coordinates visualization model, which provides a possibility to feel the flow intuitively. The rest of the paper is organized as follows: Sect. 2 discusses the related works; Sect. 3 presents a detailed explanation of the whole framework; Sect. 4 provides a specific cylindrical coordinates visualization model; Sect. 5 shows our experiment and performance evaluation; And finally, Sect. 6 offers a concluding remark and discusses future work in this area.

2   Related Works

Parallel processing models such as Hadoop [13] and Spark [14] are usually used as competent model. Lee and Lee [15] has given a measurement and analysis scheme for scalable Internet traffic based on Hadoop which can handle millions of megabytes of Libpcap file. Rettig et al. [16] devised an online anomaly detection pipeline building on Kafka queues and Spark Streaming while satisfying the generality and scalability requirements, which is useful for interactive jobs or continuous query processing programs. As a core component in Spark, Spark streaming can process RDD-based data in parallel. Compared with Hadoop, Spark streaming is more suitable for real-time computing. Zhang et al. [17] proposed a Spark based analysis model to identify abnormal packets and compute statistics for the detection model on the number of abnormal packets. In terms of visualization, Seo et al. [18] make an in-depth investigation on the issue of botnet detection and present a new security visualization tool for visualizing botnet behaviors on DNS traffic. The tools make it possible to intuitively recognize symptoms of botnet activities by only visualizing DNS traffic.

3   Visualization Defense Framework

3.1   Kafka and Spark-Streaming

Kafka is a distributed, multi-partition, multi-copy message queue system. It is a good choice when there need a log data analysis. Kafka can handle log collection, ETL, message processing, streaming and other related work on a single platform. Moreover, Kafka has real-time data processing capabilities with high throughput and low latency in online systems or offline systems.

Spark-Streaming is based on the Spark core, it has scalability, high throughput, automatic fault tolerance and other characteristics. Spark-Streaming can support Kafka and other data sources with advanced functions like map, reduce, join, window. The results can be written to a file system, database, or other real-time presentation system.

3.2   Architecture of the Framework

In this framework, Fig. 1, wireshark monitors the backbone network traffic in real time. Wireshark is a packet capture software and we set a cycle time for wireshark to general JSON files which contains detailed packet analysis in the files. From the producer of Kafka, the detailed data of packets in JSON files are analyzed and pushed into the DDoS topic and waiting for the next step.

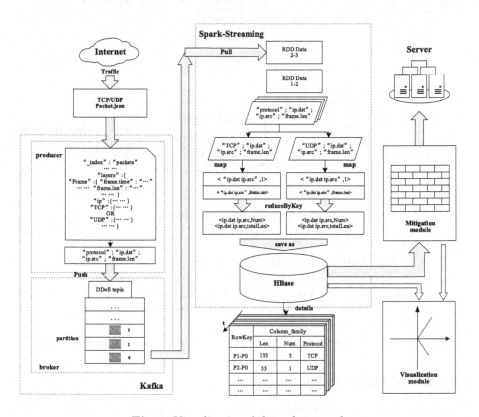

Fig. 1. Visualization defense framework

Spark-Streaming receives real-time data and divides the data into multiple batches by cycle, then dispatching each batch to spark core to calculate. According to Fig. 1, Spark-Streaming dispatch each batch into RDD Data 0-1, RDD Data 1-2, RDD Data 2-3 and so on. Being calculated by spark, every RDD Data is saved as a specific pattern in Hbase with timestamp.

According to Fig. 2 and Algorithm 1, the data format pulled from Kafka by Spark-Streaming is like (protocol_ip.src_ip.dst_size). In Spark-Streaming, this kind of data format will be extracted as the form of <K, V>. for example < "tcp", "p1_p0, 45">, here K = "tcp" and V = "p1_p0, 45". Filtering the Dstream twice in Spark and we can get TCP_Dstream and UDP_Dstream.

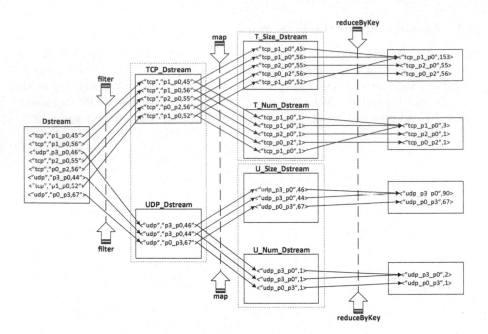

Fig. 2. Spark-streaming RDD

Then, doing map at TCP_Dstream twice, we get T_Size_Dstream and T_Num_Dstream. From T_Size_Dstream, K = "tcp_p1_p0" and V = 45, p1 means the source IP address and P0 means the destination IP address. After Reduce-ByKey, we can get the total size of the TCP packets in a cycle time and the total number of the TCP packets.

The calculation processing of UDP packets is same as the TCP packets.

3.3 Traffic Visualization Module

The traffic visualization module includes two aspects: On the one hand, the original, unmitigated traffic data will be visualized and be watched by users. On the other hand, after the mitigation module, the mitigated traffic data will also be visualized and compared at the same time. This visualization module is used to reflect traffic in real time and help users to perceive network traffic intuitively.

The design of visualization is as follows: first, visualization module retrieve traffic data in real time from Hbase and get the corresponding parameters and threshold by simple calculation. Then, via Open GL, the visualization features will be draw in a specific cylindrical coordinates timely and clearly. Finally, users will get the instantaneous state of the network and know whether there is a DDoS attack through the significant abnormal features in the cylindrical coordinates. In Sect. 4, the paper offers a detailed description of visualization model and Sect. 5 shows the related experiment.

Algorithm 1. Data Extraction

```
/*Spark-Streaming received the message generated by the producer < protocol, p.dst, p.src, size
>*/
tcpModel = InputDStream(Stream   msg).Filter{/* Filter the input stream data */
    Tuple = ParseTuple(msg) /* Extract data from the data stream */
if    (Tuple.protocol == "tcp"){ then
        /* Filter out TCP packets */
    return       Tuple
        }
end if
}.map(Tuple   T){
    TupleSize = (T.protocol_T.DST_T.SRC, T.Size) /* Returns the new tuple in order to count
the size of the packet */
    TupleNum = (T.protocol_T.DST_T.SRC, 1) /* Returns the new tuple in order to count the
number of the packet */
return       TupleSize and TupleNum
}.reduceByKey(Tuple   newTuple){
/* The elements with the same key values are added and returned */
return       (TupleSize.Size+ TupleSize.Size) and (TupleNum_2+TupleNum_2)
}
udpModel = InputDStream(Stream   msg).Filter{ /* Filter the input stream data */
    Tuple = ParseTuple(msg) /* Extract data from the data stream */
if    (Tuple.protocol == "udp") { then
    /* Filter out UDP packets */
    return       Tuple
        }
end if
}.map(Tuple   T){
    TupleSize = (T.protocol_T.DST_T.SRC, T.Size) /* Returns the new tuple in order to count
the size of the packet */
    TupleNum = (T.protocol_T.DST_T.SRC, 1) /* Returns the new tuple in order to count the
number of the packet */
return       TupleSize and TupleNum
}.reduceByKey(Tuple   newTuple){
/* The elements with the same key values are added and returned */
return       (TupleSize.Size+ TupleSize.Size) and (TupleNum_2+TupleNum_2)
}
```

4 Cylindrical Coordinates Visualization Model

4.1 Visualization Parameters

Table 1 offers the symbols of visualization parameters and the description of them. This table includes all the parameters needed in the visualization model. The following are parameters detailed description and calculation method.

P indicates the type of protocol. This visualization mechanism only considers the TCP / IP transport layer protocol. Therefore, P = "TCP" or "UDP".

C indicates the different colors. There are five different colors in the output pictures. They are, respectively, blue, green, cyan, yellow and red.

Δt Means a unit of time. In this framework, Δt is the cycle time of Kafka and Spark-Streaming.

n, S_k and $m_k(\Delta t)$ was given by Spark-Streaming in Sect. 3.

μ is calculated as follows:

for example, the server received a packet from IP address = 218.76.29.42. The IP address will be divided into 4 parts:

$$IP_1 = 218; IP_2 = 72; IP_3 = 29; IP_4 = 42$$

Table 1. Symbols of visualization parameters.

Symbol	Description
P	Indicates the type of protocol
C	Indicates the different colors
Δt	Means a unit of time
n	The total number of packets in Δt
S_k	Indicates the total size of a certain IP address k
$m_k(\Delta t)$	The number of packets from a certain IP address k in an unit of time
μ	A unique number from a certain IP address
R_k	Indicates the frequency of packets from a certain IP address k
R_o	Threshold value of DDoS attack traffic
α_r	An angle that indicates a received IP address in the first
	And second quadrants of the coordinate system
α_s	An angle that indicates a sent IP address in the third and
	Fourth quadrants of the coordinate system
L	Indicates the average length of all packets in an unit of time
$\rho_{(R)}$	Coefficient of R
$\varphi_{(R)}$	Coefficient of R
$\gamma_{(L)}$	Coefficient of L
$\tau_{(L)}$	Coefficient of L

Then, we can get a unique number by Eq. (1)

$$\mu = IP_1 * 2^{24} + IP_2 * 2^{16} + IP_3 * 2^8 + IP_4 * 2^0 \tag{1}$$

R_k indicates the frequency of packets from a certain IP address k. That means $R_k = m_k (\Delta t)$.

R_0 is a threshold value of DDoS attack traffic. When $R_k > R_0$, the visualization module will mark the abnormal network traffic by red color. Equation (2) gives the calculation method.

$$R_0 = \rho_{(R)} * \sum_{k=1}^{n} \frac{m_k(\Delta t)}{n} + \varphi_{(R)} \tag{2}$$

α_r is an angle that indicates a received IP address in the first and second quadrants of the coordinate system. Equation (3) gives the calculation method.

$$\alpha_r = \mu * \frac{\pi}{2^{32}}$$

$$\alpha_r = \sum_{i=1}^{4} \frac{IP_i * \pi}{2^{8n}} \tag{3}$$

α_s is an angle that indicates a sent IP address in the third and fourth quadrants of the coordinate system. Equation (4) gives the calculation method.

$$\alpha_s = \alpha_r + \pi \tag{4}$$

L indicates the average length of all packets in a unit of time. Equation (5) gives the calculation method.

$$L = \gamma_{(L)} * \frac{\sum_{k=1}^{n} S_k}{n} + \tau_{(L)} \tag{5}$$

$\rho_{(R)}, \varphi_{(R)}, \gamma_{(L)}, \tau_{(L)}$ are coefficients. By changing the size of the coefficient, it makes the image easier to observe and understand.

4.2 Visualization Mechanism

According to Fig. 3(a), in the first and second quadrants of the coordinate system, there is a received IP address(1) with a unique angle α_r. The size of R_1 indicates the frequency of packets from this received IP address. In the third and fourth quadrants of the coordinate system, there is a sent IP address(2) with a unique angle α_s. The size of R_2 indicates the frequency of packets from this sent IP address.

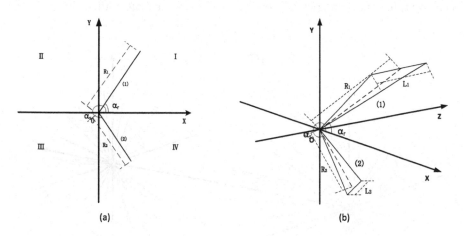

Fig. 3. Visualization step 1

According to Fig. 3, from the plane coordinate system to the three-dimensional coordinate system, the image can show another important parameters L. L_1 means the average length of all received packets in an unit of time and L_2 means the average length of all sent packets in an unit of time. In this way, there generated a unique triangle in the three-dimensional space.

In Fig. 4 there added another three-dimensional coordinate. The left one is used to visualize packets of TCP protocol and the right one is UDP protocol.

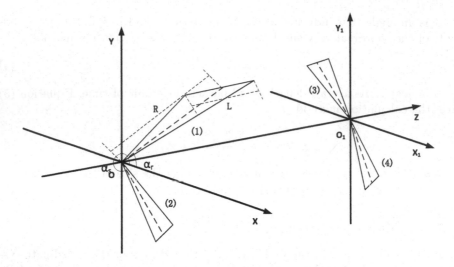

Fig. 4. Visualization step 2

According to Fig. 5, when the visualization module running, there are lots of triangles in the image with different colors. Triangles with shadows are the received traffic and triangles without shadows are the sent traffic.

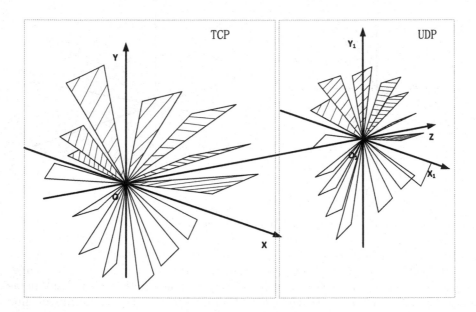

Fig. 5. Visualization step 3

5   Performance Evaluation

In this section, we present experimental results demonstrating the benefits of this Visualization model. First we explain the experimental tools and then we shows the experimental results.

5.1   Experimental Tools

In this experiment, we use two kind of tools to perform the framework. One is BoNeSi and the other is OpenGL. These two tools are described below:

BoNeSi (the DDoS Botnet Simulator) is a Tool to simulate Botnet Traffic in a test bed environment on the wire. It is designed to study the effect of DDoS attacks. BoNeSi generates UDP and TCP flooding attacks from a defined botnet size (different IP addresses). BoNeSi is highly configurable and rates, data volume, source IP addresses, URLs and other parameters can be configured. BoNeSi is a network traffic generator for different protocol types. The attributes of the created packets and connections can be controlled by several parameters like send rate or payload size or they are determined by chance. It spoofs the source ip addresses even when generating tcp traffic [19].

OpenGL is a low-level graphics library specification. It makes available to the programmer a small set of geometric primitives-points, lines, polygons, image, and bitmaps. OpenGL provides a set of commands that allow the specification of geometric objects in two or three dimensions, using the provided primitives, together with commands that control how these objects are rendered (drawn) [20].

5.2   Visualzing DDoS Flooding Traffic

According to Fig. 6, these are the experimental results. We can see that the server received TCP traffic is painted blue, sent TCP traffic is green; the server received UDP traffic is painted cyan and sent UDP traffic is yellow. The abnormal traffic is painted red remarkably.

Furthermore, with the increasing flooding traffic with short length and high frequency, the average length of packets, L, become short and the frequency of packets, R, become long and the suspicious IP address is painted red. Finally, after the mitigation module, most of the abnormal traffic is mitigated and there are little DDoS attack traffic showed from pictures.

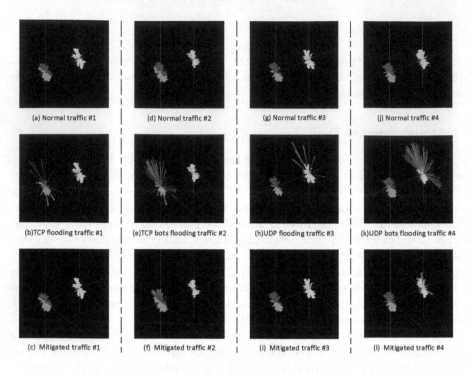

Fig. 6. Experiment results (Color figure online)

6   Conclusions and Future Work

Through the simulation experiment analysis, we can see that by extracting the feature of the front network traffic, it can effectively show the overall trend and real-time status of the traffic, and can obviously distinguish DDoS attack traffic and normal traffic, and get the expected effect.

In the future applications of DDoS attack traffic visualization, DDoS attack traffic can be improved by constantly improving the characteristics of DDoS attack traffic, which can be timely and intuitive to reflect the TCP/IP application layer DDoS attacks, such as CC attacks, DNS query flood and so on.

Acknowledgments. This work is partially supported by the Planned Science and Technology Project of Hunan Province, China (NO.2015JC3044), the National Natural Science Foundation of China (NO.61272147), and the National Science Fund for Young Scholars (NO.61309009).

References

1. YunDun 2015 H2 Report for State and Trends of the Internet DDoS Attacks (2015). Accessed 4 May 2017

2. Xmarkx. DDoS attacks in 2014: Smarter, bigger, faster, stronger (2014). https://greekorio.wordpress.com/2014/04/21/ddos-attacks-in-2014-smarter-bigger-faster-stronger/. Accessed 9 Nov 2015
3. Bogdanoski, M., Shuminoski, T., Risteski, A.: Analysis of the SYN flood DoS attack. Int. J. Comput. Netw. Inf. Secur. **5**(8), 1–11 (2013)
4. Bhandari, N.H.: Survey on DDoS attacks and its detection & defence approaches. Int. J. Sci. Mod. Eng. (IJISME) **1**(3), 2319–6386 (2013)
5. Tao, Y., Yu, S.: DDoS attack detection at local area networks using information theoretical metrics. In: 2013 12th IEEE International Conference on Trust, Security and Privacy in Computing and Communications, pp. 233–240 (2013)
6. Challa, R.K., Rai, A.: Survey on recent DDoS mitigation techniques and comparative analysis. In: 2016 Second International Conference on Computational Intelligence & Communication Technology, pp. 96–101 (2016)
7. Bhuyan, M.H., Kashyap, H.J., Bhattacharyya, D.K., Kalita, J.K.: Detecting distributed denial of service attacks: methods, tools and future directions. Comput. J. **57**(4), 537–556 (2014)
8. Krunal, P.: Security survey for cloud computing: threats & existing IDS/IPS techniques. International Conference on Control, Communication and Computer Technology, pp. 88–92 (2013)
9. Zargar, S.T., Joshi, J., Tipper, D.: A survey of defense mechanisms against distributed denial of service (DDoS) flooding attacks. Commun. Surv. Tutorials IEEE **15**(4), 2046–2069 (2013)
10. Gupta, S., Kumar, P., Abraham, A.: A profile based network intrusion detection and prevention system for securing cloud environment. Int. J. Distrib. Sens. Netw. **2013**(1), 8–10 (2013)
11. Yi, F., Shui, Y., Zhou, W., Hai, J., Bonti, A.: Source-based filtering scheme against DDoS attacks. Int. J. Datab. Theory Appl. **1**(1), 9–20 (2011)
12. Gavaskar, S., Surendiran, R., Ramaraj, E.: Three counter defense mechanism for TCP SYN flooding attacks. Int. J. Comput. Appl. **6**(6), 12–15 (2010)
13. Choi, J., Chang, C., Yim, K., Kim, J., Kim, P.: Intelligent reconfigurable method of cloud computing resources for multimedia data delivery. Informatica **24**(3), 381–394 (2013)
14. Zaharia, M., Das, T., Li, H., Hunter, T., Shenker, S., Stoica, I.: Discretized streams: fault-tolerant streaming computation at scale. In: Proceedings of the Twenty-Fourth ACM Symposium on Operating Systems Principles, pp. 423–438 (2013)
15. Lee, Y., Lee, Y.: Toward scalable internet traffic measurement and analysis with hadoop. ACM SIGCOMM Comput. Commun. Rev. **43**(1), 5–13 (2013)
16. Rettig, L., Khayati, M., Cudre-Mauroux, P., Piorkowski, M.: Online anomaly detection over big data streams. In: 2015 IEEE International Conference on Big Data (Big Data) (2015)
17. Zhang, J., Zhang, Y., Liu, P., He, J.: A spark-based DDoS attack detection model in cloud services. In: Bao, F., Chen, L., Deng, R.H., Wang, G. (eds.) ISPEC 2016. LNCS, vol. 10060, pp. 48–64. Springer, Cham (2016). doi:10.1007/978-3-319-49151-6_4
18. Han, S.C., Seo, I., Lee, H.: Cylindrical coordinates security visualization for multiple domain command and control botnet detection. Comput. Secur. **46**, 141–153 (2014)
19. https://www.openhub.net/p/bonesi. Accessed 20 Feb 2017
20. http://www.docin.com/p-1631407325.html. Accessed 8 May 2017

A Research and Analysis Method of Open Source Threat Intelligence Data

Ruyue Liu[1], Ziping Zhao[1], Chengjun Sun[2], Xiaoyu Yang[3],
Xiaoli Gong[2], and Jin Zhang[2(✉)]

[1] Computer and Information Engineering College, Tianjin Normal University,
Tianjin 300071, China
[2] College of Computer and Control Engineering,
Nankai University, Tianjin 300071, China
nkzhangjin@nankai.edu.cn
[3] Institute of Information Engineering, Chinese Academy of Sciences,
Beijing 100093, China

Abstract. As the form of cyber threats becomes more complex, which leads to a widespread concern about how to promote network security active defense system by using the exploding cyber threat intelligence. Basing on the content analysis method, introduces the precision, recall rate and timely rate on the basis of the change of time dimension, and analyzes the threat intelligence provider from three aspects. The validity of this method is verified by the test of massive source of threat data, which improves the efficiency of CIF analysis and makes it easy to analyze and extract the threat intelligence information quickly.

Keywords: Threat intelligence · Cyber security · CIF

1 Introduction

With the development of computer and network technology, there have been more and more ways threatening network security, in addition to hacker attacks, viruses, backdoors, loopholes and other traditional network attacks, APT (Advanced Persistent Threat) attack has gradually been wide concerned. In recent years, all kinds of attacks show a faster development trend than the defense. In the current severe network security situation, there are shortcomings about traditional post-passive defense network security protection which cannot meet the needs of network security issues. In order to ensure the safe operation of the network, active defense system is gradually established, all kinds of network security devices monitor the network at random, which generates a large number of security event information when used.

With the explosive growth of network security data, the network threat intelligence quietly became a new hotspot in the field of information security. In short, threat intelligence includes spam, phishing, malware, captured packets, IP, URLs, domains, and attributes associated with them, when associated these information together, there can be system information and value intelligence for information security management. [1] The correct use of cyber threat intelligence helps the defender to detect attacks before an attack occurs, or even ideally before an attack occurs. In order to play the

© Springer Nature Singapore Pte Ltd. 2017
B. Zou et al. (Eds.): ICPCSEE 2017, Part I, CCIS 727, pp. 352–363, 2017.
DOI: 10.1007/978-981-10-6385-5_30

important role of network threat intelligence in information security defense, it is necessary to analyze the information of different information sources, therefore many network threat intelligence tools and standards appeared in succession [2]. Such as the Internet threat intelligence management system CIF as an open source tool. Although CIF has made some efforts to create threat intelligence in the open source and business frontier, the collection and using of threat intelligence is still at the edge of stage [3].

Therefore, basing on the deep analysis of the CIF of the open source network threat intelligence management system, there proposes a method to extract valuable information from massive network threat intelligence. Basing on the change of its time dimension, to extract the threat intelligence that is worthy of attention, and put forward a new threat model - the outbreak model. At the same time, the quantitative evaluation of the threat intelligence providers was carried out, and the evaluation results were evaluated.

The value in using of threat intelligence depends on the quality of its supply mechanism. The threat intelligence is not the bulk of the data, on the contrary, the proliferation of spam often lead to really useful information being submerged so that it is difficult to find them from the huge collection of information in the collection of effective intelligence [4]. At present, the analysis of the contents of mass information is mainly focused on classification algorithms, correlation analysis, cluster analysis, etc., which are powerless [5].

The paper is organized as follows: Sect. 2 provides a general introduction to the tools and techniques for the application of cyber threat intelligence. Section 3 deals with the processing and design of the data set in detail; Sect. 4 focuses on the actual analysis of the data analysis process conclusions; validation and evaluation of the results of data analysis in Sect. 5 to explain; Sect. 6 makes a summary and expectation for this research.

2   Related Technology and Tools

The tool of a threat or threat intelligence platform is popular in addressing the threat intelligence gathering, storage, and sharing issues. Its main function is to collect open source intelligence and business intelligence, and store them in a safe, flexible and easy way. Such well-known open-source information libraries includes CIF (Collective Intelligence Framework), CRITs (Collaborative Research into Threats), Mantis, MISP (Malware Information Sharing Platform) [6]. They obtain threat intelligence from open or private source to collect IP, domain names, malware, email, logs and other standardizing or non-standardizing information.

Among these open source tools, CIF application is more extensive, CIF (Collective intelligence framework) [7] is an open source network threat intelligence management system. CIF can combine known malicious information together from different sources and use it to integrate, process and generate new threat intelligence.

2.1   CIF Module

CIF is C/S model including four parts which are Feed/Data, CIF Server, CIF Client and Daily Feeds.

(1) Feed/Data

Feed/Data is the source of CIF information, that is, the publisher of the threat intelligence. CIF pre-sets a portion of the public threat source, and it can also add some public threat data or private data to the CIF via feed/Data in the form of XML (RSS), JSON, or a file (e.g. CSV).

(2) CIF Server

The CIF Server mainly extracts and analyzes the online or local files, extracts the threat intelligence from the crawled data, performs the secondary processing, and uses the ElasticSearch format to store the new threat intelligence.

Figure 1 shows the core components of CIF Server, including cif_smrt, cif_worker, cif_starman, cif_router and ElasticSearch five parts.

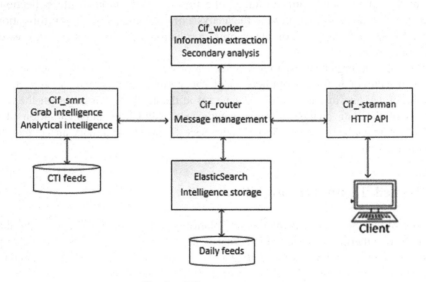

Fig. 1. CIF core components.

(3) Daily Feeds

Threat intelligence data generated from CIF can be used for dnsSinkHole, Firewall, IDS and so on.

(4) CIF Client

CIF is a command-line client tool, and users who use CIF can access the CIF server through a browser, a CIF client, or a CIF-API for quick query.

2.2 CIF Threat Intelligence Structure

The threat intelligence generated by CIF is stored in EslasticSearch in json format. Table 1 shows the information described in an information table, including Observable, ObservableAddress and ObservableIPAddress. Each attribute contains many attribute information. The bold attribute in the table is much more important attribute information in this paper.

Table 1. CIF threat intelligence attributes

Describe content	Attribute
Observable	lang、id、**provider**、group、tlp、**confidence**、**tags**、description、adata、**observable**、otype、**timestamp** (reporttime、firsttime、lasttime)、related、altid、altid_data
ObservableAddress	portlist、protocol、application、cc、rdata、rtype
ObservableIPAddress	asn、asn_desc、rir、peers、prefix、citycode、longitude、latitude、geolocation、timezone、subdivision、mask

These attributes are not required for a threat intelligence. One of the simplest threat intelligence in CIFs include observable, provider, tags, timestamps, and confidence. It considers that the most important information in the threat intelligence is observable, provider, timestamp and other original information, in which observable is reported threats, such as IP, URI, email address, Hash, etc.; provider provides threat intelligence; timestamp includes reporttime, firsttime and lasttime, each threat intelligence requires at least one timestamp, up to three.

In addition to observable, provider, timestamp and other original information, tags and confidence are also important attributes, both are the results after a CIF preliminary analysis of threat intelligence.

CIF describes an observable by using tags, and an observable can have one or more tags. Tags are defined when information is ingested by CIF. The default labels for CIF include: botnet, exploit, feodo, gozi, hijacked, malware, phishing, rdata, scanner, suspicious, whitelist, zeus.

A threat intelligence generated by CIF can be understood like this, the partial attribute information of an intelligence is as follows:

{ "reporttime": "2016-04-21T10:07:17Z",
"tags": ["scanner"],
"provider": "1d4.us",
"observable": "115.29.5.198",
"confidence": 65, ...}

The threat intelligence indicates that CIF has a 65% certainty (which means less certainty) that115.29.5.198 reported by 1d4.us at 2016-04-21T10: 07: 17Z is a scanner.

3 Data Processing Method Design

3.1 Data Preprocessing

Due to the diversity of access to open source intelligence data, it cannot be directly applied in the analysis. The preprocessing is as follows:

(1) Extracting the study object: each record of known threat intelligence text is a JSON object. The observable, provider and timestamp information of the original threat intelligence is the most important attribute of CIF threat intelligence. In addition, tags and confidence are also valuable for the evaluation information given by CIF.
(2) Removing unreasonable data: the data is stored and processed in units of days. Due to network problems, server failure and firewall and other factors, the unreasonable data should be removed, and the rules are as follows:

Rule (1) records the date set to the data as $D\{d_1, d_2, \ldots \ldots, d_n\}$, $d_i(i = 1, 2, \ldots \ldots, n)$, the day of the data file size is recorded as ds_i, about $\forall d_i \in D$,

If $ds_i < ds_{down}$, it is considered that the data file obtained on the day d_i is not reasonable, which will be removed from the D.

If $ds_i \geq ds_{up}$, it is considered that the data file obtained on the day d_i is reasonable.

If $ds_{down} \leq ds_i < ds_{up}$, judgment is made by judging the angle of the image, specific as follows:

Take the data $d_i(i = 1, 2, \ldots \ldots, n)$ as the horizontal axis, d_i get the corresponding day data file size ds_i as the vertical axis, and draw a daily change in the size of the data file for a given period of time. For each data point in the graph, it is reasonable to judge the data size of the day by calculating the angle between the two points.

Set the coordinates of a data point $(x_i, y_i)(i = 1, 2, \ldots \ldots, n)$.

When i = 1 or n, $y_i \geq ds_{down}$ which means, it is considered that the amount of data obtained on the day is reasonable;

When i = 2, ... n − 1, the left adjacent points are denoted as (x_{i-1}, y_{i-1}), the right adjacent points are denoted as (x_{i+1}, y_{i+1}), if $y_{i-1} > y_i$ and $y_{i+1} > y_i$, calculate the angle according to formula (1).

$$\cos\theta = \frac{(x_{i-1} - x_i)^2 + (y_{i-1} - y_i)^2 + (x_i - x_{i+1})^2 + (y_i - y_{i+1})^2 + (x_{i-1} - x_{i+1})^2 + (y_{i-1} - y_{i+1})^2}{2\sqrt{(x_{i-1} - x)^2 + (y_{i-1} - y)^2}\sqrt{(x_i - x_{i+1})^2 + (y_i - y_{i+1})^2}}$$

(1)

If $\cos\theta_{thre} < \cos\theta < 1$, θ_{thre} is the angle of the boundary, it is considered that the amount of data d_i obtained on the day is unreasonable.

3.2 A Collection of Observable Collections Basing on Content Extraction

Massive threat intelligence regardless of the way, the number of results after the association is still unsuitable for research, and the cost of verifying all the threat intelligence is still unacceptable. In order to reduce the time cost of data analysis and improve the efficiency of analysis, we must mention the most important threat intelligence, so this paper presents a content analysis method.

The main idea of the content-based analysis method is to detect changes in the number of reports per observable daily, that is, in the time dimension to analyze each observable daily reported number of changes in the number of cases, to determine whether the number of reported observable significant malicious object dissemination SEIQR model, that is, "appearance - burst - disappearance" the overall trend. If the observable is reported to show the overall trend of " appearance - burst - disappearance ", the observable will be placed in the observable collection O_{value} of concern; otherwise the observable is not concerned.

Figure 2 is a list of observable selected in this paper, and it uses Kibana to generate a change in the observable dimension in the time dimension, which shows an observable overall trend of "outbreak - outbreak - disappearance".

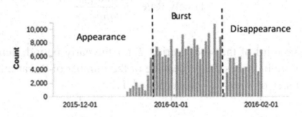

Fig. 2. Showing the change in the number of time-dimensions of an observable daily report.

3.3 Provider Evaluation System

In the reliability analysis of open source information, in addition to information content reliability evaluation index, it needs to evaluate the reliability of information sources as well [12], that is provider. The Precision, Recall, Timely Rate three indicators evaluate the provider. The definition of three indicators and the specific calculation methods are as follows:

First define the set P as the set of all providers; the set O_{value} as a collection of observable which is worthy to be noticed; T is the set of dates for all the threat intelligence obtained; O_{p_i} is the observable set involved in the threat intelligence reported by each provider $(\forall p_i \in P)$ in the time period T; O_{valuei} is the observable collection of the observable collection O_{value} which is of concern in the threat intelligence reported in the time period T.

(1) **Precision:** For each provider, $\forall p_i \in P$, the precision is defined as the number of observable sets O_{value} that are involved in the threat intelligence reported by the p_i in the time period T, and the number of observable sets O_{value} in the threat intelligence reported during this period of time. The percentage of the total number is:

$$\text{Precision} = \frac{|O_{valuei}|}{|O_{p_i}|} \qquad (2)$$

(2) **Recall:** For each provider, $\forall p_i \in P$, the recall rate is defined as the number of observable sets O_{value} that are involved in the threat intelligence reported by the p_i in the time period T, and the number of observable sets O_{value} in the threat intelligence reported during this period of time. The percentage of the total number is:

$$Recall = \frac{|O_{valuei}|}{|O_{value}|} \tag{3}$$

(3) **Timely Rate:** For each element $p_i(i = 1, 2, \ldots\ldots, n)$ in the set and each element $o_i(i = 1, 2, \ldots\ldots, n)$ in the set O_{value}, it is possible to obtain the total number $count_{total}$ of the threat information over a period of time T and the date $t_i(t_i \in T)$, the day p_i reports the $count_i$ of the number of threat information about o_i.

$$Timely\ Rate = \frac{count_{early}}{count_{total}} \tag{4}$$

Define the one third of the study period T for the early period, denoted as T_{early}.

The Timely Rate indicates the percentage of the number of threat information T_{early} and the total number of $count_{total}$ in the period T.

3.4   Improvement of Confidence Assessment of CIF Threat Intelligence

Reference Sect. 2.2 for a set of observable observations O_{value}, and accuracy of each provider calculated Precision, Recall, define a score to analyze the threat intelligence for a period of time for real-time recommendation.

For a given threat intelligence ti_i, the score $score_i$ is calculated by using the following formula:

$$Score_i = o(ti_i) + p(ti_i) + conf(ti_i)/100 \tag{5}$$

Then

$$o(ti_i) = \begin{cases} 1 & ti_i \text{ 中的observable} \in O_{value} \\ 0 & others \end{cases} \tag{6}$$

calculating $p(ti_i)$ according to ti_i's provider.

$$p(ti_i) = \log(Precision + 1) + \log(Recall + 1) \\ + \log(count(Timely\ Rate > 50\%) + 1) \tag{7}$$

$conf(ti_i)$ is the confidence value of the threat intelligence ti_i in CIF.

Calculate its Score value for the threat intelligence for over a period of time (for example, one hour), it argues that the higher the Score value of a threat intelligence is, the more the threat intelligence is worthy of attention.

4   Data Analysis and Conclusions

4.1   Obtain the Dataset

This paper uses CIF, a threat intelligence information management system to obtain threat intelligence, installs the v2 version of CIF on a server with an IP address of 103.243.182.220, and places the server in Hong Kong to reduce external interference. The data source used to obtain the data is the public threat data source which is preset in CIF, that is the public source. Server configuration is as shown in Table 2:

Table 2.   CIF server configuration.

	Server configuration
CPU	8 Cores
RAM	16.0 G
Operating system	Ubuntu 14.04 LTS
Hard disk	80.0 G

Turn on the main CIF server, and the threat intelligence from multiple websites is collected 24 h a day. The new threat intelligence generated by CIF will be automatically stored in ElasticSearch after preliminary analysis and processing of the acquired threat intelligence.

The specific data obtained in Table 3, access to data for a total of 126 days, a total of 49.3 G, a total of 49404115 of the threat intelligence, the data can be divided into three consecutive time periods, 2015.12 18-2016.02.03, 2016.01.20-2016.03.30, and 2016.05.17-2016.07.05.

Table 3.   Get the data table

Period of time	Day	Size of files	The amount of threat intelligence
2015.12.18-2016.02.03	45	18.1 G	16,444,886
2016.02.20-2016.03.30	32	7.8 G	8,335,293
2016.05.17-2016.07.05	49	23.4 G	24,623,936
Total	126	49.3 G	49,404,115

The data of 2015.12.18-2016.02.03 was selected as training set,the data of 2016.01.20-2016.03.30 and 2016.05.17-2016.07.05 were selected as testing set.

4.2   Data Obtained After Processing

In the data preprocessing phase, the training set data is filtered and its threat intelligence is processed as a json object containing only observable, provider, and report time. After the field is extracted, the size of the data file to be processed can be reduced to about one-tenth of the size of the original file, and the amount of data to be processed is greatly reduced.

Judge and remove the unreasonable data. If $ds_{down} = 5\,M$, which is equal to the minimum value of the daily data file size for each time period, when calculated by the calculation, $ds_{up} = ds_{average}$, which is equal to the average of the daily data file size for each time period, and after calculation, the three data sets were 36.42 M, 26.32 M, 50.18 M.

(x_i, y_i) is defined as follows: $x_i = (d_i - d_1 + 1) \times 2$, $y_i = ds_i$ for example, in the 2015.12.18-2016.02.03 time period, 2015.12.23 the date of the field after the extraction of data file size is 12.45 M, then the point of the coordinates are (12,12.45). Where $\cos \theta_{thre}$ takes 0.98, then remove the unreasonable data.

4.3   Observed Collections Basing on Content Extraction

Threat intelligence involves a large number of observable. In order to extract the observable data from the massive data, and find the observable which shows the general trend of "appear - burst - disappear" in the number of reports. The following operations are carried out to remove the unreasonable data:

(1)   Remove the observable with a total number of days less than 10 reported. Since the object of study is the data acquired in a period of time, it is considered that the observable can not be reflected in the observable period of the observable, which is less than 10 days. Therefore, data during the study period will not be studied.
(2)   Remove observable reports that are reported to be stable during the reporting period. This paper considers that the number of days reported only 1 or 2 per day during the study period is excessive or within the study period, the number of times the observable reported daily fluctuates within a range of not more than 4 is stable.
(3)   Make trend judgments on the remaining observable during the study period. The observable presents an overall trend of "appearance - eruption - disappearance", the observable is considered to be of interest to the observable.

In the 44 days data after removal of unreasonable data within 2015.12.18-2016.02.03, extracted the observable collection worthy of attention during the time period, which include 1134 elements.

4.4   Evaluation Provider

Through the analysis of all observables reported by the provider during that time, and according to the formulas (2) and (3) to calculate the recall rate and precision of each provider as shown in Table 4, in precision, the observable set with less than 10 total days reported is removed from the O_{p_i}.

Table 4. Precision and recall of each provider

Provider	Precision	Recall
zeustracker.abuse.ch	0.5988%	1.0582%
vxvault.net	1.2048%	0.5291%

(continued)

Table 4. (*continued*)

Provider	Precision	Recall
sslbl.abuse.ch	0.2475%	0.0882%
spamhaus.org	1.8703%	26.9841%
txt.proxyspy.net	2.7097%	5.5556%
phishtank.com	0.6045%	0.6173%
palevotracker.abuse.ch	0.0000%	0.0000%
packetmail.net	4.7151%	4.2328%
openphish.com	1.1937%	12.1693%
www.us.openbl.org	5.5633%	3.5273%
nothink.org	1.5192%	1.4991%
mirc.com	1.1374%	1.0582%
malwaredomains.com	0.6544%	35.5379%
malwareurls.joxeankoret.com	1.0907%	1.6755%
feodotracker.abuse.ch	0.0000%	0.0000%
rules.emergingthreats.net	0.7802%	1.4991%
dragonresearchgroup.org	3.9499%	3.6155%
'danger.rulez.sk'	0.8953%	0.8818%
botscout.com	4.8331%	7.4074%
blocklist.de	0.8833%	17.4603%
osint.bambenekconsulting.com	1.9048%	8.8183%
autoshun.org	3.3058%	1.4109%
'anti-phishing-email-reply'	\	0.0000%
alexa.com	0.7982%	38.1834%

5   Validation and Evaluation

5.1   Validation Test Set

Verification of the experimental test set uses the data of 2016.02.20-2016.03.30 and 2016.05.17-2016.07.05 for two time periods. The results of the data preprocessing are described in Sects. 3.1, 3.2 and 3.3, the number of elements in the two stages of 1042,1585, for all providers to assess, ranking the rate of recall and precision, and then comparing the results, there are providers equaling to each time period evaluation results. Three times period recall rate ranking change is not large, more than 3 in the following, only individual provider ranking change is greater than 5. Compared to the recall rate, the precision of the ranking of the majority of the following 8, but the ranking change is greater than 10 the number of providers still exist, the number is four.

5.2   Other Verification

In order to verify the reliability of the proposed outbreak model, the data before and after the treatment were sampled. Randomly extract the original data in the 600, and 600 data of the outbreak model after processing. The data will be placed on the

Threatbook platform for detection, and final test results are divided into two categories, one is "dynamic IP" "white list" "DIC" as label White list, and the other is included Blacklist about "botnets", "suspicious networks "," phishing sites "," malware ", which has been verified in unprocessed 600 data, of which 102 are blacklisted data, 17% of the total. The processed data, of which are 518 blacklists, accounting for 86.3% of the total.

6 Summary and Outlook

This paper analyzes each module and its data structure of CIF, and bases on the massive open source threat intelligence generated by CIF, the data of threat intelligence is preprocessed based on the characteristics of threat intelligence, and the content of the threat intelligence is extracted from it. And analyzing the trend of the time dimension of the information, so as to propose an evaluation method for the open source threat intelligence provider and improve the confidence evaluation of the CIF with reference to the evaluation result to provide the basis for the real-time analysis of the recommended threat intelligence.

Due to the period of obtaining data is limited, it is not complete in analyzing the trend of time dimension change. In later period it needs to obtain continuous data of longer time period to facilitate time series analysis. In addition, the improvement of the threat analysis method is also a work that needs to be done at a later stage. In the future, strengthening the real-time analysis and processing capabilities of threat intelligence, so as to apply the threat intelligence analysis to the actual defense system construction, real-time understanding and even forecast network threats.

Acknowledgment. This paper is supported by two projects:

1. The National Science Foundation of China (Grant No: 61103074), the International Scientific Cooperation Foundation of Tianjin(Grant No: 14RcGFGX000847).
2. The Research Plan in Application Foundation and Advanced Technologies in Tianjin (14JCQ NJC00700), the Open Project of the State Key Laboratory of Computer Architecture, Institute of Computing Technology, Chinese Academy of Sciences (CARCH201604).

References

1. Fan, J.: The threat intelligence in the big data age. J. Libr. Inf. Serv. **6**, 15–20 (2016)
2. Farnham, G., Leune, K.: Tools and Standards for Cyber Threat Intelligence Projects. SANS Institute InfoSec Reading Room (2013)
3. Paul Poputa-Clean.: Automated Defense – Using Threat Intelligence to Augment. https://www.sans.org/reading-room/whitepapers/threats/automated-defense-threat-intelligence-augment-35692
4. Li, J.: Overview of the technologies of threat intelligence sensing, sharing and analysis in cyber space. J. Chin. J. Netw. Inf. Secur. **2**(2), 17–29 (2016)
5. Mi, Y., Mi, C., Liu, W.: Research advance on related technology of massive data mining process. J. Front. Comput. Sci. Technol. **9**(6), 641–659 (2015)

6. Paul Poputa-Clean.: Automated Defense-Using Threat Intelligence to Augment Security. SANS Institute InfoSec Reading Room (2015)
7. CSIRT Gadgets Foundation. Collective Intelligence Framework. http://csirtgadgets.org/collective-intelligence-framework
8. RSA-sponsored SBIC.: When Advanced Persistent Threats Go Mainstream. http://www.emc.com/collateral/industry-overview/sbic-rpt.pdf.2011.08.02
9. Shackleford, D.: Who's Using Cyberthreat Intelligence and How?. SANS Institute InfoSec Reading Room (2015)
10. Elasticsearch.elasticsearch-definitive-guide.   https://github.com/elastic/elasticsearch-definitive guide/blob/master/010_Intro/05_What_is_it.asciidoc
11. Li, T., Liu, Z., Zhou, Y.: Application-driven big data mining. J. ZTE Technol. J. 22(02), 49–52 (2016)
12. Li, G., Hua, B.: Relationship between big data analysis and intelligence analysis. J. Libr. Sci. China 40(213), 14–22 (2014)
13. Mishra, B.K., Jha, N.: SEIQRS model for the transmission of malicious objects in computer network. J. Appl. Math. Model. 34(3), 710–715 (2010)
14. Zhao, X., Zhou, A.: Research on the fundamental value and reliability assessment of open source intelligence. J. Intell. 30(10), 16–20 (2011)
15. Chen, M., Mao, S., Liu, Y.: Big data: a survey. J. Mob. Netw. Appl. 19(2), 171–209 (2014)
16. Wu, X., Kumar, V., Quinlan, J.R., et al.: Top 10 algorithms in data mining. J. Knowl. Inf. Syst. 14(1), 1–37 (2008)
17. Dong, C., Chen, L., Wen, Z.: When private set intersection meets big data: an efficient and scalable protocol. In: Proceedings of the 2013 ACM SIGSAC Conference on Computer & communications security, pp. 789–800. ACM (2013)
18. Chaiken, R., Jenkins, B., Larson, P.Å., et al.: SCOPE: easy and efficient parallel processing of massive data sets. J. Proc. VLDB Endow. 1(2), 1265–1276 (2008)
19. Agrawal, R., Kadadi, A., Dai, X., et al.: Challenges and opportunities with big data visualization. In: Proceedings of the 7th International Conference on Management of computational and collective intElligence in Digital EcoSystems, pp. 169–173. ACM (2015)
20. Pan, A.: Research on A Rule-Based Approach to Network Security Event Correlation. Huazhong University of Science and Technology, Wuhan (2007)
21. Haass, J.C., Ahn, G.J., Grimmelmann, F.: Actra: a case study for threat intelligence sharing. In: Proceedings of the 2nd ACM Workshop on Information Sharing and Collaborative Security, pp. 23–26. ACM (2015)
22. Burger, E.W., Goodman, M.D., Kampanakis, P., et al.: Taxonomy model for cyber threat intelligence information exchange technologies. In: Proceedings of the 2014 ACM Workshop on Information Sharing & Collaborative Security, pp. 51–60. ACM (2014)

An Improved Data Packet Capture Method
Based on Multicore Platform

Xian Zhang[1,2], Xiaoning Peng[1,2(✉)], and Jia Liu[1,2]

[1] School of Computer Science and Engineering,
Huaihua University, Huaihua 418008, Hunan, China
zhangxian314@sina.com
[2] Key Laboratory of Intelligent Control Technology for Wuling-Mountain
Ecological Agriculture in Hunan Province, Huaihua 418008, Hunan, China

Abstract. The load balancing strategy of RSS used in the PF_RING capture method does not work well on multi-core processor platforms to achieve the disadvantage of the load balancing on the processor cores. This paper presents a packet load balancing method based on FD and RSS. The basic idea of this method is to capture the packet with the 5 tuple filter matching, and then can not be classified packets and flow oriented filter matching, and finally can not be classified packets matching RSS. Design of experiments to test the packet capture performance and load balancing performance which the packet capture method of PF_RING using the combination of load balancing strategy based on FD+RSS and RSS, the results show that the data packet stream load balancing method based on FD+RSS can improve the performance of data packet capture and load balancing among multiple cores.

Keywords: Load balancing · Multicore processor · Packet capture performance · Flow director filters

1 Introduction

With the development of the current processor, the multi-core processor technology has become the mainstream, and its high performance and parallel characteristics provide a new impetus for the development of network monitoring applications. However, the application of network monitoring does not make full use of the advantages of multi-core processor architecture in the kernel layer [1]. An important prerequisite for a network monitoring application is the capture of data packets, but the current packet capture technology is not a good use of parallelism of multi-core [2], which makes the technology of multi threading or multi queue in multi-core platform does not improve the performance of packet capture [3, 4]. The main reasons are: (1) thread resource

Fund Project: Project supported by Key Laboratory of Intelligent Control Technology for Wuling-Mountain Ecological Agriculture in Hunan Province (ZNKZ2015-5); The Research was supported in part by the grants from the Huaihua University Project (HHUY2017-13); The Research was supported in part by the grants from the Huaihua University Project (HHUY2016-05).

B. Zou et al. (Eds.): ICPCSEE 2017, Part I, CCIS 727, pp. 364–372, 2017.
DOI: 10.1007/978-981-10-6385-5_31

contention with multiple queues on a network interface card; (2) there is unnecessary duplication of data packets in the kernel transmission; (3) there is an inappropriate scheduling and interrupt balancing problem between CPU cores [5].

In order to make full use of the parallel architecture and network interface card of the modern multi-core processor, Luca Deri proposed a new packet capture technology of TNAPI(Threaded NAPI) which a multi thread packet receiving queue (RX) with polling mechanism, with the TNAPI mechanism of packet capture method firstly using hardware strategies (such as Receive Side Scaling (RSS)) to balance the network packets flow to the receiving queue, and each receive queue mapping to a specific processor [6, 7]. The packets load and the interrupt can be balanced to several processor cores by distributing the data packets to several receive queues, it makes better use of the inherent parallelism of multi-core architectures in the modern computer systems to improve its system performance [8].

In the paper of the performance evaluation of packet capture methods based on multi core platform [9], has been tested the performance of several typical packet capture methods on multi-core processor platform, these packet capture methods using the load balancing of RSS to distribute the received packets to several receive queues on multi-core platform, then each received queue is mapped to a processor core. (In the actual test case), it also can't balance the load of data packet very well, most of the captured packets will be processed by a few processors, and some processors are idle, it affect the performance of the systems seriously. Another drawback of load balancing method based on RSS, which can not change the packet flow balancing strategy in real time [10], and also affects the performance of multi-core system.

In this paper, we based on the basis of the load balancing strategy of RSS, improving the packet flow load balance method, using a load balancing strategy based on the combination of FD+RSS, combining RF_RING packet capture function library to improve on multi-core processor platform to capture the performance of the packet and analysis.

2   Method Introduction

2.1   Load Balancing Method Based on RSS

Receive Side Scaling (RSS) [11] is a packet flow load balancing method, which is divided into several different packet description queue, and then use the software to assign each queue corresponds to a processor core. At the same time, RSS is able to transfer packets in sequence to ensure that the data packets for each individual TCP connection can be processed by the same CPU processor, In order to avoid the overhead of inter core communication in different processor cores [12]. A basic idea of RSS strategy is for each of the received data packet generates a value based on the header of data package using hash function, to ensure the effective load balance to the CPU processor.

The sequence diagram for processing Data packets base on RSS is shown in Fig. 1.

The packet in the network card uses a hash function to produce a 32-bit hash result value based on the packet header information. then the hash mask is applied to the next

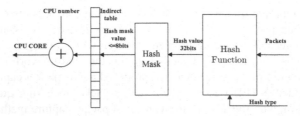

Fig. 1. Sequence diagram for processing data packets base on RSS

step to produce the hash result which to decide to the number of CPU processor cores in the indirect table, at the end of the packet is assigned to a specific CPU kernel queue according to the hash result value in the indirect table.

One of the key points of RSS hash function is the hash algorithm, Toeplitz hash algorithm is used in RSS. Toeplitz hash uses a random secret key K, the secret key K is 320 bits (40 bytes) for the IPv6, the secret key K is 128 bits (16 bytes) for the IPv4, the secret key K access in the random register of RSS. Given a input packet group containing N bytes, the byte stream is defined as follows:

input[0] input[1] input[2] ... input[n-1]

The leftmost byte is input[0], and the rightmost byte is input[n-1]. According to the sequence diagram above, the pseudo code of the input packets byte stream using the Toeplitz hash processing is defined as follows.

ComputeHash(input[], N)

For hash-input input[] of length N bytes (8N bits) and a random secret key K of 320 bits

Result = 0;
For each bit b in input[] {
if (b == 1) then Result ^= (left-most 32 bits of K);
shift K left 1 bit position;
}
return Result;

The value returned is the last result of the hash function (32 bit).

When the data packets are mapped to each processor core, the process show that the RSS processing data packets are as indicated in the sequence diagram. An example of a specific RSS indirect table is described in Fig. 2.

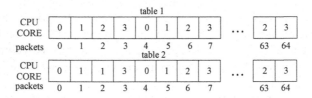

Fig. 2. Sample diagram of RSS indirect table

In the above of picture, table 1 shows that the packets is evenly distributed to each CPU core at the time of initialization, after a period of time the processor core began to appear unbalanced. When the host protocol stack is monitored for unbalanced conditions, an attempt is made to rebalance the load by calculating a new indirect table. Table 2 shows a new indirect form, it moves some packets from CPU2 to CPU1, leading to the load imbalance.

2.2 Load Balancing Method Based on FD+RSS

With the packet flow balancing strategy of RSS is not well balanced in certain conditions to the CPU cores, Another serious drawback to RSS is that the RSS algorithm takes a lot of CPU cycles and can't change the packet flow balance strategy in real time. Based on the above reasons, this paper proposed a load balancing method based on FD (Flow Director Filters) and RSS. Flow chart of load balancing method based on FD +RSS is shown in Fig. 3.

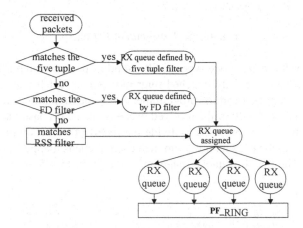

Fig. 3. Flow chart of balance method base on FD+RSS

From the flow chart of the load balancing method base on FD+RSS can be seen, the incoming packet first matches the five tuple filter to formulate the packet to the specific RX receive queue. The five tuple includes protocol type, source IP address, destination IP address, source port number, destination port number. The five tuple filter is defined as <id, protocol, ip port src/dst, ip src/dst, target RX queue id>. The specific rules for the five tuple filter are: the IP packet will match the five fields of the filter, if the packet is a pipeline packet or a piece of data packet does not match the five tuple filter.

When a packet cannot be matched with a five tuple filter, the next match filter (FD) is used to classify the packets flow, so flow director filters is an extension of the five tuple filter. The flow director filters identifies the captured data packets into a specific set of streams, and passes them to a specified queue, and finally each packet queue is mapped to a specific processor core. The basic rule of flow director filters is:

all of the IP packets can be classified using a flow director filter, encapsulation of data packets (IPv4 and IPv6 combination of packets) and incomplete packets will not be able to use flow director filters to classify. The flow director filters performs packet classification based on the following fields: VLAN header field, IP source address and destination address, source port number, destination port number, variable 2 byte data in the first 64 bytes of the protocol packet header. A block diagram of the flow director filters is shown in Fig. 4.

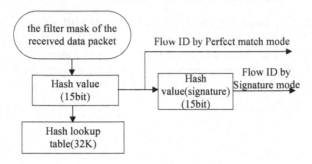

Fig. 4. Block diagram of FD filter

Flow director filters have two types of filter modes: perfect match filters modes and signature filters modes. A fully matched filter modes is a match between the masked field of the received packet and the programmed filter; the Signature filters modes is the matching of the masked field of the received packet with a hash based label of the filter. These two matched filters are used to divide the received packets into several receive queues according to the packets flow, and then each queue is mapped to a specific CPU processor core. The detail of queue mapping is shown in Fig. 5.

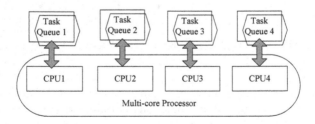

Fig. 5. Queue mapping chart of CPU

3 Experimental Evaluation

3.1 Experimental Design

The method of load balancing based on FD+RSS is used to classify the received packets, then, the classified packet queue is mapped to the corresponding CPU processor core to balance the load of each processor core, and make the data packets belong to the same flow distribution in the same processor cores in the queue, In order to improve the performance of the whole network monitoring application system, a better cache utilization and unnecessary inter processor communication overhead of CPU processors are brought forward. In this section, two experiments were designed to test the performance of packet capture and the load balance of processor core which using the packet capture method of PF_RING based on FD+RSS on multi-core processor platform.

The specific configuration of experimental platform in experimental environment is: Intel dual core processor PC, 10 Gbit Ethernet card, 10 Gbit switch, 4GDDR memory. Linux operating system, ixgbe network card driver, packet capture software program of PF_RING+TNAPI, Linux system comes with the contract module pktgen and patch are installed on PC,

Details of the basic configuration in the experimental platform are shown in Table 1.

Table 1. Configuration information of experimental platform

Configuration information	Specifications
CPU frequency	Intel Core(TM)2 Duo CPU E7200 2.53 GHz
Memory	DDR 4 GB
PCI bus	PCI-X 133 MHz 64b
Network card	RTL8168/8111 PCI-E Gigabit Ethernet NIC
Operating system	ubuntu-9.10-i386
Kernel version	Linux-2.6.31-20
NIC driver	intel e1000-8.0.16(ixgb-2.0.6)

In the experiment, 2 sets of PC are used as the data packet generator to generate enough data packets flow, and the other PC is used as the data packets receiver.

In experiment one, we tested the performance of packet capture that the packet capture method of PF_RING based on RSS and FD+RSS respectively, which in the case of packet size from 64 bytes to 1500 bytes in the 1000 Mbit packet traffic environment. In experiment two, the packet capture ability of the packet capture method of PF_RING based on RSS and FD+RSS was tested, when the packet is sent to 64 btyes, the sending speed rate is from low to high (0–2000 kpps). In experiment three, we tested the load balance of the two CPU processor cores that the packet capture method of PF_RING based on RSS and FD+RSS, which in the case of packet size from 64 bytes to 1500 bytes in the 1000 Mbit packet traffic environment.

3.2 Test the Ability of Packet Capture

In order to test the performance of packet capture that the packet capture method of PF_RING based on RSS and FD+RSS, the experiment first uses Linux's own contract module pktgen to generate packet size of 64 bytes to 1500 bytes packets in the context of 1000 Mbit traffic, respectively to test the rate of their capture packets. Figure 6 shows the performance comparison of the packet capture that the packet capture method of PF_RING based on RSS and FD+RSS.

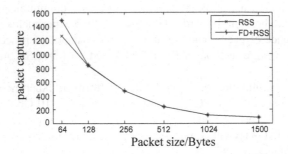

Fig. 6. The chart of the performance of packet capture

We can see from the test results that the packet capture rate can be improved by using small packets, which using data packet load balancing method based on FD+RSS than RSS. In the case of large packets, the two load balancing methods are similar.

In order to test their maximum capability of packet capture, The experiment was designed to send 64 btyes packets with the packet rate from low to high (0–2000 kpps), which tested the capabilities of packet capture of packet capture method of PF_RING based on RSS and FD+RSS. Figure 7 shows the comparison of the send rate and packet capture rate that the packet capture method of PF_RING based on RSS and FD +RSS.

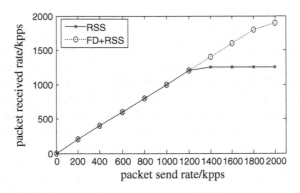

Fig. 7. Comparison of contract rate and packet capture rate

It can be seen from the test results that the performance of packet capture that packet balancing method based on FD+RSS is almost the same as based on RSS, however, the performance of packet capture based on FD+RSS is higher when the packet transmission rate is high.

3.3   Test the Load Balancing Between CPU Core

An important feature of the load balancing method based on FD+RSS is to balance the captured packets to the appropriate CPU processor core which to balance the load of each processor, in order to avoid the emergence of some processor core is busy and some processor is idle. Therefore the experimental design to test the load balance of the two CPU processor cores that the packet capture method of PF_RING based on RSS and FD+RSS, which in the case of packet size from 64 bytes to 1500 bytes in the 1000 Mbit packet traffic environment. Figure 8 shows that the load balance of the processor core that the two load balancing methods based on RSS and FD+RSS.

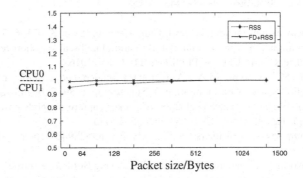

Fig. 8. Load balancing of processor core

It can be seen from the tested results that the load balance between processors of the load balancing method based on FD+RSS is better than which based on RSS during the transmission of small packets, but the performance of the two types of load balancing strategies is similar in the case of sending large packets.

4   Conclusions

This paper firstly introduces the previous packet capture method of PF_RING that using the load balancing method based on RSS. In order to effectively balance the load of packets to the core of each CPU processor and improve the ability to capture packets in the high speed environment, in this paper proposed a load balancing method based on the combination of FD and RSS. The performance of Packet Capture and Load Balancing that the packet capture method of PF_RING based on RSS and FD+RSS is tested through the designed experiments. It can be concluded from the experimental results that the load balancing method based on FD+RSS can improve the ability of data packet capture and load balancing among multiple cores. And it has a certain application value.

References

1. Yang, J.-X., Tan, G.-Z., Wang, R.-S.: Some key issues and their research progress in multicore software. Acta Electron. Sin. **9**, 2140–2146 (2010)
2. Wang, M.-Z., Zhao, G.-H., Tang, Y.: Technology research of efficient flow management based on multi-core network processor. J. Chin. Comput. Syst. **12**, 2591–2594 (2012)
3. Rajeswari, G., Nithya, B.: Implementing intrusion detection system for multicore processor. In: Advances in Recent Technologies in Communication and Computing (2009)
4. Vu, T.-T., Derbel, B.: Parallel branch-and-bound in multi-core multi-CPU multi-GPU heterogeneous environments. Future Gener. Comput. Syst. **56**, 95–109 (2016)
5. Wu, Z., Huang, Z., Gu, N., Zhang, X.: A User level real time task library on multicore. J. Chin. Comput. Syst. **36**(7), 1438–1443 (2015)
6. Yu, L.I.U., Hong, A.N., Sun, S.U.N., Jun-shi, C.H.E.N.: Improving multi-core system throughput with memory load balance. J. Chin. Comput. Syst. **3**, 671–675 (2014)
7. González-Domínguez, J., Liu, Y., Schmidt, B.: Parallel and scalable short-read alignment on multi-core clusters using UPC++. PLoS One **11**, 1–15 (2016)
8. Dashtbozorgi, M., Azgomi, M.A.: A high-performance and scalable multi-core aware software solution for network monitoring. J. Supercomput. **59**, 720–743 (2012)
9. Zhang, X., Li, W.: Performance evaluation of packet capture methods based on multi-core platform. Appl. Res. Comput. **28**(7), 2632–2635 (2011)
10. Tcpreplay. http://www.icewalkers.com/Linux/Software/528940/Tcpreplay.html. 27 July 2010
11. Scalable Networking: Eliminating the Receive Processing Bottleneck Introducing RSS. http://down-load.microsoft.com/download/5/d/6/5d6eaf2b-7ddf-476b-93dc7cf0072878e6/ndis_rss.doc,2004-4-14
12. Dashtbozorgi, M., Azgomi, M.A.: A high-performance software solution for packet capture and transmission. In: 2009 2nd IEEE International Conference on Computer Science and Information Technology (2009)

Research on Linux Kernel Version Diversity for Precise Memory Analysis

Shuhui Zhang[1,2]([✉]), Xiangxu Meng[1], Lianhai Wang[2], and Guangqi Liu[2]

[1] School of Computer Science and Technology, Shandong University,
Shunhua Road, Jinan, China
zhangshh@sdas.org, mxx@sdu.edu.cn
[2] Shandong Provincial Key Laboratory of Computer Networks,
Shandong Computer Science Center (National Supercomputer Center in Jinan),
Keyuan Road, Jinan, China
{wanglh,liuguangqi}@sdas.org

Abstract. The diversity of Linux versions brings challenges to Linux memory analysis, which is an established technique in security and forensic investigations. During memory forensics, kernel data structures are essential information. Existing solutions obtain this information by analyzing debugging information or by decompiling kernel functions to handle a certain range of versions. In this paper, by collecting and analyzing a number of Linux versions, we characterize the properties of different Linux kernel versions and how struct offsets change between versions. Furthermore, the Linux kernel provides over 10,000 configurable features, which leads to different kernel structure layouts for the same kernel version. To deal with this problem, we propose a method of identifying kernel struct layout based on brute-force matching. By examining the relationships between kernel structures, common features are extracted and exploited for brute-force matching. The experimental results show that the proposed technology can deduce structure member offsets accurately and efficiently.

Keywords: Memory analysis · Linux memory forensics · Brute-force matching

1 Introduction

Memory analysis has become critical in digital forensics because it provides insight into the state of a system that should not be represented by traditional media analysis. By analyzing a memory image, we can obtain details of volatile data, such as running processes, loaded modules, network connections, open files, and so on. Fundamentally, memory analysis is concerned with interpreting the seemingly unstructured raw memory data collected from a live system as meaningful and actionable information. Gathering and analyzing the information contained in a physical memory dump requires the following information.

© Springer Nature Singapore Pte Ltd. 2017
B. Zou et al. (Eds.): ICPCSEE 2017, Part I, CCIS 727, pp. 373–385, 2017.
DOI: 10.1007/978-981-10-6385-5_32

(1) Specific knowledge of the kernel version: the numbering for Linux kernel versions is indicated by a.b.c, where the number "a" denotes the kernel version, the number "b" denotes the major revision of the kernel, and the number "c" denotes the minor revision of the kernel.

(2) Struct members and memory layout: a complete understanding of the data structures and algorithms used by the original operating system (OS) must be understood. This contains information about memory offsets for struct members and their types. A memory framework must have the same layout information to interpret the contents of the memory. For example, in Windows systems, the _EPROCESS_ struct is used to track per process information. Similarly, the Linux kernels process descriptor structure, *task_struct*, contains all the information that links a process to its open files, memory mappings, signal handlers, network activity, and more.

(3) Global variables: almost every OS kernel contains many important constants required for analysis. For example, in Linux kernels, the *init_task* is a constant pointer to the beginning of the process linked list, and is required for listing processes by following the list. In Windows systems, *PsActiveProcessHead* is the head of the active process list [1].

(4) Function addresses: functions can be used to uncover the semantics of kernel objects that are important during memory analysis.

For a Windows system that has only a small number of officially released versions, the structure layout is mostly stable across major and minor kernel versions. The location of a global variable is determined statically by the compiler at compiling time and changes dramatically even in the same version. Modern memory analysis frameworks obtain the above information across different OS versions by using a version-specific memory layout template mechanism. For example, in Volatility (The Volatility Foundation, 2014) and Rekall (The Rekall Team), this information is called a profile [2,3].

However, under Linux systems, where new kernel versions are released frequently and distributions ship their own custom kernels, it is not feasible to write a tool that has knowledge of all kernel variations. Even in instances where an investigator has the complete source code for the kernel being investigated, there may not be enough information to accurately model the required structures. While there are numerous problems to be faced when taking a source code-based approach, the largest problem is conditional compilation during the kernel building process. Inclusion or exclusion of a kernel configuration option can cause the insertion or removal of a large number of members in key structures that are required for processing an associated memory image. For example, inside the Linux kernels struct *task_struct*, at least forty of the more than one hundred members depend on user-controlled compile time compilation options. This creates greatly different in-memory layouts of the structure between kernel versions.

The contributions of this paper towards these issues are outlined below.

(1) We describe the diversity of Linux kernel versions. Struct members, memory layout, global variables, and function addresses vary widely between differ-

ent kernel versions. Linux kernels provide more than 10,000 configurable features, which also leads to different kernel structure layouts for the same kernel version. Determining Linux kernel struct layout has become a recognized problem.
(2) We propose a method of identifying Linux kernel struct layout based on brute-force matching. Existing solutions require debugging information or source code to handle a range of versions. By exploring the relationships between kernel structures, common features are extracted and are exploited for the brute-force matching. Experiments show that this method can effectively determine offsets for critical elements over a large range.

The remainder of this paper is organized as follows. Section 2 introduces some background information. The proposed techniques based on brute-force matching are described in Sect. 3. In Sect. 4, we evaluate the effectiveness and performance of the proposed method. The final section contains our conclusions and discusses ideas for future research.

2   Linux Kernel Version Diversity

The Linux kernel was created in 1991 by Linus Torvalds for his personal computer with no cross-platform intentions [4]. Linux rapidly attracted developers and users who adapted code from other free software projects. To satisfy the variety of requirements, Linux is available for different platforms such as desktops, servers, mainframes, and on various embedded devices such as routers, wireless access points, PBXes, set-top boxes, FTA receivers, smart TVs, PVRs and NAS appliances. The Android OS for tablet computers, smartphones and smartwatches is also based on the Linux kernel. This OS is mainly used for operations like web hosting, software development, device embedding and so on.
    So far, there are more than 300 Linux distributions. Some common distributions are the following:

(1) Ubuntu is a community-developed, Linux-based OS for laptops, desktops and servers. It contains abundant applications, such as a web browser, presentation, document and spreadsheet software, instant messaging and much more. The Kubuntu, Edubuntu, Xubuntu and Dubuntu OSs were developed based on Ubuntu. Ubuntu releases its new version in April and October each year and every distribution includes 32-bit and 64-bit versions.
(2) CentOS is a community-supported enterprise-class distribution that aims to be functionally compatible with Red Hat Enterprise Linux. The first CentOS release was in May 2004, and a new CentOS Linux version is relased approximately every two years. Each CentOS Linux version is periodically updated to support newer hardware and provides a predictable and reproducible Linux environment.
(3) The Fedora Project is sponsored by Red Hat and integrates the features of SELinux, Xen virtualization and other enterprise functions. Its primary goal is to share the benefits of a common codebase for end users and developers while reducing unnecessary maintenance.

(4) OpenSUSE was created by Novell on 4 November 2003, using KDE as the default desktop environment and also providing the GNOME desktop. OpenSUSE provides YaST Software Management as a graphical user interface and Zypper as a command line interface.

(5) Other Linux distributions include Debian, PCLinuxOS, Redhat, Turbolinux, RedFlage, Kylin and so on.

The distribution version number is independent of kernel version. Even for the same kernel, the kernel struct and memory layout of different distributions may vary.

2.1 Characterizing Kernel Version Variability

Figure 1 shows sample offsets for four valuable struct members in the *task_struct* structure.

- The *task_struct.pid* is a positive integer that uniquely identifies a process. The identifier shared by the threads is the *pid* of the thread group leader, which is stored in the *tgid* field of the process descriptors. The *sys_getpid()* function returns the value of *tgid* for the current process instead of *pid*. The offset betwwen *pid* and *tgid* is four.
- The *task_struct.comm* field stores up to 16 ASCII characters for the executable filename of the process.
- The *task_struct.mm* points to the *mm_struct* structure, which saves all the information related to the process address space.
- The *task_struct.files* points to the *files_struct* structure, which describes the opening file table related to the process.
- The *task_struct.tasks* points to the *list_head* structure, which points to a doubly-linked list and can be used to enumerate the currently running processes.

As can be seen from Fig. 1, struct offsets change dramatically between each build. Even when the kernel version and the major revision are the same, the offsets can still be different.

Other than the structures memory layout, the global constants are also significant for memory analysis. Figure 2 shows the virtual addresses of several critical global constants. For a 64-bit system, all the virtual addresses are encoded as 0xffffffff00000000. For convenience, the offset presented in Fig. 2 is from 0 to 31 bits of the virtual address.

- The *init_task* function is the first process, called the *swapper*, and can be used to list the running processes.
- The *init_mm* function is the location of the initial *mm_struct* in the system, and can also be used for enumerating all *mm_struct* structures.
- The *init_files* function points to the *files_struct* structure in the swapper process. According to the *init_files* value, subtraction can be used to obtain the offset of the files variable in *task_struct*.

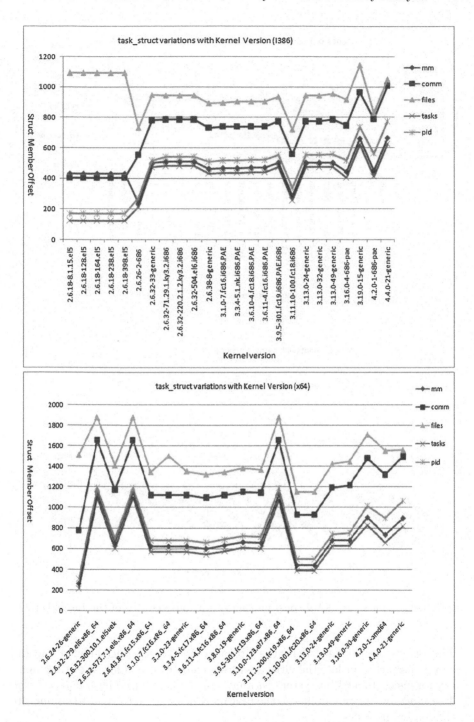

Fig. 1. Offsets for some critical struct members across various versions of the Linux kernel.

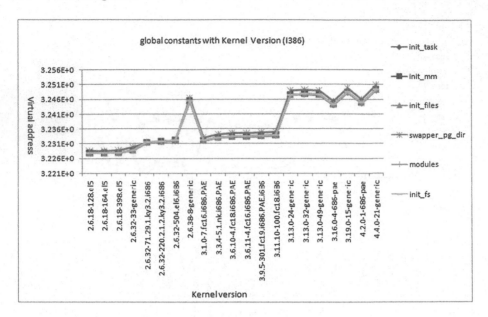

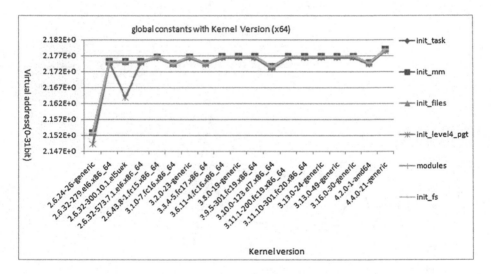

Fig. 2. Offsets for some global constants across various versions of the Linux kernel.

- The *swapper_pg_dir* function is the virtual address of the initial page table and can be used to translate the virtual address to a physical address during memory forensics. In a 64-bit system, *swapper_pg_dir* is replaced by *init_level4_pgt*.
- The *modules* function points to the first module and can be used to enumerate the loaded modules.

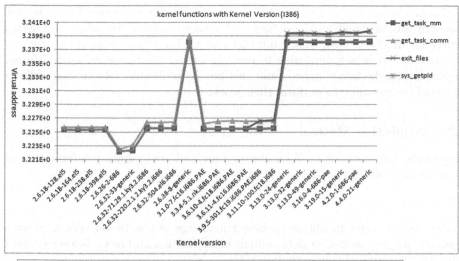

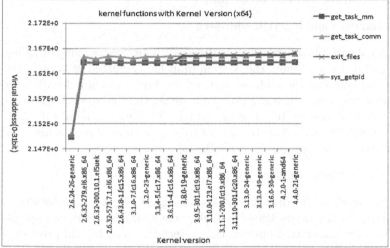

Fig. 3. Offsets for some kernel functions across various versions of the Linux kernel.

It can be seen that there is only a slight difference between addresses for a global constant in the same kernel. However, the addresses for a certain global constant in various kernels are different.

Next, the addresses used for comparison are the kernel functions shown in Fig. 3.

- The *get_task_mm* function is used to return the *mm_struct* address for a specified process.
- The *get_task_comm* function is used to obtain the name of specified process.
- The *exit_files* function is used to close the files for a specified process.
- The *sys_getpid* function can be used to obtain the *tgid* value.

From the figures above, we can see that even for the same Linux kernel, the addresses of struct member offsets, addresses of global constants, and kernel functions are diverse. All these differences have effects on the accuracy of memory analysis. It is not unreliable to directly define the above values by simply depending on the kernel version and this breaks tools that rely on exact offsets of structure members to perform their work.

2.2   Acquisition Method

Nowadays, the methods of obtaining system and configuration information in existing solutions are described below [6, 7].

(1) Kernel version

There are two ways to obtain specific knowledge of the kernel version from a memory image: *vmcoreinfo_data* content identification and *linux_banner* content identification from the memory image [8].

(2) Global variables and function addresses

To obtain global variable and function addresses, /boot/System.map or kallsyms file can be useful. Under Linux, the System.map file is created at kernel compilation time and contains the address, type of symbol, and name for all defined symbols within the kernel. Investigators can obtain the file from the target system locally [9] or remotely [10]. The method for obtaining symbol information should ideally not affect the OS. In view of this, a method of recovering symbol information from an image was proposed in [8].

(3) Struct members and memory layout

Linux memory analysis requires precise knowledge of struct layout information in the memory. As the Linux kernel is highly configurable, users can specify a large number of configuration options that affect the kernel build process. For example, the definition of *task_struct* is shown in Fig. 4. As can be seen, some struct members may be included only when certain configuration parameters are set. For example, *sched_info* only exists when the CONFIG_SCHEDSTATS macro or the CONFIG_TASK_DELAY_ACCT macro is set. That is to say, the structure elements can vary significantly even within the same kernel version and it is not accurate to specify struct member information and memory layout. In [11], the research goal was creating a system that would be able to fully analyze Linux 2.6 memory dumps without relying on specific knowledge of kernel versions and distributions. Dynamic reconstruction of kernel data structures is realized by analyzing the memory dump for instructions that reference needed kernel structure members. The ability to dynamically recreate C structures used within the kernel allows for a large amount of information to be obtained and processed.

```
struct task_struct {
        volatile long state;      /* -1 unrunnable, 0 runnable, >0 stopped */
        void *stack;
        atomic_t usage;
        unsigned int flags;       /* per process flags, defined below */
        unsigned int ptrace;
#ifdef CONFIG_SMP
        struct llist_node wake_entry;
        int on_cpu;
        struct task_struct *last_wakee;
        unsigned long wakee_flips;
        unsigned long wakee_flip_decay_ts;
        int wake_cpu;
#endif
        int on_rq;
        int prio, static_prio, normal_prio;
        unsigned int rt_priority;
... .
#if defined(CONFIG_SCHEDSTATS) || defined(CONFIG_TASK_DELAY_ACCT)
        struct sched_info sched_info;
#endif
        struct list_head tasks;
#ifdef CONFIG_SMP
        struct plist_node pushable_tasks;
        struct rb_node pushable_dl_tasks;
#endif
        struct mm_struct *mm, *active_mm;
...
```

Fig. 4. Partial members of the *task_struct* structure.

3 Identifying Linux Kernel Struct Layout Based on Brute-Force Matching

As can be seen from Sect. 2.2, existing methods of obtaining the kernel version, global variables and function addresses are effective. However, the prerequisites for obtaining structure layout are the config file and related function codes. By decompiling the memory segment, offsets of the elements within structures may be attained. The source codes for different kernel versions are different, which means that the analysis method may vary depending on the version in the target system.

To avoid complex and expensive analysis of kernel source code, we propose a method of recovering the kernels data structure. This method is based on brute-force matching and fully exploits the relationships between kernel data structures. We use *task_struct* as an example to describe our method. The relationships among *task_struct* (TS), *mm_struct* (MS), *files_struct* (FILES), *fs_struct* (FS), *nsproxy* (NS), and *signal_struct* (SGS) structures are shown in Fig. 5. It can be seen that TS has six pointers, one each to FS, FILES, MS, TS, NS,

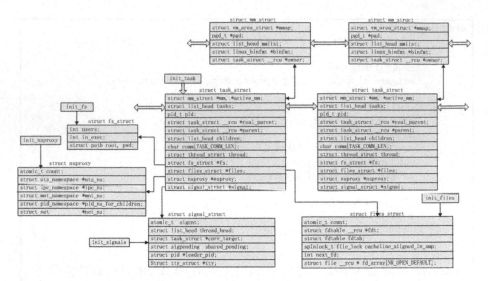

Fig. 5. Relationships among several structures.

and SGS, while MS has two pointers to TS and MS. The analysis algorithm is outlined below.

1. The *init_task* variable from the System.map file represents the TS structure of swapper process. From the address of *init_task*, we can obtain the offsets of the *files*, *fs*, *nsproxy*, *signal*, and *comm* members by searching for the value of the *init_files*, *init_fs*, *init_signals*, *init_nsproxy* and *swapper* strings.
2. The values of tasks point to tasks elements for another process. We obtain the content of the address from the offset of the *comm* variable. If the content is a character string, the offset of the tasks is valid. Otherwise, we keep scanning the TS structure.
3. According to the features of the TS structure, the *tgid* variable is close to the *pid* variable (see Eq. (1)). Furthermore, both variables are between the *comm* and *tasks* variables (see Eq. (2)), and their values are equal in most cases. Based on these features, we can obtain the offset of the *pid* variable.

$$Offsettgid = Offsetpid + 4. \tag{1}$$

$$Offsettasks < Offsetpid < Offsetcomm. \tag{2}$$

4. The offset of the *mm* variable needs to be determined from the second process. We obtain the beginning of second process from step 2. As is shown in Fig. 6, we choose a candidate MS structure address from the TS structure and traverse the memory content from the candidate address to find the TS structure address. If the TS structure address is found, then we also find the offset of the *mm* variable.

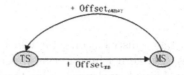

Fig. 6. Relationships between TS and MS structures.

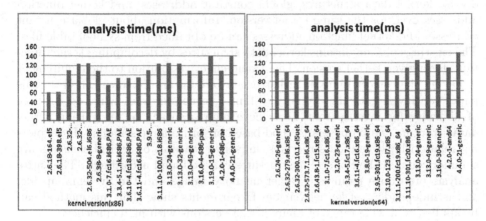

Fig. 7. Analysis time for RAMAnalyzer.

Using the algorithm described above, the offsets of the *mm, fs, files, nsproxy, signal, pid, comm,* and *tasks* variables in *task_struct* can all be obtained. The search conditions used in the algorithm are universal to all Linux versions. The efficiency of this algorithm is described in Sect. 4.

4   Experiments and Results

Based on the techniques described above, we expand our Linux memory analysis tool (RAMAnalyzer) to acquire the offsets of struct members. In this section, we present our experimental results. An experiment was performed to test the effectiveness of RAMAnalyzer for 38 Linux kernels. All of our experiments were performed on a host machine with an Intel Core i5-2500 CPU, 3.24 GB memory, and a 32-bit Windows 7 OS.

Our analysis targets are memory images that cover versions 2.6.18 to 4.4.0 for both 32-bit and 64-bit systems. The preconditioning for our analysis procedure is kernel symbol information of the target system. Using the techniques described in Sect. 3, RAMAnalyzer can deduce the offset of several elements in the *task_struct* structure, including the *mm, fs, files, nsproxy, signal, pid, comm,* and *tasks* variables. About 90–140 ms were required for the acquisition process. Detailed execution times are presented in Fig. 7.

The experimental results show that RAMAnalyzer can accurately and efficiently deduce the offsets of structure members.

5 Conclusion

There are more than 300 Linux distributions, for which kernel data structures, global constant addresses, and kernel function addresses vary greatly between kernel builds. In particular, Linux kernel struct layouts vary wildly by kernel configuration as well as kernel version. That is, different configurations can generate diverse kernel data structures, global constant addresses, and kernel function addresses even for the same kernel version. Information about global constant addresses and kernel function addresses can be obtained from symbol table files. The existing solution obtains kernel data structures by decompiling kernel functions for a given kernel version. There is no effective solution to the requirement of having the correct Linux kernel struct layout for an unknown kernel. In this paper, we investigate the diversity of Linux kernel versions, which is a source of difficulty for investigators. To achieve high accuracy in recognizing kernel data structure, we propose a method based on brute-force scanning. By exploring relationships between structures, universal features of critical elements are summarized and their offsets in structures can be found accordingly. The entire analysis procedure requires a target memory image and the System.map file, and is independent of kernel version. Experiments show that this method can be applied to a wide range of Linux kernel versions with high efficiency.

The techniques proposed in this paper provide forensics research with a starting point for delving into Linux memory forensics. We have implemented this function and have added it to our Linux memory forensics tool named RAMAnalyzer. Furthermore, these techniques can be conveniently embedded into other forensics frameworks as an analysis fundamental. In future research, we will study internal kernel functions and kernel structures for kernel malware analysis.

Acknowledgments. This work is supported by the National Natural Science Foundation of China (Grant Nos. 61572297, and 61602281), the Shandong Provincial Natural Science Foundation of China (Grant Nos. ZR2016YL014, ZR2016YL011, ZR2014FM003, and ZY2015YL018), the Shandong Provincial Outstanding Research Award Fund for Young Scientists of China (Grant Nos. BS2015DX006, and BS2014DX007), and the Shandong Academy of Sciences Youth Fund Project, China (Grant Nos. 2015QN003).

References

1. Cohen, M.I.: Characterization of the windows kernel version variability for accurate memory analysis. Digit. Invest. **12**, 28–49 (2015)
2. The rekall profile repository. https://github.com/google/rekall-profiles
3. The volatility framework. http://www.volatilityfoundation.org/
4. Linux Kernel. https://en.wikipedia.org/wiki/Linux_kernel
5. Linux memory forensics. http://www.drdobbs.com/linux-memory-forensics/199101801
6. Memoryze. https://www.mandiant.com/resources/download/memoryze

7. Moonsols windows memory toolkit. http://www.moonsols.com/windows-memory-toolkit

8. Zhang, S., Meng, X., Wang, L.: An adaptive approach for Linux memory analysis based on kernel code reconstruction. EURASIP J. Inf. Secur. **14**, 1–13 (2016)

9. Ligh, M.H., Case, A., Levy, J., Walters, A.: The Art Of Memory Forensics: Detecting Malware and Threats in Windows, Linux, and Mac Memory. Wiley Publishing, Indianapolis (2014)

10. Socala, A., Cohen, M.: Automatic profile generation for live Linux memory analysis. Digit. Invest. **16**, 11–24 (2016)

11. Case, A., Marziale, L., Richard, G.G.: Dynamic recreation of kernel data structures for live forensics. Digit. Invest. **7**, 32–40 (2010)

Unsupervised Anomaly Detection for Network Flow Using Immune Network Based K-means Clustering

Yuanquan Shi[1,2](✉), Xiaoning Peng[1], Renfa Li[2], and Yu Zhang[3]

[1] School of Computer Science and Engineering,
Huaihua University, Huaihua 418008, China
syuanquan@163.com
[2] College of Computer Science and Electronic Engineering,
Hunan University, Changsha 410082, China
[3] College of Information Science and Technology,
Hainan Normal University, Haikou 571158, China

Abstract. To detect effectively unknown anomalous attack behaviors of network traffic, an Unsupervised Anomaly Detection approach for network flow using Immune Network based K-means clustering (UADINK) is proposed. In UADINK, artificial immune network based K-means clustering algorithm (aiNet_KMC) is introduced to cluster network flow, i.e. extracting abstract internal images from network flows and obtaining an optimizing parameter K of K-means by aiNet model, and network flows are clustered by K-means algorithm. The cluster labeling algorithm (clusLA) and the network flow anomaly detection algorithm (NFAD) are introduced to detect anomalous attack behaviors of network flows, where the clusLA algorithm is used for labeling whether each cluster belongs to malicious, and the labeled clusters are regarded as detectors to identify anomaly network flows by NFAD. To evaluate the effectiveness of UADINK, the ISCX 2012 IDS dataset is considered as the simulating experimental dataset. Compared with the NDM based K-means anomaly detection approach, the results show that UADINK is a radical anomaly detection approach in order to detect anomalies of network flows.

Keywords: Unsupervised anomaly detection · Artificial immune network · K-means · Clustering · Network flow

1 Introduction

With the rapidly development of Internet technologies, and the application of mobile internet platforms and smart devices, network security problem has become severe society focus, such as virus infection, intrusion attack, illegal access and network phishing etc. The objects of network attacks mainly include network nodes, terminal computers, and smart devices of Internet environment, especially smartphone which provides network access admission and payment capability. To evaluate effectively network security, many security strategies are employed, such as private protection, firewall mechanism, virus defense, intrusion detection and risk evaluation etc.

© Springer Nature Singapore Pte Ltd. 2017
B. Zou et al. (Eds.): ICPCSEE 2017, Part I, CCIS 727, pp. 386–399, 2017.
DOI: 10.1007/978-981-10-6385-5_33

Network flow is reviewed as a description method of network behaviors based on the connections of network terminals, and records high-level description of network connections, but network flow isn't an actual network packet [1, 2]. Network flow is defined as an unidirectional or bidirectional sequence of packets that transmits between network terminals using a particular protocol (e.g. TCP) with common features [3], for example duration time, IP address, port number, and the transferred packets etc. In flow based network security field, anomaly detection is an important network intrusion detection strategy [4, 5], and it can detect unknown malicious attacks from network traffic. In traditional anomaly detection, administrators firstly need define normal profiles of the protected network system in order to effectively distinguish malicious and benign behaviors of network traffic, namely detecting anomalies by the supervised method. However, some researchers have proposed and improved anomaly detection approaches with misuse detection technique to raise detection rates (DR) of known malicious attacks and decrease false alarm rates (FAR) of unknown attacks [6]. Compared with the packet based anomaly detection, the flow based anomaly detection analyzes network security problems by using network flows, and it can help to solve the problems that are data reduction and processing time [7, 8]. Some inherent rules of network flows can be analyzed by common features of sending/receiving protocol packets of network flows and their similarity, especially TCP flow that is a typical flow based on handshake mechanism of TCP protocol to build logical connection between any two network terminals. Importantly, the flow based anomaly detection has become a research hotspot that can be seen as an effective complement of packet inspection [1, 2, 9].

Clustering is viewed as an unsupervised classification method based on the similarity of data samples [10], and it doesn't need the labeled data samples during training phase. Many researchers think that clustering holds the internal homogeneity and the external separation, i.e. samples of same clusters hold similar pattern. The representative clustering techniques [11, 12] relates to K-means clustering, hierarchical clustering, partition clustering, and evolutionary clustering algorithms. As one of the most difficult and challenging problems in machine learning fields, many evolutionary clustering algorithms are proposed successively to analyze the unsupervised nature problem that its data spatial distribution is unknown [11, 13–16], such as artificial immune system, genetic algorithm, artificial neural network etc. Many clustering algorithms are proposed for solving anomaly detection problems of network flows [5, 17]. Portnoy et al. [18] proposed a variant of single-linkage clustering based on distance to classify data instances. Leung and Leckie [19] proposed the density-based and grid-based high dimensional clustering algorithm which is used for unsupervised anomaly detection of large datasets. Petrovic et al. [20] combined the Davies-Bouldin index of the clustering and the centroid diameters of the clusters to detect massive network anomaly attacks. Syarif et al. [21] investigated the performances of five different clustering algorithms, namely, k-means, improved k-means, k-mediods, expectation maximization (EM) and the distance-based outlier detection algorithms. In the clustering based anomaly detection of network flows, some researchers have obtained remarkable outcomes. Erman et al. [22] proposed a semi-supervised clustering method used for classifying network flows. Münz et al. [23] proposed a flow based anomaly detection scheme based on K-means clustering algorithm. And its cluster centroids can be used for detecting anomaly behaviors from on-line monitoring data. Ahmed and

Mahmood [24] used X-means clustering to detect collective anomaly flows such as DOS attacks. X-means clustering provides an effective strategy to select its parameter k. Sheikhan and Jadidi [7] proposed artificial neural network-based NIDS that is used for detecting network flow anomaly attacks. This system distinguishes benign and malicious flows using multi-layer perceptron neural classifier, and optimizes the interconnection weights of neural anomaly detector using gravitational search algorithm. Winter et al. [25] proposed a network intrusion detection system to analyze anomaly flows, and used the unsupervised learning method of One-Class Support Vector Machines to distinguish whether network flow is malicious or not. Therefore, the advantages of the clustering based anomaly detection approach mainly include: (1) obtain anomaly detectors by self-learning mode; (2) extract common features rapidly from big dataset; (3) detect new and unknown anomalous attacks to cope with the changeable network traffic.

In this paper, we propose an Unsupervised Anomaly Detection approach for network flows using artificial Immune Network (aiNet) based K-means clustering (UADINK). The mainly contributions of this paper are: (1) propose an unsupervised anomaly detection framework to detect anomalies of network flows; (2) propose an aiNet based K-mean clustering (aiNet_KMC) algorithm to cluster effectively network flows; (3) give a cluster labeling algorithm (clusLA) to distinguish benign and malicious cluster centroids of network flows. In UADINK, we use aiNet model [16] to obtain a reasonable parameter K of K-means algorithm, and devise the clusLA algorithm to raise DRs and decrease FARs. And the ISCX 2012 IDS dataset [26] is adopted for detecting anomalies of network flows in this paper.

The remainder of this paper is organized as follows. The proposed UADINK approach for network flows based on aiNet model and K-means algorithm is described in Sect. 2. Section 3 illustrates the performance evaluations of UADINK on the ISCX IDS dataset. Finally, the conclusion is given in the last section.

2 The Proposed UADINK Approach

The proposed UADINK approach is an unsupervised anomaly detection approach for network flows using immune network based K-means clustering algorithm. In UADINK, it provides an automatic mechanics to detect anomalous network flows, so it doesn't need the labeled data samples to cluster network flows. The flowchart of UADINK is shown in Fig. 1. Known from Fig. 1, UADINK mainly includes five aspects: (1) Capture network flows. Network flows can be captured by the real world or generated by replaying network packets of the benchmark dataset. (2) Preprocess network flows. To effectively distinguish anomalous attacks of network flows, the network flows need be preprocessed by selecting the typical features of every network flow which can easy to identify network behaviors. (3) Obtain the optimizing K. The aiNet model [16] is introduced for clustering the given training dataset, the learned cluster number is used as the optimizing K of K-means algorithm. (4) Cluster network flows. The K-means algorithm [23] is used for clustering the given training dataset to obtain the optimizing cluster centroids for network flows, and each clustering centroid represents one class of network flows. (5) Label clusters of network flows. The obtained cluster centroids need be labeled as abnormal/normal network flows so that UADINK can

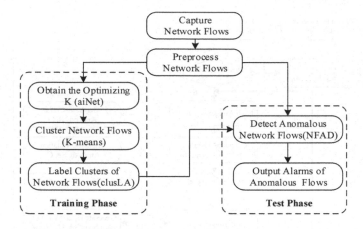

Fig. 1. The flowchart of UADINK.

detect effectively anomalous network flows. The used relevant models and algorithms of UADINK are described at the following subsections, namely artificial immune network (aiNet), the aiNet based K-mean clustering (aiNet_KMC) algorithm, cluster labeling algorithm (clusLA) and network flow anomaly detection algorithm (NFAD).

2.1 Artificial Immune Network

Artificial immune network is one of important theories of artificial immune system inspired by biological immune system, and it holds some merits of artificial immune system, such as self-learning, self-adaption, self-organization and immune memory etc. [27–30]. According to immune network theory [31], the binding between idiotopes (molecular portions of an antibody) located on B cells and paratopes (other molecular portions of an antibody) located on B cells has a stimulation effect for B cells, and the interaction of B cells within a network will produce to a stable memory structure and account for the retainment of memory cells. The artificial immune network (aiNet) model [16] is firstly used for analyzing and filtering the crude dataset, and an internal image of all data points in dataset, namely a refined relationship map, is constructed by immune evolution mechanisms. Therefore, the aiNet model can be used to refine some important characters from complex information data. At present, its application areas mainly include clustering data, pattern recognition, and data compression etc.

The aiNet model is shown in Fig. 2. The aim of the aiNet model is how to find out optimal memory antibodies of each antigen ag_j by using immune evolution strategies which include clonal proliferation, immune selection, immune mutation, and antibody suppression etc. For each antigen ag_j, the aiNet model can generate a memory antibody subset M_j, and the memory antibody set M aggregate and storage optimal antibodies after all antigens are travelled. For each iterative running of the aiNet model, all antibodies of M will be suppressed to avoid similar antibodies entering next generation. If the iterative stop criterion is satisfied, the optimizing memory antibody set M will be outputted as the results of the specific application, for instance, obtaining the cluster number and cluster centroids of clustering algorithms. Therefore, the design of immune

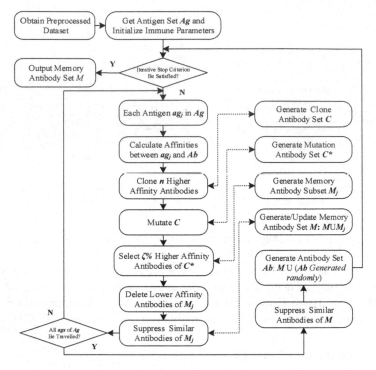

Fig. 2. The flowchart of aiNet model.

strategies is a key step to enhance the evolution learning capabilities of the aiNet model, and the corresponding immune evolution strategies of the aiNet model have been introduced in Reference [16].

2.2   The aiNet_KMC Algorithm

The purpose of K-means algorithm [32] is to partition a dataset into K sub datasets, namely executing automatically clustering operation for the given dataset. The basic idea of K-means algorithm consists of four aspects [33] : (1) define the cluster number K; (2) choose randomly K data samples as initial cluster centroids from dataset; (3) partition K clusters for the given dataset according to the nearest theory between any two different data samples of dataset; (4) obtain K optimal cluster centroids to keep having disjoint clusters for dataset. The K-means algorithm is a simple method to solve the clustering problems because of its easily programming and saving computing cost. But the accuracy of the K-means algorithm is dependent on K and the initial values of cluster centroids, and the value of K is uncertain. To find an optimal value of K, the aiNet model is introduced into K-means algorithm in this section. The aiNet based K-mean clustering (aiNet_KMC) algorithm is described as Algorithm 1.

In the aiNet_KMC algorithm, K is a positive integer number that is generated by aiNet model. The K optimal cluster centroids can be obtained by the aiNet_KMC clustering algorithm when $C_1, C_2, ..., C_K$ are stabilized during the clustering of Ds

according to the mean-squared error criterion. Once a data d is added to the specific cluster, the centroid of this cluster needs be updated because it is the arithmetic mean of the data samples in this cluster. For the distance between cluster centroids and data sample of dataset, it can be calculated by distance functions [23], such as Euclidean distance, Mahalanobis distance and Hamming distance etc. The squared Euclidean distance function is adopted in this paper, and is defined as Eq. (1), where m represents how many quantitative features that the vectors x or y has.

$$d(x, y) = \sum_{i=1}^{m} (x_i - y_i)^2 \tag{1}$$

Algorithm 1 The aiNet_KMC Algorithm
1.   **Input:** A preprocessed dataset Ds
2.   **Output:** The cluster number K and cluster centroids C
3.   **Start**
4.   Obtain an optimal value of the parameter K by the aiNet model
5.   Select K data samples randomly from Ds as initial cluster centroids $C_1, C_2, ..., C_K$
6.   **While** unstabilized for $C_1, C_2, ..., C_K$ **Do**
7.   **For** each data sample d of Ds **Do**
8.   Compute the Euclidean distance $Dist$ between d and C_i (i=1...K)
9.   Find a cluster centroid C_j ($1 \leq j \leq K$) which is closest to d
10.   Assign d to C_j, and update the j-th cluster centroid
11.   **End For**
12.   **End While**
13.   **End**

2.3   The ClusLA Algorithm

Algorithm 2 The clusLA Algorithm
1.   **Input:** Cluster centroids C, Training dataset Trd, Cluster indices idx for each data point of Trd, Anomaly recognition threshold Rt about normal/abnormal clusters
2.   **Output:** Label set Nal of C that represent normal or abnormal clusters
3.   **Start**
4.   Determine size of Nal and the scale of Trd
5.   **For** each data point of Trd **Do**
6.   Calculate the statistic value $Sc(i)$ that this data point belongs to the i-th specific cluster of C according to C and idx
7.   **End For**
8.   **For** each cluster of C **Do**
9.   Calculate the percent ratio Pr of the recognized data points, which belong to specific cluster of C, in training dataset Trd;
10.   **If** (Pr of each different cluster is not less than Rt) **Then**
11.   Label this cluster as normal cluster and storage its cluster number in Nal
12.   **Else**
13.   Label this cluster as abnormal cluster and storage its cluster number in Nal
14.   **End If**
15.   **End For**
16.   **End**

The cluster labeling algorithm (clusLA) is used for labeling normal and abnormal cluster centroids to identify malicious flows and benign flows. In clusLA, the accuracy of the labeled clusters will influence the detection performance of UADINK. The clusLA algorithm is described as Algorithm 2. It mainly include two important modules, the first module is used for accumulating the matched times of each cluster centroid in C with data point in Trd. The second module is used for calculating the percent ratio Pr of all similar data points recognized by each cluster in training dataset Trd, and then each cluster centroid of C is labeled as normal or abnormal cluster according to the percent ratio Pr and the anomaly recognition threshold Rt.

The recognition threshold Rt is an important parameter of the clusLA algorithm, and is used for labeling normal/abnormal clusters. A reasonable value of Rt can increase DRs and decrease FARs of anomaly detection system. For choosing Rt, there are two strategies to obtain the value of Rt. The first strategy is that the ratio, which is 10% of all samples of training dataset, may be considered as the value of Rt. The second strategy is that the ratio between the existing real attacks and the total amount samples in training dataset also may be considered as the value of Rt. We adopt the first strategy for partitioning normal clusters and anomaly clusters in this paper, and Rt is set to 0.05.

2.4   The NFAD Algorithm

In this section, the network flow anomaly detection (NFAD) algorithm is adopted for detecting anomalies of network flows. Administrators can obtain anomaly alarms from network flows according to the results which the cluster centroids of C match data points of test dataset or the preprocessed flow data from Internet, and then some security strategies can be deployed timely in network environment. The NFAD algorithm is described as Algorithm 3.

Algorithm 3 The NFAD Algorithm
1.   **Input:** Cluster centroids C, Label set *Nal*, Test dataset *Ted*
2.   **Output**: Alarmed network flows
3.   **Start**
4.   Initialize anomaly detection parameters
5.   **For** each datum of *Ted* **Do**
6.   Calculate distance between each cluster centroid of C and this data point
7.   Find a cluster centroid with the minimum distance *dist*
8.   **If** (its cluster number is equal to abnormal cluster number in *Nal*) **Then**
9.   Alarm and Output this data point, namely abnormal network flow
10.   **End If**
11.   **End For**
12.   **End**

---

3   Simulation and Experimental Results

3.1   Dataset Description

To demonstrate the effectiveness of UADINK in this section, we adopt the ISCX 2012 IDS dataset [26] as benchmark dataset to evaluate UADINK for detecting anomaly behaviors of network flows. The ISCX 2012 IDS dataset consists of seven days' capturing with overall 2,450,324 network flows generated by a controlled test bed, and designed by the Information Security Centre of Excellence (ISCX) at the University of New Brunswick. According to the activity characters of anomaly behaviors in the complex network environment, the Saturday sub dataset (4.22 GB) of the ISCX 2012 IDS dataset is considered on our evaluation experiments, and it only contains a small number of Bruteforce attacks. The brief statistics of Saturday sub dataset is listed in Table 1, where the total number of network flows in this sub dataset is equal to 133,193, the number of attack flows is equal to 2,086, and the attack ratio of Saturday sub dataset is equal to 1.566%.

Table 1. Brief statistics for Saturday sub dataset of ISCX 2012 IDS dataset.

Feature	Value	Feature	Value
Flows	133,193	Source bytes	311,170,654
Normal flows	131,107	Destination bytes	4,148,028,048
Attack flows	2,086	Source packets	2,340,671
TCP flows	95,117	Destination packets	3,599,451
UDP flows	37,966	Source IPs	44
ICMP flows	81	Destination IPs	2,610

The attack flow statistics for different percent of network flows of Saturday sub dataset is listed in Table 2. Known from Table 2, there isn't any attack if the percent of network flows exceed the former 50% of Saturday sub dataset, i.e. network flows from the $66,598^{th}$ flow to the $133,193^{th}$ flow. For the former 10% of this sub dataset, there are 163 attack flows in it. And the increased rapidly number of attack flows is found at the range from 10% to 20% of Saturday sub dataset, namely the number of its attack flows is equal to 1371 and its attack ratio reach to 10.29%. Table 3 describes the attack distributions of the network flows from 10% to 20% of the Saturday sub dataset. The network flows listed by Table 3 are partitioned 10 flow intervals according to the time sequence that network flows are generated, each flow interval contains 1% network flows of Saturday sub dataset, and the number of attack flows in each flow interval is listed by Table 3. In network flows range from 10% to 11% of Saturday sub dataset, for example, its flow interval is [13,321, 14,651], i.e. existing 1331 network flows between the $13,321^{th}$ flow and the $14,651^{th}$ flow, where the number of attack flows is equal to 136. According to the characteristics analysis for network flows of Saturday sub dataset, it can satisfy the experimental requirement of anomaly detection approach. To verify the proposed UADINK, we select network flows of any one flow interval listed in Table 3 as training samples or test samples to conduct the simulation experiments.

Table 2. Attack statistics for different percent flows of Saturday sub dataset.

Network flows & percent	Attacks of	Network flows & percent	Attacks of
13320 (10%)	163	79916 (60%)	2086
26639 (20%)	1534	93236 (70%)	2086
39958 (30%)	1769	106555 (80%)	2086
53278 (40%)	2047	119874 (90%)	2086
66597 (50%)	2086	133,193 (100%)	2086

3.2 Dataset Preprocessing

Dataset preprocessing is regarded as an important part of machine learning method. Preprocessing operation can decrease the running costs both times and spaces by feature selection and dimension reduction for the given experimental dataset. For the common features of network flows in the ISCX 2012 IDS dataset, we extracted 10 typical features of Saturday sub dataset listed by Table 4 to analyze anomalous behaviors of network flows according to the empirical method of the existed References [7, 34], and the selected features relate to source/destination TCP flags, source/destination port, protocol name, the transferred total source/destination packets/bytes etc. For the network flow preprocessing, it can not only improve the detection precision but also save the running costs of anomaly detection system.

As an important part of the preprocessing operation, the numerical operation for the selected features of network flows is necessary. For each extracted feature, its minimum value and maximum value are given by Table 4. According to the definitions and specifications of IP, TCP, UDP and ICMP protocols, the relevant default values of their numeric ranges are assigned, such as protocol name, source/destination port, and source/destination TCP flags. For example, due to only having 6 flag bits in TCP header, the fifth feature, namely 'SourceTCPFlags', is set to [0, 63]. For the rest features listed in Table 4, their maximum values are arbitrary, but they don't less than the real values of any selected network flows. Take the third feature as an example, its numeric range is set to [0, 190,000] because the largest value of the transferred destination packets in any flows is not greater than 180,150.

Table 3. Attack distribution for 10%–20% flows of Saturday sub dataset.

No.	Flow interval	Range from	Attacks of
1	[13321, 14651]	10%–11%	136
2	[14652, 15983]	11%–12%	154
3	[15984, 17315]	12%–13%	117
4	[17316, 18647]	13%–14%	161
5	[18648, 19979]	14%–15%	110
6	[19980, 21311]	15%–16%	120
7	[21312, 22643]	16%–17%	130
8	[22644, 23975]	17%–18%	187
9	[23976, 25307]	18%–19%	183
10	[25308, 26639]	19%–20%	73

Table 4. Flow features description of Saturday sub dataset.

Feature name	Description	Minimum value	Maximum value
TotalSourceBytes	Transferred source octets	0	7,000,000
TotalDestinationBytes	Transferred destination octets	0	300,000,000
TotalDestinationPackets	Transferred destination packets	0	190,000
TotalSourcePackets	Transferred source packets	0	95,000
SourceTCPFlags	Source TCP flags	0	63
DestinationTCPFlags	Destination TCP flags	0	63
ProtocolName	IP protocol number	0	255
SourcePort	Source port number	0	65,535
DestinationPort	Destination port number	0	65,535
Duration	Duration of flow (in seconds)	0	90,000

3.3 Evaluation Matrices

Anomaly detection is regarded as a two-class problem. Therefore, the results of anomaly detection for network traffic may be classified as benign or malicious behaviors. In this section, three metrics are adopted to evaluate the performance of UADINK [35]: (1) Accuracy Rate (AR). It represents the portion of all test samples of network traffic which are only clustered correctly, and its definition is given in Eq. (2). (2) Detection Rate (DR). It represents the portion of malicious attack behaviors which can be recognized correctly from test samples, and its definition is given in Eq. (3). (3) False Alarm Rate (FAR). It represents the portion of benign behaviors which are recognized as malicious attack behaviors from test samples, and its definition is given in Eq. (4). In Eqs. (2)–(4), TP (True Positive) represents the cumulative number of malicious attack behaviors that these behaviors are labeled as real attack behaviors in test samples, FP (False Positive) represents the cumulative number of malicious attack behaviors that these behaviors are labeled as benign behaviors in test samples, TN (True Negative) represents the cumulative number of benign behaviors that these behaviors are labeled as normal network behaviors in test samples, and FN (False Negative) represents the cumulative number of benign behaviors that these behaviors are labeled as malicious attack behaviors in test samples.

$$AR = \frac{TP + TN}{TP + FP + TN + FN} \tag{2}$$

$$DR = \frac{TP}{TP + FP} \tag{3}$$

$$FAR = \frac{FP}{TN + FP} \tag{4}$$

3.4 Performance Evaluation of UADINK

According to the aim of the proposed UADINK approach, clustering algorithm is regarded as a key part of anomaly detection for network flows. In this section, we discuss the effects of two different clustering algorithms, namely NDM based K-means [23] and UADINK, for detecting anomaly network flows. After clustering network flows, the obtained cluster centroids are labeled as normal clusters and abnormal clusters, and then are used for detecting anomaly network flows. Table 5 shows the experimental results of two different anomaly detection approaches based on the afore-mentioned clustering algorithms. Our experimental strategy considers two aspects: training phase and test phase. In training phase, the detection capability of the obtained normal and anomalous clusters is evaluated after these clusters are generated by the corresponding clustering algorithms and the partition method of normal/anomalous clusters.

Known from Table 5, the accuracy rates (ARs) and the false alarm rates (FARs) of the proposed UADINK approach are superior to that of the NDM based K-means

Table 5. Performance comparisons for different anomaly detection approaches.

Detection approaches	Training phase				Test phase			
	Data of no.	AR (%)	DR (%)	FAR (%)	Data of no.	AR (%)	DR (%)	FAR (%)
UADINK	1	78.44	100	24.02	2	75.83	100	27.33
	2	86.64	94.16	14.35	3	80.63	90.60	20.33
	3	72.97	91.45	28.81	4	75.60	91.30	26.56
	4	83.18	89.44	17.68	5	81.38	92.73	19.64
	5	72.67	73.64	27.41	6	72.52	75.00	27.72
	6	87.46	77.50	11.55	7	83.26	76.15	15.97
	7	91.74	87.69	7.82	8	90.92	82.35	7.69
	8	85.74	87.17	14.50	9	85.59	83.61	14.10
	9	88.44	95.63	12.71	10	82.58	95.89	18.19
NDM based K-means clustering	1	68.60	100	34.98	2	63.81	100	40.92
	2	62.01	100	42.95	3	56.98	100	47.16
	3	80.56	0	11.69	4	78.75	0	10.42
	4	54.50	100	51.75	5	62.84	100	40.51
	5	64.79	100	38.38	6	65.62	100	37.79
	6	66.59	100	36.72	7	59.61	100	44.76
	7	58.93	100	45.51	8	53.15	100	54.50
	8	60.44	100	46.03	9	60.89	100	45.34
	9	62.46	100	43.52	10	74.32	100	27.16

approach. The ARs of UADINK in training phase and test phase are vary from 72.67% to 91.74% and from 72.52% to 90.92%, respectively, and contrastively the ARs of the NDM based K-means approach are respectively vary from 54.50% to 80.56% and from 53.15% to 78.75%. Compared with the NDM based K-means approach, the majority of ARs of UADINK in training phase and test phase are 80% above. The FARs of UADINK in training phase and test phase are vary from 28.81% to 7.82% and from 27.72% to 7.69%, respectively; and the FARs of the NDM based K-means approach are respectively vary from 51.75% to 11.69% and from 54.50% to 10.42%. The average FARs of UADINK in training phase and test phase are respectively 17.65% and 19.73%, and contrastively the average values of FARs for the NDM based K-means approach are 39.06% and 38.73%, respectively. The average FARs of UADINK in training phase and test phase are exceeding to 21.41% and 19% of the average of the NDM based K-means approach, respectively. For the detector rates (DRs), the DRs of the NDM based K-means approach all are equal to 100% except the third group training and test data. Compared with the DRs of the NDM based K-means approach, the DRs of UADINK is inferior to it. But the average DRs of UADINK in training phase and test phase can reach respectively to 88.52% and 87.51%, and the maximum values of DRs of UADINK in training phase and test phase all are 100%.

According to the analysis of the experimental results listed in Table 5, it shows that the proposed UADAIN approach is superior to the NDM based K-means approaches. The proposed UADINK approach to detect anomalies of network flows consists of aiNet_KMC, clusLA and NFAD. Therefore, UADINK can find the optimizing cluster number and clustering centroids by using the aiNet model and K-means algorithm. Although the DRs of UADINK is not enough ideal, this approach is radical to detect anomalous flows from network traffic when it compares with the NDM based K-means approach.

4   Conclusions and Future Works

This paper has proposed an unsupervised anomaly detection approach using artificial immune network based K-means clustering (UADINK) to detect anomalous behaviors of network flows. To demonstrate the effectiveness of UADINK, we firstly describe two different clustering algorithms, namely the NDM based K-means and the proposed aiNet_KMC algorithm, to classify network flows of training dataset, respectively. In the course of experiments, the ISCX 2012 IDS dataset is adopted to evaluate the proposed UADINK approach. The results show that UADINK is a radical anomaly detection approach for network flows after comparing with the existing anomaly detection approach using the NDM based K-means clustering. In our future works, we will further improve the proposed UADINK approach that relates to unsupervised clustering algorithms, automatic detection algorithms and the running costs etc. To keep higher detection rates and decrease false alarm rates, we try to adopt novel immune optimizing mechanisms and parallel computing strategies to improve UADINK for detecting anomalies of network flows.

Acknowledgements. This work was funded by the National Natural Science Foundation of China under Grant No. 61462025, the China Postdoctoral Science Foundation under Grant No. 2014M562102, Hunan Provincial Natural Science Foundation of China under Grant No. 2015JJ2112, and the Scientific Research Fund of Hunan Provincial Education Department of China under Grant No. 12B099.

References

1. Li, B., et al.: A survey of network flow applications. J. Netw. Comput. Appl. **36**(2), 567–581 (2013)
2. Sperotto, A., et al.: An overview of IP flow-based intrusion detection. IEEE Commun. Surv. Tutor. **12**(3), 343–356 (2010)
3. Moore, A., Denis, Z., Crogan, M.: Discriminators for use in flow-based classification. Department of Computer Science, Queen Mary and Westfield College (2005)
4. Tan, Z., et al.: Detection of denial-of-service attacks based on computer vision techniques. IEEE Trans. Comput. **64**(9), 2519–2533 (2015)
5. Ahmed, M., Mahmood, A., Hu, J.: A survey of network anomaly detection techniques. J. Netw. Comput. Appl. **60**, 19–31 (2016)
6. Buczak, A., Guven, E.: A survey of data mining and machine learning methods for cyber security intrusion detection. IEEE Commun. Surv. Tutor. **18**(2), 1153–1176 (2016)
7. Sheikhan, M., Jadidi, Z.: Flow-based anomaly detection in high-speed links using modified GSA-optimized neural network. Neural Comput. Appl. **24**(3–4), 599–611 (2014)
8. Jadidi, Z., et al.: Flow-Based Anomaly Detection Using Neural Network Optimized with GSA Algorithm, pp. 76–81(2013)
9. Sperotto, A., et al.: A labeled data set for flow-based intrusion detection. In: International Workshop on IP Operations and Management, pp. 39–50 (2009)
10. Rodriguez, A., Laio, A.: Clustering by fast search and find of density peaks. Science **344** (6191), 1492–1496 (2014)
11. Jain, A., Murty, M., Flynn, P.: Data clustering: a review. ACM Comput. Surv. **31**(3), 264–323 (1999)
12. Xu, R., Wunsch, D.: Survey of clustering algorithms. IEEE Trans. Neural Netw. **16**(3), 645–678 (2005)
13. Hruschka, E., et al.: A survey of evolutionary algorithms for clustering. IEEE Trans. Syst. Man Cybern. Part C **39**(2), 133–155 (2009)
14. Nanda, S., Panda, G.: A survey on nature inspired metaheuristic algorithms for partitional clustering. Swarm Evol. Comput. **16**, 1–18 (2014)
15. He, H., Tan, Y.: A two-stage genetic algorithm for automatic clustering. Neurocomputing **81**, 49–59 (2012)
16. de Castro, L., Von Zuben, F.: aiNet: an artificial immune network for data analysis. Data Mining: Heuristic Approach **2001**(1), 231–259 (2001)
17. Agrawal, S., Agrawal, J.: Survey on anomaly detection using data mining techniques. Procedia Comput. Sci. **60**, 708–713 (2015)
18. Portnoy, L., et al.: Intrusion detection with unlabeled data using clustering. In: Proceedings of ACM CSS Workshop on Data Mining Applied to Security (DMSA) (2001)
19. Leung, K., Leckie, C.: Unsupervised anomaly detection in network intrusion detection using clusters. In: Proceedings of the Twenty-Eighth Australasian Conference on Computer Science, pp. 333–342 (2005)

20. Petrovic, S., et al.: Labelling clusters in an intrusion detection system using a combination of clustering evaluation techniques. In: Proceedings of the 39th Annual Hawaii International Conference on System Sciences (HICSS 2006), p. 129 (2006)

21. Syarif, I., Prugel-Bennett, A., Wills, G.: Unsupervised clustering approach for network anomaly detection. In: Benlamri, R. (ed.) NDT 2012. CCIS, vol. 293, pp. 135–145. Springer, Heidelberg (2012). doi:10.1007/978-3-642-30507-8_13

22. Erman, J., et al.: Offline/realtime traffic classification using semi-supervised learning. Perform. Eval. **64**(9–12), 1194–1213 (2007)

23. Münz, G., Li, S., Carle, G.: Traffic anomaly detection using k-means clustering. In: GI/ITG Workshop MMBnet (2007)

24. Ahmed, M., Mahmood, A.: Network traffic analysis based on collective anomaly detection. In: 9th IEEE Conference on Industrial Electronics and Applications, pp. 1141–1146 (2014)

25. Winter, P., Hermann, E., Zeilinger, M.: Inductive intrusion detection in flow-based network data using one-class support vector machines. In: 4th IFIP International Conference on New Technologies, Mobility and Security (NTMS), pp. 1–5. IEEE (2011)

26. Shiravi, A., et al.: Toward developing a systematic approach to generate benchmark datasets for intrusion detection. Comput. Secur. **31**(3), 357–374 (2012)

27. Timmis, J., et al.: Theoretical advances in artificial immune systems. Theoret. Comput. Sci. **403**(1), 11–32 (2008)

28. Shi, Y., et al.: An immunity-based time series prediction approach and its application for network security situation. Intel. Serv. Robot. **8**(1), 1–22 (2015)

29. Shi, Y., et al.: Network security situation prediction approach based on clonal selection and SCGM (1, 1) c model. J. Internet Technol. **17**(3), 421–429 (2016)

30. Shi, Y., et al.: An immunity-based IOT environment security situation awareness model. J. Comput. Commun. **5**(7), 182–197 (2017)

31. Jerne, N.: Towards a network theory of the immune system. In: Annales d'immunologie (1974)

32. MacQueen, J.: Some methods for classification and analysis of multivariate observations. In: Proceedings of the Fifth Berkeley Symposium on Mathematical Statistics and Probability, Oakland, CA, USA (1967)

33. Gaddam, S., Phoha, V., Balagani, K.: K-Means+ ID3: a novel method for supervised anomaly detection by cascading K-Means clustering and ID3 decision tree learning methods. IEEE Trans. Knowl. Data Eng. **19**(3), 345–354 (2007)

34. Li, W., et al.: Efficient application identification and the temporal and spatial stability of classification schema. Comput. Netw. **53**(6), 790–809 (2009)

35. Maloof, M.: Machine Learning and Data Mining for Computer Security: Methods and Applications. Springer, New York (2005). doi:10.1007/1-84628-253-5

The Research on Cascading Failure of Farey Network

Xiujuan Ma$^{(\boxtimes)}$ and Fuxiang Ma

School of Computer Science, Qinghai Normal University, Xining 810008, China
qhnumaxiujuan@163.com

Abstract. Cascading failure is an important part of the dynamics in complex network. In this paper, we research the cascading failure of Farey network which is scale-free network with fractal properties by data analysis. According to the analyses, we obtain an iterative expression of the failure nodes' number at the certain time step, anther iterative expression of the time which is needed for the network being global collapse, and the sequence of nodes failure in the Farey network. By theoretical derivation, we get an approximate solutions of perturbation threshold R to make the network achieve the global collapse. If the nodes number of the Farey network is lager, simulate results are closer to theoretical value. By the simulation, we obtain the cascading failure process of the Farey network after suffering deliberate attack and random attack. Simulation results show that, the failure nodes of Farey network increase gradually as R raises, until the network is global collapse. Moreover, Farey network shows stronger robustness for random attack. *abstract* environment.

Keywords: Farey network · Cascading failure · Couple map lattices · Perturbation threshold

1 Introduction

Data science age, data processing could reveal dynamical character and statistical feature of complex system. In the area of network science, researcher could obtain the dynamical properties of systems by data analysis, such as synchronization, diffusion, cascading failure and so on.

Watts and Strogatz proposed the small-word network model [35]. Soon after Barabási and his students found that the distribution of the degree sequence in many real networks followed power law [4] by data statistics. Because of their investigations, a lot of scholars are dedicated to the research of the structure and dynamics of complex networks [1,2,5,10,12,17,19,20,24,27–30,33]. In their research, many researchers use the data statistics method and obtain the dynamical properties of complex network. So the data statistics and analysis is an important method to study the cascading failure of complex network. Cascading failures have been discovered in various real complex networks [6,7,9,14–16,18,22,25,26,31,32,40,41]. Once large-scale cascading failure happens in complex network, it often has highly destructive power and influence [13]. A typical example of cascading failure is in electrical power grids [1].

© Springer Nature Singapore Pte Ltd. 2017
B. Zou et al. (Eds.): ICPCSEE 2017, Part I, CCIS 727, pp. 400–411, 2017.
DOI: 10.1007/978-981-10-6385-5_34

In the previous work, some researchers have investigated the relationship between the cascading failure phenomenons and the topologies of complex networks [3,23,34,37–39,43]. They have studied the cascading failure in various networks' topologies, including ER random network, small-word network scale-free network and interdependent network though data analysis. Their researches have shown that different network models have different properties after being attacked. Scale-free networks are robust to random attack but vulnerable to intentional attack on the hubs [34,39]. The front results of cascading failure are based on classic network models. In 2010, Buldyrev and his team found the cascading failure in interdependent networks and developed a framework for study the cascading failure in interdependent network [6]. Later, the cascading failure of interdependent networks have aroused extensive interest among scholars and there are quantity of interesting results of great theoretical and practical meanings [7–9,14–16,18,22,31,32,40,41].

Fractal geometry and patterns have been found to be ubiquitous, such as coastlines, trees, frost crystals, Romanesco broccoli, and more. Therefore, it is significant to study the cascading failure of fractal networks. In the field of complex networks, fractal structures are shared by many complex systems. In the past decades, the self-similar properties of the fractal networks have attracted many researchers. Since we can obtain exact solution or approximate solution in a finite fractal structure. It is important how the fractal structure affects the dynamical process in complex network. Cascading failure is one of the important research branches of dynamical processes in complex networks. The study of cascading failure in fractal network will help us deepen our understanding of cascading failure in complex systems. So far, there are a few results of cascading failure in fractal networks.

Among fractal network models, Farey network has many interesting properties: they are minimally 3-colorable, uniquely Hamiltonian, maximally outer planar and perfect [36]. Farey graphs were first introduced by Matula and Kornerup [21] and further studied by Colbourn [11], Zhang and Comellas [42] introduced a simple generation method for Farey graph family, and studied relevant topological properties: order, size, degree distribution and correlation, clustering, transitivity, diameter and average distance.

In this paper, using the method of data analysis, we investigate the cascading failure of coupled map lattices (CML) models in a classic Farey network with fractal properties which is proposed by Zhang. We analyze the cascading failure processes in Farey network, and gain the critical factors affecting the sequence of collapse nodes and recurrence relations. And the derivation method is also applicable to other networks. Our results are of great practical importance to understand the avalanches of the fractal networks and have potential applications in enhancing the robustness of real network system with fractal structure.

Coupled map lattices have been widely investigated in the past decades to model the rich space-time dynamical behaviors of complex systems. Some researchers have investigated dynamical process on CML's in small-world and scale-free topologies. Wang and Xu proposed a cascading failure model based on

CML theory [34]. They investigated the cascading failure in the CML model with different coupling topologies, including global coupling, small-world coupling and scale-free coupling. They also derived the theoretical perturbation threshold R of globally coupled network. However, there are few reference about cascading failure of fractal networks base on CML model. Besides, in the previous research work, the study of collapse time and collapse sequence of nodes is absent. In this paper, we give the collapse sequence, collapse time step of nodes in the Farey network and the theoretical deduction of the perturbation threshold R in Farey network.

2   The Model

2.1   Farey Network Model

Zhang and Comellas introduced an iterative construction method for a family of Farey graphs [42] with fractal properties.

Definition 2.1. Let F_t be the Farey network. The graph $F_t = (V, E), t \geq 0$, with vertex set V and edge set E, is constructed as following:

For $t = 0$, F_0 has two vertices and an edge connecting them.

For $t \geq 1$, F_t is obtained from F_{t-1} by adding to every edge introduced at step $t - 1$ a new vertex adjacent to the end vertices of this edge.

Therefore, at $t = 0$, F_0 is K_2; at $t = 1$, the Farey network F_1 is K_3; at $t = 2$, the Farey network is composed of five nodes and seven edges, etc. Notice that the Farey network $F_t(t > 1)$, can also be constructed recursively from two copies of F_{t-1}, by identifying two initial nodes (one from each copy of F_{t-1}) and adding a new edge between the other two initial nodes; as shown in Fig. 1.

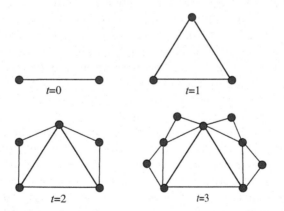

Fig. 1. The evolution of the Farey networks from time steps 0 to 3.

According to the construction of Farey networks, we can gain their relevant properties exactly. The numbers of new vertices and edges added at step t are denoted by V_t and E_t, respectively. At each generation t $(t \geq 1)$, the number of new nodes is $V_t = 2^{t-1}$ and the number of new edges is $E_t = 2^t$. Then, the nodes number (the number of set V) of Farey network F_t is $V = \sum\limits_{i=0}^{t} V_i = 2^t + 1$ and the edges number (the number of set E) of Farey network F_t is $E = \sum\limits_{i=0}^{t} E_i = 2^{t+1} + 1$.

2.2 Cascading Failure Model Base on Coupled Map Lattices (CML)

We consider a CML of N nodes described as follows:

$$x_i(t+1) = \left| (1 - \varepsilon)f(x_i(t)) + \varepsilon \sum_{j=1,j\neq i}^{N} a_{i,j} f(x_i(t))/k(i) \right|, \quad i = 1, 2, \ldots, N; \quad (1)$$

where $x_i(t)$ is the state variable of the ith node at the time step t. The connecting information among the N nodes is given by the adjacency matrix $A = (a_{ij})_{N \times N}$. If there is an edge joining node i and node j, then $a_{ij} = a_{ji} = 1$; otherwise, $a_{ij} = a_{ji} = 0$. Suppose that there is at most one edge connecting two different nodes and no loop in the network. The degree of node i is denoted by $k(i)$. Using $\varepsilon \in (0,1)$ denotes the coupling strength. The function f is used to define the local dynamics, which is chosen as the chaotic logistic map $f(x) = 4x(1-x)$. Absolute value sign in Eq. (1) can guarantee that each state is always non-negative.

In order to show the cascading failure which is triggered by an initial shock on a single node, we add an external perturbation $R \geq 1$ to node c at the time step m as follows:

$$x_c(m) = \left| (1 - \varepsilon)f(x_c(m-1)) + \varepsilon \sum_{j=1,j\neq c}^{N} a_{c,j} f(x_j(t))/k(c) \right| + R. \quad (2)$$

Since $R \geq 1$, node c is failed at the mth time step and we have $x_c(t) \equiv 0$ for all $t > m$. According to Eq. (1), the states of nodes adjacent to c will be affected by $x_c(m)$ at the time step $(m+1)$. The states of neighbor nodes of c may also be larger than 1 and this maybe lead to new failures.

3 The Numerical Analysis

3.1 The Recursion Relations of Cascading Failure in Farey Network

Farey Network is one of the fractal networks with self-similar properties. We construct a Farey network and analyze its cascading failure based on the CML model. In simulations, initial states of all the nodes are chosen randomly from the interval $(0, 1)$. A perturbation $R \geq 1$ is added to node c at the time step 6. We adopt two different attacking strategies: random attack and deliberate attack. In the random attack case, an initial shock is added to a node which is chosen randomly in Farey network. In the deliberate attack case, an initial shock is added to the node with largest degree in Farey network.

If the attacked node is selected randomly, we have the following results:

Let node i be failure node at time step t, $n_i(t)$ be the set of normal neighbor nodes of node i at time step t, f_t be the set of failure nodes which are added at time step t in the network. Using N_t denotes the set of all failure nodes at tth time in the network. T is the total number of time steps $(T > t)$, and t_{star} is the time step beginning to attack the network, t_{now} is the current moment and t_{end} is the time step the network being global collapse. Let $d(c, j)$ be the length of shortest path between c and j, c is the attacked node. So we obtain the following recursive expressions:

(i) $f_{t+1} = \bigcup\limits_{i \in f_t} n_t(i)$,

(ii) $N_{t+1} = N_t \cup f_{t+1}$,

(iii) $t_{end} = t_{start} + \max\{\mathrm{d(c, j)}, \ j \neq c, \ j = 1, 2, \ldots N\}$.

(iv) The smaller shortest path length between attacked node c and other node j $(j \neq c, \ j \in \{1, 2, \ldots N\})$ is, the easier node j is to become failure node.

In order to explain the recursive expressions, we give an example in Appendix.

3.2   The Perturbation Threshold of Cascading Failure in Farey Network

An external perturbation R is added to node c at time step m in Farey network, which maybe trigger some new failure nodes or lead to a global collapse. The approximate value of perturbation threshold can be estimated as follows: The node i is failed at time step $(m + 1)$, which means that:

$$x_i(m+1) = \left| (1-\varepsilon)f(x_i(m)) + \varepsilon \sum_{j=1, j\neq i}^{N} a_{i,j}f(x_j(m))/k(i) \right| \geq 1, \ i = 1, 2, \cdots, N, i \neq c.$$

(3)

The neighbor set of node i is denoted by Γ_i, so we have:

$$x_i(m+1) = \left| f(x_i(m)) + \frac{\varepsilon}{k(i)} \left(\sum_{j \in \Gamma_i} f(x_j(m)) - k(i)f(x_i(m)) \right) \right| \geq 1, i = 1, 2, \cdots, N, i \neq c$$

(4)

That is

$$f(x_i(m)) + \frac{\varepsilon}{k(i)} \left(\sum_{j \in \Gamma_i} f(x_j(m)) - k(i)f(x_i(m)) \right) \geq 1 (i = 1, 2, \cdots, N, i \neq c); \quad (5)$$

or

$$f(x_i(m)) + \frac{\varepsilon}{k(i)}\left(\sum_{j \in \Gamma_i} f(x_j(m)) - k(i)f(x_i(m))\right) \leq -1 (i = 1, 2, \cdots, N, i \neq c). \quad (6)$$

Let $\bar{x}_c(m)$ be the state value of node c at the time step m without external perturbation R, so we have $0 < \bar{x}_c(m) < 1$. Let $x_c(m)$ be the state value of node c at the time step m with external perturbation R, then we have $x_c(m) \geq 1$, so

$$f(x_c(m)) = 4 \times x_c(m) \times (1 - x_c(m)) \leq 0. \quad (7)$$

For the left of expression (5) and expression (7), we have,

$$\begin{aligned}
&f(x_i(m)) + \tfrac{\varepsilon}{k(i)}(\sum_{j \in \Gamma_i} f(x_j(m)) - k(i)f(x_i(m))) \\
&= f(x_i(m)) + \tfrac{\varepsilon}{k(i)}\{\sum_{j \in \Gamma_i} [f(x_j(m)) - k(i)f(x_i(m))] \\
&\quad + [f(\bar{x}_c(m)) - f(x_c(m))] + [f(x_c(m)) - f(\bar{x}_c(m))]\} \\
&= f(x_i(m)) + \tfrac{\varepsilon}{k(i)}\sum_{j \in \Gamma_i} [f(x_j(m)) - k(i)f(x_i(m))] \\
&\quad + \tfrac{\varepsilon}{k(i)}[f(\bar{x}_c(m)) - f(x_c(m))] + \tfrac{\varepsilon}{k(i)}[f(x_c(m)) - f(\bar{x}_c(m))] \\
&\geq 1 + \tfrac{\varepsilon}{k(i)}[f(\bar{x}_c(m)) - f(x_c(m))] + \tfrac{\varepsilon}{k(i)}[f(x_c(m)) - f(\bar{x}_c(m))] \\
&\geq 1 + \tfrac{\varepsilon}{k(i)}[f(x_c(m)) - f(\bar{x}_c(m))] \\
&\geq 1 + \tfrac{\varepsilon}{k(i)}f(x_c(m)).
\end{aligned}$$

According to expression (6), we have

$$1 + \frac{\varepsilon}{k(i)}f(x_c(m)) \leq -1. \quad (8)$$

Therefore expression (8) holds if

$$1 + \tfrac{\varepsilon}{k(i)}f(x_c(m)) \leq 1 - \tfrac{\varepsilon}{k(i)}4R(R-1) \leq -1.$$

Solving the follow inequation:

$$1 - \frac{\varepsilon}{k(i)}4R(R-1) \leq -1,$$

we get:

$$R \geq \frac{1}{2}\left(1 + \sqrt{\frac{2k(i)}{\varepsilon}}\right). \quad (9)$$

In equation (9), if the $k(i) = \langle k \rangle$, where $\langle k \rangle$ is the average degree of farey network, then we get the approximate solution of external perturbation R as follow,

$$R_{threshold} \simeq \frac{1}{2}\left(1 + \sqrt{\frac{2\langle k \rangle}{\varepsilon}}\right). \quad (10)$$

3.3   The Simulation Results of Cascading Failure in Farey Network

(1) The random attack and deliberate attack in Farey network

We adopt random attack and deliberate attack respectively in simulation experiment, and analyse the simulation data. In the simulation, let T be the total
diffusion time steps of cascading failure. In Fig. 2, the curves show the relationships between the number of failure nodes I and the external perturbation R
in random attack and deliberate attack for $T = 10(N = 1025), \varepsilon = 0.4$, respectively. The simulation results indicate that the Farey network shows stronger
robustness in the face of random attack than deliberate attack. When the Farey
network is suffered random attack, we needed larger external perturbation R
leading it to be global collapse.

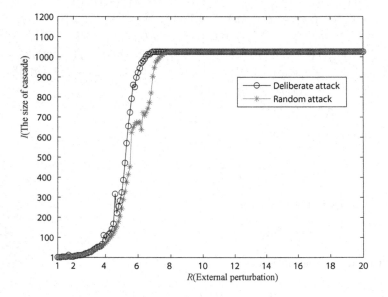

Fig. 2. The curves between the number of failure nodes I and external perturbation
R. Red circles show the curve of deliberate attack and blue stars show the curve of
random attack. (Color figure online)

(2) The comparement of simulation results and theoretical value for perturbation threshold R

The Fig. 3 plots the theory values and simulation results of perturbation threshold R. We find that simulation result is much closer to theoretical value if the size
of the network is larger, which indicates that the theory analysed of perturbation
threshold is effective.

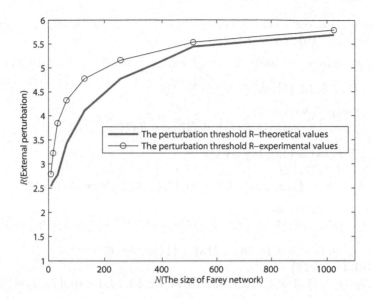

Fig. 3. The theoretical values and experimental values of perturbation threshold R. Blue stars show the theoretical values of the perturbation threshold R, red circles show the simulation results of the perturbation threshold R. (Color figure online)

4   Conclusion

In this paper, we investigate the cascading failure of CML model in a classic Farey network with fractal properties. We obtain the cascading failure processes of Farey network and the critical factors affecting the sequence of collapse nodes and recurrence relations. we gain the recursive expressions of failure nodes in certain time step, collapse time step of nodes and collapse sequence of nodes in the Farey network. Moreover, we obtain the approximate solutions of perturbation threshold R to make the network be global collapse, which is $R \simeq \frac{1}{2}(1 + \sqrt{\frac{2\langle k \rangle}{\varepsilon}})$. Meanwhile, the simulations results are in good agreement with the theoretical threshold. By analyzing the simulation, we find that in the face of deliberate attack the Farey network is shown stronger robustness. The theoretical deduction of the perturbation threshold R in Farey network.

Acknowledgments. This research is supported by the National Natural Science Foundation of China (No. 61603206) and the Nature Science Foundation from Qinghai Province (No. 2017-ZJ-949Q).

Appendix

In order to explain the recursive expressions in Subsect. 3.1, we give the following example. In Fig. 4, there are 17 nodes in the network ($N = 17$). We add a perturbation to node 6 at time step 5, i.e. $t_{start} = 5$. Start the recursive as follows:

At initial moment, there is only one failure node at time step 5, and it is node 6 , i.e. $f_5 = \{6\}$, $N_t = \{6\}$, $I_5 = |N_5| = 1$.

At time step 6, we have $f_6 = \bigcup_{i \in f_5} n_5(i) = n_5(6) = \{3, 4, 10, 11\}$. $N_6 = N_5 \cup f_6 = \{3, 4, 6, 10, 11\}$, $I_6 = |N_6| = 5$.

At time step 7, we get
$$f_7 = \bigcup_{i \in f_6} n_6(i) = n_6(3) \cup n_6(4) \cup n_6(10) \cup n_6(11)$$
$$= \{1, 2, 5, 8, 15\} \cup \{1, 7, 12\} \cup \emptyset \cup \emptyset$$
$$= \{1, 2, 5, 7, 8, 12, 15\}$$
$$N_7 = N_6 \cup f_7 = \{1, 2, 3, 4, 5, 6, 7, 8, 10, 11, 12, 15\}, I_7 = |N_7| = 12.$$

At time step 8, we obtain
$$f_8 = \bigcup_{i \in f_7} n_7(i) = n_7(1) \cup n_7(2) \cup n_7(5) \cup n_7(7) \cup n_7(8) \cup n_7(12) \cup n_7(15)$$
$$= \{13\} \cup \{9, 17\} \cup \{9, 14, 16\} \cup \{13\} \cup \{14\} \cup \emptyset \cup \emptyset$$
$$= \{9, 13, 14, 16, 17\}$$
$$N_8 = N_7 \cup f_8 = \{1, 2, 3, 4, 5, 6, 7, 8, 9, 10, 11, 12, 13, 14, 15, 16, 17\}, I_8 = |N_8| = 17.$$

If $I_t = N$, then the recursive is over.

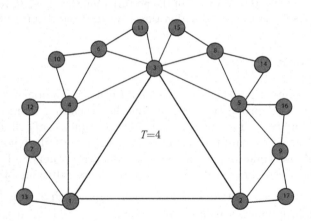

Fig. 4. The farey network of T = 4, the number of nodes is 17. The nodes are ordered with their order joined in the network.

(2) In the front example, suppose the attacked node is node 6. According to Fig. 4, obviously we have

$$d(6, 3) = d(6, 4) = d(6, 10) = d(6, 11) = 1;$$
$$d(6, 1) = d(6, 2) = d(6, 5) = d(6, 7) = d(6, 8) = d(6, 12) = d(6, 15) = 2;$$
$$d(6, 9) = d(6, 13) = d(6, 14) = d(6, 17) = 3.$$

So, $t_{end} = t_{start} + max\{d(6, i), i \neq 6, i = 1, 2, \ldots N\} = t_{start} + max\{1, 2, 3\} = 5 + 3 = 8$. It means that the network F_4 is complete collapse at time step 8.

(3) The sequence of failure nodes

In the above example, when attack node 6, the first group of failure nodes are 3, 4, 10, 11. All the length of the shortest pathes between node 6 and nodes 3, 4, 10, 11 are 1. All the length of the shortest pathes between node 6 and nodes 9, 13, 14, 17 are 3, thus nodes 9, 13, 14, 17 are the last group of failure nodes.

References

1. Albert, R., Albert, I., Nakarado, G.L.: Structural vulnerability of the north American power grid. Phys. Rev. E **69**(2 Pt 2), 025103 (2004)
2. Albert, R., Jeong, H.: Internet: diameter of the world-wide web. Nature **401**(6), 130–131 (1999)
3. Bao, Z.J., Cao, Y.J.: Cascading failures in local-world evolving networks. J. Zhejiang Univ.-SCIENCE A **9**(10), 1336–1340 (2008)
4. Barabsi, A., Albert, R.: Emergence of scaling in random networks. Science **286**(5439), 509–512 (1999)
5. Berlow, E.L., Neutel, A.M., Cohen, J.E., Ruiter, P.C.D., Bo, E., Emmerson, M., Fox, J.W., Jones, J.I., Kokkoris, G.D., Logofet, D.O.: Interaction strengths in food webs: issues and opportunities. J. Anim. Ecol. **73**(3), 585–598 (2004)
6. Buldyrev, S.V., Parshani, R., Paul, G., Stanley, H.E., Havlin, S.: Catastrophic cascade of failures in interdependent networks. Nature **464**(7291), 1025 (2010)
7. Buldyrev, S., Kadish, B., Shere, N., Aharon, M., Cwilich, G.: Cascades of failures in various models of interdependent networks. In: APS Meeting (2012)
8. Cheng, Z., Cao, J.: Cascade of failures in interdependent networks coupled by different type networks. Phys. A Stat. Mech. Appl. **430**, 193–200 (2015)
9. Cho, W., Goh, K.I., Kim, I.M.: Correlated couplings and robustness of coupled networks. Physics (2010)
10. Christiansen, M.H., Kirby, S.: Language evolution: consensus and controversies. Trends Cogn. Sci. **7**(7), 300–307 (2003)
11. Colbourn, C.J.: Farey series and maximal outerplanar graphs. SIAM J. Algebraic Discrete Methods **3**(2), 187–189 (1982)
12. Colizza, V., Barrat, A., Barthlemy, M., Vespignani, A.: The role of the airline transportation network in the prediction and predictability of global epidemics. Proc. Nat. Acad. Sci. **103**(7), 2015–2020 (2006)
13. Dombrowsky, W.R.: Again and again: is a disaster what we call a 'disaster'? Int. J. Mass Emerg. Disasters **13**(3), 241–254 (1995)
14. Dong, G., Gao, J., Du, R., Tian, L., Stanley, H.E., Havlin, S.: Robustness of network of networks under targeted attack. Phys. Rev. E Stat. Nonlinear Soft Matter Phys. **87**(5), 052804 (2013)
15. Dong, G., Tian, L., Du, R., Fu, M., Stanley, H.E.: Analysis of percolation behaviors of clustered networks with partial support-dependence relations. Physica A Stat. Mech. Appl. **394**(2), 370–378 (2014)
16. Dong, G., Tian, L., Zhou, D., Du, R., Xiao, J., Stanley, H.E.: Robustness of n interdependent networks with partial support-dependence relationship. EPL **102**(102), 68004 (2013)
17. Dorogovtsev, S.N., Mendes, J.F.F.: Evolution of networks with aging of sites. Phys. Rev. E Stat. Phys. Plasmas Fluids Related Interdisc. Topics **62**(2 Pt A), 1842 (2000)

18. Gao, J., Buldyrev, S.V., Havlin, S., Stanley, H.E.: Robustness of a network formed by n interdependent networks with a one-to-one correspondence of dependent nodes. Phys. Rev. E Stat. Nonlinear Soft Matter Phys. 85(6 Pt 2), 066134 (2012)
19. Guimer, R., Amaral, L.A.N.: Modeling the world-wide airport network. Eur. Phys. J. B 38(2), 381–385 (2004)
20. Guimer, R., Mossa, S., Turtschi, A., Amaral, L.A.N., Wachter, K.W.: The world-wide air transportation network: anomalous centrality, community structure, and cities' global roles. Proc. Nat. Acad. Sci. 102(22), 7794–7799 (2005)
21. Hardy, G.H., Wright, E.M.: An Introduction to the Theory of Numbers. Institute of Physics Publishing, Bristol (1979)
22. Havlin, S.: Cascade of failures in coupled network systems with multiple support-dependence relations. Phys. Rev. E: Stat., Nonlin, Soft Matter Phys. 83(2), 1127–1134 (2011)
23. Huang, L., Lai, Y.C., Chen, G.: Understanding and preventing cascading break-down in complex clustered networks. Phys. Rev. E Stat. Nonlinear Soft Matter Phys. 78(3 Pt 2), 036116 (2008)
24. Kinney, R., Crucitti, P., Albert, R., Latora, V.: Modeling cascading failures in the north american power grid. Eur. Phys. J. B 46(1), 101–107 (2005)
25. Moreno, Y., Gomez, J.P.A.: Instability of scale-free networks under node-breaking avalanches. Europhys. Lett. 58, 630–636 (2002)
26. Motter, A.E., Lai, Y.C.: Cascade-based attacks on complex networks. Phys. Rev. E Stat. Nonlinear Soft Matter Phys. 66(2), 065102 (2002)
27. Newman, M.E.J., Forrest, S., Balthrop, J.: Email networks and the spread of com-puter viruses. Phys. Rev. E 66, 1162–1167 (2002)
28. Redner, S.: How popular is your paper? An empirical study of the citation distri-bution. Eur. Phys. J. B 4(2), 131–134 (1998)
29. Roopnarine, P.: Extinction cascades and catastrophe in ancient food webs. Paleo-biology 32(1), 1–19 (2006)
30. Schneider, J.J., Hirtreiter, C.: The impact of election results on the member num-bers of the large parties in Bavaria and Germany. Int. J. Mod. Phys. C 16(8), 1165–1215 (2011)
31. Shao, S., Huang, X., Stanley, H.E., Havlin, S.: Robustness of a partially interde-pendent network formed of clustered networks. Phys. Rev. E 89(3), 032812 (2013)
32. Su, Z., Li, L., Peng, H., Kurths, J., Xiao, J., Yang, Y.: Robustness of interrelated traffic networks to cascading failures. Sci. Rep. 4, 5413 (2014)
33. Wang, J., De, W.P.: Properties of evolving e-mail networks. Phys. Rev. E Stat. Nonlinear Soft Matter Phys. 70(2), 066121 (2004)
34. Wang, X.F., Xu, J.: Cascading failures in coupled map lattices. Phys. Rev. E Stat. Nonlinear Soft Matter Phys. 70(5 Pt 2), 056113 (2004)
35. Watts, D.J., Strogatz, S.H.: Collective dynamics of 'small-world' networks. Nature 393(6684), 440 (1998)
36. Weisstein, E.W.: Farey sequence. MathWorld-A Wolfram Web Resource (2014). http://mathworld.wolfram.com/FareySequence.html
37. Wu, J.J., Gao, Z.Y., Sun, H.J.: Cascade and breakdown in scale-free networks with community structure. Phys. Rev. E Stat. Nonlinear Soft Matter Phys. 74(6 Pt 2), 066111 (2006)
38. Xia, Y., Fan, J., Hill, D.: Cascading failure in wattscstrogatz small-world networks. Phys. A Stat. Mech. Appl. 389(6), 1281–1285 (2010)
39. Xu, J., Wang, X.F.: Cascading failures in scale-free coupled map lattices. IEEE Int. Symp. Circ. Syst. 4, 3395–3398 (2005)

40. Yağan, O., Gligor, V.: Analysis of complex contagions in random multiplex networks. Phys. Rev. E Stat. Nonlinear Soft Matter Phys. **86**(3 Pt 2), 036103 (2012)
41. Yuan, X., Shao, S., Stanley, H.E., Havlin, S.: The influence of the broadness of the degree distribution on network's robustness: comparing localized attack and random attack. Phys. Rev. E **92**, 032122 (2015)
42. Zhang, Z., Comellas, F.: Farey graphs as models for complex networks. Theoret. Comput. Sci. **412**(8–10), 865–875 (2011)
43. Zheng, J.F., Gao, Z.Y., Zhao, X.M.: Clustering and congestion effects on cascading failures of scale-free networks. EPL **79**(5), 58002 (2007)

Optimal Task Recommendation for Spatial Crowdsourcing with Privacy Control

Dan Lu, Qilong Han, Hongbin Zhao$^{(\boxtimes)}$, and Kejia Zhang

College of Computer Science and Technology, Harbin Engineering University,
Harbin 150001, China
{ludan,hanqilong,kejiazhang}@hrbeu.edu.cn,
hqlwxm2002@163.com

Abstract. *Spatial Crowdsourcing (SC)* is a transformative platform that engages a crowd of mobile users (i.e., workers) in collecting and analyzing environmental, social and other spatio-temporal information. However, current solutions ignore the preference of each worker's remuneration and acceptable distance, and the lack of error analysis after privacy control lead to undesirable task recommendation. In this paper, we introduce an optimization framework for task recommendation while protecting participant privacy. We propose a Generalization mechanism based on Bisecting k-means and an efficient algorithm considering the generalization error to maximization the reward of *SC* server. Both numerical evaluations and performance analysis are conducted to show the effectiveness and efficiency of the propose framework.

Keywords: Spatial crowdsourcing · Task recommendation · Privacy control · Reward

1 Introduction

Spatial Crowdsourcing (SC) is the combination of Crowdsourcing and spatio-temporal features, individuals with mobile devices that perform the tasks by physically traveling to specified locations. *SC* engages individuals in the act of collecting and analyzing environmental, social and other information for which spatio-temporal features are relevant [1]. *SC* has numerous applications in domains such as environmental sensing, journalism, crisis response and urban planning. Existing commercial *SC* applications include traffic monitoring, ride sharing, Ali crowdsourcing and wireless coverage mapping. Nonetheless, *SC* is still in its infancy, and there are many undergoing research exploring applications.

With *SC*, task requesters outsource their spatio-temporal tasks to a set of workers that usually gain a small reward for providing services. *SC* is feasible only if workers and tasks are matched effectively (i.e., tasks are completed timely and workers do not travel long distances). [1, 11, 12] mentions that workers prefer to do tasks nearing themselves. To the extent, we must take into account the locations of the workers. Typically, requesters and workers register with a centralized *spatial crowdsourcing server (SC-server)*. *SC-server* receives the task information, the locations where workers submit and choose some appropriate workers to recommend. However, the

© Springer Nature Singapore Pte Ltd. 2017
B. Zou et al. (Eds.): ICPCSEE 2017, Part I, CCIS 727, pp. 412–424, 2017.
DOI: 10.1007/978-981-10-6385-5_35

SC-server may not be trusted (e.g., easy to be attacked) and disclosing individual locations has serious privacy implications [2–5]. Some research have also paid attention to the security of the *SC-server* itself [14–16]. According to [1], there are two categories of *SC*, based on how workers are matched to tasks. In *Worker Selected Tasks (WST)* mode, the *SC-server* publishes the spatial tasks online, and workers autonomously choose tasks in their vicinity without coordinating with the *SC-server*. In *Server Assigned Tasks (SAT)* mode, online workers send their location to the *SC-server* which assigns tasks to nearby workers.

WST is simpler, and does not require workers to share their location to the *SC-server*. However, the assignment is often sub-optimal, as workers do not have a global system view. *SAT* mode requires complex matching algorithm at the *SC-server*, but the best-suited workers are selected for a task. This requires the *SC-server* to know the workers' location. In our work, we consider the *SAT* mode, but we also provide location privacy protection for the workers. Several solutions [2, 13] have been proposed to protect user location privacy in Crowdsourcing applications. Such schemes include data generalization [18], k-means [8], mix zones [19], differential privacy [10], entity resolution approaches to preserve user privacy [1], but ignoring the impact of location privacy protection on the task recommendation results. Others use the predefined generalization method to blur the workers' location, the *SC-server* assigns a series of tasks to workers, workers choose the most beneficial task to complete, the *SC-server* thus benefit the most [6]. This approach takes the maximization the reward of *SC-server* into account but ignores different workers' reward preference.

According to [7] location privacy protection methods are broadly divided into three categories: anonymous, false location and hidden areas. In the anonymous method, the k-means is the typical and classical algorithm mostly used for position clustering in location privacy protection. For the k-means algorithm, it is guaranteed that worker A is indistinguishable from other k-1 workers in one area. In crowdsourcing application, it's not applicable since we cannot determine the k value in advance.

In this paper, we proposed a framework for protecting privacy of worker location while enabling effective task recommendation in *SC* systems. Specifically, our proposed framework contains two main components that may operate in parallel: *location generalization* which obtains workers' original location and releases them in noisy way, and *optimal task recommendation* which aggregates workers' preference with workers' noisy location needed for tasks recommendation.

Location Generalization. In our framework, the *SC-server* only has access to data sanitized according to the generalization algorithm. Workers submit their location to a *Safe Third-Party (STP)* that provides Internet connectivity. As opposed to the *SC-server*, the STP signs a contract with its subscribers, stipulating the terms and conditions of location disclosure. The STP collects user locations and releases them to *SC-servers* in noisy form. In this paper, iterate several times based on 2-means to obtain n clusters so that the distance between the cluster center and all points in this class are not greater than the threshold Y.

Optimal Task Recommendation. In this paper, we assume that the probability of a worker completing a task is only relevant to the task distance and the reward. Due to the personalization, we use the questionnaire survey to measure workers' distance preference and reward preference. After receiving the noisy data from the *STP*, the *SC-server* takes into account generalization errors and assigns tasks to workers, maximizing the reward of the *SC-server*. In summary, the main contributions of this paper are as follows.

- We take into account the impact of location generalization error when the *SC-server* assigns tasks.
- We considered personalized demands (e.g., task distance and the reward preference)
- We conduct both numerical evaluations and performance analysis to show the effectiveness and efficiency of our proposed framework.

The remainder of this paper is organized as follows. We present our framework in Sect. 2. Next we represent the Y-threshold clustering algorithm based on bisecting cluster algorithm in Sect. 3. Section 4 details the optimal task recommendation solutions. Experimental results are presented in Sect. 5, and conclusions in Sect. 6.

2   The Proposed Framework

In this section, we describe the basic system model for task recommendation in *SC* platform and design goals.

2.1   System Model

We consider the problem of privacy-preserving *SC* task recommendation in the *SAT* mode. Figure 1 shows the basic model of the proposed framework. Workers send their locations to a *Safe Third-Party (STP)* which collects updates and releases a set of noisy location data. The noisy data is accessed by the *SC-server*, which also receives tasks from a number of requesters. For simplicity, we focus on the single *SC-server*.

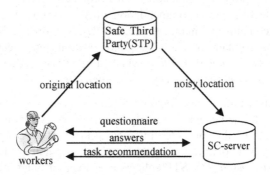

Fig. 1. Basic system model for task recommendation in SC

When the *SC-server* receives a task t, it determine a series of workers that ensures the requires of the task according to the noisy data and the workers' reward preference.

Due to the uncertain nature of the noisy data, this is a challenging process which will be detailed later in Sects. 3 and 4.

Assumption 1. Whether a worker completes a task is only related to the distance between worker and task and the reward is given.

Definition 1. Locations and Task

- Denote by $L = \{l : l = 1, 2, \ldots, |L|\}$ the set of all workers' location.
- Denote by $L' = \{l' : l' = 1, 2, \ldots, |L'|\}$ the set of all generalized workers' location.
- Denote by $t = \{l_t, Total, Need\}$ l_t represents the location of the task, Total represents the total amount of the task and Need represents the requires of the task.

Definition 2. p_d and Reward

- p_d represents the probability that the worker complete the task in terms of distance between task and workers.
- Denote by $R = \{r : r = 1, 2, \ldots, |r|\}$ the set of all workers' reward preferences.

2.2 Questionnaire

Recommended tasks are based on statistics including both workers' distance preferences and the reward preferences. The *SC-server* constructs a statistics query which asks two questions: (1) "How far can you accept the task? Selected from five level {1000, 1500, 2000, 2500, farther}" and (2) "The minimum reward that you can accept?". Both questions expect numerical answers. The answer from each worker k is an array $[d_k, r_k]$ that consists of two parts, each corresponding to a question. Therefore, we can aggregate answers to these two questions from workers, and calculate p_d and obtain the set of workers' reward preferences.

[17] considers that the effect of distance on the completed probability is linear or exponential. In this paper, we assume that the effect is staged considering five levels: (1) d ≤ 1000, (2) 1000 < d ≤ 1500, (3) 1500 < d ≤ 2000, (4) 2000 < d ≤ 2500, (5) d > 2500. According to the questionnaire survey results, we statistic the number of each level. (e.g., Assuming there are 10 workers, their answers form an array [1000, 1000, 1000, 1500, 1500, 2000, 2000, 2500, 2500, farther], the worker whose answer is farther is willing to do the task regardless the five level, so we let each level's count +1, but the worker whose answer is 1500 is willing to do tasks at level1 or level2, so we let level1's count +1, level2's count +1. Table 1 shows the general probability result.

Table 1. The completed probability in terms of distance

Distance between worker and task	p_d
d <= 1000	1
1000 < d <= 1500	0.7
1500 < d <= 2000	0.5
2000 < d <= 2500	0.3
d > 2500	0.1

2.3   Design Goals

We aim to provide good privacy, utility, and efficiency in the proposed framework. Since location generalization and task recommendation differ in nature, we describe their design goals separately.

Goals for Location Generalization. We have two design goals for location generalization: privacy and utility.

- Privacy. Worker locations are needed for task recommendation, which may be leveraged by the *SC-server* to uniquely identify an individual worker. To reduce the risk of being identified, the STP performs the generalization algorithm. The *SC-server* can only access the noisy data.
- Utility. Utility represents the accuracy of the original and generalized data. We set different threshold Y to illustrate the generalization error. In our experiment, we set Y = 500 and the utility is great.

Goals for Task Recommendation. We have two design goals for task recommendation: accuracy and scalability.

- Accuracy. In task recommendation process, we take into account the generalization error, whereby ensure task is completed while maximizing the reward of the *SC-server*.
- Scalability. The system should be scalable to a large population of workers, which means that the proposed framework should be highly efficient.

3   The Generalization Method

The first step in the proposed framework consists of location generalization (at the *STP* side) to be later used for task recommendation at the *SC-server*. Generalizing the location is an essential step, because it determines how private and accurate is the released data, which is turn affects p_d, the result of task recommendation and the maximum reward that the *SC-server* obtain. In this section, we extend the bisecting k-means algorithm proposed in [9] to address the specific requirements of the *SC* framework.

3.1   Definition

Definition 3. Clusting

- Denote by $C = \{c: c = 1, 2, \ldots, |C|\}$ the set of all clusters.
- Denote by $C_t = \{c_t: c_t = 1, 2, \ldots, |C_t|\}$ the set of all cluster centers.
- Denote by $c_i = \{l: l = 1, 2, \ldots, |c_i.size|\}$ the set of cluster c_i.

Definition 4. Distance between A and B

$$\Delta = PI/180$$
$$c = sin(LatA * \Delta)sin(LatB * \Delta) + cos(LatA * \Delta)cos(LatB * \Delta)cos((MlonA - MlongB) * \Delta)$$
$$distance(A, B) = R_{EARTH} * Arccos(c) * \Delta$$

where *LatA* is the latitude of location A while *LatB* is the latitude of location B, *MlonA* is the longitude of the location A, while *MlongB* is the longitude of location B. R_{EARTH} is the radius of the earth.

The bisecting k-means algorithm is a direct expansion of the basic k-means algorithm, which is based on a simple idea: To get k clusters, first splitting all the sets into two clusters, selecting one of them to keep on splitting, keep going until there are k clusters. Since we cannot determine the k value in advance, we cannot use it directly.

In this paper, we cannot put the k value as the end flag of the bisecting k-means algorithm, taking into account the balance between privacy protection and data availability. We present a bisecting clustering algorithm with threshold definition. The first step is initialized, just like the bisecting k-means algorithm do. Next, after each bisection, calculate whether or not the following condition is satisfied.

$$distance(l, c_i) < Y, l \in c_i$$

Ensure the distances from all points to its cluster center are no greater than the threshold Y.

3.2 Trade-Offs Among Utility and Privacy

The generalization method specifies how to generalize an original location into a noisy data. There are two conflicting design goals in this section: utility and privacy. These two goals cannot be optimized simultaneously. First, suppose that privacy is optimized, which means that a greater degree of generalization. It's impossible for the task recommendation process to assign tasks accurately as the utility. Second, consider the case that utility is optimized. In other words, the noisy data is close to the original location which may disclosure workers' privacy. Therefore, in practice, we have to find a good trade-off among these two goals.

In our work, we fix the threshold Y to control the degree of generalization to effect the privacy protection. When Y increases, it executes a greater degree of generalization which means stronger privacy protection, while weaker utility. On the other hand, Y decreases leading to a smaller degree of generalization, stronger utility. Table 2 shows the effect of the threshold on Δp_d.

Inference1. When the threshold Y = 500, the error of Δp_d is no more than 0.2

Proof:

(1) $distance(l, l_t) \leq 1000, distance(l, l') \leq 500$

$$\begin{cases} distance(l', l_t) \leq 1000 & 50\% \\ 1000 < distance(l', l_t) \leq 1500 & 50\% \end{cases} \rightarrow \Delta p_d = 1 - (1*0.5 + 0.7*0.5) = 0.15$$

(2) $1000 < distance(l, l_t) \leq 1500, \quad distance(l, l') \leq 500$

$\quad\quad 1000 < distance(l', l_t) \leq 1500 \quad 100\% \rightarrow \Delta p_d = 0.7 - 0.5 = 0.2$

The last three cases are calculated in the same way. Then, we get the $max\{\Delta p_d\} = 0.2$.

Table 2. The effect of the threshold on Δp_d

Threshold Y	$max\{\Delta p_d\}$
0	0
200	0.12
400	0.16
500	0.2
600	0.24
800	0.32
1000	0.4

3.3   The Bisecting Clustering Algorithm with Threshold

Given a set of all workers' location and the threshold Y, the generalization algorithm must balance two conflicting requirements: determine the set of clusters that (1) satisfies the above conditions (the distance between all the points and its cluster centers are no more than threshold Y, and (2) the number of the clusters is small. The input to the algorithm is the set of all workers' location as well as the threshold Y.

The algorithm chooses as initial cluster c_1 that covers all the points, and determines its Y value. As long as the algorithm execution, it bisect one of the clusters cannot satisfied with the conditions. Clusters are bisected once at a time. The algorithm stops when the distance between all the workers' location and its cluster center are no more than threshold Y.

Algorithm 1 *The Bisecting Cluster Algorithm With Threshold (BCT)*

1.Input: The set of all workers' location(L), Threshold Y
2.Output: The set of clusters(C)
3.$C = \{c_1\}$that points are all in cluster c1
4.Init$c_1 = \{l: l = 1, 2, \ldots, |l|\}$
5.Calculate the distance between all cluster center and the points covered
6.If exists $distance(l, c_i) > Y$
7.Bisectingc_i, Goto Line 5
8.If all $distance(l, c_i) > Y$
9.End, return C

The pseudo code of the algorithm is depicted in Algorithm 1. After each bisection, calculate the distance between all cluster center and the points covered until all clusters satisfy the above conditions. After obtained the clustering result, we use cluster's center to represent all the points in each cluster.

4 Optimization Model for Task Recommendation

In the precious sections, we have assumed that the *SC-server* only have the accesses to the noisy points. In this section, we describe how to use these noisy points for task recommendation while achieving design goals described in Sect. 2.3. Table 3 summarizes the notations used in this section.

Table 3. Summary of notations

Symbol	Definition
R'	The reward of the task t published by the *SC-server*
N	The expectation of the number of task completed (worst case)
N'	The expectation of the number of task completed (best case)

Definition 4. *Reward Attractiveness (RA)*: Both workers and the *SC-server* can earn some money when the tasks are completed successfully (i.e., the actual number of task completed is more than the need of the task). According to the Assumption 1, Whether a worker completes a task is only related to the distance between worker and task and the reward is given. The effect of the reward can be characterized by the *reward attractiveness (RA)*, which can be calculated as R', the reward of the task t which is published by the *SC-server*, minus r which represent worker' s reward preference, then divided by r, i.e., $RA(t|i) = \frac{R'-r_i}{r_i}$.

4.1 Optimization Problem Formulation

In our framework, the workers first submit their location to the *STP*. After generalization, the *STP* releases the workers' location to the *SC-server* in a noisy form. Based on this noisy data, the server selects N workers and sends the task to them. Then these workers decide whether to accept this task and complete it. Then return the result back to the server.

Since the worker knows his own location, he can easily make the decision whether to accept a task. This is not true for the *SC-server* as it can only accessed the noisy data provided by the *STP*. To increase the probability that the task can be completed, the server must take into the generalization threshold Y.

Assume that the server already has prior knowledge on the workers' *RA*, total revenue of the task and the required workers. From the perspective of the server, its reward (i.e., revenue) depends on the revenue paid to workers and the number of workers who actually completed the task. We use the expected revenue of the task to quantify it. Since the server does not know the exact location of the worker, it considers the maximum distance between task and workers first, and then uses the minimum distance to recalculate among these people to find out the maximum expected of the number who accomplished the task. The expected revenue R of a task t as follows:

$$\sum_{i=0}^{N}(p_d + RA) \geq Need \begin{cases} distance(l', l_t)' = distance(l', l_t) + Y \\ P_d = Sort(\{p_d\}) \end{cases}$$

$$E[R(t)] = Total - R' * \sum_{i=0}^{N'}(p_d + RA) \begin{cases} distance(l', l_t)'' = distance(l', l_t) - Y \\ N' = \sum_{i=0}^{N}(p_d + RA) \end{cases}$$

where l' are the points of workers that can be obtained from the *STP*. Let R' denote the revenue that the *SC-server* released. The server need to select R' that maximize the expected revenue given a set of workers' noisy locations.

$$R^* = \max_{0 < R' < \frac{Total}{Need}} E[R(t)]$$

4.2 Optimal Task Recommendation Algorithm

Given task t and the set of workers' noisy location, the algorithm takes into account the threshold Y for task recommendation. The input to the algorithm is task t, which consisting of location of the task with its total revenue and its required and the set of workers noisy location as well as the RA. The algorithm chooses as initial $R' = 0$, and stops when $R' \geq \frac{Total}{Need}$.

Algorithm 2 Optimal Task Recommendation Algorithm

1. Input: task t, the set of workers' noisy locations (L'), workers' reward preferences(R)
2. Output:R^*, N
3. Init $R' = 0$
4. $distance(l', l_t)' = distance(l', l_t) + Y$, calculate $\{p_d\}$
5. Sort($\{p_d\}$), get the N that satisfied $\sum_{i=0}^{N}(p_d + RA) \geq Need$
6. $distance(l', l_t)'' = distance(l', l_t) - Y$, calculate $\{p_d\}$
7. Sort($\{p_d\}$), calculate $E[R(t)] = Total - R' * \sum_{i=0}^{N'}(p_d + RA)$
8. $R' = R' + 1$, Goto Line 4
9. Find the $max_{0 < R' < \frac{Total}{Need}} E[R(t)]$

The pseudo code of the optimal task recommendation algorithm is depicted in Algorithm 2. In line 4, we first consider the worst case in that the generalization points are closer to the task than the original points, and then calculate the N that satisfied $\sum_{i=0}^{N}(p_d + RA) \geq Need$. Among these N workers, we recalculate the p_d using the best case in that the generalization points are farther than the original points (Line 6). In line 9, we find the $max_{0 < R' < \frac{Total}{Need}} E[R(t)]$.

5 Performance Evaluation

We use a real-world dataset which contains the check-in history of users in a location-based social network of a city in South of China. We assume that users are the workers of the *SC* system, and their locations are those of the check-in points. We also selected two check-in points as two tasks that will be accomplished by some workers. For our experiments, we use 1000 users' check-in points at 3 pm as the original location dataset. Table 4 presents the detail of the dataset.

Table 4. Dataset characteristics

Parameter	Value
Task	{[118.777224, 32.062331], [118.780503, 32.087687]}
Total	3000
Threshold Y	0, 200, 400, **500**, 600, 800, 1000
Need	30, 35,40, 45, **50**, 55, 60
R	Randomly (30 ~ 50)
Average Error	$Average\ Error = \frac{\sum_{i=0}^{|L|} distance(l'-l)}{|L|}$

5.1 Evaluation of Generalization Algorithm

Firstly we evaluate the performance of the *bisecting clustering algorithm with threshold (BCT)* for generalization from Sect. 3.3. To this end, we compare our generalization algorithm (Algorithm 1) with a baseline algorithm. The algorithm use a predefined hierarchy method to generalized the original data. For instance, location information represented by the latitude and longitude with a total of 6 decimal digits can be generalized by keeping 6-a decimal digits for level-a generalization. In our experiment we only considered level 1 generalization due to the limit of data accuracy. Figure 2 shows the average error of generalization. Our BCT algorithm generally perform better in terms of minimizing the Average Error, when the threshold Y within 500. Moreover, even if the threshold value Y = 1000, the average error is less than 400.

5.2 Evaluation of Task Recommendation Algorithm

5.2.1 The Effect of Varying Y

We evaluate the performance of our optimal task recommendation algorithm on two different tasks by varying the threshold Y. Figure 3 shows the result. A higher Y yields lower R^* and higher revenue (R') (workers are more willing to accept tasks). (Total = 3000, Need = 50)

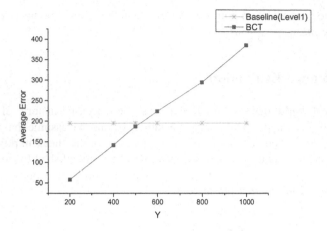

Fig. 2. Comparison of average error

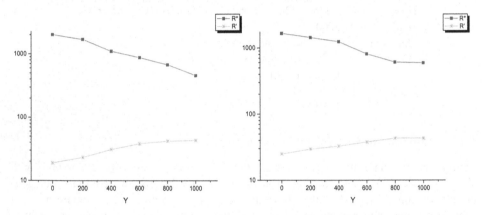

Fig. 3. Performance of task recommendation algorithm when varying the threshold *Y*

Interestingly, the *SC-server* and the workers are game-theory relationship. When Y increases, workers get higher privacy protection and higher reward at same time, while the *SC-server* gets lower reward. When Y decreases, workers get lower privacy protection and lower reward while the *SC-server* gets higher reward.

5.2.2   The Effect of Varying *Need*

We evaluate the performance of our optimal task recommendation algorithm on two different tasks by varying the number of workers required to complete the task. We experiment on two tasks which are randomly selected from the check-in points. Figure 4 shows the results. As expected, higher *Need* yields higher overhead as more workers are required to perform a task. That is to say, as the *Need* gets larger, the R^* drops down, while the R' rises up. (Total = 3000, Y = 500)

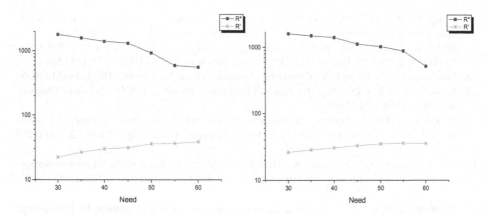

Fig. 4. Performance of task recommendation algorithm when varying the *Need*

6 Conclusion

We have considered the privacy issues in task recommendation for spatial crowdsourcing. We have proposed a task recommendation framework which recommends spatial crowdsourcing task without violating worker privacy. The proposed framework is comprised of two components: location generalization and optimal task recommendation. In the location generalization component, we have developed a bisecting cluster algorithm with threshold definition, guaranteeing the utility while providing strong privacy protection. In the optimal task recommendation component, we take into account the generalization error. We have evaluated the effectiveness and efficiency of the proposed framework. For future work, we intend to incorporate other popular task recommendation algorithms. We also plan to consider a worker may accept more than one task at a time.

Acknowledgments. This article is partly supported by the National Natural Science Foundation of China under Grant No. 61370084, and the China Numerical Tank Project.

References

1. Kazemi, L., Shahabi, C.: GeoCrowd: enabling query answering with spatial crowdsourcing. In: ACM SIGSPATIAL GIS, pp. 189–198 (2012)
2. Gruteser, M., Grunwald, D.: Anonymous usage of location-based services through spatial and temporal cloaking. In: MobiSys (2003)
3. Mokbel, M.F., Chow, C.-Y., Aref, W.G.: The new Casper: query processing for location services without compromising privacy. In: Proceedings of Very Large Data Bases, pp. 763–774 (2006)
4. Ghinita, G., Kalnis, P., Khoshgozaran, A., Shahabi, C., Tan, K.-L.: Private queries in location based services: anonymizers are not necessary. In: SIGMOD, pp. 121–132 (2008)
5. Cormode, G., Procopiuc, C., Srivastava, D., Shen, E., Yu, T.: Differentially private spatial decompositions. In: ICDE, pp. 20–31 (2012)

6. Gong, Y., Wei, L., Guo, Y., Zhang, C., Fang, Y.: Optimal task recommendation for mobile crowdsourcing with privacy control. IEEE Internet Things J. **3**(5), 745–756 (2016)
7. Wang, L., Meng, X.-F.: Location privacy preservation in big data era: a survey. Ruan Jian XueBao/J. Softw. **25**(4), 693–712 (2014). http://www.jos.org.cn/1000-9825/4551.html
8. Sun, J.G., Liu, J., Zhao, L.Y.: Clustering algorithms research. J. Softw. **19**(1), 48–61 (2008)
9. Savaresi, S., Boley, D.: On performance of bisecting k-means and PDDP. In: Proceedings of the 1th SIAMICDM (2001)
10. Dwork, C.: Differential privacy. In: Bugliesi, M., Preneel, B., Sassone, V., Wegener, I. (eds.) ICALP 2006. LNCS, vol. 4052, pp. 1–12. Springer, Heidelberg (2006). doi:10.1007/11787006_1
11. Alt, F., Shirazi, A.S., Schmidt, A., Kramer, U., Nawaz, Z.: Locationbased crowdsourcing: extending crowdsourcing to the real world. In: 6th Nordic Conference on Human-Computer Interaction, pp. 13–22 (2010)
12. Musthag, M., Ganesan, D.: Labor dynamics in a mobile micro-task market. In: Proceedings of ACM SIGCHI (2013)
13. Mokbel, M.F., Chow, C.-Y., Aref, W.G.: The new casper: query processing for location services without compromising privacy. In: Proceedings of the 32nd International Conference on Very Large Data Bases, pp. 763–774. VLDB Endowment (2006)
14. Guha, S., Reznichenko, A., Tang, K., Haddadi, H., Francis, P.: Serving ads from localhost for performance, privacy, and profit. In: HotNets (2009)
15. Fredrikson, M., Livshits, B.: Repriv: Re-imagining content personalization and in-browser privacy. In: 2011 IEEE Symposium on Security and Privacy (SP), pp. 131–146. IEEE (2011)
16. Chakraborty, S., Raghavan, K.R., Johnson, M.P., Srivastava, M.B.: A framework for context-aware privacy of sensor data on mobile systems. In: Proceedings of the 14th Workshop on Mobile Computing Systems and Applications, p. 11. ACM (2013)
17. To, H., Ghinita, G., Fan, L., Shahabi, C.: Differentially private location protection for worker datasets in spatial crowdsourcing. IEEE TMC **16**, 934–949 (2016)
18. Fawaz, K., Shin, K.G.: Location privacy protection for smartphone users. In: CCS, pp. 239–250 (2014)
19. Gao, S., Ma, J.F., Shi, W., Zhan, G., Sun, C.: TrPF: a trajectory privacy-preserving framework for participatory sensing. IEEE Trans. Inf. Forensics Secur. **8**(6), 874–887 (2013)

An Information-Aware Privacy-Preserving Accelerometer Data Sharing

Mingming Lu$^{(\boxtimes)}$, Yihan Guo, Dan Meng, Cuncai Li, and Yin Zhao

Central South University, Changsha, Hunan, China
mingminglu@csu.edu.cn

Abstract. In the age of big data, plenty of valuable data have been shared to enhance scientific innovation, which, however, may disclose unexpected privacy leakage. Although numerous privacy preservation techniques have been proposed to conceal sensitive information, it is usually at the cost of the application utility reduction. In this paper, we present a data sharing scheme, which balances the application utility and privacy leakage for specific data sharing. To illustrate our scheme, smartphones' acceleration data have been adopted as an illustrative example. Experimental study has shown that sampling frequency play dominant roles in reducing privacy leakage with much less reduction on utility.

Keywords: Data sharing · Privacy preservation · Activity recognition · Identity privacy · Mutual information · Visualization

1 Introduction

As accelerometer sensors are ubiquitously embedded into smartphones and wearable devices [8], accelerometer-based human activity recognition [21] has become a fascinating research topic. Most recently, some pioneer institutions [18] begin to share a large amount of accelerometer sensor data collected from volunteers, which has given a strong impetus to the development of activity recognition techniques and context-aware services [7]. The growing sophistication of activity recognition techniques substantially contributes to the emergency of various applications, such as sport tracker, advanced game play and healthcare monitoring [24].

Existing approaches to protecting privacy are largely focused at removing or obfuscating common sensitive information, such as location, identification, gender and age [6, 25], but little attention is paid to preserve the complicated privacy derived from individual gait characters implied in accelerometer data. More importantly, most smartphone users have been unaware of such unobtrusive private information. As Christin et al. indicated that people often think raw accelerometer readings may appear less threatening in revealing private information, but this hypothesis is not always true [15].

In this paper, we consider a specific privacy protection for specific usage requirements of the data set, which aims to enable data providers and users to reach a consensus through balancing data utility and privacy leakage. To discover the main factors that affect the accuracy of activity recognition and identity authentication, we conduct numerous experiments and found that sampling frequency play dominant roles in distinguish activity recognition and identity authentication. The sampling level,

© Springer Nature Singapore Pte Ltd. 2017
B. Zou et al. (Eds.): ICPCSEE 2017, Part I, CCIS 727, pp. 425–432, 2017.
DOI: 10.1007/978-981-10-6385-5_36

which maintain the enough activity information but hide many identity details, have been identified. Based on these explorations, we develop a sharing scheme for smartphones' acceleration data. Finally, we design a visualization tool, which shows each feature extracted from different sample rate containing the quantity of information about activity and identity. Besides, the tool provides variety of interaction means to help users analyze the combination of features and sampling rate, and find optimization proposal.

2 Related Work

Due to the embedded accelerometer, modern mobile devices have been utilized to infer human activity, identity, and other attributes. Through accelerometer, devices' dynamic acceleration can be easily tracked [21]. To use these efficiently, lots of researchers have brought out masses of achievements. Bao & Intille investigated an application called human activity identification [10]. Bao and Intille brought out a new method Activity Recognition from User-Annotated Acceleration Data [5]. Győrbíró et al. presented a real-time accelerometer based activity recognition system [10]. Kwapisz et al. brought activity recognition through cell phone accelerometers [9].

Aside from activity recognition, accelerometer data is also found useful in identify recognition. Ailisto et al. [11] proposed human gait identification through accelerometer sensors. Researches on natural gait cycles from accelerometer data for authenticate identity [12] was extracted. It is worth mentioning that with the popularity of machine learning, the extraction of statistical gait features is leading the mainstream approaches of identity recognition [23, 24].

Normally, traditional privacy preservation methods, such as data anonymization [27] and data perturbation [26], do some processes of either encrypting or revising personally identifiable information from sharing data sets. For instance, AnonySense [1] and LOCATE [2] adopted anonymous-based technique to protect location information of data contributors; Pickle [3] and PoolView [4] made significant efforts on permutation-based technique to preserve privacy in time-series data. Wang proposed a novel ambiguity technique to against both presence leakage and association leakage for privacy-preserving data sharing in cloud computing [6]. SensorSafe [16] and Ipshield [17] are newly collaborative privacy-enforcing frameworks for protecting diverse private information captured by more than one smartphone sensors.

Focusing on the privacy-preserving problem on social media, Dasgupta et al. [28] suggested current information visualization techniques commonly faced privacy leakage risks for assuming unrestricted access to data, and then discussed how to allow users to explore the data without breaching privacy by taking an example of Parallel Coordinates. Tarameshloo et al. [29] designed a novel tool to visualize location-based social network (LBSN) data both before and after anonymization, which could help privacy experts understand the effects of the anonymization techniques on LBSN data utility.

3 Data Description

3.1 Feature of Human's Gait Cycle

Gait cycle is the time interval of the duplicative actions during human's walking. Thus, it is in other normal actions, such as running, climbing stairs and so on. Thang et al. make it a figure to explain the details of gait cycle.

From Fig. 1, we could find that the gait cycle starts while the right foot touching the ground and end at the next touching of right foot. It is a half process of the cycle when the left foot touches the ground.

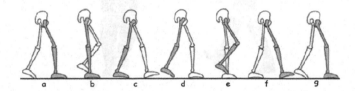

Fig. 1. The details of human's gait cycle

Congenital genetic characteristics and acquired environmental characteristics related to it, human's gait activity is quite complex. which includes coordination of movement between various parts of the body. A large number of medical studies show that muscle growth, bone structure, height, weight, and many other aspects all could be different. Besides, considering acquired environmental impact, the intensity of each person's gait, the degree of tilt of the body as well as the angle of the lower limb joints will be different. Thus, the gait could be considered individual characteristics.

Gait differences means different gait patterns during walking, such as the max distance from the ground and the distance of each step while walking. All those patterns are related to height and age, but as for a certain one, it is stable at a time.

3.2 Accelerometer Data Description

Accelerometer is an electromechanical instrument that measures the accelerations of an object in physical space. Today's smartphones and wearable devices commonly come with embedded accelerometer sensors. These sensors can continuous gather accelerometer data which is enabling the emergence of various personal and group sensing applications. The key items of accelerometer data generally include timestamp, sampling rate, sensor placement and acceleration values. For an observed person, accelerometer sensors can be positioned in different places of human body (e.g. head, forearm, wrist, waist and foot), and record acceleration values in three orthogonal directions with a constant sampling rate. Figure 2 shows the three axial directions when an accelerometer sensor is located in the waist of a walking person. In the view of human activity characteristics, the accelerometer data collected from different body locations are suitable for the requirements of different practical applications. For example, the acceleration data captured from wrist can achieve accurate gesture recognition [16], and the acceleration data collected from waist-placement has the best performance for activity recognition because the waist is close to the center of a whole human body [17]. Moreover, the acceleration data reported by data contributors often contain other meta-data, including personal information and labeled contextual information (e.g. walking and running).

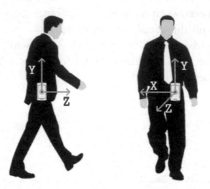

Fig. 2. Three acceleration directions relative to a walking person

We select a subset of HASC corpus dataset as our experimental data, which has been universally utilized for activity and identity recognition based on smartphone accelerometer data [31–33]. The data is contributed by 100 volunteers, among these people, meal and female are accounted for half of the proportion with height distributed between 147 cm and 187 and weight mainly distributed in 45 kg to 80 kg. The data consists of meta-data and acceleration-data. The meta-data includes several personal information (e.g. personID, gender, weight, height) and labeled activities. The acceleration-data contains four fields, namely timestamp and accelerations on three axial directions. The sampling rate of the data is 100 Hz, which is higher than common sampling rates, such as 50, 32 and 20 Hz [19]. Shown as Fig. 2, the accelerations of the data are obtained from waist-placement, the x-axis captures horizontal movement of a person, the y-axis tracks the upward and downward motion, and the z-axis records the forward movement. Six basic daily activities are labeled in the data, namely stay, walk, jogging, skip, stair-up, stair-down, which are all correspond to the movements of human body. Each activity for one volunteer lasts about 100 s, and the timestamp of the data is to the millisecond.

4   Privacy-Preserving Accelerometer Data Sharing

4.1   Accelerometer Data Sharing Scenario

The typical scenario of accelerometer data sharing is that data contributors report their raw accelerometer data to credible third parties, like government agencies and research communities. Data consumers submit their proposals to the third parties for requiring accelerometer data. After a fully scrutiny on proposals, data managers would process raw accelerometer data by data mask techniques, for instance, replacing sensitive information with a generic value such as 'X', or substituting the average value for the original value. Then, the data query API of processed data would be shred to approved data consumers.

However, there is a failure to comprehensively protect the personal privacy in the accelerometer data after proposal scrutiny and data masking. With regard to the fact, data managers encounter difficulty to judge the true intention of data consumers purely by the summited proposals. Specifically, some malicious data users are skilled at concealing their actual plans or intentions, for example, they can assertively claim as normal data users that the data will be used to train a robust activity recognition model. In contrast, they may utilize the data to crack the personal information of data contributors or save contributors' gait biological characteristics for illegal purposes. Hence, it's indispensable to provide a privacy-preserving process to data managers before sharing accelerometer data.

4.2 Factor Analysis for Accelerometer Data Privacy-Preserving

What factors should be considered in a privacy-preserving process on accelerometer data? We focus our attention on statistical machine learning techniques which are recently widely used for analyzing accelerometer data. Numerous studies show that these techniques have high accuracy and performance both in activity recognition [21, 22] and identify recognition [13, 14]. These studies have also concluded that two key factors, namely, algorithm model, data sampling rate, have significant influence on the recognition accuracy [20]. Because data managers cannot know what algorithm models will be used by data consumers, consequently, the sampling rate factor becomes our focus.

We conduct experiments to observe how these factors affect the accuracy of activity and identify recognition based on machine learning techniques. Two acceleration datasets were selected from the raw experimental data, one for activity recognition which consisted of the acceleration data of 100 volunteers' six activities, namely stay, walk, jogging,

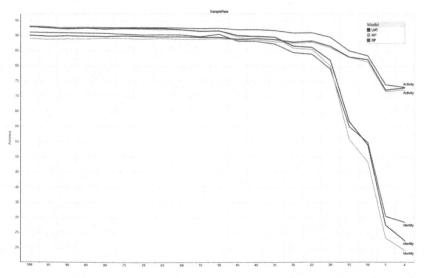

Fig. 3. The relationship between precision and the sampling rate under different feature data sets with different models

skip, stair-up and stair-down; the another for identity recognition which only contained the acceleration data of 100 subjects' walk, because of walking style is known as the most common and stable human activity for distinguishing between individuals [11].

The sampling rate of the original experiment data is 100 Hz. To observe the influence of the sampling rate on activity recognition and identity recognition, respectively, we drop the sampling rate from 100 down to 4 Hz step by step. Figure 3 shows the recognition accuracy for different sampling rates with different classification models. Both the accuracy of activity and identity recognition stabilize between 100 Hz to 35 Hz, and decrease with lower sampling rates, and the accuracy of identity recognition declines much faster than activity recognition. This experiment results show that the accuracy of activity and identity recognition will simultaneously decrease with sampling rate decreasing, but different amplitude and fall time vary with different features.

5   Experimental Study

To estimate the variation of the accuracy of the activity recognition and identity recognition along with different sampling rate, the sampling rate varies from 100 Hz to 4 Hz, using the RF model for activity recognition, using the RF, LMT and MP models for identity recognition, consistent with the previous analysis of HASC data. Figure 4 shows the effect of the sampling rate on the recognition accuracy. It can be seen that the accuracy of the activity recognition decreases from 45 Hz to 4 Hz, and the identity recognition's accuracy declines much faster. At 4 Hz, compared with the previous accuracy, it decreased slowly. At this time, the activity recognition accuracy is 93.7708%, the identity recognition's accuracy using RF is up to 52.7536%, a difference of 41.0172% with activity recognition.

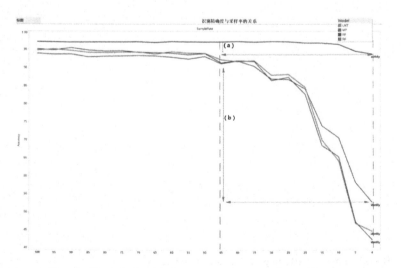

Fig. 4. The relationship between accuracy and sampling rate

6 Conclusion

This work proposed an information-aware privacy protection scheme for mobile phone acceleration data through in-depth analysis of activity recognition and identity recognition for mobile phone acceleration data set. The main factors influencing the accuracy of the two recognitions are discussed experimentally. In the future, we will study the influence of feature on accelerometer data sharing.

Acknowledgement. This work was partially supported by Project Nos. 61232001, 61173169, and 60903222 supported by National Science Foundation of China, No. 2016JJ2149 supported by Science Foundation of Hunan, and also supported by the Major Science & Technology Research Program for Strategic Emerging Industry of Hunan (Grant No. 2012GK4054).

References

1. Kapadia, A., Triandopoulos, N., Cornelius, C., Peebles, D., Kotz, D.: AnonySense: opportunistic and privacy-preserving context collection. In: Indulska, J., Patterson, D.J., Rodden, T., Ott, M. (eds.) Pervasive 2008. LNCS, vol. 5013, pp. 280–297. Springer, Heidelberg (2008). doi:10.1007/978-3-540-79576-6_17
2. Boutsis, I., Kalogeraki, V.: Privacy preservation for participatory sensing data. In: IEEE International Conference on Pervasive Computing and Communications, pp. 103–113 (2013)
3. Liu, B., Jiang, Y., Sha, F., et al.: Cloud-enabled privacy-preserving collaborative learning for mobile sensing. In: Proceedings of the 10th ACM Conference on Embedded Network Sensor Systems, pp. 57–70. ACM (2012)
4. Ganti, R.K., Pham, N., Tsai, Y.E., et al.: PoolView: stream privacy for grassroots participatory sensing. In: Proceedings of the 6th ACM Conference on Embedded Network Sensor Systems, pp. 281–294. ACM (2008)
5. Bao, L., Intille, S.S.: Activity recognition from user-annotated acceleration data. In: Ferscha, A., Mattern, F. (eds.) Pervasive 2004. LNCS, vol. 3001, pp. 1–17. Springer, Heidelberg (2004). doi:10.1007/978-3-540-24646-6_1
6. Wang, H.: Privacy-preserving data sharing in cloud computing. J. Comput. Sci. Technol. **25**(3), 401–414 (2010)
7. Hull, R., Kumar, B., Lieuwen, D., et al.: Enabling context-aware and privacy-conscious user data sharing. In: Proceedings of the 2004 IEEE International Conference on Mobile Data Management, 2004, pp. 187–198. IEEE (2004)
8. Lane, N.D., Miluzzo, E., Lu, H., et al.: A survey of mobile phone sensing. IEEE Commun. Mag. **48**(9), 140–150 (2010)
9. Kwapisz, J.R., Weiss, G.M., Moore, S.A.: Activity recognition using cell phone accelerometers. ACM Sigkdd Explor. Newsl. **12**(2), 74–82 (2010)
10. Győrbíró, N., Fábián, Á., Hományi, G.: An activity recognition system for mobile phones. Mob. Netw. Appl. **14**(1), 82–91 (2009)
11. Ailisto, H.J, Lindholm, M., Mantyjarvi, J., et al.: Identifying people from gait pattern with accelerometers. In: International Society for Optics and Photonics, Defense and Security, pp. 7–14 (2005)
12. Gafurov, D., Helkala, K., Søndrol, T.: Biometric gait authentication using accelerometer sensor. J. comput. **1**(7), 51–59 (2006)

13. Kwapisz, J.R., Weiss, G.M., Moore, S.A.: Cell phone-based biometric identification. In: 2010 Fourth IEEE International Conference on Biometrics: Theory Applications and Systems (BTAS), pp. 1–7. IEEE (2010)
14. Derawi, M., Bours, P.: Gait and activity recognition using commercial phones. Comput. secur. **39**, 137–144 (2013)
15. Christin, D., Reinhardt, A., Kanhere, S.S., et al.: A survey on privacy in mobile participatory sensing applications. J. Syst. and Softw. **84**(11), 1928–1946 (2011)
16. Cho, I.Y., Sunwoo, J., Son, Y.K., Oh, M.-H., Lee, C.-H.: Development of a single 3-axis accelerometer sensor based wearable gesture recognition band. In: Indulska, J., Ma, J., Yang, Laurence T., Ungerer, T., Cao, J. (eds.) UIC 2007. LNCS, vol. 4611, pp. 43–52. Springer, Heidelberg (2007). doi:10.1007/978-3-540-73549-6_5
17. Yang, C.C., Hsu, Y.L.: A review of accelerometry-based wearable motion detectors for physical activity monitoring. Sensors **10**(8), 7772–7788 (2010)
18. Kawaguchi, N., Ogawa, N., Iwasaki, Y., et al.: HASC challenge: gathering large scale human activity corpus for the real-world activity understandings. In: Proceedings of the 2nd Augmented Human International Conference, p. 27. ACM (2011)
19. Lockhart, J.W., Weiss, G.M.: The benefits of personalized smartphone-based activity recognition models. In: SDM, pp. 614–622 (2014)
20. Shoaib, M., Bosch, S., Incel, O.D., et al.: A survey of online activity recognition using mobile phones. Sensors **15**(1), 2059–2085 (2015)
21. Kwapisz, J.R., Weiss, G.M., Moore, S.A.: Activity recognition using cell phone accelerometers. ACM SigKDD Explor. Newsl. **12**(2), 74–82 (2011)
22. Incel, O.D., Kose, M., Ersoy, C.: A review and taxonomy of activity recognition on mobile phones. BioNanoScience **3**(2), 145–171 (2013)
23. Guo, B., Nixon, M.S.: Gait feature subset selection by mutual information. IEEE Trans. Syst. Man Cybern. Part A Syst. Hum. **39**(1), 36–46 (2009)
24. Avci, A., Bosch, S., Marin-Perianu, M., et al.: Activity recognition using inertial sensing for healthcare, wellbeing and sports applications: a survey. In: 2010 23rd International Conference on Architecture of Computing Systems (ARCS), pp. 1–10. VDE (2010)
25. Parameswaran, R., Blough, D.M.: Privacy preserving collaborative filtering using data obfuscation. In: IEEE International Conference on Granular Computing, 2007. GRC 2007, p. 380. IEEE (2007)
26. Agrawal, R., Srikant, R.: Privacy-preserving data mining. In: ACM Sigmod Record. vol. 29 (2), pp. 439–450. ACM (2000)
27. Sweeney, L.: k-anonymity: a model for protecting privacy. Int. J. Uncertain. Fuzziness Knowl.-Based Syst. **10**(05), 557–570 (2002)
28. Aritra, D., Min, C., Robert, K.: Measuring privacy and utility in privacy-preserving visualization. Comput. Graph. Forum **32**(32), 35–47 (2013)
29. Tarameshloo, E., et al.: Using visualization to explore original and anonymized LBSN data. In: Eurographics/IEEE VGTC Conference on Visualization Eurographics Association, pp. 291–300 (2016)
30. Wang, Y., Gou, L., Xu, A., et al.: VeilMe: an interactive visualization tool for privacy configuration of using personality traits. In: ACM Conference on Human Factors in Computing Systems. pp. 817–826. ACM (2015)
31. Kawaguchi, N., Ogawa, N., Iwasaki, Y., et al.: HASC Challenge: gathering large scale human activity corpus for the real-world activity understandings. In: Proceedings of the 2nd Augmented Human International Conference, p. 27. ACM (2011)
32. Lockhart, J.W., Weiss, G.M.: The benefits of personalized smartphone-based activity recognition models. In: SDM, pp. 614–622 (2014)
33. Gjoreski, H., Kozina, S., Gams, M., et al.: Competitive live evaluations of activity-recognition systems. IEEE Pervasive Comput. **14**(1), 70–77 (2015)

A Range-Threshold Based Medical Image Classification Algorithm for Crowdsourcing Platform

Shengnan Zhao, Haiwei Pan[✉], Xiaoqin Xie, Zhiqiang Zhang,
and Xiaoning Feng

College of Computer Science and Technology, Harbin Engineering University,
Harbin 150001, China
panhaiwei@hrbeu.edu.cn

Abstract. Medical images are important for medical research and clinical diagnosis. The research of medical images includes image acquisition, processing, analysis and other related research fields. Crowdsourcing is attracting growing interests in recent years as an effective tool. It can harness human intelligence to solve problems that computers cannot perform well, such as sentiment analysis and image recognition. Crowdsourcing can achieve higher accuracies in medical image classification, but it cannot be widely used for its low efficiency and the monetary cost. We adopt a hybrid approach which combines computer's algorithm and crowdsourcing system for image classification. Medical image classification algorithms have a high error rate near the threshold. And it is not significant by improving these classification algorithms to achieve a higher accuracy. To address the problem, we propose a hybrid framework, which can achieve a higher accuracy significantly than only use classification algorithms. At the same time, it only processes the images that classification algorithms perform not well, so it has a lower monetary cost. In the framework, we device an effective algorithm to generate a range-threshold that assign images to crowdsourcing or classification algorithm. Experimental results show that our method can improve the accuracy of medical images classification and reduce the crowdsourcing monetary cost.

Keywords: Medical image · Range-threshold based · Crowdsourcing · Image classification

1 Introduction

Medical images are important for medical research and clinical diagnosis [1]. The research of medical images includes image acquisition, processing, analysis and other related research fields. As a large branch of image processing, medical image processing is thriving [2]. As the medical images contain an abundance of images and medical information, data mining technology for medical images becomes a hot spot in interdisciplinary research about medical and computer. Computer technology has been widely cited in the medical field. The involvement of multimedia technology and the application of medical image technology provide favorable conditions for further

© Springer Nature Singapore Pte Ltd. 2017
B. Zou et al. (Eds.): ICPCSEE 2017, Part I, CCIS 727, pp. 433–446, 2017.
DOI: 10.1007/978-981-10-6385-5_37

medical research. The information, which is obtained through data mining, will also contribute to the improvement of medical diagnosis, treatment and analysis [3].

To improve the level of medical care, using the computer to deal with the images is clearly unparalleled superiority. At present, it is commonly used in classification methods which includes neural network, Naive Bayesian, genetic algorithm and decision tree algorithm. In the field of medical image research, classification methods are usually used to divide medical images into several categories, for example, [4] proposed a hierarchical nonparametric Bayesian approach for medical images gene expressions classification. [5] used decision tree to detect tumor and classify brain MRI images.

Medical image classification algorithms have a high error rate near the threshold. And it is not significant by improving these classification algorithms to achieve a higher accuracy.

Crowdsourcing has attracted a great deal of interests as a platform for leveraging human intelligence [6]. It can harness human intelligence to solve problems that computers cannot perform well, such as sentiment analysis and image recognition. Using crowdsourcing for medical image classification can achieve the desired results. Existing solutions will publish all medical images on the crowdsourcing platform. Since the way of using human intelligence is wasteful, prior works [7–9] showed how to build hybrid human-machine approaches.

To address these problems, a hybrid machine-crowdsourcing framework (HMC), which combined with crowdsourcing and machine algorithms, is proposed in this paper. The framework can achieve a higher accuracy significantly than only use classification algorithms. At the same time, it only processes the images that classification algorithms perform not well, so it has a lower monetary cost.

In the framework, we device an effective algorithm to generate a range-threshold that assign images to crowdsourcing or classification algorithm.

Experimental results show that our method can improve the accuracy of medical images classification and reduce the crowdsourcing monetary cost.

The remainder of this paper is organized by follows. The second section introduces problem formulation. We define the medical image classification problem and describe the process of our framework (HMC). In the third section, we propose a Range threshold calculation method (RTC) and RT-Based classification method (RBC). The forth section is experimental results and analysis. Finally, we make a conclusion about our work.

2 Problem Description

Existing crowdsourcing method is to publish all the medical images on the crowdsourcing platform. Then, waiting for the platform to returned classification results. But in the "Big Data" era, with the explosive growth of data, all kinds of image information capacity are constantly expanding.

In the above context, the drawbacks of directly using crowdsourcing platform are obvious, such as a large number of workers involved in the completion of crowdsourcing work. Meanwhile, it will cost too much time to wait for the crowdsourcing results.

The Fig. 1 shows the difference between the crowdsourcing image classification method and our method. The details of the framework will be presented in Sect. 4.

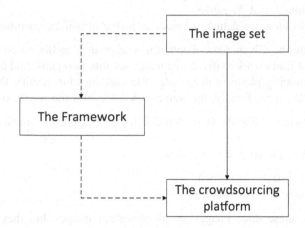

Fig. 1. The difference between crowdsourcing method and HMC

Therefore, this paper proposed an efficient framework HMC, which firstly use the classification algorithm to pre-classify the image collection. This pre-classify will pick out the most suitable ways for the image set. In order to better describe the problem, we introduce three concepts as below.

Crowdsourcing Tasks Cost. To motivate all the workers to complete the task, crowdsourcing platform will provide some money for the works as a reward.

Supposing each task (or medical image) cost c, n tasks need c * n; supposing all tasks cost C, maximum number of tasks is C/c.

Range Threshold. In this paper, *range-threshold*, extended from the single-threshold by algorithm, is denoted as RT(rt1, rt2), where rt1 is the left boundary and rt2 is the right boundary. Crowdsourcing task cost and the number of *Ambiguous State* images influence the *range threshold*.

Picture(I$_x$, I$_y$) denotes the image set whose feature value is in the range R(r$_1$, r$_2$), where I$_x$ is the image of left boundary and rt2 is the image of right boundary.

For example, supposing R(0.5, 0.7), Picture$_R$(I$_4$, I$_9$). The feature value of I$_4$ is 0.5, and the feature value of I$_9$ is 0.7. There are 6 elements in the Picture$_R$(I$_4$, I$_9$).

Image State. Through the traditional medical image classification algorithm, every image is classified as normal or abnormal image. As the traditional classification algorithm has a low accuracy near the threshold, in this paper, it divides the images into two states and four categories according to the classification algorithm.

After the medical image is processed by the algorithm, if the feature value is much smaller (or greater) than the threshold, image classification results are more accurate, we call it a steady state. In this state, if the image is classified as normal, it belongs to *S-Positive*. And if the image is classified as abnormal, it belongs to *S-Negative*.

If the feature value of the image is close to the threshold, we call the image Ambiguous State, as image classification results are not accurate. In this state, if the image is classified as normal, it belongs to *A-Positive*. And if the image is classified as abnormal, it belongs to *A-Negative*.

Based on this idea, we define our medical image classification problem as below.

Problem Definition. Given a set of medical images that needed to be classified, our goal is to use the framework to divide the image set into two parts, and then one part is sent to crowdsourcing platform to re-judge the classification results, the other part's results remain the same. Finally, the framework integrates the results of these parts.

Example 1. Assuming that the algorithm C is used to map every medical image i to a value v within (0, 1) according to an eigenvalue. And we suppose the threshold t is 0.88, the classification rules are as follows:

(1) if $v > t$, the image i is normal;
(2) if $v < t$, the image i is abnormal;

In Fig. 2(a), these three images are all abnormal images, but they are classified incorrectly as normal images by the image classification algorithm.

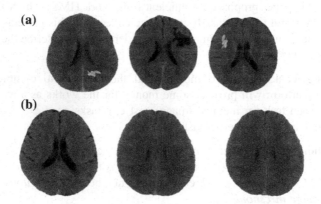

Fig. 2. (a) The feature value of images is 0.89 (b) The feature value of images is 0.87

Figure 2 shows that some real results are contrary to the results of the classifications. It also shows that the accuracy of the classification algorithm is not ideal around the threshold.

In Fig. 2(b), these three images are all normal images, but they are classified incorrectly as abnormal images by the image classification algorithm. The main reason for classifying incorrectly is that the feature values of these 6 images are very close to the threshold (0.88).

3   The HMC Framework

We propose a hybrid medical images classification framework in this section. Our framework takes as input a set of unclassified medical images and decide to classify these images either by using the classification or through the crowdsourcing platform. As shown in Fig. 3, the framework mainly consists of three parts. Pre-classification stage will preprocess the unclassified medical images. The pre-classification algorithm will classify all the images at first. Algorithm evaluation stage is to evaluate the performance of the classification algorithm and give a range-threshold to *the image selector*. The classification stage is to decide that the images classified either by the algorithm or using the crowdsourcing platform. In this stage, *the image selector* gets the range threshold from *the pre-classification algorithm evaluator*. And then, *the image selector* will send the images in the range threshold to the crowdsourcing platform to re-classify the images. And it will also send other images to the final results set. The detailed process of the machine-crowdsourcing framework (HMC) are given. On this basis, we propose a range threshold calculation method (RTC) and RT-Based classification method (RBC).

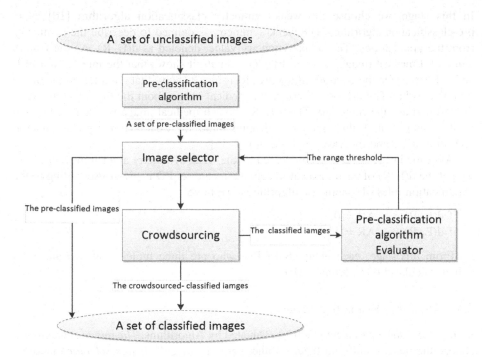

Fig. 3. The HMC Framework

3.1   The Concept of the Framework

In this paper, the untreated medical image set is denoted as $D(MI_1, MI_2, MI_3 \ldots \ldots MI_n)$, MIi denotes a medical image.

The medical image set processed by the classification algorithm is denoted as $PD(I_1, I_2, I_3 \ldots\ldots I_n)$, I_i denotes a medical image, which is processed by the algorithm. Each element is a triad form, $I_i(WR_i, AR_i, CR_i)$, where i denotes image number, WR denotes the feature value ($d(L, R)$) calculated by the classification algorithm, AR denotes the result calculated by the classification algorithm,

if AR = 0, the image is normal;
if AR = 1, the image is abnormal.
CR denotes the result determined by the crowdsourcing platform.
if CR = 0, the image is normal;
if CR = 1, the image is abnormal.

For example, $I_1(0.92, 0, 0)$ denotes that the image I_1, its feature value is 0.92, and it's a normal medical image.

The sample image set is denoted as $SD(S_1, S_2, \ldots, S_m)$. The m elements in SD are randomly selected from PD.

3.2   Pre-classification Stage

In this stage, we choose the weak symmetry classification algorithm [10] as ta pre-classification algorithm. The weak symmetry is defined to describe the symmetry from the granularities. The weak symmetry value, denoted as $d(L, R)$. As the feature value of a medical image, if the d $d(L, R)$ is small, it shows that the intersection of L and R is small. On the contrary, if the $d(L, R)$ is big, it shows that the intersection of the two parts is big. The $d(L, R)$ describes the medical image about the difference between the left part and the right part. The $d(L, R)$ of each medical image is calculated by the weak symmetry algorithm. To get the feature value, the weak symmetry classification algorithm use some theories proposed in [11–15].

After the image set D is preprocessed by the weak symmetry algorithm (Step 1), as we get the $d(L, R)$ of each medical image, the image set PD is given. According to the classification rules of symmetric algorithms, we have

if $d(L, R) > e$, AR = 0;
if $d(L, R) < e$, AR = 1.

From this stage, each elements of PD, who are three tuples, has two attributes values calculated ($d(L, R)$ and AR).

3.3   Algorithm Evaluation Stage

In this stage, some parameters will be evaluated by Algorithms. *The image selector* will choose the images and send them to other parts. Firstly, *the image selector* randomly select m elements from the images set PD. These m elements constitute the images set SD. And the parameters will be evaluated based on the data on SD. Then, we will utilize the crowdsourcing platform to classify the medical images from the images set SD. We assumed that the crowdsourcing works are professionals or professionally trained, so they can give the completely correct answers. Up to this step, each image's three attributes are known.

Next, *the pre-classification evaluator* compares the results of machine classification algorithms(AR) with the results of crowdsourcing (CR). As the AR is completely correct, for each medical image $I_i(WR_i, AR_i, CR_i)$, if AR_i is not equal to CR_i, the image I_i is classified incorrectly by the machine classification algorithms. For example, supposing $I_7(0.89, 0, 1)$.

In this method, *the pre-classification evaluator* can figure out all the incorrectly classified images in the sample collection SD. We take These incorrectly classified images constitute a set W.

Finally, *the pre-classification evaluator* counts the number of the incorrectly classified images in the unit interval and use the statistical tool to draw the error distribution curve (EDC).

The Error Distribution Curve (EDC). *The pre-classification evaluator* will count the images in images set W and use the statistical tool to draw a curve to describe the distribution of incorrectly classified images. The EDC shows the errors produced by the machine classification algorithms.

As the images around the single-threshold have the feature values similarly, the images, whose WR are around the single-threshold, have the high probability to be classified incorrectly. On the contrary, the images, whose WR are away from the single-threshold, have the high probability to be classified correctly.

According to this discovery, this paper chooses the images whose WR are close to the single-threshold (provided by the weak symmetry algorithm). These images will be sent to the crowdsourcing platform to be re-judged. In this way, the accuracy of image classification will be improved for we reduced the error rate near the single-threshold. And this error rate cannot be avoided by improving the traditional classification algorithm.

We fit the statistical results into a curve of approximate normal distribution. According to the nature of the normal distribution curve, the area covered by the normal distribution curve is 1.

The 3 Sigma principle of the normal distribution is shown as Fig. 4. The area of normal distribution also represents the ratio. The principle supposes the area surrounded by the curve and the x-axis is 1. Three important intervals occupy the different ratio as shown below.

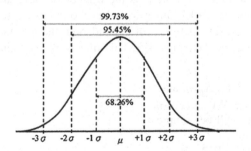

Fig. 4. The 3 Sigma principle

For Example. Supposing there are w images in the EDC, the interval $(u - \sigma, u + \sigma)$ will include about 68%w images classified incorrectly by the algorithm, and the interval $(u - 3\sigma, u + 3\sigma)$ will include about 99%w images classified incorrectly by the algorithm. Ideally, if we send all the images whose WR are in the range of $(u - 3\sigma, u + 3\sigma)$, we can almost avoid all classification errors by the algorithm. As it will cost too much time and money if there are large number of images sent to the crowdsourcing platform. We should combine some parameters to determine the number of images that need to be classified by crowdsourcing works. This paper proposed a way combined some parameters to calculate the interval. And the interval is called the range threshold (RT).

Range threshold calculation method:

Supposing there are m images randomly selected from the image set ID, and these m images constitute the image set SD. After the workers classify all the images from SD, we find that w images of the SD are classified incorrectly by the classification algorithm. Using the feature values of the wrong image set W distribution characteristics, we get the EDC and the curve obeys $N(u, \sigma^2)$. The crowdsourcing cost C and the cost c of each task (image) are given by employers. So, we can complete no more than C/c tasks.

Let $C/c = t$, $R_1 = (u - \sigma, u + \sigma)$, $R_2 = (u - 2\sigma, u + 2\sigma)$, $R_3 = (u - 3\sigma, u + 3\sigma)$ and we count the total numbers of classified images in three intervals from the images set ID (Table 1).

Table 1. The numbers of images in R

Intervals	R1	R2	R3
Numbers	k1	k2	k3

Since R2 and R3 cover more than 90% image classified incorrectly, we will not extend R2 and R3 when $t > k2$ (or $t > k3$).

If $k2 > t > k1$, R1 will be extended.

The images set Picture(Ix, Iy) presents the images whose feature value are in the range of R1. So there are (t-k1) images to be chosen to be sent to the platform. According to symmetry, this method will expand both the left and right borders. We will add (t-k1) pictures to Picture(Ix, Iy). This makes RT contain t images. And the image set of RT is

Picture$(I_{x-\lceil(t-k1)/2\rceil}, I_{y+\lceil(t-k1)/2\rceil})$, and
RT = $(WR_{x-\lceil(t-k1)/2\rceil}, WR_{y+\lceil(t-k1)/2\rceil})$.
If $k1 > t > k1$, R1 will be compress.
Similarly,
RT = $(WR_{x+\lfloor(t-k1)/2\rfloor}, WR_{y-\lfloor(t-k1)/2\rfloor})$.

Algorithm1 shows the description of RTC method.

Algorithm 1. RTC method
Input: Pre-processed image set $PD(I_1,I_2,..I_n)$, sample image set $SD(S_1,S_2,...S_m)$,the single-threshold e(=0.88),the crowdsourcing cost C and each task cost c.
Output: A: the range threshold $RT(rt_1,rt_2)$.
1. Count the numbers of medical images whose AR is not equal to CR
2. Draw the error distribution curve(EDC) and get the EDC parameters: u, σ.
3. Count the numbers of medical images in PD whose WR are in the range of R_1 (R_2or R_3), and regard as k_1,k_2,k_3.
4. Let $t=\frac{c}{c}$
5. If t>k3, RT=R1
    If k3>t>k2, RT=R2
    If k2>t>k1, RT=R2
    $$Picture(I_{x-\lceil(t-k1)/2\rceil},I_{y+\lceil(t-k1)/2\rceil})$$
    $$RT=(WR_{x-\lceil(t-k1)/2\rceil}, WR_{y+\lceil(t-k1)/2\rceil})$$
    If k1>t
    $$Picture(I_{x+\lfloor(t-k1)/2\rfloor},I_{y-\lfloor(t-k1)/2\rfloor})$$
    $$RT=(WR_{x+\lfloor(t-k1)/2\rfloor}, WR_{y-\lfloor(t-k1)/2\rfloor})$$
6. RETURN RT

3.4   The Classification Stage

In this section, the RT-Based classification method (RBC) is proposed to classify the medical images into 4 classes, so that the images can be chosen to crowdsourcing platform. Algorithm 2 shows the description of RBC method.

Algorithm 2. RBC method
Input: Pre-processed image set $PD(I_1,I_2,..I_n)$, range threshold $RT(rt_1,rt_2)$.
Output: A: the classification results F(P,N),where P denote the normal images set, N denotes the abnormal images set.
1. For each image I_i in the image set PD.
2. If WR<rt_1, $I_i \in$ S-Negative
    If WR>rt_2, $I_i \in$ S-Positive
    If rt_1<WR<rt_2, I_i will be re-judged by crowdsourcing works:
        If CR=0, $I_i \in$ A-Negative
        If CR=1, $I_i \in$ A-Positive
3. P=S-Positive ∪ A-Positive
4. N=S-Negative ∪ A-Negative
5. RETURN F(P,N);

4   Experimental Results and Analysis

This section will discuss the experiment results of our methods. To evaluate the performance of our proposed methods, we conducted experiments on real-world datasets. We show that our proposed methods

4.1   Experimental Setup

Dataset: We used a real dataset to evaluate our method. There are 1500 (n = 1500) brain CT images. We randomly select 500 (m − 500) images as a sample dataset (SD).

The Experimental Hardware Environment: Intel(R) Core(TM)2 Quad CPU Q8400 @2.66 GHz, 2.0 GB RAM, Microsoft Windows 10 operating system. The experimental software development environment is Visual Studio 2015, Origin Lab 9.0.

The Classification Algorithm: The weak symmetry classification algorithm is chosen in the pretreatment stage [10].

Single-Threshold: Our framework needs a single-threshold to classify the images by machine algorithm. The weak symmetry classification algorithm is verified to have the highest accuracy by using above dataset when the single-threshold is 0.88.

4.2   The Pretreatment of the Medical Images Dataset

In the framework, all the unclassified medical images will be pretreatment. We calculate the feature values (WR) of each image (Step 1). The results are shown partly as Fig. 5.

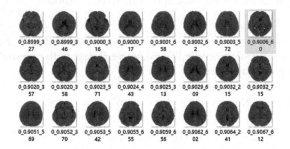

Fig. 5. The pretreatment results

And then, each image is represented in triad form $I_i(WR_i, AR_i, CR_i)$. For example, $I_7(0.77, 0, null)$ presents (1) the feature value is 0.77; (2) it is classified as an normal image by the algorithm; (3) the image have not sent to the crowdsourcing platform.

The preprocessed images set is formalized as $PD(I_1, I_2, \ldots I_{1500})$.

4.3   Algorithm-Machine Interaction Evaluation Parameter Stage

In the section, our framework randomly selects 500 images from PD (Step 3). The sample set of 500 images is denoted as $SD(S_1, S_2, \dots S_{500})$. We send all these images to the crowdsourcing platform to classify images by professional workers (Step 4). Each image is as a crowdsourcing task as Fig. 6 shown.

1.Is the following image normal?

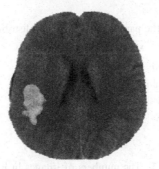

O Normal image O Abnormal image

Fig. 6. A classification crowdsourcing task

Next, we get 500 classification results by works (Step 5). The image, whose algorithm results is not equal to the crowdsourcing works' results (AR $\neq$ CR), is added to the error distribution. We count the error results distribution (Step 6). Figure 7 shows the error distribution Histogram.

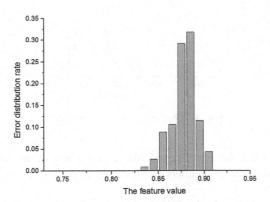

Fig. 7. The error distribution histogram

Next, we fit the normal distribution curve and get the probability density function (Table 3).

$$y = y_0 + A * e^{-\frac{1}{2} \times \frac{(x-x_c)^2}{w^2}}$$

The Fig. 8 shows that the error has a high probability when it is near the single-threshold. Two parameters (u and σ) are given by the probability density function.

We use the formula to fit the EDC. From the Table 2, two parameters (u = 0.88 and σ = 0.01) are given.

Table 2. The parameters of normal distribution

	Value	Standard error
y0	0.00487	0.0045
xc	0.88027	6.13088E-4
w	0.01088	6.49985E-4
A	0.32744	0.01631

Table 3. The numbers of images in R

Intervals	(0.87, 0.89)	(0.86, 0.90)	(0.85, 0.91)
Numbers	273	521	1083

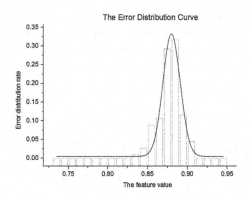

Fig. 8. The error distribution curve

Then, the range-threshold will be calculated according to the parameters.

Now, we count the total numbers of classified images in three intervals from the images set PD.

C-number denotes the numbers of the images which are corrected the results by crowdsourcing works. For example, if C-Num = 119, it shows that the number of images which are classified incorrectly by classification algorithms is 119.

T-Number denotes the total number of images which are classified incorrectly by classification algorithms. In this paper, the T-number is 183.

C-Rate is short for the correction rate. The C-Rate can describe the performance of our framework.

C-Rate = C-Num/T-Num

According to the RT-Based classification method, the crowdsourcing cost influent the range threshold. We give the cost according to different circumstances. It is shown as Table 4.

Table 4. The influence of deferent values of t

Condition	t < k1	t = k1	k1 < t < k2	t = k2	K2 < t < k3	t = k3	t = n
t	100	273	400	521	700	1083	1500
RT	(0.885, 0.874)	(0.871, 0.889)	(0.866, 0.892)	(0.861, 0.900)	(0.856, 0.906)	(0.850, 0.909)	(0.703, 0.935)
C-Num	59	119	142	151	160	167	183
C-Rate	32.2%	65.0%	77.5%	82.5%	87.4%	91.2%	100%

Table 4 shows that (1) the crowdsourcing cost influence the RT; (2) classified incorrectly images have a higher probability around the single-threshold (u); (3) we can expand TR to correct more misclassifications; (4) the correction rate is close to the expectation (the 3-sigma principle).

4.4   The Classification Stage

In the section, our framework randomly selects 500 images from PD (Step 3). The sample set of 500 images is denoted as $SD(S_1, S_2, ...S_{500})$. We send all these images to the crowdsourcing platform to classify images by professional workers (Step 4). Each image is as a crowdsourcing task as Fig. 6 shown.

5   Conclusion

In this paper, we mainly focus on the medical images classification method. According to the features of medical images and inspired by domain knowledge, we propose a medical image classification method based on the range-threshold and the RT-Based classification method. Firstly, medical images are pre-processed. We can get a method to calculate the feature values of medical images. Secondly, the sample medical image dataset is selected to be used to evaluate the classification algorithm and calculate the parameters. Last but not the least, we propose the RT-Based classification method to divide the images set into 2 parts. One part is sent to the crowdsourcing platform, the other is sent to the final results set. The experimental results show that our framework can classify medical images efficiently and perform well in cost and accuracy. This method can assist doctors to diagnose illnesses and provide better service to doctors.

Acknowledgement. The paper is partly supported by the National Natural Science Foundation of China under Grant Nos. 61672181, 61370084, 61272184 and 61202090, Natural Science Foundation of Heilongjiang Province under Grant No. F2016005.

References

1. Li, J., Zou, Z., Gao, H.: Mining frequent subgraphs over uncertain graph databases under probabilistic semantics. VLDB J. **21**, 753–777 (2012)
2. Wang, R.: Image Understanding, pp. 25–30. National Defense University of Science and Technology Press, Changsha (1995)
3. Luo, S.: Image guidance technology and application. World Med. Equip. **7**(5), 22–27 (2001)
4. Guo, J., Ma, Z.: The significance of medical image processing in the development of medical research. J. Psychiatry **6**(2), 42–43 (2012)
5. Elguebaly, T., Bouguila, N.: A hierarchical nonparametric Bayesian approach for medical images and gene expressions classification. Soft. Comput. **19**(1), 189–204 (2015)
6. Naik, J., Patel, S.: Tumor detection and classification using decision tree in brain MRI. Int. J. Comput. Sci. Netw. Secur. (IJCSNS) **14**(6), 87 (2014)
7. Wang, J., Li, G., Kraska, T.: Leveraging transitive relations for crowdsourced joins. In: SIGMOD 2013, vol. 108, no. 1–2, pp. 133–147 (2013)
8. Demartini, G., Difallah, D.E., Cudré-Mauroux, P.: ZenCrowd: leveraging probabilistic reasoning and crowdsourcing techniques for large-scale entity linking. In: WWW, pp. 469–478 (2012). Huang, C.W., Lin, K.P., Wu, M.C., et al.: Intuitionistic fuzzy c-means clustering algorithm with neighborhood attraction in segmenting medical image. Soft Comput. **19**(2), 459–470 (2015)
9. Wais, P., Lingamneni, S., Cook, D., Fennell, J., Goldenberg, B., Lubarov, D., Marin, D., Simons, H.: Towards building a high-quality workforce with mechanical turk. In: Proceedings of Computational Social Science and the Wisdom of Crowds (NIPS), pp. 1–5 (2010)
10. Wang, J., Kraska, T., Franklin, M.J., Feng, J.: CrowdER: crowdsourcing entity resolution. PVLDB **5**(11), 1483–1494 (2012). Menon, N., Ramakrishnan, R.: Brain tumor segmentation in MRI images using unsupervised artificial bee colony algorithm and FCM clustering. In: 2015 International Conference on Communications and Signal Processing (ICCSP), pp. 0006–0009. IEEE (2015)
11. Rong, J., Pan, H., et al.: Medical image multi-stage cassification algorithm based on the theory of symmetric. Chin. J. Comput. **38**(9), 1810–1821 (2015)
12. Pan, H., Li, P., Li, Q., Han, Q., Feng, X., Gao, L.: Brain CT image similarity retreval method based on uncertain location graph. IEEE J. Biomed. Health Inform. **18**(2), 574–584 (2013)
13. Quddus, A., Basir, O.: Semantic image retrieval in magnetic resonance brain volumes. IEEE Trans. Inf Technol. Biomed. **16**(3), 348–355 (2012)
14. Ruppert, G.C.S., Teverovskiy, L., et al.: A new symmetry-based method for mid-sagittal plane extraction in neuroimages. In: Proceedings of the 2011 IEEE International Symposium on Biomedical Imaging, Chicago, USA, pp. 285–288 (2011)
15. Kropatsch, W.G., Torres, F., Ramachandran, G.: Detection of brain tumors based on automatic symmetry analysis. In: Proceedings of the 18th Computer Vision Winter Workshop, Hernstein, Austria, pp. 4–6 (2013)
16. Tuzikov, A.V., Colliot, O., Bloch, I.: Evaluation of symmetry plane in 3D MR brains images. Pattern Recogn. Lett. **24**(14), 2219–2223 (2003)

A New Method for Medical Image Retrieval Based on Markov Random Field

Tiaodi Wang, Haiwei Pan$^{(\boxtimes)}$, Xiaoqin Xie, Zhiqiang Zhang, and Xiaoning Feng

College of Computer Science and Technology, Harbin Engineering University,
Harbin 150001, China
panhaiwei@hrbeu.edu.cn

Abstract. The development of medical images acquisition and storage technology has led to the rapid growth of the relevant data. Retrieval of similar medical images can effectively help doctors to diagnose diseases more accurately. But because of the particularity of medical images, traditional content-based image retrieval (CBIR) method such as bag-of-words (BOW) cannot be applied to medical images. For example, when retrieving a diseased image, we should not only consider the similar characteristics but also need to consider the type of lesion. And for medical images, images with the same lesion may have different image features, similar images may have different types of lesions. In this paper, a Markov random field (MRF) is structured, and an approximate belief propagation algorithm is used to retrieval images. An adjust-ranking step after initial retrieval is incorporated to further improve the retrieval performance. This paper uses the real brain CT images. The experimental results show that the proposed method can significantly improve the retrieval accuracy and has good efficiency.

Keywords: Medical image retrieval · Markov random field · Belief propagation

1 Introduction

Over the past decades, the medical imaging technology such as computed tomography (CT) and magnetic resonance imaging (MRI) help doctors to diagnose diseases with the medical images. Hence, a vast amount of brain CT images are generated from hospitals every year and computerized methods are needed to solve this problem of the increasing amount of data.

A medical image not only contains information about the picture itself, but also indicates a series of treatments for a particular patient. By using image-to-image comparison doctors will find similar images that come from different patients. These patients may have a high probability of getting same disease because they have the similar pathological characteristics in their images. According to this domain knowledge, we believe that finding similar images from the medical image set will significantly assist doctors in finding patients who may get the same disease. And it will also

© Springer Nature Singapore Pte Ltd. 2017
B. Zou et al. (Eds.): ICPCSEE 2017, Part I, CCIS 727, pp. 447–461, 2017.
DOI: 10.1007/978-981-10-6385-5_38

help doctors to make diagnoses by acquiring information from the previous diagnoses and results.

However, it is a challenging problem to find a suitable target image by searching a large number of image set. Now the most common method for image retrieval is content based image retrieval (CBIR).

The use of CBIR technology to determine the similarity among images is mainly divided into two steps. Firstly, the image feature vector is got by the feature extraction method (e.g., histogram, Haralick, Haar Wavelets, and Daubechies Wavelets et al.). Secondly, metric distance (e.g., City-Block, Euclidean, Chebyshev, Jeffrey Divergence, and Canberra et al.) is adopted to get the distance among the image feature vectors. The distance is used to represent the degree of similarity.

But because of the particularity of medical images, medical image similarity retrieval is different from general image similarity retrieval [1, 2].

Medical image similarity retrieval requires high accuracy, little changes of the size or the location of the object can lead to different results. For example, given several medical images which are shown in Fig. 1. Figure 1(a) is two images of cerebral hemorrhage. Although they are the same pathological, they have very different image features. The image on the left in Fig. 1(b) is cerebral hemorrhage, and the image on the right in Fig. 1(b) is normal. Although they are different types of pathological, but their image features are largely same.

a   b

Fig. 1. Sample images. (a) Cerebral hemorrhage images. (b) The image on the left is cerebral hemorrhage, and the image on the right is normal brain.

Because of the special nature of medical images, the result is not satisfactory when the CBIR is used in medical image retrieval. But some other work can be done to boost the retrieval performance, such as improving feature extraction method or using the matching relationships among image set images. To address the feature representation problem, Tian et al. [3] proposed an edge oriented difference histogram (EODH) descriptor, which was combined with color-SFIT descriptors in their image retrieval system. In [4], Li et al. proposed an image retrieval framework by concatenating image global features-based similarities with multiple learnt ranking features.

In this paper, on the one hand, the relationships between the query image and the image set images are obtained. On the other hand, the relationships among the image set images are obtained. At the same time, two relationships are considered to get the final result. Our work is similar to method introduced in [5]. As to be explained in the following sections in more detail, the key difference between our work and [5] is that the construction way of Markov random field (MRF) is different.

The main contributions of this paper are the following:

By integrating the relationships between the query image and the image set images and the relationships among image set images, we use a MRF based principled probabilistic framework to boost the image retrieval performance.

An approximate belief propagation (BP) algorithm is presented to effectively compute the beliefs of image set images which represent the similarity between query image and image set images. The value of belief is larger, the similarity is higher.

An adjust-ranking scheme is proposed to further improve the retrieval accuracy by utilizing both relationships between the query image and the image set images and the beliefs of the image set images.

The rest of the paper is organized as follows. Related work is introduced in Sect. 2. The MRF based medical image retrieval method is explained in detail in Sect. 3. Experimental results are presented in Sect. 4. Finally this paper is concluded in Sect. 5.

2 Related Work

Up to now, there are two categories of image retrieval methods: (1) description-based image retrieval, which is performed using object retrieval based on image descriptions (such as keywords et al.), and (2) content-based image retrieval (CBIR), which supports retrieval based on the image content [6].

Tao et al. suggested a local point-indexed representation for all images, where the exponential similarity measure was used to calculate the relevance of two local descriptors [7]. This local representation can facilitate the retrieval tasks with very large vocabularies.

With the successes of deep learning techniques in computer vision and machine learning areas, they also provided many encouraging results for the content-based image retrieval, especially for effective feature representations [8].

In order to avoid the false rejection problem introduced from the vocabulary quantization in traditional BOW based methods, Avrithis [9] proposed a dimensionality-recursive quantization method that can be used for image retrieval efficiently. In [10], Zheng et al. proposed a Bayesian based merging scheme, which reduced the correlation between vocabularies by calculating the cardinality ratio of intersection and union sets from indexed features list. In [11], Qin et al. presented a probabilistic framework to measure the similarity between features where a new cost function was also proposed in this paper to score the similarity between images. Inspired by this work, Yang et al. also proposed an adaptive similarity function to replace Euclidean distance or Gaussian kernel which was used by nearest neighbor (NN) search for image retrieval [12].

Different from previously described basic image retrieval methods that concentrate on querying using a single image, multiple-queries based approaches tackle the problem of finding all images of an object by using multiple images. By combining multiple queries in a principled manner, Arandjelovic and Zisserman [13] described a framework where the query images are obtained by using textual Google image search. In [14], Fernando and Tuytelaars designed a set of mid-level patterns based on the images retrieved by the Google image search engine. These patterns were used for improving the retrieval performance. Most recent studies used the matching relationships among dataset images to boost the retrieval performance. Based on the initially retrieved images, a re-ranking method was proposed by Shen et al. [15] where a voting-based approach was used to evaluate the similarity measure between K-nearest neighbors (K-NN) of the query. Chen et al. [16] applied a ranking consistency verification procedure to refine the existing ranking list for retrieval.

Lu et al. [5] presented a MRF based principled probabilistic framework to solve the problem of image retrieval. By integrating the similarity between the query image and the dataset images in feature space and the relationships among images in the dataset, the proposed framework can overcome the retrieval failures caused by the differences of image conditions.

Inspired by [5], a MRF based medical image retrieval method is introduction in this paper.

3   MRF-Based Retrieval Method

A system architecture of our approach is shown in Fig. 2. In the offline process, features of image set images are extracted and saved. In the online process, features of the query image are also extracted. All features are used to construct the MRF. Retrieval in MRF is achieved by BP to get the beliefs of all image set images. At last, an image adjust-ranking scheme is used to improve the retrieval accuracy. In this paper, MRF constructing and retrieval are the cores of this paper, which will be introduced in detail in the following sections.

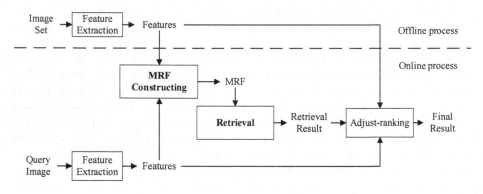

Fig. 2. The system architecture

3.1   Feature Extraction

For feature extraction, we choose Histogram of Oriented Gradients (HOG) [17]. HOG operates by first computing the image gradients in local cells in the horizontal and vertical directions of an image. The orientation for each pixel is binned into evenly divided orientation channels spreading from 0° to 180° or 0° to 360°. Since HOG is a local-region based descriptor, it tolerates some geometric and photometric variations.

Compared to other features such as color histograms, HOG features are less sensitive to illumination change.

To extract image features by HOG: the Gamma method is used to standardize the color space of the grayscale image; the gradient of each pixel of the image is calculated; the image is divided into $m * m$ scale cells; the gradient histogram of each cell is got to describe cell and can be described as n-dimensional vector according to the number of orientation bins set; k number of cells are composed of a block, so the vector dimension of block is $n * k$, and the block is normalized; Block moves in image with s as step for traversal. So the image can be described by multiple connected block vector.

3.2   Multi-strategy MRF Constructing Method (MMRFC)

Given a query object image I_q and an image dataset V, which is represented by an undirected graph $G(V, E)$, the number of V is $|V|$. Each image set images can be denoted by an element v of the vertex set V, and E which connect two elements v represent the relationships among image set images.

Let $X = \{x_1, x_2, \ldots x_{|V|}\}$ represent a set of variables in set V, where x_i indicates image set image I_i. Suppose $Y = \{y_1, y_2, \ldots y_{|V|}\}$ are the observations of the set V. In MRF, variable $\psi_{uv}(x_u, x_v)$ is used to indicate the relationship between x_u and x_v, and $\phi_v(x_v, y_v)$ is a joint compatibility function that measures the relationship between x_v and its observation y_v at vertex v. In this paper, an example which indicates medical images MRF is shown in Fig. 3.

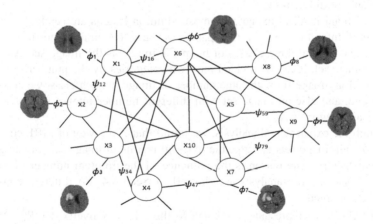

Fig. 3. Subgraph of MRF for medical images

S which indicates correlation coefficient between vectors is used to measure the relationship between image pair I_u and I_v

$$S_{uv}(I_u, I_v) = \frac{cov(U, V)}{\sigma_U \sigma_V} = \frac{\sum_{i=1}^{n}(U_i - \overline{U})(V_i - \overline{V})}{\sqrt{\sum_{i=1}^{n}(U_i - \overline{U})^2}\sqrt{\sum_{i=1}^{n}(V_i - \overline{V})^2}} \tag{1}$$

where $U = (U_1, U_2, \ldots U_n)$ represents the feature vector of image I_u by HOG and n is the vector dimension, $V = (V_1, V_2, \ldots V_n)$ represents the feature vector of image I_v by HOG and n is the vector dimension, $cov(U, V)$ represents the covariance of U and V, σ_U represents the standard deviation of vector U, σ_V represents the standard deviation of vector V, $\overline{U}$ represents the mean value of vector U, $\overline{V}$ represents the mean value of vector V. From Eq. (1), we can see that the size of $S_{uv}(I_u, I_v)$ is between 0 and 1.

The main focuses of constructing Markov random field are to get the variables $\psi_{uv}(x_u, x_v)$ and $\phi_v(x_v, y_v)$, as shown in Fig. 3. In this paper, $\phi_v(x_v, y_v)$ which indicates the relationship between x_v and y_v can be defined by

$$\phi_v(x_v, y_v) = S_{vq}(I_v, I_q) \tag{2}$$

where I_q indicates the query image, I_v indicates the image set image I_v. $\psi_{uv}(x_u, x_v)$ means the relationship between x_u and x_v and can be defined by

$$\psi_{uv}(x_u, x_v) = \begin{cases} 0, & u = v \\ \beta S_{uv}(I_u, I_v), & T_u = T_V \\ (1 - \beta)S_{uv}(I_u, I_v), & T_u \neq T_v \end{cases} \tag{3}$$

where T_u indicates the lesion class of I_u, T_v indicates the lesion class of I_v, I_u and I_v indicate image set images, β is a factor in the range of [0,1] which indicates the degree of impact of different classes. According to experience, the value of β should be larger than 0.5 which indicates the degree of relationship among same class images is larger than that among different classes.

For real brain medical images, the most common lesions are cerebral hemorrhage and cerebral infarction. So in this paper, we use cerebral hemorrhage, cerebral infarction, and normal three classes of brain CT image as the image set. And further, for a real image set, it can be divided into two categories by the number of images in each class. If an image set with same number of image in each class, we classify it as category one, and else an image set with different number images in each class, we classify it as category two.

Through the experimental analysis, we found that if the way of MRF constructing (MRFC) for all image sets is same, the retrieval results in image sets of category two are not satisfactory. The reason is the influence of the different number of images in each class, and specific analysis are showed in Sect. 3.4. So different construction methods are adopted.

The MRF can be represented as A and R, the relativity matrix $A \in R^{|V|*|V|}$, whose element is $A_{uv} = \psi_{uv}(x_v, y_v)$, the relativity list $R \in R^{|V|}$, whose vth element is $R_v = \phi_v (x_v, y_v)$.

In this paper, for image set of category two, in order to eliminate the influence of the number and improve the retrieval effect: the value of β is taken as 1, and what is more important, A need to be normalized. But for image set of category one, the step of normalizing A is not necessary since without the influence of the number.

So, to construct MRF, after given an image set, the image set should be judged to determine the value of β. If the image set is category two, the value of β is 1, else the value of β is not 1. Specific analysis of the value of β is introduction in Sect. 4.

The overall approach of MMRFC is summarized in Algorithm 1:

> **Algorithm 1.** MMRFC Algorithm
> **Input:** features of image set, features of query image I_q, the value of β
> **Output:** A: relativity matrix, R: relativity list
> 1.   BEGIN
> 2.   FOR ($u \leftarrow 1$ to $|V|$)   // $|V|$ is the number of image in image set
> 3.   FOR ($v \leftarrow 1$ to $|V|$)
> 4. $A_{uv} = \psi_{uv}(x_v, y_v)$ // get relativity matrix by Eq. (3)
> 5.   END FOR
> 6.   END FOR
> 7.   IF (β = 1)   // determine the value of β
> 8.   FOR ($u \leftarrow 1$ to $|V|$)
> 9.   FOR ($v \leftarrow 1$ to $|V|$)
> 10. $A_{uv} \leftarrow A_{uv} / \sum_{n=1}^{|V|} A_{uv}$ // normalized
> 11.   END FOR
> 12.   END FOR
> 13.   END IF
> 14.   FOR ($v \leftarrow 1$ to $|V|$)
> 15. $R_v = \phi_v(x_v, y_v)$ // get relativity list by Eq. (2)
> 16.   END FOR
> 17.   RETURN A and R
> 18. END

For Algorithm 1, the implementation of MRFC can be divided into two parts: get A and get R. In order to get A, for each image set image, the image set needs to be traversed once, and if the value of β is 1, A need to be normalized. A is a matrix, the process of normalization is implemented by row as unit. The computation complexity of getting A is depended on the size of image set, which is $O(n^2)$, n is the number of image set images. R is a list which indicates the relationship between query image and image set images, the computation complexity of getting R is $O(n)$. Further analysis, the process of getting A can be completed in the offline process, the computation complexity of MMRFC Algorithm in the online process can be $O(n)$.

3.3 Retrieval Method Based on MMRF (RMMRF)

In Fig. 3, by taking Bayes rule, the similarity between query image and image set images can be indicated as the marginal probability $P(X|Y)$

$$P(X|Y) = \frac{P(Y|X)P(X)}{P(Y)} = \frac{P(X, Y)}{P(Y)} \qquad (4)$$

since the denominator $P(Y)$ is just a normalizing constant and remains unchanged, we can disregard it during the inference procedure and focus on the joint probability $P(X, Y)$.

In the framework of Markov random field, the joint probability of X and Y can be expressed as [18].

$$P(X, Y) = \frac{1}{Z} \prod_{u,v \in V} \psi_{uv}(x_u, x_v) \prod_{v \in V} \phi_v(x_v, y_v) \tag{5}$$

Z is a normalization constant.

Based on Eq. (5), the inference of the marginal probabilities for Markov random field can be achieved by using belief propagation (BP) method [19].

In the rest of this paper, the marginal probabilities are called as "beliefs", and the belief of image I_v is denoted as $b_v(x_v)$. To infer $b_v(x_v)$, Let the variable $m_{uv}(x_v)$ be a "message" from vertex u to vertex v, then the belief $b_v(x_v)$ at vertex v is proportional to the product of the joint compatibility function at vertex v and all the messages coming into vertex v:

$$b_v(x_v) \propto k\phi_v(x_v, y_v) \prod_{u \in N(v)} m_{uv}(x_v) \tag{6}$$

where k is a normalization constant and $N(v)$ means the neighbors of the vertices v.

In BP method, the message from vertex u to vertex v is iteratively updated according to

$$m_{uv}(x_v) \propto \sum_{x_u} \phi_u(x_u, y_u) \psi_{uv}(x_u, x_v) \prod_{k \in N(v) \backslash v} m_{ku}(x_u) \tag{7}$$

where $N(u)\backslash v$ means all the neighboring vertices of vertex u except v.

By taking Eq. (6) into Eq. (7), the message update rule can be rewritten as

$$m_{uv}(x_v) \propto \sum_{x_u} \psi_{uv}(x_u, x_v) \frac{\phi_u(x_u, y_u) \prod_{k \in N(u)} m_{ku}(x_u)}{m_{vu}(x_u)} = \sum_{x_u} \psi_{uv}(x_u, x_v) \frac{b_u(x_u)}{m_{vu}(x_u)} \tag{8}$$

Equations (6) and (8) can be used directly to compute the beliefs [18]. But in large-scale loopy graphs, the efficiency is low. For MRF, feedback messages are inevitable due to graph loops, where as they carry litter information [21]. Thus, we can drop denominator of Eq. (8) and obtain an approximation of message update rule:

$$m_{uv}(x_v) \propto \psi_{uv}(x_u, x_v) \cdot b_u(x_u) \tag{9}$$

In order to effectively calculate the belief for each vertex and avoid overflow, Eq. (6) can be modified by using summation instead of multiplication which leads to an approximation to the belief:

$$b_v(x_v) \propto \alpha\phi_v(x_v, y_v) + (1 - \alpha) \sum_{u \in N(v)} m_{uv}(x_v) \tag{10}$$

where α is a decay factor between 0 and 1.

Generally, the belief $b_v(x_v)$ of each vertex $v \in V$ can be obtained by iteratively updating Eqs. (9) and (10). A scheme is applied to update messages in a parallel manner, the following variables are defined to facilitate the computation of beliefs: vector $B \in R^{|V|}$ whose vth element is the belief of b_v, relativity list $R \in R^{|V|}$ whose vth element is $\phi_v(x_v, y_v)$ and relativity matrix $A \in R^{|V|*|V|}$, whose element is $\psi_{uv}(x_u, x_v)$. Thus, the beliefs of image set images can be calculated by the following matrix represent

$$B^{k+1} = (1 - \alpha)R + \alpha AB^k \tag{11}$$

Prior to belief propagation, B is initialized by $B = R$ and updated according to Eq. (11) by N times, where N determines how far a message can be propagated in the graph $G(V, E)$. So, the overall approach of RMMRF is summarized in Algorithm 2:

Algorithm 2. RMMRF Algorithm
Input: A: relativity matrix, R: relativity list α: decay factor, N: iterations,
Output: B: beliefs of vertex set V
1. BEGIN
2. $B^0 = R$ // initialization
3. FOR ($n \leftarrow 1$ to N)
4. $B^n = \alpha AB^{n-1} + (1 - \alpha)R$ // belief propagation by Eq. (11)
5. END FOR
6. RETURN B
7. END

After belief propagation, the similarity between query image and image set images can be indicated as beliefs B, the value of belief is larger, the similarity is larger. The computation complexity of RMMRF is O(n), n is the number of iteration.

3.4 An Optimal Method for RMMRF (ORMMRF)

In MRF, the retrieval results is obtained by ranking beliefs. The first result is the most similar image whose value of belief is largest. Suppose an image set with the number of images in a class is much larger than the number in other classes. According to Algorithm 2, the final larger beliefs of images will only be the images in this specific class in despite of the different query images. So the retrieval effect in image set of category two is not ideal. In order to solve this problem, the value of β is taken as 1 and A is normalized for image set of category two.

To further improve the retrieval results, an adjust-ranking process is applied by considering not only beliefs but also the relationships between query image and image set images by using features, so after Eq. (11), the adjust-ranking process is implemented by

$$S = \gamma R + (1 - \gamma)B^N \tag{12}$$

where B^N are the beliefs after n iterations of Eq. (11), vector $R \in R^{|V|}$ whose vth element is $\phi_v(x_v, y_v)$, and γ is a control factor between 0 and 1. $S \in R^{|V|}$ whose vth element indicates the similarity between query image I_q and image set image I_v.

But for Eq. (11), we find that if the relationships among images are relatively large, with the increase in the number of updates, the elements in B will increase rapidly and be far greater than the elements in R. The result is that the Eq. (12) will not make sense. Therefore, the normalization of B should be processed after each iteration to avoid this problem.

The overall approach of ORMMRF is summarized in Algorithm 3:

Algorithm 3. ORMMRF Algorithm
Input: A: relativity matrix, R: relativity list, α: decay factor, γ:control factor, N: iterations
Output: S: the value of similarity
1.   BEGIN
2. $B^0 = R$ //initialization
3.   FOR ($n \leftarrow 1$ to N)
4. $B^n = \alpha R + (1 - \alpha)AB^{n-1}$
5. $B^n = B^n / \sum B^n$ //use normalization to improve retrieval accuracy
6.   END FOR
7. $S = \gamma R + (1 - \gamma)B^N$ //adjust-ranking
8.   RETURN S
9.   END

The values of S are the final retrieval results, the degree of similarity of image set images can be ranked by the size of the corresponding elements in S. The computation complexity of ORMMRF is O(n), n is the number of iteration.

4   Experiments

4.1   Image Sets and Evaluation Metrics

In the experiments, two real brain CT images sets are used to evaluate our proposed method. The image sets consist of three classes of images which are cerebral hemorrhage images, cerebral infarction images, and normal images. In image set one, the number of three classes of images is same 800. In image set two, the number of cerebral hemorrhage images is 400, the number of cerebral infarction images is 800, and the number of normal images is 1200. For image set two, the value of β is 1.

To evaluate retrieval performance, three metrics are used to evaluate the retrieval performance: mean average precision (MAP), precision at K ($P@K$) for $K = 10$ and $K = 20$. Mean average precision is the mean of the average precision scores for each query image and is calculated by

$$MAP = \frac{1}{N}\sum_{i=1}^{N} \int_0^1 p_i(r)dr \qquad (13)$$

where N is the number of queries, and $p_i(r)$ is the precision at recall r for query image i. The precision at K is the ratio between relevant retrieved images and total retrieval images at a cut-off rank K.

4.2 Impact of Parameters

For HOG feature extraction, we choose the implementation with the number of orientation bins set to 9. Standardize the medical image size of 256 * 256, select the size range of 16 * 16 as a cell, each cell is a 9 dimensional vector, then select the 2 * 2 cell composed of a block and normalize the block, so each block can be a 4 * 9 = 36 dimensional vector. For traversal, horizontal and vertical direction can be moved 15 times, a total of 15 * 15 = 225 block can be obtained, so each medical image can be expressed by the vector 225 * 36 = 8100.

As a result of the different values of β according to different image sets, we conduct experiments in two cases: $\beta = 1$ and $\beta \neq 1$.

To evaluate the effects of the parameters α and γ, we set $N = 10$ in advance, and for the case of $\beta \neq 1$, we set $\beta = 0.7$ in advance. The MAP scores are measured on two cases using different values of α and γ, as shown in Figs. 4 and 5. From these figures, we can see that the MAP increases with the increase of γ while $\gamma < 0.7$, decreases with the increase of γ while $\gamma > 0.7$, with $\alpha = 0.2$, the value of MAP is the largest. After getting the values of α and γ, the impact of parameter β is shown in Fig. 6. The MAP increases with the increase of β while $\beta < 0.8$, decreases with the increase of β while $\beta > 0.8$. So in this paper, we set $\beta = 0.8$ for the case of $\beta \neq 1$, and $\alpha = 0.2$, $\gamma = 0.7$.

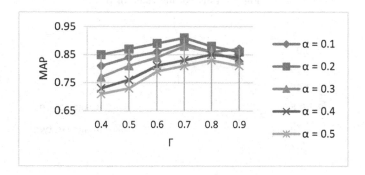

Fig. 4. Effects of α and γ with $\beta \neq 1$

For the number of iteration N, we also measure the MAP scores using different values of N, as shown in Fig. 7. From this figure, we can see that usually a higher MAP score is achieved by increasing the number of iteration N. Obviously, increasing N also increases the computational complexity. Thus, there is a tradeoff between the accuracy and efficiency. We set $N = 10$ in this paper.

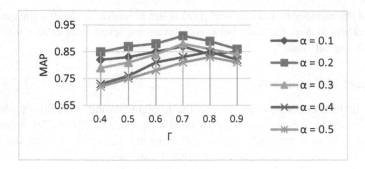

Fig. 5. Effects of α and γ with β = 1

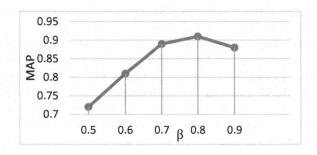

Fig. 6. Effect of the value of β

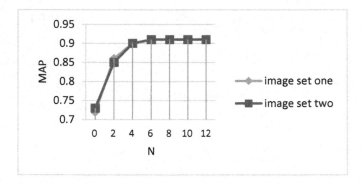

Fig. 7. Effect of iteration number N

4.3 Retrieval Results

With the same intention for image retrieval, we compare our method to the method of Zhang et al. [21].

Tables 1 and 2 show the performance comparison in terms of MAP, $P@10$, and $P@20$ for both image set one and image set two. Figure 8 shows two samples of retrieval results on image set two. One of the retrieval images is cerebral hemorrhage, and the other is cerebral infarction. We can see that the retrieval effect is accepted.

Table 1. Comparison on image set one

Method	Image set one		
	MAP	P@10	P@20
Method in [21]	0.81	0.89	0.83
Our method	0.91	0.92	0.86

Table 2. Comparison on image set two

Method	Image set two		
	MAP	P@10	P@20
Method in [21]	0.82	0.88	0.84
Our method	0.91	0.89	0.87

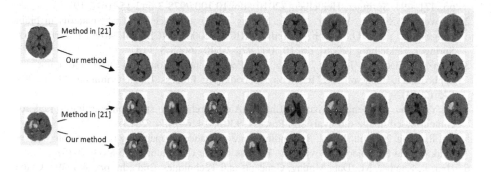

Fig. 8. Top 9 of the retrieval results with different methods

4.4 Retrieval Time

Given a query image, the time for online retrieval process mainly consist of three parts: feature extraction, retrieval in MRF, the top-k images are showed. We implement our method in Matlab on a laptop with Intel i7 2.2 GHz CPU and 16 GB RAM. On average for one query image, the feature extraction 0.12 s. On the image set one, the retrieval in MRF takes 0.14 s, and on the image set two it completes in 0.16 s. The time for showing top-k images is 0.56 s. Note that our current implementation is not optimized and further speed up can be achieved using parallel computing and/or C++ implementation.

5 Conclusions

In this paper, we address the problem of medical image retrieval by structuring MRF, an approximate belief propagation algorithm is used to retrieval images, and an adjust-ranking step after initial retrieval is incorporated to further improve the retrieval performance. The experimental results show that our method has a high precision for medical image retrieval with little time cost. This method can help doctors to diagnose diseases more accurately.

Acknowledgements. The paper is supported by the National Natural Science Foundation of China under Grant Nos. 61672181 and 51679058, Natural Science Foundation of Heilongjiang Province under Grant No. F2016005.

References

1. Müller, H., Deserno, T.M.: Content-based medical image retrieval. In: Deserno, T. (ed.) Biomedical Image Processing. Biological and Medical Physics, Biomedical Engineering, pp. 471–494. Springer, Heidelberg (2010). doi:10.1007/978-3-642-15816-2_19
2. Haiwei, P., Xie, X., Wei, Z., Li, J.: Mining image sequence similarity patterns in brain images. In: Yang, Q., Webb, G. (eds.) PRICAI 2006. LNCS, vol. 4099, pp. 965–969. Springer, Heidelberg (2006). doi:10.1007/978-3-540-36668-3_115
3. Tian, X., Jiao, L., Liu, X., Zhang, X.: Feature integration of EODH and color-sift: application to image retrieval based on codebook. Sig. Process. Image Commun. **29**(4), 530–545 (2014)
4. Li, Y., Zhou, C., Geng, B., Xu, C., Liu, H.: A comprehensive study on learning to rank for content-based image retrieval. Sig. Process. **93**(6), 1426–1434 (2013)
5. Lu, P., Peng, X., Zhu, X., Li, R.: Finding more relevance: propagating similarity on Markov random field for object retrieval. Sig. Process. Image Commun. **32**, 54–68 (2015)
6. Han, J., Kamber, M.: Data Mining: Concepts and Techniques, 2nd edn, pp. 396–399. China Machine Press, Beijing (2006)
7. Tao, R., Gavves, E., Snoek, C., Smeulders, A.: Locality in generic instance search from one example. In: IEEE Conference on Computer Vision and Pattern Recognition (CVPR), pp. 2099–2106 (2014)
8. Wan, J., Wang, D., Hoi, S.C.H., Wu, P., Zhu, J., Zhang, Y., Li, J.: Deep learning for content-based image retrieval: a comprehensive study. In: ACM Multimedia (2014)
9. Avrithis, Y.: Quantize and conquer: a dimensionality-recursive solution to clustering, vector quantization, and image retrieval. In: IEEE International Conference on Computer Vision (ICCV), pp. 3024–3031 (2013)
10. Zheng, L., Wang, S., Zhou, W., Tian, Q.: Bayes merging of multiple vocabularies for scalable image retrieval. In: IEEE Conference on Computer Vision and Pattern Recognition (CVPR), pp. 1963–1970 (2014)
11. Qin, D., Wengert, C., Van Gool, L.: Query adaptive similarity for large scale object retrieval. In: IEEE Conference on Computer Vision and Pattern Recognition (CVPR), pp. 1610–1617 (2013)
12. Yang, H., Bai, X., Zhou, J., Ren, P., Zhang, Z., Cheng, J.: Adaptive object retrieval with kernel reconstructive hashing. In: IEEE Conference on Computer Vision and Pattern Recognition (CVPR), pp. 1955–1962 (2014)

13. Arandjelovic, R., Zisserman, A.: Multiple queries for large scale specific object retrieval. In: British Machine Vision Conference (2012)
14. Fernando, B., Tuytelaars, T.: Mining multiple queries for image retrieval: on-the-fly learning of an object-specific mid-level representation. In: IEEE International Conference on Computer Vision (ICCV), pp. 2544–2551 (2013)
15. Shen, X., Lin, Z., Brandt, J., Avidan, S., Wu, Y.: Object retrieval and localization with spatially-constrained similarity measure and K-NN re-ranking. In: Proceedings of the IEEE Conference on Computer Vision and Pattern Recognition, pp. 3013–3020 (2012)
16. Chen, Y., Li, X., Dick, A., Hill, R.: Ranking consistency for image matching and object retrieval. Pattern Recogn. **47**(3), 1349–1360 (2014)
17. Dalal, N., Triggs, B.: Histograms of oriented gradients for human detection. In: IEEE Conference on Computer Vision and Pattern Recognition, vol. 1, pp. 886–893 (2005)
18. Li, S.Z.: Markov Random Field Modeling in Image Analysis. Advances in Pattern Recognition. Springer, London (2009)
19. Yedidia, J.S., Freeman, W.T., Weiss, Y.: Understanding belief propagation and its generalizations. Technical report (2001)
20. DiMaio, F., Shavlik, J.: Belief propagation in large, highly connected graphs for 3D part-based object recognition. In: Sixth International Conference on Data Mining, ICDM 2006, pp. 845–850 (2006)
21. Zhang, F., Song, Y.: Dictionary pruning with visual word significance for medical image retrieval. Neurocomputing **177**(C), 75–88 (2016)

Three-Dimensional Reconstruction of Wood Carving Cultural Relics Based on CT Tomography Data

Guiling Zhao[1,2], Zongji Deng[1], Jun Shen[1(✉)], Zhaowen Qiu[1(✉)], and Jing Huang[2]

[1] The Northeast Forestry University, Harbin, Heilongjiang, China
15776998153@163.com, 3256073977@qq.com
[2] The Northeast Agricultural University, Harbin, Heilongjiang, China

Abstract. Three-dimensional reconstruction of wood carving based on CT tomography data is important. In this paper, we propose a novel 3D variational framework for this task, which includes two procedures. First, a fitting approach is applied to a sequence of wood carving images acquired by CT scanner. The regions of interest (ROIs) can be obtained for the second segmentation after fitting. Second, we utilise a 3D TV (total variation) L1 variational model to directly segment the 3D volume of the ROIs. In addition, the TV-L1 model can smooth the volume and gives a clear 3D volume rendering result. By introducing a dual auxiliary variable, the fast primal-dual method is developed to improve the computational efficiency of the 3D TV-L1 model. Extensive experimental results are conducted to demonstrate the performance of the proposed framework.

Keywords: 3D wood carving segmentation · CT · Total variation · 3D

1 Introduction

CT scanning can obtain high precision 3D model without affecting the cultural relics [1]. Wood carving is a form of sculpture. On the basis of relief, the hollow out part of its background, generally can be divided into two: one is on the basis of relief, the hollow out part of its background, there can be sorted single-sided and double-sided carving. The two is a form between carvings and reliefs, also known as intaglio, hollow carving, or relief [2]. Because of wood carving Heritage structure is complex, the general effect of 3D scanner is not ideal. The author has used GOSAN scanning Wood carving crafts chromatic aberration of the wood part is little. It is needed to deal with by sticking point. But some precious cultural relics are not allowed to touch, even not stuck point. Using CT scanning can get the internal structure of cultural relics clearly. As the camphor wood carving scanning for example in this experiment, we not only get the same height of 3D model as the original, but also see material texture of wood inside clearly. Please note that the first paragraph of a section or subsection is not indented.

© Springer Nature Singapore Pte Ltd. 2017
B. Zou et al. (Eds.): ICPCSEE 2017, Part I, CCIS 727, pp. 462–471, 2017.
DOI: 10.1007/978-981-10-6385-5_39

2   The Two-Procedure Variational Segmentation Method

In this section, a fitting approach is first introduced by the 3D TV segmentation model and its fast primal dual implementation. In addition, the 3D discrete differential derivatives based on a finite difference scheme are also given (Fig. 1).

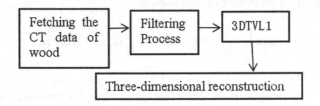

Fig. 1. The flow chart of 3D reconstruction of wood carving

2.1   Fitting the CT Data

There are many kinds of methods for nonlinear. The classical contains median filter, mean value filter and bilateral filter. In the median filter, the noise component is very difficult to choose, it will not affect the output results. But the speed is very slow. While the mean value filter is the average calculation for noise, output is influenced, Gaussian low-pass filter also produces the edge fuzzy phenomenon when eliminating noise smooth image. Bilateral filter is a compromise deal based on image space proximity and pixel value similarity. Considering spatial information and gray level similarity, it can be preserved the edge and eliminated the noise, and has the character for simple, non-iterative part [3, 4]. This paper uses bilateral filter for de-noising.

2.2   TV-L1 Segmentation

In this section, we first introduce the edge-weighted TV-L1 model for 2D image segmentation. Then we present 3D TV-L1 segmentation model and show how to implement the model using fast primal-dual method based on a finite difference scheme.

2.2.1   2D TV-L1
In [6] Bresson et al. Proposed the following edge-weighted total variation norm:

$$TV_g(u) = \int_\Omega g|\nabla u|dxdy \tag{1}$$

where g is an edge detection function, which can be chosen as $g = \exp\left(-\alpha|\nabla f|^\beta\right)$. In [6] the author suggests that $\alpha = 10$ and $\beta = 0.55$ suits well for different segmentation tasks. Theoretically, if u is a characteristic function 1_C (i.e. 1_C is a binary image and C denotes its boundary), this equation is equivalent to geodesic active contour (GAC) segmentation model proposed in [4]. However, the advantage of this model over GAC is that it can become a convex functional if variable u is relaxed to interval [0, 1]. This means one can find the global minimizer of Eq. (1) regardless of the initializations of u. The final segmentation can be then extracted from the optimized u by choosing a level set between [0, 1].

In used above edge-weighted TV norm (5) together with the L1 data fitting term for 2D image segmentation. When u is the characteristic function 1_C, the edge-weighted TV-L1 minimization problem can expressed as

$$\min_{u}\left\{ \lambda \int_{\Omega} g|\nabla 1_C|dxdy + \int_{\Omega} |1_C - f|dxdy \right\} \qquad (2)$$

where the variable $f \in [0, 1]$ is an input 2D image and λ is a smoothing parameter, controlling the smoothness of the output image u. Note that minimization problem is not strictly convex, indicating that more than one global minimum may exist. In addition, model (2) is computationally expensive due to two L1 norms used in the formulation. Moreover, it cannot be used to segment volume data. The three issues can be addressed in the next section.

2.2.2   3D TV-L1 Segmentation Model and Fast Primal-Dual Projection

We now extend (2) to the 3D case, which can be easily done by changing its integrated region from xy domain to xyz domain.

$$\min_{u \in [0,1]}\left\{ \lambda \int_{\Omega} g|\nabla u|dxdydx + \int_{\Omega} |u - f|dxdydx \right\} \qquad (3)$$

Note that u is no longer a characteristic function in (3) because it varies continuously between [0, 1]. Next, we introduce an auxiliary variable v to decouple (3) and reformulate it to become a strictly convex problem as follows

$$\min_{u \in [0,1],v}\left\{ \lambda \int_{\Omega} g|\nabla u|d\Omega + \int_{\Omega} |v|d\Omega + \frac{1}{2\theta} \int_{\Omega} |u + v - f|^2 d\Omega \right\} \qquad (4)$$

where $d\Omega = dxdydz$ and θ is a positive penalty parameter. Mathematically, if $\theta \to 0$, model (4) is exact the TV-L1 model (3). However, practice showed that the algorithm is robust even when θ is large. Note that the problem (4) has now become a convex optimization problem with respect to two variables u and v, meaning that it allows to compute the global minimizers of them. More importantly, (4) can be solved efficiently with a fast algorithm. To do so, an alternating minimization technique is performed on the two variables sequentially.

1. We solve the Eq. (4) for u by fixing v. The resulting energy functional to be minimized is

$$\min_{u \in [0,1]}\left\{ \lambda \int_{\Omega} g|\nabla u|d\Omega + \frac{1}{2\theta} \int_{\Omega} |u + v - f|^2 d\Omega \right\} \qquad (5)$$

This minimization problem is in fact exactly the ROF model [5, 6]. The only difference to the original ROF model lies in the edge-weighted TV norm. We can solve this problem using the fast primal-dual projection method proposed in [7] with an extrapolation step as follows

$$\begin{cases} \xi^{k+1} = \Pr oj_K\left(\xi^k - \sigma\lambda\nabla U^k\right) \\ u^{k+1} = f - v - \lambda\theta\nabla \cdot \xi^{k+1} \\ U^{k+1} = u^{k+1} + \alpha\left(u^{k+1} - u^k\right) \end{cases} \tag{6}$$

where ξ is the dual variable and σ is a time step which ensures that the scheme remains stable. In 3D $\sigma \leq 1/6$ dimension case to guarantee convergence of the algorithm. $\alpha \in (0, 2)$ is an over or under-relaxation factor and in the paper, we set $\alpha = 1$. $\Pr oj_K$ denote the back projection on set $K(i.e.\{x, y, z || \xi(x, y, z)| \leq 1\})$ and it is defined as $\Pr oj_K(\xi) = \frac{\xi}{\max(|\xi|, g)}$.

2. We solve the Eq. (4) for v by fixing u. The resulting energy functional to be minimized is

$$\min_v \left\{ \int_\Omega |v| d\Omega + \frac{1}{2\theta} |u + v - f|^2 d\Omega \right\} \tag{7}$$

Which can be easily solved by the one dimensional analytical shrinkage scheme below.

$$v = \begin{cases} f - u - \theta & \text{if} \quad f - u \geq \theta \\ f - u + \theta & \text{if} \quad f - u \leq -\theta \\ 0 & \text{if} \quad |f - u| \leq \theta \end{cases} \tag{8}$$

2.2.3 Discretization

In order to implement the fast primal-dual algorithm for the TV-L1 model, the 3D discrete first order gradient and divergence differential operators are required. In this paper, a forward-back word finite difference scheme is used for these two operators. Specifically, let $\Omega \to R^{MNL}$ denotes the 3D grid space of the size MNL. The forward discrete gradient of a volume u at a voxel i, j, k reads

$$\nabla u_{i,j,k}^+ = \left(\partial_x u_{i,j,k}^+, \partial_y u_{i,j,k}^+, \partial_z u_{i,j,k}^+ \right)$$

Where

$$\partial_x u_{i,j,k}^+$$
$$\partial_x u_{i,j,k}^+$$
$$\partial_x u_{i,j,k}^+$$

The 3D divergence is computed with a backward scheme: for the dual vector $\zeta = \left(\xi^1, \xi^2, \xi^3\right)$, we have

$$(div\xi)_{i,j,k} = \begin{cases} p_{i,j}^1 & j=1 \\ p_{i,j}^1 - p_{i,j-1}^1 & 1<j<N \\ -p_{i,j-1}^1 & j=N \end{cases} + \begin{cases} p_{i,j}^2 & i=1 \\ p_{i,j}^2 - p_{i-1,j}^2 & 1<i<M \\ -p_{i-1,j}^2 & i=M \end{cases} + \begin{cases} p_{i,j}^2 & i=1 \\ p_{i,j}^2 - p_{i-1,j}^2 & 1<i<M \\ -p_{i-1,j}^2 & i=M \end{cases}$$

Having defined all necessary discrete quantities, the numerical optimization for two sub-problems (5) and (8) can be implemented practically. For clarity, we present the flow chart of the primal-dual projection for the 3D TV-L1 model is shown in Table 1. Note that there is no need to exactly solve this sub-optimization problem (1). On eiteration $(k = 1)$ of this scheme is sufficient to make the entire algorithm converge. Note that as the input f has been stretched into [0, 1], the constraint $u \in [0, 1]$ can be omitted. To obtain the final segmentation, a level set of u is selected as $u = 0.5$.

Table 1. Primal-dual projection algorithm for TV-L1 model

Steps	Method
1.	Initialization: linear stretch f to [0,1], $v^0 = f$, $\xi^0 = 0$ $v^0 = f$ and $u^0 = 0$
2	Repeat
3	Compute dual variable ξ using (10)
4	Compute primal variable u using (10);
5	Compute extrapolating variable U using (10);
6	Compute v using (12);
7	Until the stopping criterion is satisfied.

3 Three-Dimensional Reconstruction

Three-dimensional reconstruction of wood carving [8] refers to using computer image processing technology, rebuilding the wood carving or wood carving part which is gained by industry CT into three-dimensional model. Researchers can observe reconstruction models in different angles and keep in the data.

The 3D reconstruction of wood carving relic image is divided into two types: one is the volume rendering method; the other is a surface rendering method. Volume rendering is described as a series of "according to 3D scalar data, two-dimensional pictures can be generated" technology [9], surface rendering is one of the important minus for image 3D reconstruction. It can re-reduce inspected objects' 3D model and can show as surface form.

The standardized DICOM3.0 data which comes from the computed tomography images can use ripe arithmetic to make 3D reconstruction. For example, VTK based MC algorithm can achieve wood carving's 3D reconstruction.

3.1   Marching Cube

MC algorithm is a classical algorithm of surface rendering, and the basic principle is the formation of contour plane [8, 10, 11] specifically, an individual data is divided into a plurality of small cubes; Each cube has 8 vertices; the vertices are in the internal or external of objects. Observe all cubes and find out the intersection of the cube and the subject's surface. If a cube has vertices in inside and outside of the object, we can determine the intersection. The subsurface triangles can be found from a predefined state table. Because a cube has 8 vertices, and each only has two possible (in the internal or external of the object), it total has 28 = 256 state. However, considering the reflection rotation, the 256 state can be reduced to 15 states, as is shown in Fig. 2. According to the triangle in Fig. 2, the objects contour plane can be drawn.

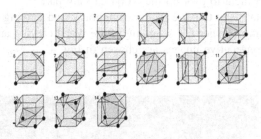

Fig. 2. The principle of MC

3.2   Ray Projection Algorithm

Beside the MC algorithm, ray projection algorithm is also widely used in 3D visualization body data. Figure 3 illustrates the principle of the algorithm: a ray comes from the line of sight, and it passes through the 2D image plane projects onto the 3D body data. The pixel value of the 2D image is determined by the color value and the transparency of each unit cell which is passed through the line. This phenomenon can be simulated by physical model.

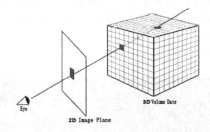

Fig. 3. The principle of volume rendering

VTK (Visualization Tool-kit) is a set of open source tool for visualization which launched by the United States Kit ware company in 1998. Its main function is image processing, computer graphics and 3D visualization [12, 13]. Use DICOM data to make VTK three-dimensional reconstruction. In the algorithm and data structure point of view, it can be seen as reading data and graphics display. VTK can classify many commonly used visualization algorithms of the 3D reconstruction, such as MC and ray projection method which can be directly used by users, and bring convenience to the user [14–20].

4 The Two-Procedure Variational Segmentation Method

In order to verify the effectiveness of the method, on the Window7 operating system, image segmentation experiments are made by using MATLAB tools. Experiments were performed to extract the selected part of CT wood carving crafts image data is chosen in the experiment to pick up the wood carving part.

The experimental data: the size of the camphor Wood carving is 25 cm * 25 cm, and no lacquer. Using 16 slice CT scanning of the wood carving artifacts, 320 images are sequenced, 8 consecutive images are chosen in Fig. 4.

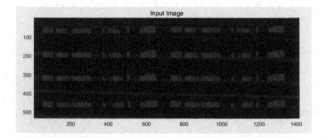

Fig. 4. The original noise CT wood engraved image artifacts

According to observation, the acquisition of CT wood carving relic image exists a lot of pseudo contour points and gray level information distribution is not uniform. Using the total variation filter to remove noise and preserve edge for CT wood carving, and chart 5 image results are gained. It is clear that the gray area of the wood carving become uniform and the organization boundary is clear and smooth.

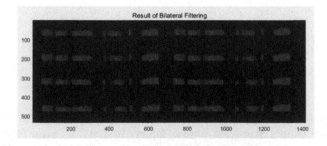

Fig. 5. The result of bilateral filtering

Figure 6 is the segmentation result for Fig. 5 using the 3DTVL1. Using 3DTVL1 algorithm can separate and wood carving and non-wood Wood carving.

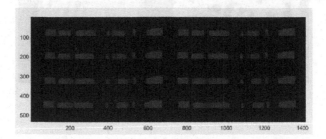

Fig. 6. The result by using 3DTVL1 segmentation

Finally, using the medical imaging system which is independent researched and based on VTK Library to build body rendering and surface rendering 3D reconstruction. The 3D reconstruction results are shown as Figs. 7 and 8 [21, 22].

The SS2 3D printer which is more accuracy, fast printing is used in this experiment.

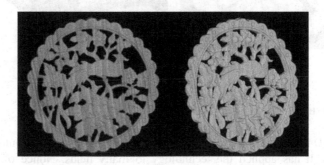

Fig. 7. The result of volume rendering

5   3D Printing

The 3D printer contains many kinds. The SS2 3D printer which is used in this experiment is more accurate and faster in printing.

The 3D printing result is shown below. Time of the 3D reconstruction is 10 min, and print time is 1 h. The 3D model of the wood carving is shown in Fig. 9. The Print specific parameters are as follows Table 1.

Fig. 8. The result of surface rendering

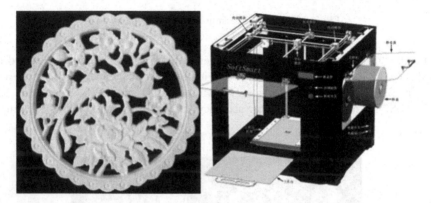

Fig. 9. The result of 3D printing and the 3D printer

6 Conclusion

At present, 3D reconstruction has been a hot area of research in computer technology and graphics. It has applied the technology to many fields. Nonce of a traditional Segmentation algorithm is used to get ideal division effect. In this paper, we use Pretreatment method and Segmentation comprehensively. In the pretreatment process, ideal image effects are gained. Then Segmentation of the image can obtain ideal segmentation results. At last we use classic reconstruction algorithm further get 3D reconstruction model. 3D reconstruction results are shown high accuracy. It breaks the laser to scan the need for cultural relics engraved with poor accuracy and point defects.

With the development of 3D printing materials, this technique will be applied to the restoration and reproduction of ancient carved relics.

References

1. Qiu, Z.W., Zhang, T.W.: Key techniques on cultural relic 3D reconstruction. Acta Electronica Sinica **32**(12), 2423–2427 (2008)

2. Lin, J.J.: Adhering to the tradition and achieve perfect realm-on the art characteristic of putian wood carving. Sculpture **3**, 70–71 (2013)
3. Yu, H.C., Zhao, L., Wang, H.X.: Image denoising using trivariate shrinkage filter in the wavelet domain and joint bilateral filter in the spatial domain. IEEE Trans. Image Process. **18**(10), 2364–2369 (2009)
4. Cai, C., Ding, M.Y., Zhou, C.P., Zhang, T.X.: Bilateral filtering in the wavelet domain. Acta Electronica Sinica **32**(1), 128–131 (2004)
5. Nikos, P., Rachid, D.: Active regions: a new paradigm to deal with frame partition problems in computer vision. J. Vis. Commun. Image Represent. **13**(1–2), 249–268 (2002)
6. Rousson, M., Deriche, R.: A variational framework for active and adaptative segmentation of vector valued images. In: Workshop on Motion and Video Computing, vol. 4, pp. 56–61 (2002)
7. Herbulot, A., Jehan-Besson, S., Barlaud, M., Aubert, G.: Shape gradient for multi-modal image segmentation using joint intensity distributions. In: International Workshop on Image Analysis for Multimedia Interactive Services, vol. 4, no. 4, pp. 2729–2732 (2004)
8. Li, C., Huang, R., Ding, Z., Gatenby, C., Metaxas, D., Gore, J.: A variational level set approach to segmentation and bias correction of images with intensity inhomogeneity. In: Metaxas, D., Axel, L., Fichtinger, G., Székely, G. (eds.) MICCAI 2008. LNCS, vol. 5242, pp. 1083–1091. Springer, Heidelberg (2008). doi:10.1007/978-3-540-85990-1_130
9. Li, C.M., Kao, C.Y., Gore, J.C., Ding, Z.H.: Minimization of region-scalable fitting energy for image segmentation. IEEE Trans. Image Process. **17**(10), 1940–1949 (2008)
10. Duan, L.M., Liu, Y.B., Wu, Z.F.: Method of reconstructing 3-D CAD model based on industrial computed tomography. Comput. Integr. Manuf. Syst. **15**(3), 479–486 (2009)
11. Rizzo, G., Castiglioni, I., Russo, G.: Data rebinning and reconstruction in 3-D PET/CT oncological studies: a Monte Carlo evaluation. IEEE Trans. Nucl. Sci. **53**(1), 139–146 (2006)
12. Li, Z., Jiang, S., Yang, X.: ZSU-E-T-279: realization of three-dimensional conformal dose planning in prostate brachytherapy. Med. Phys. **41**(6), 288 (2014)
13. Du, J.-J., Yang, X.-Y., Du, Y.-J.: From medical images to finite grids system. In: Proceedings of Annual International Conference of the IEEE Engineering in Medicine and Biology Society, vol. 2, pp. 1630–1633 (2005)
14. Wolf, I., Vetter, M., Wegner, I.: The medical imaging interaction toolkit. Med. Image Anal. **9**(6), 594–604 (2005)
15. Zhou, Y., Zou, C.H., Zhang, C.Y.: Evaluation of bone structure of hip joint using three-dimensional visualization system. Chin. J. Tissue Eng. Res. **18**(4), 601–606 (2014)
16. Lu, X.Q., Ren, X.Y., Jia, D.Z.: Image reading, writing and display for DICOM files based on ITK, VTK and MFC. J. Clin. Rehabil. Tissue Eng. Res. **15**(13), 2416–2420 (2011)
17. Myong, H.K., Ahn, J.K.: Development of a post-processing program for flow analysis based on the object-oriented programming concept. Trans. KSME, B **32**(1), 62–69 (2008)
18. Yan, R.G., Guo, X.D., Xu, C.Q.: Visualization toolkit-based three-dimensional model of the colorectal segment. J. Clin. Rehabil. Tissue Eng. Res. **14**(52), 9807–9811 (2010)
19. Yu, W.W., XI, P., He, F.: Study and realization for 3D reconstruction of medical models based on VTK and MFC **30**(4), 125–130 (2009)
20. Hong, T., Pan, Z.F., Lin, L.B., Yang, L.L., Shen, Q.Q.: Application and realization of 3D reconstruction of medical images of VTK **20**(4), 127–130 (2011)
21. Hou, H.L., Wang, M.Q., Ren, S.Q.: Research for improved surface rendering and volume rendering algorithms based on ICT data field. Comput. Eng. Appl. **50**(23), 172–175 (2014)
22. Niklas, M., Bartz, J.A., Akselrod, M.S.: Ion track reconstruction in 3D using alumina-based fluorescent nuclear track detectors. Phys. Med. Biol. **58**(18), N251–N266 (2013)

Text Feature Extraction and Classification Based on Convolutional Neural Network (CNN)

Taohong Zhang[1,2](✉) (iD), Cunfang Li[1,2] (iD), Nuan Cao[1] (iD),
Rui Ma[1,2] (iD), ShaoHua Zhang[1] (iD), and Nan Ma[3] (iD)

[1] University of Science and Technology Beijing, Beijing, China
watersword zth@163.com
[2] Beijing Key Laboratory of Knowledge Engineering for Materials Science,
Beijing, China
[3] Robotics Institute of Beijing Union University, Beijing, China

Abstract. With the high-speed development of the Internet, a growing number of Internet users like giving their subjective comments in the BBS, blog and shopping website. These comments contains critics' attitudes, emotions, views and other information. Using these information reasonablely can help understand the social public opinion and make a timely response and help dealer to improve quality and service of products and make consumers know merchandise. This paper mainly discusses using convolutional neural network (CNN) for the operation of the text feature extraction. The concrete realization are discussed. Then combining with other text classifier make class operation. The experiment result shows the effectiveness of the method which is proposed in this paper.

Keywords: Convolutional neural network (CNN) · Text feature extraction · Class operation

1 Introduction

With the rapid development of web, the Internet has been integrated into the various fields of our daily life. The Internet influence our life all the time. The ways of the information of the Internet are varied such as BBS, microblogging, blog and so on. People like to express their views through the network and publish commercial information. So the information of the Internet contains a large number information of expressing users view, mood and emotion.

The main method of the traditional text sentiment analysis is the method which based on rules and based on statistics. Rule-based methods are mainly concentrated rule-making. This part of the job may require a lot of manual operation. When the language phenomenon is a lot and very complicated in the text, process of the rule-making becomes very difficult. The current commonly used method is based on the statistical method. The process of based on statistical methods is learning characteristics of samples in the first place. Then according to the features of extraction

© Springer Nature Singapore Pte Ltd. 2017
B. Zou et al. (Eds.): ICPCSEE 2017, Part I, CCIS 727, pp. 472–485, 2017.
DOI: 10.1007/978-981-10-6385-5_40

operate the subsequent text mining. Text emotional classification can be divided based on the lexical level, sentence level, document level and so on. The length of text emotional classification which is based on the sentence level is moderate, so this paper mainly research on text emotional classification based on the sentence level.

The first to use Convolutional Neural Network model (CNN) is the papers such as the references [1, 2] whose writer is Collobert in the field of natural language processing Organization of the Text. Maas [3] and others use the training corpus which has emotional tagging and has marked with good training corpus. The training corpus combine supervised learning and unsupervised to build a language model and obtain the certain effect. Tomas Mikolov [4] and others expanded continuous Skip-Gram which is in continuous word bag model in 2013. Then word2vec which is a computing word vector deep learning tool was released.

Kalchbrenner et al. [5] and other researchers introduced the dynamic CNN model and the accuracy of binary classification and multi-classification at Stanford Sentiment Treebank test set had obtained the good effect in 2014. From these studies we can see that neural network model has great potential in the field of natural language processing. Kim [6] and others constructed CNN language model in 2014. After the training model, they tested the test data. Accuracy of the result of classification don't lose to the accuracy of traditional text emotional classification. Zhang et al. [7] and other researchers split each word in the sentence into Single character. Then Character vector was generated into CNN and finally they had obtained good result. Facebook team [8] did something that deep CNN model was applied to natural language processing tasks in 2016. And through a number of different text corpus they had tested and finally they had achieved good effect.

2   CNN Model

CNN uses Convolutional layer to learn local features. In natural language processing, the CNN model has achieved very good results in many ways, such as grammar analysis, search for words, sentence modeling, and other traditional natural language tasks. CNN model that Kim's proposed has achieved good results, but for the text emotional classification, the traditional CNN model in the Pooling layer will lose useful features, so this paper presents MaxTop-k-Pooling to retain more features. In order to make full use of the extracted features, this paper replaces the last softmax classifier of CNN model with other commonly used text emotional classifiers to obtain better results.

The CNN model used in this article executes a convolutional using three convolutional windows of different length. Besides, in order to keep more effective features, we come up with the MaxTop-k-Pooling operation in the pooling layer (Fig. 1).

The input shown in the graphic is a matrix of 8 * 6, indicating that this text include 8 words and each word has a vector of 6 dimension. After the input layer followed the convolutional layer, we use two different convolutional windows to operate the convolution, one is 2 * 6, another is 3 * 6, and each has 5 filters, we make k in the MaxTop-k-Pooling equals 3 to get the most important features, at last, sent the features to the connection layer firstly, then to the classifier to proceed classification operation. To collect more significant features, we use three different kinds of convolutional

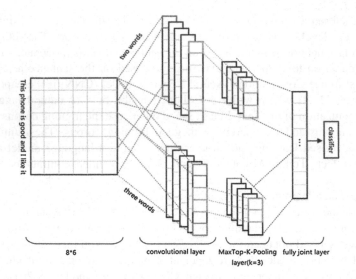

Fig. 1. CNN model of convolutional granularity based on Word mode

windows (sliding windows), what is more, we also use SVM, Naive Bayesian and Random Forests to compare with each other and get the final result. There are totally five layer in the CNN model: finding layer, input layer, convolutional layer, pooling layer, connection layer and classifier layer, we did not show the finding layer in the graphic, the main purpose of this layer is that vectoring the word in the sentence which was input so that we can matrix the sentences. Convolutional layer operation and MaxTop-k-Pooling layer operation are quite important operations in this model. We will explain these two operations respectively below.

2.1   Convolutional Layer Operation

Before the comment statements was generated into CNN model, in order to facilitate the input of CNN model, every word need to be divided into same length. For example, a sentence named s include N words. Let $s = w_1, w_2, \ldots, w_i, \ldots, w_N$. If the length of sentences is less than the regulation length of sentence which are intercepted named l (n < l), at the end of the sentences we fill specific placeholder and the length of the sentences is equal to l. If the length of the sentences is greater than l, we need intercept the sentences. Every word in the sentences which have expanded replace corresponding word vectors. Now, s is expressed as:

$$s = [v_{w_1}, \ldots, v_{w_n}, v_{z_1}, \ldots, v_{z_m}] \tag{1}$$

In formula (1), z_i is the specific placeholder, w_i is the word in the sentences, m is the number of the placeholder, n is the length of the sentences, l is equal to m plus n, $v_{w_i} \in \mathbb{R}^d$ is d-dimensional word vector, i is the ith word in the sentences. If convolutional windows is $w \in \mathbb{R}^{h \times d}$, h is the size of convolutional windows and the number of covering consecutive words in a convolutional operation, d is the dimension of the word vector, so a convolutional window in a convolutional computation will produce a feature which is named c_i, c_i is shown in formula (2).

$$c_i = f(w \cdot s_{i:i+h-1} + b) \qquad (2)$$

In formula (2), w and b is super-parameter in CNN model, f is the nonlinear activation function.

The activation function commonly used is ReLU, tanh, sigmoid and so on. $s_{i:i+m-1}$ is a matrix between ith and $(i + m-1)$th corresponding word vectors. As we all know, $s_{i:i+h-1}$ can be expressed as the formula (3). v_i is every word vector. $\oplus$ is a joint mark. Convolutional kernel that the size of the convolutional window is $d \times h$ can generate $(l - h + 1)$ number of a feature which is named c. The characteristic pattern c is described as the follows in formula (4).

$$s_{i:i+h-1} = v_i \oplus v_{i+1} \oplus v_{i+2} \cdots \oplus v_{i+h-1} \qquad (3)$$

$$c = [c_1, c_2, \ldots, c_{l-h+1}] \qquad (4)$$

In formula (4), Convolutional operation is similar to feature extraction methods based on N-Gram in traditional natural language processing and $c \in \mathbb{R}^{l-h+1}$. The size of the convolutional window in the model is h that is amount to N. In order to extract features from the different perspective and not increase the computational complexity, Kim's model used three different convolutional window.

So this paper also use three different convolutional window. Convolutional operation can extract local semantic characteristics from the sentence. Furthermore, a number of different size of convolutional operation can extract features from many different angles. As a result, multiple semantic combination can be extracted from many different angles.

2.2 MaxTop-k-Pooling Layer Operation

CNN's pooling layer can be regarded feature selection process aiming at the result of convolutional operation in the process of natural language processing. Pooling operation can retain more useful features. CNN's pooling layer include Max-Pooling and Average-Pooling which are commonly used. Max-Pooling is the largest value in the characteristic pattern c. We can think that this maximum value is the most effective feature. Average-Pooling is the average value in the characteristic pattern c. However, common NLP tasks use Max-Pooling way. Moreover, Zhang Ye [56] also thought that Max-Pooling is more suitable in text categorization. So this paper is based on Max-Pooling to do research. The eigenvalue $\hat{c}$ in the Max-Pooling is expressed as

$$\hat{c} = \max\{c_i\}, \ \forall i = 1, \ldots, (l - h + 1) \qquad (5)$$

From formula (5), we can see that the length of characteristic vector is no longer associated with the length of the sentence after Max-Pooling operation. At the moment, the effective characteristics of each sentence are unified. When CNN is applied to NLP, convolutional operation is amount to extract local N-Gram characteristics from the sentence. However, Max-Pooling can be regarded as to merge feature. So $\hat{c}$ is the most effective feature. Due to the convolutional windows of same size have a lot, this window after convolutional and pooling operation can be expressed as the follows in formula (6) in the corresponding eigenvectors of the pooling layer.

$$h = \hat{c}^0 \oplus \hat{c}^1 \oplus \ldots \oplus \hat{c}^t \tag{6}$$

In formula (6), pooling operation extract the most powerful features. This is a very good characteristic in the study of image. In the field of text can be a very general characteristic, because sequence and location of the features are important in the text. Max-Pooling discard the two features. Sometimes some strong features in the sentences will appear many times such as TF · IDF.

TF represents the number of occurrences of some characteristics. The more it appears, the stronger it becomes. However, Max-Pooling extract the strongest feature. If there are multiple important features that can only take one. It has lost some more important features. The characteristics missed may be important for the expression of text emotional intensity and language translation operations, which may affect the later effect.

This paper proposes MaxTop-k-Pooling operation in order to retain the first k characteristics of the most important characteristics in the eigenvalues and retain sequence in the original text. Through this operation can retain more effective information. There are t convolutional windows which are $d \times h$ dimension. The corresponding eigenvector $\hat{c}$ in these convolutional windows is calculated as:

$$\hat{C} = \max^{(k)}(c_i), \ \forall k < t, \ \forall i < (l - h + 1) \tag{7}$$

In formula (7), k is the first k characteristics of the most important characteristics in the eigenvalues. This value is less than the total number of features. The minimum of k is one. When k s equal to one, this operation is MaxTop-Pooling operation. Every convolutional window will produce several $\hat{c}$ matrix. The value of MaxTop-k which is extracted will make full join then put it into activation function (ReLU) and finally put it into softmax classifier to train.

3   The Training of CNN Model and the Choice of Parameters

Corpus used in this paper are random selected the part of reviews from Yelp2016 and Douban. Including: binary classification of corpus is the part of the corresponding on Yelp2016 and Douban which are "very poor" and "strongly recommended". The corresponding training corpus in both Chinese and English each have article forty thousand, which is the derogatory sense and good article 20000, article 8000 test data including derogatory sense and good each 4000. The five classification include five categories "very poor", "poor", "not too bad", "recommended", "strongly recommended" on Douban reviews and Yelp2016 corpus. For five classification, we randomly select 20000 comments of 100000 data and the corpus of equilibrium from corpus of Douban and Yelp2016 in each category, and then we randomly select 4000 data consisting of 20000 corpus as test data. Choose the parameter when macro average F_1 is maximum.

3.1 The Choice of Convolutional Window and the Number of Convolutional Unit

For the training of CNN Model, the more important structure in CNN is Convolutional layer and MaxTop-k-Pooling layer, so in this section this paper focus on the parameters of two structure. The value of dropout in CNN model choice 0.5 in this paper with reference to Kim's model, meanwhile, ReLU is chosen as the activation function in each layer in this paper. Iterations use EarlyStopping function in keras to detect when the training accuracy rate and loss rate are not to change, the training of model will terminated early to prevent over-fitting. In the training phase we use stochastic gradient descent method to update the model parameters.

This paper discusses the relationship between the size of convolutional window, the number of convolutional window and classification effect. This article uses the volume of data is not very large. So this article chooses three kinds of different lengths of convolutional window. In experiment, the value of k is one in MaxTop-k-Pooling layer. The dimension of the word vector is 50. The number of convolutional window is 100. Mini-batch is 50. Select different lengths of convolutional window to test. The relationship between the size of convolutional window and the results of final classification is shown in Fig. 2. The index of value evaluation use macro average F_1 (Macro_F_1).

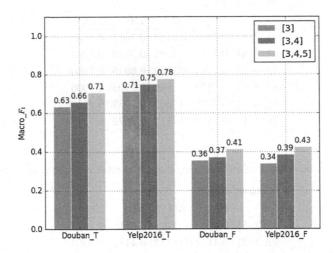

Fig. 2. The contrast of different convolutional sizes and convolutional number

In Fig. 2, "[3]" means covering three words in a convolution. "[3, 4]" have two different sizes of convolutional operation and a convolution covers three and four words. "[3–5]" have three different sizes of convolutional operation and a convolution covers three, four and five words. Douban_T and Yelp2016_T represent to extract the reviews of Douban and we classify the two types which are very poor and strongly recommended in the reviews of Yelp2016. Douban_F and Yelp2016_F represent to express the reviews of Douban and we are classify the five types which are very poor, poor, not too bad, recommended and strongly recommended in the reviews of Yelp2016. In Fig. 2, when the parameter of the types of convolutional window increase, Macro_F_1 will increase. The data of convolutional window and convolutional number is shown in Tables 1, 2, 3 and 4.

Table 1. Convolutional window and convolutional number of Douban binary classification

Window	Number			
	50	100	150	200
[1, 2, 3]	0.6898	0.6980	0.7061	0.6975
[2, 3, 4]	0.6866	0.7008	0.6991	0.7031
[3, 4, 5]	0.7065	0.6963	0.7104	0.6816
[4, 5, 6]	0.6953	0.6890	**0.7164**	0.7086
[5, 6, 7]	0.7098	0.6998	0.6909	0.7015
[6, 7, 8]	0.6944	0.7039	0.7135	0.6937

Table 2. Convolutional window and convolutional number of Yelp2016 binary classification

Window	Number			
	50	100	150	200
[1, 2, 3]	0.7398	0.7453	0.7613	0.7565
[2, 3, 4]	0.7656	0.7598	0.7929	0.7703
[3, 4, 5]	0.7781	0.7802	**0.8016**	0.7806
[4, 5, 6]	0.7533	0.7956	0.7903	0.7906
[5, 6, 7]	0.7208	0.7602	0.7939	0.7625
[6, 7, 8]	0.7568	0.7812	0.7672	0.7737

Table 3. Convolutional window and convolutional number of Douban five classification

Window	Number			
	50	100	150	200
[1, 2, 3]	0.4063	0.4198	0.4355	0.4173
[2, 3, 4]	0.4122	0.4233	**0.4459**	0.4060
[3, 4, 5]	0.4174	0.4053	0.4276	0.3920
[4, 5, 6]	0.3932	0.4301	0.4388	0.4272
[5, 6, 7]	0.4087	0.3929	0.4428	0.4366
[6, 7, 8]	0.4139	0.4176	0.4208	0.4029

Table 4. Convolutional window and convolutional number of Yelp2016 five classification

Window	Number			
	50	100	150	200
[1, 2, 3]	0.3992	0.4318	0.4422	0.4538
[2, 3, 4]	0.4157	0.4353	0.4196	**0.4633**
[3, 4, 5]	0.4236	0.4255	0.4337	0.4588
[4, 5, 6]	0.4066	0.4175	0.4592	0.4426
[5, 6, 7]	0.4202	0.4263	0.4459	0.4561
[6, 7, 8]	0.4083	0.4325	0.4281	0.4372

The influence of the result of classification is show in table above. As seen, In Table 1, when the convolutional window is [4–6] and convolutional number is 150, the value of Macro_F_1 is maximum. In Table 2, when the convolutional window is [3–5] and convolutional number is 150, the value of Macro_F_1 is maximum. In Table 3, when the convolutional window is [2–4] and convolutional number is 150, the value of Macro_F_1 is maximum. In Table 4, when the convolutional window is [2–4] and convolutional number is 200, the value of Macro_F_1 is maximum. Using these parameters set the parameters of corpus of corresponding model below.

3.2   The Choice of the Dimensions of Word Vectors

For the influence of the dimensions of different word vectors on the classification effect in the CNN model, the dimensions of word vectors are tested from 10 dimensions up to 200 dimensions in steps of 10. The convolutional size and number in the CNN model are determined by the values setting obtained from the previous results. The value of k is still 1 in the MaxTop-k-Pooling layer. Specific results are shown below.

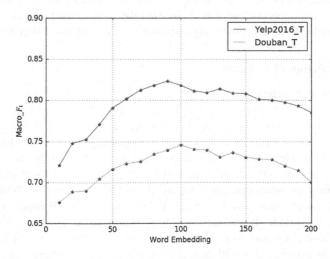

Fig. 3. The influence of the dimensions of word vectors on binary classification

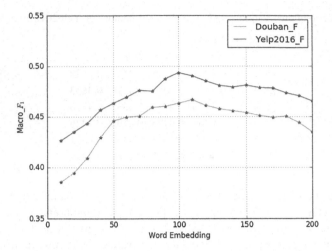

Fig. 4. The influence of the dimensions of word vectors on multiclass classification

The influence of the dimensions of word vectors on classification effect is shown in the figure above when classifying these four kinds of corpus. It can be seen from the figure that the value of overall Macro_F_1 raises with the increase of the dimensions of word vectors, but when the word vector increases to a certain dimension, the classification effect begin to decline or become stable. It is not like that the greater the dimensions of word vectors are, the better results are which was proposed by Mikolov in 2013. The reason is that corpus is not enough. High dimensional word vectors can increase the computational complexity of model in the process of training. Due to the limit of the number of the corpus, it makes higher dimensional word vectors cannot give full play to its advantages. In Fig. 3, when the dimensions of word vectors are 100 and 90, classification effect is best. In Fig. 4, when the dimensions of word vectors are 110 and 100, classification effect is best. We find the most suitable dimensions of word vectors, so we began to do fine-tuning operations aiming at word vectors in the later.

3.3   The Selection of K Value

The following parameters bases on the above analysis and we can find the most appropriate value of k by bringing the different values of k. When the value of k is 1, it is Max-Pooling operation. According to experiments, we find that the classification effect begins to deteriorate when the k value exceeds 6. So we choose the range of k value from 1 to 6, the experimental results are shown in Figs. 5 and 6.

In Figs. 5 and 6, when k is 3 on Douban binary classification and k is 4 on Yelp2016 binary classification, the value of Macro_F_1 is maximum. When k is 4 on Douban five classification and k is 5 on Yelp2016 five classification, the value of Macro_F_1 is maximum. It can be seen from the results Macro_F_1 value is also increasing with the k value raising, when the k value reaches a certain value, on the contrary, the Macro_F_1 value will decline. The reason for this phenomenon is that the noise and computation are increasing with the k value raising in the model, so the value of k can not be selected too small and too big, if it is too small, it will lose important information, in contrary, it will increase the noise and the computation.

Therefore, from the above experimental results, we can get the main settings of several super parameters in the CNN model, as shown in Tables 5 and 6, the sliding window in the table is the size of the convolutional window.

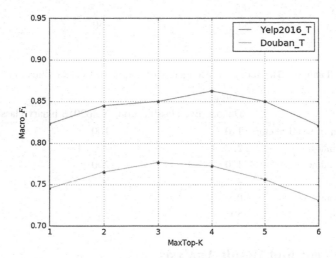

Fig. 5 The selection of k value on binary classification

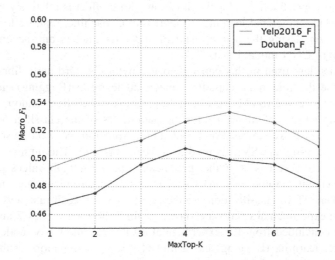

Fig. 6. The selection of k value on multiclass classification

Table 5. The setting of parameters of corpus of binary classification

Attribute	Corpus	
	Douban binary classification	Yelp2016 binary classification
Dimension of word vectors	100	90
Sliding window	4, 5, 6	3, 4, 5
Convolutional number	150	150
k	3	4
Dropout rate	0.5	0.5
Mini-batch	50	50

Table 6. The setting of parameters of corpus of five classification

Attribute	Corpus	
	Douban five classification	Yelp2016 binary classification
Dimension of word vectors	110	100
Sliding window	2, 3, 4	2, 3, 4
Sliding window	150	200
k	4	5
Dropout rate	0.5	0.5
Mini-batch	50	50

4   Experiment and Result Analysis

In general, the final classifier is softmax in CNN model. In this paper, Softmax classifier is replaced the other text classifiers in machine learning. By comparing different classifiers, this paper finally find a classification whose effect is relatively good. These classifiers include integrated learning, support vector machine, random forests, Naïve Bayes and so on. Using the above result of super parameters and according to the different corpus, CNN model is set.

These classifiers used in this paper is to use python's third-party library sklearn. This paper use the four main classifier: integrated learning (Bagging) classifier and support vector machine (SVM) classifier, Naïve Bayes classifier, random forest classifier. The paper uses default values for the parameters of the classifier to train. The penalty factor C is 1. Kernel function is radial basis function. We need to select a weak classifier for Bagging and SVM is used as a weak classifier. The number of decision trees with random forest is 50. Other parameters use the default values provided by sklearn. Five classification use the macro accuracy rate, macro recall rate, and macro F_1 value to evaluate. Two classification use the accuracy rate, recall rate, and F_1 value to evaluate. The experimental results are shown in the following Tables 7 and 8.

It can be concluded from the experimental results that the CNN model combined with ensemble learning (Bagging) has achieved good classification results and relatively stable in corpus of Douban binary classification. In order to keep the same

Table 7. Comparison of feature extraction algorithms of corpus of Douban binary classification

Classifier	Evaluating indicator		
	Accuracy rate(P)	Macro recall rate (Macro_R)	Macro F_1 (Macro_F_1)
Bagging	**0.8436**	**0.7967**	**0.8195**
SVM	0.8066	0.7853	0.7958
Naive Bayes	0.7831	0.7932	0.7881
Random forests	0.7869	0.7966	0.7917

Table 8. Comparison of feature extraction algorithms of corpus of Douban five classification

Classifier	Evaluating indicator		
	Macro accuracy rate (Macro_P)	Macro recall rate (Macro_R)	Macro F_1 (Macro_F_1)
Bagging	0.5023	0.5394	0.5201
SVM	0.4686	0.5302	0.4975
Naive Bayes	0.5129	0.5013	0.5070
Random forests	**0.5443**	**0.5166**	**0.5301**

classifier, this paper will choose Bagging classifier. So it will be combined with the Bagging classifier below.

The following tables are a summary of the commonly used models together for experimental comparison. CNN uses softmax classifier. The main parameters of the model are set according to the values obtained from the upper experiments (Tables 9, 10, 11 and 12).

Table 9. Corpus of Douban binary classification

Classifier	Evaluating indicator		
	Accuracy rate (P)	Recall rate (R)	F_1
CNN	**0.7869**	**0.7928**	**0.7898**
CNNB	**0.8436**	**0.7967**	**0.8195**
DCNN	0.7655	0.7193	0.7416
TF-IDF+Bagging	0.7461	0.6885	0.7161

Table 10. Corpus of Yelp2016 binary classification

Classifier	Evaluating indicator		
	Accuracy rate (P)	Recall rate (R)	F_1
CNN	**0.8726**	**0.8653**	**0.8689**
CNNB	**0.8748**	**0.9167**	**0.8953**
DCNN	0.8491	0.7732	0.8093
TF-IDF+Bagging	0.7933	0.7602	0.7764

Table 11. Corpus of Douban five classification

Classifier	Evaluating indicator		
	Macro_P	Macro_R	Macro_F_1
CNN	**0.4988**	**0.5373**	**0.5173**
CNNB	**0.5443**	**0.5166**	**0.5301**
DCNN	0.5103	0.4986	0.5043
TF-IDF+Bagging	0.3968	0.4296	0.4125

Table 12. Corpus of Yelp2016 five classification

Classifier	Evaluating indicator		
	Macro_P	Macro_R	Macro_F_1
CNN	**0.5275**	**0.5603**	**0.5434**
CNNB	**0.5193**	**0.6076**	**0.5600**
DCNN	0.5043	0.5288	0.5162
TF-IDF+Bagging	0.4022	0.4186	0.4102

CNNB uses Bagging classifier. DCNN is a model proposed by Kalchbrenner and the parameters of this model are the same as those used by Kalchbrenner. TF-IDF+ Bagging represents the use of traditional text classification methods. This paper uses the word frequency (TF) and inverse document frequency (IDF) to extract features and then the extracted features are sent to Bagging for classification.

The experimental results show that the CNNB model proposed in this paper can effectively improve the Macro_F_1 value. The F_1 value of CNNB for Douban binary classification reaches 0.8195. The F_1 value of CNNB for Yelp2016 binary classification reaches 0.8195. It can be seen that the validity of the model proposed in this paper. It can be seen from the experimental results that the effect of Chinese corpus classification is not good. Therefore, there is a certain degree of difficulty in the study of Chinese Text Sentiment Classification and need to further enhance.

Acknowledgments. This paper is sponsored by Natural Science Foundation of China (61300078).

References

1. Collobert, R., Weston, J.: A unified architecture for natural language processing: deep neural networks with multitask learning. In: Proceedings of 25th International Conference on Machine Learning, pp. 160–167. ACM (2008)
2. Collobert, R., Weston, J., Bottou, L., et al.: Natural language processing (almost) from scratch. J. Mach. Learn. Res. **12**(August), 2493–2537 (2011)
3. Maas, A.L., Daly, R.E., Pham, P.T., et al.: Learning word vectors for sentiment analysis. In: Proceedings of 49th Annual Meeting of the Association for Computational Linguistics: Human Language Technologies, vol. 1, pp. 142–150. Association for Computational Linguistics (2011)

4. Mikolov, T., Sutskever, I., Chen, K., et al.: Distributed representations of words and phrases and their compositionality. In: Advances in Neural Information Processing Systems, 3111–3119 (2013)
5. Kalchbrenner, N., Grefenstette, E., Blunsom, P.: A convolutional neural network for modelling sentences. arXiv preprint arXiv:1404.2188 (2014)
6. Kim, Y.: Convolutional neural networks for sentence classification. arXiv preprint arXiv:1408.5882 (2014)
7. Zhang, X., Zhao, J., LeCun, Y.: Character-level convolutional networks for text classification. In: Advances in Neural Information Processing Systems, pp. 649–657 (2015)
8. Conneau, A., Schwenk, H., Barrault, L., et al.: Very deep convolutional networks for natural language processing. arXiv preprint arXiv:1606.01781 (2016)

Predicting Big-Five Personality for Micro-blog Based on Robust Multi-task Learning

Shuguang Huang[1], Jinghua Zheng[1(✉)], Di Xue[2], and Nan Zhao[3]

[1] Hefei Electronic Engineering Institute, Hefei 230037, China
zhengjhl001@163.com
[2] PLA University of Science and Technology, Nanjing, China
[3] CAS Institute of Psychology, Beijing, China

Abstract. Personality prediction on social network has become a hot topic. At present, most studies are using single-task classification/regression machine learning. However, this method ignores the potential association between multiple tasks. Also an ideal prediction result is difficult to achieve based on the small scale training data, since it is not easy to get a lot of social network data with personality label samples. In this paper, a robust multi-task learning method (RMTL) is proposed to predict Big-Five personality on Micro-blog. We aim to learn five tasks simultaneously by extracting and utilizing appropriate shared information among multiple tasks as well as identifying irrelevant tasks. For a set of Sina Micro-blog users' information and personality labeled data retrieved by questionnaire, we validate the RMTL method by comparing it with 4 single-task learning methods and the mere multi-task learning. Our experiment demonstrates that the proposed RMTL can improve the precision rate, recall rate of the prediction and F value.

Keywords: Sina Micro-blog · Personality prediction · Multi-task learning · Prediction accuracy

1 Introduction

The study of personality, or what makes a person unique, is of central importance in psychology. Personality is closely related to human behavior. By researching the correlation between music preferences and personality, Rentforw and Gosling [1] testified they were related to each other, which can be used to make music personalized recommendation. Individuals can also better understand existing job opportunities by researching the relations of the personality traits to career exploration behavior [2]. Studying personality can help employers predict employees' job performance in their team [3]. Among personality-related research, Big-Five model is the most frequently used one, which proposes five dimensional traits: neuroticism (N), agreeableness (A), extraversion (E), conscientiousness (C) and openness (O). Each dimension describes a person's personality from a different aspect. Personality prediction on social network is a hot topic recently due to the rapid development of information technology. These researchers use information on Facebook, Twitter, YouTube, LinkedIn, Instagram, Sina Micro-blog or Renren network to predict personality. In China, Sina Micro-blog

© Springer Nature Singapore Pte Ltd. 2017
B. Zou et al. (Eds.): ICPCSEE 2017, Part I, CCIS 727, pp. 486–499, 2017.
DOI: 10.1007/978-981-10-6385-5_41

becomes one of the most popular internet services. Much research has been done to identify the relationship between social network and personality and the researchers got some findings.

Many researchers showed that different online behavior comes from different personality traits [4]. Michal et al. proved that the behavior data in social network can predict the user's personality traits [5]. Until now, most of researchers use machine learning to predict the Big-Five personality of social network users. Wald R et al. used linear regression, RepTree and decision table algorithm to predict the Big-Five personality traits, and they could predict 74.5% Facebook users [6]. In the same year, they predicted the Big-Five personality traits of Twitter users by linear regression, multi-layer perceptron, random forest and SVM [7]. Ortigosa et al. used Naive Bayesian and C4.5 algorithm to predict the Big-Five personality traits of Facebook users, and the three classification accuracy was about 70% [8]. These data characteristics from the social network include tweet characteristics, linguistic characteristics [9] and semantic characteristics. The tweet characteristics include gender, age, the number of fans and the number of friends etc. [4–12]. The semantic characteristics include emotions and attitude, opinions and so on [13, 14].

At present, personality prediction on social network is still plagued by a number of problems. There are three problems: (1) Based on the results, the five dimensions of personality are related. If a person's agreeableness has a higher score, then his/her conscientiousness will be higher; (2) There are unrelated tasks. For example, there is no obvious relationship between neuroticism and conscientiousness, agreeableness and openness.

In recent years, most research used single-task classification/regression machine learning. The result of these single-task modules had no significant difference by training same dataset. These modules are two shortcomings. Firstly, not only the correct rate of the prediction is relatively low, but also the module generalization ability is relatively weak. Their results make slight difference by training the same dataset. On the other hand, every kind of personality prediction is looked as a task and has no association with others. However personality is a set of individual differences affected by the development of an individual, such as values, attitudes, social relationships and so on. Personality generalizes the characteristics of a person, so each one of the personality dimensions can be influenced by the others. But these methods ignore the task relatedness. Secondly, it is so hard to get the dataset of Micro-blog users' Big-Five personality. Thus it can easily result in overfitting due to the small scale training data.

So we predict Big-Five personality using multi-task learning (MTL) to learn more than one task simultaneously. It can not only utilize appropriate shared task learning but can also avoid model overfitting by learning multiple related tasks simultaneously.

But most multi-task learning formulations assume that all tasks are relevant, which is however not the case in many real-world applications, such as Big-Five personality. So we use Robust Multi-Task Learning (RMTL) to solve this problem. Robust multi-task learning is aimed at identifying irrelevant tasks when learning from multiple tasks. The correlation matrix W between task and feature is the kernel of the method. In order to solve it, first, we transform it into optimal solution by minimizing the penalized empirical loss, i.e., where L(.) is the empirical loss on the training set, is the regularization terms, and correlation matrix W is estimated from the training samples.

Secondly, W is decomposed into two components, i.e., P which captures task-relatedness and Q which detects irrelevant tasks. And the mixed norm such as trace norm and L_{12} norm is introduced as constraints of regular terms. Finally, the optimal solution of the incidence matrix W is obtained by solving the constrained optimization problem.

In our research, we obtain the data of 994 Micro-blog users from Sina Micro-blog network and use the robust multi-task learning to create a Big-Five personality prediction model. With Sina Micro-blog users' data, we validate the robust multi-task learning method. First, we utilize appropriate shared task learning by learning multiple related tasks simultaneously, our model avoids overfitting and successfully improve the predictive performance from correct rate, precision rate and recall rate by comparing with 4 single task learning methods. Secondly, by detecting irrelevant tasks in Big-Five personality, our model outperform the simple multi-task learning from this there indicators.

The rest of this paper is organized as follows: In Sect. 2, we'll introduce some related work including multi-task learning and robust multi-task learning. In Sect. 3, we create the Micro-blog user Big-Five personality prediction model using the proposed method. Experiment results will be presented in Sect. 4 followed by a discussion about the results. At last, we will make a conclusion in Sect. 5.

2   Related Work

Recently, most of the personality prediction based on social network use the traditional single-task classification or regression algorithm in machine learning. In this paper, we use Robust multi-task learning to predict social network user personality in order to avoid overfitting caused by insufficient data, and to improve prediction performance by sharing the relatedness between multiple tasks.

A. Multi-task learning

In real-world, most problems are related, and we deal with multiple related classification/regression tasks. In 1997, Rich Caruana first proposed multi-task learning method which aims at learning the target task by learning multiple tasks that relate the target task. Being differences between multiple tasks, we use their characteristics to solve the small-scale problem caused by learning independently. Thus accurate knowledge is provided for the target task. Now many studies have proved this point [15–18], so the multi-task learning method has become a hot topic in many research fields.

From the task relevance, Multi-task learning has two approaches:

(1) Mining tasks that have same characteristics from data samples. Argyriou et al. [18] divided training data features into many sub tasks which share the same low dimensional subspace to share the potential information between features. Xue et al. [23] proposed an efficient LSVM-MTL model which was nonlinear classifier by using linear SVMs and multi-task learning algorithm.

(2) Mining related tasks from target tasks. Bai et al. [24] proposed a multi-task learning model to predict social media user personality based on relatedness between multiple personality. At last they proved the validity of this method.

Many different multi-task learning methods are proposed by scholars based on hypothesis of different models which are proposed in multi-task learning, i.e., shared variables, shared subspaces, and shared model parameters, and so forth. All in all, multi-task learning can improve the learning performance by assisting the target tasks with the valuable information sharing.

There are two ways to realize multi-task learning method:

(1) Constrained learning by regularization term (penalty term).

Evgeniou and Massimiliano [25] proposed an approach to multi-task learning based on the minimization of regularization functionals similar to existing ones. For T tasks, their parameter model W was

$$W = \arg\min_{W} \left(L(Y, W^T X) + \sum_{t=1}^{T} \sum_{i=1}^{m} \xi_{it} + \frac{\lambda_1}{T} \sum_{t=1}^{T} \|v_t\|^2 + \lambda_2 \|w_0\|^2 \right)$$

where $t \in \{1, \cdots, T\}$, L(.) is the empirical loss on the training set, λ_1 and λ_2 are positive regulation parameters and ξ_{it} are slack variables measuring the error that each of the final models w_t makes on the data. And $w_t = v_t + w_0$, where the vectors v_t are small when the tasks are similar to each other and all close to w_0. Thus they successfully transfer single task SVM method to multi-task SVMs method. At last, their experimental results show the multi-task learning methods largely outperforms single task leaning using SVMs.

(2) Probabilistic modeling of covariance matrix.

Probabilistic modeling of covariance matrix can learn correlation between tasks by modeling correlation and independence between multiple tasks simultaneously. Zhang and Yeung [19] proposed a new Bayesian extension model for solving problems during covariance matrix estimation.

B. Robust Multi-task learning

Multi-task learning methods strongly assume that multiple tasks are all relevant. But it is not the case, such as Big-Five personality, because this hypothesis ignores the irrelevant tasks in multiple tasks. Robust multi-task learning (Robust Multi-Task Learning) method [19, 26–29] is aimed at identifying irrelevant tasks. The way of most robust multi-task learning methods is dividing multiple tasks into two categories, i.e., relevant tasks and irrelevant tasks. They decompose the parameter matrix W into two components which are used to capture task-relatedness and detected irrelevant tasks respectively.

In Chen et al. [27] research, the parameter matrix W is decomposed into two components as

$$W = P + Q$$

where P is low rank structure matrix and used to capture task-relatedness, Q is group sparse structure matrix and used to detect irrelevant tasks. So the regularization term is decomposed into two parts as

$$W = \arg \min_{W=P+Q} \left(L(Y, (P+Q)^T X) + \rho_1 \|P\|_{2,1} + \rho_2 \|Q\|_{1,2} \right)$$

where constraint is learned by different regularization terms.

Xu et al. [28] proposed a robust multi-task regression method, which is used to deal with the sparse gross errors caused by the high dimensional data with two regularization term. They regression model is as follows

$$Y = X\Theta^* + W + G^*$$

where $Y \in R^{n \times q}$ is the response matrix, $X \in R^{n \times p}$ is the covariance matrix, $\Theta^* \in R^{p \times q}$ is unknown linear relationship between the predicted values and the response values, $W \in R^{n \times q}$ is the noise matrix, G^* is sparse gross errors matrix. They use Frobenius norm, L_1 norm and L_2 norm to model from error, and improve the robustness of the model.

Gong et al. [29] divided the weight matrix into two parts, and use Lasso to deal with relevant tasks and group Lasso to deal with irrelevant tasks.

Another method is modeling the task covariance matrix. Yu et al. [26] proposed a robust framework for Bayesian multi-task learning based on t process which can well detect irrelevant tasks. Their experiments proved that their method improves overall system performance.

The nonparametric method is often used in modeling with the last one which makes very high computational complexity. Therefore, in this paper, we use the first one to solve the robust multi-task learning.

3 Our Method

3.1 Big-Five Personality Prediction Based on MTL

Big-Five personality prediction problem needs to learn the equivalent of 5 tasks. For every task belonging to the space X × Y, its training dataset is denoted as

$$\{(x_{t1}, y_{t1}), (x_{t2}, y_{t2}), \cdots, (x_{tm}, y_{tm})\}$$

where X_{ti} is a dimension feature vector for task t, and Y_{ti} is the true value for this point for task t. So for T = 5 tasks, the training dataset is denoted as

$$\{\{(x_{11}, y_{11}), \cdots, (x_{1n}, y_{1n})\}, \cdots, \{(x_{T1}, y_{T1}), \cdots, (x_{Tn}, y_{Tn})\}\}$$

So Sina Microblog user personality prediction is aimed at learning 5 functions, Fig. 1 below shows it.

$$f_i, f_2, \cdots, f_T, f_t(x_{it}) = X_t^T w_t \approx y_{it} \qquad (1)$$

where t = 1, 2, ..., T, and each function represents a predictive model for each of the Big-Five personality.

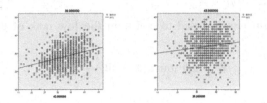

Fig. 1. The relationship between different dimensions of Big-Five personality. The relationship between agreeableness and conscientiousness is mapped to the left one. The relationship between agreeableness and openness is mapped to the right one.

For each personality prediction task, the learning goal is to solve the optimization algorithm, as shown in formula 2.

$$w_t = \arg \min L(X_t, y_t, w_t) + \lambda \Omega(w_t) \qquad (2)$$

where $w_t \in R^n$ is the model parameter, $L(\cdot, \cdot)$ is the loss function, $\Omega(w_t)$ is the regularization term of the parameter w_t, which is used to avoid overfitting so as to minimize the training error in single-task learning, and λ is the regularization parameter which is aimed at balancing the loss function and regularization term.

In this paper, we use the robust multi-task learning to predict the Sina Micro-blog user Big-Five personality, which is equivalent to learn five tasks in parallel.

So the Big-Five personality prediction is aimed at formula 3.

$$f(X) = X^T W \approx Y \qquad (3)$$

where W is the model parameter matrix, in which its row is the characteristic vector of each task, it's column is the attribute of a certain characteristic. We use the loss function and the regularization for modeling to mine the relationship between the columns or the rows of W. Then we realize multiple tasks learning in parallel to avoid the overfitting, and improve the generalization performance of the model.

The Big-Five personality predictions on Sina Micro-blog based on robust multi-task learning method as shown in formula 4.

$$W = \arg \min_W L(X, Y, W) + \lambda \Omega(W) \qquad (4)$$

where $W = [w_1, w_2 \cdots, w_t] \in R^{n \times t}$, $\Omega(W)$ is regularization term which is used to constrain the feature relatedness, λ is the regularization term coefficient which is used to balance the loss function and regularization term. Different assumptions on task relatedness lead to different regularization terms. Such as, L_1 norm is often used to minimization the matrix W nonzero row problem [30] which called matrix group sparse representation, as follows.

$$W = \arg \min_{W} L(X, Y, W) + \lambda \|W\|_1 \tag{5}$$

where $\|W\|_1 = \sum_{i=1}^{T} \sum_{j=1}^{d} |W_{ij}|$.

The kernel norm is often used to constrain the low rank of the matrix [31], as follows.

$$W - \arg \min_{W} L(X, Y, W) + \lambda \|W\|_* \tag{6}$$

where $\|W\|_* = \sum_{i=1}^{\min(d,T)} \sigma_i(W)$, σ_i is the ith singular values of matrix W.

3.2 Big-Five Personality Prediction Based on RMTL

Big-Five personality can fully describe one person's personality. Sina Micro-blog users' Big-Five personality data indicate that there is a correlation between scores of personality, as is shown in Fig. 1. The higher the score of agreeableness, the higher the conscientiousness. But there is no correlation between agreeableness and openness. This means that there is not only relevant but also irrelevant between the five personalities. Therefore, it is difficult to use the general multi-task learning method to improve the prediction performance. On the contrary, it may lead to worse results.

It is for this reason that we use robust multi-task learning which can identify irrelevant tasks to predict social network user Big-Five personality. The parameter model is decomposed into a structure term and an anomaly term. Then the regularization term is correspondingly divided into two. One is used to share the relevant information; the other is used to identify the irrelevant tasks. Thus we can avoid not only the overfitting due to the small scale training data by share multiple tasks attribute data, but also mutual influence between unrelated tasks.

So for incidence matrix $W = [w_1, w_2, \cdots w_t] \in R^{d \times t}$ of the T tasks, $W = P + Q$.

where $P = [p_1, p_2, \cdots p_t] \in R^{d \times t}$ which is a low-rank structure, and $Q = [q_1, q_2, \cdots q_t] \in R^{d \times t}$ is a group-sparse matrix.

In the field of multi-task learning, there are many prior work model relationships among tasks using different regularizations. In this paper, we choose the trace norm regulation term to encourage the desirable low-rank structure in the matrix P. And the $L_{1,2}$ norm regularization term induce the desirable group sparse structure in the matrix Q. Mathematically, the robust multi-task learning is formulated as

$$W = \min_{W=P+Q} \sum_{i=1}^{t} \|W_i^T X_i - Y_i\|_F^2 + \rho_1 \|P\|_* + \rho_2 \|Q\|_{1,2} \tag{7}$$

where ρ_1 and ρ_2 are trade-off parameters, the first item represents the averaged least square loss of the T tasks. The trace norm of matrix P represents as

$$\|P\|_* = \sum_{i=1}^{r} \sigma_i(P) \tag{8}$$

where r is the rank of matrix $P, \sigma_i(.)$ is the singular value of matrix P. The trace norm can realize the sparse representation of matrix, so we can mine the relevant of multiple tasks. The $L_{1,2}$ norm of matrix Q represents as

$$\|Q\|_{12} = \sum_{i=1}^{d} \|q_i\|_2 \tag{9}$$

That is to say, the $L_{1,2}$ norm of matrix Q is sum of the L_2 norm of the matrix Q column vector. This norm can achieve sparsity on variable groups, so it can identify the irrelevant tasks. Thus we choose the trace norm and $L_{1,2}$ norm for restraining the structure matrix and outlier matrix(irrelevant tasks) to transform the multi-task learning to an optimization problem with regular constraints.

3.3   Model Solving

Solution of the optimization algorithm for formula $\min f(x) + h(x)$ is often implemented via the proximal gradient method. For formula 7, we assume the first item is f (x), and the other is h(x). So the iterative formula is obtained by the proximal gradient method, as follows:

$$W_{k+1} = \text{Pr } ox_{\gamma_k,h}(W_k - \gamma_k \nabla f(W_k)) \tag{10}$$

where $W_k - \gamma_k \nabla f(W_k)$ is gradient descent, γ_k specifies the step size which can be appropriately determined by iteratively increasing its value. Pr $ox_{\gamma_k,h}(\cdot)$ is the proximal operator which will project a point in space on to the convex set h. γ_k is the projection parameter, i.e., the step size, which can obtained by linear search.

We now apply the accelerated proximal gradient method [27] to solve the robust multi-task learning in Eq. (7). This method increases extrapolation in the process of searching step size, as follows.

$$\begin{aligned} Z_{k+1} &= W_k + \theta_k(W_k - W_{k-1}) \\ W_{k+1} &= \text{Pr } ox_{\gamma_k,g}(Z_{k+1} - \gamma_k \nabla f(Z_{k+1})) \end{aligned} \tag{11}$$

where $\theta_k \in [0, 1)$ is a extrapolation parameter, and typically selects $\theta_k = \frac{k}{k+3}$.

The accelerated proximal gradient algorithm is as:

The Big-Five personality traits prediction model on RMTL

Input: X_i: training data matrix from the ith task

y_i: label vector on personality from the ith task

1. Initialization of γ_k and $\beta \in (0,1)$.

2. $\gamma = \gamma_k$

3. do

4. Set $Z = prox_{\gamma h}(W_{k+1} - \gamma \nabla f(W_k))$

5. break if

$$l(W_{k+1}) \leq l(\hat{W}_k) + \langle \nabla l(\hat{W}_k), W_{k+1} - \hat{W}_k \rangle + \frac{\gamma_k}{2} \left\| W_{k+1} - \hat{W}_k \right\|_F^2$$

6. updating step $\gamma = \beta \gamma$

7. while $\gamma_{k+1} = \gamma$, $W_{k+1} = Z$.

This method uses a linear combination of previous two points as the search point in every iteration, instead of only using the latest point. The AGM has the convergence speed of $O(1/k^2)$.

4 Experiment and Comparison

4.1 Data Acquisition

Our experiment will use the dataset of Sina Micro-blog Users including Micro-blog feature data as well as personality trait score from Chinese Academy of Sciences Institute of Psychology. The personality trait score comes from the Big-Five Experiment, i.e. the Berkeley 44 Personality Questionnaire (http://www.ocf.berkeley.edu/~johnlab/index.htm/). There are totally 1604 users which are active users.

First, we preprocess the collected dataset.

Preprocessing the Micro-blog data feature mainly is transforming the non-numerical data into numerical data. For example, the gender is converted to 0 and 1. Calculate user nickname and self description length. The regional information may be converted to numerical values, etc.

Preprocessing the text of Micro-blog mainly focus on the two points, as follows.

(1) Filtering all Micro-blog text which do not meet the requirement, i.e., whose content is empty, just only hyperlinks, or pictures, or video contents.
(2) Extraction of Micro-blog text features. First, we integrate all the micro-blog text from same user. Then we extract 102 dimension text features using the TextMind system (http://ccpl.psych.ac.cn/textmind/), including first person, second person, third person, emotional words, positive/negative emotional words, mental words, @ numbers, emoticon and so on.

At last, we got 994 dataset of users including Micro-blog and Big-Five personality data, in which there are 391 men, and their average age is 24.6 years old, and they are from 19 provinces and cities around the nation. The Big-Five personality scores of this dataset is the normal distribution. Therefore the practical dataset is infective.

We extracted 114 features, including 12 Micro-blog characteristics and 102 text features. The 12 Micro-blog characteristic include gender, location, the number of followers, friends, and states, etc. The 102 text feature, which were extracted using the TextMind system, include the number of first, second, third person singular/plural pronoun, the rate of positive/negative emotional words, the number of Micro-blog emoticons in the Micro-blog content, and so on.

4.2 Result

Prediction of users' Big-Five personality traits with their public information on Microblog is the goal of our experiment. Our training dataset is 70 percent of 114 dimensions of 994 Micro-blog users, and the rest is used to test. We use the robust multi-task learning (RMTL) without any assumptions, and will learn five tasks, i.e., the five personality traits simultaneously. We choice the least square loss and mixed-norm (trace norm and $L_{1,2}$ norm) to model which can automatically mine the relevant and identify the irrelevancy between multiple tasks.

There are two important parameters, i.e., ρ_1 and ρ_2 in our experiment. The first one is used to control lower rank of the structure, and the later is used to constrain the group sparse. Those are obtained by training model. When $\rho_1 = 400$ and $\rho_2 = -20$, the model performance is the best.

In our experiments, 4 typical classification/regression models including Naive Bayesian (NB), logistic regression (LR) and random forest (RF) and RepTree algorithm are selected to compare with the RMTL method mentioned in this paper. Meanwhile, we used multi-task learning method (MTL) which assumes that all of the tasks are relevant to predict Big-Five personality on the same dataset with the least square loss and Lasso norm, as follows.

$$W = \arg \min_W L(X, Y, W) + \lambda \Omega(W)$$

$$= \arg \min_W \sum_{i=1}^{t} \left\| W_i^T X_i - Y_i \right\|_F^2 + \lambda \|W\|_1$$

By 5-folds cross validation, our experimental results are showed in Figs. 2, 3 and 4 from precision rate, recall rate and the F1 value. The result demonstrates the usage of RMTL can improve the Big-Five personality prediction performance compared with the 4 single-task learning methods and MTL method.

4.3 Discussion

It is very difficult to get the Micro-blog users Big-Five personality sample dataset. In our experiment, there are 114 dimensions characteristic data from 994 Sina Micro-blog users.

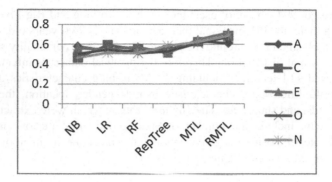

Fig. 2. The precision rate result of the six methods

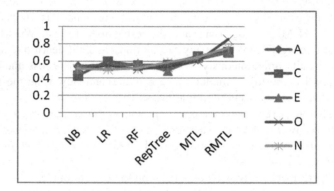

Fig. 3. The recall rate result of the six methods

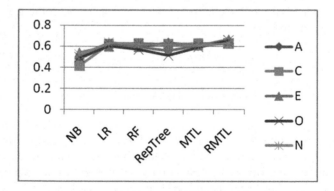

Fig. 4. The F value result of the six methods

Being the small training dataset, it is so easy to happen overfitting, the performance of the model may not be desired. There is a certain relevance between the 5 personality traits, yet the relevance will not used to the full by the traditional single-task learning method. However, the multi-task learning methods make up for these defects. On the other hand, there is relatedness between 5 Big-Five personality traits, thus the simple multi-task learning method which assumes that all tasks are relevant. Thus this method cannot direct the irrelevant tasks, so that it cannot get the best result.

Base on this, we use robust multi-task learning method to predict the user personality on Sina Micro-blog. And using the regularization term, this method fully utilize the relatedness between the five personalities, such as the positive correlation between agreeableness and conscientiousness. At the same time, it can direct the irrelevant tasks, such as the relationship between agreeableness and openness. From Figs. 2, 3 and 4, the result about the precision rate, the recall rate and the F value can shows our model got the better result and the better generalization performance.

5 Conclusion

In this paper, we analyzed personality traits characteristics on Micro-blog users and created the Big-Five personality traits prediction model based on the robust multi-task learning method. Experiment was conducted to demonstrate our models performance well. At present, most personality prediction research use single-task learning method. However, these methods ignore the task relatedness. We firstly predict the Big-Five personality of Micro-blog users using the RMTL method which can improve the generalization of prediction model.

There are wide areas and great application value to study personality traits based Micro-blog users' public information. Different social media has different data structure, which leads to different prediction models. In the future, we will do further work on multi-task modeling and feature extraction. And we will look for a better multi-task model and features in a larger scale to improve the accuracy of prediction models.

Acknowledgement. The author would like to thank the anonymous reviewers for their valuable comments. This work is supported by the National Natural Science Foundation of China (Grant Number: 61602491). We also would like to thank the Institute of Psychology, Chinese Academy of Sciences for generously supporting our research.

References

1. Rentforw, P.J., Gosling, S.D.: The do re mi's everyday life: the structure and personality correlates of music preferences. J. Pers. Soc. Psychol. **84**(6), 1236–21256 (2003)
2. Li, Y., Guan, Y., Wang, F., et al.: Big-five personality and BIS/BAS traits as predictors of career exploration: the mediation role of career adaptability. J. Vocat. Behav. **89**(4), 39–45 (2015)
3. Neuman, G.A., Wagner, S.H., Christiansen, N.D.: The relationship between work-team personality composition and the job performance of teams. Group Org. Manag. **24**(1), 28–45 (1999)
4. Schrammel, J., Koffel, C., Tscheligi, M.: Personality traits, usage patterns and information disclosure in online communities. In: Proceedings of 23rd British HCI Group Annual Conference on People and Computer, London, UK, pp. 169–174 (2009)

5. Kosinski, M., Stillwell, D., Graepel, T.: Private traits and attributes are predictable from digital records of human behavior. Proc. Natl. Acad. Sci. **110**, 5802–5805 (2013)
6. Wald, R., Khoshgoftaar, T., Summer, C.: Machine prediction of personality from Facebook profiles. In: Proceedings of 2012 IEEE 13th International Conference on Information Reuse and Integration, Las Vegas, USA, pp. 109–115 (2012)
7. Wald, R., Khoshgoftaar, T.M., Napolitano, A., et al.: Using Twitter content to predict psychopathy. In: Proceedings of 2012 11th International Conference (ICMLA) on Machine Learning and Applications, USA, pp. 394–401 (2012)
8. Ortigosa, A., Carro, R., Quiroga, J.: Predicting user personality by mining social interactions in Facebook. J. Comput. Syst. Sci. **80**, 57–71 (2013)
9. Mairesse, F., Walker, M.A., Mehl, M.R., et al.: Using linguistic cues for the automatic recognition of personality in conversation and text. J. Artif. Intell. Res. **30**(1), 457–500 (2007)
10. Goldberg, L.R., Johnson, J.A., Eber, H.W., et al.: The international personality item pool and future of public-domain personality measures. J. Res. Pers. **40**(1), 84–96 (2006)
11. Yu, Z., Dit, Y.Y.: Semi-supervised multi-task regression. In: Proceedings of European Conference on Machine Learning and Knowledge Discovery in Databases: Part II (2009)
12. Bachrach, Y., Kosinski, M., Graepel, T., et al.: Personality and patterns of Facebook usage. In: Proceedings of 3rd Annual ACM Web Science Conference, New York, USA, pp. 24–32 (2012)
13. Iacobelli, F., Gill, A.J., Nowson, S., Oberlander, J.: Large scale personality classification of bloggers. In: D'Mello, S., Graesser, A., Schuller, B., Martin, J.C. (eds.) ACII 2011. LNCS, vol. 6975, pp. 568–577. Springer, Heidelberg (2011). doi:10.1007/978-3-642-24571-8_71
14. Nowson, S., Oberlander, J.: Identifying more bloggers: towards large scale personality classification of personal weblogs. In: International AAAI Conference on Web and Social Media, Colorado, USA (2007)
15. Caruana, R.: Multitask learning. Mach. Learn. **28**(1), 41–75 (1997)
16. Argyriou, A., Evgeniou, T., Massimiliano, P.: Convex multi-task feature learning. Mach. Learn. **73**(3), 243–272 (2008)
17. David, S.B., Borbely, R.S.: A notion of task relatedness yielding provable multiple-task learning guarantees. Mach. Learn. **73**, 273–287 (2008)
18. Argyiou A, Evgeniou T, Pontil M.: Multi-task feature learning. Advances in Neural Information Processing Systems.MIT Press. 41–48(2006)
19. Zhang, Y., Yeung, D.: Multi-task learning using generalized t process. J. Mach. Learn. Res. Proc. Track **9**, 964–971 (2010)
20. Anveshi, C., Huzefa, R.: Classifying protein sequences using regularized Multi-task Learning. EEE/ACM Trans. Comput. Biol. Bioinform. **11**(6), 1087–1098 (2014)
21. Zhang, J., Ghahramani, Z., Yang, Y.: Learning multiple related tasks latent independent component analysis. In: Advances in Neural Information Systems (2005)
22. Olshausen, B., Field, D.: Emergence of simple-cell receptive field properties by learning a sparse code for natural images. Nature **381**, 607–609 (1996)
23. Xue, M., Ou, W., Weiming, H., Peter, O.: Nonlinear classification via linear SVMs and multi-task learning, pp. 1955–1958. ACM (2014). http://dx.doi.org/10.1145/2661829. 2662068
24. Bai, S., Yuan, S., Cheng, L., et al.: Application of multi-task regression in social media mining. J. Harbin Inst. Technol. **9**, 100–110 (2014)
25. Evgeniou, T., Massimiliano, P.: Regularized multi-task learning. In: Proceedings of Knowledge Discovery and Data Mining, pp. 109–117 (2004)
26. Yu, S., Volker, T., Yu, K.: Robust multi-task learning with t-processes. In: Proceedings of 24th International Conference on Machine Learning, pp. 1103–1110 (2007)

27. Chen, J., Zhou, J., Ye, J.: Integrating low-rank and group-sparse structures for robust multi-task learning. In: Proceedings of 10th ACM SIGKDD International Conference on Knowledge Discovery and Data Mining (2011)
28. Xu, H., Leng, C.: Robust multi-task regression with grossly corrupted observations. In: Proceedings of 15th International Conference on Artificial Intelligence and Statistics (AISTATS), La Palma, Canary Islands, pp. 1341–1349 (2012)
29. Gong, P., Ye, J., Zhang, C.: Robust multi-task feature learning. In: KDD 2012. ACM, Beijing, vol. 8, pp. 895–903 (2012)
30. Tibshirani, R.: Regression shrinkage and selection via the lasso. J. Roy. Stat. Soc. **58**, 267–288 (1996)
31. Ji, S., Ye, J.: An accelerate gradient method for trace norm minimization. In: Proceedings of 26th Annual International Conference on Machine Learning, pp. 457–464 (2009)

Automatic Malware Detection Using Deep Learning Based on Static Analysis

Liu Liu[✉] and Baosheng Wang

School of Computer, National University of Defense Technology,
Changsha 410073, China
hotmailliuliu@163.com

Abstract. Malware detection is an important challenge in the field of information security. The paper proposes a novel method using deep learning based on static analysis. Deep learning has stronger nonlinear expression ability than shallow learning, so it has received much attention from scholar and manufacturers. We use static analysis to extract the malware features are mapped into the input of deep learning. The experiments show that the method is suitable for detecting malware.

Keywords: Ensemble learning · K-means · Negative correlation · Neural network

1 Introduction

With the development of information technology, big data is changing the status of information security. Traditional security vendors based on signature have been difficult to adapt to the requirements of massive data. The signature not only needs to build a large characteristic database, but also have to be manually updated in real time [1, 2]. Malware writers often use obfuscation techniques to avoid anti-virus software, such as encryption, compression, metamorphic and so on [3]. Compared with the signature technique, anomaly detection can discover new malware, but it has a higher rate of false positive. Obviously, the anomaly detection could cause a large number of false alarms in big data.

In order to automate the detection of malware, many experts have used machine learning to analyze malware, and have achieved good results [4, 5]. But machine learning which is applied to the malware analysis is still facing two challenges: (1) how to extract features of malware [6]; (2) how to choose suitable model for different problems [7].

In order to solve the above problems, the paper proposes an automatic malware detection system. It consists of two parts: the feature extraction and deep learning model. The input of deep learning is the format of matrix or vector by mapping the normative features of malware. We adopt gray scale image and OpCode n-gram to obtain high quality features. Deep learning model is constructed by using Keras v0.3.0. In the experiments, we use the malware set, which contains 4,755 samples from 10

© Springer Nature Singapore Pte Ltd. 2017
B. Zou et al. (Eds.): ICPCSEE 2017, Part I, CCIS 727, pp. 500–507, 2017.
DOI: 10.1007/978-981-10-6385-5_42

malware families. The result shows that the feature extraction method and the deep learning are effective.

The remainder of paper is organized follows. In Sect. 2, we outline the feature extraction methods. In Sect. 3, deep learning model of the internal structure is described. In Sect. 4, we validate our system by multiple experiments. Finally, we conclude the paper and discuss future work.

2 Related Work

In the paper, Annachhatre et al. [8] tried to change the dynamical analysis on the high level. It got all the values which were written into destination registers or memories in execution traces. They aimed to analyze a detailed profile of program and more accurately understand the purpose of functions in each program. Then, the paper designed a two-step procedure to calculate the similarity of two traces based on the sequence values. The paper showed two applications based on the procedure. The first application grouped the samples by clustering technique and the second application dug out the code reuse.

Kang et al. [9] put forward a fast detection system which could find the malware and assign the malware to the corresponding family. In this paper, they used Bloom filter to generate the signature of each malicious family and compress the feature space for reducing the amount of computation. The features were abstracted by using control flow graphs (CFGs). For finding the common features amongst the same family, CFGs should abide by the following rules: (1) the same number of blocks; (2) the same directed edges; (3) the same sequence of instructions within each block. In the experiments, the average similarity ratio was 0.82315 in the same family and 0.01970 for the different families.

Gonzalez and Vazquez [10] proposed a method to classify malware effectively. The method extracted the features of malware by counting the times of each library in the malware. For authenticating their approach, they got the large numbers of malware from Kaspersky, F-Secure, Norton and McAfee. To analyze the common characteristics and difference of malware family, the paper obtained the amount of API calls per DLL and the imported functions of the executable program. Then, the paper employed Euclidean Distance and Multilayer Perceptron to classify the malware variants.

In recent years, malware variants multiply rapidly by automated tools. Therefore, some variants maybe have similar characteristics belong to a same family. To distinguish different malware families, Han et al. [11] proposed a new method to classify malware families. The method transformed malicious code into a visualized image. Compared with our method, the paper continued to convert the visualized image into an entropy graphs. The paper used a modified histogram similarity measuring method to measure the graph similarities [12]. From the results of their examples, their method got a higher accuracy rate. However, the efficiency of this method is quite low.

3 Our Method

3.1 Features Extraction

In the paper, the feature extraction method is composed of gray scale image and OpCode 3-gram. The Fig. 1 shows the overview of the extraction method. Firstly, the malicious code is a binary file which is converted into a gray-scale image by IDA Pro. Secondly, n-gram method is used to abstract the opcode feature which is derived from a binary file. Then, we combine the image and the opcode feature. Finally, the integrated feature is mapped into a vector which is the input of deep learning.

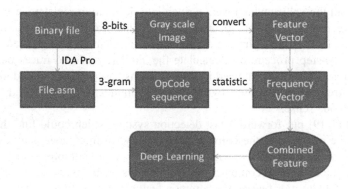

Fig. 1. The overview of feature extraction

The Fig. 2 shows the transformation of grayscale which is expressed into 256 grades, and the different grade denotes the different light between black to white [13, 14]. 0 grade stands for black, and 255 denotes white. If the image is described by the format of gray scale, the image is called gray scale image. To convert a binary file into a gray scale image, we put 8 bits binary into integer, which lies between 0 and 255. The image is mapped into one-dimensional vector. Figure 2 shows the process of converting a binary file into a gray scale image. The size of gray scale images are the same. Due to the size of binary files is different, so the size of their images is also different. The gray image does not change the texture feature by scaling down. Therefore, we are able to adjust the size of all images for a fixed size which is equal to 1000 byte in the paper.

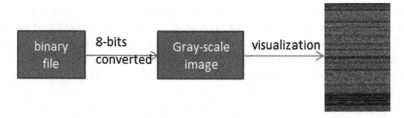

Fig. 2. Malware visualization

N-gram has been widely validated is that it distill the effective feature of malware [15]. The Fig. 2 shows the process. The operation codes are taken as a sequence, and they are arranged in their corresponding position. Form the Fig. 3, we can obtain an opcode sequence which is expressed into a set {mov, mov, xor, mov, call, mov, test, jz, call, pop, add}. In this paper, we adopt that n is equal to 3. After a lot of experiments, we found that the results will be better when n belongs to the range of 2 to 4. If n is more than 5, the results of experiments is more and more poor. The set includes 11 operation codes so that we get 9 characteristic components. We employ n-gram model to extract the feature of operation code. Firstly, we use IDA pro to translate the software into assembly file. Then, the frequency of each 3-gram sequence is the characteristic of operation code.

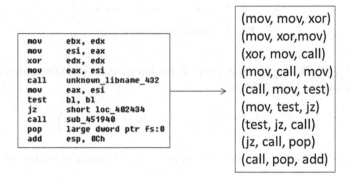

Fig. 3. OpCode 3-gram

3.2 Deep Learning Structure

In Fig. 4, we construct a deep feedback network consisting of seven layers, where the model contains the input layer and the output layer [16]. In the input layer, 3600 nodes represent that the feature contains 3600 attributes. The active function of each hidden layer is tanh-function. Each layer contains different number of neuron nodes, which is based on our experience. The output layer predicts the result using the softmax-function.

Formal Description: Let x be the input of the deep learning model. $y^{(l)}$ and $y^{(l+1)}$ are the input and output of $(l+1)$ layer, respectively. $W^{(l)}$ is a vector of parameters, which are the weights of layer. The weight represents that each component of input affect the prediction [17]. b^l is the bias, meanwhile f and F respectively express the activation functions of hidden and output layer. We give the formula of the output in $(l+1)$ layer.

$$y^{(l+1)} = f^{(l+1)}\left(y^{(l)} \cdot W^{(l+1)} + b^{(l+1)}\right) \tag{1}$$

Dropout: We employ the dropout way, which has been proved to be an effective in preventing the overfitting of deep learning and combining different neural network

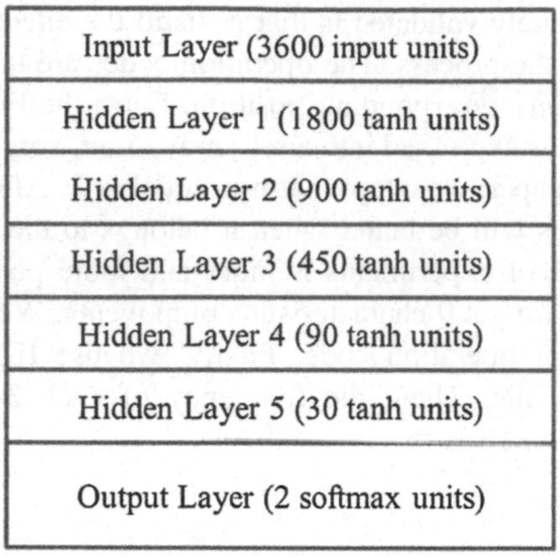

| Input Layer (3600 input units) |
| Hidden Layer 1 (1800 tanh units) |
| Hidden Layer 2 (900 tanh units) |
| Hidden Layer 3 (450 tanh units) |
| Hidden Layer 4 (90 tanh units) |
| Hidden Layer 5 (30 tanh units) |
| Output Layer (2 softmax units) |

Fig. 4. Deep learning structure

architectures [18]. During training, Dropout randomly removes some neurons. We have to modify formula (1) as follows:

$$y^{(l+1)} = f^{(l+1)}\left(y^{(l)} \cdot r^{(l+1)} \cdot W^{(l+1)} + b^{(l+1)}\right) \qquad (2)$$

where $r^{(l+1)}$ is the vector with $\left|y^{(l)}\right|$ component. The elements of vector are equal to 0 or 1 (Fig. 5).

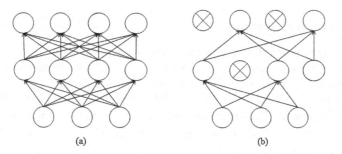

(a)   (b)

Fig. 5. (a) Standard Neural Network; (b) Dropout Model

4   Experiment

Dataset: In the last two weeks, we capture the malware using Kingsoft, ESET NOD32, and Anubis from our campus network. The malware contains 4,755 samples, which are from 10 malware families. Benign software is collected under 32-bit windows 7. It includes 2,100 samples from the clients of campus.

Algorithms: seven classical algorithms are from the sckit-learn of python 2.7, such as Random Forest (RF), SVM (kernel = 'poly'), Logistic Regression (Logistic), Gradient Boosting Classifier (Boosting), GaussianNB (GNB), Decision Tree Regressor (DT) and Multilayer perceptrons (MLP). It should be noted that MLP is the deep learning structure mentioned earlier.

Detection Accuracy: The Fig. 6 shows the detection accuracy of algorithms. MLP has the best performance, and its detection accuracy has reached 0.985. Boosting Classifier and Random Forest belong to the ensemble learning, so their accuracy also is higher than other algorithms. The picture also depicts the accuracy comparison of the algorithms, which represent 3-gram and combined method. The combined method integrates gray scale image and 3-gram. The average accuracy of the latter is 2.3% higher than that of former. The multilayer perceptron is still the best, and SVM performance is the most improved.

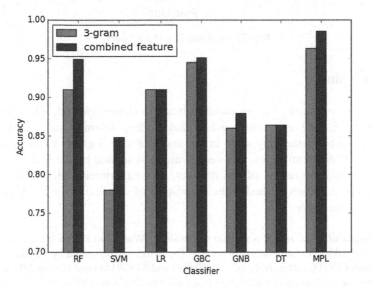

Fig. 6. Algorithm accuracy comparison

The detection performance often is evaluated in terms of precision and recall. Therefore, we use the Fig. 7 to depict the Precision-Recall rate curves of seven different algorithms using integrated features. As is shown in the diagram, the P-R curve of Multilayer perceptrons encases the other curves means that it has the best performance. The curve of SVM is located on the innermost side to indicate that its detection performance is the worst.

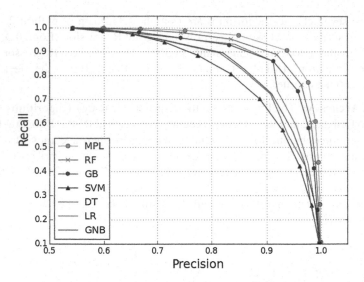

Fig. 7. Precision-Recall rate curve

5   Conclusion

In this paper, we employ an automatic malware detection method, which consists of two parts: a feature extraction method and a deep learning model. The feature extraction combines the gray scale image and OpCode 3-gram, which describe the malware from different angles. The deep learning is trained by the feature of training set, and predicts the results of test data set. The experiments prove our method is effective. The detection method is the foundation of our future work, which captures the unknown malware.

Acknowledgment. The author is grateful to Baosheng Wang and Bo Yu for the guidance and advice, and thanks to the support of the project. The work was supported by Science Foundation of China under (NSFC) (Nos. 61472437, 61303264 and 61379148) and National Basic Research and Development Program of China (973 Program, No. 2012CB315906)..

References

1. Gonzalez, L.E., Vazquez, R.A.: Malware classification using Euclide - an distance and artificial neural networks. In: 2013 12th Mexican International Conference on Artificial Intelligence (MICAI), pp. 103–108. IEEE (2013)
2. Zhong, Y., Yamaki, H., Takakura, H.: A malware classification method based on similarity of function structure. In: 2012 IEEE/IPSJ 12th International Symposium on Applications and the Internet (SAINT), pp. 256–261. IEEE (2012)
3. Tsyganok, K., Tumoyan, E., Babenko, L., et al.: Classification of poly-morphic and metamorphic malware samples based on their behavior. In: Proceedings of 5th International Conference on Security of Information and Networks, pp. 111–116. ACM (2012)

4. Sahu, M.K., Ahirwar, M., Shukla, P.K.: Improved malware detection technique using ensemble based classifier and graph theory. In: 2015 IEEE International Conference on Computational Intelligence & Communication Technology (CICT), pp. 150–154. IEEE (2015)
5. Lin, C.T., Wang, N.J., Xiao, H., et al.: Feature selection and extraction for malware classification. J. Inf. Sci. Eng. **31**(3), 965–992 (2015)
6. Liu, L.I.U., Wang, B.S., Bo, Y.U., Zhong, Q.X.: Automatic malware classification and new malware detection using machine learning. Frontiers **1** (2016)
7. Pascanu, R., Stokes, J.W., Sanossian, H., et al.: Malware classification with recurrent networks. In: 2015 IEEE International Conference on Acoustics, Speech and Signal Processing (ICASSP), pp. 1916–1920. IEEE (2015)
8. Annachhatre, C., Austin, T.H., Stamp, M.: Hidden Markov models for malware classification. J. Comput. Virol. Hack. Tech. **11**(2), 59–73 (2015)
9. Kang, B., Kim, H.S., Kim, T., et al.: Fast malware family detection method using control flow graphs. In: Proceedings of 2011 ACM Symposium on Research in Applied Computation, pp. 287–292. ACM (2011)
10. Gonzalez, L.E., Vazquez, R.A.: Malware classification using Euclidean distance and artificial neural networks. In: 2013 12th Mexican International Conference on Artificial Intelligence (MICAI), pp. 103–108. IEEE (2013)
11. Han, K.S., Lim, J.H., Kang, B., et al.: Malware analysis using visualized images and entropy graphs. Int. J. Inf. Secur. **14**(1), 1–14 (2015)
12. Strelkov, V.V.: A new similarity measure for histogram comparison and its application in time series analysis. Pattern Recogn. Lett. **29**(13), 1768–1774 (2008)
13. Han, K.S., Lim, J.H., Im, E.G.: Malware analysis method using visualization of binary files. In: Proceedings of 2013 Research in Adaptive and Convergent Systems, pp. 317–321. ACM (2013)
14. Kancherla, K., Mukkamala, S.: Image visualization based malware detection. In: 2013 IEEE Symposium on Computational Intelligence in Cyber Security (CICS), pp. 40–44. IEEE (2013)
15. Jain, S., Meena, Y.K.: Byte level n–gram analysis for malware detection. In: Venugopal, K.R., Patnaik, L.M. (eds.) Computer Networks and Intelligent Computing. CCIS, vol. 157, pp. 51–59. Springer, Heidelberg (2011). doi:10.1007/978-3-642-22786-8_6
16. Saxe, J., Berlin, K.: Deep neural network based malware detection using two dimensional binary program features. In: 2015 10th International Conference on Malicious and Unwanted Software (MALWARE), pp. 11–20. IEEE (2015)
17. LeCun, Y., Bengio, Y., Hinton, G.: Deep learning. Nature **521**(7553), 436–444 (2015)
18. Srivastava, N., Hinton, G.E., Krizhevsky, A., et al.: Dropout: a simple way to prevent neural networks from overfitting. J. Mach. Learn. Res. **15**(1), 1929–1958 (2014)

High-Level Multi-difference Cues for Image Saliency Detection

Jianwei Sun[1,2], Junfeng Wu[1,2], Hong Yu[1,2(✉)], Meiling Zhang[1,2], Qiang Luo[1,2], and Juanjuan Sun[1,2]

[1] College of Information Engineering, Dalian Ocean University,
Dalian 116023, China
yuhong@dlou.edu.cn
[2] Key Laboratory of Marine Information Technology of Liaoning Province,
Dalian 116023, China

Abstract. Salient detection approaches mainly use single local cues or global cues as its inputs features to detect salient objects, which are sensitive to complex background, so the effect of detection were not satisfactory. In this paper, we investigate the traits of saliency detection and observed the two following facts: Firstly, high-level saliency cues achieve better saliency detection results than low-level saliency cues. Secondly, multi-difference cues achieve better saliency detection results than single difference cues. Based on deeply analysis, we proposed an image saliency detection algorithm through high level multi-difference cues (HMDS). By using multi-difference, not only HMDS could remove the non-salient region effectively, but also it could enhance the pixel value of salient region at the same time. In order to evaluate the performance of HMDS, the proposed method is compared with seven state-of-the-art algorithms on five popular datasets. The final experimental results show that the proposed method performs effectiveness, and will have a perfect application prospect.

Keywords: High-level · Saliency detection · Multi-difference · Saliency map · Salient region

1 Introduction

Human vision is one of the important manners to observe and understand the world, about 80% information obtained from the outside world is obtained by human vision. Image is one of valuable information for human vision. When observe one image, people are not interested in all the objects in the pictures, some salient things in the image may get more attention. For instance, when human observe an image as Fig. 1 (a), most people's attention may focus on the "balloon" in the image, instead of the sky, trees or other background elements. That is to say, people are more concerned about the salient objects in the image. It is therefore most necessary for us to detect salient objects in the image, which could be solve many computer vision problems, such as object detection [1–3], image retrieval [4, 5], action recognition [6, 7]. So saliency detection becomes a research hotspot in computer vision and many researchers proposed lots of

© Springer Nature Singapore Pte Ltd. 2017
B. Zou et al. (Eds.): ICPCSEE 2017, Part I, CCIS 727, pp. 508–519, 2017.
DOI: 10.1007/978-981-10-6385-5_43

(a) original image   (b) our saliency map

Fig. 1. What are you interesting in this image?

effective methods to detect salient objects in the images in order to fit the visual characteristics of human eyes [8–10].

The saliency detection algorithms can be categorized as low-level cues and high-level saliency cues. The low-level algorithms could produce pixel-level saliency map [11] and high level algorithms could region-level saliency map [12]. Compared with pixel-level saliency map, the region-level saliency map which is detected based on higher level cues has greater difference. Furthermore, the region-level saliency map is influenced by the results of region segmentation. In order to improve the results of region segmentation, Felzenszwalb and Achanta proposed super-pixel algorithms respectively [13, 14]. Jiang and Margolin applied super-pixel in saliency detection and get better performance. However, there are two important factors which could influence the saliency detection results, one is interference from background, the other is difference between salient regions and background in the image. In order to improve the effectiveness of saliency detection, we should solve the following two problems:

(1) How to reduce the interference from background during the process of saliency detection?
(2) How to enlarge the contrast between salient regions and background?

Through deep research, we found that some work should be done image pattern and semantics information in the image. In fact, the pattern distance among image patches in PCA space should get a sparse and precise saliency map, and Gaussian prior weighting can be used to farther reduce the luminance in the background. Generally speaking, people pay more attention to the object in image than background. Besides, compared with other regions in the image, the object-level region can describe objects in the image better. By utilizing object-level region, the difference between object and background could be made greater. Based on the consideration above, we proposed a new saliency detection method which uses high-level multi-difference cue (HMDS).

To evaluate the performance of our algorithm, some experiments are made to compare the proposed method with seven state-of-the-art methods on 5 most popular datasets [15]. Simulation experimental results show that HMDS is efficient.

The rest of this paper is organized as follow: The related work is discussed in Sect. 2. We elaborate the principle of our method and describe HMDS algorithm in Sect. 3. Experiment results are presented in Sect. 4. Finally, our work is summarized.

2 Related Work

With the development of research, more and more methods are proposed to detect salient objects in the image. Achanta et al. proposed a pixel-level saliency detection algorithm based on global pixel difference [16]. And they got a better result by using the proposed evaluation method based on ground truth image. Ming-Ming Cheng proposed pixel-level saliency map method based on histogram contrast (HC) to measure saliency [8]. The algorithm can detect the salient region quickly and gets good results in the MSRA dataset. However, it is observed that the saliency detection should pay more attention to the region and object [8, 14]. So researchers proposed some region-level saliency map detection methods. Goferman et al. combined pixel-level saliency map and region-level saliency map together and used high-level factors to further optimize the result [17]. The quality of saliency map is better than the pixel level algorithm, however, there are still lots of noises in the background and the energy distribution of foreground is not well-distributed. Jiang proposed super-pixel segmentation based CBS algorithm to find object-level information [12]. It can distinguish salient region in the image and the gray value distribution in salient region is more uniform. But the residue of background is relatively obvious. Margolin think that images have different patch distributions [18], which can affect the accuracy of saliency detection. They proposed a pattern and color difference based saliency detection algorithm. The background of saliency map is quite clear but the luminance foreground is reduced at the same time.

Through studying the work mentioned above, we find that the object-level difference and difference of image patches distribution can affect the performance of saliency detection. Therefore, a new method is proposed to solve the problems.

3 High-Level Multi-difference Cues for Saliency Detection

High-level difference means that the difference is more abstract than the low-level difference, which can represent the difference of pixel better. Therefore high-level difference contain more effective information. According to information theory, the more effective information one thing has, the smaller information entropy, which means a smaller indetermination. So the saliency of region or image can be detected more easily using high-level difference. Based on this, we would like to use high-level difference to get a better saliency detection result rather than low-level difference. A high-level difference is found by contrast method. For example, Compared with pixel-level difference, the difference based on image patches is a higher level difference. Image patch could represent larger area than pixel, so the image patch could contain more information than pixel. Generally, the difference is greater between patches than pixels, which means that use patch-level difference has more possibility to get a good saliency results.

But compared with object-level difference, the patch-level difference may not have enough semantic information because object-level contains complete objects. Therefore using object-level difference, the saliency results can conform to human vision better. In conclusion, the object-level difference is a high-level difference. But it is difficult to calculate the object-level difference. On the one hand, it is difficult to define the range of object region [12]. The performance of object-level difference is based on the accuracy of object region. On another hand, object detection is effected by background. Because the distribution of image information is effected by the background, the same object in different background may have different saliency value. So we should evaluate the pattern of image and calculate the difference in the pattern to reduce the effect of background [18]. Based on the discussion above, we summarized four principles:

(1) Different images have different patch distributions and their salient patches should have far away the principal direction of distribution.
(2) The color of image patch may affect its saliency
(3) The salient object is always different from the background and a region with closed boundary may contain objects
(4) There may have high-level factors to optimize the saliency result.

Based on these principles, we propose a algorithm for saliency detection, which considered the advantages of pattern-level difference and object-level difference.

3.1 Saliency Detection Based on Pattern Distance

Image patch based saliency detection needs to calculate the pattern distance of image patches, which calculates from the patches to principal direction.

The pattern distance of image patches in PCA space should get a sparse and accurate saliency map. We use PCA method to calculate pattern distance.

When the distribution of patches in different image is very similar, the performance of pattern distance is poor. The color distance in LAB color space can be used to enhance the accuracy of the results (Fig. 2(c)). The distance is defined as formula (1) [18].

$$PD(p_x) = P(p_x) \cdot C(p_x) \tag{1}$$

where p_x is xth patch of image; $P(p_x)$ is the pattern distance of image. $C(p_x)$ is the color distance of image.

Gaussian weighting is generated by the map of pattern-level difference and used to give a center prior to reduce the luminance of background by the 4th principle and compared with others weighting methods, the Gaussian weighting is matching with human attention that human is more interesting in the center of the image than the surrounding (center-surrounding model). The saliency result made by pattern distance is defined as formula (2).

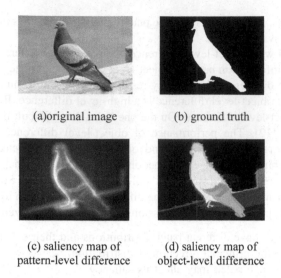

(a)original image   (b) ground truth

(c) saliency map of   (d) saliency map of
pattern-level difference   object-level difference

Fig. 2. The comparison of object-level and image patch saliency results

$$PS(p_x) = PD(p_x) \cdot G(p_x) \qquad (2)$$

where $G(p_x)$ is Gaussian weighting, which is generated by $PD(p_x)$.

3.2   Saliency Detection Based on Object-Level Difference

Based on the 3th principle, it is likely that the region with closed boundary contains objects. Therefore, the key point of object-level saliency detection is to find close boundary regions of image and calculate their color difference to get the saliency results (Fig. 2(d)).

Super-pixel algorithm is a classical image segmentation algorithm [13]. We use super-pixel algorithm to get the region surrounded with close boundary by iteration. The object-level difference-$OD(r_i)$ between these regions is calculated as formula (3)

$$OD(r_i) = \sum_{j=1}^{k} \alpha d_{color}(r_i, r_j) \quad (i \neq j, 1 \leq i \leq n, k < n) \qquad (3)$$

where r_i and r_j is the close boundary region, which likely has object. r_j is one neighbor of r_i. α is weighting of distance. d_{color} is the color distance between two regions of image.

Then the saliency value of region is calculated as (4),

$$OS(r_i) = -w_i \ln(1 - OD(r_i)) \qquad (4)$$

where w_i is the area weighting of region r_i.

The greater the color difference of the region is, the greater the saliency value of the region is.

The Gaussian pyramid is applied to fitting the center-surrounding model as above.

3.3 A Method with Complementary Advantages

Both pattern-level and object-level difference are high-level difference, they have distinct characteristic respectively. The pattern distance based method can get an accuracy result but its luminance of foreground is not high enough, because the pattern-level difference is calculated in PCA space where the information is compressed, therefore the saliency map is sparse and accurate. The object-level based method has high contrast between foreground and background but the luminance of background is not low enough. The reason is that compared with pixel and image patch, the object-level difference is calculated in object-level region, which has larger regions and the semantic information is clearer, therefore the salient object which use object-level can get greater difference but more detail information is lost in larger regions. Focus on the two problems, we use collaborative and priority of weight thinking to combine the advantage of two methods. So that not only the objects are contained in salient region but also the background is removed effectively (Fig. 3).

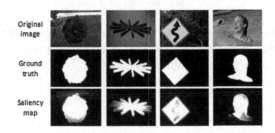

Fig. 3. A method of complementary advantages **results**

The pattern-level and object-level saliency results should be normalized respectively (as formula 5 and 6) to unify the standard of difference at first.

$$PS^* = PS./\max(\max(PS)) \tag{5}$$

$$OS^* = OS./\max(\max(OS)) \tag{6}$$

These two kind of saliency results are processed based on a collaboration thinking as (7),

$$HS_{tmp1} = PS^* \cdot OS^* \tag{7}$$

where '·' is the multiplication by bit, the salience value could become greater only if both the value of PS^* and OS^* become greater at the same time, otherwise, the salience value could become smaller, which could reflect the collaborative thinking.

The new saliency map is calculated based on the priority of weight thinking.

$$HS_{tmp2}^{*i} = \begin{cases} w_1 \cdot HS_{tmp1}^{*i} & if \ \left(HS_{tmp1}^{*i} \geq m \right) \\ HS_{tmp1}^{*i} & if \ \left(HS_{tmp1}^{*i} < m \right) \end{cases} \tag{8}$$

where HS_{tmp2}^{*i} is the one saliency value of Saliency map. m is the mean value of HS_{tmp1}^{*}. Because these pattern-level and object-level salience results are generated by Gaussian weighting. In this paper, we define weighting coefficient as $w_1 = 2$.

Finally, based on priority weighting, we selectively superpose these saliency values as formula 9, which not only could reduce the noise of saliency map but also optimize its luminance.

$$HS = w_2 \cdot HS_{tmp2} + w_3 \cdot PS^* \tag{9}$$

where w_2, w_3 are the weightings of HS_{tmp2} and PS^*. In this paper, we define weighting coefficient as $w_2 = 1$, $w_3 = \frac{1}{2}$. We suggest that these weightings should enhance the performance of salience result, at the same time, prevent the saliency value not to big. The relationship of these weightings are defined as (10):

$$w_1 \cdot w_2 \cdot w_3 = 1 \tag{10}$$

The saliency result is generated by pattern-level difference and object-level difference by using the complementary method. In conclusion, we suggest the saliency result is generated by high-level multi-difference.

4 Experiment

A series of experiments are designed to evaluate the performance of the proposed algorithm (HMDS). The experiments are implemented with MATLAB 2012b in windows7. The configuration of computer is the Intel(R) Core i3-4150 CPU 3.5 GHz and 4.00 GB memory.

Dataset: We have evaluate our algorithm on the 5 most popular dataset for saliency detection, which all have original images and corresponding ground truth image of them:

(1) **MSRA** [19]: It has 5000 images and the ground truth image was labeled by 9 different person, which is the biggest dataset for saliency detection study nowadays.
(2) **ASD** [16]: It has 1000 images from MSRA, which a more fine about ground truth image.
(3) **SED1** [20]: 100 images and every ground truth image has a single object.
(4) **SED2** [20]: 100 images and every ground truth image has two objects.
(5) **SOD** [21]: 300 image and every ground truth image was made by 7 different users.

Baseline: We have compared our algorithm with other 7 mainly popular algorithms: PCAS [18], CBS [12], FT [16], HC [8], LC [11], RC [8], SR [16]. The results of different algorithm are all provided by authors which are running available codes or soft wares. We run them and get the results on the same software and hardware.

Parameters and Evaluation Metrics: We evaluate all methods using precision-recall curves which are obtained by segmenting the saliency maps with a threshold sliding from 0 to 255, and compare these binary results with their ground truth so as to test their accuracy. Considering that the high precision and recall are both important for a good result, we use F-measure curves to make the overall performance measurement [1]. In order to make the evaluation more intuitive, The AUC (Area-Under-The-Curve) is used to evaluate the accuracy of these algorithms on the 5 datasets and the Mean AUC is used to evaluate the overall robustness and accuracy of them. Finally we also calculate the running time of our algorithm.

Validation of the HMDS: Figure 4 presents an intuitive comparison of HMDS with the most famous current methods about saliency detection on the 5 image datasets. From these images, we can see that HMDS map generally have greater luminance of saliency region and at the same time, the luminance of background also has been reduced lower than other baseline algorithms. Therefore, HMDS results are consistently more accurate than those of other methods according to human vision.

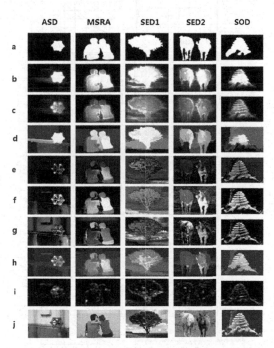

a: Ground truth; b: HMDS; c: PCAS; d: CBS; e: FT; f: HC; g:LC h: RC; i: SR; j: Original image;

Fig. 4. Visual comparisons with state of the art methods

As shown in Fig. 5, the proposed algorithm has higher precision and recall values on the 5 datasets. Besides, we can find that these F-measure values of HMDS are higher in a wider range than others on the 5 datasets. Especially, HMDS has a remarkable advantage when the threshold value is smaller than 0.5, to some extent, which can prove that HMDS can get a more accuracy saliency result using a more loose condition than other algorithms.

Aiming to compare these results more concisely, we calculate the AUC and mean AUC values of these 8 kinds of algorithms as Figs. 6 and 7. From the Fig. 6 we can easily find that HMDS get the highest AUC value in every test dataset and the Fig. 7 is

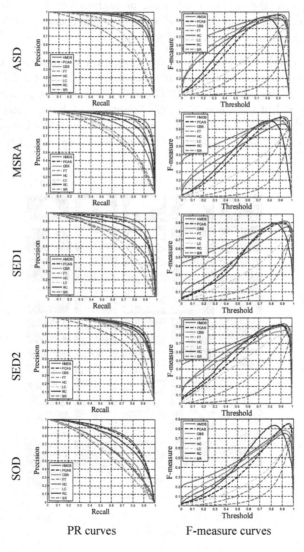

PR curves   F-measure curves

Fig. 5. The PR curves and F-measure curves of 8 algorithms on 5 datasets

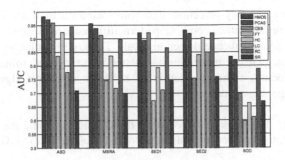

Fig. 6. AUC result

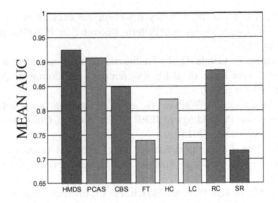

Fig. 7. Mean AUC result

the comprehensive evaluation result of these algorithm in all the test-datasets and it can prove that our algorithm has the most extensive robustness compared with other 7 baseline algorithms.

In summary, through series experiments we can get the conclusion that the algorithm proposed in the paper has great accuracy and robustness for the image saliency detection question.

Running Time: HMDS focus on accuracy and robustness. In general, the more accurate the result is, the more time we need to process, while we use a parallel computing method to speed up our algorithm. It takes on average 3.961 s to generating a saliency map on the 5 dataset via Matlab with a PC equipped with Intel(R) Core i3-4150 CPU 3.5 GHz and 4.00 GB memory.

5 Conclusion

In this paper, we introduce saliency detection and its development. Specific in how to make the saliency region more prominent, at the same time the background be removed more clearly.

We proposed a saliency detection algorithm based on high-level multi-difference cues (HMDS). The algorithm combine pattern difference and object-level difference together to enhance the salience detection result. Through simulation experiments, we get the conclusion that our algorithm has better accuracy and robustness. In future work, we will focus on the application of saliency detection algorithm in image retrieval and image segmentation.

Acknowledgement. This work is supported by the General Scientific Research Program of Universities in Dalian City Scientific and Technological Projects (2015A11GX022).

References

1. Qin, Y., Lu, H., Xu, Y., et al:. Saliency detection via cellular automata. In: 2015 IEEE Conference on Computer Vision and Pattern Recognition (CVPR), pp. 110–119. IEEE Computer Society (2015)
2. Frintrop, S., Werner, T., García, G.M.: Traditional saliency reloaded: a good old model in new shape. In: Proceedings of IEEE Conference on Computer Vision and Pattern Recognition, pp. 82–90 (2015)
3. Mauthner, T., Possegger, H., Waltner, G., et al.: Encoding based saliency detection for videos and images. In: Proceedings of IEEE Conference on Computer Vision and Pattern Recognition, pp. 2494–2502 (2015)
4. Wan, S., Jin, P., Yue, L.: An approach for image retrieval based on visual saliency. In: International Conference on Image Analysis and Signal Processing, IASP 2009, pp. 172–175. IEEE (2009)
5. Chen, T., Cheng, M.M., Tan, P., et al.: Sketch2Photo: internet image montage. ACM Trans. Graph. (TOG) **28**(5), 124 (2009)
6. Vig, E., Dorr, M., Cox, D.: Space-variant descriptor sampling for action recognition based on saliency and eye movements. In: Fitzgibbon, A., Lazebnik, S., Perona, P., Sato, Y., Schmid, C. (eds.) ECCV 2012. LNCS, vol. 7578, pp. 84–97. Springer, Heidelberg (2012). doi:10.1007/978-3-642-33786-4_7
7. Rapantzikos, K., Avrithis, Y., Kollias, S.: Dense saliency-based spatiotemporal feature points for action recognition. In: IEEE Conference on Computer Vision and Pattern Recognition, CVPR 2009, pp. 1454–1461. IEEE (2009)
8. Cheng, M., Mitra, N.J., Huang, X., et al.: Global contrast based salient region detection. IEEE Trans. Pattern Anal. Mach. Intell. **37**(3), 569–582 (2015)
9. Itti, L., Koch, C., Niebur, E.: A model of saliency-based visual attention for rapid scene analysis. IEEE Trans. Pattern Anal. Mach. Intell. **11**, 1254–1259 (1998)
10. Borji, A., Itti, L.: State-of-the-art in visual attention modeling. IEEE Trans. Pattern Anal. Mach. Intell. **35**(1), 185–207 (2013)
11. Zhai, Y., Shah, M.: Visual attention detection in video sequences using spatiotemporal cues. In: Proceedings of 14th Annual ACM International Conference on Multimedia, pp. 815–824. ACM (2006)
12. Jiang, H., Wang, J., Yuan, Z., et al.: Automatic salient object segmentation based on context and shape prior. In: BMVC, vol. 6, no. 7, p. 9 (2011)
13. Felzenszwalb, P.F., Huttenlocher, D.P.: Efficient graph-based image segmentation. Int. J. Comput. Vis. **59**(2), 167–181 (2004)
14. Achanta, R., Shaji, A., Smith, K., et al:. SLIC superpixels (2010)

15. Borji, A., Sihite, D.N., Itti, L.: Salient object detection: a benchmark. In: Fitzgibbon, A., Lazebnik, S., Perona, P., Sato, Y., Schmid, C. (eds.) ECCV 2012. LNCS, pp. 414–429. Springer, Heidelberg (2012). doi:10.1007/978-3-642-33709-3_30
16. Achanta, R., Hemami, S., Estrada, F., et al.: Frequency-tuned salient region detection. In: IEEE Conference on Computer Vision and Pattern Recognition, CVPR 2009, pp. 1597–1604. IEEE (2009)
17. Goferman, S., Zelnik-Manor, L., Tal, A.: Context-aware saliency detection. IEEE Trans. Pattern Anal. Mach. Intell. 34(10), 1915–1926 (2012)
18. Margolin, R., Tal, A., Zelnik-Manor, L.: What makes a patch distinct? In: IEEE Conference on Computer Vision and Pattern Recognition, (CVPR), pp. 1139–1146. IEEE (2013)
19. Liu, T., Yuan, Z., Sun, J., et al.: Learning to detect a salient object. IEEE Trans. Pattern Anal. Mach. Intell. 33(2), 353–367 (2011)
20. Alpert, S., Galun, M., Brandt, A., et al.: Image segmentation by probabilistic bottom-up aggregation and cue integration. IEEE Trans. Pattern Anal. Mach. Intell. 34(2), 315–327 (2012)
21. Movahedi, V., Elder, J.: Design and perceptual validation of performance measures for salient object segmentation. In: CVPRW, vol. 5, pp. 49–56 (2010)
22. Hou, X., Zhang, L.: Saliency detection: a spectral residual approach. In: IEEE Conference on Computer Vision and Pattern Recognition, CVPR 2007, pp. 1–8. IEEE (2007)

Nonlinear Dimensionality Reduction via Homeomorphic Tangent Space and Compactness

Shaoqun Zhang$^{(\boxtimes)}$ and Wanyun Xie

Sichuan University, Chengdu 610024, China
zsqhndn@126.com

Abstract. This paper establishes the geometric framework of manifold learning. After summarizing the requirements of the classical manifold learning methods, we construct the smooth homeomorphism between the manifold and its tangent space. Then we propose a new algorithm via homeomorphic tangent space (LHTS). We also present another algorithm via compactness (CSLI) by analyzing the topological properties of manifolds. We illustrate our algorithm on the completed manifold and non-completed manifold. We also address several theoretical issues for further research and improvements.

Keywords: Manifold learning · Smooth homeomorphism · Homeomorphic tangent space · Compactness

1 Introduction

With the quick advancement and extensive application of information technology, data with high dimension and complex structure occurs very quickly. High dimension not only makes the data hard to understand, and makes traditional machine learning and data mining techniques less effective. Dimension reduction is one of the important techniques to deal with high-dimensional data. The methods of dimensionality reduction can be divided into two categories, one is linear dimensionality reduction, such as Principal Component Analysis (PCA), Nonnegative Matrix Factorization (NMF) and Multi-dimensional Scaling Algorithm (MDS) [1, 2], etc.; the other is nonlinear dimensionality reduction, such as Kernel Principal Component Analysis (KPCA) [1, 2], Isomorphic Mapping Algorithm (Isomap) [3], Local Linear Embedding (LLE) [4] and Local Tangent Space Arrangement (LTSA) [5].

Manifold learning is a recent popular approach to non-linear dimensionality reduction. Algorithms for this task are based on the idea that the dimensionality of data sets is only artificially high, and the data points are actually samples from a low-dimensional manifold that is embedded in a high-dimensional space. Manifold learning algorithms attempt to find the low-dimensional representation of the data. Manifold learning can be broadly divided into two categories, one is global optimization algorithm, such as Isomorphic Mapping (Isomap); the other is local optimization algorithm, such as Local Linear Embedding (LLE), Local Tangent Space Arrangement (LTSA).

© Springer Nature Singapore Pte Ltd. 2017
B. Zou et al. (Eds.): ICPCSEE 2017, Part I, CCIS 727, pp. 520–533, 2017.
DOI: 10.1007/978-981-10-6385-5_44

The research on manifold learning has attracted many scholars just in recent years and many problems are still unsolved. The contributions of this paper are as follows:

- Establish the geometric framework of manifold learning based on the theories of fundamental mathematics;
- Summarize the classical manifold learning methods, present the new algorithm via homeomorphic tangent space and another algorithm via compactness;
- We give some synthetic examples to compare our proposed algorithms with Isomap, LLE and LTSA, and the experimental results show the efficiency of the proposed new methods in this paper.

2 Mathematical Framework of Manifold Learning

Firstly, we introduce several important concepts and symbols about manifolds [6–9].

Definition 1. (M, σ) is a Hausdorff space, if for any $x \in M$, there exists an open neighborhood $U_x \in \sigma$ that satisfies U_x is homeomorphic to an open subset of Euclidean space R^d, Then we call (M, σ) **topological manifold of dimension** d or **topological** d**-manifold**, abbreviated as M. And (U_x, φ_x) is called **the coordinate chart** of M, φ_x is a local homeomorphism.

Definition 2. Let M be a topological manifold of dimension d, if there is a C^∞ structure $\Sigma = \{(U_\alpha, \varphi_\alpha), \alpha \in I\}$ which is the maximum covering structure, then we call (M, Σ) d**-dimensional smooth manifold.**

Definition 3. Let M, N are two manifolds and there is a mapping $f: M \rightarrow N$. If f is a bijection, f and f^{-1} are continuous, then we name f as homeomorphism. Specially, if M, N are smooth, then we call f is a **smooth homeomorphism** or **diffeomorphism.**

Definition 4. Let M, N are non-empty sets, $f: M \rightarrow N$ is a linear map. If for any $x \in M$, the equation

$$\|f(x)\|_N = \|x\|_M$$

is held, then f is called an **operator preserving norm.**

Definition 5. For any $x \in M$, the set of all tangent vectors of x is called **the tangent space of M at the point x**, noted by $T_x(M)$. And the dimension of $T_x(M)$ is d or $dim(T_x(M)) = d$.

After introducing the above concepts, we redescribe the manifold learning. Assuming that the data set is distributed over a smooth manifold M in a high-dimensional Euclidean space R^m and for $\forall x \in M$, $\exists U_x$ homeomorphic to an open subset of Euclidean space $R^d (d < m)$, then there exists a mapping

$$f: M \rightarrow N \subset R^d$$

so that N is a d-dimensional geometric structure in R^d. The geometry N is the result of dimension reduction. Obviously, M and N must be smooth and maintain the same topology. The mapping f is the relationship between M and N. To analysis the mapping f, we quote a theorem in [10].

Theorem 1. $f: M \to N$ is a homeomorphism. As a result, any property of M that is entirely expressed in terms of the topology of M (that is, in terms of the open sets of M) yields, via the correspondence f, the corresponding property for the space N. And such a property of M is called topological property of M.

By this theorem we can conclude that f is a smooth homeomorphism at least.

We summarize the requirements of mapping f in the classical manifold learning algorithms. Isomap requires f must be an operator preserving norm. Let $\|\cdot\|_M$ be the geodesic distance on M and $\|\cdot\|_N$ be the Euclidean distance on R^d. In this way, the distance relationship between any two points of M is projected onto N by f. Since M and N are equivalent structures, if M is a convex set in R^m, then N is also a convex set in R^d [3, 11]. Convex set has good connectivity, so Isomap can get an effective embedding result. On the contrary, if M is not a convex set, such as the manifold with "holes", the effect will be much worse. LLE, LTSA HLLE, and other local algorithms only need mapping f is a smooth homeomorphism [4, 5, 12]. As a condition, the operator preserving norm is much stronger than the homeomorphism. In general, the requirements of the stronger conditions are more strict, so the results under the stronger conditions will be more effect. But the stronger conditions are difficult enough to be meet, if there is a little difference, the result is worse.

Remark: The sample set is the discretization result of the ideal smooth manifold. So for convenience, we use the symbol M to denote both a smooth manifold and a sample set $M = \{x_i\}_{i=1}^l, x_i \in R^m$.

3 The Algorithm via Homeomorphic Tangent Space

3.1 Theories of Homeomorphic Tangent Space

At first, we introduce an important conclusion about the tangent space [7, 8].

Theorem 2. Let M be a d-dimensional smooth manifold. For $\forall x \in M$, $T_x(M)$ is its corresponding tangent space, then $T_x(M)$ is a d-dimensional real linear space.

We define the 2-norm $\|\cdot\|_2$ on $T_x(M)$, then $T_x(M)$ is equivalent to the Euclidean space R^d. So N is a subset of $T_x(M)$. Therefore, the original description of manifold learning can be transformed as follows, for a certain point $x \in M$, the corresponding open neighborhood which is homeomorphic to R^d is noted by U_x and the corresponding tangent space is $T_x(M)$, then there exists a smooth homeomorphism $F: M \to N \subset T_x(M)$.

We use $V_x = \{v_x^1, \ldots, v_x^d\}$ as a set of bases of the tangent space $T_x(M)$. The tangent vector $v_y \in N$ has uniquely certain coordinates $(\theta_x^1, \ldots, \theta_x^d)$ in $T_x(M)$. And a certain point $y \in M$ is uniquely determined corresponding to these coordinates by F. In other words, for $\forall y \in M$, the point y is 1–1 corresponding to the coordinates $(\theta_x^1, \ldots, \theta_x^d)$, which are the global coordinates in $N \subset T_x(M)$. In fact, we cannot get the mathematical expression of the global mapping F, we can only get the local homeomorphism—the coordinate chart (U_x, φ_x). So we try to find the local coordinates. Then we introduce the following theorem.

Theorem 3. Let M be a d-dimensional smooth manifold, $f: M \to N$ is a smooth homeomorphism, and for a certain point $x \in M$, there is a corresponding tangent space $T_x(M)$. Then for $\forall y \in M$, there is an uniquely corresponding tangent vector $v_y \in T_x(M)$ s.t.

$$F(y) = v_y.$$

In particular, if $y \in U_x$, then there exists the homeomorphism φ_x s.t.

$$\varphi_x(y) = v_y.$$

Theorem 3 ensures that the local coordinates of any point within the neighborhood of x are available.

3.2   Find the Local Coordinates

Let the k-neighborhood points of the sample point x includes itself be $= \{x_i\}_{i=1}^{k}(k > d)$. Since $\varphi_x: U_x \to R^d$, then $X - x1_k^T$ are the k tangent vectors in $T_x(M)$. We use the singular value decomposition of the matrix $X - x1_k^T$ to get the largest d singular values and the corresponding right singular vectors. Obviously, these d right singular vectors are a set of bases of $T_x(M)$, noted by V_x [12].

Thus we can obtain the corresponding coordinates of $\{x_i\}_{i=1}^{k}$ in $T_x(M)$:

$$\vartheta_x = \left\{ \left(\theta_{xi}^1, \ldots, \theta_{xi}^d \right)^{\mathrm{T}} \right\}_{i=1}^{k} \tag{1}$$

This set of coordinates represents the original topology of set $\{x_i\}_{i=1}^{k}$. The topological relationship is local, so the coordinates ϑ_x can only be called local coordinates.

We can also make the assumption to proof it. If there are two certain points x and y, U_x and U_y are the corresponding neighborhoods and $U_x \cap U_y \neq \emptyset$, V_x and V_y are the bases of $T_x(M)$ and $T_y(M)$ respectively, then there are a coordinate transformation matrix P_{xy} s.t.

$$P_{xy}V_x = V_y. \tag{2}$$

Then for $\forall z \in U_x \cap U_y$, there are corresponding coordinates ϑ_x and ϑ_y under the bases V_x and V_y. If (1) can represent the global coordinates of N, then

$$P_{xy}\vartheta_x = \vartheta_y \tag{3}$$

must be held. Thus the manifold M is a linear hyperplane. Obviously, this conclusion is not true in fact. So the coordinates ϑ_x can only be called local coordinates.

Now we have successfully projected the local structure on M to the local coordinates in R^d by the homeomorphism. The next step we have to do is to project the local coordinates onto the global coordinates.

3.3   Get the Global Coordinates

Now, there are many ways to solve the problem of projecting the local coordinates onto the global coordinates, such as Neural Network, Ensemble Learning [1, 2] etc. In this section, we propose a linear optimal method.

For a certain point $x \in M$, V_x and ϑ_x are the corresponding bases of the tangent space and the local coordinates respectively. For $\forall y \in M$, we have corresponding V_y, ϑ_y and the coordinate transformation matrix P_{yx}, which satisfies $V_x = P_{yx} V_y$.

We note

$$\vartheta_x^y = P_{yx} \vartheta_y \tag{4}$$

Then ϑ_x^y represent the coordinates of k-neighborhood points $\{y_i\}_{i=1}^k$ under the bases V_x. For convenience, (4) is called the unity of coordinates. We use the unity of coordinates to calculate all the points of M. Then we can get the local coordinates $\bigcup_{y \in M} \vartheta_x^y$ under the bases V_x. Obviously, there are $k \cdot l$ local coordinate vectors in the set $\bigcup_{y \in M} \vartheta_x^y$. Let $\left\{ \left(\theta_x^1, \ldots, \theta_x^d \right)_x^T, x \in M \right\}$ be the set of global coordinates of N. K_x is noted as the number of occurrences of x in all neighborhoods. And set $\left\{ \left(\theta_{xj}^1, \ldots, \theta_{xj}^d \right), j = 1, \ldots, K_x \right\}$ represent all local coordinates of x corresponding to $\left(\theta_x^1, \ldots, \theta_x^d \right)$. Then there exists the set of mappings $\{ g_i \}_{i=1}^l$ s.t.

$$\left(\theta_x^1, \ldots, \theta_x^d \right) = g_i \left(\theta_{xj}^1, \ldots, \theta_{xj}^d \right) + \epsilon_i, \forall x \in M \tag{5}$$

In Sect. 3.2, we mentioned (3) is not held. It is worth noting that for $\forall \left(\theta_{xi}^1, \ldots, \theta_{xi}^d \right) \in \vartheta_x$, corresponding $\left(\theta_{yi}^1, \ldots, \theta_{yi}^d \right) \in \vartheta_y$, there is

$$\frac{P_{xy} \left(\theta_{xi}^1, \ldots, \theta_{xi}^d \right)^T}{\left(\theta_{yi}^1, \ldots, \theta_{yi}^d \right)^T} = \gamma \tag{6}$$

The ratio γ should be a constant vector. (6) shows that the local coordinates of x obtained by the unity of coordinates are linearly related. This means that the elements of $\{ g_i \}_{i=1}^l$ are allowed to be linear. This is a very important conclusion. Thus, we can use the classical manifold learning algorithms LLE and LTSA to solve the problem from the perspective of optimization [4, 5]. It is worth mentioning that, if we use the optimization method in LTSA, its arrangement matrix A should be defined as follows:

$$A = \sum_{i=1}^{l} \left(S_i (I - V_x)(I - V_x)^{\mathrm{T}} S_i^{\mathrm{T}} \right) \tag{7}$$

And $S_i \in R^{l \cdot k}$ is the corresponding selection matrix.

Finally, we organize the steps of the algorithm via homeomorphic tangent space (LHTS):

- Firstly, we prove the equivalence relation between low-dimensional Euclidean space R^d and tangent space $T_x(M)$, $\forall x \in M$;
- Secondly, project the neighborhood U_x into $T_x(M)$ by the local homeomorphism φ_x;
- Thirdly, construct the bases of $T_x(M)$ and obtain the local coordinates of U_x;
- Finally, integrate the local coordinates to get the global coordinates.

3.4   Experimental Results

Then we design an experiment to verify the superiority of the LHTS algorithm. In the experiment, we compare LTSA with LHTS. We extract $l = 2000$ points from "S"-shaped manifold in the interval $[0, 2\pi] \times [0, 1] \times [-1, 1]$. The original sample set and the ideal dimension reduction result are shown in Figs. 1 and 2. We take the number of neighbors $k = 20$ and the experimental results of LTSA and LHTS are shown in Figs. 3 and 4.

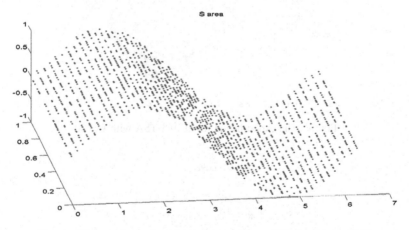

Fig. 1. The original sample set extracted from "S"-shaped manifold

With the comparison between Figs. 3 and 4, we find that after a coordinate transformation, Fig. 3 is basically the same to Fig. 4. And the computational complexity of LHTS is basically the same as the LTSA. But there are some differences:

- Almost the whole LTSA algorithm is present on the optimized theory; this paper pays more attention to the basic theory of mathematics and adds many important conclusions;

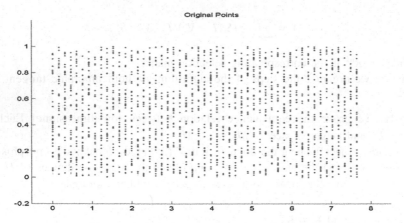

Fig. 2. The ideal result of dimension reduction

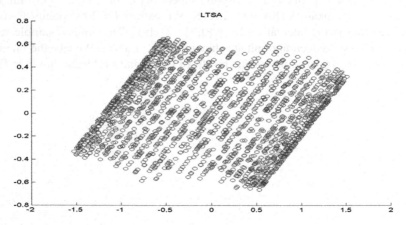

Fig. 3. The experimental result of LTSA with $k = 20$

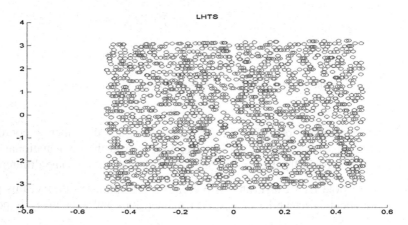

Fig. 4. The experimental result of LHTS with $k = 20$

- The local coordinates ϑ_x are different;
- The proof of the linear relationship between the local coordinates is also a slight difference.

4 The Algorithm via Compactness

Isomap suppose that the mapping $f: M \rightarrow N$ is an operator preserving norm and the sample set M (manifold) is a convex set in R^m. As mentioned in Sect. 2, the convex is a stronger condition which is too difficult to be meet. And the large computational complexity of Isomap is another shortcoming [13]. L-Isomap is an improved dimensionality reduction algorithm. It takes advantage of landmark points and L-MDS which greatly reduces the computational complexity. But the number and the landmark points are random, so it is very unstable.

In this chapter, we present an improved manifold learning algorithm via L-Isomap and the compactness of manifolds. In our algorithm, we maintain the assumption that the mapping $f: M \rightarrow N$ is an operator preserving norm. And we also suppose that M is a bounded closed set. Bounded closure is much weaker than convex [9] and ubiquitous. In order to deeply analysis the bounded closure, we should introduce the compactness.

4.1 Compactness and Topology

Definition 6. The set M in the distance space X is compact, if every open covering of M has a finite sub-covering. The set M is also called **a compact set** [8, 9].

The compactness is a topological property, if set M is a compact set, then there is a finite covering structure. We take Euclidean space X for example, if M is a compact set on X, there are a finite sequence $\{\alpha_i\}_{i=1}^n \subset M$ and its neighborhood set $\{U_i\}_{i=1}^n$, satisfying $M \subset \prod_{i=1}^n U_i$.

Theorem 4. Compact set in finite-dimensional Euclidean space is equivalent to bounded closed set [8].

The theorem shows that the manifold M in which the sample set is located is a compact set in R^m. According to the Definition 6, there are a finite sequence $\{\alpha_i\}_{i=1}^n$ and its neighborhood set $\{U_i\}_{i=1}^n$ can cover M.

Then we try to analysis the operator preserving norm f to get the properties of N. First, f is homeomorphism. According to Theorem 1, N is a compact set in R^d. So there are a finite sequence $\{\beta_i\}_{i=1}^n$ and neighborhood set $\{\tilde{U}_i\}_{i=1}^n$ can cover N. And then, f is bijection. So the relation between $\{\alpha_i\}_{i=1}^n$ and $\{\beta_i\}_{i=1}^n$ is 1–1 correspondence. After that, f is preserving norm. So the set $\{U_i\}_{i=1}^n$ is 1–1 corresponding to the set $\{\tilde{U}_i\}_{i=1}^n$.

In this case, we can replace all points of M by the landmark points set $\{\alpha_i\}_{i=1}^n$, which greatly reduces the computational complexity. It also shows that the landmarks cannot be randomly selected. Now we need to select the appropriate finite sequence $\{\alpha_i\}_{i=1}^n$ and the neighborhood set $\{U_i\}_{i=1}^n$.

4.2   Compact Selection

In the theoretical analysis, the selection of $\{\alpha_i\}_{i=1}^n$ and $\{U_i\}_{i=1}^n$ is various, but in the actual data experiment, it must be representative and robustness. We introduce the selection method as follows.

Firstly, we find two points from M:

$$dist(x_i, x_j) = \max dist(x_t, x_s), t, s = 1, \ldots, N$$

$dist(\cdot, \cdot)$ is 2-norm in Euclidean space. Let x_i (or x_j) be the first landmark point α_1, and compute its nearest k points to make up the neighborhood U_1. Then select second landmark point α_2 and its neighborhood U_2 from the remaining sample set M/U_1 by the same way like (4.2.1). And so on, we can collect all landmark points $\{\alpha_i\}_{i=1}^n$ and the neighborhood set $\{U_i\}_{i=1}^n$. This method of selection can ensure that the finite sequence $\{\alpha_i\}_{i=1}^n$ is uniformly dispersed and the neighborhoods are partially over-lapped. We name this method as Compact Selection. It is easy to calculate that the computational complexity of the Compact Selection is $O(l)$.

After that, we combine the Compact Selection with the L-Isomap algorithm [13]. The new algorithm is called CSLI.

As we all know, the computational complexity of the Isomap algorithm is very large. So we should compute the computational complexity of CSLI. The steps and the computational complexity of CSLI are as follows [3, 11, 13–15]:

- Select k-nearest neighborhood and construct the undirected graph G. Its computational complexity is $O(ml^2)$;
- Compact Selection. Its computational complexity is $O(l)$;
- Calculate the geodesic distance matrix between the landmark points and the sample points. It's $O(n^2 l)$;
- L-MDS. It is $O(nl \log l)$.

4.3   Experimental Results

We design two experiments to verify the stability and validity of the algorithm CSLI. In the experiment, we compare CSLI with Isomap, L-Isomap and LLE on the completed manifold and non-completed manifold. The first experiment still use the experimental data in Sect. 3.4. The number of neighbors $k = 20$, the number of landmark points $n = 100$. Then the experimental results of Isomap, L-Isomap, CSLI, and LLE are shown in Figs. 5, 6, 7 and 8.

We can see that the results of Figs. 5 and 7 are basically the same. It shows that the CSLI algorithm can maintain the faithfulness of the Isomap algorithm. Figure 6 differs to Fig. 5. It shows that L-Isomap reduces the computational complexity, but it cannot guarantee a great effect of dimensionality reduction. Figure 8 is the result of LLE. There are two obvious twisted lines (red lines in Fig. 8) where the principal curvatures of "S" manifold are larger. It is shown that the LLE algorithm is more sensitive to the curvature of the manifold.

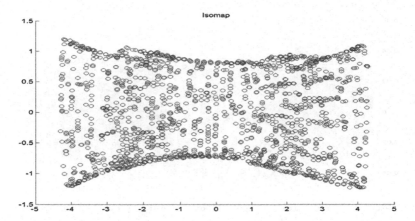

Fig. 5. The experimental result of Isomap with $k = 20$

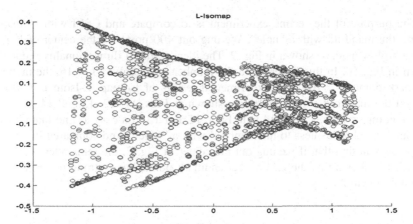

Fig. 6. The experimental result of L-Isomap with $k = 20$ and $n = 100$

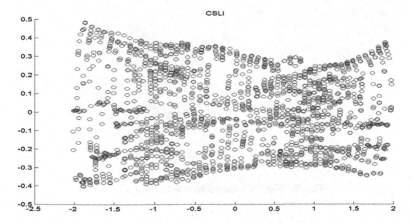

Fig. 7. The experimental result of CSLI with $k = 20$ and $n = 100$

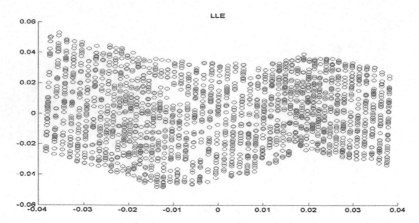

Fig. 8. The experimental result of LLE with $k = 20$

The purpose of the second experiment is to compare and CSLI with Isomap and LLEon the manifold with a "hole". We dug out 400 points in the center of Fig. 1 to form a hollow hole, as shown in Fig. 9. The ideal result of dimensionality reduction is shown in Fig. 10. In this experiment, the number of neighbors $k = 16$, the number of landmark points $n = 100$. The experimental results of Isomap, L-Isomap, CSLI, and LLE on the manifold with a "hole" are shown in Figs. 11, 12, 13 and 14.

We compare Figs. 13 and 14 with Figs. 11 and 12. It's easy to find that CSLI and LLE have good robustness to the manifold with a "hole". And compared Fig. 13 with Fig. 7, we can find that if we dug out the center of Fig. 7, the remainder is similar to Fig. 13. It's shown that the grid of the landmark points has a strong representation of the whole manifold.

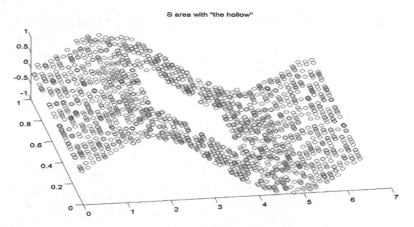

Fig. 9. The "S"-shaped manifold with a "hole"

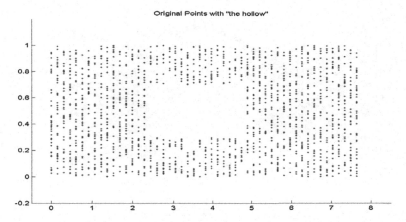

Fig. 10. The ideal result of dimensionality reduction

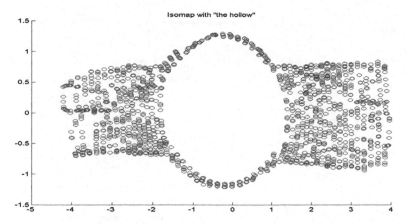

Fig. 11. The experimental result of Isomap with $k = 16$

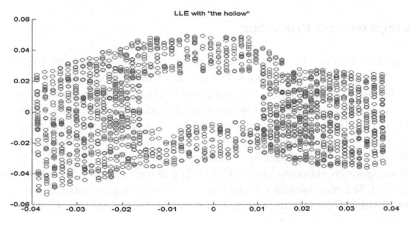

Fig. 12. The experimental result of L-Isomap, $k = 16$, $n = 100$

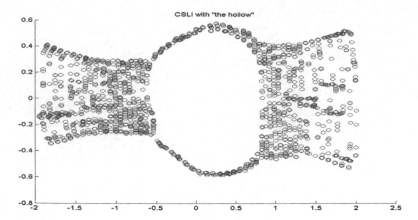

Fig. 13. The experimental result of CSLI with $k = 16$ and $n = 100$

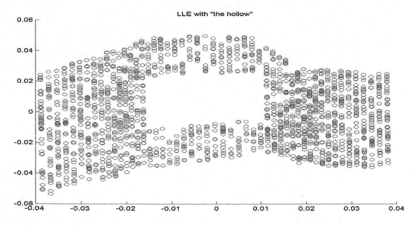

Fig. 14. The experimental result of LLE with $k = 16$

5   Summary and Promotion

This paper establishes the geometric framework of manifold learning based on the theories of fundamental mathematics. In this framework, we summarizes the requirements of some classical manifold learning methods about the mapping f, and construct the smooth homeomorphism between the manifold and its tangent space. We obtain the conclusion that the local coordinates obtained by the unity of coordinates are linearly related. Then we proposed the algorithm via homeomorphic tangent space (LHTS).

After analyzing the compactness of manifold, this paper also propose a dimension reduction algorithm via compactness (CSLI). The algorithm has achieved three points as follows, CSLI can maintain the faithfulness of the Isomap; compared with Isomap, the computational complexity of CSLI is greatly reduced; the manifold M is only bounded closed. Experimental results show that CSLI has some robustness to the manifold with a "hole".

The second experiment is based on which the "hole" is smaller and the region of the simple set has good connectivity. If the "hole" is too large to break the region, then the CSLI algorithm will not show a strong superiority. This is because the original topological structure of manifold is destructed. The dimensionality reduction for manifolds with multi-connected components is a difficult problem to be solved.

The most important purpose of this paper is to try to build the link between fundamental mathematics and computer science, put some good theorems and conclusions in mathematics into the application, in order to get some better algorithms and create greater value.

References

1. Theodoridis, S.: Konstantinos Koutroumbas: Pattern Recognition, 4th edn. Machinery Industry Press, Beijing (2009)
2. Zhou, Z.: Machine Learning, pp. 53–63, 97–113, 171–173. Tsinghua University Press, Beijing (2016)
3. Tenenbaum, J.B., De Silva, V.: A globel geometric framework for nonlinear dimensionality reduction. Science **290**(5500), 2319–2323 (2000)
4. Roweis, S., Saul, L.: Nonlinear dimensionality reduction by locally linear embedding. Science **290**(5500), 2323–2326 (2000)
5. Zhang, Z., Zha, H.: Principal manifolds and nonlinear dimensionality reduction via tangent space alignment. SIAM J. Sci. Comput. **26**(1), 313–338 (2004)
6. Chen, W.: Introduction to Differential Geometry, pp. 37–94. Higher Education Press, Beijing (2013)
7. John, M.L.: Introduction to Smooth Manifolds, pp. 2–122. Springer, Heidelberg (2012)
8. Peter, D.L.: Functional Analysis, pp. 43–51, 233–244. Higher Education Press, Beijing (2007)
9. Jiang, Z., Sun, S.: Functional Analysis, pp. 19–25. Higher Education Press, Beijing (2005)
10. Munkres, J.R.: Topology, 2nd edn, pp. 75–118, 148–186, 200–227, 322–340. Prentice Hall, Upper Saddle River (2000)
11. Zha, H., Zhang, Z.: Continuum Isomap for manifold learnings. Comput. Stat. Data Anal. **52**, 184–200 (2007)
12. Donoho, D., Grimes, C.: Hessian eigenmaps: new tools for nonlinear dimensianality reduction. Proc. Natl. Acad. Sci., 5591–5596 (2003)
13. Tenenbaum, J.B., Silva, V.D.: Global versus local methods in nonlinear dimensionality reduction. In: Proceeding of Advances in Neural Information Processing Systems 15. MIT Press, Cambridge (2003)
14. Orsenigo, C.: An improved set covering problem for Isomap supervised landmark selecion. Pattern Recogn. Lett. **49**, 131–137 (2014)
15. Samko, O., Marshall, A.D., Rosin, P.L.: Selection of the optimal parameter value for the Isomap algorithm. Pattern Recogn. Lett. **27**, 968–979 (2006)

Selective Image Matting with Scalable Variance and Model Rectification

Xiao Chen, Fazhi He$^{(\boxtimes)}$, Yiteng Pan, and Haojun Ai

School of Computer Science, Wuhan University, Wuhan 430072, China
fzhe@whu.edu.cn

Abstract. Bayesian Matting has four limitations. Firstly, Bayesian matting makes strong assumption that the texture distribution of nature image satisfies Gaussian distribution with fixed variance. This assumption will fail for complex texture distribution. In order to extract the nature images with complex texture distribution, we design an information entropy approach to estimate the scalable variance. Secondly, when the opacity is near the boundary of the value range, Bayesian matting method may be failure because of the error computation of opacity. Therefore, a rectification approach is proposed to adjust the computation model and keep the opacity within the valid value range. Thirdly, Bayesian matting is a local sample method which may miss some valid samples of matting. We propose a selection function to integrate valid global sample matting result into above matting framework as a supplement to the local sample matting result. The proposed function is compose of three criteria, that is, the similarity of results, the overlapping degree of samples, and the similarity of neighborhood. Fourthly, in order to obtain a smooth and vivid matte, the result is further refined by considering correlation between neighbouring pixels. Finally, We use online benchmark for image matting to evaluate the proposed method with both qualitative observation and quantitative analysis. The experiments show that our method achieves a competitive advantages over other methods.

Keywords: Image matting · Information entropy · Model rectification · Constraint optimization · Global sample · Selection function

1 Introduction

Images matting and images segmentation are important areas of computer graphics and computer vision [1–4].

Both images matting and images segmentation try to separate one part from an image [5], where the separated part is foreground and the remained part is

F. He—This paper is supported by the National Science Foundation of China (Grant No. 61472289) and the Key Technology R&D Program of Hubei Provence (2014BAA153).

© Springer Nature Singapore Pte Ltd. 2017
B. Zou et al. (Eds.): ICPCSEE 2017, Part I, CCIS 727, pp. 534–548, 2017.
DOI: 10.1007/978-981-10-6385-5_45

background. The difference between matting and segmentation is: the result of segmentation is binary; the result of matting is a continuous opacity which is distributed among $[0\tilde{1}]$.

Porter and Duff defined the digital analog of the matting, the alpha channel, and showed how synthetic images with alpha could be useful in creating complex digital images [6]. Any image I can be decomposed into a linear combination of opacity with foreground image F and background image B:

$$I = \alpha F + (1 - \alpha) \tag{1}$$

In order to solve the problem, it is necessary to add constraints by user interaction. There are two types of interactive matting methods: (1) Trimap based matting, which mainly corresponds to the sample matting; (2) Simple stroke based matting, which mainly corresponds to the propagation matting [7].

Bayesian matting [8] method uses maximum a-posteriori probability (MAP) to establish the mathematical model for matting which has a good balance between matting speed and effectiveness. An outstanding effect can also be obtained even if the foreground and background texture are non-smooth. However, this method still has three deficiencies: the sample method in Bayesian matting is local sample method, which may miss some valid samples; Bayesian matting assumes that the matting likelihood function is a Gaussian distribution with fixed variance, which is not suitable for the texture distribution of natural images; The solving process is an unconstrained optimization, in which the result of which may be unreliable in some situations.

In order to over the limitations, we present a selective image matting with scalable variance and model rectification. The paper is organized as follows. Section 2 introduces the related work. Our approach and new contributions are presented in Sect. 3. Experiments are discussed in Sect. 4, and we conclude the paper in Sect. 5.

2   Related Work

Image Matting method can be divided into two categories: sample-based and propagation-based methods.

Sampling-based methods is based on the similarity and continuity of image. It assumes that both the F value and B value of each unknown pixel can be estimated from the known pixels [9–13, 15–20].

Propagation-based matting method adopted the field theory [21]. It assumes that α graph is a field and the opacity of each pixel are correlated with its neighborhood. Thus, the solution process of α is transformed into a solution process of field. Propagation-based matting method includes [7, 22–28].

In recent years, there is a new way for propagation-based matting, which uses learning for the matting process. Learning based matting method includes [29, 31–36].

3 Proposed Method

3.1 Scalable Variance

The variance of the Gaussian distribution σ_c^2 describes the dispersion degree of data distribution. If the texture distribution of image is complex, the variance should be large, whereas the variance of distribution model should be small. In order to make $L(C|F, B, \alpha)$ likelihood function to fit the texture distribution of natural images, we use the image complexity to describe the variance.

Image complexity has different definitions in different applications. In the applications of image analysis, it indicates the complexity of an image texture. In communication theory, Shannon proposed the concept of information entropy to describe the uncertainty of the source [37]. Information entropy can be used to measure the complexity of a system [38].

$$H = -\frac{\sum_{i=1}^{k} n_i * \log(n_i/N)}{N} \tag{2}$$

In this paper, the theory of information entropy is adopted to describe the image complexity as Eq. 2.

The fixed variance in the Gaussian likelihood function of Bayes matting framework is re-designed and replaced with a scalable variance.

The complexity H is used to model the scalable variance in Eq. 3.

$$\sigma'^2_c = \sigma_c^2 * (1 + H)$$
$$while\ H = -\frac{\sum_{i=1}^{k} n_i * \log(n_i/N)}{N} \tag{3}$$

where, k is the number of clusters; n_i is the number of pixels in each cluster; N is the total number of pixels; The k is determined by the number of histogram peaks; σ_c^2 is the original variance.

As shown in Fig. 1, image complexity changes with k.

Figure 1a contains the ground, leaves and flowers, so the texture is complex. And the corresponding histogram has 8 peaks. So $k = 8$ and the complexity H is 0.7532. The new variance is bigger than the original variance.

Figure 1b contains only the ground, the texture is simple, and the histogram has only one peak. So $k = 1$ and the complexity H is 0. The new variance is equal to the original variance. The Eq. 3 replaced the variance in the standard bayesian matting.

A new expression of the likelihood function based on the complexity of image texture with information entropy is described as Eq. 4

$$L(C|F, B, \alpha) = \frac{-\|C - \alpha F - (1 - \alpha)B\|^2}{\sigma'^2_c} \tag{4}$$

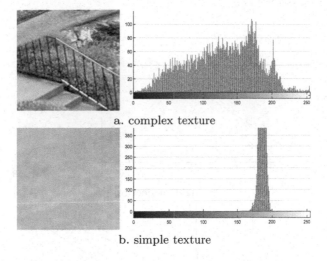

a. complex texture

b. simple texture

Fig. 1. Histogram of different textures distribution

3.2 Model Rectification

Error in Projection. Bayesian matting method obtains α by projection. Sometimes even if the samples are valid, the result of projection may be fall out of FB.

As shown in Fig. 2a, C' falls in FB. So F, B is valid sample pair for C.

As shown in Fig. 2b, $C1'$ and $C2'$ are the projection points of $C1$ and $C2$. Both $C1'$ and $C2'$ fall out of FB.

- However $C1'$ is close enough to B and the error can be tolerated. So F, B can be still considered as a valid sample pair for $C1$.
- While $C2'$ is far away from B that the error is big. So F, B is invalid sample pair for $C2$.

Error in Selection. Bayesian matting force $\alpha = 1$ $(\alpha > 1)$ or $\alpha = 0$ $(\alpha < 0)$. However this way will change the fitness term of posterior probability.

- As shown in Fig. 3a. A is a reasonable result because α_A is in the $[0,1]$, α_B is an error result because B is more greater than 1. When Bayes matting forced $\alpha_B = 1$, the posterior probability has been changed to B' or B". If B' is the new posterior probability while $\alpha_B = 1$, the selection strategy selected B as the final result. But unfortunately, A is the real right result.
- As shown in Fig. 3b. A, B both are reasonable results because the alpha is in $[0, 1]$. B is the better result.

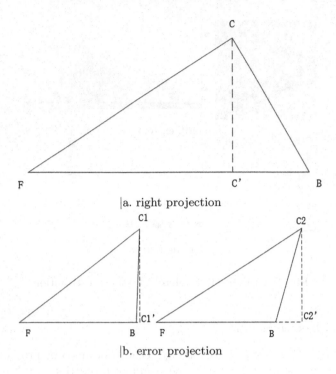

|a. right projection

|b. error projection

Fig. 2. Projection

Model Rectification by Regularization Strategy. The matting result may be error under the above situation. Therefore, this paper present rectification method. The main idea of the method is shown in Fig. 4.

- Fig. 4a is an error result. A large white area is appeared in the narrow region.
- Fig. 4b is the ground truth.
- Fig. 4c is our result rectified by our method.

Our correction to color estimation and optimization model is modeled by following Eq. 5

$$
\begin{aligned}
&\arg\max_{F,B,\alpha} P(F,B,\alpha|C) \\
&= \arg\max_{F,B,\alpha} P(C|F,B,\alpha)P(F)P(B)P(\alpha)/P(C) \\
&= \arg\max_{F,B,\alpha} L(C|F,B,\alpha) + L(F) + L(B) + L(\alpha) \\
&s.t.\, F : 0 \le F \le 1, B : 0 \le B \le 1, \alpha : 0 \le \alpha \le 1
\end{aligned}
\tag{5}
$$

Different from the basic optimization model in bayesian matting, our model 5 introduces inequality constraint, which can correct the optimization parameters of F, B and α.

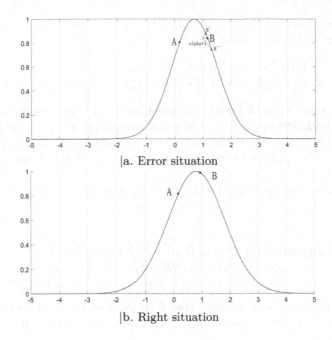

|a. Error situation

|b. Right situation

Fig. 3. Error select result

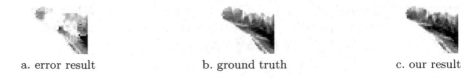

a. error result   b. ground truth   c. our result

Fig. 4. Error result

One way of solving constrained optimization problems is to convert it to unconstrained optimization problems. Augmented Lagrange multiplier method can integrate all the equality constraints, inequality constraints and functions, combination into an objective function equation by Lagrange multipliers.

- It has been proved that the augmented Lagrangian multiplier parameters can be customized. It isn't affecting the accuracy of result, but only affecting the iterative speed [39].
- From view of the model, regularization and objective function with constraints are equivalent [40]. The essence of regularization is to give a certain constraint on optimization parameters, and add a penalty term to constrain the optimization space. From the view of model solution, regularization provides the possibility of a unique solution.

Therefore, based on the regularization theory, this paper constructs the constraint items and penalty terms 6:

$$\lambda_{f1} \|F - 1\| + \lambda_{b1} \|B - 1\| + \lambda_{\alpha 1} \|\alpha - 1\|$$
$$+ \tfrac{1}{2}\lambda_{f2}\|F - 1\|^2 + \tfrac{1}{2}\lambda_{b2}\|B - 1\|^2 + \tfrac{1}{2}\lambda_{\alpha 2}\|\alpha - 1\|^2 \tag{6}$$

where, λ_{f1} is the ratio of foreground pixel in an effective sample pixel; λ_{b1} is the ratio of background pixel in an effective sample pixel; $\lambda_{\alpha 1}$ is the mean value of α in neighborhood; λ_{f2} is the variance of foreground clustering result; λ_{b2} is the variance of background clustering result; and $\lambda_{\alpha 2}$ is the mean value of α in neighborhood.

Finally, we add the Eq. 6 to the basic optimization of the original Bayes matting, and deduct the corrective color estimation and optimization model as shown in 7.

$$\arg\max L(\cdot) = L(C|F, B, \alpha) + L(F) + L(B) + L(\alpha)$$
$$+\lambda_{f1} \|F - 1\| + \lambda_{b1} \|B - 1\| + \lambda_{\alpha 1} \|\alpha - 1\|$$
$$+\tfrac{1}{2}\lambda_{f2}\|F - 1\|^2 + \tfrac{1}{2}\lambda_{b2}\|B - 1\|^2 + \tfrac{1}{2}\lambda_{\alpha 2}\|\alpha - 1\|^2 \tag{7}$$

Equation 7 is the regularization transformation of Eq. 5, where the "regularization" means to have "restriction", "constrain" or "correction" on the optimized parameters.

After correction, the solution process is shown as follows.

$$\begin{bmatrix} \Sigma_F^{-1} + I\alpha^2/\sigma_c'^2 - \lambda_{f2} & I\alpha(1-\alpha)/\sigma_c'^2 \\ I\alpha(1-\alpha)/\sigma_c'^2 & \Sigma_B^{-1} + I(1-\alpha)^2/\sigma_c'^2 - \lambda_{b2} \end{bmatrix} \begin{bmatrix} F \\ B \end{bmatrix}$$
$$= \begin{bmatrix} \Sigma_F^{-1}\overline{F} + C\alpha/\sigma_c'^2 - \lambda_{f1} - \lambda_{f2} \\ \Sigma_F^{-1}\overline{B} + C(1-\alpha)/\sigma_c'^2 - \lambda_{b1} - \lambda_{b2} \end{bmatrix}$$
$$\alpha = \frac{(C - B)(F - B)/\sigma_c'^2 - \lambda_{\alpha 1} - \lambda_{\alpha 2}}{\|F - B\|^2/\sigma_c'^2 - \lambda_{\alpha 2}}$$

3.3   Global-Supplement Sample

Local Sample vs. Global Sample. As a typical local sample method, Bayesian matting assumes that image is continuous in local space.

- If the natural image satisfy the assumption, good samples can be collected in the neighbor area.
- However, in some situation, the image patches are not continuous, and good samples cannot be collected in the neighbor area. In this situation, Bayesian matting fail to get the right result.

In this paper, we propose a selection strategy to integrate valid global sample matting result into the matting framework as a supplement to the local sample matting result. The proposed method has following steps as follows.

Step 1. The proposed method mattes the input image to get two types of initial results waiting for selection:

– the initial local result, which is got from above matting method untill Sect. 3.2 of this paper;
– the initial global results is got from global samples methods.

Step 2. Based on our selection strategy, the better α result is chosen as the output result of this Sect. 3.3.

Selection Strategy. A comprehensive selection strategy is proposed, which compares the initial global result with the initial local result as follows.

Firstly, the similarity of the results is the distance between the results and the observed values. The lower distance indicates the better result. We used color cost function 8 to evaluate the cost.

$$\xi(C, F, B) = \|C - \alpha_c F_c - (1 - \alpha_c)B_c\|$$

$$K_c = \begin{cases} 1 & \xi(C_g, F_g, B_g) < \xi(C_l, F_l, B_l) \\ 0 & otherwise \end{cases} \tag{8}$$

where, C is unknown pixel; α is opacity; F is foreground sample; B is background sample; subscript G represents the global result and subscript L represents the local result.

Secondly, the overlapping degree of samples is a distance between foreground samples and background samples in color space. The larger distance indicates that the result samples are more reliable. We used equation Eq. 9 to evaluate the overlapping degree.

$$\xi(F, B) = \|F_c - B_c\|$$

$$\eta_c = \begin{cases} 1 & \xi(F_g, B_g) > \xi(F_l, B_l) \\ 0 & otherwise \end{cases} \tag{9}$$

where, C is unknown pixel; F is foreground sample; B is background sample; subscript g represents the global result and subscript l represents the local result.

Thirdly, different from exiting researches, this paper presents Eq. 10 to consider the neighborhood similarity, which is the opacity difference between the result opacity and the known neighborhood opacity.

Equation 10 comes from a fact of observation that the opacity changes with the gradient in most situation. A weak gradient corresponds with smooth image texture or vice versa. Therefore, the smaller difference between unknown pixel and its neighborhood opacities is, the more reliable result is. We used Eq. 10 to measure the neighborhood similarity.

$$D = \sum_{i \in A_k} \|\alpha_c - \alpha_i\|$$
$$R_c = (Gradient(C) \leq \varsigma) \times (D_g < D_l) + (Gradient(C) > \varsigma) \times (D_g > D_l) \tag{10}$$

where, A is the neighborhood region; Gradient(C) is the gradient of the C; ς is the threshold to determine the strength of the gradient; subscript g represents the global result and subscript l represents the local result.

Finally, based on above equations, we propose a final comprehensive selection Eq. 11, which can be applied to select a better result.

$$O_c = K_c \times \eta_c \times R_c \qquad (11)$$

If $O_c = 1$, we choose the global sample result. Otherwise, we choose the local sample result.

3.4 Smooth Process

The sample-based matting methods calculate each unknown pixel separately. The matte opacity may not be smooth enough. On the other hand, the propagation-based matting methods fully take use of the relationship among adjacent pixels. The result is very smooth. Therefore, it is necessary to further process the sample-based matting result through the smooth technique from propagation-based matting methods.

In this paper, a propagation-based matting method, the CF method [22], is applied to establish the cost function for smooth purpose. Cost function consists of data item $\bar{\alpha}$, the trust value f and matting Laplacian matrix. We solve the cost function to obtain the final $\hat{\alpha}$.

$$f = \exp(-\frac{\|C - \alpha F - (1 - \alpha)B\|}{2}) \qquad (12)$$

$$\hat{\alpha} = \arg\min \hat{\alpha}^T L \hat{\alpha} + \lambda(\hat{\alpha} - \bar{\alpha})^T D(\hat{\alpha} - \bar{\alpha}) + \gamma(\hat{\alpha} - \bar{\alpha})^T T(\hat{\alpha} - \bar{\alpha}) \qquad (13)$$

where, L is matting Laplacian matrix in CF matting. $\hat{\alpha}$ is a vector of opacity to be solved after smoothness; λ is a larger weighting value, we set $\lambda = 800$; $\bar{\alpha}$ is composed of α that calculated through the matting method; D is a diagonal matrix which the elements on the main diagonal is 0 or 1; γ is a constant, we set $\gamma = 0.1$; T is a diagonal matrix, and elements on the main diagonal consists 0 and trust value. The known foreground pixel and background pixel are marked as 0, the unknown pixel is marked as the corresponding of trust value f.

Confidence value f represents the credibility of currently solved α, the closer the previously obtained of α, F and B to the actual result, the greater the value of f, otherwise the smaller.

After smooth process in Sect. 3.4, we have a more vivid matting result.

4   Experimental Research

Rhemann proposed an evaluation method in ICCV paper [41] and provided the benchmark test set of matting effect on the website www.alphamatting.com. It has been widely used for matting evaluation. Our experiments use dataset from www.alphamatting.com. We also use the evaluation methods, Sum of absolute differences (SAD) and mean squared error (MSE), from ICCV paper [41].

In Sect. 4.1, the local details comparison is conducted. Section 4.2 makes an overall evaluation.

4.1   Comparison on Local Details

Figure 5 is the experimental result of wind image.

- As shown in Fig. 5d, the experimental results of CF method is not good at the long and narrow area. The reason is that, in the box region, the unknown area is treated as the background if it is surrounded by the background.
- As shown in Fig. 5e, the result of weighted color matting is worse than CF method.
- As shown in Fig. 5f, Bayesian matting can get good result in the narrow area, however, but it cannot get fine result in the continuous narrow area.
- As shown in Fig. 5g, our method obtain the fine result in the box area.

In Fig. 5, compared to other methods, out result is most close to ground truth.

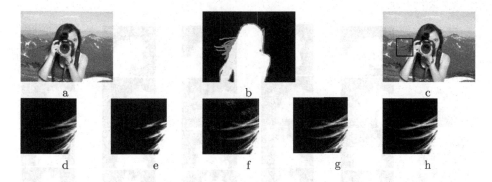

Fig. 5. Details comparison on woman results. (a) input image (b) trimap (c) details to be compared of (d) closed form (e) weighted color matting (f) bayes matting (g) our method (h) ground truth

Figure 6 is the experimental result of dog image. The background texture is complex and changeable. In the box region, there are semi-transparent region and long and narrow structure. These are difficult problems in matting.

- As shown in Fig. 6d, CF method cannot treat the narrow area in box region. It was incorrectly treated as the foreground.
- As shown in Fig. 6e, weighted color matting can get partial result in the narrow area but it lost too much details.
- As shown in Fig. 6f, Bayesian matting get the good result in semitransparent area but it get the error result in narrow area.
- As shown in Fig. 6g, our method can obtain the detailed result in the narrow and semitransparent area.

In Fig. 6, compared to other methods, out result is most close to ground truth.

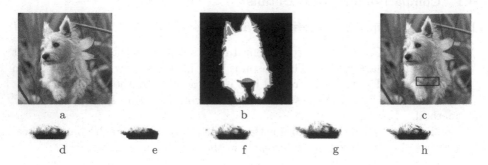

Fig. 6. Details comparison on dog results. (a) input image (b) trimap (c) details to be compared of (d) closed form (e) weighted color matting (f) bayes matting (g) our method (h) ground truth

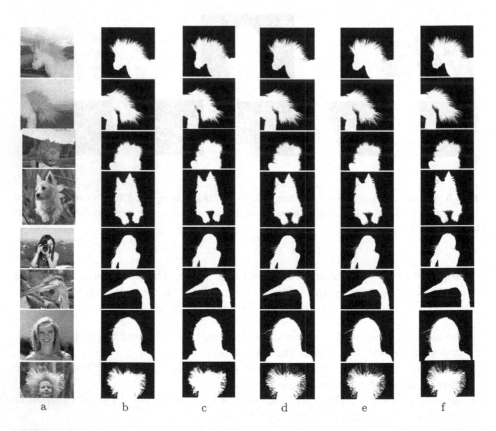

Fig. 7. Matting result comparison (a) original image (b) closed form (c) weighted color matting (d) bayesian matting (e) our method (f) ground truth

4.2   Comparison with Other Matting Method

Figure 7 is the result of matting. As shown in Fig. 7, our results are better than other results in vision. Our results keep more details and perform better in the narrow area.

Figure 8 is the SAD evaluation and MSE evaluation respectively. As shown in Fig. 8, our results have minor degree of error than other results.

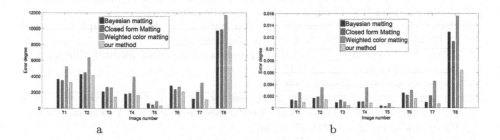

a   b

Fig. 8. (a) SAD evaluation result, (b) MSE evaluation result

The Figs. 7 and 8 illustrate that our method is better than others.

5   Conclusions and Prospect

This paper present a selective image matting with scalable variance and model rectification. Main contributions of this article includes follows. First of all, we construct a scalable variance of Gaussian distribution based on information entropy to effectively deal with the complex texture of natural images. Secondly, in terms of model solution, we propose a rectification method for color estimation optimization model. Thirdly, we present a comprehensive selection strategy to choose a global sample result to complement the local sample result. Numerous experiments on benchmark dataset demonstrate that our method achieves a competitive performance over other methods.

In future work, we will continue our research, such as how to get more valid samples and how to reach the balance between region and precision. We want to accelerate and optimize our algorithm with advanced computing techniques, such as parallel computing [42–46] and intelligent computing [47,48]. We also want to extend the idea to other related areas, such as image processing [49–51], video processing [52–54], CAD&Graphics [55–62].

References

1. Yao, G.: A survey on pre-processing in image matting. J. Comput. Sci. Technol. **32**(1), 122–138 (2017)

2. Pitié, F.: Rank reduced alternative matting laplacian. In: IEEE International Conference on European Conference on Visual Media Production, CVMP 2016, p. 7. ACM (2016)
3. Liu, B., Wang, C., Zhu, L.: A new automatic human body slice segmentation method based on closed-form solution matting and distance transform. J. Med. Imaging Health Inform. **7**(1), 247–250 (2017)
4. Carey, P., Bennett, S., Lasenby, J., Purnell, T.: Aerodynamic analysis via foreground segmentation. Electron. Imaging **2017**(16), 10–14 (2017)
5. Shen, Y., Lin, X., Xie, Z.: A survey on interactive alpha matting technique. J. Comput.-Aided Des. Comput. Graph., 511–519 (2014)
6. Porter, T., Duff, T.: Compositing digital images. In: ACM Siggraph Computer Graphics, New York, USA, vol. 18, pp. 253–259 (1984)
7. Wang, J., Cohen, M.F.: An iterative optimization approach for unified image segmentation and matting. In: IEEE International Conference on Computer Vision, vol. 2, pp. 936–943, October 2005
8. Chuang, Y., Curless, B., Salesin, D.H., Szeliski, R.: A Bayesian approach to digital matting. In: IEEE International Conference on Computer Vision and Pattern Recognition, vol. 2, pp. 264–271 (2001)
9. Berman, A., Dadourian, A., Vlahos, P.: Method for removing from an image the background surrounding a selected object. US Patent 6,134,346, 17 Oct 2000
10. Ruzon, M.A., Tomasi, C.: Alpha estimation in natural images. In: IEEE International Conference on Computer Vision and Pattern Recognition, vol. 1, pp. 18–25 (2000)
11. Hillman, P., Hannah, J., Renshaw, D.: Alpha channel estimation in high resolution images and image sequences. In: IEEE International Conference on Computer Vision and Pattern Recognition, vol. 1, pp. 1063–1068 (2001)
12. Rhemann, C., Rother, C., Gelautz, M.: Improving color modeling for alpha matting. In: BMVC, vol. 1, p. 3. Citeseer (2008)
13. Gastal, E.S., Oliveira, M.M.: Shared sampling for real-time alpha matting. In: Computer Graphics Forum, vol. 29, pp. 575–584. Wiley Online Library (2010)
14. Wu, Y., He, F.: Color sampling based on fuzzy connectedness for image matting. J. Comput.-Aided Des. Comput. Graph., 1194–1200 (2010)
15. He, K., Rhemann, C., Rother, C.: A global sampling method for alpha matting. In: IEEE International Conference on Computer Vision and Pattern Recognition, pp. 2049–2056 (2011)
16. Rajan, D.: Weighted color and texture sample selection for image matting. In: IEEE Conference on Computer Vision and Pattern Recognition, pp. 718–725 (2012)
17. Sun, W., Luo, S., Wu, L.: A saliency-based sampling method for image matting. In: International Conference on Fuzzy Systems and Knowledge Discovery, pp. 1686–1689 (2012)
18. Shahrian, E., Rajan, D., Price, B., Cohen, S.: Improving image matting using comprehensive sampling sets. In: IEEE International Conference on Computer Vision and Pattern Recognition, pp. 636–643 (2013)
19. Tan, G., Chen, H., Qi, J.: A novel image matting method using sparse manual clicks. Multimed. Tools Appl. **75**, 10213 (2016)
20. Yan, X., Hao, Z., Huang, H.: Alpha matting with image pixel correlation. Int. J. Mach. Learn. Cyber. **8**, 1–7 (2016)
21. Sun, J., Jia, J., Tang, C., Shum, H.: Poisson matting. ACM Trans. Graph. **23**, 315–321 (2004)
22. Levin, A., Lischinski, D., Weiss, Y.: A closed-form solution to natural image matting. IEEE Trans. Pattern Anal. Mach. Intell. **30**, 228–242 (2008)

23. Lee, P., Wu, Y.: Nonlocal matting. In: Proceedings of the IEEE International Conference on Computer Vision and Pattern Recognition, pp. 2193–2200 (2011)
24. Chen, Q., Li, D., Tang, C.: KNN matting. In: IEEE International Conference on Computer Vision and Pattern Recognition, pp. 869–876 (2012)
25. Chen, X., Zou, D., Zhou, S., Zhao, Q.: Image matting with local and nonlocal smooth priors. In: IEEE International Conference on Computer Vision and Pattern Recognition, pp. 1902–1907 (2013)
26. Kim, B.K., Jin, M., Song, W.J.: Local and nonlocal color line models for image matting. IEICE Trans. Fundam. Electron. Commun. Comput. Sci. **E97.A**(8), 1814–1819 (2014)
27. Xiao, C., Liu, M., Xiao, D.: Fast closed-form matting using a hierarchical data structure. IEEE Trans. Circuits Syst. Video Technol. **24**(1), 49–62 (2014)
28. Fiss, J., Curless, B., Szeliski, R.: Light field layer matting. In: IEEE Conference on Computer Vision and Pattern Recognition, pp. 623–631 (2015)
29. Zheng, Y., Kambhamettu, C.: Learning based digital matting. In: Proceedings of the IEEE 12th International Conference on Computer Vision, pp. 889–896 (2009)
30. Wang, J.: Image matting with transductive inference. In: Gagalowicz, A., Philips, W. (eds.) MIRAGE 2011. LNCS, vol. 6930, pp. 239–250. Springer, Heidelberg (2011). doi:10.1007/978-3-642-24136-9_21
31. Zhang, Z., Zhu, Q., Xie, Y.: Learning based alpha matting using support vector regression. In: IEEE International Conference on Image Processing, pp. 2109–2112 (2012)
32. Yoon, S.M., Yoon, G.J.: Alpha matting using compressive sensing. Electron. Lett. **48**(3), 153–155 (2012)
33. Johnson, J., Varnousfaderani, E.S., Cholakkal, H.: Sparse coding for alpha matting. IEEE Trans. Image Process. **25**, 3032–3043 (2016)
34. Feng, X., Liang, X., Zhang, Z.: A cluster sampling method for image matting via sparse coding. In: Leibe, B., Matas, J., Sebe, N., Welling, M. (eds.) ECCV 2016. LNCS, vol. 9906, pp. 204–219. Springer, Cham (2016). doi:10.1007/978-3-319-46475-6_13
35. Shen, X., Tao, X., Gao, H., Zhou, C., Jia, J.: Deep automatic portrait matting. In: Leibe, B., Matas, J., Sebe, N., Welling, M. (eds.) ECCV 2016. LNCS, vol. 9905, pp. 92–107. Springer, Cham (2016). doi:10.1007/978-3-319-46448-0_6
36. Cho, D., Tai, Y.-W., Kweon, I.: Natural image matting using deep convolutional neural networks. In: Leibe, B., Matas, J., Sebe, N., Welling, M. (eds.) ECCV 2016. LNCS, vol. 9906, pp. 626–643. Springer, Cham (2016). doi:10.1007/978-3-319-46475-6_39
37. Shannon, C.E.: A mathematical theory of communication. Bell Syst. Tech. J. **27**, 623–656 (1948)
38. Biaynicki-Birula, I., Mycielski, J.: Uncertainty relations for information entropy in wave mechanics. Commun. Math. Phys. **44**(2), 129–132 (1975)
39. DiPillo, G., Grippo, L.: A new class of augmented lagrangians in nonlinear programming. SIAM J. Control Optim. **17**(5), 618–628 (1979)
40. Wu, M., Schölkopf, B.: Transductive classification via local learning regularization. In: AISTATS, pp. 628–635 (2007)
41. Rhemann, C., Rother, C., Wang, J., Gelautz, M.: A perceptually motivated online benchmark for image matting. In: IEEE International Conference on Computer Vision and Pattern Recognition, pp. 1826–1833 (2009)
42. Zhou, Y., He, F., Qiu, Y.: Optimization of parallel iterated local search algorithms on graphics processing unit. J. Supercomput. **72**(6), 2394–2416 (2016)

43. Krajcevski, P., Pratapa, S., Manocha, D.: GPU-decodable supercompressed textures. ACM Trans. Graph. **35**(6), 230 (2016)
44. Li, R., Hou, Q., Zhou, K.: Efficient GPU path rendering using scanline rasterization. ACM Trans. Graph. **35**(6), 228 (2016)
45. Wang, H., Yang, Y.: Descent methods for elastic body simulation on the GPU. ACM Trans. Graph. (TOG) **35**(6), 212 (2016)
46. Yan, X., He, F., Hou, N.: An efficient particle swarm optimization for large scale hardware/software co-design system. Int. J. Coop. Inf. Syst. (2017). doi:10.1142/S0218843017410015
47. Yan, X., He, F., Chen, Y., Yuan, Z.: An efficient improved particle swarm optimization based on prey behavior of fish schooling. J. Adv. Mech. Des. Syst. Manuf. **9**(4), JAMDSM0048 (2015)
48. Wang, D., Tan, D., Liu, L.: Particle swarm optimization algorithm: an overview. Soft. Comput. **21**, 1–22 (2017)
49. Huang, Z., He, F., Cai, X.: Efficient random saliency map detection. Sci. China Ser. F: Inf Sci. **54**(6), 1207–1217 (2011)
50. Ni, B., He, F., Pan, Y., Yuan, Z.: Using shapes correlation for active contour segmentation of uterine fibroid ultrasound images in computer-aided therapy. Appl. Math.-J. Chin. Univ. **31**(1), 37–52 (2016)
51. Yu, H., He, F., Pan, Y., Chen, X.: An efficient similarity-based level set model for medical image segmentation. J. Adv. Mech. Des. Syst. Manuf. **10**(8), JAMDSM0100 (2016)
52. Li, K., He, F., Chen, X.: Real-time object tracking via compressive feature selection. Front. Comput. Sci. **10**(4), 689–701 (2016)
53. Sun, J., He, F., Chen, Y., Chen, X.: A multiple template approach for robust tracking of fast motion target. Appl. Math.-J. Chin. Univ. **31**(2), 177–197 (2016)
54. Li, K., He, F., Yu, H., Chen, X.: A correlative classiers approach based on particle filter and sample set for tracking occluded target. Appl. Math.-J. Chin. Univ. (2017). doi:10.1007/s11766-017-3466-8
55. Jing, S., He, F., Li, X., Cai, X.: A method for topological entity correspondence in a replicated collaborative CAD system. Comput. Ind. **60**(7), 467–475 (2009)
56. Li, X., He, F., Cai, X., Zhang, D.: A method for topological entity matching in the integration of heterogeneous CAD systems. Integr. Comput. Aided Eng. **20**(1), 15–30 (2013)
57. Cheng, Y., He, F., Cai, X., Zhang, D.: A group Undo/Redo method in 3D collaborative modeling systems with performance evaluation. J. Netw. Comput. Appl. **36**(6), 1512–1522 (2013)
58. Wu, Y., He, F., Zhang, D., Li, X.: Service-oriented feature-based data exchange for cloud-based design and manufacturing. IEEE Trans. Serv. Comput. (2015). doi:10.1109/TSC.2015.2501981
59. Zhang, D., He, F., Han, S., Li, X.: Quantitative optimization of interoperability during feature-based data exchange. Integr. Comput. Aided Eng. **23**(1), 31–51 (2016)
60. Chen, Y., He, F., Wu, Y., Hou, N.: A local start search algorithm to compute exact Hausdorff Distance for arbitrary point sets. Pattern Recogn. **67**, 139–148 (2017)
61. Zhang, D., He, F., Wu, Y., Han, S.: An efficient approach to directly compute the exact Hausdorff Distance for 3D point sets. Integr. Comput.-Aided Eng. (2017). doi:10.3233/ICA-170544
62. Wu, Y., He, F., Han, S.: Collaborative CAD synchronization based on symmetric and consistent modeling procedure. Symmetry **9**(4), 59 (2017)

Side-Channel Attacks Based on Collaborative Learning

Biao Liu[1]([✉]), Zhao Ding[2], Yang Pan[1], Jiali Li[2], and Huamin Feng[1,2]

[1] Beijing Electronic Science and Technology Institute, Beijing, China
liubiao@besti.edu.cn
[2] Xidian University, Xi'an 100070, China

Abstract. Side-channel attacks based on supervised learning require that the attacker have complete control over the cryptographic device and obtain a large number of labeled power traces. However, in real life, this requirement is usually not met. In this paper, an attack algorithm based on collaborative learning is proposed. The algorithm only needs to use a small number of labeled power traces to cooperate with the unlabeled power trace to realize the attack to cryptographic device. By experimenting with the DPA contest V4 dataset, the results show that the algorithm can improve the accuracy by about 20% compared with the pure supervised learning in the case of using only 10 labeled power traces.

Keywords: Side-channel attacks · Supervised learning · Collaborative learning · Power trace

1 Introduction

In order to protect information security in the national defense, financial and network security and many other areas, encryption device has been widely used in people's lives. These devices will leak the key-related physical information when executing the cryptographic algorithm. With using of these information, side-channel attacks narrows the key search space by combining the mathematical properties of cryptographic algorithms and the physical properties of cryptographic equipment. Then we can restore the whole key [1].

The side-channel attacks based on supervised learning makes full use of the power trace of all information, which is considered to be the strongest attack method [2]. The method is divided into two stages: the first stage is the model training stage. A large number of the labeled power traces are used to training model; the second stage is the attack stage, where the model is used to determine the unknown power trace of the key. The earliest supervised learning attacks include template attacks [2], random model attacks [3, 4]. With the development of machine learning technology in recent years, the methods based on support vector machine (SVM) [5], random forest (RF) [6] and linear discriminant (LDA) [7] have been appeared. Compared to template attacks, the machine learning attack algorithm loosens the hypothesis of the Gaussian distribution of noise, which with a wider applicability.

However, side-channel attacks based on supervised learning need to assume that the attacker has completely control the cryptographic device and require a large number

© Springer Nature Singapore Pte Ltd. 2017
B. Zou et al. (Eds.): ICPCSEE 2017, Part I, CCIS 727, pp. 549–557, 2017.
DOI: 10.1007/978-981-10-6385-5_46

of labeled power trace to be training models. But this assumption is often unable to set up. More common is that an attacker can easily get a few labeled power traces, as well as a large number of unlabeled power trace. For example, the devices with known key between relatives and friends can be collected to get labeled traces. At the same time, the electromagnetic leakage of the cryptographic equipment with unknown key can be collected to get a large number of unlabeled power traces secretly. In this paper, an attack algorithm based on collaborative learning is proposed. This algorithm makes full use of the information of a large number of unlabeled power traces, and realizes the attack on the cryptographic device when the labeled power trace is small. Similar to our work, Lerman et al. [8] also proposed a semi-supervised template attacks method which only requires two tagged power traces to find the feature points, and then through the clustering of unlabeled power to achieve the attack. This method makes good use of the priori knowledge in side-channel domain, which is very different from the idea of collaborative learning in this paper.

2   Collaborative Learning

2.1   Collaborative Learning Algorithm

The earliest collaborative learning algorithm is the Co-training algorithm proposed by Blum and Mitchell [9]. The algorithm is implemented on two fully redundant "views" of the same dataset. In the course of learning, each classifier selects the unlabeled samples with high confidence for each other, and then adds the samples and their predictive labels to the labeled sample set, re-training the classifier until the conditions are fulfilled. Co-training algorithm can take advantage of the information of unlabeled samples to improve the performance of classifiers in the case of fewer labeled samples. However, in realistic tasks, the conditions for fully redundant views tend to be more difficult to meet.

Against to this problem, Goldman and Zhou [10] proposed an algorithm that does not require sufficient redundancy. Instead, they use different types of classifiers to learn, but the algorithm introduces the restrictions on the classifier and also increases the overhead of the system. In order to further improve the collaborative learning algorithm, Zhou and Li [11] proposed a Tri-training algorithm, which does not require a sufficiently redundant view and a different type of classifier. The core idea of Tri-training algorithm is to handle the confidence implicitly to achieve the purpose of selecting unlabeled samples. The Tri-training algorithm uses three classifiers to train. In the process of algorithm iteration, a classifier is selected as the main classifier and the rest is used as the auxiliary classifier in turn. If two auxiliary classifiers are the same for a certain sample, the sample is added to the labeled sample set. The collaborative learning algorithm itself is designed for multi-view data, and subsequent studies have found that the algorithm does not require data to have multiple views, and only the weak classifier algorithm with differences is still valid [12].

Because the Co-training algorithm makes it easy to assign the wrong label to the sample in the process of selecting unlabeled samples, the performance of the initial state classifier in the collaborative learning is usually weak. However, the Tri-training algorithm overcomes this deficiency by implicitly managing the confidence. Thus, the two algorithms can be combined by introducing the method of voting for unlabeled samples into the Co-training algorithm and using the thresholds to select the appropriate samples. This combination not only enhances the ability of the algorithm to select unlabeled samples, but also simplifies the Tri-training algorithm complex training process. This paper will focus on the method of the combination in Sect. 3.3.

2.2   Base Classifier-Support Vector Machine

SVM algorithm is used as the base classifier for collaborative learning in this paper. The SVM algorithm is proposed by Vapnik et al. [13, 14]. It is a kind of machine learning algorithm based on the theory of statistical learning. Its standard output is discrete classification, which belongs to hard classification. But in many cases we not only need to know the sample classification results, but also know the probability of a certain class that samples belongs to. At this point, soft classification is required. In the process of collaborative learning, this paper uses Platt [15] method to map the output of SVM decision function into class probability by sigmoid function model. The objective function is as follows:

$$p(y = c|f) = \frac{1}{1 + exp(Af + B)} \tag{1}$$

In the formula (1), f is the function value of the corresponding sample, c is the class of the sample, and A, B is the morphological parameter of the Sigmoid function. The parameter values are obtained by solving the maximum likelihood problem of the following equation:

$$\min_{A,B} F(Z) = - \sum_{i=1}^{n} (t_i \log(p_i) + (1 - t_i) \log(1 - p_i)) \tag{2}$$

where $p_i = 1/exp(Af_i + B)$ is determined by:

$$t_i = \begin{cases} \frac{N_+ + 1}{N_+ + 2}, y_i = 1 \\ \frac{1}{N_- + 2}, y_i = -1 \end{cases}, i = 1, \cdots, n \tag{3}$$

where N_+, N_- are positive and negative samples respectively.

3   Side-Channel Attacks Algorithm Based on Collaborative Learning

In this paper, a collaborative learning (Co-training SCA) algorithm based on co-learning is proposed in the case of a small number of labeled samples and a large number of unlabeled samples. The CoSCA algorithm uses implicit methods to process confidence in the process of selecting unlabeled samples, and sets the thresholds when selecting unlabeled samples, which reducing noise introduction when base classifier is weaker in initial state. Thereby we will improve the classifier performance.

In the model training process, CoSCA uses the collected power trace as the training data. The intermediate value in the process of encrypting the algorithm is used as the label of the training sample. This intermediate value is the function value after operation between the plaintext and the key. Take AES-256 as an example, the first round of the algorithm contains the first 16 bytes of the complete key, each byte is sent to a single S-box for operation, and then the S-box outputs the result. CoSCA uses each bit of the S-box output byte as a training sample (intermediate value) to train, and then uses the "divide and rule" strategy to attack each bit in turn. Finally, we find all the key bits and combine them, then restore the integrity key.

3.1   Attack Masked AES Algorithm

In order to resist side-channel attacks, cryptographic devices typically use a loop mask, clock scrambling, and power balance to protect the device. This article uses the DPA contest V4 dataset [16]. The information in the dataset is collected from the Atmel ATMega-163 smart card, which uses the cyclic mask technology of the literature [17] to resist side-channel attacks. The smart card generates a random number before executing the encryption algorithm and uses the random number to select a mask from an array containing 16 bytes of mask. At this point the output of the first S box is:

$$S(P_i \oplus K_i) \oplus Mask_{(i+offset+1)mod16} \tag{4}$$

where S represents a permutation function, $Mask$ represents byte mask array, and P_i and K_i denote the input plaintext and subkey respectively. Because the byte mask array is known to attack the cryptographic device, this article first uses the method [18] to attack the mask with a success rate of 81%.

3.2   Feature Point Selection

The power trace contains more power information and the valuable feature points are usually selected before training the model. Selecting a feature point can either reduce the dimensionality of the vector to reduce the calculated pressure and filter out the irrelevant features in the power trace. Therefor the introduction of noise to the classifier is avoided. In this paper, we use the calculated Pearson correlation coefficient method [19] to carry out feature selection. The method is based on the formula as shown below:

$$\rho_i = \frac{Cov(t_i, l)}{\sqrt{Var(t_i) \cdot Var(l)}} \tag{5}$$

In the formula, Cov represents the covariance, Var represents the variance, t_i is the power trace vector, which consists of the values of the i position in all training power traces, and l is the labeler vector consisting of the tag values of all training power traces. The largest m values are selected from the Pearson correlation coefficient

$\rho_i, i = 1, \ldots, n$, and its corresponding power trace value is used as the characteristic of model training. The characteristic number is expressed as FP_{num}.

3.3 CoSCA Algorithm

The CoSCA algorithm first assigns the corresponding tag to the power trace according to the known key, forming the labeled sample set. The unlabeled set of samples consists of the power trace of the unknown key. Subsequently, CoSCA uses radial basis function SVM and linear SVM as base classifier for collaborative learning. In the process of model training, different types of SVM algorithms are used to predict unlabeled samples, and the class probability of unlabeled samples is obtained by Sigmoid function model. If the probability of class which is predicted from unlabeled sample is greater than (or less than) the threshold and in the meantime, the result of the two classifiers is consistent, it is guaranteed that the confidence of prediction is higher. Then we add the selected sample and it's predict label into the training set, update the training set, and re-training classifier. Cycle the above steps until the termination condition is met. CoSCA combines the Tri-training algorithm to implicitly handle confidence, and meanwhile sets a threshold before selecting unlabeled samples, which reduces the introduction of noise in the weak state of the base classifier in the initial state. The algorithm is described in Fig. 1 as follows:

Given:

A set TL of labeled training power traces;

A set TU of unlabeled power traces;

Create a pool TU';

For $i = 0 : k$ **do**

$SVM(TL) \rightarrow HSVM_{rbf}$;

$SVM(TL) \rightarrow HSVM_{linear}$;

Allow $HSVM_{rbf}$ to label trace u from TU', $u_1 = \{rbf_{pre}, rbf_{proba}\}$;

Allow $HSVM_{linear}$ to label trace u from $TU', u_2 = \{linear_{pre}, linear_{proba}\}$;

If $rbf_{proba} \geq \omega_1, linear_{proba} \geq \omega_1$ and $rbf_{pre} = linear_{pre}$:

$TL = TL + u$;

If $rbf_{proba} \leq \omega_2, linear_{proba} \leq \omega_2$ and $rbf_{pre} = linear_{pre}$:

$TL = TL + u$;

end for

Fig. 1. Pseudo code of CoSCA

During training, the initial classifiers are $HSVM_{rbf}$, $HSVM_{linear}$, Uses base classifier to predict the class of unlabeled sample TU'. Select a sample u that is consistent with the predicted result and whose class probability is greater than the threshold ω_1 (or less than the threshold ω_2) and add them to the training set. After K iterations, the CoSCA model H_{CoSCA} is obtained by using the integrated method.

4 Experimental Results and Analysis

4.1 Experimental Setup

In order to verify the validity of the CoSCA algorithm, this paper uses the DPA contest V4 data set [16] and selects one of the 10 power traces as the experimental sample. At the same time, the SVM attack algorithm in [20] is taken as a reference, denoted by SVM_{rbf} and SVM_{linear} respectively.

The CoSCA algorithm is based on Scikit-Learn. By calculating the Pearson correlation coefficient, the 200 largest feature points of the correlation coefficient are selected as the characteristics of the training sample, that is the number of features $FP_{num} = 200$. In the training process, a total of 1000 samples were randomly selected from 10,000 power traces as test set, and 5000 samples were randomly selected as the unlabeled sample set TU'. In the experiment, the number of iterations of the CoSCA algorithm is set to k = 5, and the threshold of the probability of the sample class is set to $\omega_1 = 0.9$ and $\omega_2 = 0.1$ by experimental comparison. The number of random samples is 5, 10, 20, 30, 40, 50, 100. In each case, the algorithm runs independently 10 times in the data set, and the final result is the average of 10 times run results.

4.2 Experimental Results

Table 1 shows the average correct rate of the algorithm under different number of labeled samples, and Table 2 shows the accuracy of the algorithm to run 10 times when using 5 labeled samples.

Table 1. The accuracy under different number of labeled samples

Number of labeled samples	SVM_{rbf} [20]	SVM_{linear}[20]	CoSCA
5	44.77%	45.57%	73.89%
10	61.86%	61.94%	80.04%
20	65.17%	72.12%	83.92%
30	71.21%	72.85%	85.37%
40	74.86%	76.67%	91.33%
50	73.97%	77.64%	92.05%
100	77.79%	79.25%	92.97%

When using 5 labeled samples, the accuracy for the algorithm to run 10 times is shown in Table 2.

Table 2. The accuracy when the number of labeled samples is 5

Run times	Algorithm		
	SVM_{rbf} [20]	SVM_{linear} [20]	CoSCA
1	52.2%	63.9%	61.3%
2	42.2%	46.8%	93.0%
3	38.3%	33.2%	91.1%
4	48.5%	45.4%	94.0%
5	47.4%	46.0%	93.6%
6	50.9%	51.1%	50.9%
7	32.1%	34.4%	32.2%
8	39.0%	39.8%	38.6%
9	48.7%	47.2%	92.9%
10	48.4%	47.9%	93.2%
Average value	**44.7%**	**45.57%**	**73.89%**

The change trend of CoSCA in the test sample set is shown in Fig. 2 under the number of different labeled samples.

Observing the above experimental results, we can see that with the increase of the number of labeled samples, the classification accuracy of various algorithms in the test set is basically increasing trend. This reflects the role of labeled samples for the correct classification of the model.

The results of the CoSCA algorithm are superior to the simple SVM attack algorithm in the case of fewer samples. It is shown that the CoSCA algorithm can make full use of the information of unlabeled samples and improve the accuracy of model

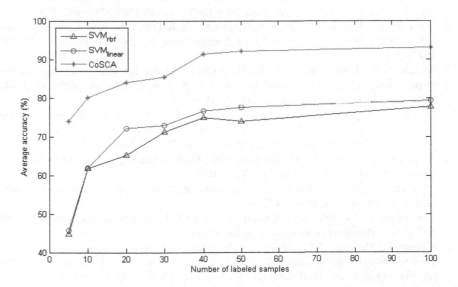

Fig. 2. The change trend of the accuracy under different labeled samples

prediction in the case of fewer labeled samples. When the number of labeled samples is very small (5), the predictive accuracy of CoSCA algorithm in most cases is significantly improved compared with using single algorithm. It also proves that when the initial classifier performance is weaker, the algorithm still picks out the correct sample from the unlabeled sample set. and then gradually improves the classifier performance in the iterative process.

5   Summaries and Outlook

In this paper, a side channel attacks algorithm CoSCA based on collaborative learning is proposed. The performance of the algorithm is better than the attack algorithm based on supervised learning in the case that the attacker has a small number of labeled power traces and a large number of unlabeled power traces. However, when the number of labeled samples is small, the prediction result of the algorithm is not completely stable due to the poor performance of the initial classifier. How to make the algorithm more stable is the goal of the next step. In addition, it was found that the threshold of the probability of the sample category had a greater effect on the final result. How to set the threshold is also the next step in the research work.

Acknowledgment. This work has been supported by the Fundamental Research Funds for the Central Universities (No. 328201507).

References

1. Standaert, F.-X., Koeune, F., Schindler, W.: How to compare profiled side-channel attacks? In: Abdalla, M., Pointcheval, D., Fouque, P.-A., Vergnaud, D. (eds.) ACNS 2009. LNCS, vol. 5536, pp. 485–498. Springer, Heidelberg (2009). doi:10.1007/978-3-642-01957-9_30
2. Chari, S., Rao, J.R., Rohatgi, P.: Template attacks. In: Kaliski, B.S., Koç, K., Paar, C. (eds.) CHES 2002. LNCS, vol. 2523, pp. 13–28. Springer, Heidelberg (2003). doi:10.1007/3-540-36400-5_3
3. Schindler, W., Lemke, K., Paar, C.: A stochastic model for differential side channel cryptanalysis. In: Rao, J.R., Sunar, B. (eds.) CHES 2005. LNCS, vol. 3659, pp. 30–46. Springer, Heidelberg (2005). doi:10.1007/11545262_3
4. Lemke-Rust, K., Paar, C.: Analyzing side channel leakage of masked implementations with stochastic methods. In: Biskup, J., López, J. (eds.) ESORICS 2007. LNCS, vol. 4734, pp. 454–468. Springer, Heidelberg (2007). doi:10.1007/978-3-540-74835-9_30
5. Hospodar, G., Gierlichs, B., Mulder, E.D., et al.: Machine learning in side-channel analysis: a first study. J. Crypt. Eng. 1(4), 293–302 (2011)
6. Patel, H., Baldwin, R.O.: Random forest profiling attack on advanced encryption standard. In. J. Appl. Crypt. 3(2), 181–194 (2014)
7. Karsmakers, P., Gierlichs, B., Pelckmans, K., et al.: Side channel attacks on cryptographic devices as a classification problem. Esat.kuleuven.be
8. Lerman, L., Medeiros, S.F., Veshchikov, N., Meuter, C., Bontempi, G., Markowitch, O.: Semi-supervised template attack. In: Prouff, E. (ed.) COSADE 2013. LNCS, vol. 7864, pp. 184–199. Springer, Heidelberg (2013). doi:10.1007/978-3-642-40026-1_12

9. Blum, A., Mitchell, T.: Combining labeled and unlabeled data with co-training. In: Colt, pp. 92–100 (1998)
10. Goldman, S.A., Zhou, Y.: Enhancing supervised learning with unlabeled data. In: Proceedings of the Seventeenth International Conference on Machine Learning, pp. 327–334 (2000)
11. Zhou, Z., Li, M.: Tri-training: exploiting unlabeled data using three classifiers. IEEE Trans. Knowl. Data Eng. 17(11), 1529–1541 (2005)
12. Zhou, Z.: Disagreement-based semi-supervised learning. Acta Autom. Sin. 39(11), 1871–1878 (2013)
13. Vapnik, V.: The nature of statistical learning theory. IEEE Trans. Neural Netw. 8(6), 1564 (1995)
14. Cortes, C., Vapnik, V.: Support-vector networks. Mach. Learn. 20, 273–297 (1995)
15. Platt, J.C.: Probabilistic outputs for support vector machines and comparisons to regularized likelihood methods. Adv. Large Margin Classif. 10(4), 61–74 (2000)
16. Description of the masked AES of the DPA contest v4. http://www.dpacontest.org/v4/data/rsm/aes-rsm.pdf
17. Nassar, M., Souissi, Y., Guilley, S., et al.: RSM: a small and fast countermeasure for AES, secure against 1st and 2nd-order zero-offset SCAs. In: Design, Automation & Test in Europe Conference & Exhibition. IEEE, pp. 1173–1178 (2012)
18. Liu, B.: Correlation power attack on aes cipher chip with rotating masking. J. Huazhong Univ. Sci. Technol. (Nat. Sci. Ed.) 11, 112–116 (2014)
19. Rechberger, C., Oswald, E.: Practical template attacks. In: Lim, C.H., Yung, M. (eds.) WISA 2004. LNCS, vol. 3325, pp. 440–456. Springer, Heidelberg (2005). doi:10.1007/978-3-540-31815-6_35
20. Lerman, L., Bontempi, G., Labelowitch, O.: A machine learning approach against a masked AES. J. Crypt. Eng. 5, 123–139 (2015). International Conference Smart Card Research and Advanced Applications, Cardis

Research on Hydrological Time Series Prediction Based on Combined Model

Yi Cheng$^{(\boxtimes)}$, Yuansheng Lou, Feng Ye, and Ling Li

College of Computer and Information, Hohai University, Nanjing, China
18260624431@163.com

Abstract. Water level prediction of river runoff is an important part of hydrological forecasting. The change of water level not only has the trend and seasonal characteristics, but also contains the noise factors. And the water level prediction ability of a single model is limited. Since the traditional ARIMA (Autoregressive Integrated Moving Average) model is not accurate enough to predict nonlinear time series, and the WNN (Wavelet Neural Network) model requires a large training set, we proposed a new combined neural network prediction model which combines the WNN model with the ARIMA model on the basis of wavelet decomposition. The combined model fit the wavelet transform sequences whose frequency are high with the WNN, and the scale transform sequence which has low frequency is fitted by the ARIMA model, and then the prediction results of the above are reconstructed by wavelet transform. The daily average water level data of the Liuhe hydrological station in the Chu River Basin of Nanjing are used to forecast the average water level of one day ahead. The combined model is compared with other single models with MATLAB, and the experimental results show that the accuracy of the combined model is improved by 7% compared with the traditional wavelet network under the appropriate wavelet decomposition function and the combined model parameters.

Keywords: Combined model · Autoregressive Integrated Moving Average · Prediction · Wavelet neural network · Hydrological time series

1 Introduction

Artificial neural network (ANN) has been widely used in hydrological forecasting [1]. Zhang and Benveniste proposed the wavelet neural network (WNN) in 1992 [2]. The wavelet network is not easy to fall into the local minimum, the fitting precision is higher than the BP neural network, and the required parameters and the number of the neurons are smaller than the latter. However, as the input dimension increases, the training samples of the network will increase exponentially and the convergence rate of the network is greatly decreased [3]. Box and Jenkins proposed a time series prediction method of ARIMA model in 1972 [4]. The theory of ARIMA model is complete, but with the increase in the forecast period, the forecast error will gradually increase [5].

The individual optimization of these methods can't overcome the limitation of single method. Therefore, the combined forecasting method based on multiple models

© Springer Nature Singapore Pte Ltd. 2017
B. Zou et al. (Eds.): ICPCSEE 2017, Part I, CCIS 727, pp. 558–572, 2017.
DOI: 10.1007/978-981-10-6385-5_47

has become the direction of time series prediction. In order to predict precisely, the combined forecasting method use the weight coefficient to combine two or more different forecasting methods, which is proved to be theoretically feasible [6]. In literature [7–9], the prediction error of ARIMA model is modeled by BP neural network, but it does not consider the local convergence of BP network. The literature [10, 11] combines the ARMA model with the neural network, ignoring the unstable decomposition sequence is not suitable for ARMA model.

The combined model above has achieved some effect in predicting the time series, but it does not take into account the applicability of the method and principle, such as the influence of time series characteristics on the method selection, the method of data preprocessing, the coupling way of combined model, etc., so there are some limitations. Based on the idea of combinatorial model, this paper presents a scientific combined forecasting model of hydrological time series. Firstly, the hydrological time series is decomposed by wavelet transform, and then the wavelet transform and the wavelet transform are used to model the ARIMA transform and the wavelet transform respectively. Then, the sub-model output is reconstructed by wavelet transform, and the predicted value is obtained. The coupling mode of the combined model is different from the traditional combination model which is optimized by the weight coefficient, and it embeds the two models into the same wavelet analysis framework without solving the objective optimal function. Compared with the single ARIMA model and the WNN model, the predicted accuracy of the combined model is higher than that of the single model, and the optimal effect of the combined model on the hydrological time series is proved.

2 Models and Methods

2.1 Wavelet Neural Network

The basic idea of wavelet neural network is to replace the Sigmoid family transfer function in BP neural network by wavelet basis function in wavelet analysis. WNN has the characteristics of wavelet transform and BP neural network. Compared with the BP neural network, the convergence speed of the wavelet network is faster, and it can avoid the local optimization, and also has the advantages of time-frequency local analysis [12]. The wavelet network has strong network fault-tolerant ability, and only a subset of the coefficients is involved in the computation and updating of the single point weights.

The WNN consists of input layer, hidden layer and output layer. The input layer and output layer weights are replaced by the scale factor and the translation factor of the wavelet function, respectively. Similar to the BP neural network, the signal propagates in the forward direction, the error back propagation. The function in the hidden layer nodes is a wavelet function. The concrete structure of the WNN is shown in Fig. 1:

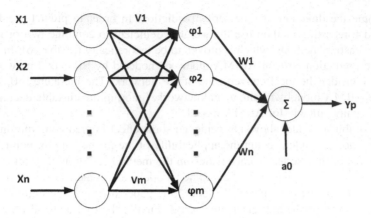

Fig. 1. Structure of wavelet neural network

2.2   Autoregressive Integrated Moving Average

Hydrological time series usually show some tendency and periodicity, which are more obvious in the low frequency scale transformation series generated by wavelet decomposition. After the differential operation of the low frequency sequence, we can get the difference sequence which satisfies the requirement of the stability. Therefore, the ARIMA model is used to simulate and predict the sequence that is obtained from scale transformation. The definition of d-order difference operation is as follows:

$$x_t = \sum_{i=1}^{d} (-1)^{i+1} C_d^i x_{t-1} + \nabla^d x_t. \tag{1}$$

Assuming that the order of autoregressive model is p, the order of the moving regression model is q, and the order of differential operation is d, The ARIMA model is denoted as $\mathrm{ARIMA}(p, d, q)$, and the ARIMA model is defined as follows:

$$\Phi(B)\nabla^d y_t = \Theta(B)\varepsilon_t \tag{2}$$

$$\mathrm{E}(\varepsilon_t) = 0, \mathrm{Var}(\varepsilon_t) = \delta_\varepsilon^2, E(\varepsilon_t \varepsilon_s) = 0, s \neq t \tag{3}$$

$$\mathrm{E}y_s \varepsilon_t = 0, \forall s < t \tag{4}$$

Among them: $\nabla^d = (1 - \mathrm{B})^d$, $\Phi(B) = 1 - \emptyset_1 B - \dots - \emptyset_p B^p$, $\Theta(B) = 1 - \theta_1 B - \dots - \theta_q B^q$.

In general, if the sequence has a linear trend, $d = 1$; If the sequence has a nonlinear trend, $d = 2$ or $d = 3$; If there is a fixed period in the sequence, the difference operation is carried out with the length of the periodic length.

2.3   Model Performance Evaluation

In this paper, the Mean Square Error (MSE) and the determination coefficient R^2 are used as the evaluation factors of the model. The formula of MSE and R^2 is as follows:

$$MSE = \frac{\sum_{i=1}^{N} (Q_i - \hat{Q}_i)^2}{N} \tag{5}$$

$$R^2 = 1 - \frac{\sum_{i=1}^{N} (Q_i - \hat{Q}_i)^2}{\sum_{i=1}^{N} (\hat{Q}_i - \bar{Q})^2} \tag{6}$$

In the formula, R^2, MSE, N, Q_i, $\hat{Q}$, $\bar{Q}$ represent the coefficient of determination, mean square error, the number of observations, the observed data, the predicted value and the mean value of the observed data, respectively. MSE is used to determine the accuracy of the prediction model, which is usually a positive value. When the observed values and the predicted values are well fitted, the MSE is close to 0, and when the error is large, the MSE value is also large. The higher the R^2 value (upper limit is 1), the higher the explanatory degree of the independent variable to the dependent variable, and The percentage of changes caused by the independent variable is higher in the total change.

2.4   Data Preprocessing Method

The water level time series fluctuates frequently and the difference is large, so it is necessary to normalize the time series to reduce the model processing error. After the training of the network, it is necessary to carry out the anti-normalization and restore the data to the actual value. The normalized and anti-normalized formulas used are shown in Eqs. (7) and (8), respectively:

$$n(t) = \frac{x(t) - \min(x(t))}{\max(x(t)) - \min(x(t))} \tag{7}$$

where $x(t)$ is the sample data and $n(t)$ is the input sample.

$$O(t) = \min(x(t)) + o(t) * \max(x(t)) - \min(x(t)) \tag{8}$$

where $o(t)$ is the output sample, $x(t)$ is the sample data, $O(t)$ is the output predicted value.

3 The Proposed Combination Model

3.1 The Network Structure of the Combined Forecasting Model

The core strategy of the combined model is "divide and conquer". According to the randomness of high frequency sequence, the WNN network is used to predict the high frequency sequence, which can effectively improve the accuracy and efficiency of prediction. For the low frequency sequence containing the trend and the periodicity, the difference processing is performed, and the future value of the sequence is predicted by the ARIMA model. The modeling flow of the combined model is shown in Fig. 2.

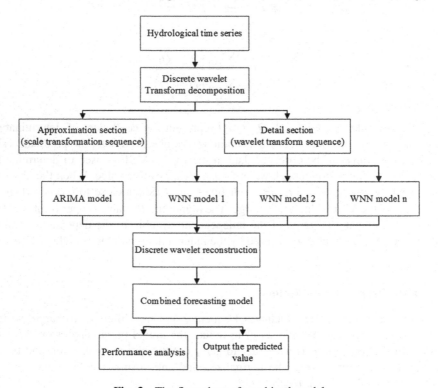

Fig. 2. The flow chart of combined model

The concrete steps of the combined model are as follows:

Step 1: Normalized water level time series data;
Step 2: The normalized sequence is decomposed into a scale transformation sequence and several wavelet transform sequences by wavelet analysis;
Step 3: The ARIMA model is applied to the scale transformation sequence to predict the sequence;
Step 4: The WNN model is used to fit the wavelet transform sequence, and the prediction value is obtained;
Step 5: The predicted sequence of values in step 3 and 4 is reconstructed by wavelet to obtain the normalized water level prediction sequence;

Step 6: The normalized water level time series is anti-normalized to obtain the predicted value of the original sequence.

3.2 Wavelet Decomposition Function

Wavelet is a waveform that effectively limits the duration and has a mean of zero. The literature [13–16] have proved the effect of wavelet decomposition in hydrological time series. The wavelet decomposition of the signal is helpful to give full play to the advantage of wavelet de-noising, and it is also a necessary step for the division of labor. For the selection of wavelet functions, there is no direct formula available, usually in accordance with the self-similarity principle and the length of the wavelet support to choose. Self-similarity principle refers to the choice of wavelet and the signal has a certain similarity, then the transformation of energy is more concentrated, and the amount of calculation can be reduced effectively [17]. In most applications, the wavelet with the support length of 5–9 is chosen, because it will produce a boundary problem if the support is too long, and if the support is too short, the vanishing moment is too low to concentrate the energy of signal.

The water-level fluctuation of Liuhe site is similar to that of Daubechies wavelet (db wavelet), and there are often some potential or obvious jump characteristics in hydrological time series while db wavelet is suitable for extracting time series jump characteristics. The db wavelet is a compactly supported orthogonal wavelet, and the support of the wavelet function $\psi(t)$ and the scale function $\varphi(t)$ is $2N - 1$, and the vanishing moment of $\psi(t)$ is N. The regularity of dbN wavelet is better, and the vanishing moment increases with the increase of order, and the frequency domain localization ability is also stronger, but the amount of calculation will increase too. Therefore, the db10 wavelet is chosen as the wavelet basis for the decomposition of the original time series, and the level of decomposition is 6.

3.3 ARIMA Sub-module Modeling

The ARIMA model of the wavelet transform sequence is established as follows:

Step 1: Test the stationarity of the original sequence. If passed, go to the next step, otherwise, continue to make a difference to the sequence until the differential sequence satisfies the stationarity test;

Step 2: Determine the order of the model, based on the AIC information criterion, limit the scope of p and q, and traverse the (p, q) in order to find the minimum AIC (p, q);

Step 3: Apply the $\text{ARIMA}(p, d, q)$ model to the d-order differential prediction;

Step 4: The differential sequence of prediction is reduced, then return the actual predicted result.

In this paper, we utilized the ADF test (Auger Dickey-Fuller test) in the time series to carry out the unit root test, and the test utilized the daftest () function in the MATLAB software. If the ADF test passed to reject the null hypothesis $H_0{:}\delta = 0$, the test result is 1, otherwise, the result is 0. The ADF test of the scale transformation sequence shows that the sequence fluctuation is obvious and the sequence stability is achieved after the third order difference.

The MATLAB pseudo program code segment of ARIMA module is as follows:

```
Input: the scale transformation sequence: a.
Output: the prediction sequence of the Input.
while (adftest(da)==0)// Stationarity tests
r11<- autocorr(a);       //the autocorrelation
r12<- parcorr(a);        //the partial autocorrelation
da <- diff(a);//Differences and Approximate Derivatives
//According the AIC traverse the i=0:3,j=0:3 group
spec<- garchset('R',i,'M',j,'Display','off');
[coeffX,errorsX,LLFX] <- garchfit(spec,dc);
num <- garchcount(coeffX);
[aic,bic] <- aicbic(LLFX,num,n);// the Minimum AIC value
garchset('R',r,'M',m,'Display','off');// D order differ-
ential prediction;
[coeffX,errorsX,LLFX]<- garchfit(spec2,da);
[sigmaForecast,w_Forecast]<- garchpred(coeffX,da,10);
x_pred<-a(end)+cumsum(w_Forecast);// Reductive difference
and Return forecast results
end
```

3.4   WNN Sub-module Modeling

The WNN's input dimensions and wavelet function are important factors that affect the network performance. Since the research of this paper is to compare the prediction accuracy of the combined model with the traditional model, the focus of the study is not the long-term forecast of the runoff water level, so the forecast period of the water level is 1 day. Through the autocorrelation evaluation of wavelet coefficients, it is found that the wavelet transform coefficient of the forecast date has a high correlation with the coefficients of 5 days before the date. Therefore, the input matrix with hysteresis period of 5 days is used as the input of the corresponding wavelet network. That is, the input of the wavelet network is the variables $\{X_{t-1}, X_{t-2}, X_{t-3}, X_{t-4}, X_{t-5}\}$, and the output of the wavelet network is X_t. The wavelet function used in this paper is Morlet wavelet, and its basis function and derivative function are as follows:

$$\psi(x) = \exp\left(-\frac{x^2}{2}\right) * cos1.75x \tag{9}$$

$$\psi'(x) = -(xcos1.75x + 1.75sin1.75x) * \exp\left(-\frac{x^2}{2}\right) \tag{10}$$

Prediction steps of wavelet neural network model are as follows:

Step 1: design the structure of wavelet network. Such as the input dimension of the network, the number of hidden layers and the dimension of the output vector;

Step 2: sample handling. The sample is divided into training sample and test sample. The training sample is used to train the network, and the test sample is used to test the network accuracy;

Step 3: Wavelet Neural Network Learning. Set the maximum number of iterations, calculate the neural network output according to the formula, and perform the neural network learning according to the matrix form of the error back propagation learning algorithm;

Step 4: Determine whether the error is less than the target value, if it is greater than the target value, return to step 3, if less than the target value, stop training.

The MATLAB pseudo code of the WNN sub-module is shown below:

```
Input: the detail data after wavelet transform : x, target
data of the WNN model: target, time interval: t, the number
of the wavlet function nodes: H, the output nodes: J;
Output: the predicted value of the Input data series;
whi<-rand(I,H);// weight between input and hidden layler
whj<-rand(H,J);// weight between hidden and output layler
//Calculate of net input:
for-loop{xpl<-xpl+whi(i,h)*x(i);ixp(h)<-xpl+c(h);};
oxp(h)<-fai((ixp(h)-b(h))/a(h)); //The Gauss function
for-loop{ixjpl<-ixjpl+whj(h,j)*oxp(h);ixjp(j)<-ixjpl+d(
j);} ;
for-loop{oxjp(i)<-fnn(ixjp(i));};//the output of network
ERROR<-1/2*sumsqr(oxjp-target);
Err_NetOut<-[Err_NetOut ERROR]; //the error of output
detaj(j)<- -(oxjp(j)-target(j))*oxjp(j)*(1-oxjp(j));
//the gradient of network:
for-loop{detawhj(h,j)<-eta*detaj(j)*oxp(h);};
detab2<-eta*detaj;
sum<-detaj(j)*whj(h,j)*diffai((ixp(h)-b(h))/a(h))/a(h)+
sum;// First order Gauss derivative function:diffai()
detah(h)<-sum;
detawhi(i,h)<-eta*detah(h)*x(i);
detab1<- eta*detah;detab<- -eta*detah;
detaa(h)<- -eta*detah(h)*((ixp(h)-b(h))/a(h));
whj<-whj+(1+aerfa)*detawhj;// adjust of weight value
whi<-whi+(1+aerfa)*detawhi;
end
```

3.5   Integration of ARIMA Model and WNN Model

The ARIMA sub module deals with the scale transform sequences obtained by discrete wavelet decomposition, and the predict sequence values of the scale transformation sequence are obtained. The WNN sub module is responsible for processing the wavelet transform sequence, and obtains the predictive sequence values of the wavelet transform sequence. The predicted sequence values of the combined model can be obtained by discrete wavelet reconstruction of the predicted sequence values. The discrete wavelet is reconstructed by the single reconstruction function wrcoef() in MATLAB.

The concrete integration process of the two models is as follows:

Step 1: use the single reconstruction function to reconstruct the predicted sequence which is output by the ARIMA model;
Step2: use the single reconstruction function to reconstruct the predicted sequence which is output by the WNN model;
Step3: The reconstructed signals are obtained by superimposing the low and high frequency reconstruction signals obtained by steps 1 and 2. That is, the combined model predictive values.

4   Experiment and Discussion

The data set used in this paper is the daily average water level data of Liuhe hydrological station in Chuhe River from January 1, 2010 to December 31, 2015, and the number of the data is 2129. The data set is divided into two parts: the first part is the training data set, accounted for 95% of the total data, which is used to train to adjust the network weights and coefficients; the second part is used to test the network performance test data set, accounted for 95% of the total data. The time series curve of the original data is shown in Fig. 3.

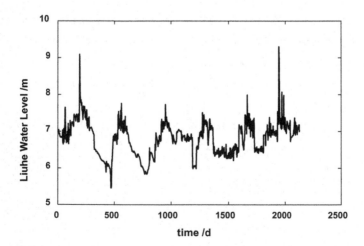

Fig. 3.   Daily average water level curve of Liuhe station in 2010–2015

It can be seen from Fig. 3, the Liuhe site's water level has obvious nonlinear characteristics of the overall while maintaining the $[5.0, 9.5]$ water level fluctuation interval. On a large time scale, there is a cycle of "up - down - up - down". And the time series need to be normalized to the $[0, 1]$ interval, the obtained sequence is shown in Fig. 4. The normalized sequence is decomposed into 6 layers with Daubechies10 wavelet, as shown in Fig. 5.

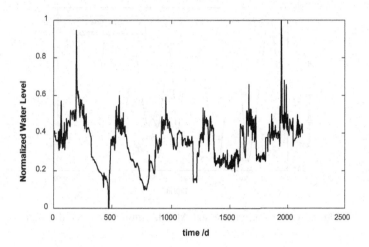

Fig. 4. Normalized curve of original water level time series

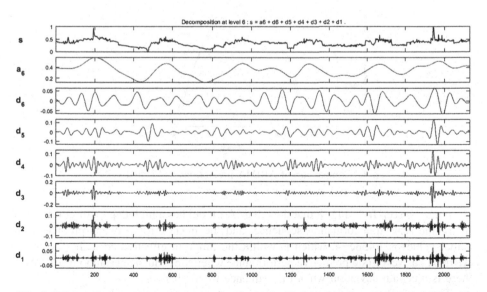

Fig. 5. The scale transformation sequence a6 and the wavelet transform sequences d1 to d6

The ARIMA model is used to simulate and predict the scale transform sequence a6. The third-order difference of time series can pass the ADF stationary test. The order of the minimum AIC obtained by using the ACI information criterion: p = 1, q = 1, therefore, the ARIMA (1, 3, 1) model is established. The fitting and prediction effect of ARIMA model on the original sequence is shown in Figs. 6 and 7:

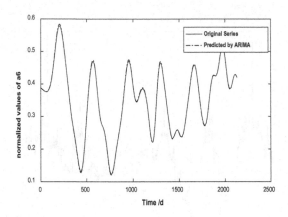

Fig. 6. Original sequence and ARIMA fitting sequence diagram

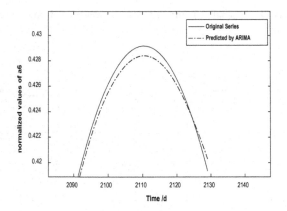

Fig. 7. Comparison of ARIMA predictions and actual values

It can be seen from Fig. 8 that the ARIMA model has a high degree of goodness of fit for the scale transformation sequence, and the mean error between predicted and actual values is 1.458e−4, which is within the range of effective error. It's clear that ARIMA has very good predictive performance when the sequence is smooth. However, there is a certain deviation when the sequence trend changes.

The wavelet transform sequence of d1 to d6 are trained through wavelet network. Via duplicate test, and when the number of input nodes are 5, the number of hidden

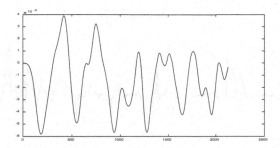

Fig. 8. Predictive error curves for the ARIMA model

layer nodes are 13 and the number of output nodes is 1, the WNN can get good results. The input matrix of 2018 × 5 and the output matrix of 2018 × 1 are set up for training samples. And the test samples are built with 101 × 5 input matrix and an output matrix of 101 × 1.

Due to the limited space, this paper only shows the network performance analysis of d1 sequence. The MSE curve of training data is shown in Fig. 9. It can be seen that the WNN network achieves the minimum MSE of the order of magnitude e−5 after 12 iterations, so the network has very good convergence speed and precision. The error histogram shows that the error is normal distributed between −0.01084−+0.01006. The fitting results of the wavelet transform and the real data are shown in Fig. 10.

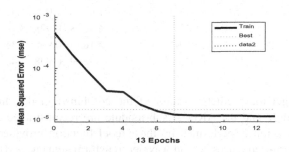

Fig. 9. The MSE curve of training data

As can be seen from Table 1, the MSE and R2 of training data and test data are on the same order of magnitude, and this indicates that the prediction accuracy of the wavelet neural network is stable, and the mean square error is in the order of e−5 and below, and the coefficient value is 8.3 and above. With the improvement of the wavelet detail level, the number of iterative times of the compact wavelet network is also increasing, which means that the more detailed the fitting details are, the more iterations are needed to achieve the target accuracy. The training data and the test data also improve the prediction accuracy with the improvement of the detail level, and the minimum MSE is controlled on the e−9 scale eventually. This indicates that WNN model is very effective to predict the wavelet transform sequence.

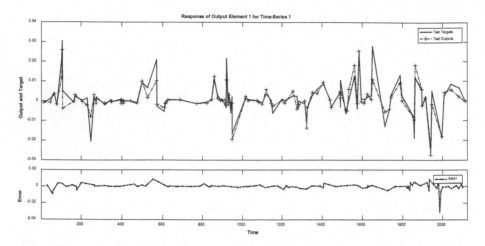

Fig. 10. Target output and actual output and error of detail1

Table 1. WNN model performance of wavelet transform sequences

Decomposition detail level	WNN architecture	Training data		Testing data	
		MSE	R^2	MSE	R^2
Detail 1	5-13-1	1.3e−5	0.865	1.4e−5	0.81
Detail 2	5-13-1	1.1e−5	0.941	1.9e−5	0.941
Detail 3	5-13-1	1.6e−6	1.0	3.9e−6	0.980
Detail 4	5-13-1	4.9e−8	0.998	8.1e−8	0.998
Detail 5	5-13-1,	2.1e−9	0.998	1.3e−9	0.998
Detail 6	5-13-1	1.2e−9	0.998	1.4e−9	0.998

In order to get the predicted values of the combined model, the results of the wavelet sub-module and the ARIMA sub-module are reconstructed by wavelet. The wavelet reconstruction is performed by the wrcoef function provided by MATLAB, The scale transform sequences a6 and wavelet transform sequences d1, d2, d3, d4, d5, d6 are reconstructed respectively, after reconstruction, the outputs of the composite model are obtained by superimposing them. And in order to judge the performance of the combined model, the predicted results of ARIMA-WNN model are compared with the WNN model and the traditional hydrological time series forecasting model ARIMA. The fitting curve of the three prediction sequences and the original observation sequence are shown in Fig. 11.

The figure shows that both of the predicted results of the combined model and WNN model are very close to the original sequence, while the predicted values of the ARIMA model have obvious offset to the actual value. The results of the MSE and R^2 values of the predicted model are shown in Table 2. The table shows that the MSE of the combined model of ARIMA and WNN is less than that of WNN and ARIMA. The error of the combined model is close to that of the single WNN, but it is still better than

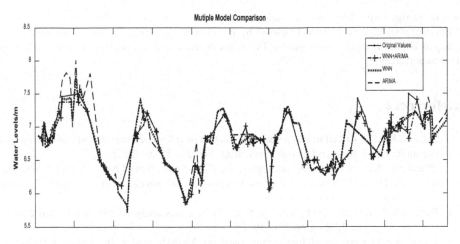

Fig. 11. Fitting curve of combined model, WNN and ARIMA model

Table 2. Comparison of model performance

Model type	Model parameters	MSE	R^2
ARIMA + WNN	(1,3,1) & (5-13-1)	0.00511	0.9770
WNN	(5-13-1)	0.00787	0.9530
ARIMA	(1,3,1)	0.03419	0.8808

the WNN, and the accuracy of the two is higher than the ARIMA model by an order of magnitude. The determination coefficient R^2 show a strong correlation between the input variables and predicted values.

5 Conclusion and Prospect

In this paper, the WNN model and ARIMA model is combined to predict the accuracy of hydrological time series. The model is applied to predict the average daily water level of Liuhe station on the Chu River Basin, and compared with the single WNN model and ARIMA model. The predicted results shows that the prediction accuracy of the combined model is higher than that of the single model, which shows that the combination model is effective. In this paper, the wavelet preprocessing- double model parallel processing – prediction pattern is different to the traditional combined model which is for the goal of the optimal weight coefficient. The combined model can integrate their advantages of the two single models, and effectively improve the pre-diction accuracy of the two single models. Due to the uncertainty and complexity of time series, it is necessary to combine the specific application with specific analysis, and select the appropriate processing methods in the field. In addition, the prepro-cessing of the data has an important influence on the accurate prediction of the model. And the wavelet decomposition is indeed good for data de-noising and time frequency analysis.

Acknowledgement. This work is supported by (1) National Natural Science Foundation of China (61300122); (2) A Project Funded by the Priority Academic Program Development of Jiangsu Higher Education Institutions; (3) Water Science and Technology Project of Jiangsu Province (2013025)

References

1. Rahimi, K.A.: Artificial neural network estimation of reference evapo transpiration from pan evaporation in a semi-arid environment. Irrig. Sci. **27**, 35–39 (2008)
2. Zhang, Q., Benveniste, A.: Wavelet networks. IEEE Trans. Neural Netw. **3**, 889–898 (1992)
3. Chen, Z., Feng, T.: Research progress and prospect of wavelet neural network. J. Ocean Univ. Qingdao **04**, 663–668 (1999)
4. Box, G.E.P., Jenkins, G.M., Reinsel, G.C.: Time series analysis: forecasting and control. J. Oper. Res. Soc. **22**(2), 199–201 (1976)
5. Zheng, P.: The combined forecasting based on ARIMA model. In: Yanshan University (2009)
6. Bates, J.M., Granger, C.W.J.: The combination of forecasts. J. Oper. Res. Soc. **20**(4), 451–468 (1969)
7. Mo, D.: The application of ARIMA and BP neural network mixture model in the GDP estimation of Guangxi. J. Guangxi Univ. Financ. Econ. **24**(6), 31–34 (2011)
8. Zhai, J., Cao, J.: Combined forecasting model based on time series ARIMA and BP neural network. Stat. Decis. **4**, 29–32 (2016)
9. Dang, G.: Research on combined forecast method in short-term load forecasting. In: Hunan University (2011)
10. Zhang, P., Li, X.: Based on wavelet neural network and ARMA model of CPI fluctuation range forecast. Inf. Technol. **2**, 131–135 (2016)
11. Zheng, J.: The research on forecasting of financial time series based on wavelet analysis and neural network. In: Xiamen university (2009)
12. Yang, Q., Zhang, J., Li, W., et al.: Prediction of wind speed and wind power generation based on Wavelet Neural Network. Power Grid Technol. **17**, 44–48 (2009)
13. Wei, S., Yang, H., Song, J., et al.: A wavelet-neural network hybrid modelling approach for estimating and predicting river monthly flows. Hydrol. Sci. J. **58**(2), 374–389 (2013)
14. Falamarzi, Y., Palizdan, N., Huang, Y.F., et al.: Estimating evapotranspiration from temperature and wind speed data using artificial and wavelet neural networks (WNNs). Agric. Water Manag. **140**(3), 26–36 (2014)
15. Zhu, Y., Li, S., Fan, Q., et al.: Wavelet neural network model based complex hydrological time series prediction. J. Shandong Univ. (Eng. Sci.) **04**, 119–124 (2011)
16. Nourani, V., Özgür, K., Komasi, M.: Two hybrid artificial intelligence approaches for modeling rainfall–runoff process. J. Hydrol. **402**(1–2), 41–59 (2011)
17. Gao, Z., Yu, X.: Wavelet Analysis and Application in MATLAB, 3rd edn. National Defense Industry Press, Beijing (2007)

A Cross-View Model for Tourism Demand Forecasting with Artificial Intelligence Method

Siming Han[1], Yanhui Guo[1,2], Han Cao[1(✉)], Qian Feng[1], and Yifei Li[1]

[1] Shaanxi Normal University, Xi'an 710119, Shaanxi, China
caohan@snnu.edu.cn
[2] Shandong Women's University, Ji'nan 250300, Shandong, China

Abstract. Forecasting always plays a vital role in modern economic and industrial fields, and tourism demand forecasting is an important part of intelligent tourism. This paper proposes a simple method for data modeling and a combined cross-view model, which is easy to implement but very effective. The method presented in this paper is commonly used for BPNN and SVR algorithms. A real tourism data set of Small Wild Goose Pagoda is used to verify the feasibility of the proposed method, with the analysis of the impact of year, season, and week on tourism demand forecasting. Comparative experiments suggest that the proposed model shows better accuracy than contrast methods.

Keywords: Cross-view · BPNN · SVR · ARIMA

1 Introduction

With the rapid development of the national economy and the escalation of people's living standard, the tourism demand of people has become greater and greater. The number of tourists shows a blowout growth. This brings great opportunities to the tourism industry, but also ushered in unprecedented challenges. In recent years, it is often seen that more and more safety accidents have occurred as a result of the crowded tourists and overload of tourism spots. So, tourism demand forecasting is significantly necessary for tour managers to make a decision and tourists to develop a travel plan.

As early as 1960s, foreign scholars began to study tourism demand forecasting. Lots of forecasting models were presented, such as initial time series models, subsequent econometric models, and artificial intelligence models. There are a large number of research results both in theory and in methodology. The following will be introduced from three aspects of the previous work done.

Classical time series forecasting methods based on exponential smoothing include AutoRegressive (AR), Moving Average (MA) and so on. Autoregressive Moving Average (ARMA) and Autoregressive Integrated Moving Average (ARIMA) [1] are more popular than others recently.

Gustavsson employs ARMA model [2] to forecast tourism demand of different areas. Lim takes advantage of ARMA to forecast monthly and seasonal tourism flows [3, 4]. Pai makes use of ARIMA [5] to forecast tourism demand. For the seasonal feature, lots of seasonal ARIMA models are presented by researchers, such as papers [6–8]. However, the result of classical time series forecasting relies on the accuracy of

© Springer Nature Singapore Pte Ltd. 2017
B. Zou et al. (Eds.): ICPCSEE 2017, Part I, CCIS 727, pp. 573–582, 2017.
DOI: 10.1007/978-981-10-6385-5_48

figures, and is sensitive to dirty data [9]. Also, it is suitable for linear data, but not for nonlinear data.

The econometric models are widely used in the tourism demand forecasting. The advantages of these models are to reflect the causal relationship between the influencing factors and to explain the changes of tourism demand from the view of economic. The methods of econometric models for tourism demand forecasting include the error correction model (ECM), the vector autoregressive (VAR), the autoregressive distributed lag model (ADLM), and so on. Kulendran and Witt [10] uses an ECM model to predict the UK outbound travel to six major destinations, which suggests that ECM is more suitable for long-term forecasting. Song et al. [11] employs ADLM model for forecasting inbound tourism demand in Hongkong. The almost ideal demand system model (AIDS) is also used for tourism demand forecasting recently, such as the papers [12, 13]. The econometric models mainly explain the change of tourist flow by economic factors. So, they need additional economic variables to support it. Meanwhile, as the impacts of each site are not the same, the model cannot be used as a common method.

Recently, artificial intelligence methods are widely used in all areas of life. Law [14, 15] applies a feed-forward neural network (NN) model and a backward propagation neural network (BPNN) to forecast Japanese tourism demand. Fernandes et al. [16] highlights the usefulness of the ANNs applied for tourism demand forecasting, and proposes a new ANN [17] using the years and months in the input. Lin [18] employs three methods to forecast Taiwan's visitors. Chen et al. [19] develops a new model based on empirical mode decomposition and ANN to forecast tourism demand. Claveria and Torra [20] evaluates the performance of ANN.

In addition to the above methods, a statistical method is also widely used in time series prediction named support vector regression (SVR). SVR has a better generalization owing to the structural risk minimization principle, and overcomes the shortcomings of over-fitting and lots of parameters. Chen and Wang [21] and Hong et al. [22] adopts SVR with genetic algorithms in forecasting tourism demand. Pai et al. [23] uses least-squares SVR combined with fuzzy c-means to forecast tourism demand.

However, these methods focus on improving models, but ignoring data modeling. Researchers used to combine a variety of factors into a model to improve accuracy. This paper proposes a new data modeling method by using a cross model to forecast tourism demand, and then obtain a weight by learning algorithm. The method can be used in many models, such as ANNs and SVR.

The rest of the paper is as follows. Section 2 presents the methodology, and the problem analysis is introduced in Sect. 3. Section 4 shows the experiment results. Finally, we conclude the paper in Sect. 5.

2 Methodology

2.1 Artificial Neural Networks (ANNs)

Artificial Neural Networks, named ANNs or NNs for short, is a kind of artificial intelligence method to simulate the human brain. ANNs is divided into forward and backward networks. At present, the back propagation neural network (BPNN) is more popular. We take BPNN for example to describe the principle of ANNs.

BPNN is based on a large collection of neural units. Each neural unit is connected with many others, and links can be enforcing or inhibitory in their effects on the activation state of connected neural units. There may be a threshold function or limiting function on each connection and on the unit itself, such that the signal must surpass the limit before propagating to other neurons.

Take a three layers network for instance, given one training example (x, y), forward propagation:

$$a^{(1)} = x \Rightarrow z^{(2)} = \ominus^{(1)} a^{(1)} \Rightarrow a^{(2)} = g(z^{(2)}) \left(\text{add } a_0^{(2)} \right) \Rightarrow z^{(3)} = \ominus^{(2)} a^{(2)} \Rightarrow$$
$$a^{(3)} = g(z^{(3)}) \left(\text{add } a_0^{(3)} \right) \Rightarrow z^{(4)} = \ominus^{(3)} a^{(3)} \Rightarrow a^{(4)} = g\left(z^{(4)} \right) \tag{1}$$

The total loss is as following:

$$E = \frac{1}{2} \sum_i (y_i - a_i)^2 \tag{2}$$

Each layer will update their weight like the following:

$$\ominus_{ij}^{(l+1)} = \ominus_{ij}^{(l)} + \Delta \ominus_{ij}^{(l)} = \ominus_{ij}^{(l)} - \alpha \frac{\partial E(\ominus)}{\partial \ominus_{ij}^{(l)}} \tag{3}$$

The residual function of the *i-th* node in the *l-th* layer is as the below:

$$\delta_i^{(l)} = \sum_j^{N^{(l+1)}} \left(\delta_j^{(l+1)} \right) \cdot g'(z_i) \tag{4}$$

The update function of BPNN is as following:

$$\ominus_{ji}' = \ominus_{ji}' - \alpha \cdot \delta_i^{(l+1)} \cdot a_j^{(l)} \tag{5}$$

2.2   Support Vector Regression (SVR)

SVR, originally proposed by Vapnik, uses linear model to implement nonlinear class boundaries by mapping the input vectors x into the high-dimensional feature space. In the new space, a non-linear model is converted into a linear model. Thus, SVR is interested in the solution of the maximum margin hyperplane in the new space. Then, the classification/regression problem is transformed into a quadratic programming problem, which is easy to be solved by an optimization program.

Given a data (x_i, y_i), i = 1, 2, ..., n, $x_i = R^n$, $y_i \in R$, where x_i is then-dimensional input, that is historical stock data in this paper, y_i is the output, that is predicted value, n is the number of training set. The form of SVR can be represented as the following equations:

$$f(x) = w^T \varphi(x) + b \quad \varphi : R^n \to F, w \epsilon F \tag{6}$$

where w is an element in the high-dimensional spaces, $\varphi(x)$ is the non-linear mapping function, bis the threshold. Then, the above regression problem is to seek to minimize the risk of f, by minimizing the structural risk, as the following.

$$R_{reg} = \frac{1}{2} \| w \|^2 + R_{emp} = \frac{1}{2} \| w \|^2 + C \times \frac{1}{n} \sum_{i=1}^{n} |y_i - f(x_i)|_\varepsilon \tag{7}$$

Where $\| w \|^2$ is the description function, R_{emp} is the empirical risk, C is a penalty parameter, $|y_i - f(x_i)|_\varepsilon$ is an insensitive parameter. The parameters of C and ε affect the precision of SVR significantly. The value of ε is defined as following.

$$|y_i - f(x_i)|_\varepsilon = \begin{cases} 0, |y_i - f(x_i)| \le \varepsilon \\ |y_i - f(x_i)| - \varepsilon, else \end{cases} \tag{8}$$

By introducing nonnegative slack variable ξ and Lagrange function, using Karush-Kuhn-Tucker (KKT) condition, the function $f(x)$ is described as following:

$$f(x) = \sum_{i=1}^{n} (a_i - a_i^*) k(x_i, x) + b \tag{9}$$

Where a_i, a_i^* are Lagrange multiplier, $k(x_i, x)$ is defined as the kernel function. There are some different kernels for generating the inner products to construct machines in the input space. Different kernels will have different nonlinear decision surfaces to obtain different results. We should choose the kernel according to the feature of inputs. Generally, common example of the kernel function is Gaussian radial basis function (RBF). We also choose RBF as our kernel in this paper, as following:

$$f(x) = \sum_{i=1}^{n} (a_i - a_i^*) \exp(- \| x_i - x \|^2 / 2\sigma^2) + b \tag{10}$$

Where x_i is the element of training set, x is the element of test set, σ is a parameter of RBF.

3   Problem Analysis

In order to accurately forecast tourism demand, we need to take the impacts of seasons, weeks, and years into account. The papers [6] suggested that the tourism demand is seasonal.

3.1   Analysis of Annual Variation

We take the small wild goose pagoda as an example to analyze the travel data from 2011 to 2014, as is shown in Fig. 1. Taking the year as the research object, the traffic has the following characteristics:

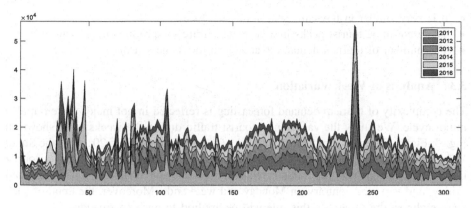

Fig. 1. Annual variation of the tourism demand from 2011 to 2016.

- Passenger traffic shows an upward trend year by year.
- The trend of tourist flow is similar to that in other years, and tourism demand has annual periodicity.
- Festival will bring great changes to the flow of tourists.

Figure 1 shows that it is very different from 20 to 50 days. The main reason for this phenomenon is that the date of the Spring Festival in China is not the same day in solar calendar. In addition to this, the other shows a strong consistency.

3.2   Analysis of Season Variation

We also take the small wild goose pagoda as an example to analyze the travel data from 2011 to 2016. For the purpose of analyzing the factor of season, we use the average of 35 days to replace the value of the day. Figure 2 illustrates the impact of season on tourism flow. We can draw the following conclusions:

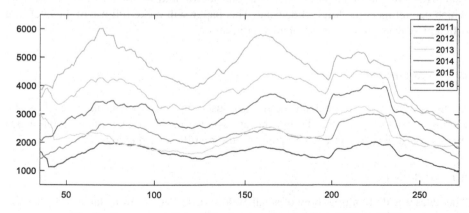

Fig. 2. Season variation of the tourism demand from 2011 to 2016.

- It is very similar in the same season of each year.
- There are three tourist peaks in a year, which are in spring, summer and fall.
- The number of tourism demand is arising in the same season.

3.3   Analysis of Week Variation

The complexity of tourism demand forecasting is reflected in not major cycle but also minor cycle. We study the variation of tourist traffic during the weeks, as is shown in Fig. 3. Figure 3 depicts the average number of weeks from 2011 to 2016. As the Small Wild Goose Pagoda closes on Tuesday, the chart does not show Tuesday's tourist traffic. We can conclude that traffic is relatively not busy on Wednesday, Thursday and Friday, and is on the contrary on Monday and weekends. Moreover, the trend of each year is almost the same. So, this rule will be applied to our experiments.

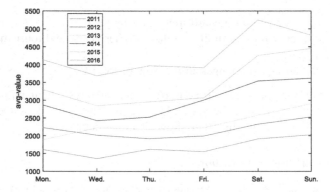

Fig. 3. Week variation of the tourism demand from 2011 to 2016.

In summary, tourism demand is affected by different scales, e.g., week, season, and year. However, most methods only take one factor into account to model the data, which cannot meet the forecast demand. This paper will make full use of the cycles in different scales to improve the prediction accuracy.

4   Experiment

In this section, we first explain the datasets used in our experiments as well as our proposed model. Next, we discuss the cross-view model with artificial intelligence method on different datasets, and compare the results with previous works.

4.1   Data Set

The dataset is daily tourist flow of small Wild Goose Pagoda from January 1, 2011 to December 31, 2016, provided by Xi'an Museum.

4.2   Proposed Model

According to our analysis in Sect. 3, tourism demand there are a small weekly cycle and two large cycles for a season or a year. We propose a simple data modeling approach, called the cross-view model, which considers not only the small cycles but also the large cycles. The modeling method is shown in Fig. 4. The Small Wild Goose Pagoda shuts on Tuesday, which is absent in the Fig. 4.

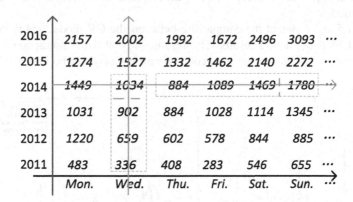

Fig. 4. Schematic diagram of prediction model.

First, we establish the forecasting model with artificial intelligence method in the vertical axis. BPNN and SVR are used in our experiments. We use passenger traffic data of Monday in 2011–2013's to forecast data for Monday, 2014. The training set is constructed as the following:

$$\{(x_{2011}, x_{2012}, x_{2013}) \rightarrow x_{2014}; (x_{2012}, x_{2013}, x_{2014}) \rightarrow x_{2015}\}$$

The test set is constructed like $\{(x_{2013}, x_{2014}, x_{2015}) \rightarrow x_{2016}\}$.

Then, we build the forecasting model with BPNN and SVR in the horizontal axis. All the data make up a time series. The form of the training set is like $\{(x_{t-3}, x_{t-2}, x_{t-1}) \rightarrow x_t\}$. Also, we employ the former data to forecast the tourism demand in 2016.

Finally, a linear fitting method is adopted to cross the results. We take the results obtained by the vertical axis and the horizontal axis as an input, and true value as output. The final form is as follows:

$$f(x) = \alpha f(x_{z-3}, x_{z-2}, x_{z-1}) + \beta f(x_{t-3}, x_{t-2}, x_{t-1}) \tag{11}$$

where α and β denote the weight of the model, z denote yearly cycle, t denote daily cycle.

4.3   Experiment and Results

To demonstrate the effectiveness of the proposed model, we evaluate the performance of the methods, such as ARIMA, V-BPNN (BPNN in the vertical axis), H-BPNN (BPNN in the horizontal axis), V-SVR (SVR in the vertical axis), H-SVR (SVR in the horizontal axis), cross-view BPNN, cross-view SVR. In the experiments, we adopt RMSE, MAPE, MAD to measure the accuracy of the models. In our experiments, we set $\alpha = 0.3632$, $\beta = 0.936$ for CV-BPNN model, and set $\alpha = 0.5683$, $\beta = 0.6367$ for CV-SVR model.

Figures 5 and 6 depict the comparison between the CV model and the traditional BPNN and SVR prediction results. The results show that the prediction accuracy is improved greatly. However, the prediction of outliers needs to be improved.

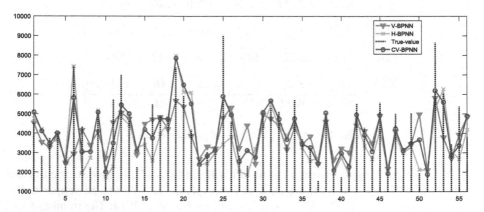

Fig. 5. Comparison of the forecasting results for BPNN.

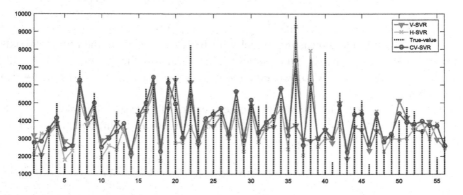

Fig. 6. Comparison of the forecasting results for SVR.

In Table 1, we report the comparison results on the seven methods. As is shown in the table, our cross-view model (CV-BPNN, CV-SVR) achieves the best performance in all cases. H-BPNN is worse than ARIMA for RMSE measure. However, CV-BPNN

Table 1. Comparison of models for RMSE, MAPE, MAD.

MODEL	RMSE	MAPE	MAD
ARIMA	1488.0	0.4054	1098.21
V-BPNN	2.0056e+03	0.2995	1.4153e+03
H-BPNN	1.5021e+03	0.2260	1.0098e+03
CV-BPNN	918.8014	0.2116	722.9307
V-SVR	1.5685e+03	0.2457	1.0542e+03
H-SVR	1.4863e+03	0.2289	998.6130
CV-SVR	1.0191e+03	0.1855	749.1628

is better than ARIMA, V-BPNN, H-BPNN. SVR is more effective for MAPE measure, and CV-BPNN is more effective for RMSE and MAD measure.

5 Conclusion and Future Work

The proposed model is suitable to all problems of tourism demand forecasting by cross-view modeling method. As the promising results shown in the experiments, the proposed model is simple and effective. Also, it shows that the modeling of data is good for improving the prediction accuracy. In the future, we will conduct a deeper analysis on the impacts of different dimensions of input. Meanwhile, we will take multivariate analysis and prediction into account.

Acknowledgements. Thanks to the support by the National Natural Science Foundation of China (No. 41271387)

References

1. Box, G.E.P., Pierce, D.A.: Distribution of residual autocorrelations in autoregressive-integrated moving average time series models. J. Am. Stat. Assoc. **65**(332), 1509–1526 (1970)
2. Gustavsson, P., Nordström, J.: The impact of seasonal unit roots and vector ARMA modelling on forecasting monthly tourism flows. Tour. Econ. **7**(2), 117–133 (2001)
3. Lim, C., McAleer, M.: Monthly seasonal variations: Asian tourism to Australia. Ann. Tour. Res. **28**(1), 68–82 (2001)
4. Lim, C., McAleer, M.: A seasonal analysis of Asian tourist arrivals to Australia. Appl. Econ. **32**(4), 499–509 (2000)
5. Pai, P.F., Hong, W.C.: An improved neural network model in forecasting arrivals. Ann. Tour. Res. **32**(4), 1138–1141 (2005)
6. Goh, C., Law, R.: Modeling and forecasting tourism demand for arrivals with stochastic nonstationary seasonality and intervention. Tour. Manag. **23**(5), 499–510 (2002)
7. Lei, K., Chen, Y.: The forecasting for China inbound toursim based on BP neural network and ARIMA combined model. Tour. Tribune **22**(4), 20–25 (2007)
8. Papatheodorou, A., Song, H.: International tourism forecasts: time-series analysis of world and regional data. Tour. Econ. **11**(1), 11–23 (2005)

9. Zheng, Y.: The research for short-term load forecasting about Phase Space Reconstruction and Support Vector Machine. Southwest Jiaotong University, Chengdu (2008)
10. Kulendran, N., Witt, S.F.: Leading indicator tourism forecasts. Tour. Manag. 24(5), 503–510 (2003)
11. Song, H., Li, G., Witt, S.F., et al.: Tourism demand modelling and forecasting: how should demand be measured? Tour. Econ. 16(1), 63–81 (2010)
12. Durbarry, R., Sinclair, M.T.: Market shares analysis: the case of French tourism demand. Ann. Tour. Res. 30(4), 927–941 (2003)
13. De Mello, M.M., Nell, K.S.: The forecasting ability of a cointegrated VAR system of the UK tourism demand for France, Spain and Portugal. Empir. Econ. 30(2), 277–308 (2005)
14. Law, R., Au, N.: A neural network model to forecast Japanese demand for travel to Hong Kong. Tour. Manag. 20(1), 89–97 (1999)
15. Law, R.: Back-propagation learning in improving the accuracy of neural network-based tourism demand forecasting. Tour. Manag. 21(4), 331–340 (2000)
16. Fernandes, P., Teixeira, J., Ferreira, J., Azevedo, S.: Modelling tourism demand: a comparative study between artificial neural networks and the Box—Jenkins methodology. Rom. J. Econ. Forecast. 5(3), 30–50 (2008)
17. Fernandes, P.O., Teixeira, J.P.: New approach of the ANN methodology for forecasting time series: use of time index. In: International Conference on Tourism Development and Management (2009)
18. Lin, C.J., Chen, H.F., Lee, T.S.: Forecasting tourism demand using time series, artificial neural networks and multivariate adaptive regression splines: evidence from Taiwan. Int. J. Bus. Adm. 2(2), 14 (2011)
19. Chen, C.F., Lai, M.C., Yeh, C.C.: Forecasting tourism demand based on empirical mode decomposition and neural network. Knowl.-Based Syst. 26, 281–287 (2012)
20. Claveria, O., Torra, S.: Forecasting tourism demand to Catalonia: neural networks vs. time series models. Econ. Model. 36, 220–228 (2014)
21. Chen, K.Y., Wang, C.H.: Support vector regression with genetic algorithms in forecasting tourism demand. Tour. Manag. 28(1), 215–226 (2007)
22. Hong, W.C., Dong, Y., Chen, L.Y., et al.: SVR with hybrid chaotic genetic algorithms for tourism demand forecasting. Appl. Soft Comput. 11(2), 1881–1890 (2011)
23. Pai, P.F., Hung, K.C., Lin, K.P.: Tourism demand forecasting using novel hybrid system. Expert Syst. Appl. 41(8), 3691–3702 (2014)

Computational Intensity Prediction Model of Vector Data Overlay with Random Forest Method

Qian Wang[1], Han Cao[1(✉)], and Yan-Hui Guo[1,2]

[1] School of Computer Science, Shaanxi Normal University, Xi'an 710119,
Shaanxi, China
{wang_qian_555, caohan}@snnu.edu.cn,
guoyanhui03@163.com
[2] Shandong Women's University, Ji'nan 250300, Shandong, China

Abstract. Spatial analysis is the core of geographic information system (GIS), of which, spatial overlay of vector data is a major job. Computational intensity of the spatial overlay has a direct effect on the overall performance of the GIS. High precision modeling for the computational intensity and analysis of the vector data overlay has been a challenging task. Thus, the paper proposes a novel approach, which utilizes self-learning and self-training features of optimized random forest algorithm to the vector data overlay analysis. Simulation experiments show that the proposed model is superior to non-optimized random forest algorithm and support vector machine model with higher prediction precision and is also capable of eliminate redundant computational intensity features.

Keywords: Random forest · Space analysis · Vector data · Computational intensity · Machine learning

1 Introduction

With the increasing scale of spatial data, the analysis and processing of geospatial data is facing the dual challenges of data expansion and the rapid growth of model analysis and calculation. Cloud computing unlimited expansion of storage capacity and strong computing power, although it can meet the requirements of mass data storage, large data parallel processing, persistent online services, etc. But for spatial analysis, especially the analysis of vector data overlay there must be a way to effectively predict its intensity, in order to achieve a balanced allocation of parallel efficiency maximization and limited computing resources.

The calculation types of vector data superposition analysis are diversity and complexity [1]. The calculation types are often relevant to logical intersection, logical sum and logic difference and related to spatial topological relation judgment, geometric information and attribute information processing. The superposition of vector data is not only relevant to the amount of data, but also the spatial data form, distribution and consequently closely related. Computational intensity is the amount of computation

© Springer Nature Singapore Pte Ltd. 2017
B. Zou et al. (Eds.): ICPCSEE 2017, Part I, CCIS 727, pp. 583–593, 2017.
DOI: 10.1007/978-981-10-6385-5_49

that is spent on a particular calculation task. It can be predicted by the calculation time, which is determined by the characteristics of the input and output data of the calculation task and the characteristics of the algorithm [2]. The expected execution time can be used as the computational intensity when the metric of the vector data overlay analysis is predicted [3]. Nevertheless, the computational time of vector space analysis is highly predictive and computational time is difficult to predict. Quite a few scholars have assumed the computational intensity metric model of some spatial computing types, but the lack of mature computational intensity modeling theory leads to the inefficiency and accuracy of the model [4]. Breiman proposed a combination of random forest classifiers that can handle high-dimensional and non-linear samples and have been widely used in many fields [5]. The random forest is adept at learning the modeling of the high-dimensional data feature set with interaction, which can deal with a large number of input variables through a large number of decision trees, with its decision-making ability to improve the prediction accuracy. Moreover, it has good tolerance to outliers and noises, and is not easy to be fitted.

Therefore, this paper presents a computational intensity metric model for vector data superposition analysis based on the computational intensity feature set of high dimensionality and interaction, which is based on the modeling method of random forest self-learning and self-training without empirical function model. Experiments show that the method has better effectiveness and accuracy.

2   Vector Data Overlay Analysis

Overlapping analysis is one of the core parts of spatial analysis of geographic information system (GIS), which is one of the methods used to extract spatial implicit information in geographic information system (GIS), and plays an important role in geographic information system. It is the distinction between geographic information systems and other information systems of the most essential feature [7]. Therefore, since the birth of geographic information system, overlay analysis has been the focus of many researchers.

In principle, overlay analysis is based on a certain mathematical model of the calculation of the characteristics of the new elements, which often involve the logic, logical and logical differences and other operations. According to the different operating elements, the overlay analysis can be divided into polygons superposition, line and polygon superposition, polygon and polygon superposition; according to the different forms of operation, stacking analysis can be divided into layer erasure, identification stacking, intersection operation, symmetry the difference analysis, the layer merge and the revision update simultaneously carry on the certain operation to the attribute [8] (Table 1).

Using vector to represent geographical objects with high accuracy, and have very low storage space, but its calculation process is more complex, also has the certain difficulty on algorithm design [9]. Vector data superposition analysis is diverse and complex, and its computational strength is not only related to the quantity of data, but also the shape and distribution of spatial data.

Table 1. Vector data overlay classification

Form	Type					
	Point and point	Point and line	Point and polygon	Line and line	Line and polygon	Polygon and polygon
Erase			√		√	√
Identify			√		√	√
Intersect	√	√	√	√	√	√
Symmetric difference						√
Union						√
Update			√		√	√

In order to solve the problem of vector data overlay analysis of the computational intensity, there are many scholars assume part of the space calculation type according to the experience of computational intensity measurement function model [10], for calculating the strength of measurement provides a rapid modeling approach. But to achieve the effective measurement calculation strength, further research must be an effective computational intensity modeling method, based on high dimensional and interaction to the calculation of strength characteristics to establish effective computational intensity measurement model.

3 Random Forest Model

Random forest is a classifier that trains and predicts samples from multiple trees [11]. The classifier was first proposed by Leo Breiman and Adele Cutler. In simple terms, a random forest is made up of multiple decision trees. For each tree, its training set is returned from the total training set. This means that some samples of the total training concentration may be repeated in the training set of a tree, and may never appear in a tree's training concentration.

Definition 1. A random forest is a classifier composed of multiple decision trees $\{h(X, \theta_k), k = 1, \ldots K\}$, where $\{\theta_k\}$ is an independent and distributed random variable, WW indicates the number of decision trees in the random forest. In the case of a given variable X, each decision tree determines the most sorted result by voting.

The traditional structure of the random forest is as follows:

1. For a given training sample, random sampling can be repeated to form new sub-sample data.
2. For the sub sample data contained in the M feature variables, using random sampling method to extract $m(m < M)$ features. Then split the node with the best splitting method to construct a complete decision tree.
3. Repeat steps 1 and 2 to obtain K decision trees to form a random forest.
4. Each decision tree voted to select the optimal classification forecast.

$$H(x) = \arg \max_Y (\sum_{k=1}^{k} I(h(X, \theta_k)) = Y) \tag{1}$$

In the formula (1), $H(x)$ represents the combined classification model, $h(\cdot)$ is the decision tree model, Y is the output variable, $I(\cdot)$ is the explicit function, $\arg \max_Y$ is the value of Y when the maximum value of $\sum_{k=1}^{k} I(h(X, \theta_k)) = Y$ is taken.

Definition 2. Given a set of classifiers $\{h_1(X), h_2(X), \ldots, h_k(X)\}$, each classifier's training set is obtained from the original data set (X, Y) random sampling, which can be obtained margin function:

$$mg(X, Y) = av_k I(h(X) = Y) - \max_{j \neq Y} av_k I(h(X) = j) \tag{2}$$

where $I(\cdot)$ is an exponential function and $av_k(\cdot)$ is an average.

The marginal function is used to measure the extent to which the average number of correct categories exceeds the average number of errors. The higher the value, the higher the confidence of the classifier.

The generalization error of the classifier can be written as:

$$PE = P_{x,y}(mg(X, Y) < 0) \tag{3}$$

When the decision tree classifier is large enough, $h_k(X) = h(X, \theta_k)$ obeys the law of strong Numbers. Subscript X, Y indicates the definition space of probability coverage. It can be shown that when the number of decision trees increases, all sequences $\theta_1 \ldots PE$ converge almost everywhere:

$$P_{X,Y}(P_\theta(h(X, \theta) = y) = y - \max_{j \neq Y} P_\theta(h(X, \theta) = j) < 0)m \tag{4}$$

where $h(X, \theta_k)$ is the output for X and θ, θ is the immediate vector of the single decision tree. Therefore, in the random forest, the number of decision trees k and the number of features m affect the performance of the classifier, but there is no over-fitting problem.

Given that random forest has the advantages of being able to handle high dimensional data and not prone to over-fitting, we model it for vector data overlay analysis.

4 Computational Intensity Model of Vector Data Overlay Analysis with Random Forest

With the increase of the size of the vector data and the increase of the dimension of the data attribute, the traditional random forest algorithm cannot effectively extract the feature and coordinate the optimal way of the decision tree node, and reduce the efficiency of dealing with massive scale vector data and cannot be done efficiently and quickly classification prediction. In order to overcome the shortcomings of the traditional random forest algorithm in the computational intensity of vector data space, we improve the traditional random forest and optimize the branching criteria to ensure that the nodes are split in an optimal way.

4.1 Feature Extraction and Pretreatment

Feature extraction is the basis and prerequisite for the collection of data samples and the establishment of computational intensity metrics. Firstly, the process of vector data superposition analysis is decomposed and analyzed according to the principle of vector data superposition and analysis, and the factors that may affect the computation intensity are extracted to the maximum, and the calculated feature sets of vector data superposition analysis are obtained. Then, Pearson correlation analysis is then performed on the characteristics of the candidate sets of the intensity signatures. Only the candidate features that are significantly related to the computational intensity are initially selected as the computational intensity characteristics of the vector data superposition analysis. This ensures the validity and reliability of the calculation intensity characteristics.

4.2 Random Forest Samples Collected and Sampled

According to the feature set selected in the previous step, we collect the data sample of vector data overlay analysis. By using the bootstrap random re-sampling method, the sample samples are sampled several times to obtain the training sample set and the test sample set of the set of vector data superposition analysis to ensure the randomness of the training set and the test sample to reduce the noise intensity of the random forest.

4.3 Construct the Computational Intensity Metric Model of Vector Data Superposition Analysis

After acquiring the training samples, we need to train each set of training sets, and finally grow into a random forest containing multiple computational intensity decision trees. It is important that the nodes of each decision tree be split in the best way to split. The traditional decision tree's branching goodness criterion is based on the sum of squares of deviation from mean.

Suppose there are P independent variables $X = (X_1, X_2, \ldots X_p)$ and continuous variable $\left\{ (\vec{x}_n, y_n) \right\}$. If the sample number of a node t is $\left\{ (\vec{x}_n, y_n) \right\}$ and the number of samples of the node is N(t), then the sum of the squared difference in the node is:

$$SS(t) = \sum_{\vec{x} \in t} [y_n - \bar{y}(t)]^2 \tag{5}$$

$$\bar{y}(t) = \frac{1}{N(t)} \sum_{\vec{x} \in t} y_n \tag{6}$$

where $\bar{y}(t)$ is the average of the node variables.

Now suppose that the branch s divides the node t into the left child node t_L and the right child node t_R, then the maximum branch of $\phi(s,t) = SS(t) - SS(t_L) - SS(t_R)$ is the optimal branch.

The traditional branching goodness criterion is often not well applied to the random forest model of vector data, so the criterion of branching preference based on residual error sum of square is proposed. This criterion allows each decision tree node to split in the best possible way. In fact, the least squares solution of the regression analysis is to minimize the sum of squares of residuals. According to this criterion, the splitting step is performed for each node using the following method:

1. From the root node of the decision tree, all the samples are here, where the tree has not yet grown, and the square of the residual of the tree is the sum of the residuals of the regression. Selecting the feature of the feature candidate set as the splitting feature, and then selecting the value of the feature in all the samples at the current node as the split points.
2. According to the characteristics of each split and the splitting point, we compute the sum of the computational intensity residuals in the two sub nodes after splitting, and use it as the judging criterion of the node branching. Starting from the root node, the current node for each division, according to the degree of branching can be found the strength calculation and prediction results of residual square sum and minimum feature split point, which will be the optimal way to split the input node.

$$t_L(s,t) = \{X|X_s \le t\} \text{ and } t_R(s,t) = \{X|X_s > t\} \tag{7}$$

$$\phi(s,t) = \min_{s,t}[\min \sum_{x_i \in t_L} (y_i - \widehat{y}(t))^2 + \min \sum_{x_i \in t_R} (y_i - \widehat{y}(t))^2] \tag{8}$$

where t_L and t_R are two sub nodes after node t splitting.

3. Starting from the root node, according to the criterion, the splitting process is carried out recursively, which makes the decision tree grow. All of the calculation of the strength of your decision tree after training and complete growth, the calculation of the intensity of the random forest model.

5 Experimental Results

5.1 Model Comparison

This paper obtains the computation time by using the vector data stacking experiment in ARCGIS10.2 environment. The data used in this paper is derived from a 1: 5000 vector map in 2015, including electronic navigation map, road network, administrative boundary and zoning (including township boundary), contour line, POI (including transportation, culture, education, nature, tourism, Medical) and other information, a total of 74 layers. Data categories include point layer, line layer, area layer, a total of 700 data. The data contains various overlay forms of vector data overlay analysis. As shown in Table 2.

Table 2. Vector overlay analysis of experimental data

SN	ILT	INE	INA	OLT	ONE	ONA	TO	T
1	3	10	5	3	10604	5	1	1.86
2	3	224	43	3	818	10	3	0.51
...	...	...	...	...	...	...	...	...
300	2	219	6	2	1043	9	3	1.17
301	2	11848	5	3	185	15	2	1.75
...	...	...	...	...	...	...	...	...
700	1	13495	41	1	672	5	3	1.66

In Table 2, the first column is the sample number. Starting from the second column followed by the input layer type, the number of input layer elements, the number of input layer properties, type of operation layer, layer number of elements operation, operating the layer number of attributes, superposition type and the computing time.

In this paper, we study the correlation characteristics of the computational feature sets, and only select the candidate characteristics which are significantly related to the calculated intensity as the calculated intensity characteristics of the vector data superposition analysis. SVM has been proved to be superior to other models in prediction [12]. So this paper chooses the support vector machine model to do the comparison. The optimized random forest is compared with SVM and the traditional random forest, as shown in Fig. 1.

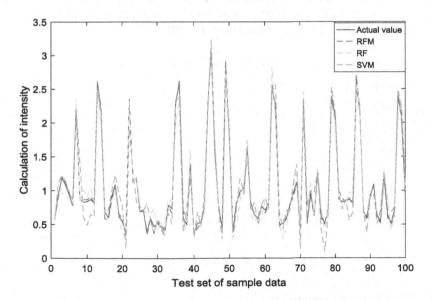

Fig. 1. Comparison of three models (Color figure online)

In Fig. 1, the purple line represents the actual value of computational intensity data of the test set. The red line represents the predicted value of the optimized random forest model. The orange line represents the predicted value of the traditional random forest model. The green line represents the predictive value of the vector machine model. It is obvious that the results of different models are obviously different, and the optimized random forest model is the closest to the real value, and the SVM model is the worst.

5.2   Parameter Optimization

In the random forest bootstrap sampling process, about 37% of the data is not in the sample. These data are known as OOB (Out of Bag). Estimation of model performance using bag data is called OOB estimation. For each decision tree, we can obtain an OOB error estimate, and then estimate the generalization error of the random forest by averaging the OOB error estimates of all decision trees. Breiman experiments show that OOB is unbiased estimation.

In this paper, the OOB error is estimated as the index to select the number of decision trees K and the number of feature selection m, when the OOB estimates at least, K and m optimal. Because the experimental data use the Pearson coefficient to remove irrelevant features, the number of feature selection m = 2. The number of decision trees K and OOB curve change as shown in Fig. 2. It can be seen from the figure that with the increase of K value, the OOB error is reduced. In the beginning, the trend of decline is faster and then gradually weakened. When k is in the range of 20–80, the trend of decline is weakened, and gradually tends to be gentle. With the increase of the K value, the running time of the algorithm is gradually increasing. Considering the OOB error and time factor, k = 80 is the best.

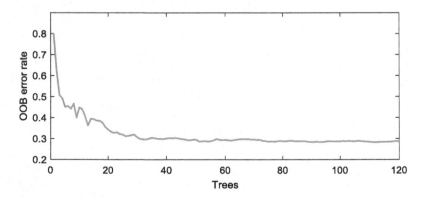

Fig. 2. Random forest decision tree number k and OOB error

5.3   Evaluation Indicators

This article uses the following indicators to assess the performance of the model:

- Root mean square error (RMSE):

$$RMSE = \sqrt{\frac{1}{N}\sum_{i=1}^{N}(\hat{y} - y_i)^2} \tag{9}$$

It is the square root of the sum of the squared and observed times n of the observed and true values. In actual measurements, the number of observations n is always finite, and the truth value can only be replaced by the most reliable (best) value. The square root error is very sensitive to the large or small error in a set of measurements. Therefore, the root mean square error can well reflect the precision of the measurement.

- Mean absolute error (MAE):

$$MAE = \frac{1}{N}\sum_{i=1}^{N}|\hat{y} - y_i| \tag{10}$$

The mean absolute error is the average of the absolute values of the deviation between all the individual observations and the arithmetic mean. Compared with the average error, the average absolute error due to the deviation is absolute value; there will be no positive and negative offset the situation. Thus, the mean absolute error can better reflect the actual situation of the predicted error.

- Mean absolute percentage error (MAPE):

$$MAPE = \left(\frac{1}{N}\sum_{i=1}^{N}\frac{|\hat{y} - y_i|}{y_i}\right) \times 100\% \tag{11}$$

The average absolute percentage error is a predictive error indicator expressed as a percentage of the relative number. The average absolute percentage error is the average of the absolute value of the ratio between the predicted and actual values divided by the actual value.

In the above three index formulas, $\hat{y}$ is the test value of the corresponding sample; y_i is the expected value of the corresponding sample; N is the number of test samples.

In Table 3, SVM is a support vector machine model, RF is a traditional random forest model, and RFM is an optimized random forest model. It can be seen from Table 3 that the RMSE, MAE and MAPE of the optimized random forest model are the smallest among the three models in the vector data superposition analysis and computation intensity prediction results. Therefore, it is shown that the optimized random forest model has higher prediction accuracy than the other two models in the calculation of vector data superimposed and calculated to effectively remove the computational intensity characteristics of redundancy.

Table 3. Comparison of three model evaluation indicators

	SVM	RF	RFM
RMSE	0.3487	0.2740	0.1366
MAE	0.3083	0.1568	0.0703
MAPE	0.2749	0.1852	0.0697

6 Conclusion

This paper first proposes a novel approach to integrate the optimized random forest algorithm and the vector data overlay analysis characteristics so that a model of computation intensity can be built through machine self-learning and self-training without requiring any experience, and the prediction accuracy is improved. So two conclusions can be drawled as follow:

1. The precision of the computational intensity of a vector data spatial overlay analysis can be improved through the appropriate optimization algorithm.
2. Before the modeling of correlation analysis and optimization of random forest algorithm can more effective place to computational intensity characteristics of redundancy. The computational intensity of vector data spatial overlay analysis predicts that such problems are valid and therefore a promising method.

Acknowledgements. Thanks to the support by the National Natural Science Foundation of China (No. 41271387)

References

1. Chang, K.T.: Geographic Information System. Wiley, Hoboken (2006)
2. Wang, S., Armstrong, M.P.: A theoretical approach to the use of cyberinfrastructure in geographical analysis. Int. J. Geogr. Inf. Sci. **23**(2), 169–193 (2009)
3. Shook, E., Wang, S., Tang, W.: A communication-aware framework for parallel spatially explicit agent-based models. Int. J. Geogr. Inf. Sci. **27**(11), 2160–2181 (2013)
4. Zhao, Y., Padmanabhan, A., Wang, S.: A parallel computing approach to viewshed analysis of large terrain data using graphics processing units. Int. J. Geogr. Inf. Sci. **27**(2), 363–384 (2013)
5. Breiman, L.: Random forests. Mach. Learn. **45**(1), 5–32 (2001)
6. Nguyen, T.T., Huang, J.Z., Nguyen, T.T.: Unbiased feature selection in learning random forests for high-dimensional data. Sci. World J. **2015** (2015)
7. Goodchild, M.F.: A spatial analytical perspective on geographical information systems. Int. J. Geogr. Inf. Syst. **1**(4), 327–334 (1987)
8. XueLian, H., ChuanYong, Y., JingZu, L.: Research and application of vector data overlay analysis based on ArcGIS engine. Urban Geotech. Invest. Surv. **3**, 014 (2010)
9. Armstrong, M.P., Densham, P.J.: Domain decomposition for parallel processing of spatial problems. Comput. Environ. Urban Syst. **16**(6), 497–513 (1992)

10. Zhao, Y., Padmanabhan, A., Wang, S.: A parallel computing approach to viewshed analysis of large terrain data using graphics processing units. Int. J. Geogr. Information Sci. 27(2), 363–384 (2013)
11. Du, S., Wang, X., Feng, C.C., et al.: Classifying natural-language spatial relation terms with random forest algorithm. Int. J. Geogr. Inf. Sci. 31(3), 542–568 (2017)
12. Ding, S., Chen, L.: Intelligent optimization methods for high-dimensional data classification for support vector machines. Intell. Inf. Manag. 2(6), 354–364 (2010)
13. Delmelle, E., Dony, C., Casas, I., et al.: Visualizing the impact of space-time uncertainties on dengue fever patterns. Int. J. Geogr. Inf. Sci. 28(5), 1107–1127 (2014)
14. Wang, S.: A CyberGIS framework for the synthesis of cyberinfrastructure, GIS, and spatial analysis. Ann. Assoc. Am. Geogr. 100(3), 535–557 (2010)
15. Yang, C., Goodchild, M., Huang, Q., et al.: Spatial cloud computing: how can the geospatial sciences use and help shape cloud computing? Int. J. Digit. Earth 4(4), 305–329 (2011)
16. Fang, K.N., Wu, J.B., Zhu, J.P., et al.: A review of technologies on random forests. Stat. Inf. Forum 26(3), 32–38 (2011)
17. Boulesteix, A.L., Janitza, S., Kruppa, J., et al.: Overview of random forest methodology and practical guidance with emphasis on computational biology and bioinformatics. Wiley Interdisc. Rev.: Data Mining Knowl. Discov. 2(6), 493–507 (2002)
18. Cheng, G., Jing, N., Chen, L.: A theoretical approach to domain decomposition for parallelization of Digital Terrain Analysis. Ann. GIS 19(1), 45–52 (2013)
19. Nicodemus, K.K.: Letter to the Editor: On the stability and ranking of predictors from random forest variable importance measure. Brief. Bioinf. 12(4), 369–373 (2011)

An Implementation and Improvement of Convolutional Neural Networks on HSA Platform

Zhenshan Bao$^{(\boxtimes)}$, Qi Luo, and Wenbo Zhang

Faculty of Information Technology, Beijing University of Technology,
Beijing 100124, China
baozhenshan@bjut.edu.cn

Abstract. Nowadays, the most heterogeneous architectures were made up by the various IP modules of different hardware vendors, but this model is less efficiently. In order to solve this problem, AMD joint other hardware vendors proposed heterogeneous system architecture (HSA) specification. On the one hand, the HSA could help developers to accelerate the design process and programming. On the other hand, it improved the system performance and reduced the power. In this paper we presented the implementation of a framework for accelerating training and classification of arbitrary Convolutional Neural Networks (CNNs) on the HSA, on the basis of implementation, we presented tow accelerated methods that are Online update weights and letting CPU to participate in calculation. Experimental results showed that the implementation of CNNs on HSA 4 to 10 times faster than on the CPU.

Keywords: Heterogeneous computing · Heterogeneous system architecture · Convolutional neural network · Batch update weights

1 Introduction

For a long time, the demand of parallel computing was endless, these domains included atmosphere [1], astronomy [2] and real-time tracking system [3], They required high processing speed, otherwise the result would appear deviation or lose its significance.

In order to improve the speed of parallel computing, improvement measures were generally had the following three aspects:

- We could take advantage of the hardware devices to improve processing speed faster, such as faster supercomputer, larger memory and faster I/O devices. It was the fundamental way to improve the capacity of parallel computing.
- We could use more optimization program design method and function library, such as introducing multi-threaded, parallel processing methods in the program.
- We could use optimization software, which is a simple effective method and relatively low cost.

Nowadays, the way of using the GPU parallel computing involves the three options above, such as CUDA (Compute Unified Device Architecture), OpenCL (Open

© Springer Nature Singapore Pte Ltd. 2017
B. Zou et al. (Eds.): ICPCSEE 2017, Part I, CCIS 727, pp. 594–604, 2017.
DOI: 10.1007/978-981-10-6385-5_50

Computing Language), etcetera. CUDA is a new infrastructure, it is a synergetic solutions based on the hardware and software. OpenCL is a programming framework for heterogeneous platforms [5]. The speedup of CUDA was recognized by most people, but we must depend on the particular NVIDIA GPU to run our benchmark, and OpenCL supports most of the GPU, but whenever there is a new GPU to be used for heterogeneous computing, OpenCL need do a lot of improvement.

Furthermore, mobile devices become more and more widespread, in order to optimize the operation rate of equipment based on the hardware, on a chip system is the best choice, CUDA was not so suitable for this kind of architecture. Therefore, in order to improve the CUDA and OpenCL is insufficient, In June 2012, AMD joint ARM, Imagination, mediatek, Texas instruments together formed a nonprofit organization "heterogeneous system architecture Foundation" (HSA Foundation), then attract the LG, samsung and qualcomm, etc. After two years of efforts, the HSA foundation had finally completed his first important mission, approved the release 1.0 Final the official version of heterogeneous system architecture specification [6]. Compared with the CUDA system architecture, the CPU and GPU controlled the respective memory area, when the CPU and GPU transmit data, they need to make a copy of their data in memory to another memory, this approach reduces the working efficiency of the platform. CPU unit and GPU unit in the APU has implemented unified memory addressing which is one of the characteristics of HSA architecture, the technology is called HUMA (heterogeneous unified access memory). The HSA is suitable for mobile devices, because the CPU and GPU were gathered in a nucleus, it improved the system performance and reduced the power.

In this paper we presented the implementation of a framework for accelerating training and classification of arbitrary Convolutional Neural Networks (CNNs) on the HSA, on the basis of implementation, the application field was handwriting recognition [7]. We presented tow accelerated methods that are letting CPU to participate in calculation and convergence speedup, the basic hardware was AMD APU core chip (A10-7850) based on HSA architecture. We had used the programming to assess their operation ability in HSA software architecture and optimize it.

The rest of this paper is organized as follows. In Sect. 2, highlights the differences between this paper and related work. Then an overall about algorithm design and implementation of strategy will be described in Sect. 3. Some preliminary experiment results are presented and discussed in Sect. 4. Section 5 summarize this work and describes the future works of implementation of strategy in HSA.

2 Related Works and Motivation

This section introduces the HSA system, related works and motivation for this study. Section 2.1 describes the features of HSA. Section 2.2 describes the Convolution neural network, Sect. 2.3 presents some existing research about our works. Section 2.4 describes the details of the problem to be solved.

2.1 HSA Essential Features

After years of development, AMD put forward the concept of the APU, the design could make the transmission channel by PCI - E bus upgrade to a more ideal internal bus speed and effectively reduced the data in the process of transmission delay. The design not fundamentally solve the problem of data sharing of CPU and GPU, but laid an important basis for the emergence of HSA.

A new APU conducted a series of upgrade on the hardware platform, the CPU had updated to steamroller, both performance and core Numbers had improved significantly than the previous generation. One of the highest specifications A10-7850k had updated to 4-core CPU and 8-core GPU, and in the middle models also have at least 4-core CPU and 6-core GPU. It was worth mentioning that the new framework of main frequency obviously less than the last two generations of products, such as the default highest frequency of A10-7850k only was 4.0 GHZ, and A10-6800K was 4.4 GHZ. But with the development of new frame and new process, the steamroller would bring us a better performance and lower power consumption.

On the one hand, HSA provided a signal mechanism working for the CPU and the accelerator. HSA signal is a 32-bit or 64-bit values, according to the target platform. HSA runtime supported signal operation which is to use a special set of runtime function, and supported HSAIL using a special signal indicator. Therefore, CPU and GPU can communicate with each other with these signals. On the other hand, HSA could join the core CPU and GPU which was called HUMA. As a result, they could be connected to each other and share the workload and Shared memory. As shown in Fig. 1, GPU and CPU have equal flexibility to be used to create and dispatch work items. This is another feature of HSA—hQ.

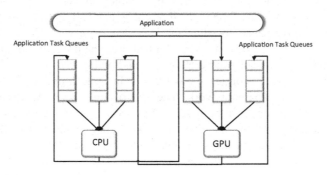

Fig. 1. HQ technology

2.2 Convolution Neural Networks

In traditional multi-layer back propagation network, it was generally need to design a feature extractor so that the data feature was extracted. The purpose was to get a set of some irrelevant information and the data variables reflecting the characteristics, rather than directly to the original data as input. There were two reasons for us don't use the original data as input. First of all, in the image recognition, the raw data is a multi-dimensional matrix, the matrix as input inevitably lead to excessive input vector and network training harder. Secondly, there are a lot of useless information in the multi-dimensional matrix, such as pure color edge area, we should remove these useless information, and finally even the matrix can be converted to vector as network input, but it still loses some of its image of two-dimensional features.

CNNs as a kind of neural network, it has important research value in the field of image recognition and speech recognition. The feature of CNNs was shared-weight, local perceptual field and down-sampling, so it was very suitable for the image and voice and data processing. Shared-weight greatly reduced the network parameters, the method of processing based on local perceptual could simplify the network structure, down-sampling could efficiently extract features. when the input was multidimensional images, these characteristics made the superiority of the convolutional neural network more prominent.

CNNs are composed of three different types of layers: convolutional layers, sub-sampling layers (optional), and fully connected layers. These layers are arranged in a feed-forward structure as shown in Fig. 2.

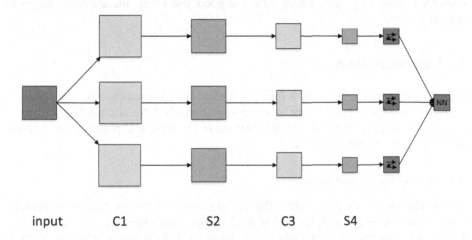

input   C1   S2   C3   S4

Fig. 2. Convolutional neural networks

2.3   Related Works

There are some existing research efforts to use CPU-GPU heterogeneous systems. Ubal et al. present Multi2Sim, an open-source, modular, and fully configurable toolset that enables ISA-level simulation of an x86 CPU and an AMD Evergreen GPU [9]. Ma et al. propose Green GPU, a holistic energy management framework for GPU-CPU heterogeneous architectures [10]. Ding et al. presents the design of HSAemu, a full system emulator for the HSA platform, and illustrates how those HSA features are implemented in the simulator [11].

Recently, the Neural Networks implementation on the heterogeneous computing platform drawn the attention of researchers. Luo et al. explored application of commodity available GPU for two kinds of ANN models [12]. Strigl et al. present the implementation of a framework for accelerating training and classification of arbitrary Convolutional Neural Networks (CNNs) on the GPU [13]. Gadea et al. describes the implementation of a systolic array for a multilayer perceptron on a Virtex XCV400 FPGA with a hardware-friendly learning algorithm [14]. Himavathi et al. present a hardware implementation of multilayer feedforward neural networks (NN) using reconfigurable field-programmable gate arrays (FPGAs) [15].

2.4 Motivation

If the resources of CPU would be used efficiently in heterogeneous systems, it could reduce the task execution time and further improve the performance of HSA. As mentioned, if the data transmission consumption between CPUs and GPUs could be reduced or removed, the total execution time can be further reduced. Nowadays, the emergence of HSA can realize our ideas. In addition, GPU and CPU are used as a combined computing unit, which makes a good use of all the available hardware resources.

3   Implementation

This section introduces the CNNs Implementation, HSA Implementation and algorithm acceleration. Section A describes the Implementation of CNNs. Section B describes the convolutional neural network to implement on HSA, Section C presents two methods to improve the runtime of CNNs.

3.1   CNNs Implementation

It can The initialization of CNN. The initialization of CNN is mainly to initialize convolution layer, the weights and bias of output layer. For example, Convolution kernels and weights were randomly initialized and All of the bias was assigned zero in DeepLearnToolbox.

The computation of feed-forward structure. This phase consists of the computation of input layer, convolution layer, Pooling layer and output layer.

- Input layer had no input value, but only one output vector, and the size of the vector is the size of the picture.
- The input of convolution layer come from the input layer or sampling layer, each map of Convolution layer has a convolution kernel with the same size.

Pooling subsampling is a resource management term that refers to the grouping together of resources (assets, equipment, personnel, effort, etc.) for the purposes of maximizing advantage and/or minimizing risk to the users. The main purpose of pooling layer was for taking a sample, in order to avoid over-fitting.

The input and output of hidden layer is as follows:

$$net_j = \sum\nolimits_{i=0}^{m} V_{ij}X_i \tag{1}$$

$$H_j = f\left(net_j\right) \tag{2}$$

The input and output of output layer is as follows:

$$net_{k=} \sum\nolimits_{i=0}^{m} V_{kj}H_j \tag{3}$$

$$O_k = f(net_k) \tag{4}$$

Weight Adjustment. This process was to minimize the residual error so that adjust the weights and bias, the error of the output layer and the error of middle layer calculation are different, the error of the output layer is the absolute value of the output value and the class value, but the error of each middle layer is the sum of weights from last layer. The residuals for the output layer is as follows:

$$\delta_i^{(n_i)} = \frac{\partial}{\partial z_i^{(n_i)}} \frac{1}{2} \left\|y - h_{W,b}(x)\right\|^2 = -\left(y_i - a_i^{(n_i)}\right) \cdot f'\left(z_i^{(n_i)}\right) \tag{5}$$

In the above formula, net_j is the j'th neuronic input of hidden layer, net_k is the k'th neuronic input of output layer, V is the weight, H is the output of hidden layer, O is the output of output layer, $f(\cdot)$ is the activation function.

3.2 Implementation on HSA Platform

Convolution neural network was the improvement of BP neural network, in addition, it was a type of feed-forward artificial neural network in which the connectivity pattern between its neurons is inspired by the organization of the animal visual cortex. The main steps are as follows:

- The network weights and other related parameters were initialized, characteristic vector data was read and transmitted to the CPU and Data from the CPU was changed to the GPU.
- We use the CPU to control training cycle and use the GPU to control the sample vector.
- We used matrix parallel algorithms to calculate the output of hidden layer and the output layer in the middle.
- Calculating the error of output layer and hidden layer on the GPU and adjusting weight with parallel algorithm.

As shown in Fig. 3, there is concrete convolution neural network in the whole process of extracting feature map. The A1, A2… An was input characteristic figure, B11, B12… B1n was convolution kernels. Convolution layer's main job is convolution

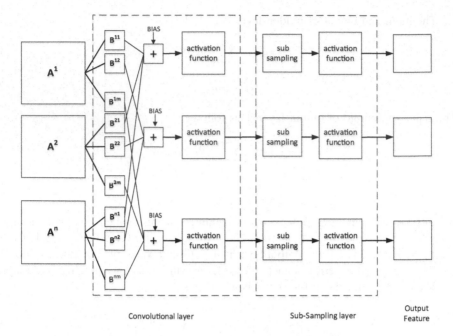

Fig. 3. The feature map of output

and nonlinear transformation, the abstraction layer is mainly for pooling operation and nonlinear transformation.

In the implementation of our algorithm, we omitted bias and weights and used the small convolution kernel. There were many differences between convolution operation and matrix multiplication, the main point of HSA heterogeneous platform is to use the GPU to do a large amount of calculation, thus, HSA heterogeneous platform for the efficiency of matrix multiplication is very high, but not particularly appropriate convolution operation. In order to accelerate the calculating process of convolution, we referred to the implementation process of caffe, converting the convolution operation to matrix operations.

3.3   Algorithm Acceleration

In order make the result output of the algorithm more quickly, we made some improvement on it. Two methods of improvement were put forward, they were letting CPU to participate in calculation and convergence speedup.

Batch Update. The network weights and bias is in each sample after prior to spread and back propagation process is updated, we call this way online updates. The other way is batch update, the network weights and bias is updated after all samples were processed. In the online update model, the error of the network is to produce a single sample, but in the batch update mode, the error of the network is the error of all samples. The various weights and bias of gradient change rate is also all samples produced by the gradient changes, through the network, the sum of each sample set by gradient changes of integrated together to obtain more accurate amount of gradient change, so it is helpful to accelerate the algorithm convergence speed. In order to optimization algorithm, we combine these two ways, it could improve the training speed and accelerate convergence speed.

Letting CPU to Participate in Calculation. CPU-GPU heterogeneous systems transfer data slowly between CPU and GPU before HSA platform. In such an environment, GPUs are assigned to the most of computing workload, and thus they provide up to 10 times the computing power of CPUs. However, in real-world applications, CPUs are often idle while GPU take all the workloads. Compared with the use of CPU to calculate, it would spend too much time on data transmission before HSA platform. Nowadays, Unified Coherent Memory enables all compute elements to have access to the same data, so it enables the application to run on the best compute element.

4   Experimental Results

4.1   Experimental Environment

Our hardware environment was given priority to AMD APU. We use A10-7850K in our test bed. A10-7850K was the first formal comprehensive products supporting hUMA technology. The operating system is Ubuntu 15.04 with a Linux kernel 3.19. We used kfd v1.4 HSA drivers and the HSA Runtime 1.0f. We used AMD CodeXL which was a comprehensive tool to obtain the GPU utilization and usage time. It turned out on a Linux system there was this command called "time" for application execution time. Final result was based on the average of multiple tests.

4.2   Experimental Results

In order to To verify that our algorithm could improve the execution efficiency, we defined the tow experiments. There are three kinds of implement included in the first experiments. And these three parts are the implement by using CPU, the implement by using online update weights and the implement by using batch update weights. Figure 4 shows the results of these three implementations.

The training cycle is 100 times, the batch update data volume is 100, the test data set size is 12001, during a training process, it need 121 times the batch update. It can be seen that after 100 iterations, the recognition rate of our convolutional neural network algorithm running on the CPU can reach 98.70%, while the recognition rate in the batch update mode is relatively low, but the difference is very small. Though there are some differences in the recognition rate, but this model training speed quickly, compared to the realization of the CPU to achieve 3.57 acceleration ratio.

The accuracy rate does not fluctuate as the number of iterations increases. As shown in Fig. 5, it is the result of three sets of tests. Respectively, real-time update, batch update of the data scale of 100 and 50. It is not difficult to see that the accuracy of the model using batch update is more obvious, the smaller size of the batch update, the higher the accuracy, with the increase in the number of iterations, the smaller the

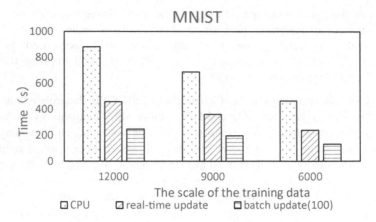

Fig. 4. Experimental results of different data sizes

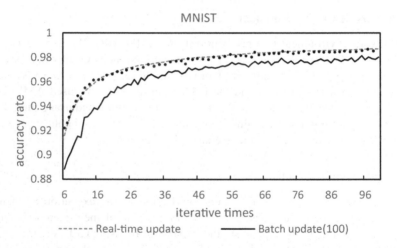

Fig. 5. The accuracy of handwritten digit recognition

fluctuation. It can be seen from the figure that the accuracy of the initial pattern update is the most stable and there is no fluctuation.

There are three kinds of implement included in the second experiments. And these three parts are the implement by using batch update weights and task allocation. As shown in Fig. 6, it can be seen that after 100 iterations, the recognition rate of these two algorithms is still 98.08%. Compared with the first implementation, the optimal allocation scheme can reach the acceleration ratio of 1.24.

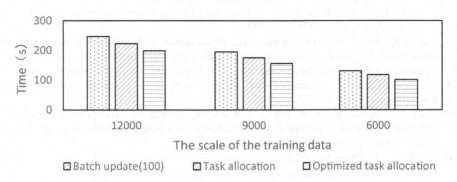

Fig. 6. The result three implement with convergence speedup

5 Conclusion and Future

This article shows that HSA work quite well for convolutional neural networks. Although convolution neural network topology structure is simple, but still need great amount of calculation. AMD HSA architecture is based on stream processors, with the support of the programming model, relative to the CPU, handwriting recognition based on convolution neural network performance played amazing advantage. Experiments show that stream processor architecture suitable for convolution neural network.

The growth of heterogeneous systems represents a solid trend in modern systems, and we believe that future work to optimize our CNNs algorithm on HSA would better effect.

Acknowledgement. This research is supported by the Natural Science Foundation of BJUT, the National Natural Science Foundation of China (Grants No. 91546111, 91646201), the Key Project of Beijing Municipal Education Commission (Grants No. KZ201610005009).

References

1. Yang, C., Wu, L.: GPU-based volume rendering for 3D electromagnetic environment on virtual globe. Int. J. Image Graph. Sig. Process. **2**(1), 53 (2010)
2. Harris, C.: GPU accelerated radio astronomy signal convolution. Exp. Astron. **22**(1), 129–141 (2008)
3. Michel, P., Chestnutt, J., Kagami, S., Nishiwaki, K.: GPU-accelerated real-time 3D tracking for humanoid locomotion and stair climbing. IEEE/RSJ International Conference on Intelligent Robots and Systems, pp. 463–469. IEEE (2007)
4. Owens, J.D., Houston, M., Luebke, D., Green, S., Stone, J.E., Phillips, J.C.: GPU computing. Proc. IEEE **96**(5), 879–899 (2008)
5. Du, P., Weber, R., Luszczek, P., Tomov, S., Peterson, G., Dongarra, J.: From CUDA to OpenCL: towards a performance-portable solution for multi-platform GPU programming ☆, ☆☆. Parallel Comput. **38**(8), 391–407 (2012)

6. Blinzer, P.: The heterogeneous system architecture: it's beyond the GPU. In: International Conference on Embedded Computer Systems: Architectures, Modeling, and Simulation, p. iii. IEEE (2014). (Maxwell, J.C.: A Treatise on Electricity and Magnetism, 3rd edn., vol. 2, pp. 68–73. Clarendon, Oxford (1892))
7. Dan, C.C., Meier, U., Gambardella, L.M., Schmidhuber, J.: Convolutional neural network committees for handwritten character classification. In: International Conference on Document Analysis and Recognition, pp. 1135–1139. IEEE Computer Society (2011)
8. Lecun, B.Y., et al.: Gradient-based learning applied to document recognition. In: Proceedings of the IEEE (2010)
9. Ubal, R., Jang, B., Mistry, P., Schaa, D., Kaeli, D.: Multi2Sim: a simulation framework for CPU-GPU computing. In: International Conference on Parallel Architectures and Compilation Techniques, pp. 335–344. ACM (2012)
10. Ma, K., Li, X., Chen, W., Zhang, C.: GreenGPU: a holistic approach to energy efficiency in GPU-CPU heterogeneous architectures. In: International Conference on Parallel Processing, pp. 48–57. IEEE (2012)
11. Ding, J.H., Hsu, W.C., Jeng, B.C., Hung, S.H., Chung, Y.C.: HSAemu: a full system emulator for HSA platforms. In: International Conference on Hardware/Software Codesign and System Synthesis, p. 26. ACM (2014)
12. Luo, Z., Liu, H., Wu, X.: Artificial neural network computation on graphic process unit. In: IEEE International Joint Conference on Neural Networks, vol. 1, pp. 622–626. IEEE (2005)
13. Strigl, D., Kofler, K., Podlipnig, S.: Performance and scalability of GPU-based convolutional neural networks. In: Euromicro International Conference on Parallel, Distributed and Network-Based Processing, pp. 317–324. IEEE (2010)
14. Gadea, R., Cerdá, J., Ballester, F., Mocholí, A.: Artificial neural network implementation on a single FPGA of a pipelined on-line backpropagation, pp. 225–230 (2000)
15. Himavathi, S., Anitha, D., Muthuramalingam, A.: Feedforward neural network implementation in FPGA using layer multiplexing for effective resource utilization. IEEE Trans. Neural Netw. 18(3), 880 (2007)

An Enhanced Transportation Mode Detection Method Based on GPS Data

Jing Liang[1], Qiuhui Zhu[1], Min Zhu[1(✉)], Mingzhao Li[2], Xiaowei Li[1],
Jianhua Wang[1], Silan You[1], and Yilan Zhang[1]

[1] College of Computer Science, Sichuan University, Chengdu, China
zhumin@scu.edu.cn
[2] School of Science, RMIT University, Melbourne, Australia

Abstract. Understanding meaningful information such as transportation mode (e.g., walking, bus) from Global Positioning System (GPS) data has great advantages in urban management and environmental protection. However, the urban traffic environment has evolved from "data poor" to "data rich", resulting in the decline in the accuracy of transportation mode detection results. In this paper, an enhanced approach for effectively detecting transportation mode with a detection model and correction method from GPS data is proposed. Specifically, we make the following contributions. First, a trajectory segmentation method is proposed to detect single-mode segments. Secondly, a Random Forest (RF)-based detection model containing several new features is introduced to enhance discrimination. Finally, a correction method is designed to improve the detection performance, which is based on the mode probability of the current segment and its adjacent segments. The results of our detection model and correction method outperform state-of-the-art research in transportation mode detection.

Keywords: GPS · Transportation mode · Random Forest

1   Introduction

A popular and innovative way based on GPS and Geographic Information System (GIS) for travel survey is increasingly substituted for the traditional travel survey. Hence, collecting and analyzing travel data becomes important, and the information of transportation modes plays a tremendous part in travel survey. It is important to understand the distribution of transportation modes in different regions and times, which not only provides a quantitative basis for the government to formulate transportation development strategies and promulgates various kinds of management policies, but also has contributed to urban planning and traffic management. Additionally, transportation mode is linked closely to environmental issues. Analyzing transportation mode information and changing people's intention by the use of more sustainable modes can be considered as a substantial approach of tackling environmental problems such as air pollution [8].

© Springer Nature Singapore Pte Ltd. 2017
B. Zou et al. (Eds.): ICPCSEE 2017, Part I, CCIS 727, pp. 605–620, 2017.
DOI: 10.1007/978-981-10-6385-5_51

The rapid development of city led to the progressively sophisticated characteristics of urban traffic. While most of previous studies [2,9,11,14] have focused on velocity and acceleration features to identify transportation modes. Furthermore, these studies take no account of correction process on the detection result.

In this context, our approach identifies transportation mode from GPS dataset in a more comprehensive way. More features related to users' behavior and motion are considered, and we propose a correction method that combines the current segment and its adjacent segment. The main contributions of this paper can be summarized as follows:

- We propose an enhanced trip segmentation method based on [17] to identify single transportation mode segments. In particular, detections of stay point and outlier handling are both introduced in our method, and we also improve the merging step of previous algorithm (Zhu et al. [17]). Subsequently, our method is compared with state-of-the-art segmentation methods by using recall and precision to show the effectiveness.
- We design a RF-based detection model including several new significant features to improve the detection accuracy, and we compare the significance of each feature.
- We propose a correction method based on the probability of each kind of mode of every segment and its adjacent segments, and the probability of each mode can be corrected apparently.

The rest of this paper is organized as follows. Section 2 reviews related works. Then, Sect. 3 presents the detection approach of transportation mode in detail. In Sect. 4, the evaluation and comparison are described. Finally, the conclusions are given in Sect. 5.

2   Related Work

In recent years, there have been a number of studies describing attempts to solve the problem of transportation mode detection. Most of these studies extracted single-mode segments from GPS trajectories by using transition point-based segmentation method, a transition point stands for a place where a user changes transportation mode. Zheng et al. [14] used a loose upper bound of velocity and acceleration to identify walk segments, where the transition point is the start or end point of each walk segment. Lin et al. [7] labeled each point at each trip with high or low velocity mode according to velocity and generated walk segments. Gong et al. [4] and Xiao et al. [11] identified transition points as the first and last points of a gap or of a walk segment that is longer than 60s, and walk segments are extracted from trips by several rules related to velocity and duration features. Nevertheless, none of these studies have addressed the impact of outliers existing in GPS dataset.

Selecting features that have a favorable effect is critical in transportation mode detection. The majority of previous studies used diversified velocity and acceleration related features [2,9,11,14]. Additionally, Zheng et al. [13] identified

Table 1. Comparison of different methods for transportation mode detection

Method	Outlier handling	Feature		Correction processing	GIS data usage
		Classification	UBF usage[a]		
Biljecki et al. [2]					✓
Gone et al. [4]					✓
Lin et al. [7]				✓	
Xiao et al. [11]		✓			
Zheng et al. [13]		✓		✓	
Zhu et al. [17]		✓		✓	
Our method	✓	✓	✓	✓	✓

[a]UBF usage is short for users' behavior features, which excludes velocity and acceleration related features and contains information of users' behavior and motion.

that heading change rate, stop rate and velocity change rate are more robust to traffic conditions. Xiao et al. [11] added low speed rate and average heading change to improve the performance of the classification model. Timeslice type and acceleration change rate are used to infer transportation modes by [17]. However, these studies have limitations that lack exploring the impact of various combination of features on the detection results, and the number of features is no more than nine. Therefore, for purpose of a better detection result, the number of features is increased, and several of them are new proposed features on the basis of the diversity of GPS data and people's behavior and motion. In this paper, we propose a detection model with new features and make a feature classification and combination evaluation.

Most existing studies have not considered the correction of detection results. Lin et al. [7], Zheng et al. [13] and Zhu et al. [17] combined with the transition probability of transportation mode and the current road segment inference probability through the classifier to improve the inference performance. However, the result of Lin et al. [7] turned out that the ratio of segments being processed is generally 1% or even less of all the segments, and three of them took no account of the influence of the adjacent segments (preceding and following) on the current segment. Hence, we introduce a transportation mode correction method by way of re-calculating the probability of each kind of mode of every segment with the transition probability and the probability of adjacent segments.

As compared in Table 1, some utilized users' behavior features and GIS infomation, while some did not. In addition, only three of previous studies took the correction process, and all of them did not give much thought to outlier handling and feature classification. Whereas our method outperforms other studies in a more comprehensive way.

3   Methodology

In this section, we first define terms used in our method, and then describe the procedure of detecting transportation mode with our method at a high level (Fig. 1). Here are some clarifications of terminologies used in our method.

- Trip: A trip contains a sequence of GPS points. Each point consists of a pair of latitude and longitude and a timestamp. A trip might have more than one transportation mode.
- Stay point: A stay point refers to a certain spatial region where a user stays for a period exceeding a threshold.
- Transition point: A transition point refers to a place where a user changes transportation mode.
- Segment: A segment denotes a portion of a trip whose transportation mode number might be no more than one.

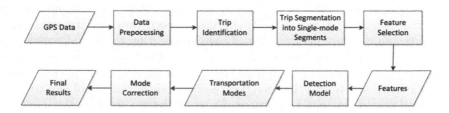

Fig. 1. Procedure of detecting transportation mode

3.1   Data Description

Two datasets are used in this paper: GPS trajectory dataset from (Microsoft Research Asia) Geolife project [13,15,16] and bus station dataset from Open-StreetMap (OSM) [1].

We use the GPS trajectory dataset to test the generality, flexibility and effectiveness of our method, since it records a broad range of users' outdoor movements. The GPS dataset was collected by 182 users from April 2007 to August 2012 in Beijing, and it contains 17,621 trajectories recorded by different GPS loggers and phones. Each point includes three components: latitude, longitude and time. The uneven time sampling rate is set to be 2 s or 5 s. 73 users out of 182 have labeled their trajectories with transportation mode.

The bus station dataset from OSM contains latitude, longitude and name of bus stations in Beijing. OSM enables user free access to current geographical information [5], and the coverage and quality of OSM data in China have been continuously improved in recent years, which makes OSM-China data more dense and applicable.

In data preprocessing, three steps are taken to deal with the problems existing in the GPS dataset. Firstly, obvious errors are amended or deleted based on the context of erroneous points. Secondly, points with no label are eliminated. Finally, trips with duration less than 5 min are also removed from the GPS dataset.

3.2   Trip Segmentation

In this section, we first identify trips from GPS dataset by stay point detection and outlier handling. Then, a transition point detection is proposed to extract single-mode segments from trips.

3.2.1   Trip Identification

Algorithm 1 gives a detailed description of trip identification procedure through stay points. We adopted the stay point detection algorithm from [6]. Line 2 to Line 12 describes the process of detecting stay point sets. Due to the effect of urban canyon, outliers may exist in the set of stay point. Therefore, we eliminated outliers in Line 13 to Line 25. First of all, random sampling and Kolmogorov-Smirnov test are performed to validate whether the collection of stay point sets follows Gaussian distribution. Secondly, three-sigma rule is employed to find outliers, and we remove outliers from GPS trajectory or the current stay point set. Finally, the central stay point of its set is calculated, which is regarded as a mark to generate a trip collection.

3.2.2   Transition Point Detection

We propose an enhanced transition point detection method based on [17] to identify segments that might have only one transportation mode, which is different from the segment merging step of [17] in two aspects. We first merge segments whose distance meets the threshold merely into its following segment instead of preceding and following segments to identify more transition points. Secondly, iterating over the items in the segment collection until the number of segments keeps unchanged is substituted for traversing once, which clearly improves the accuracy of detecting transition point.

3.3   Transportation Mode Detection

This section is organized as follows. The first part gives a detailed description of new proposed features. The second part is focused on the detection model.

3.3.1   Feature Selection

Enlightened by related studies, 29 features are extracted from GPS dataset to form one feature set including 85% percentile velocity and acceleration (85thV, 85thA), first maximum velocity and acceleration (MaxV1, MaxA1), second maximum velocity and acceleration (MaxV2, MaxA2), median velocity and acceleration (MedianV, MedianA), minimum velocity and acceleration (MinV, MinA),

---

Algorithm 1. Trip identification algorithm

Input: GPS logs T, distance threshold $dist_{thd}$, time span threshold $time_{thd}$
Output: A trip collection
1: **function** TRIPALG($T, dist_{thd}, time_{thd}$)
2:     $i = 0$
3:     **while** i $<T.length$ **do**
4:         $j = i + 1$
5:         **while** j $<T.length$ **do**
6:             **if** $p_j.time - p_i.time > time_{thd}$ **and** $p_j.dist - p_i.dist > dist_{thd}$ **then**
7:                 Generate a new stay point set SP starting from p_i, end the current
   set at p_j
8:             **else**
9:                 $j = j + 1$
10:            **end if**
11:        **end while**
12:    **end while**
13:    Apply three-sigma rule to get the outliers in every stay point set
14:    Initialize the collection of stay point sets CSP
15:    $CSP \leftarrow \phi$
16:    **for** each stay point set SP **do**
17:        **for** each $p_i \in SP$ **do**
18:            **if** p_i is an outlier **then**
19:                **if** p_i is the first point or last point in the set **then**
20:                    Remove p_i from SP
21:                **else**
22:                    Remove p_i from T
23:                **end if**
24:            **end if**
25:        **end for**
26:        Calculate the mean coordinate, departure and arrive time of central stay
   point $centralSP$ in SP
27:        $CSP \leftarrow CSP \cup centralSP$
28:    **end for**
29:    Generate a trip collection $CTRIP$ based on CSP
30: **end function**

---

average velocity and acceleration (MeanV, MeanA), the expectation of velocity and acceleration (ExpV, ExpA), the covariance of velocity and acceleration (CovV, CovA), velocity and acceleration change rate (VCR, ACR), high, medium and low velocity rate (HVR, MVR, LVR), high, medium and low acceleration rate (HAR, MAR, LAR), timeslice type (TS), bus stop rate (BSR), heading change rate (HCR) and stop rate (SR). In the feature set, seven of them are new proposed features. Moreover, these 29 features as shown in Fig. 2 are divided into three categories: velocity, acceleration and behavior.

Velocity Category. The first category contains all the features related to velocity. Among these features, *High Velocity Rate* (HVR), *Middle Velocity Rate* (MVR) and *Low Velocity Rate* (LVR) are new features.

The velocity distribution of transportation modes are somewhat distinct from each other. Figure 3(a) shows the velocity cumulative distribution for six different transportation modes. People usually keep low velocity while walking. For bus mode, it needs to stop at every station, which results in alternating low and medium velocity. While the velocity of plane is the highest. Based on

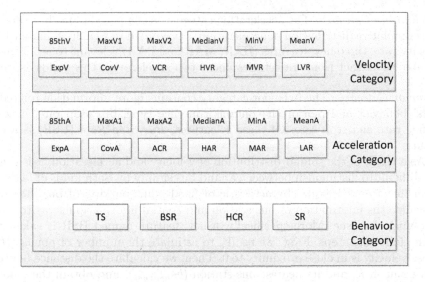

Fig. 2. Feature category

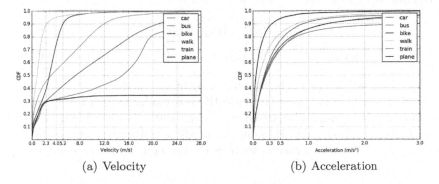

(a) Velocity   (b) Acceleration

Fig. 3. The distribution of velocity and acceleration

this phenomenon, three features related to velocity are calculated from GPS trajectories: HVR, MVR and LVR (Eqs. (1), (2) and (3)).

$$HVR = |P_h|/N; \tag{1}$$

$$MVR = |P_m|/N; \tag{2}$$

$$LVR = |P_l|/N; \tag{3}$$

where $P_h = \{p_i|p_i \in P, p_i.v \geqslant V_{hm}\}$, $P_m = \{p_j|p_j \in P, V_{ml} \leqslant p_j.v < V_{hm}\}$, $P_l = \{p_k|p_k \in P, p_k.v < V_{ml}\}$, $|P_h|$, $|P_m|$ and $|P_l|$ represent the number of elements in P_h, P_m and P_l.

Acceleration Category. The acceleration category is composed of twelve features. The cumulative distribution of acceleration is presented in Fig. 3(b). Similarly to velocity rate, three new features (HAR, MAR and LAR) relevant to acceleration are able to extract from trajectories under two acceleration thresholds as well.

Behavior Category. The behavior category contains more information related to users' behavior and motion, it contains timeslice type, bus stop rate, heading change rate and stop rate. Of these 4 features, *Bus Stop Rate* (BSR) is a new feature.

As a result of the similar distribution of velocity and acceleration, it is generally difficult to distinguish between car and bus simply by both kinds of features. According to the unique characteristic of fixed-point parking of bus, however, we utilize the information of bus station and spatial information so as to assist in solving the issue. More specifically, a new feature named BSR is calculated by the following steps. First, we use V_s to estimate the number of points ($|P_s|$) whose velocity is in close proximity to 0. Then, we calculate the distance between each point in P_s and its nearest bus station ($bs_{nearest}$), and obtain the number of points ($|P_{bs}|$) whose distance is smaller than threshold D_{bs}. Finally, BSR can be calculated by Eq. (4).

$$BSR = |P_{bs}|/|P_s|; \tag{4}$$

where $P_s = \{p_i | p_i \in P, p_i.v < V_s\}$, $P_{bs} = \{p_j | p_j \in P_s, dist(p_j, bs_{nearest}) < D_{bs})\}$.

BSR takes advantage of the characteristic of bus mode whose speed near the bus station is generally close to 0 to identify bus mode. Even if GPS trajectories may have the possibility of repeated crossover, BSR still has a strong distinction for bus mode. Additionally, BSR merely needs to calculate those points whose speed is close to 0, which greatly reduces the amount of feature computation.

3.3.2 Classification Model

In previous works, Zheng et al. [14] obtained from the experiment that the segmentation method based on transition points followed by Decision Tree (DT) model outperformed other models including Support Vector Machine (SVM) and Bayesian Net. According to this and the characteristic of multi-dimensional feature, we select DT and five kinds of DT-based combinatorial classification models including Random Forest (RF), AdaBoost, Bagging, Gradient Boost Decision Tree (GBDT) and eXtreme Gradient Boosting (XGBoost) [3] to learn the transportation mode of each segment. In the experiment section, the accuracy of six classification models will be tested, and RF is chosen to be the optimal model eventually.

3.4 Transportation Mode Correction

Since Lin et al. [7], Zheng et al. [13] and Zhu et al. [17] took no account of the influence of the adjacent segments (preceding and following) on the current segment, the detection accuracy is further improved by combining the current segment and its adjacent segments in this paper, and the predicted transportation mode of each segment is ranked by its probability value. The procedure of transportation mode correction is shown in Fig. 4.

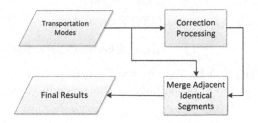

Fig. 4. Procedure of correction method

- If the maximum probability of transportation mode is greater than the threshold value for each segment in the inference sequence, we select the mode with the maximum probability as the final result of the current segment and skip to the merge procedure. Otherwise the correction processing is performed.
- In the merge procedure, we merge the current segment into its adjacent segments whose mode is identical to the mode of the current segment, so as to get the final results.
- In the correction processing procedure, if the absolute value of the difference between the first and second modes in the current segment ranked by its probability value is greater than the threshold P_{th1}, the first mode is the final result of the current segment, and the merge procedure is performed directly. Otherwise, we obtain the transportation mode corresponding to the maximum probability of the current segment (e.g. mode 1) and adjacent segments. If the modes of adjacent segments are the same (e.g. mode 2) and the maximum probability of both is greater than the threshold P_{th2}, the probabilities of mode 1 and mode 2 in the current segment are particularly close to each other in the meantime, then the mode of current segment is modified to mode 2, and skip to the merge procedure. If both of the two ways above are not satisfied, the following procedure is adopted:

1. Acquire the transition matrix $T = (t_{ij})_{6\times6}$ of transportation modes, where t_{ij} stands for the transition probability from mode i to mode j.
2. Calculate the impact factor matrix $P = (p_{ij})_{6\times6}$ of the preceding segment on the current segment according to Eq. (5).
3. Calculate the impact factor matrix $F = (f_{jk})_{6\times6}$ of the following segment on the current segment on the basis of Eq. (6).
4. Correct the transportation mode probability of the current segment as Eq. (7), and select the transportation mode with the maximum probability as

the final result of the current segment, which is the maximum value of $P_{cur}(j)$. Finally, perform the merge procedure.

$$p_{ij} = P(i) \times t_{ij} \tag{5}$$

$$f_{jk} = P(k)/t_{jk} \tag{6}$$

$$P_{cur}(j) = \max_{i,k \in M}(P(j) \times p_{ij} \times f_{jk}),$$
$$(j = \{walk, bike, bus, car, train, plane\}) \tag{7}$$

where $M = \{walk, bike, bus, car, train, plane\}$, $i, j, k \in M$, and $P(i)$, $P(j)$, $P(k)$ represent mode probability of the preceding, current and following segment, respectively.

4   Experiment

4.1   Parameter Setting

Trip Identification. The sensitivity of *minTime* and *maxRange* are analyzed and tuned in details to identify trips reasonably. The lower *minTime*, the more fake stay points will be generated which are caused by crossroads, traffic congestions, etc. If the maximum range setting is too small, part of stay points will be lost. Consequently, we make an experiment with ground truth data to find optimal parameters. Based on previous works [4,10,12], *minTime* was set to 180 s, 200 s and 30 min, respectively, and the corresponding *maxRange* was 30 m, 50 m and 200 m. The best performance goes to the combination of 30 min and 200 m.

Feature Selection. As shown in Fig. 3(a) and (b), for more than 90% of walk and bike have a velocity less than 5.2 m/s. The probability of other modes is under 60%. What is more, more than 80% of bus and car have a velocity lower than 16 m/s, and less than 55% of train and plane exhibit velocity lower than this value. Therefore, 5.2 m/s and 16 m/s are selected for the two threshold (V_{ml}, V_{hm}) separating velocity into three levels (high, medium and low). Thresholds for acceleration (A_{ml}, A_{hm}) can be chosen as 0.3 m/s^2 and 1.0 m/s^2 through Fig. 3(b) similarly to velocity.

4.2   Transition Point Evaluation

Table 2 shows the comparison by the use of both recall and precision rate between our method and other segmentation methods based on transition point, including Fixed-Length Segmentation (FLS, dividing one segment every 100 m), Fixed-Duration Segmentation (FDS, dividing one segment every 120 s) and the method of Xiao et al. [11] and Zheng et al. [14]. The recall and precision of our method is 90% and 54%, which achieves the highest precision among all the segmentation methods. Besides, the recall of our method is higher than FLS, FDS and that of the method of Zheng et al. [14]. FLS, FDS and the method of Xiao et al.

[11] are at the expense of the precision to improve the recall rate, resulting in poor performance of segment feature information and detection results. Zheng et al. [14] lacks consideration of noise in the trajectory. Therefore, our method not only maintains a high recall, but also outperforms the other four methods in precision, which achieves balance in improving the recall without significantly decreasing the precision.

Table 2. Comparison of different methods for transportation point evaluation

Measure (%)	FLS	FDS	Xiao et al. [11]	Zheng et al. [14]	Our method
Recall	87	88	91	88	90
Precision	15	20	32	41	54

4.3 Feature Evaluation

We perform an evaluation on the 29 features set from two aspects: the information gain and the accuracy of single feature classification. As shown in Fig. 5(a) and (b), LVR, MVR and HVR display higher rank in the results of both information gain and accuracy, and BSR has a better performance in terms of accuracy. From the results of two feature rankings, it is found that 16 of the top 20 features are exactly the same. As a result, these 16 features are used as a selected feature set to infer transportation mode, and the comparison result between the selected set and the entire feature set including 29 features is shown in Table 3. Table 3 shows that using the selected feature set instead of the entire feature set would not lose too much accuracy, and the evaluation value (precision, recall and F-score) of some modes would increase instead. BSR is not included in the selected feature set, so we add it to the selected feature set and re-evaluate. Compared Table 4 with the selected feature set part of Table 3, the F-score of bus is improved from 78.0% to 79.3%, the increased value of the F-score of car is 1.8%, and the overall accuracy increases by 0.7% as well.

Table 3. Feature comparison

Transportation mode	Selected feature set (%)			Entire feature set (%)			Difference (%)		
	Precision	Recall	F-score	Precision	Recall	F-score	Precision	Recall	F-score
Walk	90.4	93.2	91.8	89.2	95.7	92.3	1.2	−2.5	−0.5
Bike	91.7	88.7	90.2	93.6	86.7	90.0	−1.9	2.0	0.2
Bus	76.0	80.1	78.0	80.2	78.7	79.5	−4.2	1.4	−1.5
Car	77.9	70.1	73.8	79.7	73.0	76.2	−1.8	−2.9	−2.4
Train	77.4	72.1	74.7	80.4	72.6	76.3	−3	−0.5	−1.6
Plane	100	100	100	100	96.8	98.4	0	3.2	1.6
Accuracy(%)	85.1			86.5			−1.4		

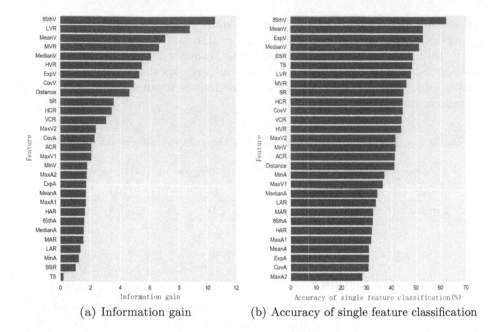

(a) Information gain   (b) Accuracy of single feature classification

Fig. 5. Feature evaluation

Table 4. The result of adding BSR into selected feature set

Transportation mode	Selected feature set + BSR (%)		
	Precision	Recall	F-score
Walk	90.0	93.8	91.9
Bike	92.7	89.1	90.8
Bus	78.3	80.4	79.3
Car	79.5	72.0	75.6
Train	78.1	73.2	75.6
Plane	100	100	100
Accuracy (%)	85.8		

4.4 Comparison Between RF Model and Other Classification Models

To acquire the optimal detection results, DT and five kinds of DT-based combinatorial machine learning classification models are considered. In the experiment, 2/3 of the datasets are divided into training set, the rest is test set, and the number of trees used in five combinatorial classification models is set to 100. As shown in Table 5, RF outperforms other classification models based on the accuracy, and all three evaluation standards (precision, recall and F-score)

Table 5. Classification model comparison

Model	Measure (%)	Transportation mode						Accuracy (%)
		Walk	Bike	Bus	Car	Train	Plane	
DT	Precision	85.9	85.9	72.0	66.3	60.7	100	78.8
	Recall	87.4	82.4	68.3	67.8	63.5	100	
	F-score	86.6	84.1	70.1	67.0	62.0	100	
RF	Precision	89.2	93.6	80.2	79.7	80.4	100	86.5
	Recall	95.7	86.7	78.7	73.0	72.6	96.8	
	F-score	92.3	90.0	79.5	76.2	76.3	98.4	
Adaboost	Precision	86.9	86.6	70.9	64.9	58.3	100	78.9
	Recall	87.4	83.9	68.5	66.4	62.4	96.9	
	F-score	87.1	85.2	69.7	65.6	60.3	98.4	
Bagging	Precision	89.0	94.0	80.0	75.1	80.4	100	85.7
	Recall	95.7	85.9	76.5	72.8	68.5	96.0	
	F-score	92.2	89.8	78.2	73.9	74.0	98.0	
GBDT	Precision	89.4	91.8	80.5	76.8	81.5	95.0	85.9
	Recall	94.9	86.7	77.2	72.8	73.6	100	
	F-score	92.0	89.2	78.8	74.7	77.3	97.4	
XGBoost	Precision	85.1	92.5	73.1	63.9	80.4	100	81.4
	Recall	94.5	74.6	73.7	70.1	39.6	100	
	F-score	89.5	82.6	73.4	66.9	53.1	100	

of RF in walk, bike, bus and car mode are obviously superior to other models. Therefore, there is no doubt that RF is the optimal classification model.

4.5   Evaluation of Transportation Mode Correction Method

The detection results are dealt by three correction methods separately, the result of corrected segment number is presented in Table 6. The method considering transition probability and adjacent segments (preceding and following) obviously achieves a higher correction performance over the other two methods, which brings 46 and 18 promotion.

Table 6. Correction method evaluation

Method	The number of corrected segment
Transition matrix	95
Transition matrix + preceding segment	123
Transition matrix + adjacent segments (preceding and following)	141

618   J. Liang et al.

4.6 Results

Table 7 presents the confusion matrix of final detection results with our detection model and correction method. The overall accuracy is 86.5%. The recall and precision of walk mode reach 95.7% and 89.2%, and bike mode with good recall and precision both higher than 85% simultaneously. Due to the similarity between bus and car mode, there is a certain degree of misinterpretation existing in these two modes. 19% of train mode is incorrectly classified as walk mode, because the train mode has a great speed similarity with walk mode when entering and leaving the station.

Table 7. Confusion matrix of final detection results

		Walk	Bike	Bus	Car	Train	Plane	Recall (%)	Segment number
Ground truth(%)	Walk	95.7	0.8	2.1	0.4	0.9	0	95.7	3537
	Bike	9.8	86.7	2.3	1.2	0	0	86.7	1047
	Bus	7.6	1.3	78.7	10.4	2.0	0	78.7	1430
	Car	7.2	1.2	14.8	73.0	3.8	0	73.0	1107
	Train	19.7	0	2.6	5.1	72.6	0	72.6	583
	Plane	3.2	0	0	0	0	96.8	96.8	63
Precision (%)		89.2	93.6	80.2	79.7	80.4	100	86.5	

4.7 Comparing with Other Transportation Mode Detection Methods

For the purpose of comparing the transportation mode detection method with other existing works, three previous methods [11,13,17] are applied to the same dataset we used in this paper. As shown in Table 8, our method achieves the

Table 8. Transportation mode detection methods comparison

Method	Measure (%)	Walk	Bike	Bus	Car	Train	Plane	Accuracy (%)
Xiao et al. [11]	Precision	88.6	77.2	75.8	74.2	68.8	88.0	81.0
	Recall	86.9	81.8	78.2	76.5	55.2	100	
	F-score	87.8	79.5	77.0	75.3	61.3	93.6	
Zheng et al. [13]	Precision	88.0	80.0	73.0	73.1	64.8	100	80.4
	Recall	91.0	76.5	75.8	66.4	60.4	85.0	
	F-score	89.5	78.2	74.4	69.6	62.5	91.9	
Zhu et al. [17]	Precision	90.0	90.7	76.8	75.3	75.7	100	84.9
	Recall	95.1	86.7	76.3	68.1	69.5	66.7	
	F-score	92.5	88.7	76.6	71.5	72.5	80.0	
Our method	Precision	89.2	93.6	80.2	79.7	80.4	100	86.5
	Recall	95.7	86.7	78.7	73.0	72.6	96.8	
	F-score	92.3	90.0	79.5	76.2	76.3	98.4	

highest accuracy among all the methods, even the value of every evaluation criteria for each transportation mode is the best, especially the F-score of 92.3% for walk and 90.0% for bike when it is compared with other methods.

5 Conclusions

In this paper, we aim to detect transportation mode from GPS data through detection model and correction method. Firstly, we design an enhanced trip segmentation method including trip identification and transition point detection, and the evaluation result shows that our method achieves a high recall without decreasing the precision. Secondly, we introduce a detection model based on RF classifier which contains several new features (IIVR, MVR, LVR, HAR, MAR, LAR and BSR) to significantly manifest a stronger identification to complicated traffic condition. A clear feature structure containing three categories is proposed as well. Results demonstrate that LVR, MVR and HVR outperforms other 23 features at information gain, and the classification accuracy of BSR, LVR and MVR show their advantages beyond others apparently. The overall accuracy is 86.5%, the recall, precision and F-score of most modes are above 80%, even exceeding 95.7% for walk mode. Additionally, we propose a correction method with the view of amending misclassification of transportation mode. The method considers not only the transition probability and the probability of current segment, but also the great influence of adjacent segments on the current segment. The comparison of corrected segment number with other two correction methods clearly verifies the effectiveness of our method. Future work can explore more GIS sources in the transportation mode detection to help discriminate similar modes. In addition, the correction method by using adjacent segments needs to be improved by further studies as well.

Acknowledgments. The research was supported by projects of Chengdu City Science and Technology Bureau (2014-HM01-00302-SF, 2015-HM01-00484-SF). We would also like to thank all of the reviewers for their valuable and constructive comments, which greatly improved the quality of this paper.

References

1. Openstreetmap. http://www.openstreetmap.org
2. Biljecki, F., Ledoux, H., Van Oosterom, P.: Transportation mode-based segmentation and classification of movement trajectories. Int. J. Geogr. Inf. Sci. **27**(2), 385–407 (2013)
3. Chen, T., He, T.: Xgboost: extreme gradient boosting. R package version 0.4-2 (2015)
4. Gong, H., Chen, C., Bialostozky, E., Lawson, C.T.: A GPS/GIS method for travel mode detection in New York city. Comput. Environ. Urban Syst. **36**(2), 131–139 (2012)
5. Haklay, M., Weber, P.: Openstreetmap: user-generated street maps. IEEE Pervasive Comput. **7**(4), 12–18 (2008)

6. Li, Q., Zheng, Y., Xie, X., Chen, Y., Liu, W., Ma, W.Y.: Mining user similarity based on location history. In: Proceedings of the 16th ACM SIGSPATIAL International Conference on Advances in Geographic Information Systems, p. 34. ACM (2008)
7. Lin, M., Hsu, W.J., Lee, Z.Q.: Detecting modes of transport from unlabelled positioning sensor data. J. Locat. Based Serv. **7**(4), 272–290 (2013)
8. Mir, H.M., Behrang, K., Isaai, M.T., Nejat, P.: The impact of outcome framing and psychological distance of air pollution consequences on transportation mode choice. Transp. Res. Part D: Transp. Environ. **46**, 328–338 (2016)
9. Rasmussen, T.K., Ingvardson, J.B., Halldórsdóttir, K., Nielsen, O.A.: Improved methods to deduct trip legs and mode from travel surveys using wearable gps devices: a case study from the greater copenhagen area. Comput. Environ. Urban Syst. **54**, 301–313 (2015)
10. Tran, L.H., Nguyen, Q.V.H., Do, N.H., Yan, Z.: Robust and hierarchical stop discovery in sparse and diverse trajectories. Technical report (2011)
11. Xiao, G., Juan, Z., Zhang, C.: Travel mode detection based on GPS track data and bayesian networks. Comput. Environ. Urban Syst. **54**, 14–22 (2015)
12. Ye, Y., Zheng, Y., Chen, Y., Feng, J., Xie, X.: Mining individual life pattern based on location history. In: 2009 Tenth International Conference on Mobile Data Management: Systems, Services and Middleware, pp. 1–10. IEEE (2009)
13. Zheng, Y., Li, Q., Chen, Y., Xie, X., Ma, W.Y.: Understanding mobility based on GPS data. In: Proceedings of the 10th International Conference on Ubiquitous Computing, pp. 312–321. ACM (2008)
14. Zheng, Y., Liu, L., Wang, L., Xie, X.: Learning transportation mode from raw gps data for geographic applications on the web. In: Proceedings of the 17th International Conference on World Wide Web, pp. 247–256. ACM (2008)
15. Zheng, Y., Xie, X., Ma, W.Y.: Geolife: a collaborative social networking service among user, location and trajectory. IEEE Data Eng. Bull. **33**(2), 32–39 (2010)
16. Zheng, Y., Zhang, L., Xie, X., Ma, W.Y.: Mining interesting locations and travel sequences from GPS trajectories. In: Proceedings of the 18th International Conference on World Wide Web, pp. 791–800. ACM (2009)
17. Zhu, Q., Zhu, M., Li, M., Fu, M., Huang, Z., Gan, Q., Zhou, Z.: Identifying transportation modes from raw GPS data. In: Che, W., et al. (eds.) ICYCSEE 2016. CCIS, vol. 623, pp. 395–409. Springer, Singapore (2016). doi:10.1007/978-981-10-2053-7_35

The Triangle Collapse Algorithm Based on Angle Error Metrics

Xiaorong Yan[✉], Yuansheng Lou, and Ling Li

College of Computer and Information, Hohai University, Nanjing 211100, China
Yanxiaorong0129@gmail.com

Abstract. To solve the problems in the mesh simplification, such as the poor features preserving and too uniform in some areas, an improved triangle collapse algorithm based on angle error metrics is proposed. First, divide the triangles into three types, and different processing methods are adopted for different types of triangles. The algorithm uses the triangle collapse based on the quadratic error metrics. In order to preserving the topological structure and geometric boundary feature of the original model, the angle error metrics is added as a new feature factor on the basis of the two factors: the long and narrow degree and the local region area. This algorithm is implemented by VC++6.0 and OpenGL. The experiment shows that the algorithm preserves the boundary feature and topology of the original model well, and the speed of simplification is faster than other algorithm.

Keywords: Angle error · Triangle collapse · Quadric error metrics · Mesh simplification

1 Introduction

Along with the development of computer graphics, computer aided design and virtual reality technology, the 3D model visualization as one of the key technology is also constantly improving, and it makes the current 3D digital model is growing, which requires a high speed computing platform and a large amount of memory to run. However,the high-precision model is not necessarily necessary, so how to use a relatively simple model instead of the original model and how to simplify the complex triangular mesh model without losing the details of the model have become the focus of research. Mesh simplification is to make the model more concise by reduce the number of points, edges, and faces of the model as much as possible while keeping the original model features, which can be divided into vertex removal, edge collapse, triangle collapse and wavelet decomposition method [1, 2]. The vertex removal is less efficient because of it requires a re-triangulation for the holes which due to the deletion of vertex. A single triangle collapse operation is equivalent to two edge collapse, and therefore triangle collapse has a higher computational efficiency.

On the triangle collapse, the Quadratic Error Metrics (QEM) which is firstly proposed by Garland and Heckbert [3] is a widely used method of error calculation. The QEM uses the sum of the squares with the distance from the vertex to the plane to measure errors and then determine the sequence of collapse. This method is simple and

© Springer Nature Singapore Pte Ltd. 2017
B. Zou et al. (Eds.): ICPCSEE 2017, Part I, CCIS 727, pp. 621–632, 2017.
DOI: 10.1007/978-981-10-6385-5_52

has a relatively fast operation speed. Hamann [4] proposed a mesh simplification method based on triangular curvature, this method removes the narrow and low curvature triangular by calculating. Isler and Green [5] realizes the semi-real-time hierarchical detail model by combining triangle collapse and edge collapse. Zhou et al. [6] applied QEM to the triangle collapse algorithm, which is faster than the Garland algorithm, but the calculation of the new vertex position is unstable during the simplification process, and the simplified mesh tends to be averaged and easy to lose feature details. Later, some scholars have taken the volumetric error, curvature analysis, the maximum area of the triangle and the normal factor as control factors of the collapse error, which is significantly better than previous methods [7, 8]. In [9], the edges of the model are classified into four categories: contour edge, transition edge, transboundary edge and general edge, in order to preserving the features of model, each kind of edge has different collapse methods.

In this paper, an improved triangle collapse algorithm based on angle error metrics is proposed to solve the problems in above mesh simplification algorithms, such as the poor features preserving and too uniform in some areas. The algorithm uses the triangle collapse based on the quadratic error metrics, and incorporates the angle error metrics as the new feature factor. The improved algorithm can better preserve the topology and geometrical boundary of the original model.

2   Related Work

2.1   Triangle Collapse

The triangle collapse simplification method is the continuation of the edge collapse. It uses a triangle as the basic element of the deletion. The process of collapse is to choose a new vertex instead of the deleted triangle, and then calculate the collapse error for the new vertex of each triangle in the model according to some error measurement method. Make the cost of collapse from small to large arrangement. Finally remove the triangle which the smallest cost, while removing all the related element and re-grid the collapse area. As shown in Fig. 1, after the triangle T_0 is collapsed, the adjacent triangles T_1, T_2, T_3 and vertices v_1, v_2, v_3 are deleted and then generate the new vertices v_0.

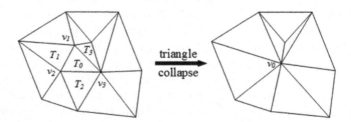

Fig. 1. Schematic diagram of triangle collapse

The basis of this study is the triangle collapse, and the key is to determine the location of the vertex v_0. The choice of the position is determined by the result of the collapse error calculation. And the smaller the error is, the better the choice of the vertex is.

Zhou et al. [6] used the QEM to measure the cost of the triangle collapse. First, set the new vertex $v_{i0} = [x_{i0}\ y_{i0}\ z_{i0}\ 1]^T$. Second, for each of the triangles T_i in the original mesh, a 4×4 symmetric error matrix Q is set, and the cost of collapse the triangle T_i can be expressed as:

$$
\begin{aligned}
\Delta(T_i) = v_{i0}^T \bar{Q} v_{i0} \\
= q_{i11}x_{i0}^2 + 2q_{i12}x_{i0}y_{i0} + 2q_{i13}x_{i0}z_{i0} + 2q_{i14}x_{i0} \\
+ q_{i22}y_{i0}^2 + 2q_{i23}y_{i0}z_{i0} + 2q_{i24}x_{i0} + q_{i33}z_{i0}^2 + 2q_{i34}z_{i0} + q_{i44} \\
+ q_{i33}z_{i0}^2 + 2q_{i34}z_{i0} \\
+ q_{i44}
\end{aligned}
\tag{1}
$$

Let the point with the smallest value of $\Delta(T_i)$ be the best position of the new vertex v_{i0}. Minimize the three vertices, the center and the midpoint of the three sides of the triangle T_i, that is, to derive the x_{i0}, y_{i0}, z_{i0} in Eq. (1):

$$
\partial\Delta(v_{i0})/\partial x_{i0} = \partial\Delta(v_{i0})/\partial y_{i0} = \partial\Delta(v_{i0})/\partial z_{i0} = 0
\tag{2}
$$

The above formula can be sorted out the following form of linear equations:

$$
\begin{bmatrix}
q_{i11} & q_{i12} & q_{i13} & q_{i14} \\
q_{i21} & q_{i22} & q_{i23} & q_{i24} \\
q_{i31} & q_{i32} & q_{i33} & q_{i34} \\
0 & 0 & 0 & 1
\end{bmatrix}
v_{i0} =
\begin{bmatrix}
0 \\ 0 \\ 0 \\ 1
\end{bmatrix}
\tag{3}
$$

If the Eq. (3) has a solution, the solution of Eqs. (3) can get new vertex position for v_{i0}:

$$
v_{i0} =
\begin{bmatrix}
q_{i11} & q_{i12} & q_{i13} & q_{i14} \\
q_{i21} & q_{i22} & q_{i23} & q_{i24} \\
q_{i31} & q_{i32} & q_{i33} & q_{i34} \\
0 & 0 & 0 & 1
\end{bmatrix}^{-1}
\begin{bmatrix}
0 \\ 0 \\ 0 \\ 1
\end{bmatrix}
\tag{4}
$$

If there is no solution to the Eqs. (3), then v_{i0} is the point choose from the middle of the three vertices, the center and the midpoint of the triangle, which makes the result of Eq. (4) the smallest.

2.2   QEM Based on Feature Preserving

QEM. Garland's QEM [3] use the point-to-plane distance as error criterion. Each triangle T_i in the original model is defined a set C_i which is the union of the triangles adjacent to the three vertices of T_i. Moreover, the vertices of each triangle in C_i contain at least one of the vertices of T_i. After collapse the triangle T_i, the distance from the new vertex v_{i0} to the plane of each triangle in C_i is obtained. Where the maximum distance is, the simpler the model is closer to the original model.

Assume that the equation of the plane which adjacent triangle of T_i is located is:

$$ax + by + cz + d = 0 \left(a^2 + b^2 + c^2 = 1 \right) \tag{5}$$

Let $p = \begin{bmatrix} a & b & c & d \end{bmatrix}^T$, and the collapse error of the triangle T_i is:

$$
\begin{aligned}
\Delta(T_i) &= |ax + by + cz + d|^2 \\
&= \left(p^T v_{i0} \right)^2 \\
&= v_{i0}^T p p^T v_{i0} \\
&= v_{i0}^T M_p v_{i0}
\end{aligned}
\tag{6}
$$

Where M_p is a 4×4 symmetric matrix and the error matrix Q in Eq. (1) is the M_p that minimizes T_i:

$$
M_p = pp^T = \begin{bmatrix}
a^2 & ab & ac & ad \\
ab & b^2 & bc & bd \\
ac & bc & b^c & cd \\
ad & bd & cd & d^2
\end{bmatrix}
\tag{7}
$$

Feature Retention. Feature preserving means that the feature factor is added into the calculation of error to control the process of model simplification. The feature factor is used to measure the weight of the collapse error. And the more obvious features of the region, the smaller the value of the weight, and the priority should be given to the greater weight of the triangle which feature is not obvious. This paper uses those two commonly used factors: the long and narrow of triangle and local region area [8].

Long and Narrow of Triangle. In order to ensure that the triangle shape in the mesh will not be too large deformation, the triangle with poor shape should be preferentially collapsed. In general, an equilateral triangle is considered to be the best shape, as shown in Fig. 2(b). The smallest angle of the triangle is close to $0°$ or $180°$ is considered the worst shape, as shown in Fig. 2(a). So the long and narrow triangle should be preferentially collapsed. Use the length of the triangle L_i (Long and Narrow Degree) as a measure of the shape of the triangle.

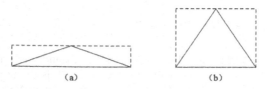

(a)   (b)

Fig. 2. The pros and cons of the shape of the triangle

Since the sum of the inner angles of triangle is $180°$, the range of the minimum internal angle $min(\alpha)$ is:

$$0 < min(\alpha) \leq \pi/3 \tag{8}$$

The long and narrow of triangle L_i can be expressed as:

$$L_i = 2(1 - \cos(min(\alpha))) \tag{9}$$

The Eq. (9) is used to describe the extent of "narrow" of a triangle. The smaller the value of L_i, the less irregular the triangle is, and this triangle should be preferentially collapsed. The introduction of this factor in the collapse can better preserve the feature of the model.

Local Region Area. The Local Region Area (LRA) is the sum of the triangles' area that changed by collapse. As shown in Fig. 3, the local region area of the triangle T_0 consists of its adjacent triangles from T_1 to T_{10}, and the sum of them is the LRA of the triangle T_0.

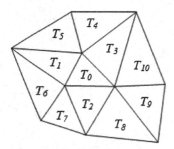

Fig. 3. Schematic diagram of the local region

In general, the features of the mesh model are more obvious in the area, and the density of the mesh is higher, where the triangle area is smaller, resulting in a relatively small area of the local region. Collapse these area will produce a significant triangle missing that is feature missing, so it should be preferred to collapse the feature is not obvious and the local region of the larger triangle. Controlling the sequence of triangle collapse by defining LRA is a very important measure of error. The set of triangle containing the three vertices of the triangle T_i is set to T_n–T_k, then the LRA of the triangle T_i can be expressed as:

$$LRA(T_i) = \sum_{i=n}^{k} S_i \tag{10}$$

3 Triangle Collapse Algorithm Based on Angle Error Metrics

In the triangle mesh, the collapsed triangle and the triangle area formed by new vertex will form a dihedral angle error. The error reflects the overall features change in the local region which is caused by collapse, and has a significant effect on the regions with different curvatures. The effective control of the angle error can avoid the phenomenon that the rotation angle is too large or even flip when the model is simplified. In this paper, the angle error metrics is introduced as a new feature factor, and an improved triangle collapse algorithm based on angle error metrics is proposed.

3.1 Angle Error Metrics

After a triangle is collapsed, the shape and position of the triangle in its local region will change. When the triangle T_0 in Fig. 4 is collapsed to the point v_{i0}, the first-order neighborhood triangles T_1 to T_{10} of T_0 are transformed into the first-order neighborhood triangles $\triangle OAB$, $\triangle OBC$, $\triangle OCD$, $\triangle ODE$, $\triangle OEF$, $\triangle OFG$ and $\triangle OAG$ of the new vertex v_{i0}, and they formed the dihedral angle.

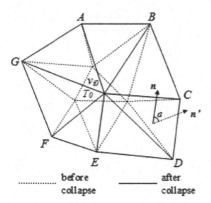

Fig. 4. Analysis of angle error

In this paper, the maximum value of these dihedral angles is taken as the angle error of the triangle, denoted as α_i. For a single adjacent triangle, the angular error can be expressed as:

$$\alpha_i = arccos\left(\frac{n_i \cdot n_i'}{|n_i| \cdot |n_i'|}\right) \tag{11}$$

Let $n_{i0}, n_{i1}, n_{i0}, n_{i1}, \cdots, n_{ik}$ be the unit normal vector of the first-order neighborhood triangle before the triangle T_i is collapsed, let $n_{i0}', n_{i1}' \cdots n_{ik}'$ be the unit normal vector of the first-order neighborhood triangle of the new vertex v_{i0} after collapse, and let k be the number of first-order neighborhood triangles of vertex v_{i0}, the angle error of the collapsed triangle T_i can be expressed as:

$$\alpha_i = \max\left\{\arccos\left(\frac{\boldsymbol{n}_{i0} \cdot \boldsymbol{n}'_{i0}}{|\boldsymbol{n}_{i0}| \cdot |\boldsymbol{n}'_{i0}|}\right), \arccos\left(\frac{\boldsymbol{n}_{i1} \cdot \boldsymbol{n}'_{i1}}{|\boldsymbol{n}_{i1}| \cdot |\boldsymbol{n}'_{i1}|}\right), \ldots, \arccos\left(\frac{\boldsymbol{n}_{ik} \cdot \boldsymbol{n}'_{ik}}{|\boldsymbol{n}_{ik}| \cdot |\boldsymbol{n}'_{ik}|}\right)\right\}$$

(12)

The angle error of collapsing is arranged from small to large. The larger the angle error represents, the larger the change of triangular normal vector which is caused by the collapse, and the feature with too large loss should be delayed or not collapsed. The smaller the error is, the better the feature of the mesh are, and can be preferentially collapsed.

The simplified method based on the angle error metrics is more effective than the distance-based metrics in QEM. The latter uses the same collapse cost for the different curvature of the region, which is easy to make high curvature area loss feature, and low curvature area is not fully simplified. After the angle error is obtained by the calculation of the normal vector changes, the high curvature region and the low curvature region can be effectively distinguished, and the feature is effectively simplified with feature preserving. The distance-based simplification produces a triangle folding due to the collapse of the adjacent triangle, resulting in a non-smooth plane that affects the feature and visual effects of model. As shown in Fig. 5, after the triangle T_0 is collapsed, the triangular $\triangle v_1 v_2 v'_0$ is turned over due to the addition of the new vertex v'_0. After adding the angle error control, these triangles that affect the features of the mesh will not be collapsed due to the high cost of collapse.

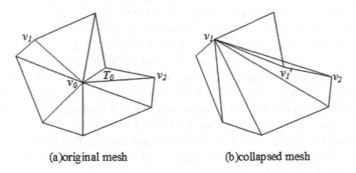

(a)original mesh   (b)collapsed mesh

Fig. 5. Structural changes due to lack of angle error

3.2   The Determination of New Vertex

Before simplifying the model, it is necessary to determine whether the triangle satisfies the collapse requirement, thus we should firstly pre-classify the triangle in the original model, and it can be more targeted to determine the location of the new vertex. The triangle mesh is divided into internal triangles and boundary triangles. Boundary triangles are also divided into vertex-boundary triangles and edge-boundary triangles [10]. To determine the type of triangle by judging whether the vertex is the boundary point, if all the adjacent points of the vertices appeared at least twice in their adjacent triangles, then the vertex is the interior point, otherwise the boundary point, as shown in Fig. 6 shows:

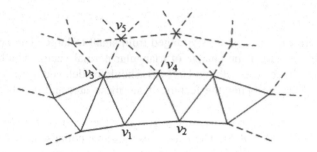

Fig. 6. Schematic diagram of triangles classification

$\triangle v_3 v_4 v_5$ do not have boundary point which means it's an internal triangle; $\triangle v_1 v_3 v_4$ has a boundary point, which is a vertex-boundary triangle; $\triangle v_1 v_2 v_4$ has two boundary points, which are an edge-boundary triangle.

We don't collapse the edge-boundary triangle, because the change of boundary will lead to the change of the outline of the model, resulting in collapse, deformation and so on. For the vertex-boundary triangle, the three vertices, centers, and midpoints of the triangle are taken into Eq. (6), and the point where the result $\Delta(T_i)$ is minimized is taken as the new vertex v_{i0}. The set of new vertices after the simplification is a subset of the original model' vertexes, so the computational complexity is small and fast, and the basic geometric features of the original model are preserved.

As for the internal triangle, we use QEM to minimize Eq. (1) and determine the new vertex position. If the Eqs. (3) are resolved, then the collapsed triangle is not coplanar with the adjacent triangle, the solution of the equations is the new vertex. If the equations is not solved, the collapsed triangle is coplanar with the adjacent triangle. The new vertex v_{i0} position refers to the method in literature [7], which takes the barycenter of the adjacent region as the position of new vertex.

Classifying the triangle mesh in the model and using different new vertex determination methods for different triangles can avoid overly useless operations and keep the boundaries of the mesh model well.

3.3   Calculation of Collapse Cost

First, according to Garland's QEM to calculate the cost of triangle collapse. Then the long and narrow of triangle L_i and the local region area $LRA(T_i)$ are taken as the feature factors. Among them, the smaller the L_i, the larger the LRA of the triangle, the less obvious the regional features, so the cost of collapse the triangle is small and should be collapsed firstly, thereby the triangle collapse error matrix is obtained:

$$Q'_i = \frac{L_i}{LRA(T_i)} Q_i \tag{13}$$

where L_i and $LRA(T_i)$ are obtained from Eqs. (9) and (10), respectively.

For the angle error metrics proposed in this paper, the smaller the error caused by the collapse, the triangle is in the area where the features are less obvious, then the area should be collapsed first, and the new collapse error matrix is obtained by combining Eq. (6):

$$Q''_i = \frac{L_i}{LRA(T_i) \cdot \alpha_i} Q_i \tag{14}$$

The new collapse error metrics which is obtained from Eq. (14) and Eq. (6) is:

$$\Delta'(T_i) = v_{i0}^T Q''_i v_{i0} \tag{15}$$

3.4 Algorithm Description

The Triangle collapse algorithm based on angle error metrics in this paper is as follows:

Step 1. Enter the original mesh model that needs to be simplified and pre-classify the triangles.

Step 2. After the classification, if it is edge-boundary, then it is not collapsed. The vertex-boundary triangle takes the vertices, centers, and midpoints of the vertices to determine the position of the new vertices. The point with the smallest result of Eq. (6) is used as the new vertex v_{i0}; If it is the internal triangle, then minimize the Eq. (1) to solve the new vertex v_{i0}, if there is no solution, we take the barycenter of the adjacent region as the new vertex v_{i0}, then update the relevant information of the local region triangles.

Step 3. For each triangle T_i, L_i and $LRA(T_i)$ are calculated by Eqs. (9) and (10).

Step 4. According to the parameters obtained in step 3, the new collapse error matrix Q''_i is calculated by Eq. (14). Combined with the new vertex v_{i0} and the Eq. (15) to calculate the new triangle error $\Delta'(T_i)$, and make the cost of collapse from small to large arrangement, so we can get a collapse error queue.

Step 5. Remove the triangle with the least cost of collapse in the error queue, perform the collapse and delete the first-order adjacent triangles, then repeat steps 2 to 4.

Step 6. Repeat step 5 until the simplification meets the final requirements.

4 Experiment

The experimental environment of this algorithm is based on 2.60 GHz Core (TM) i7 CPU, 8 GB memory PC computer, VC++6.0 in conjunction with OpenGL [11, 12], and select the pulley model as the experimental object to simplify and render.

The experimental results are shown in Fig. 7, Fig. 7(a) is the original model of the pulley, the edge of the model are region with high curvature, and the middle arc plane is a low curvature region. Figure 7(b) is a simplified model when the triangle mesh simplification method with angle error control is used herein to simplify 50% of original model. Figure 7(c) is a model which uses the improved Garland algorithm in literature [13] to simplify 50% of original model. Because of the lack of angle error control, there is no distinction between different curvature regions, and due to a large number of triangles' turning angle is too large, the model surface is not smooth, and flat region has obvious defects due to uneven, which seriously affect the visual effects.

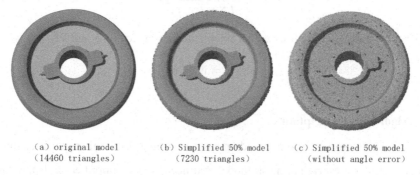

(a) original model   (b) Simplified 50% model   (c) Simplified 50% model
(14460 triangles)   (7230 triangles)   (without angle error)

Fig. 7. Pulley original model and simplified contrast

In addition, the algorithm classified the triangles before the simplification, and took different approaches for the distinguished internal triangle, edge-boundary triangle and vertex-boundary triangle. All of these measures simplified the calculation of the feature factors and reduces the complexity of the algorithm, which makes the algorithm has a high speed of simplification on the basis of preserving the feature. In this paper, the improved algorithm is compared with Garland's algorithm, and the execution speed is improved (Table 1).

Table 1. The speed comparison of Garland's algorithm and the improved algorithm

Number of remaining triangle in the model	14460	7230	3612	1446
Simplification ratio	0%	50%	75%	90%
Time(s) Garland's algorithm	–	4.58	6.03	7.12
Improved algorithm	–	4.08	5.66	6.23

Experimental results show:

(a) With the increase of collapse cost, the simplification rate increases, and the simplification effect is strengthened.
(b) In a range, the simplification is carried out at the same time with high curvature region and low curvature region.

(c) The simplified method based on angle error metrics could generate the high quality simplified model and improve the simplified efficiency.

5 Conclusion

The algorithm in this paper is based on a number of domestic and foreign models which are used to simplify. Aiming at the problem of preserving the model features of the existing algorithms, a triangle collapse algorithm based on angle error metrics is proposed. The triangles are classified before simplification, and then respectively determine the location of the new vertex so that the algorithm can reduce the amount of calculation and improve efficiency. When calculating the cost of triangle collapse, the angle error metrics is added as a new feature factor on the basis of the two factors: the long and narrow degree and the local region area. Then analyze the collapse precedence of the triangle and get the collapse error queue. The experiment shows that the algorithm preserves the boundary feature and topology of the original model well, and the speed of simplification is faster than other algorithm.

The next phase of the work is to consider the simplification of the mesh with texture and color attributes. The triangle mesh simplification will be researched to make the simplified algorithm more practical and effective.

Acknowledgement. This work is supported by (1) National Natural Science Foundation of China (61300122); (2) A Project Funded by the Priority Academic Program Development of Jiangsu Higher Education Institutions; (3) Water Science and Technology Project of Jiangsu Province (2013025).

References

1. Kalvin, A.D., Taylor, R.H.: Superfaces: polygonal mesh simplification with bounded error. IEEE Comput. Graph. Appl. **16**(3), 64–77 (1996)
2. William, J.S., Jonathan, A.Z., William, E.L.: Decimation of triangle meshes. Comput. Graph. **26**(2), 65–70 (1992)
3. Garland, M., Heckbert, PS.: Surface simplification using quadric error metrics. In: Proceedings of the 24th Annual Conference on Computer Graphics and Interactive Techniques, pp. 209–216. ACM Press, New York (1997)
4. Hamann, B.: A data reduction scheme for triangulated surface. Comput. Aided Geom. Des. **11**(2), 197–214 (1994)
5. Lau, W.H., Green, M.: Real-time multi-resolution modeling fore complex virtual environments. In: Proceedings of Virtual Reality Software and Technology 1996, Hongkong, pp. 11–19. ACM Press (1997)
6. Zhou, K., Pan, Z., Shi, J.: Mesh simplification algorithm based on triangle collapse. Chin. J. Comput. **21**(6), 506–513 (1998)
7. Li, N., Xiao, K., Li, Y.: Improved algorithm of mesh simplification based on triangle collapse. Comput. Eng. Appl. **45**(34), 192–194 (2009)
8. Liu, X., Liu, Z., Gao, P.: Edge collapse simplification based on sharp degree. J. Softw. **16**(5), 669–675 (2005)

9. Dong, F., Liu, Y., Xiao, R.: Improved QEM simplification algorithm based on features preserved. J. Comput. Appl. **28**(8), 2040–2045 (2008)
10. Zhou, Y., Zhang, C., He, P.: Feature preserving mesh simplification algorithm based on square volume measure. Chin. J. Comput. **32**(2), 203–212 (2009)
11. Yong, T., Yan, X., Li, Y.: Geometry modeling for virtual reality based on CAD data. Open Cybern. Syst. J. **9**(1), 2339–2343 (2015)
12. Ali, S.K., Rahmat, R.O.K., Khalid, F., Khalil, H.H.: Rational equation for simplifying complex surfaces. Arab. J. Sci. Eng. **39**(6), 4617–4636 (2014)
13. Mao, J., Yang, J., Zhu, B., Yang, Y.: A new mesh simplification algorithm based on quadric error metrics. In: IEEE, International Conference on Consumer Electronics - Berlin, pp. 463–466. IEEE, Berlin (2008)

Spatial-Temporal Event Detection Method with Multivariate Water Quality Data

Yingchi Mao[✉], Zhitao Li, Xiaoli Chen, and Longbao Wang

College of Computer and Information, Hohai University, Nanjing 211100, China
yingchimao@hhu.edu.cn

Abstract. With the rapid development of the society, water contamination events cause great loss if the accidents happen in the water supply system. A large number of sensor nodes of water quality are deployed in the water supply network to detect and warn the contamination events to prevent pollution from speading. If all of sensor nodes detect and transmit the water quality data when the contamination occurs, it results in the heavy communication overhead. To reduce the communication overhead, the Connected Dominated Set construction algorithm-Rule K, is adopted to select a part fo sensor nodes. Moreover, in order to improve the detection accuracy, a Spatial-Temporal Abnormal Event Detection Algorithm with Multivariate water quality data (M-STAEDA) was proposed. In M-STAEDA, first, Back Propagation neural network models are adopted to analyze the multiple water quality parameters and calculate the possible outliers. Then, M-STAEDA algorithm determines the potential contamination events through Bayesian sequential analysis to estimate the probability of a contamination event. Third, it can make decision based on the multiple event probabilities fusion. Finally, a spatial correlation model is applied to determine the spatial-temporal contamination event in the water supply networks. The experimental results indicate that the proposed M-STAEDA algorithm can obtain more accuracy with BP neural network model and improve the rate of detection and the false alarm rate, compared with the temporal event detection of Single Variate Temporal Abnormal Event Detection Algorithm (M-STAEDA).

Keywords: Spatial-temporal data · Event detection · Mutlivariate water quality data · Data analysis · Connected dominating set

1 Introduction

When a large-scale water contamination event occurs in a public water supply network, this can have a dramatic effect on the society. It is one of the most important issues to detect and warn contamination events in a public water supply network and prevent pollution from spreading. One effective way is to deploy a large-scale number of water quality sensors for contamination detection to monitor generic water quality parameters [1]. At present, there have online water quality sensors deployed in the Water Supply Networks (WSNs), which can measure many contaminants and provide an early indicator of possible pollution [26]. Contamination event detection approaches focus on identifying abnormal system behavior. Accurate detection should consider the

© Springer Nature Singapore Pte Ltd. 2017
B. Zou et al. (Eds.): ICPCSEE 2017, Part I, CCIS 727, pp. 633–645, 2017.
DOI: 10.1007/978-981-10-6385-5_53

response strategies, operation optimization, and overall system efficiency [2]. It is an important issue how to detect contamination events in the real-time and accurate way with the multivariate time series data of water quality parameters in a WSN. Due to large-scale sensor nodes deployment, the amount of data from dissimilar sensors has increased tremendously. For example, there are 128,291 hydrometric sensor nodes deployed in China. They generate the hydrologic data about 60 TB in one year. To reduce the computational cost of event detection and improve the response time, sensor nodes selection is needed to cut down the amount of data for the spatio-temporal correlation analysis. To avoid forwarding and collecting the sensed data from all of the sensors in the WSNs, Connected Dominating Set (CDS) [2, 5] is introduce to select backbone nodes and construct an effective connected structure in a WSN to reduce the amount of transmitted data and improve the response speed for contamination event detection. In this paper, a distributed and effective algorithm to construct CDS called Rule K is adopted to select backbone nodes and forward the sensed data of water quality. On the other hand, the main water quality parameters are free chlorine, electrical conductivity (EC), pH, temperature, total organic carbon (TOC), and turbidity [3]. Almost existing contamination event detection approaches are based on only one single water quality parameter, which results in the high false negative rate and low false positive rate [4]. It is difficult to detect potential contamination events in the water in the WSNs due to the number of parameters involved [5].

Therefore, a spatial-temporal event detection algorithm with multivariate data (M-STAEDA) was proposed. First, M-STAEDA adopts Rule K algorithm to select backbone nodes and forward the sensed data of water quality. Then, Back Propagation Neural Networks (BP) models are adopted to simulate the single parameter of water quality and establishes the BP model to analyze the multivariate time-series data of water quality parameters, which can identify the possible outliers. Third, the univariate event probability is updated using sequential Bayesian analysis for each water quality parameter. Forth, each water quality parameter is assigned a weight reflecting its influence on the water pollution. Univariate event probabilities are fused to give a unified multivariate event probability based on all water quality parameters. Finally, a spatial correlation model is applied to determine the spatial-temporal contamination event in the water supply networks.

The rest of this paper is organized as follows. Related work is discussed in Sect. 2. Section 3 presents the overall framework of event detection model with multivariate time-series data for water quality monitoring. The abnormal event detection algorithm with multivariate time-series data (M-TAEDA) was proposed. The performance evaluation is given in Sect. 5. The final section concludes the work.

2 Related Work

The event detection framework relies on continuously transmitted hydraulic and water quality data from the supervisory control and data acquisition system. The contamination event detection approaches based on water quality measurements consist of two phases. One is to set up the prediction model with the historical data as the training dataset. The other is to determine the water quality by comparing the predicted values with the measurements. Various approaches have been proposed for addressing the problem of contamination detection, using single or multiple-type measurements which are analyzed separately or in combination, from one or more locations in the network, using model-based or model-free approaches.

In the threshold-based approaches, the threshold values were set through statistical models. Simplicity is the main advantage since raw data can be directly processed. The abnormal event detection with two thresholds were adopted in [6]. However, the threshold-based approaches cannot obtain the spatio-temporal feature of water quality data, which results in the low accuracy and high false positives of event detection. In the pattern matching approaches, the pattern was established and verified with the water quality sensor readings to infer the contamination event [7]. Byer and Carlson assumed that the water quality parameters obey Gaussian distribution. One statistical model was established to detect the contamination event [8]. The statistical model detection methods cannot be used in the applications.

Learning-based methods can make inference of the possibility of contamination event based on the temporal and spatial correlation of water quality measurements [2, 9, 10]. It is promising to make full use of the spatial and temporal correlation to detect abnormal events. Markov Random Fields (MRFs), Bayesian Network (BN), Dynamical BN, and SVM were common models in the WSNs with the high density. MRFs were adopted to model spatial context and stochastic interaction among observable quantities [11]. BN was considered as a means for unsupervised learning and anomaly detection in gas monitoring sensor networks for underground coal mines [9]. Using a single water parameter-free chlorine, Radial Basis Function (RBF) neural network and wavelet analysis were applied to determine the contamination event occurrence with on-line data of water quality [12]. Perelman proposed one BN-based contamination event detection method to determine the event occurrence through estimating the possible locations of potential contaminants in the WSNs [13].

A multiple type measurement approach at a single location was proposed in [14], where each parameter was compared to its 3 bounds. Control charts and Kalman filters have also been proposed in [15]. When multiple types of sensors are available, these can be used to compute distance metrics to detect contamination [16]. The probability of contamination event could be computed and compared to an adaptive threshold by utilizing a sequential Bayesian rule [17]. The estimation error of event detection was computed with respect to the measurements from a moving window [18]. Moreover, the US Environmental Protection Agency provides the event detection tool - CANARY [19, 20] for time series analysis of multiple water quality parameters. Fluctuations in water distribution networks may cause significant variability, a Bayesian Belief Network was presented to infer the probability of contamination [21].

Based on the above analysis, most approaches for the contamination event detection have been discussed via using single type water quality parameter. There are multiple components to indicate the water quality in s WSN, such as free chlorine, EC, pH, temperature, TOC, and turbidity. Unfortunately, single parameter of water quality cannot reflect the real water quality in a WSN. When contamination event occurred, the observable values of multiple water quality parameters changed in a significant way. The contamination event detection methods based on single parameter may result in the low detection accuracy and high false alarms. To improve the detection accuracy and reduce the false alarms, multiple parameters of water quality should be considered to make decision with data fusion. However, the development of multi-parameter water quality models and their calibration is challenging due to the large amount of information and number of parameters involved. Thus, their application in the context of contamination event detection is complicated. In this paper, the proposed event detection approach, M-STAEDA, utilizes back propagation neural networks model to estimate the relationships between water quality parameters in a WSN. The proposed M-STAEDA can detect the potential contamination events for temporary analysis of multivariate water quality time series with Bayesian sequential analysis.

3 M-STAEDA Framework

The contamination event detection framework relies on continuously transmitted hydraulic and water quality data from the WSNs. The contamination event detection approach consists of two phases: off-line phase and on-line phase. In the off-line phase, one data-driven model – BP neural network is established and trained for temporary analysis of multivariate water quality time series and model assessment. In the on-line phase, first, it selects part of sensor nodes as the connected domination set. Second, it can obtain the predicted values based on the BP model. Then, residuals are estimated as the difference between the measured and predicted parameters' values. Adopting the sequential Bayesian analysis for each new measured parameter's value, the probability of each parameter is computed. Forth, multiple parameters of water quality are fused to provide a unified decision support about a contamination event. Finally, a spatial correlation model – Bayesian Network, is applied to determine the spatial-temporal contamination event in the water supply networks.

On-line phase includes six main steps for each incoming measured value. (1) backbone nodes selection – Rule k algorithm is applied to optimal nodes selection to avoid the heavy communication overhead. (2) data-driven model training – the relationships between water quality parameters are examined using BP models. (3) error threshold setting – calculated residuals are classified as normal or abnormal values using thresholds for each water quality parameter. (4) identification of events – the probability of an event is updated using sequential Bayesian analysis based on the residual classification. (5) information fusion – univariate event probabilities are fused to provide a unified multivariate event probability reflecting the likelihood of an event based on all water quality parameters. Each water quality parameter is assigned a weight reflecting its influence on the fusion decision. (6) spatial correlation model – Bayesian Network is adopted to model the spatial correlation and the maximum likelihood parameter estimation algorithm is used to compute the parameters of BN. Step 3 and 4 are initially trained using available data collected from the WSN. Step 1 to 6 are then repeatedly executed in real-time ways for each new incoming measured value.

4   M-STAEDA Approach

4.1   Backbone Nodes Selection

Thousands of sensor nodes are deployed in a water supply network to sense, transmit and forward massive sensed data, which brings heavy burdens to further contamination event detection in the context of real-time analysis. To avoid forwarding and collecting all of the sensed data from all sensors in the WSNs, optimal sensor nodes selection is needed to reduce the amount of transmitted data and improve the response time for event detection. CDS is one of the optimal nodes selection strategies from Graph Theory [5].

A dominating set (DS) of an undirected graph $G(V, E)$ is a subset of V' of the vertex set V such that every vertex in $V - V'$ is adjacent to a vertex in V'. A dominating set V' is connected if for any two vertices u and v, a path (u, u_1), (u_1, u_2), ..., (v_n, v), $(v_i \in V', 1 \le i \le n)$ in E exists. Rule K algorithm was proposed in our previous work [2], which is one energy efficient and distributed algorithm for CDS selection. Through selecting the appropriate communication range, a set of CDS can be constructed to meet the requirements of network connectivity and coverage. In each round of selection, the sensor nodes firstly adjust the appropriate communication range and select the nodes as nodes in the CDS. The sensor nodes in the CDS as the backbones execute the tasks of monitoring and data forwarding. After the collected data from the CDS with a WSN are analyzed, the system can detect the contamination event in the context of real-time analysis. The backbone nodes can keep monitoring several streams from backbone nodes instead of all streams, therefore, the CDS-based nodes selection strategy can reduce the amount of transmitted data and improve the response time for real-time event detection. In this paper, Rule K algorithm [2] is adopted to select the backbone nodes and construct an effective connected structure in a WSN.

4.2   Water Parameters Simulation Based on BP Model

BP model is adopted to estimate the relationships between six water quality parameters through nonlinear, weighted functions during normal operation. The development of BP model does not require any knowledge of the physical and chemical laws affecting the water quality parameters. Thus, BP model can be available to model the multivariate reactions in the WSN.

In the water supply networks, contamination events heavily rely on the multiple water quality factors making the events hard to prespecified. In the case of free chlorine, BP model is just established with the historical data from single water parameter – free chlorine to estimate the predicted value. When there is small difference between the measured and estimated free chlorine values, it has the obvious great deviations between the measured values of other water parameters and their normal states. The reason is that the measured values of free chlorine exhibit great deviations from the normal operation conditions due to the interplay among some of water quality parameters. The prediction model based on single water quality parameter would obtain the lower accuracy of estimated values. To improve the accuracy of prediction model, BP neural networks are adopted and trained for modeling the relationships between multiple variate water quality parameters in WSNs. One of the most important components in the BP model are multi-layer perceptron (MLPs). MLPs are represented by an input layer, a hidden layer, and an output layer and correlate input variables to output variables depending on the nonlinear, weighted, parameterized functions. The BP model is shown in Eq. (1):

$$f_k(x; w) = \varphi_0(w_0 + \sum_j w_{jk}\varphi(w_{0j} + \sum_i w_{ij}x_i)) \tag{1}$$

where w_{jk}, w_{ij} are weights, w_0, w_{0j} are biases, φ and φ_0 are activation function and output function, respectively, x_i is one of water quality parameters, and $f_k(x; w)$ is used to estimate the target value y. Data is passed from the input layer with p inputs $x = (x_1, \ldots, x_p)$ to hidden layer having m neurons. Each neuron in the hidden layer receives the summed weighted outputs of the preceding layer. The output layer with K targets $y = (y_1, \ldots, y_k)$ receives the summed weighted outputs of the preceding layer again. The final output is the function $f_k(x; w)$.

To avoid the interplay among the multiple water quality parameters, for each target water quality parameter, the corresponding BP model is established with the input measured time series of all predictive parameters and lagged target parameter, as formulated in Eq. (2).

$$\hat{x}_i(t) = f(x_1(t), \ldots, x_{i-1}(t), x_i(t-1), x_{i+1}(t), \ldots, x_n(t)) \tag{2}$$

where $x_i(t)$ and $\hat{x}_i(t)$ are the measured and estimated water parameters at time t respectively, n n is the number of water quality parameters, and $f(\cdot)$ is a function defined by the BP model, as in Eq. (1).

The water quality parameters include free chlorine, electrical conductivity (EC), pH, temperature, total organic carbon (TOC), and turbidity. Six BP models can be created and trained. According to Eq. (2), one model was constructed for each target water quality parameter with the input vector containing measured time series of all predictive parameters at the same time and lagged target parameter. For example, the BP model for estimating free chlorine takes the following inputs:

$$\hat{x}_{free\_chlorine}(t) = f(x_{EC}(t), x_{pH}(t), x_{temperature}(t), x_{TOC}(t), x_{turbidity}(t), x_{free\_chlorine}(t-1)) \tag{3}$$

4.3   Residual Estimation and Classification

The established BP models through training the historical data can be utilized to predict the values of water quality parameters. The next step is to calculate the residuals as the difference between the predicted and measured values of parameters and then to classify the residuals. Residuals can be calculated as shown in Eq. (4), represented as time series.

$$ER_i(t) = x_i(t) - f(\cdot) = x_i(t) - \hat{x}_i(t) \tag{4}$$

where $x_i(t)$, $\hat{x}_i(t)$, and $ER_i(t)$ represent the measured value, the estimated values, and the estimated residual for parameter i at the time interval t, respectively, $f(\cdot)$ is defined by Eq. (1). To distinguish the normal operation of a water supply network from the contamination events, the estimated residuals $ER_i(t)$ can be classified into two types: Normal and Outlier.

4.4 Sequential Bayesian Updating

In this stage, the probability of a contamination event is updated using sequential Bayesian analysis for each incoming value. In the sequential Bayesian analysis, the number of incoming values is unknown in advance. In fact, the measured values come in sequence and a decision requires to be made about the current state. The performance of the BP model is measured through a confusion matrix during the training [22]. The confusion matrix represents the model's classification of all measured values to one of four classes. True Positive (TP): the residual is classified as an outlier during an actual event; False Positive (FP): the residual is classified as an outlier during normal operation condition; True Negative (TN): the residual was classified as reasonable model error during normal operation; False Negative (FN): the residual was classified as a reasonable model error during a contamination event.

Additional common metrics, two others can be derived from the confusion matrix as described in Eq. (5).

$$RD = \frac{TP}{TP+FN} \times 100\% \quad FAR = \frac{FP}{TN+FP} \times 100\% \tag{5}$$

where RD represents the ratio of the number of the detected contamination events to the number of the actual contamination events, and FAR is the ratio of the number of the detected false alarms to the total number of the determined contamination events.

To determine the possibility of an event occurrence from previously classified states (normal or outlier), the probability of a contamination event is updated based on the classification of the residuals. The state of an event has two states: normal and outlier, as shown in Eq. (6). In the Sequential Bayesian analysis for each new incoming data, the number of measured values is not known in advance. The measured values come in sequence, and a decision is made based on the current state. With each new incoming measured value, the probability of an event is updated with sequential Bayesian rule, as shown in Eqs. (7) and (8).

$$P(E_t)=\begin{cases} P(E_t|O_t) \text{ if Residual}_t \text{ is Outlier} \\ P(E_t|\overline{O_t}) \text{ if Residual}_t \text{ is Normal} \end{cases} \tag{6}$$

$$P(E_t|O_t) = \frac{P(O|E) \times P(E_{t-1})}{P(O|E) \times P(E_{t-1})+P(O|\overline{E}) \times P(\overline{E}_{t-1})}$$
$$= \frac{RD \times P(E_{t-1})}{RD \times P(E_{t-1}) + FAR \times (1 - P(E_{t-1}))} \tag{7}$$

$$P(E_t|\overline{O}_t) = \frac{P(\overline{O}|E) \times P(E_{t-1})}{P(\overline{O}|E) \times P(E_{t-1}) + P(\overline{O}|\overline{E}) \times P(\overline{E}_{t-1})}$$
$$= \frac{(1 - RD) \times P(E_{t-1})}{(1 - RD) \times P(E_{t-1}) + (1 - FAR) \times (1 - P(E_{t-1}))} \tag{8}$$

where $P(E_t)$ is the probability of an event at time t, O_t and $\overline{O}_t$ are "Outlier" and "Normal" states at time t, respectively. $P(E_t|O_t)$ is the conditional probability of a contamination event when the residual is classified as an outlier. $P(E_t|\overline{O}_t)$ is the conditional probability of a contamination event when the residual is classified as a normal state.

4.5   Multivariate Fuse Decision

To improve the model's estimation of an event and to lower the number of false alarms, a process of multivariate fusion decision is invoked. This process ensures that only when a predetermined number of parameters indicate an event, an alarm is raised. This process reduces the number of false alarms and improves the alarms' reliability. The univariate event probabilities are fused to compute the multivariate event probability reflecting the likelihood of a contamination event based on all measured water parameters.

4.6   Spatial Correlation Model

A BN (Bayesian Network) is a directed acyclic graph. Each sensor node is regarded as a node of the BN, and the flow direction is regarded as the direction of each edge. BN can incorporate the knowledge and encode the uncertainty relationships from different sources. BN can be used to model spatial correlation of water supply networks.

A BN can model spatial relationship of sensor nodes in the network G. S_i^t is denoted as the observed value from sensor node v_i at the time interval t. One n-tuple $(S_1^t, S_2^t, \ldots S_i^t)$ can be used as one of the training data cases for learning the parameters of the BN. Thus, it can generate lots of training data cases from the historical sensed data set. The maximum likelihood parameter estimation algorithm is adopted to compute the parameters of the BN.

Assume that the historical data are time series $(t_1, t_2, \ldots t_n)$ from the interval 1 to n. A BN is used to model causal and spatial relationship between different sensor streams. As illustrated in Fig. 3, there exists causal relationship between upstream nodes and downstream nodes. The structure of BN is the same as the topological structure of water supply network.

5   Performance Evaluation

5.1   Experiments Setup

The methodology is tested on real data from CANARY [13]. The real data was collected from the SCADA system sensing hydraulic and water quality in the water supply network. The data consists of six online multivariate water quality parameters: free chlorine (mg/L), EC (mS/cm), pH, temperature (C), TOC (ppb), and turbidity (NTU).

Event Simulation: Contamination events in water supply networks heavily depend on the environmental factors, which brings harder difficult to detect the contamination event. To cope with the above difficulty, contamination events were superimposed on the normal patterns, characterized by the magnitude, direction, and duration [24], which can reflect the water parameters variation caused by the contamination events. Contamination events were assumed to be Gaussian distribution [25]. Figure 1 shows a partial time series plot of six water quality parameters during normal operation and randomly simulated events.

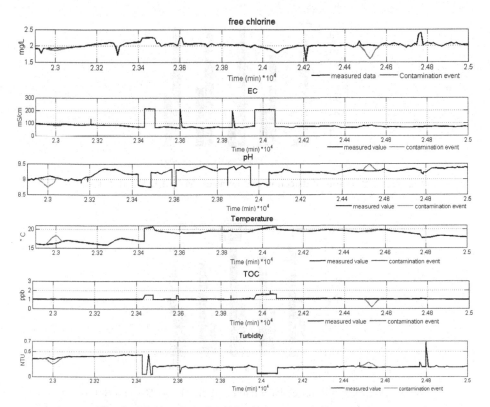

Fig. 1. Time series of multivariate water quality parameters

5.2   Experimental Results

Comparison Analysis

To improve the accuracy of event detection, M-TAEDA adopts BP models to establish the prediction model for each water quality parameter. Different water quality parameters have different predictive ability of contamination events, it should allocate the appropriate weights used to make a unified decision. Contamination events detection based on the multivariate parameters can greatly reduce the false alarms and improve the detection rate, compared with the detection based on the univariate parameter. As shown in Figs. 2 and 3, according on the results from the number of simulated contamination events varying from 20 to 100, the contamination events detection rate fall in the range between 50%–60% with S-TAEDA method, while the detection rate is greater than 75% with M-TAEDA method, which is 40% higher than that of S-TAEDA method. Meanwhile, the false alarms rate of S-TAEDA is greater than 14%, while the detection rate of M-TAEDA is lower than 9%, which is 45% lower than that of S-TAEDA method.

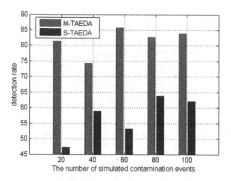

Fig. 2. Comparison of detection rate between M-TAEDA and S-TAEDA

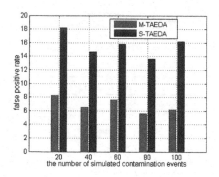

Fig. 3. Comparison of false alarm rate between M-TAEDA and S-TAEDA

Detection Delay

Different number of sensor nodes are selected in the same area to execute the proposed M-STAEDA, BN-only, and simple threshold approach. As shown in Fig. 4, with the increase of the number of sensor nodes, the average delay of the three approaches significantly increase. The average delay time of the proposed M-STAEDA is smaller than that of BN-only and simple threshold one. The results show that the communication overhead increases with the increase of the number of nodes. However, the proposed M-STAEDA approach exhibits a slower increase than other two approaches. The reason is that Rule K algorithm is adopted to select part of nodes as backbone

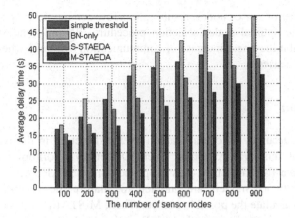

Fig. 4. Comparison of average delay time between M-STAEDA and S-STAEDA

nodes for forwarding the sensed data, which can greatly reduce the communication overhead. Therefore, the proposed M-STAEDA can achieve shorter average delay for contamination event detection. While BN-only and simple threshold approaches need to collect all of the sensed data, resulting in a large communication overhead. The response time of event detection is longer than that of the proposed M-STAEDA approach.

Scalability

The communication complexity of the proposed M-STAEDA approach depends on the local transition of the sensed values. Detecting temporal events in the IDLE state requires no communication overhead, because the nodes run the algorithm individually. Whenever, a temporal event is detected by the backbone nodes, they broadcast the corresponding messages and collect the sensed data. In TMPE state, the communication overhead and the message transmission increase when the number of sensor nodes increases. From Fig. 5, the proposed M-STAEDA approach can achieve better performance on data transimssion. The reason is that Rule K algorithm can optimally

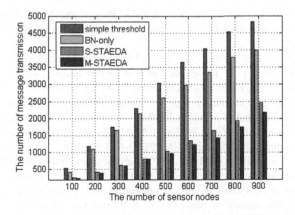

Fig. 5. Comparison of transmission overhead between M-STAEDA and S-STAEDA

reduce the number of sensor nodes, resulting in a low communication overhead. On the contrary, without optimal nodes selection procedure, BN-only and simple threshold approaches exhibit the linear increase of communication overhead when the number of sensor nodes increases.

6 Conclusion

To improve the accuracy of contamination events detection in the water supply networks, a Spatial-Temporal Abnormal Event Detection Algorithm with Multivariate water quality data (M-STAEDA) was proposed. In M-STAEDA, first, Back Propagation neural network models are adopted to analyze the multiple water quality parameters and calculate the possible outliers. Then, M-STAEDA algorithm determines the potential contamination events through Bayesian sequential analysis to estimate the probability of a contamination event. Third, it can make decision based on the multiple event probabilities fusion. Finally, a spatial correlation model is applied to determine the spatial-temporal contamination event in the water supply networks. The experimental results indicate that the proposed M-STAEDA algorithm can obtain more accuracy with BP neural network model and improve the rate of detection and the false alarm rate, compared with the temporal event detection of Single Variate Temporal Abnormal Event Detection Algorithm (M-STAEDA).

Acknowledgement. This study was supported by the National Key Technology Research and Development Program of the Ministry of Science and Technology of China under Grant Nos. 2013BAB06B04, 2016YFC0400910, 2017ZX07104-001; the Fundamental Research Funds for the Central Universities under Grant No. 2015B22214.

References

1. Heidemann, J., Stojanovic, M., Zorzi, M.: Underwater sensor networks: applications, advances and challenges. Philos. Trans. Roy. Soc. A: Math. Phys. Eng. Sci. **370**, 158–175 (2012)
2. Eliades, D.G., Lambrou, T.P., Panayiotou, C.G., Polycarpou, M.M.: Contamination event detection in water distribution systems using a model-based Approach. Procedia Eng. **89**, 1089–1096 (2014)
3. Yang, J.Y., Haught, C.R., Goodrich, A.J.: Real-time contaminant detection and classification in a drinking water pipe using conventional water quality sensors: techniques and experimental results. J. Environ. Manag. **90**(8), 2494–2506 (2009)
4. Huang, T., Ma, X., Ji, X., Tang, S.: online detecting spreading events with the spatio-temporal relationship in water distribution networks. In: Motoda, H., Wu, Z., Cao, L., Zaiane, O., Yao, M., Wang, W. (eds.) ADMA 2013. LNCS, vol. 8346, pp. 145–156. Springer, Heidelberg (2013). doi:10.1007/978-3-642-53914-5_13
5. Stotey, M.V., Gaag, B.V.D., Burns, B.P.: Advances in on-line drinking water quality monitoring and early warning systems. Water Res. **45**(2), 741–747 (2011)
6. Yim, S.J., Choi, Y.H.: Fault-tolerant event detection using two thresholds in wireless sensor networks. In: Proceedings of the 15th IEEE Pacific Rim International Symposium on Dependable Computing. Picataway, pp. 331–335. IEEE (2009)

7. Xue, W., Luo, Q., Wu, H.: Pattern-based event detection in sensor networks. Distrib. Parallel Databases **30**(1), 27–62 (2012)
8. Byer, D., Carlson, K.H.: Expanded summary: real-time detection of intentional chemical contamination in the distribution system. J. Am. Water Works Assoc. **97**(7), 130–133 (2005)
9. Wang, X.R., Lizier, J.T., Obst, O., Prokopenko, M., Wang, P.: Spatiotemporal anomaly detection in gas monitoring sensor networks. In: Verdone, R. (ed.) EWSN 2008. LNCS, vol. 4913, pp. 90–105. Springer, Heidelberg (2008). doi:10.1007/978-3-540-77690-1_6
10. Uusital, L.: Advantages and challenges of Bayesian networks in environmental modelling. Ecol. Model. **203**(3), 312–318 (2014)
11. Piao, D., Menon, P.G., Mengshoel, O.J.: Computing probabilistic optical flow using markov random fields. In: Zhang, Y.J., Tavares, J.M.R.S. (eds.) CompIMAGE 2014. LNCS, vol. 8641, pp. 241–247. Springer, Cham (2014). doi:10.1007/978-3-319-09994-1_22
12. Hou, D., Chen, Y., Zhao, H., et al.: Based on RBF neural network and wavelet analysis the water quality of anomaly detection method. Transducer Microsyst. Technol. **32**(2), 138–141 (2013)
13. Peremlan, L., Ostfeld, A.: Bayesian networks for source intrusion detection. J. Water Resour. Plan. Manag. **139**(4), 426–432 (2012)
14. Byer, D., Carlson, K.: Expanded summary: real-time detection of intentional chemical contamination in the distribution system. J. Am. Water Works Assoc. **97**, 130–133 (2005)
15. Jarrett, R., Robinson, G., O'Halloran, R.: On-line monitoring of water distribution systems: data processing and anomaly detection. In: Proceedings of Water Distribution Systems Analysis Symposium, Cincinnati, Ohio, USA (2006)
16. Kroll, D., King, K.: Laboratory and flow loop validation and testing of the operational effectiveness of an on-line security platform for the water distribution system. In: Proceedings of Water Distribution Systems Analysis Symposium, pp. 1–16 (2006)
17. Perelman, L., Arad, J., Housh, M., Ostfeld, A.: Event detection in water distribution systems from multivariate water quality time series. Environ. Sci. Technol. **46**, 8212–8219 (2012)
18. Arad, J., Housh, M., Perelman, L., Ostfeld, A.: A dynamic thresholds scheme for contaminant event detection in water distribution systems. Water Res. **47**, 1899–1908 (2013)
19. Hart, D., McKenna, S.: CANARY User's Manual, 4.1 edn., National Security Applications Dept., Sandia National Laboratories, Albuquerque, NM 87185-0735 (2009)
20. Murray, R., Haxton, T., Janke, R., Hart, W.E., Berry, J., Phillips, C.: Water quality event detection systems for drinking water contamination warning systems—development, testing, and application of CANARY. Technical report, EPA/600/R-10/036 U.S. EPA (2010)
21. Murray, S., Ghazali, M., McBean, E.A.: Real-time water quality monitoring: assessment of multisensor data using Bayesian belief networks. J. Water Res. Plan. Manag. **138**, 63–70 (2011)
22. Kong, Y.H., Jing, M.L.: Classification method based on confusion matrix and the integrated learning research. Comput. Eng. Sci **34**(6), 111–117 (2012)
23. Murray, R., Haxton, T., et al: Water quality event detection systems for drinking water contamination warning systems: development testing and application of CANARY. US Environmental Protections Agency (2010)
24. Mckenna, S.A., Klise, K.A.: Multivariate applications for detecting anomalous water quality. American Society of Civil Engineers (247) (2010)
25. Mckenna, S.A., Wilson, M., Klise, K.A.: Detecting Changes in Water Quality Data. J. Am. Water Works Assoc. **77**(1), 74–85 (2008)
26. Jayaraman, P.P., Yavari, A., Georgakopoulos, D., Morshed, A., Zaslavsky, A.: Internet of things platform for smart farming: experiences and lessons learnt. Sensors **16**(11), 1884 (2016)

Context-Aware Technology of Disabled Health Service for Intelligent Community

Yao Tan[1] and Wenbi Rao[1,2]([✉])

[1] School of Computer Science and Technology,
Wuhan University of Technology, Wuhan, China
tanyao_wuhan@163.com, wbrao@whut.edu.cn
[2] Hubei Key Laboratory of Transportation Internet of Things, Wuhan, China

Abstract. With the rapid development of Ubiquitous computing, context-aware technology as one of the core contents of the former has made a series of research progress and achievements. Context-aware technology can automatically provide the corresponding services according to the strategy given by the inference engine through sensing the related context of the environment and tasks. It is because context-aware technology can significantly improve the intelligence of computer interaction, the application of this technology to "Intelligent community" which is a new concept built on the highly developed Internet and sensor networks has become more promising and meaningful. As an important part of the intelligent community, the technology is also of great use for the disabled health service.

Based on the research of the theory of context-aware technology and the examples of international development applications, this paper designs an ontology based context-aware system framework named CA-DHS for disabled health service. The hierarchy based framework uses ontology to standardize and formalize context information and complete context modeling. This framework includes the collection and encapsulation of the original context information and constructs a formal reasoning model. Moreover, the framework provides query and subscription services as the external interface of the system. On the basis of this framework, the DHS-Protosystem (Disabled Health Service Protosystem) is designed and implemented. This paper introduces implementation of query and subscription service module.

Keywords: Context-aware computing · Disabled health service · CA-DHS · Ontology modeling · DHS-Protosystem

1 Introduction

With the continuous progress of the times and the urgent requirement of social development, China has entered a new period of rapid urbanization. "Intelligent community" is a new concept which built on the highly developed Internet and

© Springer Nature Singapore Pte Ltd. 2017
B. Zou et al. (Eds.): ICPCSEE 2017, Part I, CCIS 727, pp. 646–659, 2017.
DOI: 10.1007/978-981-10-6385-5_54

sensor networks, and it constructs the intelligent environment for the development of urban community. "Intelligent community" is the necessary prerequisite
for the construction of smart city and reflects the city's intelligence level [1].

As an important part of the intelligent community, the disabled health service is closely related to the universal medical technology and the wireless sensor network technology. Under the support of universal health care technology,
the disabled health service system can strengthen information communication,
improve the level of medical services and perfect the function of the intelligent
community.

A great deal of research work in the field of health care has been carried out
around the world, the following are the main items:

A prototype system named BlueHospital for the detection of patients with
heart disease is studied at the University of technology in Valencia, Spain. The
system uses the patient's heart rate sensor and positioning technology to collect
data and send it to the central database for doctors to make diagnosis and
decision making [2].

Researchers at the University of California at Berkeley and Intel developed
a home care sensor network system for SIDS (sudden infant death syndrome):
SleepSafe. In the system, the acceleration sensor is integrated in the baby's infant
baby cloth and all kinds of action information of babys will be sent to the base
station. If there is danger of the attitude, the system will send a text message
to parents forwardly [3].

With the development of context-aware technology, health care has begun to
introduce this technology. For example, Georgia Institute of Technology's Context Toolkit project, aims at developing a series of context-aware basic application tools [4]. And in 2010, Pam of Nokia company proposed a method for
providing context aware and fusion and designed the corresponding system [5].

At present, there have a lot of research of universal health care about medical
health care at home and abroad, but research of the context awareness technology
for the disable health service are rare.

In view of the existing problems about health state monitoring and universal
health care, the paper designs and implements the "Intelligent Community Disabled Health Service System". The key technology is context-aware computing
about user's location context and physiological state context information collected by intelligent terminal in pervasive environment. The system can monitor
the physiological state and location information of the disabled in real-time by
the intelligent terminal. The aims of the system are to provide better medical
service for the disabled within the community, and improve the treatment level
for emergency.

2   Context Representation and Modeling

Context-aware computing should be based on a reasonable context model. Context modeling is an important prerequisite for context-aware computing [6,7].
The modeling method is directly related to the inference of context information.

Context reasoning is the key to the feasibility of the framework. In summary, the construction of context modeling is the most important step in the research of context-aware system.

Contrast to other context representation, the framework uses ontology to model the context. The context ontology of health services for the disabled will be designed and a software for ontology editing and knowledge acquisition named Protégé will be used to construct the corresponding ontology.

2.1 Design of Context Ontology

The context information collected by the bottom sensor can be divided into two categories: position information and physiological state information. Physiological state context include four physiological state information: Bp (Blood pressure), Hr (Heart rate), Br (breath rate) and SpO2 (Pulse Oxygen Saturation). In order to reduce the complexity of context modeling for the disabled health service, this paper defines a seven tuple to standardize and formalize context information: DHS_Context = (userId, position, Bp, Hr, Br, SpO2, time). Based on context information and personalized service which the system needs to provide, we design five ontologies in the system, which are user, device, health, position and service. The five ontologies are described as follows:

(1) User ontology can be divided into three categories, namely, the disabled users, community healthcare workers and system administrators, and the disabled users are the main objects of system services
(2) Device ontology is composed of some sensors, location devices, PDA (Personal Digital Assistant) and servers
(3) Health ontology includes four physiological states, namely, blood pressure, heart rate, breath rate and Sp02 (Pulse Oxygen Saturation). It should be mentioned that two intermediate states (syndrome diagnostic element and function diagnostic element) are introduced when the system reason about health ontology. Syndrome diagnostic element and function diagnostic element are two indexes that can respectively reflect basic signs of the human body and whether the main physiological functions are normal.
(4) Position ontology includes two categories: indoor and outdoor
(5) Service ontology includes active service and alert. The active service include providing physiological monitoring, health tips, community medical appointments and other services for the disabled

In the context model of health services for the disabled, there are many specific relationships between the five ontologies. Figure 1 describes the relationship between the various ontologies in the system and their properties.

2.2 Implementation of Context Ontology Modeling Based on Protégé

Protégé is mainly used to create ontology in Semantic Web and it makes users only need to focus on the creation of the ontology model at the conceptual level, but not to take into account the description language of the ontology [8].

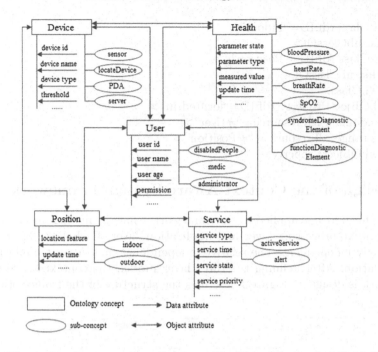

Fig. 1. Ontology structure diagram of health service for the disabled

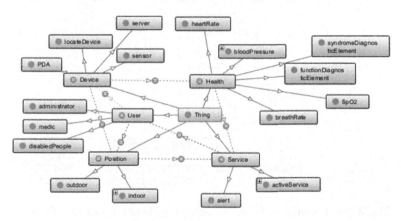

Fig. 2. Ontology structure generated by Protégé

According to the ontology structure shown in Fig. 1, we can generate the corresponding ontology class and ontology structure in Protégé. Figure 2 describes the ontology structure generated by Protégé.

As can be seen from Fig. 2, the five ontology classes have their own subclasses and especially there are various attributes between them. It is these attributes that demonstrate the advantages of using ontology to represent context. The following is a partial OWL code for describing the disabled context ontology model:

```
<owl:Class rdf:ID="Indoor">
<rdfs:subClassOf>
<owl:Class rdf:ID="Position"/>
</rdfs:subClassOf>
</owl:Class>
<owl:ObjectProperty rdf:ID="located_in">
<rdfs:domain rdf:resource="#User">
<rdfs:range rdf:resource="#Position">
</owl:ObjectProperty>
```

3 Design of the Context-Aware System Framework

A hierarchical idea is applied to the framework which is divided into perception layer, context processing layer and application layer. Each layer of the three is relatively independent. The relative independence make easy modularity and encapsulation. After defining the three layer model, the context-aware system framework is designed. Figure 3 describes the structure of the framework.

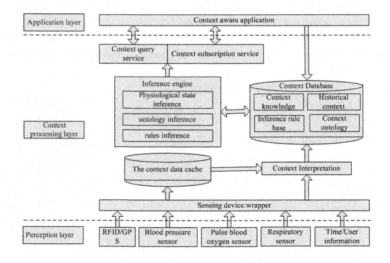

Fig. 3. Context-aware system framework of health service for disabled

In the system framework, the perception layer is responsible for collecting and encapsulating the context information. The encapsulated information will be provided by sensor wrappers to the context processing layer. In order to solve the speed mismatch problem between data collection and data usage, the recently collected data will be stored in the context data cache. The context processing layer uses the data provided by the perception layer, combining with the knowledge of the context knowledge base, to reason the high-level context. Applications in application layer can choose context subscription service or context query service according to different demand patterns.

3.1   Collection and Encapsulation of Original Context Information

Due to different types of sensors produced by different manufacturers and different formats of physiological information and location information collected by sensors, the underlying sensor data needs to be encapsulated. Encapsulation makes the upper layer of the system shield the difference between the hardware device and the data format. This paper designs a sensor wrapper to achieve encapsulation of sensor data and external system.

As the smallest unit of service, each user (the disabled) corresponds to a wrapper which encapsulates location information and physiological state information. These data is defined as data source, each of which includes some data nodes, namely, basic sensing equipment. Data source and data nodes have a unified description format. Figure 4 describes the main structure of wrapper.

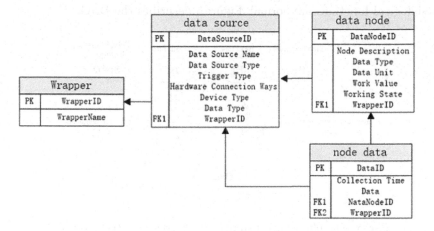

Fig. 4. Database model diagram of wrapper

Considering the speed mismatch problem between data collection and data usage, a context data cache is designed. Because the collected data is always processed sequentially, queue in data structure is suitable. However, this system need to collect data frequently so that application and release of resources in queue will take up a lot of system processing time. Therefore, cyclic queue is used to implement context data cache. It is the structure of cyclic queue that can reduce effectively application and release storage operation when context data is collected and used.

3.2   Context Reasoning

In CA-DHS, the context reasoning of health service for the disabled is divided into physiological state reasoning, ontology reasoning and rule reasoning. The physiological state reasoning is used to judge whether the disabled people are healthy according to the collected physiological state data. Ontology reasoning is a simple reasoning based on the logical relationship between ontologies. Rule reasoning is established on the various contexts and the reasoning rules defined in the rule base.

Due to the physiological state of the human body has certain uncertainty, this paper introduce bayesian network to reason physiological state context for disabled. Bayesian network is based on bayesian probability theory and is widely used in the reasoning of uncertain context [9]. A Bayesian network consists of DAG (Directed Acyclic Graph) and CPT (Conditional Probability Table). This paper, combined with domain expert guidance and medical knowledge, contructs a DAG of bayesian network for physiological state context reasoning. In the DAG, Sd (syndrome diagnostic element) reasoned from Bp and Hr is a index that can reflect basic signs of the human body. And Fd (function diagnostic element) reasoned from Hr, Br and SpO2 is a index that can reflect whether the main physiological functions are normal. Figure 5 describes the DAG.

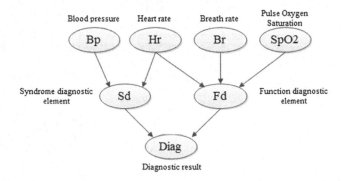

Fig. 5. The physiological state of the Bayesian network diagram

In addition, according to the expert knowledge and statistical data, this paper makes the CPT of bayesian network. The four basic physiological indexes (Bp, Hr, Br, SpO2) in the CPT have 3 values: regular, abnormal and dangerous. Sd, Fd in the CPT have 2 values: regular and dangerous. Due to limited space, this paper only uses the conditional probability table of Fd to explain the CPT of the Bayesian network. Table 1 shows a part of the CPT.

The Bayesian network constructed in this paper returns a probability value of the final diagnosis (Diag) after reasoning. This probability value reflects the health condition of the disabled. Through a large number of simulation experiments, the probability value of Diag is defined as follows:

$P(Diag) \in (0, 0.25]$ indicates that The disabled are in good condition and do nothing;

$P(Diag) \in (0.25, 0.45]$ indicates that the disabled are in poor state, and the corresponding prompts need to be made;

$P(Diag) \in (0.45, 1]$ indicates that the disabled are in danger, and help is urgently needed.

Table 1. CPT of function diagnostic element

Hr	Br	SpO$_2$	P (Fd = regular)	P (Fd = dangerous)
Regular	Regular	Regular	0.95	0.05
Abnormal	Regular	Regular	0.75	0.25
Dangerous	Regular	Regular	0.4	0.6
Regular	Abnormal	Regular	0.8	0.2
Abnormal	Abnormal	Regular	0.6	0.4
......				
Regular	Dangerous	Dangerous	0.25	0.75
Abnormal	Dangerous	Dangerous	0.1	0.9
Dangerous	Dangerous	Dangerous	0.01	0.99

In order to complete ontology reasoning and rule reasoning, this paper introduce Jena which is an open source Java semantic toolkit developed by Hewlett Packard for the Semantic Web research project. Jena provides some functions and interfaces to analysis, reason and query ontology [10].

The following two example are given to describe the ontology reasoning and rule reasoning:

(1) The system gets this information that the user Li whose ID is DP0001 is located in Room 201 of department of orthopedics. And then Jena reads this information and begins ontology reasoning. The output results after reasoning is [*Li, locate_in, communityHosipital*]. Li is in Room 201 of department of orthopedics, and the department of orthopedics is a part of the community hospital. Therefore the system judges that Li is in the community hospital.
(2) There is a rule that describes a disabled person with hypertension whose blood pressure is in abnormal state at night. The formal definition of the rules is (disabledPeople hasDisease? hypertension) & (time ? evening) & (bloodPressure hasStatus? abnormal) -> (activeService provideTo disabledPeople). The result of reasoning shows that the system will prompt the user to take the medicine on time, which is consistent with the actual situation.

3.3   Context Application Services

Context application services include context query services and context subscription services. The service module is directly connected to the application layer and is responsible for the request of the corresponding context application.

The context query services module analyzes query request, calls the corresponding query method according to the specific content and finally returns the query results in a uniform format to the context application. Figure 6 shows implementation class diagram of the query module.

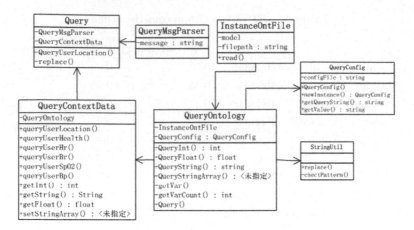

Fig. 6. Query service implementation class diagram

The specific description of the query process is as follows:

(1) The context application sends a query command to the context query service module;

(2) Create a QueryOperator object, and call the QueryContextData class to perform the query operation;

(3) The QueryContextData object gets the query type of context information by calling the typeParser method of the QueryMsgParser class;

(4) Call the getSPARQL method (this method calls the QueryConfig class to read the configuration file) of the QueryOntology class to get the SPARQL query statement;

(5) Call the getModel method of the InstanceOntFile class, and import the model instance in the SPARQL statement from the ontology instance database;

(6) The QueryContextData object creates the QueryExecution object to perform the query operation, and returns the query result to the context application.

The context subscription service module processes and saves the subscription requests sent by the application in order to send the latest, user interested context information to the application in time. The main mechanism of subscription service is to receive the subscription request from the context application, and the two is to judge whether the trigger condition of the event has been satisfied (if the condition is satisfied, the trigger message is sent to the application). Figure 7 shows implementation class diagram of the subscription module.

The specific description of the subscription process is as follows:

(1) The context application sends a subscription request to the context subscription service module;

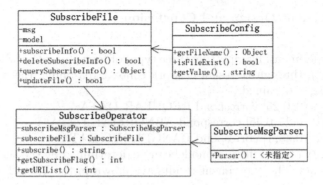

Fig. 7. Subscription service implementation class diagram

(2) After the subscription service request is accepted by the context subscription service module, the module puts the request into the message queue in the first in first out sequence, and call the SubscribeOperator class to process messages extracted from the queue;

(3) The SubscribeOperator class calls the querySubscribeInfo method of the SubscribeFile class to query subscription file library. This process needs to call the getFileName method of the SubscribeConfig class to obtain the subscription file corresponding to the context information that the application subscribes to, and then call the updateFile method of the SubscribeConfig class to update the subscription file;

(4) After the subscription result is generated, the result is returned to the context application successively.

4 Experiment

4.1 Experiment Data

Because the prototype system designed in this paper has been applied to the local nursing home, we have used the medical data of three years in the nursing home and the doctor's diagnosis as the data set to verify the validity of the prototype system. The dataset is divided into 3 types based on the diagnostic results of doctors:

(1) REQULAR_DATA: Four physiological indexes (Bp, Hr, Br, SpO2) under normal condition;

(2) ABNORMAL_DATA: Four physiological indexes (Bp, Hr, Br, SpO2) under abnormal condition;

(3) DANGEROUS_DATA: Four physiological indexes (Bp, Hr, Br, SpO2) under dangerous condition;

4.2   Experiment Design and Conclusion

Experiment 1:

In this experiment, by inputting three types of four physiological indexes (Bp, Hr, Br, SpO2), through Bayesian network context reasoning, the probability of diagnosis Diag is obtained.

$P(Diag) \in (0, 0.25]$ correspond REQULAR_DATA;

$P(Diag) \in (0.25, 0.45]$ correspond ABNORMAL_DATA;

$P(Diag) \in (0.45, 1]$ correspond DANGEROUS_DATA;

The validity of the inference model can be determined by the accuracy of the diagnosis results. The experiment results are shown in the Table 2.

Table 2. Results of experiment 1

Data type	Data size	Success number	Accuracy rate
REQULAR_DATA	308	264	0.857
ABNORMAL_DATA	500	462	0.924
DANGEROUS_DATA	285	273	0.958
Total	1093	999	0.914

The experimental data and the doctor's evaluation show that the reasoning model of this system has higher accuracy to some extent, and can solve the problem of health monitoring and early warning better.

Experiment 2:

This experiment tests the query and subscription function modules by the actual data. The data is the real data of the real user in the system.

For query service, we want to query the real-time information of physiological status and the location of the disabled. When the user is in an indoor environment, indoor RFID reader reads the user label and get his/her position through the context ontology reasoning. And then the building where the user is located is marked on the map. In addition, when the user is located in an outdoor environment, the system will automatically switch to the location information provided by the GPS receiver. Figure 8 describes the result interface of query service.

The current position description and health status of the disabled can be got by selecting and clicking on the user's location markers. In addition, the real-time curve of physiological index can also be shown in the system through selecting. For example, when oxygen saturation is chosen, the real-time state of curve, the maximum, the minimum and the average values of the recent indicators are obtained. Figure 9 describes the real-time curve of oxygen saturation.

At the time of the test, a subscription service for the system was launched when the disabled person's physiological status was abnormal. Therefore, when the human body's heart rate, respiratory frequency has a obvious change, the subscription service will be launched by the system. If the health status is judged

Fig. 8. Result interface of query service

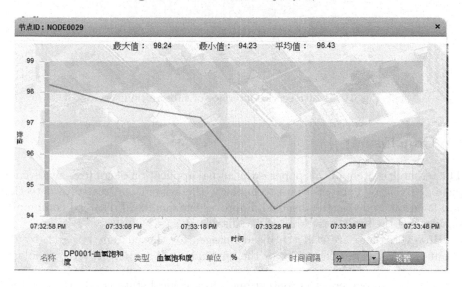

Fig. 9. Real-time curve of oxygen saturation

as "abnormal" through the reasoning of physiological state, the corresponding prompts will be sent. If the health status is judged as "dangerous", the rescue alarm will be sent. Figure 10 shows the health status and the value of the indicators, and gives the corresponding suggestions according to the current state of the user.

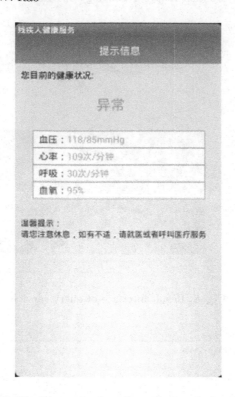

Fig. 10. Health service interface of subscription service

5   Conclusion

The research content of this paper is the actual application of context-aware computing. By using context-aware technology, the system realizes state detection and location aware for the disabled, improves the quality of disabled community service and the healthcare level of the disabled.

However, there are still the disadvantages for the current research results about the disable health service or the application of context-aware technology:

(1) The way to perceive the context and physiological state is relatively simple;
(2) Lack of information exchange and data sharing with health monitoring center;
(3) The application of context aware technology to the health service for the disabled is relatively short, and there is not yet a mature application theory for the inference of context information

References

1. Wang, J., Gao, B.: On the concept and application of intelligent community 11 (2012)
2. Cano, J.-C., Calafate, C.T., Manzoni, P.: Building a research prototype to provide pervasive services in hospiatls. In: Proceeding of 3rd International Symposium on Wireless Pervasive Computing, May 2008
3. Baker, C.R., Armijo, K., Belka, S., Benhabib, M., et al.: Wireless sensor networks for home health care. In: Proceedings of 21st International Conference on Advanced Information Networking and Applecations Workshops, 2007, vol. 2, pp. 832–837, May 2007
4. Dey, A.K., Abowd, G.D., Salber, D.: A conceptual framework and a toolkit for supporting the rapid prototyping of context-aware applications. Hum.-Comput. Interact. 0737–0024
5. Pam, R.A., Eronen, A., Leppanen, J.A.: Method and apparatus for providing context aware and fusion. Finland Nokia, May 2010
6. Zhang, L.: Research and application of ontology based context aware computing. Wuhan University of Technology, May 2011
7. Strang, T., Linnhoff-Popien, C.: A context modeling survey. In: First International Workshop on Advanced Context Modeling, Reasoning and Management, UbiComp (2004)
8. Yao, Z., Chen, X.: Problem solving evolutionary method for ontology knowledge representation with protg-2000. In: The Ninth International Conference on Electronic Business, November 2009
9. Friedman, N., Geiger, D., Goldszmidt, M.: Bayesian network classifiers. Mach. Learn. **29**, 131–163 (1997)
10. Jena 2A Semantic Web Framework for Java [EB/OL], May 2006. http://jena.sourceforge.net/index.html

DFDVis: A Visual Analytics System for Understanding the Semantics of Data Flow Diagram

Hao Xiong, Haocheng Zhang, Xiaoju Dong[✉], Lingxi Meng,
and Wenyang Zhao

Department of Computer Science and Engineering,
Shanghai Jiao Tong University, Shanghai 200240, China
xjdong@sjtu.edu.cn

Abstract. Data flow diagram (DFD), as a special kind of data, is a design artifact in both requirement analysis and structured analysis in software development. However, rigorous analysis of DFD requires a formal semantics. Formal representation of DFD and its formal semantics will help to reduce inconsistencies and confusion. The logical structure of DFD can be described using formalism of Calculus of Communicating System (CCS). With a finite number of states based on CCS, state space methods will help a lot in analysis and verification of the behavior of the systems. But the number of states of even a relatively small system is often very great that is called state explosion. In this paper, we present a visual system which combines Formal methods and visualization techniques so as to help the researchers to understand and analyze the system described by the DFD regardless of the problem of state explosion.

Keywords: Data flow diagram (DFD) · Calculus of Communicating Systems (CCS) · State space · Visualization techniques

1 Introduction

Data flow diagram (DFD) is a visual tool to depict logic models and express data transformation from input to output in a system [1]. It has been widely used in both requirement analysis and structured analysis in software development. Now it is still a very useful tool in computer field. For example, Tensorflow, a popular open source software library for machine learning developed by Google is based on DFD [2]. However, notations used in the DFD are usually graphical and different tools and practitioners interpret their notations differently that will make DFD ambiguous.

DFD formalism can help to reduce such inconsistencies and confusion. Formal methods are a kind of mathematically based techniques for the specification and verification of software and hardware systems in computer science [3]. There are a variety of formal methods [4]. The Calculus of Communicating System (CCS) is one of them which can represent DFD formally [5].

Automated tools can help to generate the formal description of DFD, get the state transition graph and even find the deadlocked states in the system [6]. When we get a

B. Zou et al. (Eds.): ICPCSEE 2017, Part I, CCIS 727, pp. 660–673, 2017.
DOI: 10.1007/978-981-10-6385-5_55

finite number of states based on CCS, state space methods will help a lot in analysis and verification of the behavior of the systems [7]. However, the number of states of even a relatively small system is often very large: state explosion [8]. Since the number of states in the state space is so large, researchers will find it difficult to deal with the problems such as how these states are organized, what the relationship between a single state and the state space is and how to measure the importance of a state.

The innovation of this paper is using well-designed visualization methods instead of simply reducing some states to solve this problem. In this paper, we combine formal methods and visualization techniques so as to make the researchers better understand and analyze the system described by the data flow diagram (Fig. 1).

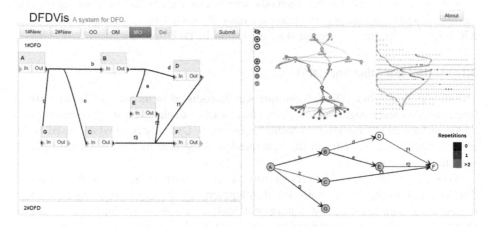

Fig. 1. An overview of the DFDVis

DFDVis is a web system which supports the drawing of the specified DFD. When the user completes the drawing, DFDVis will upload the relevant description to the server. The programs running on the server will automatically infer the semantics of DFD based on CCS, generate these semantic equations and simulate the behaviors of the system depicted in the DFD. At this stage we choose to mainly focus on a finite number of states based on CCS of the system. Since the number of states is very large in most cases, we define the creative concept of depth to classify these states to solve the problem of state positioning. The system presents the entire state space through an overview, and a view of its detail through a tree structure with dual focus [9].

Moreover, the interactive exploratory approach of this large state space will help analysts find the system deadlock problem then give improvement to solve this problem. In addition, we also define the innovative concept of task completion progress, in which each state always corresponds to a task progress of the system. There is always an action sequence from the initial state to a selected state, and each action in this action sequence corresponds to a data flow in the DFD. Schedule Graph intuitively describes the task progress of the system task which in correspondence with a state, and then reflects the importance of this state and its degree of redundancy. This will give the analysts a reference in understanding and analyzing the system design.

We first introduce the relevant background in Sect. 2, the concept of data flow diagram, formal methods and state space methods, as well as the visualization techniques we used will be introduced. Section 3 describes the design of our web system and some special methods we take to solve the problem of state explosion. In Sect. 4, two cases will demonstrate how DFDVis helps to understand and analyze system given by DFD. Section 5 concludes.

2 Background

The data flow diagram (DFD) technique is still useful even though it has been introduced for many years. Formal methods which can reduce the inconsistencies and confusion will help a lot in better understanding of DFD. Tree visualization techniques will help to solve the problem of state explosion in DFD formalism.

2.1 Data Flow Diagram (DFD)

The data flow diagram (DFD) technique was introduced in the late 1970s as a structured analysis and design method. As a visual tool to depict logic models and express data transformation from input to output in a system, DFD has been widely used in the software development soon. In the structured development method, it used to play a very important role.

Nowadays, DFD is also very useful in requirement analysis in research and development work. In the software development cycle, requirement specification, requirement understanding and requirement analysis are the most important activities. Proper Software development requires complete knowledge of the system that needs to be developed. DFD can describe and analyze system more precisely than any other artifact, especially in modeling of real time embedded system.

In addition, DFD is very important for the modernization of old legacy systems. Old systems that have become legacy systems today were modeled and documented using structured approach. When these old legacy systems demand numerous modifications and expansion in terms of functionality, the only viable solution to save legacy systems, cost effectively is to save system design and combine the old system design with the latest software development techniques. The old system design and DFD are closely related, thus the understanding of the DFD will be crucial.

DFD includes following main elements:

- →: Data stream. A data stream is a path through which data is propagated within the system. Since the data stream is data in the flow, it must have direction, and the data stream should be named with noun or noun phrase to support positioning.
- □: Data source or the end of the data stream. It can be an entity such as a human or other software system which inputs or receives data finally.
- ○: Data processing. Processing is a unit that processes data, it receives a certain input or more and then produces outputs through a specific process.

DFD includes a mechanism to model the data stream through these main elements. It supports decomposition to illustrate details of data flows and functions. However, notations used in the DFD are usually graphical and different tools and practitioners interpret their notations differently. This uncertainty makes DFD ambiguous.

Formal syntax is based on mathematical knowledge. It reduces inconsistency and helps the designers to automate the solution straightforwardly. At the same time, formalism will help in checking the syntactic and semantic inconsistencies. DFD formalism can help to reduce such inconsistencies and confusion, and will help a lot in better understanding the requirements and design.

2.2 Formalizing Data Flow Diagram

Formal methods are a particular kind of mathematically based techniques for the specification, development and verification of software and hardware systems in computer science, especially software engineering and hardware engineering. The use of formal methods for software and hardware design is motivated by the expectation that performing appropriate mathematical analysis can contribute to the reliability and robustness of a design. It can be described as the application of a fairly broad variety of theoretical computer science fundamentals, in particular formal languages, automata theory, and program semantics.

Formal methods can be applied at various points through the development process. DFD has been widely used in the software development and it also has the problem of ambiguity and uncertainty. We look forward to introducing formal methods in DFD to solve these problems. Motivation to formalize DFD is to reduce the chance of ambiguousness in requirement analysis phase. At the same time formal representation of DFD can also become beneficial in up-gradation and modernization of old legacy systems. There are some techniques that represent DFD formally or give formal semantics to the data flow diagram.

Butler and his partners discussed that logical structure of DFD can be described using formalism of Calculus of Communicating System (CCS) in their work [6]. They presented tools to extract information from data flow diagrams to support the reverse engineering. CCS provides a natural and intuitive model for concurrent systems. It focuses not on the activity within nodes of the system but rather on the interaction between nodes, which is precisely the purpose of a DFD. Moreover, they can construct a structure that consists of all states that a system can reach, and all transitions that the system can make between those states based on CCS. For any reasonable DFD, it is easy but tedious to generate corresponding CCS based logical structure. Fortunately, the Edinburgh Concurrency Workbench (CWB) can help us to generate the formal description of DFD automatically [10]. The CWB is an automated tool which caters for the manipulation and analysis of concurrent systems. When we give the semantic equations of a DFD in accordance with the rules, it is possible to analyze derivations of processes, get the transition graph of an agent and even find the deadlocked states with the workbench.

For the system described in the DFD is composed of a finite number of states based on CCS, we can achieve the goal of analysis and verification of the behavior of systems. State space methods are one of the most important approaches to computer-aided analysis and verification of the behavior of concurrent systems. In their basic form, they consist of enumerating the set of the states the system can ever reach.

Unfortunately, the number of states of even a relatively small system is often very great: state explosion. It has led many to believe that state space methods will never work well enough for practical use. The number of states of almost any system of interest is huge. During the last decade, many methods have been suggested to reduce the number of states [11]. Unfortunately, most state space reduction techniques sacrifice one or another of the advantages of state space methods.

2.3   Visualization of State Space

In the whole state space, the number of states will be very large. And the relationship between them will be very complicated. This requires us to use special visualization techniques to help to analyze the relationship and show local details.

Tree is an abstraction of reality relations, and the hierarchical data which can be represented by tree is very common in our life and work, including evolutionary tree, organization charts, file systems and websites documents [12]. As an important category in the information visualization, tree visualization is widely used in the production and scientific research. Tree visualization techniques are mostly based on nodelink-based representations. Nodelink representations depict relations between parent and child nodes by lines that connect the nodes.

Generally, there are two kinds of purposes in interaction of tree, including connecting associated object, rearranging and compacting object. In tree visualization, unfolding, scaling and enquiry are commonly used ways of interaction. Unfolding and scaling allow the user to focus on the important part, and enquiry can achieve quick localization of nodes. Moreover, linking and brushing are used to tag the interested part of the data.

Tree visualization has brought benefits to the communication of large information spaces, but they still have some drawbacks. For example, the manifestation pattern of tree visualization is restrained by the characteristic of tree structure, which incurs scalability problems when tree nodes and hierarchy increase, thus impacting the form of comparison and causing problem of node occlusion. Moreover, the context information may be lost in the process of scaling. To overcome these specific problems, the Bifocal Tree incorporates the advantages of using two smoothly integrated viewing areas [9]: one for maintaining the context and the other for displaying the region of interest, minimizing the problem of node occlusion and enhancing context perception.

Tree visualization techniques can be used to solve the problem of state explosion. Unfolding, scaling and enquiry can help to process a huge number of states, once the positioning problem of states is solved.

3   System Analysis

We develop a web system called DFDVis which combines the advantages of formal methods and visualization techniques to assist experienced researchers in understanding and analyzing system through state graph, and then avoid the problem of state explosion. Besides, DFDVis can also help users without experience of formal methods to discover deadlock in the system.

3.1   Definitions of DFD and Semantic Equations

There are a variety of definitions of DFD in the articles on software engineering. Although these definitions differ in detail, we can still extract the most general features of DFD:

- Node. A node is a process (or function), a data store, or a terminator.
- Arrow. An arrow represents a data flow. Data flow can be divided into three categories: one node to one node, one to multiple and multiple to one.
- All Nodes and arrows should be labeled.

We introduce a kind of semantics for DFD based on Calculus of Communicating Systems (CCS) in which each node of a DFD is associated with an agent and each arrow in a DFD is associated with communication between two agents.

There are four functions used to construct the logical structure of a DFD from its layout structure: sequence, or, composition, and restriction. The four functions are written as ".", "+", "|" and "\". We can use formalism of CCS to describe the logical structure of DFD. For instance, we use the DFD of Fig. 2 to show the DFD and its formalized logical structure. And we can get these semantic equations (Table 1):

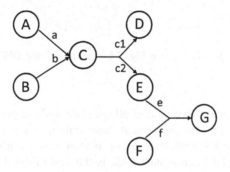

Fig. 2. A simple data flow diagram

Table 1. The semantic equations of the DFD of Fig. 2

The semantic equations
| ```
agent A=input1.'a.A;
agent B=input2.'b.B;
agent C=a.'c1.'c2.C+a.'c2.'c1.C+b.'c1.'c2.C+b.'c2.'c1.C;
agent D=c1.'output1.D;
agent F=input3.'f.F;
agent E=c2.'e.E;
agent G=e.f.'output2.G+f.e.'output2.G;
agent DFD=(A|B|C|D|F|E|G)\{a,b,c1,c2,e,f};
``` |
```

Since the syntax rules are fixed, we can make an automated tool to generate these semantic equations. These semantic equations can be used as the input of CWB. The DFD can be drawn by the users with this web system as shown in Fig. 3. When users submit the DFD, the description of this DFD's structure and a series of commands by CWB will be sent to the server. Our automated tool and CWB which is running on the server will complete all the formal analysis work and return the results, such as all the states and the exception of the system.

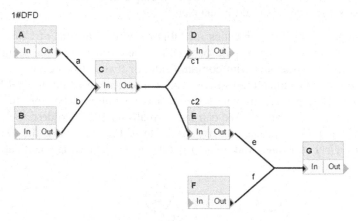

Fig. 3. The DFD in Fig. 2 is drawn with the DFDVis

3.2   State Graph

After receiving the formal expression of all the state nodes and their connections from the server, we need to process and present these state nodes appropriately. For a finite state system describing a specific task, all of its state nodes can form a directed graph. In the absence of deadlock, each state node is the end of the data stream initiated by $M(M \geq 1)$ state nodes and is the starting point of the $N(N \geq 1)$ data streams which point to the other N state nodes, and that is how the states of the state space are linked.

In the case of large number of states, which is called state explosion, there is no way to fully present all the states and their relationship. However, in our tree structure, we can always choose a state node as the core, come up and down to a limited level of relationship, and our tree structure provides the choice of the amount of the upper and lower relationship that can be displayed. On the other hand, within this limited hierarchy, not all state nodes are associated with the current state of interest. For example, since the post-state node of a previous state node of the current node in the current focus is not in the state transition path of the current state of interest, it will be folded to save space. The state node which is outside the level will be folded, so is the state node that is not on the state transition path of current node, besides, the state nodes associated with these folded state nodes will be displayed differently.

However, a large number of states coupled with the hiding of the vast majority of the states cause the requirement for method to present the relationship between two states that have to go through multiple actions and solve the problem of locating a state in a large state space.

We can classify the state in order to quickly locate a state. We use the depth to classify the state, in which the state with the same depth will be classified as the same class. And depth refers to the number of actions of the shortest action sequence from the initial state to this state. For a directed graph with loops, we can take the loop-avoid-Breadth-First-Search (BFS) algorithm and sort all the states by depth in ascending order, until traversing all the state nodes (Figs. 4 and 5).

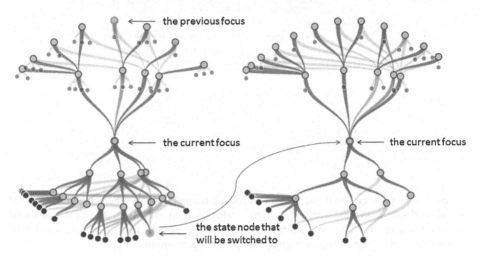

Fig. 4. The tree structure on the left shows the previous focus, the current focus, and the focus to be switched to, and once the switch is confirmed, it will become the tree structure on the right. (The circle with orange border represents the state node in the previous focus, The circle with red border represents the state node in the current focus. The purple circle represents the state node that will be switched to.) (Color figure online)

In order to avoid the disadvantages of folding in the tree structure, we can provide an overview of the entire state space in which the states will be arranged according to the category that has been classified. Moreover, each state node in this overview will be presented in the same form as the presentation in the tree, in which the state node where the focus is, the state nodes which are folded, and the hidden state nodes will be presented differently.

The state transition path is another feature that we must pay attention to when analyzing the state diagram. The state transition path indicates all intermediate states and all actions to be passed from one state to another, and each action is actually in correspondence with a data flow.

Since we want to explore the relationship between two states and their transfer path, a focus is not sufficient, thus we need to keep the focus on the previous focus in the interactive operation and present the relationship between the state node in the previous focus and the state node in the current focus. The Bifocal Tree gives us a reference (or inspiration), however, when the distance from the previous focus to the current focus exceeds the number of display layers selected by the user, it will still be folded in the

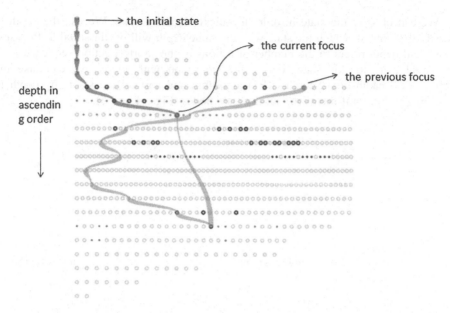

Fig. 5. This is the overview of the state space, and it intuitively displays the depth distribution of the unfolded state in the tree structure. And two state transition paths from the initial state to the current focus state (in blue color) and from the state of the previous focus to the state under the current focus (in orange color) are highlighted. (Color figure online)

tree structure, but in the overview, the previous focus can be positioned and there is a complete path display in the overview.

3.3   Schedule Graph

In addition to State Graph, Schedule Graph will also help us understand the relationship between the various processes of the system described in DFD and between each state in State Graph. Each state must be the end point of an initial state after a series of actions, and these action sequences are the processing steps of the completed task in the current system. Each processing step in Schedule Graph corresponds to a data flow in the DFD. However, in a parallel system, a process (or function) can continue to work, although the task assigned to it has been completed, thus for a specific task from the input to the given output, many actions in the action sequence are not helpful to task schedule, and the state after these actions is also independent of the specific task.

Schedule Graph presents the task completion schedule corresponding to the current state based on the analysis of the sequence actions from the initial state to the current state. Moreover, the process that has been completed, the process that has been completed for many times, the data flow that has been delivered once and the data flow that has been delivered for many times will be displayed differently. Therefore, the user can quickly perceive the progress of the current task (Fig. 6).

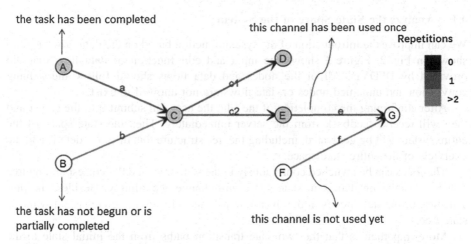

Fig. 6. The gray node indicates the task of this process (or function) has been completed, while the task which has not begun or been partially completed is represented by the white node. Meanwhile, the different colors of the data stream will represent the different number of times this channel is used. (Color figure online)

3.4 Interaction

Schedule Graph presents the task completion schedule corresponding to the current state based on the analysis of the sequence actions from the initial state to the current state. Moreover, the process that has been completed, the process that has been completed for many times, the data flow that has been delivered once and the data flow that has been delivered for many times will be displayed differently. Therefore, the user can quickly perceive the progress of the current task.

Among these multiple views, the link between State Graph and Schedule Graph is particularly important. Since Schedule Graph regards the state from the perspective of task schedule, the researchers can focus on the changes of Schedule Graph when they perform state transitions, therefore, we can get the comparison between different states of the completion situation of a specific task. If a state has a greater depth than the other state, but the completion situation of this state displayed on Schedule Graph is worse than the other state, then this state is worthy of the attention of the researcher, which may be a redundant state. In analysis of the system state space, this will help the researchers to be more quickly determine the validity of a state and its value of attention when confront with state explosion.

4   Applications

In this section we present two cases that demonstrate how DFDVis helps to understand and analyze system given by DFD.

4.1   Analyze the State Space of the System

We can improve the introduction of the system function based on the data flow diagram shown in Fig. 2. Figure 3 shows the input and edit function of data flow diagram provided by DFDVis. All of the nodes and data flows should follow the naming convention and unnamed nodes or data flows are not allowed to exist.

After confirming the completion of the edit, the user can submit it to the server and they will receive feedback from the server immediately. Then the state space of the entire system will be generated, including the tree structure that displays details and the overview of the entire state space.

The focus can be switched continuously in the state tree, and the corresponding tree structure displaying details in state space will change accordingly. Besides, the presentation of the state nodes in the overview will also change to be consistent with the state tree.

More important is that the two state transition paths from the initial state to the current focus state and from the state of the previous focus to the state under the current focus will be highlighted.

As shown in Fig. 7, the current focus state is described as:

```
('a.A|B|C|'output1.D|E|F|f.'output2.G)\{a,b,c1,c2,e,f}.
```

And the two state transition paths are also displayed differently. However, due to the folding of some state nodes in the tree structure, the state tree cannot directly show the complete path, but the user can get the full path view in the overview. As we can see in Schedule Graph, although there are 6 actions from the initial state to the current state, only 4 actions (data flows) are necessary regarding the situation of the task completion. The researchers should re-examine the current state nodes of interest, and go along the state transition path from the initial state to the current focus state to search for the state without any redundant action for completion of task, therefore, such state is indispensable, which is described as:

```
(A|B|C|'output1.D|E|F|f.'output2.G)\{a,b,c1,c2,e,f}.
```

Its depth is 5.

4.2   Deadlock Detection

A deadlock state is a state without any subsequent state, which indicates the system is stagnant. The state space of the system contains all the states that the system can reach, thus the deadlock is also one of many states in state space [13].

After finishing the automatic analysis work, DFDVis can find out such state without any subsequent state, then switch the focus to this state and generate alarm: there is deadlock in the system described in DFD. As shown in Fig. 8, the focus is directly switched to this deadlock state without any subsequent state. Given this deadlock state, the analysis can analyze the root causes of the system deadlock based on the state transition path provided by DFDVis.

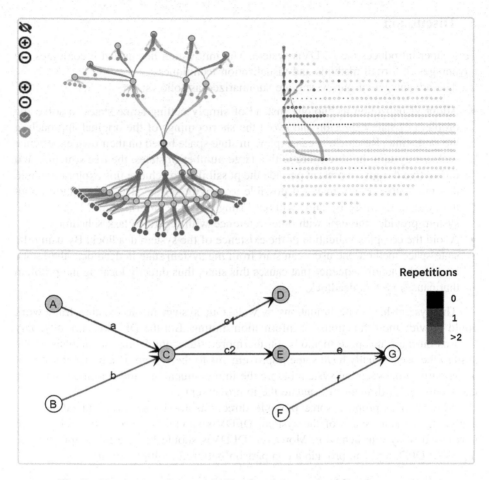

Fig. 7. Use the same input as in Fig. 3 and the result of focusing on a state.

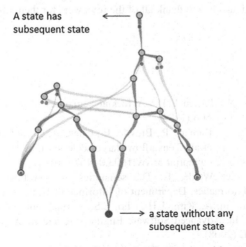

Fig. 8. The view shows that the focus is on a deadlock state without any subsequent state.

5 Discussion

This paper introduces the DFDVis system. The innovation lies in that it combines the advantages of formal methods and visualization techniques.

The advantages of our system are summarized as follows:

- Adopt visualization technology instead of simply cutting some states to solve the problem of state explosion and avoid the shortcomings of the original approach at the same time. Researchers can explore in state space based on their own experience with our system. In state space with a large number of states, the tree structure and interactive design we present provide the possibility to achieve this exploration, and this innovative design makes it possible to provide all the state information about the system described by DFD without losing any state information. Moreover, the system provides the user with other reference with regard to task schedule.
- Avoid the complex validation of the existence of the system deadlock. By using the state space method, the users can start from the system state in deadlock, then trace back to the action sequence that causes this state, thus directly locating the problem that causes system deadlock.

The system has some limitations as well. Our system in the existing framework could provide more functions for information mining. But the DFDVis currently only focuses on the state space methods. Furthermore, after completing the analysis of the system the users usually have some improvements to the system. But in the absence of a comparison between the system before the improvement and the current system, the users will find it difficult to evaluate the improvement.

Now we can propose some possible directions for future work. In addition to analyzing the state space of the system, DFDVis should support more analysis and simulation of system behavior. Moreover, DFDVis should be able to accept multiple inputs of DFD, and can provide a comparison between multiple results.

Acknowledgments. The work is supported by the National Nature Science Foundation of China (61472239, 61472238, 61100053) and CCF-Venustech Hongyan Research Initiative (2016-013). Furthermore, we would also like to thank all of the reviewers for their valuable and constructive comments.

References

1. Larsen, P.G., Plat, N., Toetenel, H.: A formal semantics of data flow diagrams. Form. Asp. Comput. **6**(6), 586–606 (1994)
2. Abadi, M., Agarwal, A., Barham, P., Brevdo, E., Chen, Z., Citro, C., Corrado, G.S., Davis, A., Dean, J., Devin, M., et al.: Tensorflow: large-scale machine learning on heterogeneous distributed systems. arXiv preprint arXiv:1603.04467 (2016)
3. Bruza, P.D., Van der Weide, T.: The semantics of data flow diagrams. Faculty of Mathematics and Informatics, Department of Informatics, University of Nijmegen (1989)
4. Jilani, A.A.A., Nadeem, A., Kim, T.H., Cho, E.S.: Formal representations of the data flow diagram: a survey. In: Advanced Software Engineering and Its Applications, ASEA 2008, pp. 153–158. IEEE (2008)

5. Milner, R.: A calculus of communicating systems. Laboratory for Foundations of Computer Science, Department of Computer Science, University of Edinburgh (1986)
6. Butler, G., Grogono, P., Shinghal, R., Tjandra, I.: Retrieving information from data flow diagrams. In: Proceedings of 2nd Working Conference on Reverse Engineering, pp. 22–29. IEEE (1995)
7. Harel, D.: Statecharts: a visual formalism for complex systems. Sci. Comput. Program. **8**(3), 231–274 (1987)
8. Valmari, A.: The state explosion problem. In: Reisig, W., Rozenberg, G. (eds.) ACPN 1996. LNCS, vol. 1491, pp. 429–528. Springer, Heidelberg (1998). doi:10.1007/3-540-65306-6_21
9. Cava, R.A., Luzzardi, P.R., Freitas, C.: The bifocal tree: a technique for the visualization of hierarchical information structures. In: Workshop on Human Factors in Computer Systems (IHC 2002) (2002)
10. Stevens, P., Moller, F.: The Edinburgh concurrency workbench user manual (version 7.1). Laboratory for Foundations of Computer Science, University of Edinburgh 7 (1999)
11. Allal, L., Belalem, G., Dhaussy, P.: Towards distributed solution to the state explosion problem. In: Satapathy, S.C., Mandal, J.K., Udgata, Siba K., Bhateja, V. (eds.) Information Systems Design and Intelligent Applications. AISC, vol. 433, pp. 541–550. Springer, New Delhi (2016). doi:10.1007/978-81-322-2755-7_56
12. Shneiderman, B.: Tree visualization with tree-maps: 2-D space-filling approach. ACM Trans. Graph. (TOG) **11**(1), 92–99 (1992)
13. Brebner, G.: A CCS-based investigation of deadlock in a multi-process electronic mail system. Form. Asp. Comput. **5**(5), 467–479 (1993)

Design and Implementation of Medical Data Management System

Jie Wang[1]([✉]), Jianqiao Liu[1], Jian Li[2], Jian Zhang[1], and Qi Lei[1]

[1] School of Information Science and Engineering, Central South University,
Changsha 410083, China
jwang@csu.edu.cn
[2] The Second Xiangya Hospital, Central South University, Changsha 410011, China

Abstract. At present, with the development of medical technology, the amount of medical data is increasing, in face of the shortcomings like the complexity of information, lack of structure, flexible changes, as well as the lack of scientific medical data management platforms at home and abroad, this paper presents a child epilepsy Data platform. This platform can extract quantitative data of structural epilepsy from the description of the complex and redundant text, which makes the data analysis and data mining more convenient and provides the basis for the future intelligent diagnosis, prediction and intelligent prognosis of epilepsy. At the same time, this paper introduces the platform database, server language and the construction of back-end and front-end of platform. The paper also introduces the technology and method of data-oriented management platform from requirement analysis, platform design to implementation stage. Through continuous testing and communication with doctors, developers finally established a data-driven system that is efficient in storing, managing and transferring and exporting medical data, which is also expandable, robust and stable.

Keywords: Medical data · Epilepsy disease · Data-driven · Medical management platform

1 Introduction

Many hospitals still use paper documents to record information about patients. The drawbacks of paper documents are obvious: it is difficult to organize, not easy to save, and so on. It is also difficult to position the document when users need to use patient information. Once the medical information needs to be exchanged between different sites, the traditional paper data is also facing great difficulties. Time of transfer and labor costs are also big problems even if barely realized the transfer. Considering the damage to paper material during the transfer, paper material is almost inoperable on data transfer. In order to solve this problem, using intelligent data platform to manage patient information and replace paper materials is almost an inevitable trend.

© Springer Nature Singapore Pte Ltd. 2017
B. Zou et al. (Eds.): ICPCSEE 2017, Part I, CCIS 727, pp. 674–688, 2017.
DOI: 10.1007/978-981-10-6385-5_56

In addition, with the emergence of emerging information technologies and application models, the global data shows an unprecedented explosive growth trend [1]. According to the International Data Corporation's Digital Universe Research Report [2], In 2011 the world created and copied more than 1.8ZB of data, and the data growth trend doubled every two years (follow the new Moore's Law), to 2020 is expected to reach 35ZB. In addition, not only the surge in the amount of data, the complexity of the data is also growing rapidly. The characters like diversity, low value density, real-time are increasing significantly. Big data is coming [3]. Big data indicates that data-oriented researches and applications have gone beyond the initial stage and have entered a new era of maturity and deepening [4].

With the emergence of various emerging data technologies, more and more areas need to be applied to the intelligent data platform for data management. As a very important category of data, the amount of medical data has generated explosive growth. By 2020, medical data will grow sharply by 35 Zetabytes, up 44 times [5] from 2009 data. As a result of an annual growth rate of 48% [6], medical data has become one of the fastest growing branches of data. More noteworthy is that with the development of advanced treatment technologies in the medical field, medical data will usher in a greater growth frenzy. However, although the data is growing, the quality of the data can not be guaranteed. More disordered and redundant data is generated in the medical field, so it is necessary to create a reasonable platform to standardize the management of the data.

Nowadays, medical data is one of the most important branches of data [7]. The technology of big data provides us with more information and methods never used before. A system that can efficiently store medical data not only can alleviate the problem of storing too much medical data, but also provides the basis and the possibility for future medical artificial intelligence and medical data mining [8].

This project is based on the needs of a hospital for children with epilepsy department to establish a required data centralized management system. This article will introduce the medical data platform of children with epilepsy and other foreign medical data platforms will be described in the following article. In addition, problems of existing common data platforms will be analyzed. The structured form and analysis of epilepsy data in the platform will be described as well. In addition, this article will introduce the background, front-end structures and database selection of this Data platform, as well as sections of testing and summary, outlook and so on.

2   Development of Medical Platforms at Abroad

In this section, we will introduce several representative medical data platforms, the problems and challenges exist in these platforms will be discussed at the same time. Then we will introduce the advantages of our medical data platform in the following passage.

2.1 Foreign Medical Platforms

IBM Watson Medical Platform. IBM Watson Medical Services have services in health care, disease management, population health, cancer, and clinical trial matching, and even intelligent tracking in life sciences research, including storage and app development [9].

Users can not only store their own data, but also to obtain specific data from 300 million patient information. Watson's superb analytical techniques, from predictive analysis, similarity analysis to natural language processing, depth quizzes, allow data-assisted decision making. It is a very powerful management platform which can erase the patient's identity while achieving the sharing of medical information, and can integrate a large number of anonymous information for clinical research, public health to provide aggregated view.

Through the direct storage of text information, as well as the handling of natural language, the platform has a meticulous but more complex features.

The People's Hospital of Peking University Data Management Platform. The people's Hospital of Peking University data management platform has established a large clinical data center and integrated patient information on this platform. Its approach is to record and store the personal information, medical records, medication status, and other information on each patient in chronological order.

The platform uses the Hadoop cluster model [10] to carry on the design, which generates medical data center of Peking University. By dividing the server into multiple blocks and connecting several large rooms in the hospital through high-speed Internet connectivity and taking advantages of redundant rooms and the distributed structure of Hadoop [11], it ensures the reliability and flexibility of the medical data platform.

The platform developed by the people's Hospital of Peking University, the data storage has a reliable and flexible features, but the internal data transfer method is also worthy of attention.

2.2 Problems and Challenges

After a full investigation of a variety of domestic and foreign data platform, we found the following problems:

The first is the structural weakness of medical data. Many medical data are derived from the patient's description or the doctor's clinical experience, so before the design of the system, we need to do a design for all the data which can be structured. But part of the medical data still can not be structured, such as image data, audio data, text description and so on. These unstructured data can be stored using an external system. In general, in data management systems, we should manage unstructured data and structural data separately, so that structural data can be easily modeled in data analysis. Unstructured data can not be discarded directly or externally, but also need to be managed in the system, because another important function of the data management system is provided to the medical staff. A lot of unstructured data such as a doctor's annotation for a particular data is meaningful to them. There is also a need for a separate system to manage unstructured data.

The second is the continued change in medical data. This problem is mainly in the frontier medical field, where the data in these areas often changes. So the system should be able to adapt to the data according to the adaptive changes, and leave enough customizable parts to ensure that the system can adapt to different changes in the configuration of the data.

The third is the transfer of data. There are many data systems that are currently mature, but the data between the systems is not compatible. Sometimes there is a need to dock two systems, and data migration may occur. Therefore, we need a method to ensure the data migration between different databases.

2.3   Advantages of This Platform

In the face of the weakness of the structure of the medical data, the difficulty of data transfer, as well as the flexible characteristics of medical data changes, the project developers designed the data platform as follows:

First, structured data. Developers extracted information from the original unstructured description of epilepsy and the doctor's advice and they designed it as a part of the constructive options and a non-structured part of the note. For example, we choose to structure the patient information, diagnosis and treatment information and personal information. And we make the personal information and medical information as options for doctors to form operation, in order to liberate medical information from previous miscellaneous text description, making the structured system easy to operate and facilitate future data analysis.

Second, content customization. Developers designed the epilepsy data platform for patient and physician needs, and designed a custom section on the data platform, such as seizures and diagnoses. Doctors can add or delete information at any time to respond to the changing characteristics of medical information.

Third, the data is easy to migrate. When we design flexible structures and custom parts, developers also designed the appropriate data interfaces and methods for the transfer of data in different system databases. This greatly reduces the energy spent on moving data, making data transfer easier and more convenient, and improves the performance of the original medical data platform which is not flexible, uneasy to operate.

Fourth, the data is easy to be analyzed. In the design process for the entire platform, the original complex and lengthy disease description and doctor's advice can be structured into a data form. And the corresponding options will be stored in the database to facilitate future analysis and operation for epilepsy Disease, which provides the foundation of intelligent prediction, diagnosis, and prognosis for epilepsy.

3   Data Quantification

In this chapter, the necessity of quantification of epilepsy information and the quantification process and the form of data in the platform are described in detail. At the same time, the role of data quantification in intelligent diagnosis, intelligent prognosis and treatment of epilepsy will be analyzed.

3.1   Data Background

According to the analysis of World Health Organization (WHO), there are about 50 [12] million epilepsy patients worldwide. In the past, the diagnosis of epilepsy most rely on the patient's description of the disease and the doctor's observation of the patient, so we lack a comprehensive and accurate diagnostic classification method for epilepsy. With the development of medical technology and the accumulation of cases, although epilepsy disease diagnosis is complex and difficult, the researchers summed up the [12] seizure classification method through the combination of different epilepsy symptoms and previous diagnostic experience. Different types of seizures were classified, and more than 20 types of syndromes [12] were classified. The emergence of this classification method combined with a variety of measurable disease characteristics provide the basis for the establishment of children epilepsy medical data platform and intelligent analysis of children epilepsy disease.

According to the understanding of ILAE epileptic seizure classification and the guidance of doctors, developers have structured the epilepsy information to convert the descriptive text information into quantifiable digital data. In addition to the basic information of patients, MRI, CT, EEG examination, medication and other detailed information are presented in the data platform as the form of options. This facilitates the future extraction and analysis of the data. And more detailed classification of information will be described in the following paper.

3.2   Structured Data

In order to make the entire data platform easy to be operated and data easy to be analyzed, developers extracted the following information from the doctor's advice and disease descriptions:

There are three main objects in the whole system, that is, the patient object, the medical record and the medication record object. Among them, there is one to many relationship between the patient's medical record and medication record. In addition to the one to many relationships, there are other forms of a one-one relationship with the patient: growth history, current history, family history, maternal history. These tables are information that a patient may have, and the patient will have only one correspondence, so there is a one to one relationship between the patient and these tables. Similarly, there are a number of subsidiary records of the diagnosis and treatment of the object: seizures, MRI, CT, EEG, prognosis. There is a one to one relationship between these objects and the medical records.

Patient Information. Patient information includes personal basic information, personal history, family history, maternal history, production history, current history. The specific information of a patient is as follows:

Table 1: Patient basic information = {Patient ID, hospital number, name, gender, age, nationality, date of birth, weight, height, head circumference, intelligence level}

Medical Records. A medical record is an object that is generated when a patient visits a hospital. Each of the medical records need to record the date of the patient's visit, the doctor's diagnosis, as well as other checks (MRI, CT, EEG) information, the patient's description of the seizure etc. The specific forms are as follows.

Table 2: Medical records = {patient ID, date, special inspection, genetic testing, metabolic examination, diagnosis}

Table 3: CT = {diagnosis and treatment record ID, check number, low density, high density, ventricles, notes}

Table 4: MRI = {diagnosis and treatment record ID, check number, cortex, date, white matter, strengthen, note}

Table 5: EEG = {medical record ID, ID EEG, EEG date, description of the same impression, with parents, monitoring of clinical seizure, ictal EEG, EEG background, during the onset of abnormal wave, seizure duration, electric activity, discharge duration, note}

Table 6: The seizures = {medical record ID, the number of seizures, the seizure, the date of first seizure, seizure of last date, the longest duration of seizure, the shortest duration of sleep onset, seizure frequency, note}

Table 7: The ID of prognosis = {medical records, clinical prognosis, prognosis of EEG, the monthly recurrence times, note}

Medication Records. Medication records are used to record the use of drug, including the name of drug, dosage, side effects, start and end date of use.

Table 8: Medical records = {patient ID, drugs, start time, end time, side effects, dosage}

3.3   Custom Data

In the face of the flexibility of medical data, developers made a special treatment on some of the changes that are flexible and need to be modified at any time. That is, we can create an option table. We can add or delete a form field at any time by flexibly modifying the options. For example, in the medical record, there is a field as a result of the diagnosis. In general, the diagnostic result is textual content, but in order to quantify the value, we must list all possible diagnostic results and add the serial numbers then use that serial number as the diagnostic result. This quantification is possible because there is a classification standard for the diagnosis of epilepsy in the international community. So whether it is only necessary to follow the standard to classify it? The answer is no. In addition, according to Epilepsia's [13] report, the epilepsy diagnostic criteria have been modified several times, indicating that direct hard-coding options are not desirable. This reflects the role of the option table, these option tables in particular save a field for all possible options, and the field is a foreign key to the table.

Another effect of the option table is multiple selection. There are some fields that also need to be quantitatively provided with some quantitative options, but the field itself can be multiple choice, such as the family's similar disease history (a person may have one or more similar diseases). One way is to save the array directly as a string to the database. However, this method violates the atomicity of the relational database, that is, the data item can no longer be divided in the structure, only the simplest unstructured value. In this case, the solution in the relational database is to create a middle table that holds the many-to-many relationship between the family history option tuple and the similar disease option tuple.

Among them, the information which is needed to establish the option table is as follows:

MRI. Analysis of magnetic resonance imaging (MRI) helps doctors to detect brain structural abnormalities [12]. This is an important method of diagnosing seizures, determining the type of seizures and epilepsy, a routine examination for patients with epilepsy. The MRI table is a subsidiary table of the medical record object. Among them, the basal nuclei needs to be changed flexibly:

Table 9: Basal nuclei = {Basal nucleus; int; 0 - normal, 1 - abnormality (exception attribute to be added)}

CT. Head CT examination is of great significance in the display of calcification or hemorrhagic disease. The CT table is an accessory table for the medical record object, where the fields are flexible. And the attribute which needs to be added or deleted is as follows.

Table 10: Background activity = {background activity; int; 0 - high voltage, 1 - low voltage, 2 - fast activity, 3 - wave, 4 - wave}

Types of Seizures. The types of seizures is the accessory table of the medical record object, where the fields are flexible, and the attributes need to be added or deleted.

Table 11: Types of Seizures = {Exercise episodes, Sensory seizures, Autonomic nervous seizures, Psychotic seizures, Simple focal attack secondary disturbance of consciousness, Onset of the beginning that is a sense of disturbance, Automatism,

Tonic - clonic seizures, Typical absence seizure, Myoclonic seizures, int, 0 - none, 1 - have}

3.4   Data Mining

A system that can store medical data efficiently can not only alleviate the problem of storing large amounts of medical data, but also provide a basis for medical data mining. "Big data" represents a new era in data exploration and utilization [14].

The large number theorem tells us that under the same conditions of the experiment, if the test is repeated several times, the frequency of the random event is similar to its probability. "Regular random events" in a large number of repeated conditions often shows almost the inevitable statistical characteristics. Therefore, the analysis of large amounts of data, we can extract the law and get behavior patterns from the corresponding data. For example, although travel patterns of people are very different, but most of us are equally predictable. This means that we can predict the likelihood of his or her future travel modes, that is, 93% of human behavior can be predicted based on the behavioral trajectory of the individual [15]. Big data show their place not only in the scientific research, they are also very valuable in other areas of society, such as the financial sector will be able to create a very high value [16] from the application of big data technology.

In the medical field, there are already many uses of big data on disease research, for example, some advanced data mining methods can use the published breast cancer patient information to obtain very accurate mortality prediction [17]. In cardiovascular disease prediction, the use of Naive Bayes, neural networks and decision trees can also achieve high accuracy results [18]. The above studies use medical data provided by government or public organizations to analyze, and the results have been able to achieve a high degree of accuracy. So if large hospitals have their own private data management systems and extract data from these systems for professional analysis, researchers will get more valuable information.

The structure of the epilepsy data provided by the platform is more likely to be analyzed compared with the previous textual epilepsy information, the researchers can get intelligent epilepsy diagnosis and prognosis by dealing with the data through clustering, neural networks, decision trees, support vector machines, DSS [19], and other methods.

4   Design of Data Platform

This chapter explains the selection of the server language, the ORM-based data storage acquisition system, and the construction process of the Observe-based reactive modular front-end, and the database selection is explained in detail. At the same time in this chapter, the function of the whole system is introduced.

4.1 Platform Construction

Django ORM Based Platform Back-End. As a universal Language, python is widely used in many fields of the science industry [20]. In the Python realm, the Django server framework is one of the most mature and popular frameworks. Django is an advanced Python Web server framework that provides a fast developing experience and simple design.

In this platform, through the inheritance of Django Model, ModelField class, the corresponding patient, medical records, medication record object can be established. And developers can define the corresponding attributes in the class, then Django generates the corresponding view in the internal interface through the Meta class. The view will be converted to the property value in the database, then developers can successfully establish the ORM model. After creating the ORM model, developers can import the object model into the database as a relational model which makes the development process simple and easy.

The relationship between data models in this data platform is shown in Fig. 1.

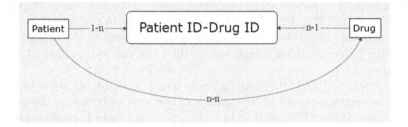

Fig. 1. Implementation of the many-to-many relationship in the relational model

In order to reduce the burden on the server as well as enhance the user experience, the system will let the part of the page rendering done by the front-end. The Web server only serves as database query and data transfer. So all the data of the server need to pass through AJAX. In addition to AJAX, the system's data-oriented features also depend on the meta interface. By using the meta interface, the details of all the fields of a model will be obtained. The information of data models and fields are sent to the front-end in the form of json data through the AJAX view. After the front-end acquires the information, it will create the corresponding forms. All processes are automatic, there is no handwritten coding. Therefore the front-end of the form can automatically adapt to the data model changes.

In general, the underlying data flows from the database, through the object relationship mapping, the data transmitted to the final front-end of the MVVM system by AJAX. All of these processes experience different deformations and changes, resulting in different levels of data performance. However, at all levels, the data is center (Fig. 2).

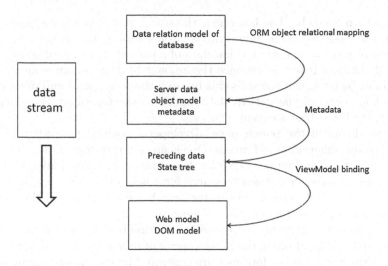

Fig. 2. Oriented data system

Observe-Based Reactive Modular Front-End. The front-end part of the system uses Vue, the framework of ViewModel part in MVVM.

In the front-end of the system, the Vue components are divided into two categories:

1. Components: They are normal components, which are belonging to parts of the functions in the page. The biggest feature of it is reuse, all external data are provided by the parent component.
2. Pages: this type of component is directly related to a route, the characteristic of it is not reusable. The data is obtained from the model layer, it refers to the normal components and use some custom code to form a page.

4.2   Introduction of Functions

Information Modification. In the patient information, there are four categories, namely the medical record details page, the medical record list, the patient information page and the medication situation page. These four pages are responsible for the display and modification of all the data in the user object. Most of the management functions of the system are in these four pages. Users can add or remove patient information, medical records, and medication information.

Information Customization. In the settings page, there is a separate classification named the option settings. Back to the beginning of the data model in the second chapter, we mentioned some of the data models have multiple choices and variable options. Developers need to provide the interface so that users can change them and the contents of the options can be arbitrarily modified. So the option setting is left for the user to modify the options of an interface.

Information Search. The last one is the search page. As a data management system, the modification of only a single patient object data is not enough, but also need to provide an entry to view data of a patient. It is unrealistic to simply make all the data listed, since once the amount of data becomes more, all the list can not be retrieved; the second is because the server load and network load will be huge once all the patient data loaded to the front-end. So providing a search entry is the best solution.

In the design of the search page, developers provided two different search options to suit different search needs. One is an ordinary search, it will be based on the patient's id and name. The other is a more sophisticated advanced search mode in which the user chooses to search for a field of the data model. Thanks to the data-oriented system design, the search data control will be based on the field of meta data and generate different types of input controls, such as the choice of gender will generate a drop-down selection box to select men or women. If users use the date of birth, the input control of date selection will also appear. All the performance on the interface are generated by the underlying meta data.

Information Display. In the website page, the patient's information can be displayed on the page. Users can view the patient's basic information, medical records, medication records and other information on the website.

Remarks System. Because of the complexity of medical data, there are many cases are still unable to quantify, such as the patient's attack situation. In order to facilitate the analysis of data in the future, the attack situation can be divided into good, bad and several other quantitative options. But the doctor may need to add some other remarks to help themselves to review the disease. Therefore, the system adds a global note system.

The implementation of the remark system is the model of the data. Since the memo system is global and need to be able to pinpoint a field of a record, the note should be composed of: the table of notes, the fields of the memo, the object id of the note and the contents of the note. According to the notes of the table, the notes of the field and the id of the object, users can find any information they want.

On the front-end page, the note input box also needs to be displayed directly on the form, so the front-end interface is designed to add or modify notes via a floating button. As long as the mouse moves to an input button, users can add, delete, or modify the comment for that field.

5 Platform Test and Evaluation

This chapter briefly introduces the detailed functions and evaluates their performance.

5.1 Platform Test

Normal Search. In the normal search page, the user input patient name or medical treatment id then diagnosis and treatment information can be displayed in the page, including EEG information, MRI information, CT information, seizure information. Click the Web page option, the user can easily browse the information of the patient at any time.

Advanced Search. Unlike normal search, in the advanced search page, the user only need to enter the target table or the primary key field of the table to get the corresponding patient information. All information about the relevant field and the patient's personal basic information will be displayed, which is more convenient compared to normal search. The target table and target fields in the advanced search are obtained directly from the meta data. Because of the existence of meta data, the contents of the search will automatically use different types of input box according to the different field types, such as the pull box.

Patient Basic Information. After entering the home page, the user can see the patient option. Clicking to add the patient, the user can add the corresponding patient personal basic information, personal history, family history, maternal history and other content to the platform. The specific form options and names are presented to the user as a drop-down list and users only need to click on the option to save the corresponding number of options to the database. There are also delete buttons next to the form, uses can just click delete button to delete.

Medical Records. After adding the basic information of the patient, the user can choose to add the patient's medical record, and a patient can add a number of corresponding medical records which include EEG, CT examination, MRI examination, seizures and other information. By selecting the save option to save the results to the database and ensure the accuracy of the diagnosis and treatment information. There are delete options next to the form, so the user can click the Delete button to delete the corresponding field content, making the modification easier. After adding the complete medical record, the user can click the display button and check the contents of the patient's medical records. If there are any questions about the contents, the user can delete and modify it, or add new medical records.

Medication Records. In addition to adding the basic information of patients, medical records and other content, the user can add the patient's medication records, including medication type, name, start and end date, dose and so on.

Information Customization. In the settings page, there is a separate classification named the option settings. Back to the beginning of the data model in the second chapter, we mentioned some of the data models have multiple choices and variable options. Developers need to provide the interface so that users can change them and the contents of the options can be arbitrarily modified. So the option setting is left for the user to modify the options of an interface.

5.2 Platform Evaluation

Platform Back-End. By applying Django ORM, developers can focus on the complex logic of data-driven systems in business. Since ORM makes SQL logic simple and clear, developers only need to obtain data from the table. Although the existence of ORM will produce time consumption, but most of the data conversion will be completed by the server, so the impact of ORM consumption can be ignored. And its simple, convenient and excellent performance can make development more efficient.

In addition, the presence of the ORM layer eliminates the dependency of the system business logic on the database, thus solving the problem of data migration. To migrate to other databases, the model can be applied to other databases as long as the corresponding database ORM driver is installed.

Developers can change the data easily since the back-end of the system is based on ORM development. If users need to modify the order of the form, users can directly modify the ORM variable definition order. If users need to add a new suffix, users can add it in the ORM while there is no need to modify other things. The custom options also make the entire platform more flexible, and if there is a new option for a data item in the later stage, it can be added directly, which adapts to the rapid changes in medical data.

System Front-End. Since the front-end uses the reactive system, the transmission in network is no longer a complete transmission of HTML pages, but the data which is get from the database by the server and then be converted and expanded. Because the amount of data is greatly reduced, developers can feel the speed is upgraded. So the network transmission time is greatly reduced.

In addition, because the front-end completes the loading of the page, and Vue architecture will reuse the existing module on the page. Vue only needs to update the part of the change, so only a small part of the changed DOM node need to be updated, which greatly reduces the time used to load pages. At the same time in the user's view, the changes of page are instantaneous.

6   Summary and Prospect

The system has solved several problems that the current majority of medical data platforms exist, such as weak structure, inflexible, and difficult to transfer shortcomings. It extracts the structured information from the original complex and redundant epilepsy disease description and doctors advice, making the data platform easy to be operated. And the system makes it convenient to extract data that can be analyzed in the future.

In managing the medical data system, the data is always the protagonist, so building a data-oriented system is also a matter of course. Due to the large scale, variability and complexity of medical data, the establishment of this system will be more convenient for the analysis and processing of data. The epilepsy condition information which has been structured will be stored in the platform database, the structured data will facilitate the researchers to carry out data mining and analysis. The use of decision tree, clustering, neural network, support vector machine and other algorithms will achieve intelligent epilepsy prediction, intelligent epilepsy diagnosis, intelligent epilepsy prognosis and other functions, which provide references for future medical artificial intelligence.

Acknowledgment. This work is supported by National Natural Science Foundation of China under Grant No. 61202495 and No. 61573379.

References

1. Lynch, C.: Big data: how do your data grow? J. Nat. **455**(7209), 28–29 (2008)
2. Gantz, J., Reinsel, D.: Digital Universe Study: Extracting Value from Chaos. IDC Go-to-Market Services (2011)
3. Hilbert, M., Lopez, P.: The worlds technological capacity to store, communicate, and compute information. J. Sci. **332**(6025), 60–65 (2011)
4. Zhiyan, F., Xunhua, G., Dajun, Z., et al.: A number of frontier issues in business management research under the background of big data. Chin. J. Manag. Sci. **16**(1), 37–39 (2013)
5. McKinsey Global Institute Analysis. ESG Research Report 2011 – North American Health Care Provider Market Size and Forecast
6. The Digital Universe Driving Data Growth in Healthcare. EMC Digital Universe (2014)
7. Mulsant, B., Servan-Schreiber, D.: Toward a new paradigm of health care: artificial intelligence and medical management. Hum. Syst. Manag. **5**(2), 137–147 (1985)
8. Chen, Z., Jiang, Y., Xu, M., Wang, H., Jiang, D.: The application and development of artificial intelligence in medical diagnosis systems
9. IBM/Nuance to Apply "WATSON" Analytics to Healthcare - ProQuest 2011
10. Rallapalli, S., Gondkar, R.R., Ketavarapu, U.P.K.: Impact of processing and analyzing healthcare big data on cloud computing environment by implementing hadoop cluster. Procedia Comput. Sci. **85**, 16–22 (2016)
11. Duan, C.M.: Design of Big Data Processing System Architecture Based on Hadoop under the Cloud Computing (2014)
12. Clinical Diagnosis and Treatment Guidelines (Epilepsy)
13. Berg, A.T., Berkovic, S.F., Brodie, M.J., et al.: Revised terminology and concepts for organization of seizures and epilepsies: report of the ILAE commission on classification and terminology, 2005–2009. Epilepsia **51**(4), 676–685 (2010)
14. Zikopoulos, P., Eaton, C.: Understanding Big Data: Analytics for Enterprise Class Hadoop and Streaming Data. McGraw-Hill Osborne Media, New York (2011)
15. Song, C., et al.: Limits of predictability in human mobility. Science **327**(5968), 1018–1021 (2010)
16. Brown, B., Chui, M., Manyika, J.: Are you ready for the era of big data. McKinsey Q. **4**(1), 24–35 (2011)

17. Delen, D., Walker, G., Kadam, A.: Predicting breast cancer survivability: a comparison of three data mining methods. Artif. Intell. Med. **34**(2), 113–127 (2005)
18. Palaniappan, S., Awang, R.: Intelligent heart disease prediction system using data mining techniques. In: IEEE/ACS International Conference on Computer Systems and Applications, AICCSA 2008. IEEE (2008)
19. Economou, G.P., Lymberopoulos, D., Karavatselou, E., Chassomeris, C.: A new concept toward computer-aided medical diagnosis-a prototype implementation addressing pulmonary diseases. IEEE Trans. Inf Technol. Biomed. **5**(1), 55–65 (2001)
20. Pedregosa, F., Varoquaux, G., Gramfort, A., et al.: Scikit-learn: machine learning in Python. J. Mach. Learn. Res. **12**, 2825–2830 (2011)

Recognition of Natural Road Sign Based on the Improved Curvature Feature

Yanqing Wang(✉), Hao Zheng, and Weiwei Chen

Key Laboratory of Trusted Cloud Computing and Big Data Analysis,
School of Information Engineering, Nanjing Xiaozhuang University,
Nanjing 211171, China
wyq0325@126.com

Abstract. To solve the recognition of road sign with an intelligent vehicle in vision-based navigation, road sign extraction and matching techniques required in outdoor scene was proposed in this paper. The method of the improved curvature based on feature extraction and binary description took the advantage of reasonable features distribution to overcome the problems of traditional features uneven distribution. Binary description method was represented to solve the real-time problem of feature matching. Through the validity and real-time performance of different algorithms are compared by experiments and indicate that the method can not only overcome negative influences from the disturb of non-targets, while spending on average only 46 ms processing each frame, but also meet the requirements of robustness, real-time, and accuracy.

Keywords: Feature matching · Feature extraction · Road sign · Curvature algorithm

1 Introduction

Positioning is one of the most basic functions of mobile robots. The earliest positioning research results are mainly through the robot's internal sensors, such as code table, inertia and other facilities which are used to achieve positioning. However, due to slippery, ground rough and uneven causes of error accumulation often resulting in inaccurate positioning results, it is not proper to navigate for a long time alone. If you are by GPS positioning or by China's Beidou navigation system positioning, positioning system will lose efficacy in a shaded area and other place with bad signal. Therefore, people begin to gradually use external sensors to assist in positioning, such as infrared and visual sensors.

Visual sensor gains attention because of its rich environmental information such as vision, texture, shape and so on. By the computer vision theory, it is known that using the visual system can accurately identify the target and determine their position. At present, there are many existing methods, and most people use the method of manually setting road signs, and then move the robot in the process of pre-set artificial road signs to match the positioning. However, with the continuous progress of research work, people gradually move the application of mobile robot scene from the room to the outside. At this time, in many specific circumstances, manually setting the road signs is

© Springer Nature Singapore Pte Ltd. 2017
B. Zou et al. (Eds.): ICPCSEE 2017, Part I, CCIS 727, pp. 689–695, 2017.
DOI: 10.1007/978-981-10-6385-5_57

unrealistic, so the natural road signs become the preferred method in outdoor environment. Because the so-called natural characteristics are exist so non-artificial set and can be used to identify different environmental scene. Compared with most of the indoor environment, the complex outdoor unstructured environment poses a considerable challenge to the positioning and navigation of the mobile robot. At the same time, compared with the indoor the change of outdoor lighting conditions and the climatic environment is also a big challenge.

The natural roadmap based on vision needs to extract its invariant point, and then use the method of feature point matching to judge and identify the landmark, so it can realize the positioning of mobile robot. Unlike the positioning method by GPS, the feature-based location method converts the original sampled data into corresponding local features in the first place, but also it can preserve rich and accurate environmental information in the case of a small amount of storage space. This method usually has good robustness to meet the relatively complex applications. But when there is vegetation distributed on both sides of the road, most of the feature points are concentrated on the trees on both sides of the road and the extracted features are similar. The feature matching and robot positioning have brought a greater impact. Therefore, How to remove the effect of vegetation on natural road sign extraction and to point the feature points as evenly as possible to the scene of the building become a natural road mark extraction in a major problem. At present, the local feature matching algorithms are Harris, SIFT, SURF, ORB, etc. [1–4] The SIFT (Scale Invariant Feature Transform) is proposed by David Lowe in 1999 and summarized in 2004. Because of its good invariant scale rotation it has been widely concerned. But its slow operation affects its application in real-time applications. Herbert Bay has proposed an improved SURF algorithm for the SIFT algorithm, which is an order of magnitude in speed, but this is not enough in some real-time applications. Until 2011 proposed ORB algorithm, it can meet most real-time applications. However, the feature points extracted by ORB algorithm are not evenly distributed in outdoor applications, so the algorithm based on curvature is used to realize more uniform feature point extraction.

2 Based on the Curvature of the Feature Point Extraction

The way that is based on the curvature of the feature point extraction is very simple to calculate. The extracted feature points have rotational invariance, and they have good robustness to the changes of noise and brightness, and the distribution of feature points is uniform. More importantly, the quantitative feature points can be extracted. The method mainly uses the change of the gray scale to judge whether it owns the corner point or not, when the small offset of the image of a point in any direction will produce a large number of changes in gray, then the point is considered as a corner. Let I_x be the horizontal gray scale change, I_y for the vertical direction of the gray scale changes, then the corner is a point which both I_x and I_y have big changes, and the edge is a point which has a relatively big change either in I_x or in I_y. Let w (x, y) be a slider centered on (x, y), and when it is sliding [u, v] in any direction, the gray scale change E (u, v) is calculated as follows:

$$E(u, v) = \sum_{x,y} w(x, y)[I(x+u, y+v) - I(x, y)]^2 \tag{1}$$

According to the Taylor series to calculate the first order partial derivatives of order n, finally, you can get a matrix.

$$M = \sum w(x, y) \begin{bmatrix} I_x^2 & I_x I_y \\ I_x I_y & I_y^2 \end{bmatrix} \tag{2}$$

According to the eigenvalues of the matrix λ1, λ2 to determine whether they are the corners, but in the actual operation, the corner response value is calculated as follows:

$$R = \det M - k(traceM)^2 \tag{3}$$

In the formula: k is the coefficient value, generally the value of k is between 0.04 and 0.06. Then, we can through R to determine whether it is the corner. If R is the maximum value in the corresponding neighborhood, then we find the location of the corner.

The original extraction algorithm is based on the characteristic points of the curvature. The way to determine whether a point is a feature point is to judge whether the calculated R is the largest in the neighborhood or not. If it is the maximum, then that point is a feature point. And the uncertainty of a point here depends mainly on the smaller eigenvalue. If we use the characteristic value to determine whether it is a feature point, just need a smaller one eigenvalue value which is greater than the preset threshold, then the point can be considered as a strong point, that is:

$$R = \begin{cases} \lambda_1 & \lambda_1 < \lambda_2 \\ \lambda_2 & \lambda_2 < \lambda_1 \end{cases} \tag{4}$$

If R is greater than a given threshold, then it is a strong corner. Because the algorithm has lower requirements on the discovery of eigenvalues, you can extract some of the features which are not particularly obvious points. As is shown in Fig. 1, when you need the uniform distribution of feature points in actual application, you can the appropriately increase the neighborhood in the algorithm, then you can get more uniform corner distribution compared with other algorithms in terms of distribution.

It can be found that the characteristic distribution is fairly uniform when the maximum suppression region is enlarged to 8 × 8.

3 Binary Features with Rotation Invariance

The binary string descriptor reduces the descriptor dimension compared with other algorithms, thereby reducing the time of description and matching. The algorithm first performs Gaussian filtering on the image to reduce the influence of noise, then selects a local block p in the image whose size is S × S pixels, and defines a τ test which is as follows:

(a) region of 3 × 3   (b) region of 8 × 8

Fig. 1. The maximum suppression region of 3 × 3 and 8 × 8.

$$\tau(p;u,v) = \begin{cases} 1, & p(u) < p(v) \\ 0, & \text{others} \end{cases} \tag{5}$$

In the formula: u, v is a two-dimensional coordinate pair like (x, y); p (u), p (v) is the brightness of the point.

The result of a number of τ tests consists of a string:

$$f(p) = \sum_{1 \le i \le n} 2^{i-1} \tau(p;x,y) \tag{6}$$

In the formula, n can take 128,256,512, etc. Taking 512 requires 512/8 = 64 B of storage space. How to effectively select the feature points has great influence on the next application. It is proved by experiments that the Gaussian distribution obeying $\frac{1}{25} S^2$ has a good resolution, and the Ham-ming distance is used to replace the traditional European distance judgment. Matching only by bitwise XOR operation is greatly reducing the time of feature matching.

BRIEF operation is simple and the memory is small which is more suitable for some real-time applications, but the algorithm also has some shortcomings: be susceptible to noise, not considering the direction of the characteristics, not have the rotation invariance, not have scale invariance.

In order to make the algorithm have the rotation invariance to reduce the influence of the rotation of the screen on the matching result during the jitter, the direction of each key point is calculated by the centroid algorithm, as shown in the following formula:

$$Mij = \sum_x \sum_y x^i y^j I(x,y) \tag{7}$$

The direction of the feature points is:

$$c_x = \frac{M_{10}}{M_{00}}, c_y = \frac{M_{01}}{M_{00}}, \theta = \tan^{-1}\left(\frac{c_y}{c_x}\right) \tag{8}$$

In the formula: (x, y) is the coordinates of the point of interest, and then the obtained θ is quantized by the scale of $2\pi/30$. Define a $2 \times n$ matrix for any n binary feature sets at position (x_i, y_i):

$$S = \begin{pmatrix} x_1 & x_2 & x_n \\ y_1 & y_2 & y_n \end{pmatrix} \tag{9}$$

$S_\theta = SR_\theta$ is obtained by the previously obtained θ correction S, and:

$$R_\theta = \begin{bmatrix} \cos\theta & \sin\theta \\ -\sin\theta & \cos\theta \end{bmatrix} \tag{10}$$

$g(p, \theta) = f(p)|(x_iy_i) \in S_\theta$ has the invariant rotation according to the obtained direction.

4   Experimental Analysis

In this experiment, we use vs-2010 and opencv2.4.10 to realize feature extraction and binary feature description method based on curvature algorithm under win7 system. Experimental hardware is AMD quad-core A6-3420 M, CPU 1.5 GHz, RAM 4 GB, video card HD 7470 M 1G. Because the realization of mobile robot positioning required for the small number of features, therefore, all the algorithms in this paper set the upper limit of the number of feature points detected for each pair of images is 200, the images used are recorded in similar video frames in the recorded video. The result of each experimental algorithm is shown in Fig. 2 and the experimental data are shown in Table 1.

From the experimental results, the distribution of feature points obtained by SIFT and SURF is relatively uniform, but the speed of it is much slower when compared to other algorithms, and does not apply to real-time applications. The distribution of feature points obtained by the fast point detection algorithm with faster velocity is more concentrated, which can not reduce the impact of trees and other features on the match.

Table 1. The experiment data of different algorithm

Feature extraction algorithm	The correct matching rate(%)	Matching time (ms)
Fast corner detection	86	68
SIFT	60	1237
SURF	58	1142
curvature	75	152
Improvement curvature	89	46

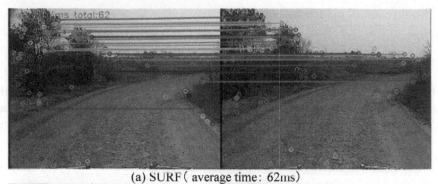

(a) SURF (average time: 62ms)

(b) ORB method (average time: 216ms)

(c) The paper method (average time: 46ms)

Fig. 2. The result of each experimental algorithm

5   Conclusion

The algorithm based on Curvature have a good balance in algorithm real - time and characteristic uniform distribution characteristics. Due to the simplification of the process, the processing time of the improved curvature-based feature extraction method is shorter; in terms of correct match rate, there is little different between the two methods relative to the original algorithm, the algorithm is more evenly distributed in this paper, reducing the impact of natural road signs such as trees on feature matching.

Therefore, the proposed algorithm can meet the requirement of effectively extracting feature points and matching in outdoor conditions and has good real-time and robustness.

Acknowledgment. This work has been supported by the Jiangsu Engineering Research Center for Networking of Elementary Education Resources (grantnumber: BM2013123).

References

1. Yanqing, W., Deyun, C., Chaoxia, S.: Vision-based road detection by monte carlo method. Sci. Alert: Inf. Technol. J. **9**(3), 481–487 (2010)
2. Yang, C., Zhang, L., Lu, H., et al.: Saliency detection via graph-based manifold ranking. pp. 3166–3173 (2013)
3. Wang, Y.Q., Zhuang, L.L., Shi, C.X.: Construction research on multi-threshold segmentation based on improved Otsu threshold method. In: 4th International Conference on Automation, Communication, Architectonics and Materials, ACAM, Advanced Materials Research, vol. 1046, pp. 425–428 (2014)
4. Wang, Y., Wang, Y., Shi, C., Shi, H.: Research on feature fusion technology of fruit and vegetable image recognition based on SVM. In: Che, W., et al. (eds.) ICYCSEE 2016. CCIS, vol. 623, pp. 591–599. Springer, Singapore (2016). doi:10.1007/978-981-10-2053-7_52

Evaluating Cities' Independent Innovation Capabilities Based on Patent Using Data Analysis Methods

Yan Zhang[1(✉)], Ping Yuan[2], and Bin Yu[3]

[1] Qingdao University of Science and Technology Library,
Qingdao 266061, China
zy@qust.edu.cn
[2] Archives Bureau, Qingdao University of Science and Technology,
Qingdao 266061, China
[3] School of Mathematics and Physics,
Qingdao University of Science and Technology, Qingdao 266061, China

Abstract. City innovation ability evaluation is becoming an important research field these days. The quantity and quality of patent that a city owns as well as the capability of managing patent institution indicate urban technological innovation ability and core competence. In this work, starting from patent analysis point of view, we construct urban technological innovation ability evaluation system. Factor analysis method is applied to 15 vice-provincial cities in China to analysis the collected patent data samples and a demonstration research is indicated upon the experimental results. Experimental results show that the littoral cities in south China have distinct advantages over the others in innovation activity, innovation capability, and innovation livability and innovation contribution. In the meantime, some old industrial bases own superiority in innovation cooperation and potential.

Keywords: Factor analysis · Cluster analysis · Cities' innovation capabilities · Patent analysis

1 Introduction

After three decades of economic growth at a relatively high rate, China is facing great challenges of economic growth pattern transformation. New patterns are needed to solve all kinds of problems such as the decrease of the energy and the pollution of the environment. As illustrated by President Xi Jinping in 2016 that China's economy, now the world's second largest, still suffers from low quality of growth. Lack of innovation ability has been the Achilles' heel for economic development. As a consequence, the driving factors behind economic growth are no longer those connected with physical capital, nor even with regional human capital. On the contrary, the new competitive conditions put a premium on social capital, formal and informal norms and rules which can foster reciprocal understanding and mutual confidence among the agents of the regional economy.

© Springer Nature Singapore Pte Ltd. 2017
B. Zou et al. (Eds.): ICPCSEE 2017, Part I, CCIS 727, pp. 696–707, 2017.
DOI: 10.1007/978-981-10-6385-5_58

However, how can we build up an innovative city is a tough problem. Basically, policy makers and contributors need to be very familiar with the city's innovative capability. On the other hand, cities on the way to innovation also have the needs to know where they are. All of these call for a comprehensive evaluation of innovative city. In the last years, the issue about the evaluation of city independent innovation has become quite debated both in economic literature and in scientific international forums. Also, social scientists have increasingly focused upon the significance of the city as a fundamental basis of economic organization and development.

In recent years, patent analysis had been recognized as an important task at the government level in some Asian countries. Public institutions in China, Japan, Korea, Singapore, and Taiwan have invested various resources in the training and performing of the task of creating visualized results for ease of various analyses [1]. For example, the Korean Intellectual Property Office plans to create 120 patent maps for different technology domains in the next 5 years [2]. Patent information is one of the most important factors in evaluating scientific innovation ability among cities. In particular, patent for invention is of particularly importance in innovation evaluation. In England, several measures of innovation suggest that the highest rates of 'visible innovation' are found in and around cities. Analysis of European Patent Office data for 1999 to 2001 shows that 67 per cent of EPO patent applications came from the 56 largest cities in England, and that 43 per cent of these applications came from just 10 cities [3]. As the Chinese practice of building innovative city goes further, it becomes more and more urgent to construct a theoretical framework and to find a strategy out for innovative city in the context of Chinese transformation. However, relatively little research has been done on city independent innovation ability evaluation based on patent analysis.

Literatures on innovation ability mostly focus on national innovation capability. Nelson compared the technology innovation's national institutional system of 20 countries and regions, which including America and Japan [4]. Riddel and Schwer applied the theoretical framework of national innovation ability to study the regional innovation ability [5]. Through empirical study on 52 states and regions in USA, they found the main influence factor of American technology innovation is knowledge stock, industrial R&D investment, high-tech human capital and the human capital quantity of obtaining bachelor degree.

Lindhult et al. studied innovation capabilities and challenges for energy smart development in medium sized European cities [6]. In their work, an assessment model was used to assess innovation capabilities in the PLEEC cities, comprising five innovation capacity dimensions, specified in a set of innovation practices which enable innovation activity and management. Fu presented her study on regional innovation capabilities [7]. Some recent research about innovation capability evaluation can be seen in reference [8–10].

In this paper, we present a new model based on multivariate statistics, allowing to synthesize cities' innovation performance by taking into consideration a set of diversified indicators in patent point of view, and to classify them into homogeneous clusters. Based on the systematic evaluation of 15 cities innovation capability by the application of factor analysis and cluster methods, along with the comparative study of the evaluation result, this paper constructs four clusters. The aim of this paper is to help cities formulating suitable innovation development strategies and assessing new business opportunities on the basis of patent information.

In the following sections, we discuss the factor analysis and cluster methods used to help evaluating cities' independent innovation ability. We design a synthesis evaluate function to rank the cities and furthermore, cluster them in order to understand the specific features of cities having similar positioning on factorial plan. We give the experimental results and finally we discuss the experimental results of the proposed algorithm and conclude the paper.

2   Methodology

Even though principal components analysis (PCA) is the default method of extraction in many popular statistical software packages, such as SPSS and SAS, Factor Analysis (FA) method is preferable to PCA which does not discriminate between shared and unique variance. In our work, the model is based on FA and kernelized fuzzy c-means cluster analysis.

2.1   Factor Analysis

Factor analysis is used to describe a number of possible lower unobserved variables called factors aspects observed, relevant variables statistical method variability. These factors cannot be directly observed and modeled as "Error" entry linear combination of observed variables, plus. Factor analysis is often confused with PCA, similar statistics. However, there are significant differences between the two. Factor analysis is a correlation (and covariance) model measures observed by checking performed between. This is the impact that may be highly correlated (either positively or negatively) the measures are likely to be affected by the same factors, but relatively irrelevant by different factors. Factor analysis is a statistical method used to describe variability among observed, correlated variables in terms of a potentially lower number of unobserved variables called factors. Such factors are not directly observable and are modeled as linear combinations of the observed variables, plus "error" terms. Factor analysis is often confused with PCA, a similar statistical procedure. However, there are significant differences between the two. Factor analyses are performed by examining the pattern of correlations (or covariances) between the observed measures. Measures that are highly correlated (either positively or negatively) are likely influenced by the same factors, while those that are relatively uncorrelated are likely influenced by different factors.

In applying Factor Analysis, the most important prerequisite is a strong correlation between the observed variables. The correlation coefficient matrix describes the relationship between the original variables which plays an important role in estimating the factor structure. The number of factors can be determined by the value of their variances. By means of coordinate transformation, there is a close relationship between the least factors related to each primitive variable so that it is easier to explain the practical meaning of the factor solution. After factor score of each sample is obtained, they can be used in a variety of analysis, for example cluster analysis etc.

There are several factor analysis extraction methods to choose from. SPSS has six (in addition to PCA; SAS and other packages have similar options): unweighted least squares, generalized least squares, maximum likelihood, principal axis factoring, alpha factoring, and image factoring.

Suppose we have a set of p observable random variables, $x_1, \ldots, x_p$ with means $\mu_1, \ldots, \mu_p$. Suppose for some unknown constants l_{ij} and k unobserved random variables F_j, where $i \in 1, \ldots, p$ and $j \in 1, \ldots, k$, where $k < p$, we have

$$x_i - \mu_i = l_{i1}F_1 + \cdots + l_{ik}F_k + \varepsilon_i. \tag{1}$$

Here, ε_i are independently distributed error terms with zero mean and finite variance, which may not be the same for all i. Let $Var(\varepsilon_i) = \psi_i$, so that we have

$$\begin{aligned} Cov(\varepsilon) &= Diag(\psi_1, \cdots, \psi_p) = \Psi \\ E(\varepsilon) &= 0 \end{aligned} \tag{2}$$

In matrix terms, we have

$$x - \mu = LF + \varepsilon. \tag{3}$$

If we have n observations, then we will have the dimensions $x_{p \times n}$, $L_{p \times k}$, and $F_{k \times n}$. Each column of x and F denote values for one particular observation, and matrix L does not vary across observations.

Also we will impose the following assumptions on F.

1. F and ε are independent.
2. $E(F) = 0$
3. $Cov(F) = I$ (to make sure that the factors are uncorrelated).

Any solution of the above set of equations following the constraints for F is defined as the factors, and L as the loading matrix.

Suppose $Cov(x - \mu) = \sum$. Then note that from the conditions just imposed on F, we have

$$Cov(x - \mu) = Cov(LF + \varepsilon). \tag{4}$$

Note that for any orthogonal matrix Q if we set $L = LQ$ and $F = Q^T F$, the criteria for being factors and factor loadings still hold. Hence a set of factors and factor loadings is identical only up to orthogonal transformation.

2.2   Kernelized Fuzzy C-Means Cluster

Cluster analysis is a collective term covering a wide variety of techniques for delineating natural groups or clusters in data sets. Cluster analysis methodology then allows to identify cities with homogeneous characteristics, by considering the multiplicity of possible determinants of cities performance as variables. Such a technique allows to identify the groups of cities that are characterized by the same distinctive elements and therefore have interior homogeneous characteristics and that present the highest heterogeneity with other groups. As a result, a homogeneous class will be described as a combination of variables.

It can be deduced that fuzzy c-means (FCM) method can yield satisfying cluster results for data of normal distribution, however this method is sensitive to isolated points and initial cluster centers. Inappropriate initial cluster centers may lead to local optimum. In order to solve the drawbacks of traditional FCM algorithm kernel function was introduced. By mapping the data into a high dimensional space FCM algorithm can cluster not only data of hypersphere but also asymmetric data and overlapping data. In the meantime the accuracy of clustering can be improved.

In KernelizedFCM (KFCM), a nonlinear map is defined as $\varphi : x \rightarrow \varphi(x) \in F$ where $x \in X$ denotes the input space, F denotes the feature spaceof the higher dimension. The objective function of KFCM can be given by

$$C_\varpi = \sum_{i=1}^{\eta} \sum_{j=1}^{N_c} \mu_{ij}^\varpi \left\| \varphi(x_i) - \varphi(n_j) \right\|^2. \tag{5}$$

where η is the size of input data, $\varpi > 1$ resents the weighting coefficient, the center of the cluster is represented by n, N_c is the number of clusters, μ_{ij} is the degree of membership of $\varphi(x_i)$ the cluster j and here

$$\left\| \varphi(x_i) - \varphi(n_j) \right\|^2 = \varphi(x_i)^T \varphi(x_i) + \varphi(n_j)^T \varphi(n_j) \\ - 2\varphi(x_i)^T \varphi(n_j). \tag{6}$$

We use the Gaussian function as the nonlinear kernel function. Hence (5) can be rewritten as

$$C_\varpi = 2 \sum_{i=1}^{\eta} \sum_{j=1}^{N_c} \mu_{ij}^\varpi \left(1 - e^{-\frac{\|x_i - n_j\|^2}{\sigma^2}} \right). \tag{7}$$

Thus we get the updating equations for finding membership value μ_{ij} and centroids n_j by minimizing the objective function with respect to μ_{ij}

$$\mu_{ij} = \frac{\left(\frac{1}{(1 - e^{-\frac{\|x_i - n_j\|^2}{\sigma^2}})} \right)^{\frac{1}{(\varpi-1)}}}{\sum_{m=1}^{\eta} \left(\frac{1}{(1 - e^{-\frac{\|x_i - n_j\|^2}{\sigma^2}})} \right)^{\frac{1}{(\varpi-1)}}}, \tag{8}$$

$$n_j = \frac{\sum_{i=1}^{\eta} \mu_{ij}^{\varpi} x_i e^{-\frac{\|x_i - n_j\|^2}{\sigma^2}}}{\sum_{i=1}^{\eta} \mu_{ij}^{\varpi} e^{-\frac{\|x_i - n_j\|^2}{\sigma^2}}}. \tag{9}$$

In Sect. 3, factor analysis was performed on the dataset in Table 5. Thereafter, in order to understand the specific features of cities having similar positioning on factorial plan, and to point out the main differences occurring among those located in different quadrants, cluster analysis was performed on the results.

3 Case Study

City innovation evaluation plays an important role in national innovation development. The aim of the paper is to evaluate city independent innovation ability of 15 vice-provincial cities in China thus to validate findings in previous studies. The evaluation results may improve its industrial structure with agglomeration scale's enlargement and enhance market competition and finally realize Small City leaping development. In order to obtain a synthetic judgment, we probe into the city scientific innovation system under the guideline of scientific, systematic, analogous, objectivity, practicality, maneuverability principles, combine the patent quantity, patent rate and patent value. Based on this, we construct a fundamental patent analysis based city independent innovation ability evaluation system, as can be seen in Table 1.

Table 1. Urban technological innovation ability evaluation guideline based on patent analysis.

Categories	Variables
Patent application (innovative vigor)	Patent Application (X1) Invention Patent Application (X3)
Patent authorization (innovation capability)	Patent Certified (X5) Invention Patent Certified (X8) Patent Authorization Rate (X7) Invention Patent Certified Annual Growth Rate (X10)
Patent validation (innovation survival ability)	Quantity of Valid Patent (X11) Quantity of Valid Invention Patent (X13) Rate of Valid Patent (X12) Rate of Valid Invention Patent (X14)
Patent development potential (innovation potential)	Patent Application Annual Growth Rate (X2) Invention Patent Application Annual Growth Rate (X4) Patent Certified Annual Growth Rate (X6) Invention Patent Certified Annual Growth Rate (X9)
Patent market value (innovation resultant)	Rate of Service Invention Patent Application (X17) Rate of Authorization Service Invention Patent (X18) Rate of Valid Service Invention Patent (X19)
Patent benefit (innovation contribution)	Quantity of Invention Patent per 100 Million Yuan (X15) Quantity of Invention Patent per 10 Thousand People (X16)

The proposed model has been tested on data extracted by statistical data of the State Intellectual Property Office of the People's Republic of China issued by National Bureau of Statistics of China. Analysis is conducted on 19 indicators, identified as able to describe some key factors of city independent innovation ability performance. Such indicators have been classified into 6 categories, according to the different aspects to whichthey are related.

The X1 and X3 indicators describe cities' patent application category;
The X5, X7, X8, X10 indicators describe cities' patent authorization category;
The X11, X12, X13, X14 indicators describe cities' patent validation category;
The X2, X4, X6, X9 indicators describe cities' patent development potential category;
The X17, X18, X19 indicators describe cities' patent market value category;
The X15, X16 indicators describe cities' patent benefit category.

In order to obtain a synthetic judgment on innovation performance and an indication of the multiple relations existing among the variables, so as to better understand the analyzed phenomenon, Factor Analysis has been performed on dataset in Table 5 in Appendix section. We first standardize the original dataset. Then correlation coefficient matrix R between variables is computed, which is measured on a scale of -1 to $+1$. The bigger the correlation coefficient is the more related the two variables are. The eigenvalue and unit eigenvector accordingly of R is computed, see Table 2.

Table 2. The eigenvalue of R and eigenvalue contribution.

Component		1	2	3	4
Initial eigenvalue	Total	9.888	4.279	1.920	1.231
	Variance %	52.040	22.523	10.104	6.481
	Accumulation %	52.040	74.563	84.667	91.148
Extraction sums of squared loadings	Total	9.888	4.279	1.920	1.231
	Variance %	52.040	22.523	10.040	6.481
	Accumulation %	52.040	74.563	84.667	91.148
Rotation sums of squared loadings	Total	7.969	4.518	3.553	1.277
	Variance %	41.945	23.779	18.02	6.722
	Accumulation %	41.945	65.723	84.426	91.148

The cumulative contribution of variances of the first four components is 91.148%, therefore they are selected as principle components. Based on the principle components, we compute factor loading matrix and perform maximum variance cross rotation on it. Component Score Coefficient Matrix and component score are obtained, as can be seen in Table 3.

Table 3. Component score coefficient matrix.

Variables	1	2	3	4
Patent Application (X1)	0.153	−0.056	−0.089	−0.012
Patent Application Annual Growth Rate (X2)	−0.104	0.273	0.025	−0.134
Invention Patent Application (X3)	0.162	−0.095	−0.017	−0.049
Invention Patent Application Annual Growth Rate (X4)	−0.079	0.271	0.105	−0.166
Patent Certified (X5)	0.142	−0.039	−0.115	0.088
Patent Certified Annual Growth Rate (X6)	−0.066	0.231	−0.015	0.076
Patent Authorization Rate (X7)	−0.001	0.059	−0.090	0.344
Invention Patent Certified (X8)	0.159	−0.096	−0.006	0.030
Invention Patent Certified Annual Growth Rate (X9)	0.072	0.066	−0.001	−0.099
Invention Patent Certified Annual Growth Rate (X10)	−0.002	−0.057	0.053	0.752
Quantity of Valid Patent (X11)	.0123	−0.007	−0.088	0.098
Rate of Valid Patent (X12)	−0.061	0.235	0.086	0.142
Quantity of Valid Invention Patent (X13)	0.156	−0.086	−0.010	−0.008
Rate of Valid Invention Patent (X14)	−0.006	0.178	0.024	−0.074
Quantity of Invention Patent per 100 Million Yuan (X15)	0.097	−0.009	0.079	−0.015
Quantity of Invention Patent per 10 Thousand People (X16)	0.124	−0.034	0.029	−0.012
Rate of Service Invention Patent Application (X17)	−0.034	0.044	0.285	0.084
Rate of Authorization Service Invention Patent (X18)	−0.066	0.078	0.312	0.037
Rate of Valid Service Invention Patent (X19)	−0.043	0.027	0.292	−0.009

We perform the Kaiser criterion and finally we get:

$$F1 = 0.153X1 - 0.104X2 + 0.162X3 + \ldots - 0.043X19 \tag{10}$$

$$F2 = -0.056X1 + 0.273X2 - 0.095X3 + \ldots + 0.027X19 \tag{11}$$

$$F3 = -0.089X1 + 0.025X2 - 0.017X3 + \ldots + 0.292X19 \tag{12}$$

$$F4 = -0.012X1 - 0.134X2 - 0.049X3 + \ldots - 0.009X19 \tag{13}$$

As can be seen that in F1 the factor loading values of Quantity of Valid Patent (X11), Quantity of Invention Patent per 10 Thousand People (X16), Patent Application (X1), Quantity of Valid Invention Patent (X13), Invention Patent Certified Annual Growth Rate (X9), Patent Certified (X5), Invention Patent Application (X3), Quantity of Invention Patent per 100 Million Yuan (X15) are relatively bigger, which means that F1 is a composite indicator indicating the quantity of patent. In other word, F1 characterizes a city's innovative vigor, innovation capability, innovation survival ability and innovation contribution.

In F2, the factor loading values of Rate of Service Invention Patent Application (X17), Rate of Authorization Service Invention Patent (X18), Rate of Valid Service Invention Patent (X19) are relatively bigger, which means that F2 is a composite indicator indicating innovation resultant.

In F3, the factor loading values of Rate of Authorization Service Invention Patent (X18), Rate of Valid Patent (X12), Invention Patent Application Annual Growth Rate (X4), Rate of Service Invention Patent Application (X17), Patent Application Annual Growth Rate (X2), Rate of Valid Service Invention Patent (X19), Patent Certified Annual Growth Rate (X6) are relatively bigger. Among them three are Service Invention Patent related factor loading values, another three are Patent Application Annual Growth Rate related factor loading values. This reveals that F3 is a composite indicator indicating innovation potential and innovation resultant.

In F4, the factor loading values of Patent Authorization Rate (X7), Invention Patent Certified Annual Growth Rate (X10) are relatively bigger which means that F4 is a composite indicator indicating patent authorization rate therefore represents a city's innovation capability.

In a word, F1, F2, F3 and F4 can express the original variables perfectly in the aspects of patent quantity, patent growth rate, rate of authorization service invention patent and patent authorization rate.

A synthesis evaluate function is designed using F1, F2, F3 and F4 and eigenvalue contribution:

$$F = 0.4195F1 + 0.2378F2 + 0.1870F3 + 0.0672F4 \qquad (14)$$

Thus, the synthesis evaluation values of the 15 cities are obtained and sorted. Also, cluster analysis is performed on the results to identify the groups of cities that are characterized by the same distinctive elements. Table 4 shows factor analysis and cluster results of innovation ability based on patent data in 15 vice-provincial cities.

As can be seen in Table 4 that the top five cities are Shenzhen, Hangzhou, Chengdu, Guangzhou and Ningbo. Among them, Shenzhen don't have high value in F2 and F3 which indicates that the performances of service invention patent and patent annual growth rate are not satisfactory. In other words, Shenzhen does not have advantage in innovation potential and innovation resultant over the others. However, it surely outperforms the others in F1 and F4 which tells why Shenzhen ranks first in the list. The values of F1, F2 and F3 for Hangzhou, Chengdu and Guangzhou are similar. They rank second, third and fourth. Ningbo ranks fifth but has a relatively high F4 value which means that it does well in Patent Authorization Rate. Nanjing, Wuhan, Jinan, Xi' an, Dalian, Qingdao and Shenyang have similar performance which rank from sixth to twelfth. Among them, Wuhan, Jinan, Qingdao have good Innovation Capability as they have larger F4 values.

We cluster the factor scores using nearest neighbor method. The 15 cities can be classified into 4 clusters according to their independent innovation ability based on patent. The first cluster is Shenzhen with a much higher score than the rests. The second cluster includes Hangzhou, Chengdu, Guangzhou and Ningbo, most of which

Table 4. Factor analysis and cluster results of innovation ability based on patent data in 15 vice-provincial cities.

City	F1		F2		F3		F4		F	Cluster result
	Value	Rank	Value	Rank	Value	Rank	Value	Rank		
Shenzhen	9.921	1	−3.665	15	−4.989	15	1.391	1	2.451	1
Hangzhou	4.531	2	−1.533	14	−2.516	14	0.929	3	1.128	2
Chengdu	4.094	3	−1.317	12	−2.443	13	0.923	4	1.010	2
Guangzhou	4.011	4	−1.347	13	−2.303	11	0.821	5	0.987	2
Ningbo	3.680	5	−1.073	11	−2.339	12	1.088	2	0.925	2
Nanjing	2.725	6	−1.023	10	−1.347	9	0.318	9	0.670	3
Wuhan	2.546	7	−0.901	9	−1.370	10	0.401	6	0.625	3
Jinan	2.260	8	−0.779	8	−1.274	8	0.346	7	0.548	3
Xi'an	2.090	9	−0.767	7	−1.067	7	0.243	12	0.511	3
Dalian	1.893	10	−0.67	6	−1.035	5	0.211	13	0.454	3
Qingdao	1.822	12	−0.615	4	−1.041	6	0.320	8	0.445	3
Shenyang	1.822	11	−0.663	5	−0.963	4	0.247	11	0.443	3
Ha'erbin	1.315	13	−0.488	3	−0.661	3	0.173	14	0.324	4
Xiamen	1.039	14	−0.326	1	−0.620	2	0.255	10	0.260	4
Changchun	0.956	15	−0.350	2	−0.487	1	0.127	15	0.235	4

are located in the south coastal areas of China. Nanjing, Wuhan, Jinan, Xi'an, Dalian, Qingdao and Shenyang are in the third cluster. The rest three cities are in the fourth cluster.

4 Conclusion

The quantity and quality of patent that a city owns as well as the capability of managing patent institution indicate urban technological innovation ability and core competence. In this work, we construct urban technological innovation ability evaluation system based on patent dataset collected from the State Intellectual Property Office of the People's Republic of China in 2010. Factor analysis and cluster method is applied to 15 vice-provincial cities in China to analysis the collected patent data samples and a demonstration research is indicated upon the experimental results. The result of the evaluation system reveals the innovation ability distribution of the cities and their advantages and disadvantages respectively. The findings of this work would be of great significance in helping cities formulating suitable innovation development strategies.

Acknowledgment. This work was supported by the Shandong Provincial Natural Science Foundation ZR2014FL021 and supported in part by Qingdao Basic Research Program 15-9-1-83-jch.

Appendix

See Table 5.

Table 5 Paten guideline data of 15 vice-provincial cities.

City	Patent Application [a]	Patent Application Annual Growth Rate [b](%)	Invention Patent Application [a]	Invention Patent Application Annual Growth Rate [b](%)	Patent Certified [a]	Patent Certified Annual Growth Rate [b](%)	Patent Authorization Rate	Invention Patent Certified [c]	Invention Patent Certified Annual Growth Rate [b](%)	Invention Patent Certified Annual Growth Rate [b](%)	Quantity of Valid Patent [c]	Rate of Valid Patent [c](%)	Quantity of Valid Invention Patent [c]	Rate of Valid Patent [c](%)	Quantity of Invention Patent per 100 Million Yuan [d]	Quantity of Invention Patent per 10 Thousand People [d]	Patent Application [d](%)	Rate of Service Invention Service Invention Patent [c](%)	Rate of Valid Service Invention Patent [c](%)
	X1	X2	X3	X4	X5	X6	X7	X8	X9	X10	X11	X12	X13	X14	X15	X16	X17	X18	X19
Guangzhou	115391	21.4	27686	28.9	72072	24.4	62.5	7850	37.6	28.4	44706	52	6469	78.9	0.7	8.1	42	39.2	47
Changchun	27480	16.5	10145	23	13744	21.3	50	2985	54.6	29.4	7876	55.5	2120	71	0.7	2.8	53.3	53.6	62.5
Wuhan	76554	24.6	20999	26.3	37383	26.3	48.8	7233	40	34.4	24969	57.9	5353	70.1	1.2	6.4	66.3	64.4	75.4
Nanjing	76130	30	31777	37.9	34541	28.8	45.4	9140	144.8	28.8	24138	68.3	7340	80.3	1.7	11.7	66.6	72.9	80.3
Hangzhou	121083	38.4	36133	45.4	75025	47.9	62	11067	115.5	30.6	57393	75.8	9493	85.8	1.9	13.9	61.6	57.5	64.8
Xi'an	63419	32.5	22224	41.8	25860	27.6	40.8	5504	43.6	24.8	18902	59.1	4759	81.3	1.7	6.1	68.3	68.5	78.8
Jinan	75016	31.6	16281	36.6	34678	37.6	46.2	3706	117.4	22.8	20394	57.7	3061	82.6	0.9	5.1	31.2	37.9	51.1
Shenyang	54447	13.8	18133	25.4	27702	10.9	50.9	4661	25.4	25.7	14508	35	3253	62.3	0.8	4.5	44.6	42.2	60
Chengdu	121596	33.1	22696	31.6	75810	39.4	62.4	5782	129.6	25.5	50705	66	4812	83.2	1.1	4.2	35.9	31.1	37.2
Dalian	68198	38.3	14464	31	24087	26.2	35.3	3552	29.6	24.6	14331	50	2699	69.8	0.6	4.6	44.9	38.9	52.2
Xiamen	29117	21.6	5807	39.9	20739	23.2	71.2	1444	97.6	24.9	13191	61.5	1318	91.3	0.8	7.4	60	58.4	65.5
Ha'erbin	36199	17.7	14660	29.6	18508	16.4	51.1	4347	35.6	29.7	11361	44.8	3401	73.6	1.1	3.4	47.6	44.5	56.2
Shenzhen	256085	28.1	116639	46.4	136519	31.5	53.3	28854	276.6	24.7	102494	73.7	29221	101.3	3.6	32.8	71.1	64.3	73.7
Qingdao	58337	22.7	12475	33.4	29813	17.1	51.1	2896	41.5	23.2	18228	49.6	2239	73.8	0.5	2.9	68.3	68.7	72.3
Ningbo	103134	47.2	11492	49.5	75391	51.6	73.1	3327	120.8	29	60621	79.6	3137	94.3	0.72	5.5	32.6	30	33.8

[a] Data source: According to statistical data of the State Intellectual Property Office of the People's Republic of China; [b] Annual Growth Rate: According to Statistical Yearbook; [c] Data source: According to the China Valid Patent Report, issued by SIPO; [d] Data source: According to CHINA STATISTICAL YEARBOOK

References

1. Liu, S.J.: Patent map – a route to a strategic intelligence of industrial competitiveness. In: Proceedings of the 1st Asia-Pacific Conference on Patent Maps, pp. 2–13, Taipei (2003)
2. Bay, Y.M.: Development and applications of patent map in Korean high-tech industry. In: Proceedings of the 1st Asia-Pacific Conference on Patent Maps, pp. 18–23, Taipei (2003)
3. Athey, G., Nathan, M., Webber, C., Mahroum, S.: Innovation and the city. Innov.: Manag. Policy Pract. **10**, 156–169 (2009)
4. Nelson, R.: National Systems of Innovation: A Comparative Study. Oxford University Press, Oxford (1999)
5. Riddel, S.: Regional innovative capacity with endogenous employment: empirical evidence from the U.S. Rev. Reg. Stud. **33**, 73–84 (2003)
6. Lindhult, E., Campillo, J., Dahlquist, E., Read, S.: Innovation capabilities and challenges for energy smart development in medium sized European cities. Energy Procedia **88**, 205–211 (2016)
7. Fu, X.L.: Foreign direct investment, absorptive capacity and regional innovation capabilities: evidence from China. Oxford Dev. Stud. **36**, 89–110 (2008)
8. Technopolis Group and Mioir: evaluation of innovation activities guidance on methods and practices. European Commission, Directorate for Regional Policy, http://ec.europa.eu/regional_policy/sources/docgener/evaluation/pdf/eval2007/innovation_activities/inno_activities_guidance_en.pdf. Accessed 21 Feb 2017

9. Xing, G.: Innovation ability evaluation research based on GIOWA operator. In: Proceedings of the International Conference on Management and Service Science (MASS), pp. 1–4, Wuhan, China(2010)
10. Zhang, D.Z., Gao, Y.Y., Peng, Z.L.: Independent innovation capability evaluation and analysis on wanjiang city-zone. In: Proceedings of the International Conference on Computer Science and Service System (CSSS), pp. 2286–2289, Nanjing, China (2011)

Research on Target Extraction Technology of Fruit and Vegetable Images in the Complex Environment

Yanqing Wang[✉] and Hao Zheng

Key Laboratory of Trusted Cloud Computing and Big Data Analysis,
School of Information Engineering, Nanjing Xiaozhuang University,
Nanjing 211171, China
wyq0325@126.com

Abstract. Considering the difficulty of fruit and vegetable images with uneven illumination and uncontrolled backgrounds, this paper proposed an image pre-process algorithm based on visual subject detection. Firstly, we can use manifold ranking to significance test and to acquire significance images, then use gradient images and position weighting to get cutting images, finally adjust the image brightness and remove image noise to complete the image preprocessing. The experiment results demonstrated the robustness and real-time of the proposed algorithm which can segment images accurately and the accuracy is higher than 91%.

Keywords: Manifold ranking · Significance images · Gradient images · Image segment

1 Introduction

With the gradual development of image recognition technology, image recognition technology has been maturing. In the image processing stage, the original image can not be directly applied to the image recognition software due to the factors such as the acquisition of the tool, the environment, the means and so on. Therefore, image pre-processing is very important and essential in order to improve the image data, suppress the adverse factors. In the actual project, the images are often collected under uncontrolled background and uncontrolled light source, and the image acquisition tool is not unified. Therefore, in the image preprocessing, the image needs to be normalized and de-noised and the brightness of it needs to be adjusted.

According to the main body of the cut image, the Image cutting can be divided into integrity zoom cut or partial cutting.

Integrity Scaling means zooming the entire image according to a certain algorithm and highlight the main part of the image. Initially, the main method is interpolation when cutting the image, and after cutting the image, it is proportional to the former, but the cutting image's clarity has been reduced. With the gradual improvement of technology, researchers focus on image cutting to maintain the visual subject of the image, the current mainstream cutting algorithm are classified as non-uniform scaling and line cutting.

© Springer Nature Singapore Pte Ltd. 2017
B. Zou et al. (Eds.): ICPCSEE 2017, Part I, CCIS 727, pp. 708–717, 2017.
DOI: 10.1007/978-981-10-6385-5_59

(1) Non-uniform scaling

Non-uniform scaling is to keep the visual subject of the image unchanged or evenly changed, and other non-interested areas of the image are non-uniform (such as stretching, twisting, etc.) to achieve the final cutting effect.

(2) Line cutting

The line cutting algorithm first looks for a pixel with a small energy value, and by changing (e.g., removing or inserting) a pixel with a small energy value, that is, the total energy of the image change is small, to achieve the purpose of changing the image size. [1] Fig. 1(a) for the classic line cutting algorithm - Forward algorithm of the cutting line diagram, as shown in the figure, the cutting line is mostly around the surfers, reasonable Fig. 1(b) shows the final result of the Forward algorithm. Figure 1(b) shows that the image after image cutting retains the image visual subject, and the surrounding area is not significantly distorted.

(a) Forward algorithm trimming line   (b) Forward algorithm cut results

Fig. 1. Forward cutting algorithm

Part of the cut is based on color, location and other factors cut part of the original image, cut the image after the original image of the sub-graph. Most of the methods used in the original image visual subject in the original image is relatively small or the original image has multiple visual subjects of the situation, the more fiercely face recognition field mostly use part of the cut. Figure 2(a) is the original image, the algorithm is based on the skin color cut, the original image is cut into a number of skin color part of the visual sub-image (such as shown in Fig. 2(b), the cut sub-image is a partial image of the original image.

As the original image of this article in the uncontrolled light source, uncontrolled background and uncontrolled shooting tool to obtain the case, so the there exits factors which would affect the image like the light is too bright or too dark, the background is not uniform, fruit and vegetable position is not uniform, the original image size is not uniform and so on.

Figure 3 is interference factor of light diagram, as shown in the figure, the two images are images of apple and picture background is the same, but because of the

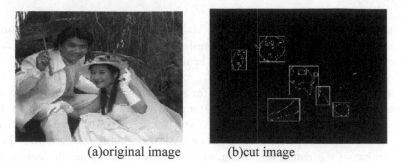

(a)original image   (b)cut image

Fig. 2. The cut schematic according to the skin color

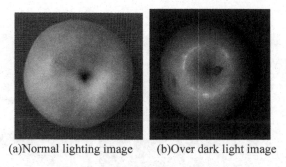

(a)Normal lighting image   (b)Over dark light image

Fig. 3. Interference factor of light diagram

different shooting time, resulting in great differences in image brightness, thus affecting the image color information and contour Information; Fig. 4 for interference factor of background diagram, as shown in the figure, two orange image background color, texture are not the same, coupled with shadow and other interference factors, for the subsequent feature extraction caused great interference. Figure 5 for interference factor of position diagram, as shown in the figure, similar to the size of fruits and vegetables in the shooting angle, the location is not the same circumstances, will produce a big difference to the smaller fruit and vegetable image for the visual body is easy Is not affected by the area of interest to lose the visual focus.

(a)Fabric background image   (b)Desktop background image

Fig. 4. Interference factor of background diagram

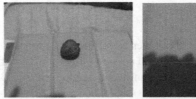

(a)The visual subject located in the center of the image
(b)The visual subject is the upper right corner

Fig. 5. Interference factor of position diagram

As shown in Fig. 5, the original image size is large, and the visual subject (fruit and vegetable) occupies less in the original image. Therefore, the partial image of the original image is preprocessed by the fruit and vegetable image preprocessing, and the sub-image of the original image is cut after the image more than 80%. As the various types of fruits and vegetables vary widely, color, shape, texture are not uniform, and the location of fruit and vegetable shooting is not uniform (to the main focus of the fruit and vegetable, there are still some part of the position of fruit and vegetable migration), cutting is not easy to use color, texture, shape and location and other factors as a basis for cutting. At present, most of the cutting algorithms use a single feature as a cutting basis, such as OTSU algorithm based on color feature [2], Markov algorithm based on texture feature and contour feature based on contour position HU invariant moments and other algorithms, these algorithms do not apply to the complex environment of fruit and vegetable image recognition segmentation process. The algorithm proposed in this paper is based on the visual subject to detect and combine many kinds of features, which greatly improves the precision of segmentation.

2 Cutting Algorithm Based on Visual Subject Detection

In 2013, Yang et al. [3] proposed a graph-based manifold prioritization detection algorithm at the CVPR International Conference. Set the image line vector $X = \{x_1, \ldots, x_n\} \in \Re^n$ for the algorithm for each pixel to define a popular value f, and set the weight $W = [w_{ij}]_{n \times n}$ and $D = diag\{d_{11}, \ldots, d_{nn}\}$ between the points, $d_{ii} = \sum_j w_{ij}$ similar to spectrum clustering algorithm. According to formula (1) to obtain f^*, which μ is balanced parameter.

$$f^* = \arg\min_f \frac{1}{2} \left(\sum_{i,j=1}^{n} w_{ij} \left\| \frac{f_i}{\sqrt{d_{ii}}} - \frac{f_j}{\sqrt{d_{jj}}} \right\|^2 + \mu \sum_{i=1}^{n} \|f_i - x_i\|^2 \right) \tag{1}$$

Since the formula (1) can also be written as the formula (2), the E in formula (2) is the unit matrix, $\alpha = 1/(1 + \mu)$, The matrix S can be obtained by formulas (1) and (2).

$$f^* = (E - \alpha S)^{-1} X \qquad (2)$$

According to the upper edge, the lower edge, the left edge and the right edge, we can get the weight W_t, W_b, W_l and W_r, and so we can get four significant graphs S_t, S_b, S_l and S_r, for a given image, the final significant degree see Eq. (3).

$$S(I) = S_t(i) \times S_b(i) \times S_l(i) \times S_r(i) \qquad (3)$$

The saliency map can reflect the image attention of human vision, and the gradient energy graph can describe the image better. Given image for a given image I, the gradient function of the image is given by Eq. (4)

$$e(I) = \left| \frac{\partial}{\partial x} I \right| + \left| \frac{\partial}{\partial y} I \right| \qquad (4)$$

Because the image library used in this algorithm has certain position information, it may add image position information when cutting. For a given image I, the distance from the upper left corner of the image to the center of the image (that is, the square of the sum of the sums of the sums) Δk is the distance from the current pixel to the center of the image. The position information values are given in Eq. (5), where μ are the control parameters, the control parameters $\mu = 40$.

$$U(I) = \mu \frac{d - \Delta k}{d} \qquad (5)$$

In this paper, we proposed the algorithm of saliency, gradient energy and position fusion. For a given image I, the cutting degree is shown in Eq. (6).

$$W(I) = S(I) + e(I) + U(I) \qquad (6)$$

3 Image Preprocess

Color space, is mainly used to illustrate colors in an acceptable way under certain criteria. The color space is the elaboration of the coordinate system and the subspace, and each color in the system has a specific value. Most color spaces are for hardware or application-oriented. In the field of image processing mostly use RGB color space, HIS color space, HSV color space, through the image color space conversion can be more convenient to extract some of the features or some kind of processing.

Figure 6(a) and (c) are too dark or too bright images, respectively, (b) and (d) for the final algorithm to adjust the image, the figure can be seen that the use of the algorithm has initially solved the non-extreme situation of too bright or too dark.

In real life, the digital image is often generated by the hardware of the acquisition device or the external noise of the collection environment during the generation and transmission, and the noise is often present in the acquired image, that is, the noise

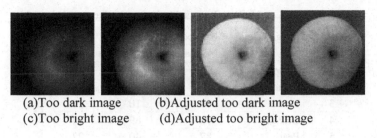

(a)Too dark image   (b)Adjusted too dark image
(c)Too bright image   (d)Adjusted too bright image

Fig. 6. Brightness adjustment comparison chart

image or the noisy image. In the field of digital image processing, the process of reducing or removing image noise is called image de-noising. This process will reduce the image interference information and improve the accuracy of the subsequent processing. Therefore, the image de-noising step is very important. There are several common methods for removing noise: the mean filter [4], which mainly uses the neighborhood mean method, which is mainly applied to the particle noise in the image obtained by scanning. The algorithm is the most basic algorithm in the de-noising algorithm. The median filter, the main function of the algorithm is to make the image of the gray value of the larger difference between the image and the value close to the surrounding pixel value, and then eliminate the isolated noise, So the algorithm is very effective for salt and pepper noise. Wavelet de-noising, the algorithm preserves most of the wavelet coefficients that contain the signal, which can preserve image detail better.

In this paper, the algorithm is based on the Gaussian de-noising algorithm with high fuzzy effect. The algorithm is based on the Gaussian de-noising algorithm with high fuzzy effect. The algorithm can better meet the research requirements. Figure 7(a) is the original image, (b) is the mean filter, (c) is the median filter image, and (d) is the Gaussian filtered image.

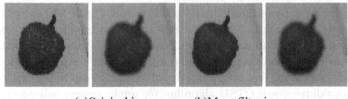

(a)Original image   (b)Mean filter image
(b)(c)Median filter image   (d)Gaussian filter image

Fig. 7. Comparison of various de-noising algorithms

4   Experimental Results and Analysis

This paper collected the total of 12 kinds of common fruit and vegetable images, the total images of 1461, as shown in Table 1. These images are collected in uncontrolled light sources and uncontrolled background conditions.

Table 1. Number of fruit and vegetable image database statistics table

Order	Image category	Number of images		
		Normal light	Over light	Dark light
1	Appple	96	47	48
2	Banana	100	72	46
3	Litchi	10	10	10
4	Orange	92	89	84
5	Pear	87	40	27
6	Orange	53	31	22
7	Nectarine	20	12	12
8	Carrot	50	34	45
9	Cucumber	49	22	21
10	Eggplant	32	10	10
11	Tomato	28	15	20
12	Winter jujube	45	38	34
Total		662	420	379
		1461		

It can be seen from the introduction that it is difficult to use a single feature for segmentation and cutting, and this paper takes the two-dimensional OTSU algorithm [5] and the adaptive Canny algorithm [6] as an example. The two-dimensional OTSU algorithm is an improved OTSU algorithm, which divides the image into foreground and background colors according to the color characteristics. Because of the large difference in the image library environment used in this paper, we can not segment the image better by using this method. The adaptive Canny algorithm is an improved algorithm of Canny algorithm. The algorithm can extract the features of image contours. However, if the image details are more, the algorithm obtains more contours and is less likely to be segmented by contours.

Figure 8 for a variety of algorithm comparison chart. Figure 8(a) shows the original images of apples, oranges and nectar; Fig. 8(b) shows the significant images of apples, oranges and nectar; Fig. 8(c) shows the gradient of apples, oranges and nectar Fig. 8(d) is the final cut of the apple, orange and nectarine image, has clearly highlighted the visual subject (red for the cutting line, the cut length and width ratio can be set, here set Fig. 8(e) shows the two-dimensional OTSU algorithm for apple, orange and nectarine (image segmentation thresholds 136, 139 and 97, respectively), which is 1:1, and is finally scaled to 64 * 64 pixel image The algorithm is mainly used to deal with the color characteristics of the image. It can be seen from (e) that the algorithm handles the complex image of the shooting environment. Figure 8(f) is the adaptation of apple, orange and nectarine Canny The thresholds of the images are 29.20 and 11.68, and the thresholds of the orange images are 24.66 and 9.86 respectively, and the thresholds of the nectarine images are 32.55 and 13.01 respectively. The algorithm mainly uses the contour features of the images to deal with (f) can be clearly seen, through the contour feature of the image is very complex, the details of the characteristics of too much, it is difficult to obtain fruit and vegetable visual subject. Compared with the algorithm image

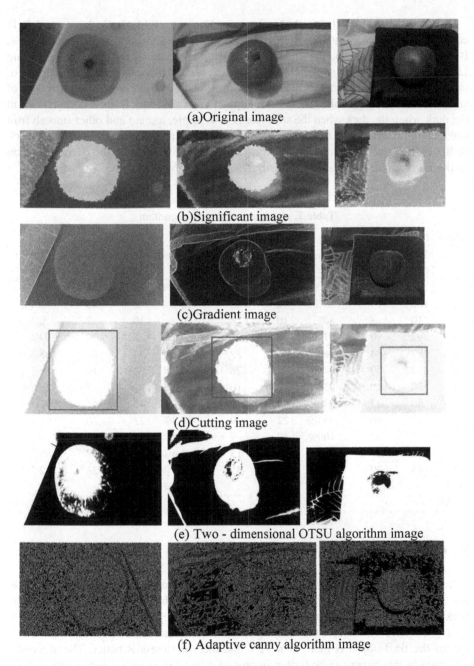

(a)Original image

(b)Significant image

(c)Gradient image

(d)Cutting image

(e) Two - dimensional OTSU algorithm image

(f) Adaptive canny algorithm image

Fig. 8. Comparison of multiple algorithms (Color figure online)

of the single feature processing, it is easier, more convenient and accurate to cut the image according to the original image than the original image. Compared with the algorithm image of the single feature processing, the algorithm described in this paper is more effective.

The correct rate of the final algorithm is shown in Table 2. Under normal light conditions, the correct rate of other fruits except lychee and jujube is more than 90%. Because litchi and jujube are small in size, they are affected by background texture Large, the correct rate is slightly lower, but also reached more than 80%. Too bright or too dark image of the correct rate is basically lower than the normal fruit image of similar fruits and vegetables, nectarine, eggplant and other dark fruits and vegetables in the dark when the dark when the split rate is low, apple, tomato and other smooth fruit and vegetable, When it is too bright, it is easy to reflect and lead to segmentation errors, so the correct rate is slightly lower. According to the overall data, the overall division of the correct rate can reach more than 91%, more robust.

Table 2. Accuracy of the cut algorithm

Order	Normal lighting image			Over light image			Over dark light image		
	Total quantity	Correct quantity	Correct rate	Total quantity	Correct quantity	Correct rate	Total quantity	Correct quantity	Correct rate
1	96	92	95.8%	47	42	89.3%	48	46	95.8%
2	100	99	99%	72	65	90.2%	46	40	86.9%
3	10	8	80%	10	8	80%	10	7	70%
4	92	90	97.8%	89	82	92.1%	84	79	94%
5	87	84	96.5%	40	37	92.5%	27	24	88.8%
6	53	50	94.3%	31	28	90.3%	22	19	86.3%
7	20	19	95%	12	10	83.3%	12	7	58.3%
8	50	47	94%	34	29	85.2%	45	41	91.1%
9	49	46	93.8%	22	19	86.3%	21	20	95.2%
10	32	32	100%	10	9	90%	10	7	70%
11	28	27	96.4%	15	13	86.6%	20	17	85%
12	45	39	86.6%	38	31	81.5%	34	28	82.3%
95%			88.8%			88.4%			
91.8%									

5 Conclusion

In this paper, a preprocessing algorithm based on visual subject detection is proposed to solve the problem that fruit images are difficult to be cut due to uneven illumination and background control in complex environment. The preprocessing of the algorithm mainly includes three steps: cutting, brightness adjustment and de-noising. The correct rate of the final object is more than 91% and the robustness is better. The algorithm proposed in this paper can be further improved when the visual subject is smaller, too bright and too dark. The algorithm described in this paper can pre-process and cut the fruit and vegetable images in complex environment, which will lay a good foundation for the follow-up feature extraction and image recognition. The algorithm described in this paper can be used in practical engineering of various fruits and vegetables identification.

Acknowledgment. This work has been supported by the Jiangsu Engineering Research Center for Networking of Elementary Education Resources (grant number: BM2013123).

References

1. Masdiyasa, I.G.S., Purnama, I.K.E., Purnomo, M.H.: Teratozoospermia classification based on the shape of sperm head using OTSU threshold and decision tree (2016)
2. Yanqing, W., Deyun, C., Chaoxia, S.: Vision-based road detection by Monte Carlo method. Sci. Alert: Inf. Technol. J. **9**(3), 481–487 (2010)
3. Yang, C., Zhang, L., Lu, H., et al.: Saliency detection via graph-based manifold ranking, pp. 3166–3173 (2013)
4. Buades, A., Coll, B., Morel, J.M.: Image data processing method by reducing image noise, and camera integrating means for implementing said method. EP Patent 1,749,278, 7 Feb 2007
5. Wang, Y.Q., Zhuang, L.L., Xia, S.C.: Construction research on multi-threshold segmentation based on improved Otsu threshold method. In: 4th International Conference on Automation, Communication, Architectonics and Materials, ACAM, Advanced Materials Research, vol. 1046, pp. 425–428 (2014)
6. Wang, Y., Wang, Y., Shi, C., Shi, H.: Research on feature fusion technology of fruit and vegetable image recognition based on SVM. In: Social Computing - 2nd International Conference of Young Computer Scientists, Engineers and Educators, ICYCSEE 2016, pp. 591–599 (2016)

Route Guidance for Visually Impaired Based on Haptic Technology and Their Spatial Cognition

Guansheng Wang[1,2], Jianghua Zheng[2(✉)], and Hong Fan[2]

[1] The Air Force Xi'an Flight Academy, Xi'an 710300, Shanxi, China
[2] College of Resource and Environment Sciences,
Xinjiang University, Urumqi 830046, Xinjiang, China
zheng_jianghua@126.com

Abstract. Haptic (vibration) information expression is an effective human-computer interaction mode and information transfer method. It makes up shortcomings of sound under certain conditions and is an important channel of information transfer for route guidance field. As a special kind of walkers, the visually impaired pedestrians have a specific type of cognition or perception of route guidance environment (including spatial orientation, distance and walking speed etc.) and they have sharp sense of touch resources form the ordinary. This work integrated application of GPS, GIS and Haptic (vibration) technology to develop more reliable mode route guidance for the visually impaired. It provides different vibration to the user under two circumstances: key nodes of roads ahead and deviation planning path. It resists environmental noise, reflects timely and has high effectiveness. Based on Google Legend phone, we developed the prototype and realized the designed functions, and verified the effectiveness of the system. We initially determined the thresholds of deviating from the path, those at road junctions and other nodes through experiments, interviews etc. And we used the thresholds for experiments testing and guiding. Then they were inspected, corrected and improved in the field practice. Finally, more reasonable thresholds were drawn out for future applications in reality.

Keywords: Visual impairment · Route guidance · Spatial cognition · Haptic (vibration)

1 Introduction

According to the "the second disabled survey data of China in 2006", statistics showed that: China has the largest number of visually impaired people in the world. The up to 12.33 million visually impaired people is about 20% of the whole world, and the number is increasing at the rate of 450000 per year [1]. More than 80% of information in everyday life is obtained through vision [2]. Unfortunately, the visually impaired people, living in a dark world, have great difficulty in life, study, work and so on because of lack of vision. As a special group in the society, they should be given more care and attention. And one of the biggest troubles is to walk safely without other's

© Springer Nature Singapore Pte Ltd. 2017
B. Zou et al. (Eds.): ICPCSEE 2017, Part I, CCIS 727, pp. 718–728, 2017.
DOI: 10.1007/978-981-10-6385-5_60

companionship. With the development of economy and technology, their troubles in traveling environment were focused.

For ordinary people, the visual sense is the most effective way to obtain environmental information, including landmarks, location, direction, distance and speed and effective route guidance information. The visually impaired people inevitably are confronted with great challenge when walking in complex and mobile environment because of the loss of visual ability. For example: how to get certain distant and spatial information; how to acquire direction through the sign; how to keep the judgment and trace of the direction and location; and how to identify the specific sign when approaching it [3, 4], etc. During their walking, if they could not only get support from general third-party assistant devices (such as a cane, guide dog, blind road, guide equipment etc.), but also obtain directions by their spatial cognition [5], they would more safe and convenient in individual walking.

The cane, as an assistant walking aid, has been widely adopted in reality. However, many restrictions in action of the blind make it has big challenge for effective applying [6]. Just because its' simple design, convenient usage, most visually impaired people are still depending on cane to grope forward. Simply from the functional point of view, guide dog is a good choice of leading the visually impaired to walk, since they could understand master's various passwords and simple intentions, and take the blind to destination successfully. However, they can't plan a path or lead the blind man to the appointed place in an unfamiliar environment. And the training of the dogs is hard, time-consuming, and high cost. These restrict the popularization of guide dogs [7]. Nowadays, many countries equipped the visual impaired with basic blind roads and voice prompt equipments, and other guide facilities. However, their limited functions cannot meet walking requirements of the blind. Their functions are even more limited in china mainland because of low usage ratio and wide occupation of the blind road (there even has open manhole covers next to the blind road, etc.).

Aiming at relieve the walking troubles of the visually impaired, many research teams from various countries are working on better guide devices except traditional cane and guide dogs. Some obtained progress make the visually impaired more convenient and safe while walking, such as the successful development of Electronic Travel Aids [8–10], Guide Robots [11–14], Smart Cane [15–18] as well as Smart Guide System [19–22], and so on. Most of these products are voice-navigation based. But this kind of products have an obvious shortcoming that is they would lose most of the functions while it was noisy around. And if the visual impaired wear headphone together, this may block the important guiding voice outside (such as the car's whistle). The visually impaired are depending on the non-visual way to groping forward. On this occasion, how to implement an effective and efficient guiding? We believe voice based on the haptic (vibration) guidance is a better solution [23, 24].

This paper focused on haptic based mobile route guidance aided mode, which integrated GPS, GIS, haptic and the specific spatial cognition of the blind, for better walking of the visually impaired. Its prototype was developing on Google Legend phone. Field experiments showed its characteristics of better resistance of noise, real-time reflection and high effectiveness, etc. It makes up the shortcomings of sound under certain conditions. Integration of vibration and voice can be a better choice for visual impaired to travel safer and more convenient. The vibration thresholds of deviating from the path were discussed. We initially determined the threshold of deviating from the path, at road junctions by experiments, interview etc. with the visually impaired students and staff of the Blind School of Urumqi. After inspecting, correcting and improving the thresholds by tests, we finally obtained more reasonable thresholds. **Sample Heading (Third Level).** Only two levels of headings should be numbered. Lower level headings remain unnumbered; they are formatted as run-in headings.

2 Design and Implementation

2.1 Design Goals and Ideas

The goal of this system is to provide the visually impaired people with different pat-terns of vibration so that it can guide them when walking in an unfamiliar envi-ronment and make them efficient, timely, and accurate walking in right directions. The basic idea is to provide various appreciable vibration mechanism (frequency and vibrating size) while the visually impaired people deviating from the path (a slight deviation from the route, serious deviation from the route) or reaching intersections, Following Fig. 1 shows the mechanism.

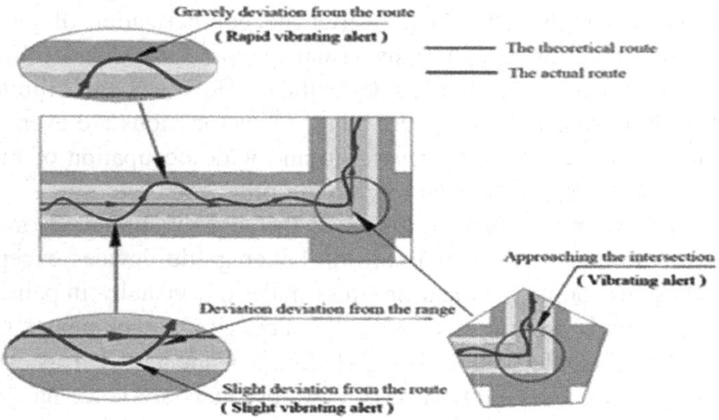

Fig. 1. Mechanism chart

2.2 Basic Framework

The basic framework mainly includes GPS module, map module, routing module, multiple vibration prompting module and man-machine interface module etc. The system provided the visually impaired people, the vibration or integration of vibration and sound after processing the collected environmental information. The users under-stood the information combined with their spatial cognition and realized posi-tioning and navigation based on them.

2.3   The Working Principle of the System

The work flow of the system is as following: (a) loads map after running main program; (b) the user inputs the starting point and the destination; (c) the optimal route computing and information display; (d) according to the route node information, the GPS module provides the latitude and longitude information real-time judge whether deviating from the route or to the next intersection nodes (the destination) when the visually impaired people are walking; (e) Haptic (vibration) module provides different patterns of vibration to route guidance.

According to the duration time of vibrating, three kinds of vibration modes were compared together, as shown in Fig. 2.

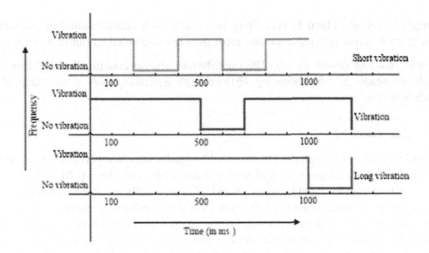

Fig. 2. Three groups of vibration mode

3   Experiment Results

3.1   Experiment 1

The organization and implementation of experiments: The tests had 30 adult participants (including 20 ordinary people from Xinjiang University, 10 visually impaired from short-term training classes of Urumqi Blind School). The experiments had been done respectively with vibration and voice for alert in a noisy environment (the school cafeteria and commercial street). The subjects were asked not to pay special attention to vibration or voice alert and do activities as usual. The mobile phones were placed in pocket or in hands. The vibration or voice volume was set to medium. In a certain period, the subjects' phones were dialed one time at random. If he/she could feel or hear it, it will be recorded as right. Then we obtained the correct rates of vibration or voice. The results as shown in Table 1.

Table 1. The correct rate and error rate

The types of the subjects	Number of people	School cafeteria (Vib & Cor)	School cafeteria (Voi & Cor)	Commercial street (Vib & Cor)	Commercial street (Voi & Cor)	The total correct rate of the vibration	The total correct rate of the voice
The ordinaiy visually impaired	20	75%	55%	65%	50%	70%	52.50%
	10	80%	60%	70%	60%	75%	60%

Note: The correct rate of the vibration is simplified as Vib & Cor; the correct rate of the voice is simplified as Voi & Cor.

Sample Heading (Third Level). Only two levels of headings should be numbered. Lower level headings remain unnumbered; they are formatted as run-in headings.

Sample Heading (Forth Level). The contribution should contain no more than four levels of headings. The following 错误!未找到引用源。 gives a summary of all heading levels.

3.2   Experiment 2

Preliminary threshold settings: There were 32 visually impaired participants (including 11 blind, 21 low vision) from short-term training classes of Urumqi Blind School, 20 males and 12 females, aged between 15 and 67 (average 30.31). There are three choice questionnaire about thresholds of deviating: (1) the threshold of slight deviation from route; (2) the threshold of serious deviation from route; (3) the preliminary remind threshold of reaching intersection nodes. The answers are set form 0.5 m to 5 m, interval is 0.5 m (For example: a. 0.5 m, b. 1 m, c. 1.5 m, d. 2 m, e. 2.5 m, f. 3 m, g. 3.5 m, h. 4 m, i. 4.5 m, j. 5 m, k. above 5 m, l. Please write yourself…etc.). The psychological thresholds are selected by visually impaired and the classification of the thresholds as shown in Fig. 3.

According to references, blind interviews, expert consultation (including orientation and mobility teachers), we preliminarily set predetermined thresholds based on spatial cognition and psychological factors for the visually impaired. We set predetermined thresholds of slight deviation as 1 or 1.5 m, serious deviation as 2.5 or 3 m, and initial distance threshold to intersection node as 4.5 or 5 m. After experiments, the initial thresholds were inspected, corrected and improved for more reasonable ones.

We had illustrated and trained the corresponding relationship between the degree of deviation from route and vibration alert firstly before tests. And it let users to feel the differences of the vibrations until mastering it. Finally, we set up a vibration mode, so that the blind can quickly tell the corresponding deviation degree and bias, and make the corresponding action.

Ten volunteers (including 4 blind, 6 having low vision) of 32 participants helped us and tried to experience it in our tests, and four of them were randomly selected (including 2 blind, 2 having low vision) to have field tests of deviation from route and the rest to experience scene tests of intersection node. Test route was shown in Fig. 4.

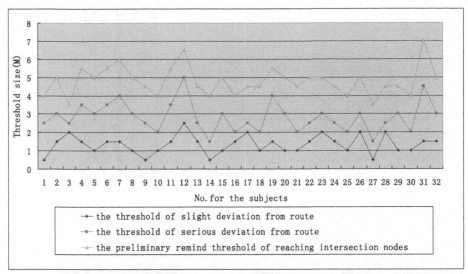

(a)The psychological thresholds are selected by visually impaired

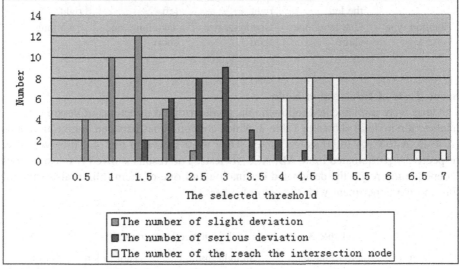

(b) The classification of the thresholds

Fig. 3. The psychological thresholds of visually impaired.

Scene Tests of Deviation from Route

The tests were implemented according to the predetermined deviation as shown in Table 2. It chose a straight line between two points A and O as the test route, as shown in Fig. 4. The results showed that the testers could wander repeatedly and independently according to the differences of vibration alert and reach the designated destination eventually.

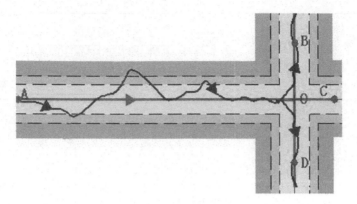

Fig. 4. Test route

Table 2. The basis of deviation from the route tests

Bias/deviation degree	Slight deviation to the left	Slight deviation to the right	Serious deviation to the left	Serious deviation to the right
Vibration type	One long, one short	One long, two short	Two long, one short	Two long

Scene Tests of Intersection Node

The tests were implemented following the predetermined directions of passing through intersection node as shown in Table 3, which contains lines between intersection nodes (A to B, A to C, A to D) as test routes. They were shown in Fig. 4. Three volunteers respectively participated route AB, AC, AD tests. The results showed two testers could independently reach the designated destination and one of them (blind) also smoothly finished the experiment with the help of others.

Table 3. The basis of the intersection node tests

Degree/turn to	Approaching the intersection node	Straight through the intersection	Turn to left	Turn to right
Vibration type	Short	Long	One long, two short	Two long, one short

4   Discussion

4.1   Discussion of the Interviews with the Blind

We drew some conclusions after having interviews with many visually impaired students and their teachers who teach orientation and mobility.

Most of the visually impaired students use cane as their walking assistant in strange environment. However, no tools are used in a familiar environment. The scope of their activities is relatively fixed, school or home community. They almost don't go out to unfamiliar environment. As to a new environment, they go there about 3–5 times with other's helps and after "mental maps" are formed in their minds, they go there individually.

The visually impaired people take obstacles (or other fixed objects, etc.) for route guidance, that is to say, they usually bring some obstacles to mind (where there is a what kind of obstacle) in familiar environment. When walking there again they can judge the position of the obstacles for position and navigation.

Not all blind people have a mobile phone. It mostly depends on the families' economic condition. With the price is also a very important factor.

There are some activity difference between Western and Oriental (China) blind. Chinese blind seldom go to unfamiliar environment, or is said not to let or not dare to. Generally, they go to unfamiliar environment with others help.

Young people with visual impairment, whose outdoor safety are considered too much by their supervisors. They usually have poor sense of distance because they are rarely allowed to go out alone which may cause the lack of experience in space.

During the developing of navigation tools for the blind, convenient, intelligent, safety and the price are important factors. The most important factor is the fitness between navigation tools and the characteristics of the blind walking activities.

4.2   Discussion of Experiment 1

The experiment 1 confirmed that vibration alert has superiority to voice navigation service in noisy environment. The total correct rate of vibration alert is 71.67% (including the ordinary and the visually impaired), and the total correct rate of voice alert is only 55%. The experiment also demonstrated the visually impaired have more sensitive touch and hearing senses that ordinary people do. In comparison with the ordinary people, the visually impaired have a slight advantage over 5% in correct rate of vibration alert and voice alert.

4.3   Discussion of Experiment 2

Experiments were conducted in relatively open areas. We took weather conditions and other factors into account. The GPS module got positioning information which met requirements of route guidance. The prototype can smoothly run and load the map. The routing module planned an optimal path according to the input of the starting position and the destination. While the user was walking, a cane was in one hand and the mobile in the other (as shown in Fig. 5). It can alarm different vibration patterns while having different degree of deviating from the planned path or reaching to an intersection node according to a preset threshold of vibrating length and time cycle. The participants can obviously feel the different vibration patterns, and most of them can judge the deviation degree and make corresponding mobility combined with the blind spatial cognition. It had reached the initial testing purposes.

Fig. 5. Training and testing of the outdoor for the blind

On the whole, the tests were practically successful and reached the initial testing purpose. The analysis of results showed that many aspects needed to be improved: (1) The corresponding relationships between the degree of deviation from route and vibration alert are more complicated, and they are not easy to remember; (2) Threshold setting should be optimized, the different people have various thresholds. It needs to do a lot of experiments and tests to obtain experience values suitable for various groups of the blind. (3) The GPS accuracy and sensitivity need to be further improved. (4) The dimension of the vibration is limited; we will consider the integration use of the multiple patterns of vibration and sound, and so on. **Sample Heading (Third Level).** Only two levels of headings should be numbered. Lower level headings remain unnumbered; they are formatted as run-in headings.

5 Conclusions and Future Work

The lack of vision caused many inconvenience to the visual impaired people. Many scholars and experts around the world have committed themselves to invent various assistant equipments to improve the life quality of the visual impaired. The effect of voice guidance might be greatly reduced or even lost when in noisy and crowded environments. However, the development of route guidance with multiple patterns of vibration makes up the shortcomings and weaknesses of sound guidance under certain conditions. The prototype realized basic functions of route guidance for the visual impairments. It can effectively provide optimal planning path between two locations with multiple patterns of vibration alert during the walking. And the field experiments demonstrated it make the blind walking efficiently, timely, and accurately. We also obtain a set of reasonable thresholds of deviating from path.

We will continue to optimize the system and service, and try to obtain more reasonable thresholds by experiment combining with spatial cognition of the blind as our future work. The integration use of the multiple patterns of vibration and sound is considered as the next generation of our prototype. It will have stronger function in various environments and be closer to the market.

Acknowledgment. This work is jointly funded by the National Natural Science Foundation, China (41361084). The authors would like to thank Jacob Ricky, PhD, who provided papers and selfless help, at the Department of Computer Science, NUI Maynooth (CS-NUIM) Ireland. We also acknowledge the staff of Urumqi Blind School and the short-term training students, who provided cooperation and supports in the field tests.

References

1. Cao, Y.H., Liu, G., Yang, H.F.: Detection of pedestrian crossing in image-based blind aid devices. Comput. Eng. Appl. **4**, 176–178 (2008)
2. He, J., Nie, M., Luo, L., Tong, S.B., Niu, J.H., Zhu, Y.S.: Auditory guidance systems for the visually impaired people. J. Biomed. Eng. **27**, 467–470 (2010)
3. Loomis, J.M., Golledge, R., Klatzky, R., Marston, J.R.: Assisting wayfinding in visually im-paired travelers. pp. 186–190. Psychology Press (30 Suppl) (2007). Author, F.: Contribution title. In: 9th International Proceedings on Proceedings, pp. 1–2. Publisher, Location (2010)
4. Koustriava, E., Papadopoulos, K.: Are there relationships among different spatial skills of individuals with blindness? Res. Dev. Disabil. **33**, 2164–2176 (2012)
5. Schmidt, S., Tinti, C., Fantino, M.: Spatial representations in blind people: the role of strategies and mobility skills. Acta Psychol. **142**, 43–50 (2013)
6. Sehellingethout, R., Bongers, R.M., Van Grinsven, R., Smitsman, A.W., Van Galen, G.P.: Improving obstacle detection by redesign of walking canes for blind persons. Ergonomics **44**, 513–526 (2001)
7. Cecelja, F., Garaj, V., Hunaiti, Z., Balachandran, W.: A navigation system for visually impaired. In: Instrumentation and Measurement Technology Conference, pp. 1690–1693. IEEE (2006)
8. Ganz, A., Gandhi, S.R., Wilson, C., Mullett, G.: INSIGHT: RFID and Bluetooth enabled automated space for the blind and visually impaired. In: 32nd Annual International IEEE EMBS Conference, New York, pp. 331–334. IEEE (2010)
9. Shang, W.Q., Jiang, W., Chu, J.: A machine vision based navigation system for the blind. In: 2011 IEEE International Conference on Computer Science and Automation Engineering (CSAE), Piscataway, vol. 3, pp. 81–85. IEEE Computer Society (2011)
10. Hernández, D.C., Jo, K.H.: Stairway segmentation using gabor filter and vanishing point. In: 2011 IEEE International Conference on Mechatronics and Automation, Piscataway, pp. 1027–1032. IEEE Computer Society (2011)
11. Yun, H.C., Park, T.H.: Auto-parking controller of omnidirectional mobile robot using image localization sensor and ultrasonic sensors. J. Inst. Control **21**(6), 571–576 (2015)
12. Yelamarthi, K., Haas, D., Nielsen, D., Mothersell, S.: RFID and GPS integrated navigation system for the visually impaired. In: 53rd IEEE International Midwest Symposium on Circuits and Systems, MWSCAS 2010, Piscataway, vol. 12, no. 8, pp. 1149–1152. IEEE (2010)
13. Chen, F., Ma, C., Ma, W.G., Zhu, H.R.: Study on mobile robot navigation based on strategy of blind man finding way. In: 2011 IEEE International Conference on Mechatronics and Automation, Piscataway, pp. 1045–1049. IEEE Computer Society (2011)

14. Capi, G., Toda, H.: A new robotic system to assist visually impaired people. In: 20th IEEE International Symposium on Robot and Human Interactive Communication, New York, pp. 259–263. IEEE (2011)
15. Hesch, J.A., Mirzaei, F.M., Mariottini, G.L., Roumeliotis, S.I.: A 3D pose estimator for the visually impaired. In: 2009 IEEE/RSJ International Conference on Intelligent Robots and Systems, Piscataway, pp. 2716–2723. IEEE Computer Society (2009)
16. Ye, C.: Navigating a portable robotic device by a 3D imaging sensor. In: IEEE SENSORS 2010 Conference, Piscataway, pp. 1005–1010. IEEE (2010)
17. Ienaga, T., Sugimura, Y., Kimuro, Y., Wada, C.: Pedestrian navigation system using tone gradient and robotic GIS. In: Miesenberger, K., Klaus, J., Zagler, W., Karshmer, A. (eds.) ICCHP 2010. LNCS, vol. 6180, pp. 239–246. Springer, Heidelberg (2010). doi:10.1007/978-3-642-14100-3_36
18. Qinghui, T., Malik, M.Y., Youngjee, H., Park, J.: A real-time localization system using RFID for visually impaired. Computer Science, pp. 958–965 (2011)
19. Chandrasekara, P., Mahaulpatha, T., Thathsara, D., Koswatta, I., Fernando, N.: Landmarks based route planning and linear path generation for mobile navigation applications. Spat. Inf. Res. **24**(3), 245–255 (2016)
20. Shaik, A.S., Hossain, G., Yeasin, M.: Design, development and performance evaluation of reconfigured mobile Android phone for people who are blind or visually impaired. In: 28th ACM International Conference on Design of Communication, pp. 159–166. ACM, New York (2010)
21. Peng, E., Peursum, P., Li, L., Venkatesh, S.: A smartphone-based obstacle sensor for the visually impaired. In: Yu, Z., Liscano, R., Chen, G., Zhang, D., Zhou, X. (eds.) UIC 2010. LNCS, vol. 6406, pp. 590–604. Springer, Heidelberg (2010). doi:10.1007/978-3-642-16355-5_45
22. Yang, R., Park, S., Mishra, S.R., Hong, Z., Newsom, C., Joo, H.: Supporting spatial awareness and independent wayfinding for pedestrians with visual impairments. In: 13th International ACM SIGACCESS Conference on Computers and Accessibility, pp. 27–34. ACM, New York (2011)
23. Jacob, R., Winstanley, A., Togher, N., Roche, R., Mooney, P.: Pedestrian navigation using the sense of touch. Comput. Environ. Urban Syst. **36**(6), 513–525 (2012)
24. Wolf, F., Feldman, P., Kuber, R.: Towards supporting situational awareness using tactile feedback. IEEE (2014)

Prediction of Passenger Flow at Sanya Airport Based on Combined Methods

Xia Liu[1(✉)], Xia Huang[1], Lei Chen[1], Zhao Qiu[2(✉)],
and Ming-rui Chen[2]

[1] Sanya Aviation and Tourism College, Sanya 572000, Hainan, China
paolo_lx@qq.com
[2] Information Science and Technology School, Hainan University,
Haikou 570228, Hainan, China
qiuzhao73@hotmail.com

Abstract. It is crucial to correctly predict the passenger flow of an air route for the construction and development of an airport. Based on the passenger flow data of Sanya Airport from 2008 to 2016, this paper respectively adopted Holt-Winter Seasonal Model, ARMA and linear regression model to predict the passenger flow of Sanya Airport from 2017 to 2018. In order to reduce the prediction error and improve the prediction accuracy at meanwhile, the combinatorial weighted method is adopted to predict the data in a combined manner. Upon verification, this method has been proved to be an effective approach to predict the airport passenger flow.

Keywords: Airport · Passenger flow prediction · Seasonal model · Regression soothing model · Linear regression · Combination

1 Introduction

It is crucial to correctly predicting the passenger flow of an airport for the capacity arrangement, future development and construction as well as the functional planning of an airport. Therefore, correct prediction of the passenger flow of an airport has become a significant subject for the airport operation management and the fundamental basis for the effective allocation of airport resources. The passenger flow can be predicted by multiple approaches. Incomplete statistics indicate there are about 300 approaches in the world, among which, about 150 approaches are comparatively mature, about 30 approaches are commonly adopted and about 10 approaches are universally used. They could be generally divided into two categories according to different standards: the linear theory and non-linear theory, the qualitative prediction and quantitative prediction; but all of them generally include the time sequence model, the grey prediction model, the expert prediction model, the exponential smoothing method, the neural network, SVM, the trend extrapolation and the regression analysis method etc.

Domestic experts have obtained certain achievements in respect of study on the airport passenger flow prediction. Yan and Zhu (2010) from Nanjing University of Aeronautics and Astronautics came up with the regression model of SVM, and compared it with BPANN and linear regression, which turned out that the regression model

B. Zou et al. (Eds.): ICPCSEE 2017, Part I, CCIS 727, pp. 729–740, 2017.
DOI: 10.1007/978-981-10-6385-5_61

of SVM would be an effective prediction method of air passenger flow for the least relative error. Qu from Xi'an Jiaotong University (2014) combined grey model with BP neural network to predict the linear variation with the grey model and predict the linear variation and make up for the prediction error with the BP neural network, which has thus solved the defects of using only one prediction model and improved the prediction accuracy of passenger throughput of airports. Chen et al. (2013) from Hefei University of Technology came up with a model in combination with the seasonal support vector regression (SSVR) and the particle swarm optimization (PSO) to predict the tourist flow volume. The prediction accuracy of this model is obviously higher than that of other methods such as SVR-PSO, SVR-GA, BPNN and ARIMA. Huang et al. (2013) from Civil Aviation Flight University of China established the multivariable linear regression prediction model in respect of airport passenger throughput and used the time sequence method to verify the predicted value so as to correctly predict the airport passenger throughput in the subsequent year. Guan (2013) from Civil Aviation University of China, according to the features of the grey prediction and the SVM, put forward the grey support vector machine in integration with these two prediction methods. In consideration of the prediction result of passenger throughput and in comparison with the prediction results of the grey prediction model, SVM and the grey support vector machine, the grey support vector machine was proved to feature high prediction accuracy, accurate and reliable prediction result. Jing (2014) from Civil Aviation Flight University of China came up with the seasonal route passenger transport demand analysis method based on the superposition trend. According to the historic data of seasonal route passenger volume, he established the model of seasonal route passenger volume variation trend and further considered the effect of seasonal change on passenger transport demand by estimating the linear trend equation parameters though OLS so as to establish the seasonal route passenger transport demand analysis model. Chen and Zeng (2014) from Civil Aviation University of China adopted multivariable linear regression model and time sequence trend extrapolation model to predict the passenger flow of the Capital Airport from 2012 to 2016. In order to improve the accuracy, they also applied the combination weighting method to predict the result in a combined manner, which thus increased the accuracy and reduced the prediction error. Tian et al. (2015) from Shenyang University of Technology proposed a network traffic prediction scheme based on empirical mode decomposition and time sequence analysis, which highlighted better prediction effect and precision. Du and Liu (2015) from East China Normal University dealt with the seasonal fluctuations in the monthly data of port container throughput. They processed the throughput of containers in a seasonal manner by using the seasonal time sequence model to improve the prediction accuracy.

Holt-Winter decomposed and studied the time sequence with linear trend, seasonal and random fluctuations, estimated the long-term trend, the increment beyond the trend and the seasonal fluctuation in combination of the exponential smoothing method, thus established the prediction model and extrapolated the prediction value. ARMA model (Auto-Regressive and Moving Average Model), as an important method to study time sequence, is composed of Auto-Regressive Model (AR model for short) and Moving Average Model (MA model for short). It is used to build model mainly based on the analysis about steady random time sequence. As it is simple in form, convenient in data fitting and easy for analyzing the data structure and the inner nature, it could make the optimal prediction and control in the sense of minimum variance. At meanwhile, it is also a short-term prediction model with high accuracy and more flexible because the order could be adjusted as the case maybe; nevertheless, it demands large amount of historic data (more than 50 in general). The linear regression, as used very extensively, is a statistical analysis method to identify the interdependent quantitative relationship of two or more variables by means of regression analysis in mathematical statistics. In the regression models, the simple regression is the simplest and steadiest, but it is not adequate to describe the behaviors in complex system. As a result, the prediction technique based on multiple regressions is more common. As the conventional mul-tivariable regression model is generally linear, the insignificant variables and the cor-relativity among the variables may result in normal equation set of the regression subject to serious ill-condition, which will thus affect the regression equation's sta-bility. Therefore, the multivariable linear regression is confronted with a basic problem of seeking the "optimal" regression equation. Neural Networks (Neural Networks, NN) is a complicated network system by extensive interconnection of numerous and simple processing units (known as the neuron), which reflects many basic characteristics of human brain. As a nonlinear dynamic learning system with extremely high complexity, the system highlights sound approximation effect, fast calculation speed and high accuracy without the necessity of establishing a mathematical model; moreover, it has strong nonlinear fitting ability. However, as the system cannot express and analyze the relationship between the input and output of the predicted system, the forecasters cannot participate in the prediction process. Meanwhile, it is hard to process mass data due to slow rate of convergence. As a result, the network obtained features weak fault tolerance and incomplete algorithm.

2 The Data Sources and Descriptive Analysis

2.1 Data Selection

The article selected the monthly passenger flow volume-about 108 data in total from Jan. 2008 to Dec. 2016 provided by Sanya Phoenix International Airport, then adopted Holt-Winters seasonal prediction mode, ARMA model and the unary linear regression respectively for modeling prediction, finally predicted and obtained the prediction result in combination of these three methods to predict the passenger flow of Sanya Airport in the subsequent two years.

2.2 The Descriptive Analysis

Eviews6.0 software was applied to make the time sequence of Sanya Airport passenger flow. Figure 1 shows the variation trend of passenger flow.

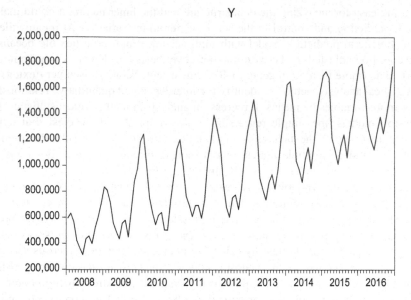

Fig. 1. Time sequence of passengers

3 Prediction Based on Holt-Winter Seasonal Model

Holt-Winter mainly decomposed the time sequence with linear trend, seasonal fluctuation and random fluctuation, then estimated the long-term trend and seasonal fluctuation respectively in combination with the exponential smoothing method; then the model was established to conduct out-of-sample prediction. This article mainly used Holt-Winter seasonal multiplication model applicable to the sequences with long-term trend and multiply seasonal change. The calculation formula of smooth sequence is:

$$\hat{y}_{t+k} = (a_t + b_t k)S_{t+k}, \quad t = s+1, s+2 \cdots, T \tag{1}$$

where a_t refers to the intercept, b_t refers to the slope, $a_t + b_t k$ refers to the long-term trend, S_t refers to the seasonal factor of the multiplicative model, and s refers to the seasonal cycle length (the months: $s = 12$ in this article)

This model needs three coefficients to give the initial value of the seasonal factor in the first year, the intercept and the slope. These three coefficients are defined as follows:

$$
\begin{aligned}
a_t &= \alpha \frac{y_t}{S_{t-s}} + (1 - \alpha)(a_{t-1} + b_{t-1}) \\
b_t &= \beta(a_t - a_{t-1}) + (1 - \beta)b_{t-1} \\
S_t &= \gamma \frac{y_t}{a_t} + (1 - \gamma)S_{t-s}
\end{aligned}
\tag{2}
$$

where $k > 0, \alpha, \beta, \gamma$ are the damping coefficients between 0 and 1. If $t = T$, the calculation formula of the prediction value is:

$$\hat{y}_{T+k} = (a_T + b_T k)S_{T+k-s} \tag{3}$$

where S_{T+k-s} is the seasonal factor in the last year of the sample data.

Use Eviews6.0 software to predict the above data, and the established result of the prediction parameters gained is shown in Table 1.

Table 1. Estimated result of the prediction parameters

Parameters:	Alpha	0.9900
	Beta	0.0000
	Gamma	0.0000
Sum of squared residuals		3.43E+11
Root mean squared error		56352.57

The result of the ample prediction is obtained according to Table 1, which is shown as in Fig. 2.

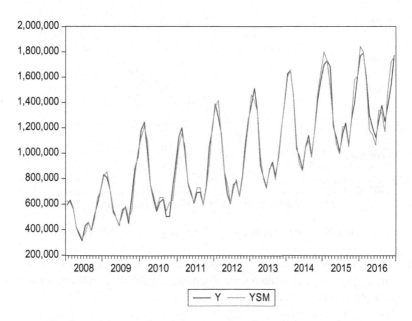

Fig. 2. Time sequence of the original and Holt-Winter sequences

Figure 2 shows that the actual passenger flow and the predicted passenger flow of Sanya Airport based on the Holt-Winter seasonal multiplicative model are almost overlapping. The calculation formula of the predicted mean absolute error is:

$$MAPE = \frac{1}{n} \sum_{i=1}^{n} \left| \frac{\hat{y}_i - y_i}{y_i} \times 100 \right| \tag{4}$$

Though calculation, we get $MAPE = 4.16\%$, indicating the prediction effect of Holt-Winter multiplicative model is comparatively ideal.

4 Prediction Based on ARMA Model

Figure 1 shows that the airport passenger flow of the sample has significant seasonal fluctuations, which requests seasonal adjustment for airport passenger flow Y before establishing ARMA model. This article selected X12 seasonal adjustment method. The time sequence of airport passenger flow upon seasonal adjustment with Eviews6.0 is shown in Fig. 3.

Fig. 3. Time sequence after eliminating seasonal fluctuations

As shown in Fig. 3, the airport passenger flow is no longer affected by seasonal change after the seasonal adjustment. Then we established ARMA model for the passenger flow sequence after adjustment.

4.1 Stationary Test of the Sequence

The unit root test was adopted to test the stationarity of the time sequence, and ADF test was also selected. Eviews6.0 software was used to conduct ADF unit root test for YSA. The result is shown as in Table 2.

Table 2. The ADF test result of unit root

Variable	ADF test value	P value	1% critical value	5% critical value	10% critical value	Test result
YSA	−4.0852	0.0089	−4.0561	−3.4524	−3.1517	steady

According to Table 2, YSA original sequence of the airport passenger flow sequence after seasonal adjustment was stable at the significant level of 0.05. Then ARMA model was established for YSA sequence.

4.2 Establishment and Identification of ARMA (p, q) Model

4.2.1 Establishment of the Algebraic Expression of ARMA (p, q) Model

$$y_t = \alpha_0 + \alpha_1 y_{t-1} + \alpha_2 y_{t-2} + \cdots + \alpha_p y_{t-p} + \varepsilon_t + \theta_1 \varepsilon_{t-1} + \theta_2 \varepsilon_{t-2} + \cdots + \theta_q \varepsilon_{t-q} \quad (5)$$

4.2.2 Identification of the Parameters p and q of the Model

Use Eviews6.0 to get the autocorrelation function ACF and partial autocorrelation function PACF map of YSA sequence after the first order difference. We made preliminary judgment to p and q, and the model made for YSA is ARMA (2,1), expressed as follows:

$$YSA_t = c + \alpha_1 YSA_{t-2} + \varepsilon_t + \beta_1 \varepsilon_{t-1} \quad (6)$$

And the established result of ARMA (2,1) modeling is as follows:

According to Table 3, under the significance level of 0.05, all the coefficients of ARMA (2,1) are all significant. The probability P value of the whole equation is 0.0000, lower than the significance level of 0.05. Thus it is also significant. The specific equation is:

$$YSA_t = 999383.4 - 0.933343 YSA_{t-2} + 1.0689\varepsilon_{t-1} \quad (7)$$

Use the model above to predict the sample data with the result shown as in Fig. 4.

Figure 4 shows that the actual passenger flow and the predicted passenger flow of Sanya Airport based on ARMA (2,1) are almost overlapping. Through calculation, the predicted mean absolute error, indicating that the prediction effect of ARMA (2,1) is comparatively ideal.

Table 3. Regression result of the ARMA (2,1) modeling

Variables	Regression coefficient	Standard deviation	T statistics	P value
C	999383.4	121539.6	8.222699	0.0000
AR(2)	0.933343	0.014391	64.85657	0.0000
MA(1)	1.068920	0.052988	20.17291	0.0000
R_a^2	0.9782			
F statistics	2363.318 (0.0000)			

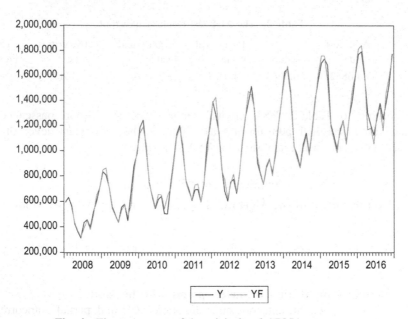

Fig. 4. Time sequence of the original and ARMA sequences

5 Prediction Based on Linear Regression Model

In Fig. 3, the airport passenger flow after adjustment shows a linear growth trend over time. Now we will establish the simple linear regression model in respect of the passenger flow over time and carry out the fitting prediction afterwards. The simple linear regression model is:

$$YSA_t = \beta_0 + \beta_1 T + \varepsilon_t \quad T = 1, 2 \cdots \tag{8}$$

among which ε_t refers to the random error, T refers to time, where January 2008 is 1, and the like. Table 4 shows the estimated result obtained by Eviews6.0.

According to Table 5, under the significance level of 0.05, the regression coefficients are all significant and the probability P value of F statistics of the general equation is 0.0000, lower than the significance level of 0.05, indicating the general equation is significant. On the other hand, the goodness of fitting $R_a^2 = 0.960357$ after adjustment, indicating that the fitting effect of the regression equation established above is very good. The regression equation obtained is:

Table 4. Estimated result of the model

Variables	Regression coefficient	Standard deviation	T statistics	P value
C	451693.7	12148.62	37.18066	0.0000
T	9852.941	193.4904	50.92211	0.0000
R_a^2	0.960357			
F statistics	2593.061 (0.0000)			

Table 5. The mean absolute errors of the methods

Prediction method	Holt-Winter seasonal model	ARMA model	Regression model	Combined prediction model
MAPE	4.16	4.19	5.40	3.67

$$YSA = 451693.7 + 9852.941T \tag{9}$$

Figure 5 shows the prediction result by means of the regression equation as established.

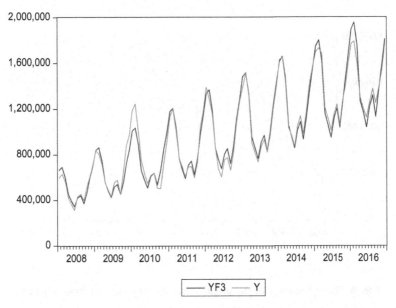

Fig. 5. Time sequence of the original and the linear regression sequences

According to Fig. 4, the actual passenger flow and the predicted passenger flow of Sanya Airport based on the regression model are almost overlapping. Through calculation, the predicted mean absolute error $MAPE = 5.40\%$, indicating that the prediction effect of the regression model is comparatively ideal.

6 Prediction Based on Combined Methods

Considering the prediction of the above three methods, the prediction error of ARMA model is the lowest while that of the linear regression model is the highest, and that of the Holt-Winter seasonal multiplicative model is between the two. As the deviation may be available by one method, this paper finally adopted the combined model, that is, make prediction by the specific value of the mean absolute error of the said two models as the weight, in which, the weight of Holt-Winter seasonal multiplicative model is 0.3023, the weight of ARMA model is 0.3052 and the weight of regression model is 0.3924. Then we can get the equation of the combined prediction:

$$Y_4 = \partial_1 Y_1 + \partial_2 Y_2 + \partial_3 Y_3 \tag{10}$$

where $\partial_1, \partial_2, \partial_3$ refer to the weight.

Predict the passenger flow of Sanya Airport in combination of the said three methods with the result shown in Fig. 6.

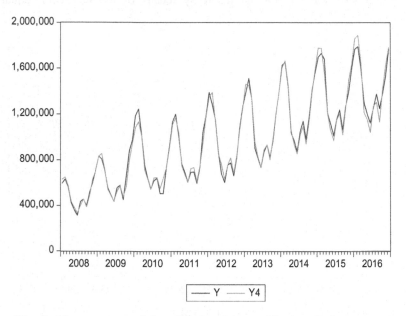

Fig. 6. Time sequence of the original and the combined method sequences

According to Fig. 6, the combined prediction features the optimal prediction result. Its average absolute error. Now, the mean absolute error of four prediction methods is shown as follows in Table 5:

From Table 5, the combined prediction features the optimal prediction effect, then the Holt-Winter seasonal multiplicative model, the ARMA model and the regression model in proper order. However, the mean absolute errors of these four prediction methods are all below 10, indicating that the overall prediction effect of the four methods is comparatively ideal. Table 6 shows the prediction values of the airport passenger flow in the future two years by the four methods and the line graph is shown in Fig. 7.

Table 6. Prediction of the passenger flow at Airport in the future

Month	Holt-winter seasonal model	ARMA model	Regression model	Combined prediction model
2017M01	2035337	2091886	2152184	2098453
2017M02	2068783	2119340	2201496	2136297
2017M03	1812981	1835060	1919429	1861495
2017M04	1340658	1320378	1394092	1355438
2017M05	1180735	1162833	1236331	1197089
2017M06	1046838	1015271	1089367	1053893
2017M07	1249058	1198306	1294786	1251512
2017M08	1310920	1253373	1366454	1315148
2017M09	1112687	1049940	1152727	1109248
2017M10	1340608	1265845	1401986	1341875
2017M11	1669629	1552549	1731652	1658233
2017M12	1916297	1773933	1995610	1903968
2018M01	2200668	2046934	2318973	2200171
2018M02	2235701	2075351	2371012	2239858
2018M03	1958283	1795960	2066284	1951121
2018M04	1447393	1293118	1500078	1420979
2018M05	1274118	1138262	1329731	1254475
2018M06	1129089	994426	1171149	1104491
2018M07	1346559	1173193	1391386	1311234
2018M08	1412589	1227784	1467770	1377836
2018M09	1198428	1028114	1237672	1161844
2018M10	1443252	1240153	1504669	1405362
2018M11	1796655	1520533	1857710	1736334
2018M12	2061170	1738143	2140007	1993511

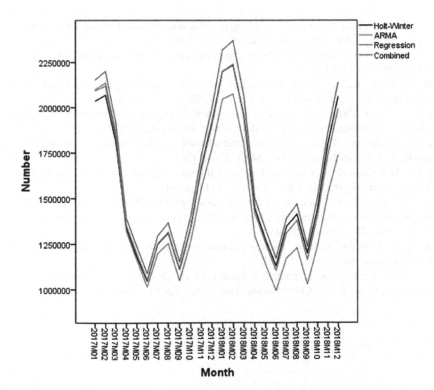

Fig. 7. Prediction of the passenger flow at airport in the future

7 Conclusion

The subsequent model improvement shall consider collecting more passenger flow data at airports and ports for analysis while adopting the grey model, the multivariable regression model, BP neural model and the grey support vector machine, etc. to improve the time sequence and increase the prediction precision.

As driven by the such three national strategies as "One Belt and One Road Initiative", strengthening our country through marine and establishing Hainan International Tourism Island, Sanya, the significant city in south Hainan Province, undertakes the historic mission of realizing such three national strategies. Meanwhile, Sanya Phoenix International Airport is an important traffic port in south China. Its passenger flow is ascending on the whole. As predicted, its passenger throughput will exceed 20 million in two years. In this sense, it is extremely urgent to rebuild and expand the airport, and also imperative to build a new airport. The overall planning of the new airport shall be taken into long-term consideration and efforts shall be made to accelerate its implementation and promotion.

Acknowledgments. This research was financially supported by A cooperative science project between colleges and local government in Sanya (2013YD64) and (2014YD52).

References

Yan, K., Zhu, J.: Research on airline passenger traffic forecast based on support vector machine regression algorithm. Enterp. Econ. **23**, 88–90 (2010)

Qu, T.: Application of combined model in airport passenger throughput forecast. Comput. Simul. **29**(4), 108–111 (2014)

Chen, C., Liang, R., Lu, W., et al.: Research on forecast model of passenger traffic based on seasonal SVR - PSO. Syst. Eng.-Theory Pract. **33**(1), 1–7 (2013)

Huang, B., Lin, J., Zhen, X.: Forecasting passenger throughput of civil transport airport based on multiple linear regression analysis. Math. Pract. Theory **43**(4), 172–178 (2013)

Guan, J.: Prediction of passenger throughput of civil aviation based on gray support vector machine. J. Dalian Jiaotong Univ. **34**(6), 41–43 (2013)

Jing, C.: Analysis method of seasonal passenger transport demand based on superposition trend. J. Civ. Aviat. Flight Univ. China 25(2), 5–7+11 (2014)

Chen, Y., Zeng, G.: Research on forecasting passenger throughput of civil aviation based on combination forecasting method: taking Capital Airport as an example. J. Civ. Aviat. Flight Univ. China **32**(2), 59–64 (2014)

Tian, Z., Li, S., Wang, Y., et al.: Plication of time series analysis in network traffic forecasting. Control Decis. **30**(5), 905–910 (2015)

Du, G., Liu, Y.: Forecast of container throughput in Shanghai port under the influence of seasonal change. J. East China Normal Univ. (Nat. Sci.) **1**, 234–239 (2015)

Research on the Copyright Protection Technology of Digital Clothing Effect Diagram

Yongqiang Chen[1(✉)] and Lihua Peng[2]

[1] School of Math and Computer, Wuhan Textile University, Wuhan 430073,
Hubei, China
chenyqwh@163.com
[2] Archive, Wuhan Textile University, Wuhan 430073, Hubei, China

Abstract. As a kind of digital clothing design works, digital clothing effect diagram shows fast and roughly designer intent expression and needs protection of digital copyright protection technology. A novel digital clothing effect diagram copyright protection scheme and some algorithms based on digital image watermark technology are proposed, used by two-dimensional chaotic encryption watermark generation algorithm to generate watermark data, watermark embedding algorithm twice using genetic algorithm to obtain the optimal embedding position and embedding strength, and watermark verification algorithm to determine the copyright ownership. Experimental results and analyses show that the proposed scheme and algorithms have the feasibility and effectiveness.

Keywords: Clothing effect diagram · Digital copyright protection · Image watermark · Photoshop

1 Introduction

Clothing design is based on the requirements of the designer to get design effect and plan diagrams, and then according to the diagrams to complete the whole producing process of a style clothing. Clothing design works include two parts of design diagram and physical clothing. As elements of the fact value of clothing design display media, these two parts are of equal importance and need copyright protection and management.

In the traditional clothing industry, according to the characteristics of design diagram and physical clothing, different copyright protection technologies are adopted. Clothing design diagram is made of paper and it's copyright can be protected by signature. Physical clothing is generally protected by patent and trademark. But with the rapid development of computer technology and software design, using computer software to complete the clothing design, or by scanning, photography and other means to get clothing image, digital clothing design works are used more and more for the garment business exchange and production [1]. It has become a kind of trend to release digital image, graphics and other media forms. With respect to the high and difficult production requirements, easy to the copyright protection of design diagram papers and physical clothing, digital clothing design works have the characteristics of low price,

© Springer Nature Singapore Pte Ltd. 2017
B. Zou et al. (Eds.): ICPCSEE 2017, Part I, CCIS 727, pp. 741–750, 2017.
DOI: 10.1007/978-981-10-6385-5_62

signed copy of judicial difficulties, convenience and tampering with the web page protection difficult, and the copyright protection problems become increasingly sharp.

Digital copyright protection technology for digital content is a series hardware and software protection technology of the intellectual property, in order to ensure the legitimate use of digital content in the entire life cycle and the balance of the value chain of digital content in the role of the interests and needs of the whole digital communication to promote the development of the market and information [2]. The usual digital copyright protection is based on the theory of cryptography. Digital encryption technology is to encrypt the file into ciphertext so that illegal users cannot read, that is, to control the file access. But with the rapid increase of computer processing ability, the method of improving system security by increasing key length has become unsafe. With the wide application of multimedia technology, multimedia data need encryption, authentication and copyright protection. Because of ignoring the properties of digital media itself, the password encryption affects the use of multimedia data.

The existing copyright protection management technology due to defects of multimedia content copyright protection is not suitable for digital clothing diagram protection between the creative design value and promotion that part of the need to publish. It is necessary to study the copyright protection technology of digital clothing design works.

2   Clothing Effect Diagram Design in Photoshop

Clothing effect diagram is a kind of accurate and quick painting form which expresses the intention of the fashion designer to require accurate and clear presentation of design awareness and performance, including line shape, material texture, color and processing technology, such as appearance of the shape and expression [3]. Clothing effect diagram is a prelude to fashion design and a general fast painting, reflecting idea of design, fashion style, charm and characteristics to prepare for the selection and production of clothing. Clothing effect diagram does not been required to express detail contents and specific performance, in which the actual parts should meet the features of manufacturing process with a clear guide to the pattern maker.

Digital clothing effect diagram is drawn by means of software which breaking the traditional hand-painted clothing rendering design and can complete more forms. Digital clothing effect diagram overcomes the drawbacks of traditional clothing design and improves the design efficiency. Computer design software for clothing design provides a wealth of resources and powerful tools. There are three main types of image processing, vector graphics and graphic layout. Clothing designers use mainly image processing and vector graphics software, such as Photoshop, CorelDraw and CorelPainter.

Photoshop is one of the famous Adobe image processing software, which is powerful, expressive, high penetration rate and recognized by the design community. It can be used to draw clothing effect diagram and deal with clothing fabric [4]. Facial features realistic performance, fabric texture simulation and rich background can be dealt with by Photoshop. Filters and other tools in Photoshop have many kinds of function in digital clothing design.

Clothing effect diagram design using Photoshop includes clothing rendering and process. The general steps are as follows.

(1) Open an existing diagram file or new a blank file, modify image size and resolution, adjust canvas size, determine whether to use ruler, grid, reference lines and other auxiliary tools.
(2) Create and edit selective area and mask by using marquee tool in existing or new layer and channel.
(3) Draw, process, regulate or match color and make special effect in part or full of image by using diagram tools, modification tools, color tools, filter tools and so on.
(4) Add a variety of graphics and text by using path tools, shape tools, text tools and so on.
(5) Combine layers and channels to complete and save the satisfactory diagram file after careful modifications.

3   Image Watermark Copyright Protection Scheme

Because of the limitation of the traditional encryption method and the existing copyright protection system for the protection and integrity authentication of multimedia content, researchers put forward a new kind of multimedia data copyright protection technology, digital watermark technology [5]. Digital watermark technology embeds some identifying information (i.e. digital watermark) directly into digital carrier (including multimedia documents, software, etc.), but does not affect the value of the original carrier and is not easy to be perceived or noticed by the perceptual systems (such as visual or auditory system). By hiding the information in the multimedia carrier, we can achieve the purpose of confirming the content creators, buyers, sending secret information or judging whether the carrier has been tampered with. According to the classification of digital watermark, watermark can be divided into text watermark, image watermark, audio watermark, video watermark and image watermark.

At present, the general digital watermark technology or product cannot take into account the copyright protection application characteristics of the digital clothing design works and the research and application of copyright protection in digital clothing work is still blank. In this paper, based on Photoshop image processing software which has the highest penetration rate in clothing effect diagram, we study the image watermark copyright protection technology of Photoshop clothing effect diagram.

3.1   Copyright Protection Guideline

In the publicity and business communication of digital clothing effect diagram, it is necessary to show the appearance of clothing and the intrinsic creative value, but also to determine the copyright owner. The main object of Photoshop is digital image which made up of pixels so that it is feasible and suitable to use image watermark technology to protect copyright of clothing effect diagram. A guideline of image watermark copyright protection in the clothing effect diagram is as follows.

Based on the design creativity and the importance of each component part of a clothing effect diagram, the whole or part of diagram area reflecting the unique characteristics of the designer and design units is selected as the main focus of the copyright protection.

In the process of designing Photoshop clothing effect diagram, image area comes from important part drawn and edited by selective area, mask, layer, channel and tools.

A certain amount of data, serial numbers, text and image signs designed and selected as copyright information is unencrypted or encrypted to watermark data by watermark generation algorithm.

The watermark data is embedded into the protection part to form the watermarked image region by spatial domain or transform domain watermark embedding algorithm.

Image regions including and no-including watermark data are combined to an image and some images are composed to clothing effect diagram protected by image watermark copyright.

In the process of the use of the clothing effect diagram, the copyright of the watermark can be determined by watermark verification algorithm.

3.2 Implementation Algorithms

An effective and feasible image watermark scheme or system generally includes watermark generation, watermark embedding and watermark verification. It should meet the requirements of the basic characteristics such as authentication, imperceptibility, robustness and security. According to the components of image watermark scheme, an image copyright protection scheme of Photoshop clothing effect diagram based on the basic characteristics and the previous research is proposed and implemented in this paper. Watermark is generated from the chaotic encryption algorithm. The genetic algorithm (GA) is used to find the optimal embedding position and intensity. The fast synergetic neural network algorithm (F-SNN) is applied to verify watermark [6, 7].

Watermark Generation Algorithm
Watermark can be meaningful or meaningless copyright information with the required amount of data. Considering the security of the watermark and the secret of embedding information, we can encode the meaningful watermark information.

The two-dimensional Logistic map system with simple coupled term is defined as:

$$\begin{cases} x_{n+1} = 4\mu_1 x_n(1 - x_n) + \gamma y_n \\ y_{n+1} = 4\mu_2 y_n(1 - y_n) + \gamma x_n \end{cases} \tag{1}$$

The dynamical behavior of this map system is controlled by control parameters of μ_1, μ_2 and γ. When $\mu_1 = \mu_2 = \mu \geq 0.89$ and $\gamma = 0.1$, the system is chaotic and can be used in digital image encryption. The encryption algorithm uses the sequences of x_i and

y_i produced by two-dimensional Logistic chaotic map to encrypt the rows and columns of a meaningful gray image respectively.

(1) Transform gray value to the binary sequence for per matrix element of meaningful gray image so that the gray matrix is converted to binary matrix.
(2) Transform the decimal fraction of x_i to binary sequence after one iterative step and select bits from the binary sequence equal to the number of binary matrix column. According to the matrix row order, do the XOR operation between one binary matrix row and bits of x_i's binary sequence till all rows of binary matrix are operated.
(3) Transform the decimal fraction of y_i to binary sequence after one iterative step and select bits from the binary sequence equal to the number of binary matrix row. According to the matrix column order, do the XOR operation between one binary matrix column and bits of y_i's binary sequence till all columns of binary matrix are operated.
(4) Revert the binary matrix to a gray value matrix and get the encrypted image.

Watermark Embedding Algorithm
Watermark embedding can be done in either spatial domain or frequency domain. Although the spatial domain watermark embedding is simple and easy to implement, it is less robust than frequency domain watermark embedding including discrete Fourier transform (DFT), discrete cosine transform (DCT), and discrete wavelet transform (DWT) [8, 9]. Watermark embedding techniques have been improved using optimization algorithms such as GA to search for location or intensity to embed the watermark [10, 11].

GA is twice used to find the optimal embedding location and intensity α_{opt} to modify the DWT coefficients of the host color image. The fitness function in GA is determined by imperceptibility and robustness of image watermark together. In watermark techniques, peak signal-to-noise ratio (PSNR) is often assessed for imperceptibility and normalized cross correlation coefficient (NCC) for robustness.

Let I and W represent the host color image with size $S \times S \times 3$ and the watermark with size $T \times T$ respectively, which S and T are even. The embedding algorithm is outlined below in detail.

(1) Do DWT to the watermark and get 4 sub-bands of low sub-band LL_w, middle sub-bands HL_w and LH_w, and high sub-band HH_w. The 4 sub-bands have same size $\frac{T}{2} \times \frac{T}{2}$.
(2) Divide 3 RGB component images of the host color image into ordinal un-overlapped $T \times T$ sub-images in spatial domain. There are $\frac{S}{T} \times \frac{S}{T} \times 3$ sub-images respectively represented by R_i, G_i and B_i $(1 \leq i \leq \frac{S}{T})$.
(3) Do DWT independently to every sub-image. The 4 sub-bands of sub-image are low sub-band LL_{tag}, middle sub-bands HL_{tag} and LH_{tag}, and high sub-bands HH_{tag}, $tag \in \{R_i, G_i, B_i\}$. The all of sub-bands have the same size $\frac{T}{2} \times \frac{T}{2}$ as the sub-bands of watermark.
(4) Select an initial value of embedding intensity α and some DWT coefficient matrixes of RGB component images to modify the corresponding sub-bands using

Eq. 2. Then do IDWT for all sub-bands and assemble RGB component images to watermarked color image I'.

$$CF'_{tag} = CF_{tag} + \alpha \cdot CF_w \tag{2}$$

Here, $CF \in \{LL, HL, LH, HH\}$.

(5) GA1: There are $\frac{S}{T} \times \frac{S}{T} \times 3$ possible corresponding sub-bands of host color image to embed one sub-band of watermark from Eq. 2. The $\frac{S}{T} \times \frac{S}{T} \times 3$ possible positions can be encoded to $\log_2(\frac{S}{T} \times \frac{S}{T} \times 3)$ bits so that the chromosome in GA has $4 \times \log_2(\frac{S}{T} \times \frac{S}{T} \times 3)$ bits together. Each chromosome represents an embedding location to embed the watermark into the host color image. Define the fitness function using PSNR between the host color image I and the watermarked color image I' as Eq. 3.

$$PSNR = 10 \lg\left(\frac{S^2 \times 255^2}{\sum\limits_{i=1}^{S}\sum\limits_{j=1}^{S}(I(i,j) - I'(i,j))^2}\right) \tag{3}$$

Create some random chromosomes into an original colony. Compute the fitness values of chromosomes and do the genetic operation until the process of GA stops and get the optimal embedding location using GA for the first time.

(6) GA2: After a certain attack to embedded image, NCC between watermark W and extracted watermark W' is computed as Eq. 4.

$$NCC = \frac{\sum\limits_{i=1}^{T}\sum\limits_{j=1}^{T} W(i,j)W'(i,j)}{\sum\limits_{i=1}^{T}\sum\limits_{j=1}^{T}(W(i,j))^2} \tag{4}$$

So, for n kinds of attack to the watermarked image with certain intensity α, the fitness function is defined.

$$f = PSNR + \frac{\lambda}{n}\sum_{k=1}^{n} NCC_k \tag{5}$$

In Eq. 5, λ is weighting factor to balance the effect and contribution of PSNR and NCC because PSNR value is usually about 40 dB and NCC values are between 0 and 1.

Like step 5, compute the fitness values of chromosomes and do the genetic operation to get the optimal embedding intensity α_{opt} using GA for the second time.

Through adapted twice GA operation to get optimal embedding location and intensity, the watermark embedding process can be completed.

Watermark Verification Algorithm

The watermark extracting is the contrary process of watermark embedding. Do DWT to the received watermarked image $\widehat{I}$ and host color image I according to the rule of watermark embedding and the final chromosomes of GA. The extracted encrypted watermark $\widehat{W}$ can be extracted as follows.

$$\widehat{CF}_w = (\widehat{CF}_{tag} - CF_w)/\alpha_{opt} \tag{6}$$

Then doing IDWT, we can decrypt the extracted encrypted watermark and get the decrypted watermark $W^{'}$ using the decryption algorithm.

The watermark identification may be taken for the process that a special existing watermark is formed and recognized in a mass of watermark patterns so that the pattern recognition method may be used in the watermark algorithm. The SNN can effectively identify the extracted watermark in our former research so that it is used in this scheme too [12].

Adding the original gray watermark image, select some gray images having the same size and similar content as the watermark image to makeup a prototype pattern set including M components. Transfer two-dimension watermark to one-dimension sequence and get the vector $v_k = [v_{k1}, v_{k2}, \cdots, v_{kN}]^T$, $k = 1, 2, \cdots, M$, $N = T^2$. A prototype pattern set may be composed of these M pattern vectors only if $M \leq N$.

According to the F-SNN model, the learning algorithm of network is the training process that adjoint vectors are calculated through prototype pattern vector.

(1) Compute prototype pattern vector v_k satisfying normal and center condition.
(2) Compute the according adjoint vector v_k^+ of prototype pattern vector v_k.

Used the F-SNN method, the watermark detecting and identification may be accomplished at one time through recognized pattern. The recognition process is showed as following.

(1) Compute the test pattern vector $q(0) = \{q_i\}$, $i = 1, 2, \cdots, N$, which satisfying normal and center condition, too.
(2) Achieve the according order parameter $\xi_k(0)$ of prototype patterns.

According to synergetic slaving principle, the pattern having the most value of order parameters will prevail in the synergetic evolution, and thus the watermark embedded in the host image carrier may be verified firstly.

4 Experiments and Analyses

As an example of copyright protection of image watermark in clothing effect diagram, an evening dress effect diagram with 1181×1890 pixels image is designed by Photoshop in Fig. 1(a) and the skirt layer is showed in Fig. 1(b). The skirt layer trimmed by square selective area with $S = 512$ pixels is showed in Fig. 1(c).

(a)effect diagram   (b)skirt layer   (c)trimmed skirt layer

Fig. 1. Evening dress image

In the experiments a gray face image as original meaningful image with size $T = 64$ pixels is selected from The ORL Database of Faces. After set control parameters $\mu = 0.9$, $\gamma = 0.1$ and initial values $x_0 = 0.1$, $y_0 = 0.11$, the face image is encrypted to watermark used the encryption algorithm. The face image and watermark are showed in Fig. 2(a) and (b) respectively. The watermark showed in Fig. 2(b) cannot be distinguished clearly by our visual organs to satisfy security requirement.

(a)face (b)watermark

Fig. 2. Watermark generation

The regular selection area with $S = 512$ pixels from the skirt layer is used as carrier image. Let the initial embedding intensity $\alpha = 0.1$ and used GADS Toolbox in the Matlab7.0, the fittest DWT coefficient matrixes of RGB component images of the carrier image are $LL_G(3,6)$, $HL_R(7,8)$, $LH_R(6,8)$, and $HH_R(8,8)$ from the chromosome [94564864] with the minimal fitness value −42.95875. Then we select 5 representative attacks to find the best embedding intensity $\alpha_{opt} = 0.0105$ in [0 0.1] range including gaussian noise with zero mean and 0.0002 variance, salt-pepper noise which zero mean and 0.0001 variance, 5×5 median filter and wiener filter, and JPEG compressing quality 90%. So the optimal watermarked image showed in Fig. 3(b) can be gotten through twice GA operation. Compared images with Fig. 3(a) and (b), any difference cannot be find out by our eyes to fit imperceptibility.

After gotten the Fig. 3(b) watermarked image that the watermark is embed in the Fig. 3(a) carrier image, it is reimported to the trimmed skirt image to compose a new PSD file format evening dress image. In the common usage of Clothing effect diagram, the watermark can be extracted from the independent layer in PSD file and decrypted by the decrypt algorithm. In the verification experiments, we accomplish a gaussian

(a)carrier image   (b)watermarked image

Fig. 3. Watermark embedding

noise attack with zero mean and 0.00005 variance of the watermarked evening dress diagram and verify the decrypted watermark to acquire the right watermark owner and diagram copyright with F-SNN algorithm shown in Fig. 4.

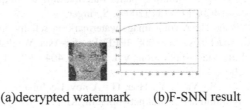

(a)decrypted watermark   (b)F-SNN result

Fig. 4. Watermark verification

5   Conclusion

With the development of modern garment industry, the proportion of digital clothing design work is gradually expanded by computer. The digital clothing work should need to be effectively protected by copyright. As a kind of digital clothing design work, digital clothing effect diagram is the expression of design intent of designer quick and rough and has important function. According to the characteristics of digital clothing effect diagram, based on the existing digital watermark technology, combined to meet the property requirements of safety, imperceptibility and robustness, a novel scheme and some algorithms for digital clothing effect copyright protection are researched in this paper are feasible and effective through some experiments and analyses. In the future, according to clothing components and style characteristics, further research will be carried out on the copyright protection technology, such as watermark generation, embedding and verification algorithms.

Acknowledgment. This work is supported by the humanities and social sciences research project No. 15Y064 of Hubei Provincial Education Department.

References

1. Xu, Q.: Digital Fashion Design and Management. China Fashion Press, Beijing (2006)
2. Yu, Y., Tang, Z.: A survey of the research on digital rights management. Chin. J. Comput. **28**(12), 1957–1968 (2005)
3. Kuang, Q.: The important role of clothing effect diagram in fashion design. J. Inner Mongolia Normal Univ. (Educ. Sci.) **25**(5), 142–144 (2012)
4. Xin, W.: Studio: Photoshop CS6 Fashion Design and Performance Techniques. Ordance Industry Press, Beijing (2013)
5. Fan, K., Mo, W., Cao, S.: Advances in digital rights management technology and application. Acta Electronica Sinica **35**(6), 1139–1147 (2007)
6. Chen, Y.: Theory and technology survey of nonlinear image watermarking. J. Comput. Appl. **33**(S1), 144–147 (2013)
7. Chen, Y., Zhang, Y., Hu, H.: A novel gray image watermarking scheme. J. Softw. **6**(5), 849–856 (2011)
8. Nasir, I., Weng, Y., Jiang, J.: Multiple spatial watermarking technique in color images. Sig. Image Video Process. **4**(2), 145–154 (2009). Springer
9. Hu, D., Wang, J., Peng, H.: A color image watermarking scheme in wavelet domain based on support vector machines. In: 2008 IEEE Pacific-Asia Workshop on Computational Intelligence and Industrial Application, vol. 01, pp. 404–407. IEEE Computer Society, Washington DC (2008)
10. Han, J., Kong, J., Lu, Y., Yang, Y., Hou, G.: A novel color image watermarking method based on genetic algorithm and neural networks. In: King, I., Wang, J., Chan, L.-W., Wang, D. (eds.) ICONIP 2006. LNCS, vol. 4234, pp. 225–233. Springer, Heidelberg (2006). doi:10.1007/11893295_26
11. Chu, S., Huang, H., Shi, Y.: Genetic watermarking for zerotree-based applications. Circuits Syst. Sig. Process **27**, 171–182 (2008)
12. Chen, Y., Hu, H., Li, X.: Extracted watermark identification using synergetic pattern recognition. In: MIPPR 2005: SAR and Multispectral Image Processing, pp. 256–264. SPIE, Bellingham (2005)

Visualization Analysis Framework for Large-Scale Software Based on Software Network

Shengbing Ren, Mengyu Jia$^{(\boxtimes)}$, Fei Huang, and Yuan Liu

School of Software, Central South University, Changsha, China
154711013@csu.edu.cn

Abstract. Large-scale software systems, which are the most sophisticated human-designed objects, play more and more important role in our daily life. Consequently effective analysis for large-scale software has become an urgent problem to be solved with the increasing issues of software security and the continuous expansion of software applications scope. For the characteristics of large scale and complex structure in large-scale software, the traditional software analysis techniques are difficult to be used. With the problem of difficulty in presentation, storage and low efficiency in the process of large-scale software analysis, the visualization analysis framework for large-scale software based on software network, named SoNet, is proposed with the combination of complex network theory and program slicing technique. Constraint logic attributes of the programs will be obtained through source code parsing. Then we will construct a global view by the theory of complex network after extracting software structure and behavior, improving user's perception of software architecture in a macro perspective. Use case slicing will be realized combined with Redis cluster, and accessibility analysis when given a keyword to be analyzed. We evaluate our prototype implementation on an open source software project named SoundSea in Github, and the results suggest that our approach can realize the analysis for large-scale software.

Keywords: Large-scale software · Software network · Visualization · Constraint logic

1 Introduction

With typically millions of lines of code, large-scale software systems are among the most sophisticated human-designed objects ever built, and they play an important role in all aspects of modern life. However defective software may produce unexpected results or behavior at run time, remaining great loss for enterprise or even human's life. After a long period of engineering practice, software developers recognized that the structure of software is closely related to its functional quality for large-scale complex systems. Nevertheless, only little is known about the actual structure and quantitative properties of large-scale software systems. As a consequence, they are at the core of an intense research and engineering activity, and much work is devoted to understanding them better.

B. Zou et al. (Eds.): ICPCSEE 2017, Part I, CCIS 727, pp. 751–763, 2017.
DOI: 10.1007/978-981-10-6385-5_63

Our solution, visualization analysis framework for large-scale software based on software network (SoNet), combines the theories of software network, constraint programming, program slicing and the tools of ANTLR, Redis, Graphviz to automatically visualize the structure and behavior of given open source program in a macro scope, with sliced view visualized and constraint logic facts related with keyword to be analyzed.

Model-driver engineering (MDE) as a paradigm for software development is gaining more and more importance. Doan [1] proposed a process for the verification of a behavioral property of a UML model enriched by OCL constraints. Gogolla [2] presented steps towards a well-founded benchmark for assessing UML and OCL validation and verification techniques. In order to systematically support modeling and formal verification for self-adaptive software, an approach which incorporated the visual UML and the strictly defined timed automata proposed by Han [3]. And network science, because of its statistical nature and grate support for exploring the feature of structure and behavior is best suited for the study of large-scale structures. Networks possibly provide the most adequate framework for the analysis of the structure of complex systems like software projects. Not novel as software network is, however, network analysis is still only rarely used in software engineering [4].

Several works make interesting proposals for solving these issues, but current state-of-the-art remains far from satisfactory. Although a drawing of a complex network is a natural entry point for describing it, it is far from sufficient in general since graph drawing is a challenge task and results are often disappointing. ANTLR (ANother Tool for Language Recognition), based on the LL (*) algorithm, is a top-down analysis tool to generate the syntax parser generator. With the characteristics of cross-platform, readable source code and advanced grammar and semantic prediction, ANTLR is very popular in the parser of XML, scripting language, Hibernate which is an open framework etc. Given a file and the corresponding g4 grammar, the lexical analyzer, grammar analyzer and walk mechanics including Listener and Visitor which is very useful for information exacting will be generated automatically.

When comes to large-ennescale software, the difficulty related to storage will appearing. Redis, as a representative of NoSQL database, has been applied into Internet Company such as Twitter, Instagram, Sina and so on with rich data types, high speed and performance in concurrent read and write [5].

As a prominent technique for program analysis, program slicing proposed by Weiser [6], leaves partial statements unrelated alone, which gives a new perspective for the analysis for large-scale software.

The core idea of SoNet is that after the directory to be analyzed being given, the system will glean enough primary information to visualize the features and to store them into Redis for searching later. And whenever a keyword is available the sliced view and constraint logic facts will be outputted for the purpose of improving the comprehension of software. We evaluate our approach on open-source project written in Java named SoundSea which is proved to be effective in the result.

2   Related Work

While there is a lot of research on large-scale software analysis, Latapy [7] applied complex networks and link streams into the analysis for large software with the contribution of dynamical aspects by defining properties directly over link streams and using these properties to describe data which are poorly captured by current approaches although they are crucial. Kumar [8] developed a tool that implements K-limited path sensitive interval domain analysis to get precise results without losing on scalability against the situation of non-relational domains is highly scalable but imprecise and relational domains may precise but doses not scale up, whereas the path segmentation can be arbitrary when paths are more than k with the configuration of k and widening parameters not given. An approach proposed by Zhou [9, 10] modeled C program based on CNets, an expansion of Petri nets with regular expression API named Jakarta - ORO being used to extracting functions, variants, operations and relations among them, whereas regular expression, as a coarse-grained parser tool, may come across bug of stack overflow and cost more time to program patterns matched with the elements that needed to be capture. Constructing event structure model after extracting behavior of program, with partial order algorithm being used to reduce the explosion of state space by compute key path directly, and defect reports, as dynamic information, were very suitable for refining program slicing proposed by Wu [11]. However they dependent on defect reports overly and depicted the characteristic of behavior only with event structure model lacking of enough comprehension of overall structural characteristics.

3   Visualization Analysis Framework

In this section we will describe key ideas of and give intuition into why and how our approach works. The scheme of visualization analysis of large-scale software based on software network proposed in this paper is shown in Fig. 1.

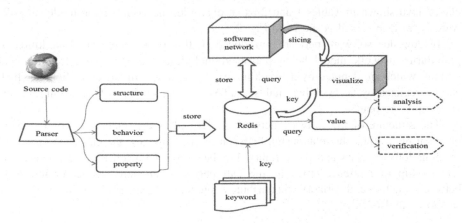

Fig. 1. Visualization analysis framework.

The code features, including structure, behavior and attribute, are extracted by parser firstly, which will generate software network related, with the source code being translated into the constraint logic expression to form the constraint logic fact library. Then the network can be stored by Redis cluster, where node in network can be denoted by the key in Redis, the property name and value in node can also be denoted by the field and value respectively, and a global view with features of structure and behavior can be depicted simultaneously. In order to ease software analysis and program validation, we will obtain the local view and the constraint logical fact library related to keyword by retrieving Redis when a keyword to be analyzed is given. The inputs of the framework are program source code and keyword to be analyzed, and the outputs are the global view related to the program source code, the local view and the constraint logical fact library related with keyword.

3.1 Modeling and Visualization

While traditional model of UML and automatic machine have encountered a great challenge in representation when comes to large-scale software, and the mechanism of package, class, function, variant in object-oriented language can prevent naming conflicts which can identify the element of source code unambiguously. A novel method concerning software modeling is proposed in this paper taking three characteristics of object-oriented language including encapsulation, inheritance and polymorphism into account.

Based on the theory of complex network, the large-scale software is modeled by characterizing four nodes including package, class, function and variant with different shape denoting vary nodes, arrows of dotted and solid line are used to depicted relationships of reference and inclusion respectively. Specially, access modifier and realization (i.e., the nodes of interface, abstract class, abstract functions and variants are regarded as not realized) are taken into account when it comes to the modeling of classes, functions and variants, which are depicted by font color in differ and fill condition of nodes respectively. Taking the code fragments of classX.java and classY.java shown in Tables 1 and 2 as an example, the visualization model of software network obtained is shown in Fig. 2.

In complex software network introduced in this paper, there are four kinds of granularity network model being portrayed including package, class, function and variant, which are denoted by network node. The approach of use case slicing, which can be realized by accessibility analysis, takes the relationships of inclusion and reference into account.

The structure of the software network is defined as the relationship of inclusion between nodes (i.e., the relationship between the package and the class, the class and the function, the class and the variant). The behavior characteristic is defined as the relationship of reference (such as the call between functions and communication between function and variant), and attribute depicts the characteristics of network node as shown in Table 3.

Table 1. classX.java.

```
package P1;
public class classX{
    public int i;
    public String str;
    public int A(double d,int
num){
        int j;
        j = (int)d+num+i;
        return j;
    }
    private String B(double d){
        int j;
        String ret_str;
        j = (int)d+i;
        ret_str =
str+String.valueOf(j);
        return ret_str;
    }
    public void C(double d){
        String count = B(d);
        System.out.println(count);
    }
}
```

Table 2. classY.java.

```
package P2;
import P1.classX;
public class classY{
    public double d;
    public int num;
    public int C(){
        classX x;
        x = new classX();
        return x.A(d,num);
    }
}
class classZ{
    private int i;
    public void A(int j){
        classY y;
        y = new classY();
        i = y.C()+j;
    }
}
```

3.2 Translating

In many software model checking techniques, the notion of constraint has been shown to be very effective, both for constructing models of programs and reasoning about them. Constraint logic programming, with the abilities of both logic programming and constraint solving, can be used to represent the execution and attribute of program.

In addition, constraint solving can be achieved by constraint logic expression with state space compressed. Besides extracting the information of structure, behavior and attribute from source code, the control flow is also available by ANTLR with its parser tree, listener mechanism and tree traversal algorithm. And constraint logic expression can be obtained easily with the information of control flow generated. The definitions of constraint logic expression are shown in Table 4.

As a summary, compound statement including condition and loop will be handled with the combination of 2 and 3 in definition regarding constraint logic command in

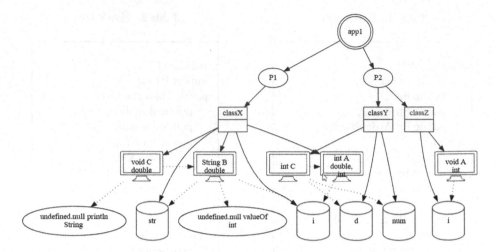

Fig. 2. The Visualization model related with Tables 1 and 2.

Table 3. Attribute depicts the characteristics of each network node.

Node	Property
Package	level, key, package, m+className+i
Class	level, key, packName, className, memFun+i, memVarName+i, memVarType+i, baseClass+i, access, isRealize
Function	level, key, packName, className, funName, funCall+i, varName+i, varType+i, access, isRealize
Var	level, packName, className, varName, varType, access, comFun+i

Table 4. Definitions of constraint logic expression.

Constraint logic expression	Definition
Instr(funName@line,sta)	Denotes that the statement is executed in function named funName at line
Instr(funName@line,ite (expr),line1,line2)	Denotes that the conditional judgement is executed in function named funName at line, if true the execution will jump into line1, and redirect to line2 otherwise
Instr(funName@line,goto (line1))	Denotes that the execution will jump into line1 when comes to the function named funName at line
phiInit(X,Y):- X = c1, Y = c2	X and Y denote member variant in class which the function belongs to, c1 and c2 is constant which maybe the value of initialized or self-defined when uninitialized by the type of member variant and programming language

Table 4, and generally simple statement can be solved by definition 1, while definition 4 can be used to deal with initial condition, such as member variant in the class which the function belongs to.

3.3 Storing

The storage of software model has encountered a great challenge with expanding of software scale. Redis, as one of the NoSQL database, is very suitable for dealing with the problem for its rich data types, high speed and performance in concurrent read and write. As identifying the element of source code uniquely by the model of software network proposed in this paper, access can be achieved efficiently, with absolute paths in software of node name being keywords, the name of attributes and behaviors being field and the value of attributes and behaviors being value storing into Redis.

Take the fragment of code in Table 1 for instance, the hash structures stored in Redis are shown in Table 5. And feature attribute including the information of node in four granularities and constraint logic attribute are the two attributes considered.

3.4 Searching and Slicing

As the difficulty in analysis and verification for large-scale software, a new perspective for analysis and comprehension of software has been provided by the theory of program slicing. And traditional algorithms regarding program slicing which excluding part of program statement and remaining specific statements related to the analysis of procedures by analyzing the dependencies between program statements is difficult to adapt to large-scale software with too small granularity considered. Based on the discovery of "small world" and "scale free" of complex network regarding object-oriented system, algorithm of reachability analysis can be used to apart from the nodes and edges unrelated with the key considered taking the relationships of inclusion and reference into account. There are two rules should be followed when slicing based on software network.

Lemma 1. Retaining the nodes and relationships of reference among the keyword and functions or variants and inclusion among them when a function or variant is given as a keyword.

Lemma 2. Retaining the nodes and relationships of reference among the same granularity where exists reference relationship among their underlying nodes and inclusion between its upper nodes when a package or class is given as a keyword. The strategy of slicing is as follows.

Algorithm 1. Slicing algorithm

Input : key denotes keyword, network stored in Redis
Output: Net denotes sliced network, Fact denotes constraint
logic facts related

```
1   Net.node <- {key};
2   Net.edge <- {};//initialize Net
3   Fact <- {};//initialize Fact
4   if (key not exists) then
5   throw an exception;
6   else if (key is package or key is class) then
7    if(key is class) then
8      Net.node.add(key.getPack);
9    end if;
10   for each i in getFunVar(key):
11     for each j in getFun(i):
12       Node <- Same(j);
13       Net.node.add(node);
14       Net.edge.add(key,node);
15     end for
16   end for
17  else if (key is variant or key is function) then
18   if (key is function) then
19     Fact.add(key.act);
20   end if;
21   Net.node.add(key.getClass);
22   Net.edge.add(key.getClass,key);
23   Net.node.add(key.getClass.getPack);
24   Net.edge.add(key.getClass.getPack ,key.getClass);
25   for each i in getFun(key):
26     Net.node.add(i);
27     Net.edge.add(i,key);
28     Fact.add(i.act);
29   end for;
30  end if
31  return Net, Fact;
```

In the strategy of slicing, the method of getFunVar(key), located in 10, denotes all the nodes of functions and variant where exists inclusion relationship between the key or its underlying nodes. And getFun(i) in 11 and 25 depicts function nodes related to given i including the function nodes where exists reference. Whereas the Same(j) in 12 searches for the node where exists inclusion relationship with j at the same granularity with given key. Given software network stored in Redis and a keyword to be analyzed firstly, and the sliced network and constraint logic facts related keyword will be outputted after the slicing strategy being employed.

Table 5. Hash structures stored in Redis.

Key	Field-value	
	Field	Value
P1.classX	key	P1.classX
P1.classX	level	class
P1.classX	packName	P1
P1.classX	className	classX
P1.classX	isRealize	true
P1.classX	access	public
P1.classX	memVarType1	int
P1.classX	memVarName1	i
P1.classX	memVarType2	String
P1.classX	memVarName2	str
P1.classX	memFun1	A(double,int)
P1.classX	memFun2	B(double)
P1.classX	memFun3	C(double)

4   Case Study

The visualization analysis framework for large-scale software based on software network is realized through ANTLR, Graphviz and Redis. And in this section we will evaluate our prototype implementation on an open source software project named SoundSea in github.

4.1   Subject Program and Platform

SoundSea, as one of the top ten java programs worthing your attention in github roaming guide in July 9, 2016, uses the iTunes search to grab the metadata that the user has requested and Pleer.com is also used to find and download corresponding song with the features of showing a lists of songs from the search mode, filtering through high, low, or VBR bitrates, playing songs within the browser and so on.

The experiment can be carried out in the platform of EclipseMARS.2, ANTLR4.4, Redis and Graphviz2.38.

4.2   Experimental Evaluation

There are four components in SoNet, including GUI, Parser, Visualization and Storage with Java Swing being used to realize GUI, ANTLR4, the theory of complex network and constraint programming being used to construct Parser, Graphviz2.38 and Redis being used to implementing the module of Visualization and Storage respectively.

The window of SoNet composed of components of menu bar, text field of input, display panels of global and sliced views and text area of error output, shown in Fig. 3. The menu item of File/Import can be clicked in GUI and file chooser is used to select directory of source code regarding open source software. And then Parser module can

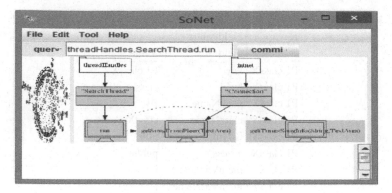

Fig. 3. GUI of SoNet.

be activated extracting the primary information of package, class, function, variant and relations among them with the information storing into Redis by calling storage module and global view displaying in the left of GUI by invocating the layout of twopi provided by Graphviz in visualization module shown in Fig. 4, where the color of red, yellow, green and blue represent the nodes of package, class, function and variant respectively. Particularly, the library functions, which is not exists in Redis, are denoted by circle, while others are denoted by ellipse.

And then the commit button can be clicked when a keyword inputted into the text field with slicing algorithm being invocated to search constraint logic facts and partial view stored in Redis with facts outputting to the file suffix by .clp and partial view displaying in the right of GUI respectively. And the nodes in different granularity correspond to vary shapes with package being tab, class being record, function and variant being self-defined shape of computer and cylinder respectively, the access correspond to different font color with public being red, private being gold, friendly

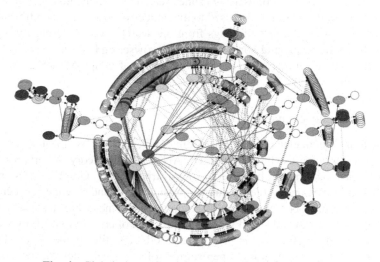

Fig. 4. Global view related SoundSea. (Color figure online)

defaulted being black and protected being blue, and the relationships of inclusion and reference are denoted by the line of solid and dotted with arrow separately, and wither realize' or not is denoted by the filled condition of node. Taken the keyword of threadHandles.SearchThread.run, threadHandles.SearchThread and threadHandles as example, the partial views are depicted in Figs. 5, 6 and 7 respectively. Especially, information of errors and exceptions are outputted in text area the down of GUI.

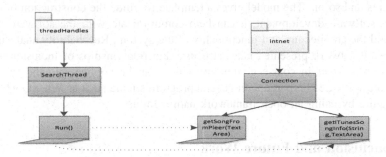

Fig. 5. The threadHandles.SearchThread.run as keyword related SoundSea.

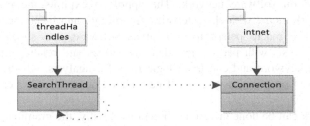

Fig. 6. The threadHandles.SearchThread as keyword related SoundSea.

Fig. 7. The threadHandles as keyword related SoundSea.

4.3 Evaluation

Compared with related work the superiorities are as follows. Firstly, the information of global structure and behavior of the given software is obtained from the global view visualized giving a macro perspective to understand the overall characteristics. The sliced partial view related to considered keyword takes reachability analysis and relationships of inclusion and reference into account with work efficiency, readability and understandability improving. And the file suffixed with .clp obtained by SoNet can also be employed as an input of ECLiPse [12] for software analysis and program verification.

Different from Guofu Zhou who extracted key information by parsing objective files using regular expression, ANTLR, with the characteristics of cross-platform, readable source code and advanced grammar and semantic prediction, can be used to extract primary information from all kinds of source code effectively including C, C++, java, XML, python, and so on. Better than general UML modeling, as unified modeling language, integrated the software structure and behavior characteristics, automatic machine can be depicted for simulation with the information of video, audio, pictures, documents and so on, The model gives a template to guide the construction of system. And for software development, it can help communicate with the software requirements and the architecture and functionality of the system. Reachability analysis based on software network presented takes structure and relationships of inclusion and reference into account in coarse granularity which can be applied into the analysis of large-scale software compared with general program slicing such as Wu XQ who sliced the program by using an open framework names Indus.

5 Conclusion and Future Work

This contribution has proposed a process for the visualization analysis of large-scale software based on software network. The approach exploits the multi-granularity software network model with characteristics of structure and behavior with the global view providing a macro perspective for given software, and slices the model by accessibility analysis with partial view denoting nodes and relationships interrelated with the given keyword and constraint logic facts for analysis. All the steps of visualization analysis proposed in paper will be realized automatically performed by our SoNet.

Future work can be done in various directions. Firstly, the grammar of g4 needs to be programmed to be able to cope with the file including xml, picture and to model the given software with elements of picture, video, audio and configuration. Then the constraint logic facts along with the rule can be useful to SMT solver for analysis and verification later. And the codes in function can also be transformed into Hoare expression for the proof of correctness in addition.

References

1. Doan, K.-H., Gogolla, M., Hilken, F.: Towards a developer-oriented process for verifying behavioral properties in UML and OCL models. In: Milazzo, P., Varró, D., Wimmer, M. (eds.) STAF 2016. LNCS, vol. 9946, pp. 207–220. Springer, Cham (2016). doi:10.1007/978-3-319-50230-4_15
2. Gogolla, M., Cabot, J.: Continuing a benchmark for UML and OCL design and analysis tools. In: Milazzo, P., Varró, D., Wimmer, M. (eds.) STAF 2016. LNCS, vol. 9946, pp. 289–302. Springer, Cham (2016). doi:10.1007/978-3-319-50230-4_22
3. Han, D.S., Yang, Q.L., Xing, J.C.: UML-based modeling and formal verification for software self-adaptation. J. Softw. **26**(4), 730–746 (2015)

4. Šubelj, L., Bajec, M.: Software systems through complex networks science: review, analysis and applications. In: Proceedings of the First International Workshop on Software Mining, pp. 9–16. ACM, Beijing (2012)
5. Liu, J., Liu, G.: Real-time storage and index strategy of massive internet small file based on Redis. J. Comput. Res. Dev. **52**(Suppl.), 148–154 (2015)
6. Weiser, M.: Program slicing. IEEE Trans. Softw. Eng. **210**(4), 352–357 (1984)
7. Latapy, M., Viard, T.: Complex networks and link streams for the empirical analysis of large software. In: Ciardo, G., Kindler, E. (eds.) PETRI NETS 2014. LNCS, vol. 8489, pp. 40–50. Springer, Cham (2014). doi:10.1007/978-3-319-07734-5_3
8. Kumar, S., Chimdyalwar, B., Shrotri, U.: Precise range analysis on large industry code. In: Proceedings of the 2013 9th Joint Meeting on Foundations of Software Engineering, pp. 675–678. ACM, Russia (2013)
9. Zhou, G., Sun, Y., Cai, Y.: CCNeter: an automatic modeling tool based on petri nets for C program. Comput. Sci. **38**(5), 96–101 (2011)
10. Zhou, G., Du, Z.: Petri nets model of implicit data and control in program code. J. Softw. **22**(12), 2905–1918 (2011)
11. Wu, X.Q., Wei, J.: Visual debugging concurrent programs with event structure. J. Softw. **25**(3), 457–471 (2014)
12. Gray, P.M.D., Runcie, T., Sleeman, D.: Reuse of constraint knowledge bases and problem solvers explored in engineering design. Artif. Intell. Eng. Des. Anal. Manuf. **29**(01), 1–18 (2015)

Author Index

Printed in the United States
By Bookmasters